# Learn the Essential *What, How & Why* of Human Anatomy & Physiology

With the **Twelfth Edition** of *Essentials of Human Anatomy & Physiology,* science educator Suzanne Keller joins bestselling author Elaine Marieb in helping learners focus on the *What, How & Why* of A&P, without getting sidetracked in details.

## 11 The Cardiovascular System

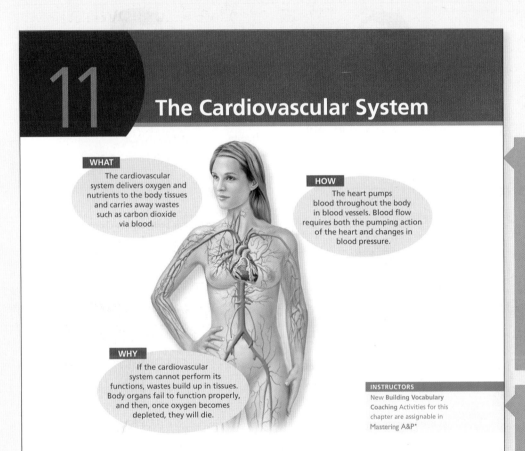

**WHAT**
The cardiovascular system delivers oxygen and nutrients to the body tissues and carries away wastes such as carbon dioxide via blood.

**HOW**
The heart pumps blood throughout the body in blood vessels. Blood flow requires both the pumping action of the heart and changes in blood pressure.

**WHY**
If the cardiovascular system cannot perform its functions, wastes build up in tissues. Body organs fail to function properly, and then, once oxygen becomes depleted, they will die.

**INSTRUCTORS**
New **Building Vocabulary Coaching** Activities for this chapter are assignable in Mastering A&P*

**NEW!** *What, How & Why* **chapter previews** introduce key examples of anatomy and physiology concepts that will be covered in the chapter. This technique helps learners hone in on *what* they are studying, *how* it functions, and *why* it is important for them to learn.

**NEW! Building Vocabulary Coaching Activities** in Pearson Mastering A&P help students learn the essential language of A&P.

When most people hear the term *cardiovascular system*, they immediately think of the heart. We have all felt our own heart "pound" from time to time when we are nervous. The crucial importance of the heart has been recognized for ages. However, the **cardiovascular system** is much more than just the heart, and from a scientific and medical standpoint, it is important to understand *why* this system is so vital to life.

Night and day, minute after minute, our trillions of cells take up nutrients and excrete wastes. Although the pace of these exchanges slows during sleep, they must go on continuously: when they stop, we die. Cells can make such exchanges only with the interstitial fluid in their immediate vicinity. Thus, some means of changing and "refreshing" these fluids is necessary to renew the nutrients and prevent pollution caused by the buildup of wastes. Like a bustling factory, the body must have a transportation system to carry its various "cargoes" back and forth. Instead of roads, railway tracks, and subways, the body's delivery routes are its hollow blood vessels.

Most simply stated, the major function of the cardiovascular system is transportation. Using blood as the transport vehicle, the system carries oxygen, nutrients, cell wastes, hormones, and many other substances vital for body homeostasis to and from the cells. The force to move the blood

382

See p. 382.

# Focus on Essential A&P Concepts

Throughout every chapter, the text's conversational writing style and straightforward explanations have been strengthened with familiar analogies and abundant mnemonic cues to help students learn and remember concepts.

**UPDATED! Exceptionally clear photos and illustrations,** including dozens of new and improved figures, present concepts and processes at the right level of detail. Many figures from the text are assignable as Art-Labeling Activities in Pearson Mastering A&P.

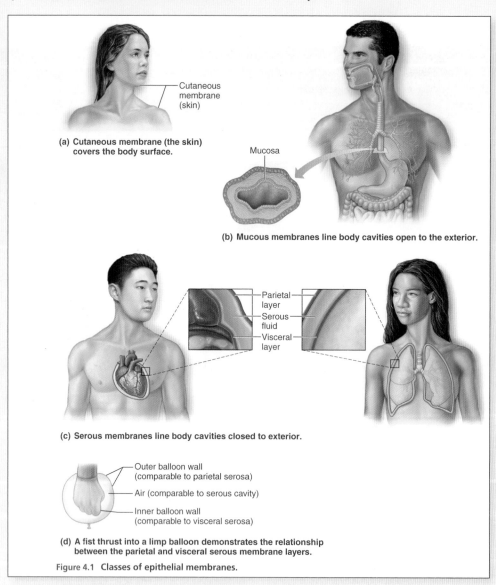

(a) Cutaneous membrane (the skin) covers the body surface.

Cutaneous membrane (skin)

Mucosa

(b) Mucous membranes line body cavities open to the exterior.

Parietal layer
Serous fluid
Visceral layer

(c) Serous membranes line body cavities closed to exterior.

Outer balloon wall (comparable to parietal serosa)

Air (comparable to serous cavity)

Inner balloon wall (comparable to visceral serosa)

(d) A fist thrust into a limp balloon demonstrates the relationship between the parietal and visceral serous membrane layers.

Figure 4.1 Classes of epithelial membranes.

See p. 137.

**Unique Concept Links** reinforce previously-learned concepts and help students make connections across body systems while learning new material.

> **CONCEPTLINK**
> The terms for the connective tissue coverings of a nerve should seem familiar: We discussed similar structures in the muscle chapter (Figure 6.1, p. 209). Names of muscle structures include the root word *mys*, whereas the root word *neuro* tells you that the structure relates to a nerve. For example, the endomysium covers one individual muscle fiber, whereas the endoneurium covers one individual neuron fiber. ←

See p. 282.

# Explore Essential Careers and Clinical Examples

To inspire and inform students who are preparing for future healthcare careers, up-to-date clinical applications are integrated in context with discussions about the human body.

**UPDATED! Homeostatic Imbalance discussions** are clinical examples that revisit the text's unique theme by describing how the loss of homeostasis leads to pathology or disease. Related assessment questions are assignable in Pearson Mastering A&P, along with Clinical Case Study coaching activities.

 **Homeostatic Imbalance 7.11**

In difficult deliveries, temporary lack of oxygen may lead to **cerebral palsy** (pawl′ze), but this is only one of the suspected causes. Cerebral palsy is a neuromuscular disability in which the voluntary muscles are poorly controlled and spastic because of brain damage. About half of its victims have seizures, are intellectually disabled, and/or have impaired hearing or vision. Cerebral palsy is the largest single cause of physical dis-

This adult patient with cerebral palsy presses a pad to communicate through a speaker.

*See p. 295.*

**Focus on Careers essays** feature conversations with working professionals and explain the relevance of anatomy and physiology course topics across a wide range of allied health careers. Featured careers include:

- Ch. 2 Pharmacy Technician
- Ch. 4 Medical Transcriptionist
- Ch. 5 Radiologic Technologist
- Ch. 8 Physical Therapy Assistant
- Ch. 10 Phlebotomy Technician
- Ch. 15 Licensed Practical Nurse

Students can visit the Pearson Mastering A&P Study Area for more information about career options that are relevant to studying anatomy and physiology.

**FOCUS ON CAREERS** Pharmacy Technician

To recognize how medications affect patients, pharmacy technicians need a thorough understanding of anatomy and physiology.

When most people get a new medication, they open up the package and toss out the little pamphlet that goes into detail about how the medication works. Not Chris Green. "I love reading the package inserts," says Green, the lead pharmacy technician at a CVS drugstore in Birmingham, Alabama. Green's enthusiasm for those details is a lifesaver for his customers. Pharmacy technicians are a vital link in the chain between doctor and patient.

Although ph...

Pharmacy technicians must have a good grasp of anatomy and physiology to understand each drug's chemical properties.

Green started working as a cashier at a drugstore ... was in high

medication that could react badly with another medication the patient is already taking. Drug interactions happen commonly when you have multiple doctors. "Sometimes, we'll get two ACE inhibitors in the same category from two different doctors [pre...

*See p. 82.*

# Continuous Learning
## Before, During, and After Class

Pearson Mastering A&P improves results by engaging students before, during, and after class.

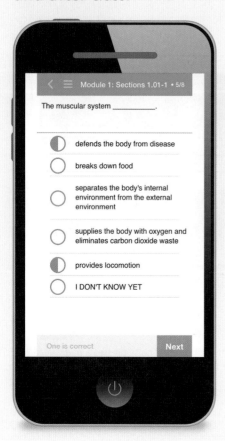

## Before Class

**Dynamic Study Modules** enable students to study more effectively on their own. With the Dynamic Study Modules mobile app, students can quickly access and learn the concepts they need to be more successful on quizzes and exams. **NEW!** Instructors can now select which questions to assign to students within each module.

Instructors can further encourage students to prepare for class by assigning **NEW! Building Vocabulary activities,** reading questions, art labeling activities, and more.

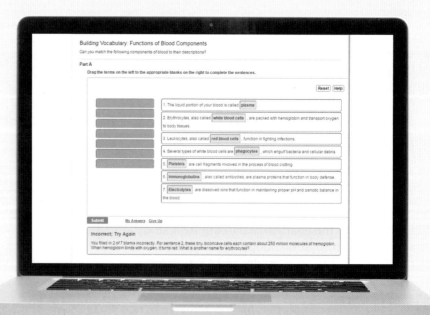

# with Pearson Mastering A&P

## During Class

**NEW! Learning Catalytics** is a "bring your own device" (laptop, smartphone, or tablet) engagement, assessment, and classroom intelligence system. Students use their device to respond to open-ended questions and then discuss answers in groups based on their responses. Visit learningcatalytics.com to learn more.

## After Class

**A wide variety of interactive coaching activities** can be assigned to students as homework, including Art-Labeling Activities, Interactive Physiology 2.0 tutorials, Clinical Case Studies, and activities featuring **A&P Flix** 3-D movie-quality animations of key physiological processes.

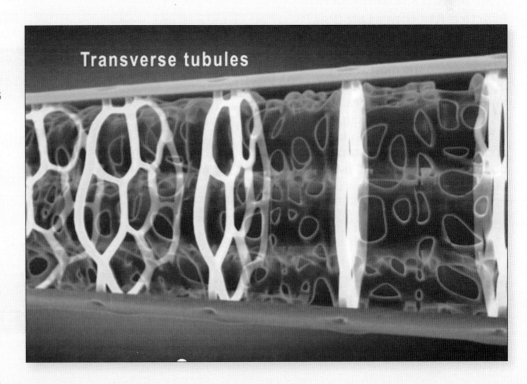

Transverse tubules

# A&P concepts come to life with Pearson Mastering A&P

Media references in the text direct learners to digital resources in the Pearson Mastering A&P Study Area, including practice tests and quizzes, flashcards, a complete glossary, and more.

## NEW! Interactive Physiology 2.0

**NEW! Interactive Physiology 2.0** helps students advance beyond memorization to a genuine understanding of complex physiological processes. Fun, interactive tutorials, games, and quizzes give students additional explanations to help them grasp difficult concepts. IP 2.0 features brand-new graphics, quicker navigation, and more robust interactivity.

**Practice Anatomy Lab (PAL™ 3.0)** is a virtual anatomy study and practice tool that gives students 24/7 access to the most widely used lab specimens, including the human cadaver, anatomical models, histology, cat, and fetal pig. PAL 3.0 is easy to use and includes built-in audio pronunciations, rotatable bones, and simulated fill-in-the-blank lab practical exams.

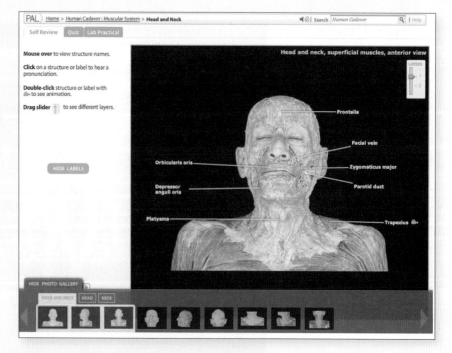

# Access the complete textbook online with the eText on Pearson Mastering A&P

**Powerful interactive and customization functions** include instructor and student note-taking, highlighting, bookmarking, search, and links to glossary terms.

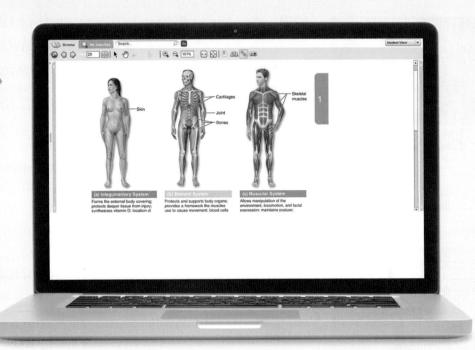

# Additional Support for Students and Instructors

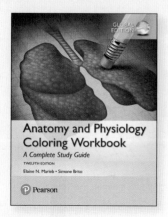

**NEW!** *Anatomy & Physiology Coloring Workbook* Twelfth Edition, Global Edition by Elaine N. Marieb and Simone Brito

The perfect companion to *Essentials of Human Anatomy & Physiology,* this engaging interactive workbook helps students get the most out of their study time. The *Twelfth Edition* includes **NEW!** crossword puzzles for every chapter, along with coloring activities, self-assessments, "At the Clinic" questions, and unique "Incredible Journey" visualization exercises that guide learners into memorable explorations of anatomical structures and physiological functions.

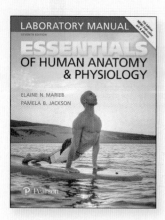

**NEW! IN FULL COLOR!** *Essentials of Human Anatomy & Physiology Laboratory Manual Seventh Edition* by Elaine N. Marieb and Pamela B. Jackson

This popular lab manual provides 27 exercises for a wide range of hands-on laboratory experiences, designed especially for a short A&P Lab course. This edition, which includes a Histology Atlas with 55 photomicrographs, features **NEW!** full-color illustrations, photos, and page design that help students navigate and learn the material faster and easier than ever before. Each concise lab exercise includes a Pre-Lab Quiz, brief background information, integrated learning objectives, student-friendly review sheets, and more.

**The Instructor Resources Area in Pearson Mastering A&P includes the following downloadable tools:**

- All of the figures, photos, and tables from the text in JPEG and PowerPoint® formats, in labelled and unlabeled versions, and with customizable labels and leader lines

- Step-edit Powerpoint slides that present multi-step process figures step-by-step

- Clicker Questions and Quiz Show Game questions that encourage class interaction

- A&PFlix™ animations bring human anatomy and physiology concepts to life

- Customizable PowerPoint® lecture outlines save valuable class prep time

- A comprehensive Instructor's Guide includes lecture outlines, classroom activities, and teaching demonstrations for each chapter.

- Test Bank provides a wide variety of customizable questions across Bloom's taxonomy levels. Includes art labeling questions, and available in Microsoft® Word and TestGen® formats.

TWELFTH EDITION
GLOBAL EDITION

# ESSENTIALS
## OF HUMAN ANATOMY
## & PHYSIOLOGY

ELAINE N. MARIEB, R.N., PH.D.,
HOLYOKE COMMUNITY COLLEGE

SUZANNE M. KELLER, PH.D.,
INDIAN HILLS COMMUNITY COLLEGE

Pearson

330 Hudson Street, NY NY 10013

*Editor-in Chief:* Serina Beauparlant
*Senior Courseware Portfolio Manager:* Lauren Harp
*Content and Design Manager:* Michele Mangelli, Mangelli
  Productions, LLC
*Managing Producer:* Nancy Tabor
*Courseware Director, Content Development:* Barbara Yien
*Courseware Sr. Analysts:* Suzanne Olivier and Alice Fugate
*Courseware Specialist:* Laura Southworth
*Editorial Coordinator:* Nicky Montalvo
*Acquisitions Editor, Global Edition:* Sourabh Maheshwari
*Senior Project Editor, Global Edition:* Amrita Naskar
*Senior Media Editor, Global Edition:* Gargi Banerjee
*Senior Manufacturing Controller, Global Edition:*
  Trudy Kimber
*Mastering Content Developer:* Cheryl Chi
*Director of Mastering Production:* Katie Foley

*Associate Producer, Science:* Kristen Sanchez
*Rich Media Content Producer:* Ziki Dekel
*Copyeditor:* Sally Peyrefitte
*Proofreader:* Betsy Dietrich
*Art and Production Coordinator:* David Novak
*Indexer:* Steele/Katigbak
*Interior Designer:* tani hasegawa and Hespenheide Design
*Cover Designer:* Lumina Datamatics Ltd.
*Illustrators:* Imagineering STA Media Services, Inc.
*Rights & Permissions Manager:* Ben Ferrini
*Photo Researcher:* Kristin Piljay
*Manufacturing Buyer:* Stacey Weinberger
*Executive Marketing Manager:* Allison Rona

*Cover Photo Credit:* Aleksandr Markin/Shutterstock

Pearson Education Limited
Edinburgh Gate
Harlow
Essex CM20 2JE
England

and Associated Companies throughout the world

Visit us on the World Wide Web at:
www.pearsonglobaleditions.com

© Pearson Education Limited 2018

ISBN 10: 1-292-21611-5
ISBN 13: 978-1-292-21611-9

British Library Cataloguing-in-Publication Data
A catalogue record for this book is available from the British Library

10 9 8 7 6 5 4 3 2 1

Typeset by iEnergizer Aptara® Ltd.

Printed and bound by Vivar in Malaysia

**Elaine Marieb** After receiving her Ph.D. in zoology from the University of Massachusetts at Amherst, Elaine N. Marieb joined the faculty of the Biological Science Division of Holyoke Community College. While teaching at Holyoke Community College, where many of her students were pursuing nursing degrees, she developed a desire to better understand the relationship between the scientific study of the human body and the clinical aspects of the nursing practice. To that end, while continuing to teach full time, Dr. Marieb pursued her nursing education, which culminated in a Master of Science degree with a clinical specialization in gerontology from the University of Massachusetts. It is this experience that has informed the development of the unique perspective and accessibility for which her publications are known.

Dr. Marieb has given generously to provide opportunities for students to further their education. She funds the E. N. Marieb Science Research Awards at Mount Holyoke College, which promotes research by undergraduate science majors, and has underwritten renovation of the biology labs in Clapp Laboratory at that college. Dr. Marieb also contributes to the University of Massachusetts at Amherst, where she generously provided funding for reconstruction and instrumentation of a cutting-edge cytology research laboratory. Recognizing the severe national shortage of nursing faculty, she underwrites the Nursing Scholars of the Future Grant Program at the university. In January 2012, Florida Gulf Coast University named a new health professions facility in her honor. The Dr. Elaine Nicpon Marieb Hall houses several specialized laboratories for the School of Nursing, made possible by Dr. Marieb's generous support.

**Suzanne Keller** Suzanne M. Keller began her teaching career while she was still in graduate school at the University of Texas Health Science Center in San Antonio, Texas. Inspired by her lifelong passion for learning, Dr. Keller quickly adopted a teaching style focused on translating challenging concepts into easily understood parts using analogies and stories from her own experiences. An Iowa native, Dr. Keller uses her expertise to teach microbiology and anatomy and physiology at Indian Hills Community College, where most of her students are studying nursing or other health science programs.

Dr. Keller values education as a way for students to express their values through the careers they pursue. She supports those endeavors both in and out of the classroom by participating in her local Lions Club, by donating money to the Indian Hills Foundation to fund scholarships, and by financially supporting service-learning trips for students. Dr. Keller also enjoys sponsoring children in need with gifts for the holidays.

Dr. Keller is a member of the Human Anatomy and Physiology Society (HAPS) and the Iowa Academy of Science. Additionally, while engaged as an author, Dr. Keller has served on multiple advisory boards for various projects at Pearson and has authored assignments for the Pearson Mastering A&P online program. When not teaching or writing, Dr. Keller enjoys reading, traveling, family gatherings, and relaxing at home under the watchful eyes of her two canine children.

# New to the Twelfth Edition

This edition has been thoroughly updated. New "What, How, Why" art opens each chapter, highlighting key concepts relating to the chapter topic. Other specific chapter-by-chapter changes include the following:

## Chapter 1: The Human Body: An Orientation
- Updated description of the integumentary system to include vitamin D production in the presence of sunlight.
- Updated definition of the term *crural* to specify the anterior leg, or shin.
- New "Critical Thinking and Clinical Application" question on blood clotting and feedback regulation.
- New "Critical Thinking and Clinical Application" question on using anatomical language to describe the location of a spinal injury and identifying the best medical imaging technique to diagnose a spinal problem.
- Updated "A Closer Look: Medical Imaging" with new discussion and images of mammogram and bone densitometry.
- New "Did You Get It?" questions throughout the chapter.

## Chapter 2: Basic Chemistry
- New example of atomic symbol and Latin derivative for potassium.
- Revised discussion of hydrogen bonds to clarify that electrons are not involved in this type of bond as they are in covalent and ionic bonds.
- New example of importance of hydrogen bond in holding DNA strands together.
- All references to *hydroxyl ion* have been corrected to *hydroxide ion*.
- New sports analogies for acids and bases: putting electrons "in the game," to represent free protons versus taking electrons "out of the game," to represent binding a proton so it cannot contribute to a shift in pH.
- New "Did You Get It?" question part on the difference in pH between solutions at pH 11 and pH 5.
- Introduced concepts of *hydrophilic* and *hydrophobic* in discussion of phospholipids to help explain the functions of the polar head and fatty acid tails in cell membranes.
- Updated "Did You Get It?" question on lipids to include both phospholipids and cholesterol as cell membrane components.
- New analogy comparing the alpha (α)-helix to a metal spring.
- New analogy comparing a beta (β)-sheet to a pleated skirt, or paper folded into a fan.
- New analogy comparing enzymes and substrates to scissors and paper, respectively.
- New description of RNA function as DNA's "molecular assistant."
- New shorthand symbols for messenger (mRNA), transfer (tRNA) and ribosomal (rRNA) added.
- New analogy comparing glucose and ATP to crude oil and gasoline; energy must be in the proper form before it can be used by cells.
- New explanation of why the terminal phosphate bonds in ATP are high energy.
- New "Critical Thinking and Clinical Application" question on sickle cell anemia.

## Chapter 3: Cells and Tissues
- New description of the principle of complementarity.
- New information about how mitochondria divide.
- New "cargo" in the form of a membrane-bound receptor protein added to pathway 2 of Figure 3.6.
- New analogy for lysosomes as "cellular stomachs."
- New art of plasma membrane and new detail of mitochondrial function including aerobic respiration in Table 3.1.
- New analogy of dust "crowd surfing" on the mucus that cilia carry from the lungs.
- New description of neuron function includes production of neurotransmitters.

- New colors used in Figure 3.14 DNA images to help students track new and old strands of DNA.
- New description of each chromosome being composed of two sister chromatids.
- New explanation of protein synthesis includes the role of the large ribosomal subunit in peptide bond formation.
- New Figure 3.18 descriptions of nuclei lining up in simple columnar and not lining up in pseudostratified epithelia.
- New description of cell shapes in different layers of stratified epithelia as "squished" and variable.
- New analogy for mucus produced by goblet cells as a "sticky trap" for dust and debris.
- New Figure 3.19 labels for osteocytes, the elastic and collagen fibers in areolar connective tissue, and the fluid matrix of blood; new art for dense fibrous connective tissue.
- New analogy for reticular tissue as "cellular bleachers" where other cells rest to monitor the body.
- Updated Figure 3.21 to include the term *neuroglia* to describe supporting cells.
- New information added to Figure 3.22: cartilage added to the connective tissue list and two major hallmarks of each of the four tissue types.
- New example of atrophy: when a broken leg is in a cast, lack of use causes muscles to atrophy during healing.
- Revised "A Closer Look: Cancer—An Intimate Enemy" and updated art.
- New "Short Answer Essay" questions on the components of the plasma membrane and their functions and on contrasting cytokinesis, interphase, and mitosis.
- New "Critical Thinking and Clinical Application" question on IV fluids and tonicity.
- New "Did You Get It?" questions throughout the chapter.

## Chapter 4: Skin and Body Membranes
- New Figure 4.1 on epithelial membranes.
- New description of sensory receptors as part of nervous system including a list of the stimuli detected.
- New text updates on Figure 4.4 on epidermal structure; included a new figure question on stratum lucidum.
- New analogy for epidermal dendritic cells as "sentries" guarding the skin.
- New photo of stage 2 decubitus ulcer added to Homeostatic Imbalance 4.2.
- New layout for Figure 4.7 combining scanning electron micrograph of hair shaft with existing art of the hair root and follicle.
- New discussion of fourth-degree burns.
- New criteria for determining whether a burn is critical, including circumferential burns, burns of the airway, and burns to the genital area.
- New images of basal cell and squamous cell carcinoma in Figure 4.11.
- New component added to ABCDE rule: now includes "Evolution," changes in a skin lesion over time.
- New "Short Answer Essay" questions on the risks of full-thickness burns, contrasting eccrine and apocrine sweat glands, and the relative severity of different skin infections.
- New "Critical Thinking and Clinical Application" question on burns.
- New "Did You Get It?" questions throughout the chapter.

## Chapter 5: The Skeletal System
- Updated description of long bones.
- New analogy comparing lubrication over articular cartilage at joints to a slick marble floor.

- Updated descriptions of red and yellow bone marrow.
- Updated descriptions of sagittal and coronal sutures.
- Updated description of the capitulum of the humerus.
- New analogy comparing the trochlea meeting the trochlear notch to a curved "tongue-in-groove" joint.
- Updated description of buttock injections to include the consequences of hitting a nerve.
- Updated description of a synovial membrane to include areolar connective tissue.
- Updated description of cartilaginous joints.
- New description of saddle joints including a reference to opposable thumbs.
- Updated list of triggers for rheumatoid arthritis.
- Discussion of the fetal skull and fontanels moved to the Developmental Aspects section.
- New analogy likening skulls of small children to "bobble heads."
- Updated review question on bones that articulate with the sphenoid to reflect only bones shown in the figures of Chapter 5.
- Updated "Short Answer Essay" question on synovial joints to include osteoarthritis.
- New "Short Answer Essay" question contrasting the foramen magnum and obturator foramen.
- New "Critical Thinking and Clinical Application" question on gouty arthritis.
- New statistics, information, and images added to "A Closer Look: Joint Ventures."
- Updated description of comminuted fractures on Table 5.2.
- Updated Figure 5.6 to include osteoblasts and osteoclasts in the descriptions of bone addition and resorption, respectively.
- Updated Systems In Sync with respect to the descriptions of relationships of cardiovascular and muscular systems to the skeletal system.

## Chapter 6: The Muscular System

- Updated descriptions of tendons and aponeuroses.
- New analogy about running to explain the difference between the contraction of skeletal muscle (fast) versus smooth muscle (slow).
- Updated description of a sarcomere to include its role as the structural and functional unit of muscle.
- Added discussion of titin to the description of a sarcomere as the elastic filament that attaches myosin to the Z disc.
- New Homeostatic Imbalance on ALS (amyotropic lateral sclerosis, or Lou Gehrig's disease).
- New "Did You Get It?" question on the roles of calcium in muscle contraction.
- Updated descriptions of cross-bridge formation and the sliding filament theory, including the role of ATP.
- New link to IP Essentials for the sliding filament theory.
- New description of flaccid versus spastic paralysis.
- New mnemonic device for *adduction*: "add" back to the body by moving toward the trunk (midline).
- New descriptions of dorsiflexion and plantar flexion with respect to the head: toes point toward the head or away, respectively.
- New girdle analogy for abdominal wall muscles "holding guts in."
- New description of the consequences of an injection being too close to, or hitting, the sciatic nerve.
- New description of tailor's muscle sitting position.
- New description of myasthenia gravis as an autoimmune disease.
- New "Short Answer Essay" question about the relationship between wrist flexors and extensors, including their locations.
- New figure question for Figure 6.20 on the origin(s) and insertion(s) of the rectus femoris depending on the action being performed.
- Updated explanation of steps in Figure 6.5.

## Chapter 7: The Nervous System

- Updated Figure 7.13 to use *superior* and *inferior* instead of *cephalad* and *caudal*.
- Updated Figure 7.24 to clarify why there are eight cervical nerves but only seven cervical vertebrae.
- New Learning Outcome on the structures and functions of neurons and neuroglia.

- Updated description of Nissl body function.
- New description clarifying the difference between a synapse and synaptic cleft.
- New analogy for a myelin sheath as the wrapping on an electrical cord.
- New explanation clarifying the differences between myelin sheaths in the CNS and PNS.
- New explanation clarifying the "short circuit" event in multiple sclerosis means that the signal may stop or "jump" to an unmyelinated neuron.
- New analogy for the structure of a unipolar cell body as a "cul-de-sac" off the "main road" that is the axon.
- Replaced references to the term *basal ganglia* with the more accurate term *basal nuclei*.
- Replaced the term *arachnoid villi* with *arachnoid granulations*.
- New statistics on stroke as the fifth leading cause of death in the United States (formerly identified as the third leading cause).
- New statistics regarding the rate of survival after a stroke.
- Replaced the phrase "mentally retarded" in the discussion of cerebral palsy with "intellectually disabled."
- New information included in "A Closer Look: The 'Terrible Three'" reflecting the role of calcium in apoptosis, two new drugs for treatment of Parkinson's disease, and the variation in dopamine levels in patients with Huntington's disease.
- New information incorporated in "A Closer Look: Tracking Down CNS Problems" to include a new dopamine imaging technique called DaTscan.

## Chapter 8: Special Senses

- New description of lacrimal caruncle.
- New description of optic disc and the resulting blind spot.
- New analogy comparing the ability to see intermediate colors (between the red, green, and blue cones) to mixing paint.
- Updated the description of cataracts.
- New example of motion detected by dynamic equilibrium: a spinning carnival ride.
- New analogy for bending of the cupula as divers' fins in water.
- New description of foliate papillae on the side of the tongue, another location for taste buds.
- New art showing the retina in Figure 8.5.
- New Figure 8.6 showing the graph of rods and cones, and which wavelengths of light are detected by each.
- Updated Figure 8.12b on maculae.
- New "Did You Get It?" questions throughout the chapter.

## Chapter 9: The Endocrine System

- Updated discussion of the mechanism of hormone action, including Figure 9.1 and its caption, to reflect that steroid hormones can act via either second messenger or direct gene activation.
- Updated explanation of how hormones alter cell activity.
- New analogy comparing second-messenger systems to delivering a letter.
- Revised coverage of endocrine glands to reflect their location in body from superior to inferior; Table 9.1 has also been revised to reflect the new order.
- Updated description explaining why a goiter forms in the absence of iodine.
- Updated description of body proportions in cretinism.
- New "Did You Get It?" question on adrenal cortex hormones.

## Chapter 10: Blood

- Updated explanation of why the normal temperature of blood is a bit higher than body temperature.
- Added definitions for the suffixes *-cytosis* and *-penia*.
- Updated the analogy comparing the shape of the eosinophil nucleus to earmuffs.
- Updated the role of monocytes to include activation of lymphocytes.
- Updated the list of locations where red marrow is found in adults.
- Updated the major anticoagulants to include warfarin.
- New description of petechiae includes comparison to a skin rash.

- Added a learning tool about blood type reminding readers that a person does not make antibodies against their own blood type antigen(s).
- Updated discussion of lack of vitamin B12 as the cause of pernicious anemia and how this relates to intrinsic factor.

## Chapter 11: The Cardiovascular System

- Updated description of pericardium.
- Revised discussion of the function of the atria to clarify that they assist with ventricular filling.
- Arteries and veins are now introduced in terms of the direction of blood flow with respect to the heart.
- New analogy comparing valve cusps filling with blood to a parachute filling with air.
- New analogy comparing the intrinsic conduction system setting heart rhythm to a drummer setting the beat for a rock band playing a song.
- New discussion of AEDs (automatic external defibrillators) included in the discussion of fibrillation.
- Reorganized section on the cardiac cycle to include five stages.
- New "Did You Get It?" question about isovolumetric contraction.
- Updated description of the effect of congestive heart failure on stroke volume.
- Updated description of pulmonary congestion.
- Updated description of pulmonary embolism.
- Discussion of fetal circulation moved to the Developmental Aspects section.
- Updated description of the blood pressure gradient to include a pressure of zero in the right atrium.
- New layout of Figure 11.8 reflecting five stages of the cardiac cycle.
- Updated Figure 11.9 description to clarify that any change in heart rate or stroke volume will also cause a change in cardiac output.
- Updated description of Figures 11.13 and 11.14 to include a statement that all vessels are bilateral unless otherwise stated in the text.
- Updated "A Closer Look" box on atherosclerosis.

## Chapter 12: The Lymphatic System and Body Defenses

- Updated Figure 12.10 on lysis by complement to reflect water flowing into the cell to cause lysis.
- Added the role of B cells in antigen presentation to Figure 12.19.
- New information added regarding discovery of lymphatics in the central nervous system.
- Updated the description of adaptive defenses as defenses that fight antigens that get past the innate defenses.
- New description of how natural killer cells kill: via *perforin* and *granzymes*.
- Updated the description of positive chemotaxis to include movement toward the stimulus.
- Revised description of interferon to clarify that interferon fights only viral pathogens, not bacteria or fungi.
- New antibody function has been listed: opsonization.
- New description of Graves' disease explaining that excess production of thyroxine is in response to antibodies that mimic TSH (thyroid-stimulating hormone).
- New descriptions of two additional types of hypersensitivities: reactions resulting in cell lysis and those forming antigen-antibody complexes.
- New example of when epinephrine is used during acute hypersensitivity: EpiPen® injection.
- New Short Answer Essay question provided on mechanisms of killing used by the immune system, including lysozyme, perforin, and granzymes, and membrane attack complex (MAC).
- Updated "A Closer Look" box on AIDS, including new title.

- Updated Table 12.1 regarding the role of nasal hairs to include filtration of airborne particles.
- Updated Table 12.3 entry for "Cytokines: Perforin and granzymes" to include natural killer (NK) cells.
- New "Did You Get It?" questions throughout the chapter.

## Chapter 13: The Respiratory System

- New information explaining neural regulation of breathing with respect to the dorsal and ventral respiratory groups of the medulla.
- New Short Answer Essay question contrasting hyperventilation and hyperpnea.
- Updated "A Closer Look" on cleanliness and asthma.
- New "Did You Get It?" questions throughout the chapter.

## Chapter 14: The Digestive System

- New illustration outlining the parietal and visceral layers of the peritoneum (Figure 14.5).
- New illustrations showing both deciduous and permanent teeth in greater detail (Figure 14.9).
- Updated description of circular folds to provide students with a visual image of a corkscrew that slows progression of food and increases surface area at the same time.
- Added detail that rennin in infants is the same enzyme used to curdle milk in cheesemaking.
- Added narcotic pain medications to the list of causes of constipation, with stool softeners as a method of treatment.
- New "Did You Get It?" question on the four types of teeth and their functions.
- Added brief discussion of nucleic acid digestion, including the source of the enzymes and the reminder that nucleotides are the building blocks.
- Revised "A Closer Look" box on obesity to update references, statistics, and methods used to determine body composition, such as DEXA, the Bod Pod, and underwater weighing.

## Chapter 15: The Urinary System

- Updated descriptions of the arterioles that connect to the glomerulus.
- Included a new learning tool describing the *in*ternal urethral sphincter as *in*voluntary.
- New "Short Answer Essay" question contrasting the homeostatic imbalances oliguria, anuria, polyuria, and nocturia.
- New "Critical Thinking and Clinical Application" question about the relationship between hypertension and impaired kidney function, and tests that are used for determining impaired kidney function.
- New information included in "A Closer Look: Renal Failure and the Artificial Kidney" about a blood test to determine the creatinine level in order to estimate the rate of glomerular filtration.
- Did You Get It?

## Chapter 16: The Reproductive System

- New explanation of the purpose of polar bodies: to reduce the chromosome number during oogenesis.
- Update of suggested age range for women to begin having regular mammograms: between 45 and 54.
- New Concept Link on chemotaxis.
- New photomicrograph showing sperm swarming an oocyte in Figure 16.16.
- New explanation of how an egg blocks additional sperm from entering; the surface sperm receptors on an oocyte are shed after the first sperm enters the cell.
- New clarification with updated definitions of *miscarriage* and *abortion*.

# Acknowledgments

Many people contributed to our efforts in the creation of this twelfth edition. We offer our profound thanks to the following reviewers, whose thoughtful critiques informed and enhanced our development of this edition:

- William Brazelle, University of South Florida
- Sheree Daniel, Trinity Valley Community College
- Trevor Day, Mount Royal University
- Camille Di Scala, Chandler-Gilbert Community College
- Pamela Boyter Jackson, Piedmont Technical College
- Roop Jayaraman, Central Michigan University
- Kimberly Kushner, Pueblo Community College
- Frances Mills, Lake Michigan College
- Diane Pelletier, Green River Community College
- Heidi Peterson, Indian Hills Community College
- Kenneth Ryan, Alexandria Technical and Community College
- Holly Sanders, Gwinnett Technical College
- Scott Schaeffer, Harford Community College
- I-Chia Shih, Leeward Community College, University of Hawaii
- K. Dale Smoak, Piedmont Technical College
- Bill Snyder, Bluegrass Community and Technical College
- Greg Tefft, Northwest State Community College
- Sandra Uyeshiro, Modesto Junior College
- Khursheed Wankadiya, Central Piedmont Community College
- Carol T. Wismer, College of Lake County

Thanks are also extended to the reviewers of the Eleventh Edition: Carmen Carpenter, South University; Steven D. Collins, Niagara College; Janie Corbitt, Central Georgia Technical College–Milledgeville Campus; Eric D. Forman, Sauk Valley Community College; Andrew Goliszek, North Carolina A&T State University; Amy Goode, Illinois Central College; Jeannette Hafey, Springfield College; Ashley Hagler, Gaston College; Frances Miles, Lake Michigan College–Napier Avenue Campus; Margaret Ott, Tyler Junior College; Heidi Peterson, Indian Hills Community College–Ottumwa Campus; Laura Ritt, Burlington County College; Holly Sanders, Gwinnett Technical College; Leba Sarkis, Aims Community College; Gustavo A. Solis, Forsyth Technical Community College; Ginny Stokes, Nash Community College; Robert Suddith, Cape Fear Community College; John F. Tarpey, City College of San Francisco; Deborah S. Temperly, Delta College; Claudia Williams, Campbell University.

The entire group from Pearson and beyond deserves our heartfelt thanks for being the best team around! They have provided support, guidance, and humor throughout the writing process, which made the process fun and was most appreciated. Special thanks to Serina Beauparlant, Editor-In-Chief, Brooke Suchomel, Sr. Acquisitions Editor, and Tiffany Mok, Program Manager who passed the reins to Lauren Harp, Sr. Portfolio Manager, for crossing the finish line. Thanks also to Nicky Montalvo, Editorial Coordinator. Thank you to the content development team—Suzanne Olivier, Alice Fugate, and Laura Southworth—for their attention to detail and collaborative minds. Thank you Patrice Fabel, Ziki Dekel, and Lauren Hill for supervising an impressive variety of media content that will benefit both students and instructors. A special thank-you to Gary Hespenheide for the book's beautiful and creative new interior and cover designs. The work of Kristin Piljay, Photo Researcher, resulted in some striking new photos for this edition; and Sally Peyrefitte, our excellent and diligent copyeditor, ensured a consistent style throughout the book. Proofreader Betsy Dietrich skillfully reviewed every page proof, and Sallie Steele provided a thorough and accurate index. Our talented art house, Imagineering STA Media Services, Inc., and compositor, Aptara, worked tirelessly to provide stunning artwork and student-friendly page layouts. Stacey Weinberger, Senior Manufacturing Buyer, Allison Rona, and Derek

Perrigo deserve special thanks for their expertise in delivering and presenting the final product to the market. A special thank you goes to David Novak, our Production and Art Coordinator, for taking on the role of two people during this edition and flawlessly handling every text and art-related production detail—David made the whole process smooth and successful. And last, but not least, Michele Mangelli—a tremendous thank-you for your skillful oversight of all aspects of the 12th edition, including assisting a first-time co-author in all things publishing . . . you're an unflappable rock star!

*Elaine N. Marieb*

*Suzanne M. Keller*

Anatomy and Physiology
Pearson Education
1301 Sansome Street
San Francisco, CA 94111

The publishers would like to thank the following for their contribution to the Global Edition:

**Contributor**

Christiane Van den Branden, Vrije Universiteit Brussel

**Reviewers**

Hemant Kumar, MBBS
Snezana Kusljic, The University of Melbourne
Eva Strandell, Halmstad University

# Contents

# 5 The Skeletal System  160

# 6 The Muscular System  207

# 7 The Nervous System  251

# 8 Special Senses  304

# 9 The Endocrine System  334

# 10 Blood   363

# 11 The Cardiovascular System   382

# 14 The Digestive System and Body Metabolism   489

# 15 The Urinary System   537

# 16 The Reproductive System  564

# Appendixes

# The Human Body: An Orientation

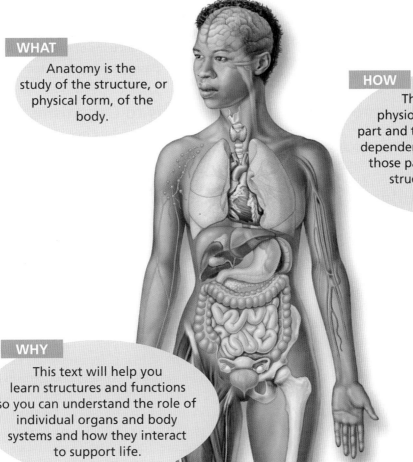

**WHAT**
Anatomy is the study of the structure, or physical form, of the body.

**HOW**
The function, or physiology, of each body part and the body as a whole is dependent on the anatomy of those parts; in other words, structure determines function.

**WHY**
This text will help you learn structures and functions so you can understand the role of individual organs and body systems and how they interact to support life.

**INSTRUCTORS**
New **Building Vocabulary Coaching** Activities for this chapter are assignable in Mastering A&P®

## An Overview of Anatomy and Physiology

→ **Learning Objectives**

☐ **Define** *anatomy* **and** *physiology*.

☐ **Explain how anatomy and physiology are related.**

Most of us are naturally curious about our bodies; we want to know what makes us tick. Infants can keep themselves happy for a long time staring at their own hands or pulling their mother's nose. Older children wonder where food goes when they swallow it, and some believe that they will grow a watermelon in their belly if they swallow the seeds. Adults become upset when their hearts pound, when they have uncontrollable hot flashes, or when they cannot keep their weight down.

Anatomy and physiology, subdivisions of biology, explore many of these topics as they describe how our bodies are put together and how they work.

### Anatomy

**Anatomy** (ah-nat′o-me) is the study of the structure and shape of the body and its parts and their relationships to one another. Whenever we look at our own body or study large body structures such as the heart or bones, we are observing *gross anatomy*;

that is, we are studying large, easily observable structures. Indeed, the term *anatomy*, derived from the Greek words meaning to cut (*tomy*) apart (*ana*), is related most closely to gross anatomical studies because in such studies, preserved animals or their organs are dissected (cut up) to be examined. *Microscopic anatomy*, in contrast, is the study of body structures that are too small to be seen with the naked eye. The cells and tissues of the body can only be seen through a microscope.

## Physiology

**Physiology** (fiz″e-ol′o-je) is the study of how the body and its parts work or function (*physio* = nature; *ology* = the study of). Like anatomy, physiology has many subdivisions. For example, *neurophysiology* explains the workings of the nervous system, and *cardiac physiology* studies the function of the heart.

## Relationship between Anatomy and Physiology

Anatomy and physiology are always inseparable. The parts of your body form a well-organized unit, and each of those parts has a job to do to make the body operate as a whole. Structure determines what functions can take place. For example, the lungs are not muscular chambers like the heart and so cannot pump blood through the body, but because the walls of their air sacs are very thin, they *can* exchange gases and provide oxygen to the body. We stress the intimate relationship between anatomy and physiology throughout this text to make your learning meaningful.

### Did You Get It?

1. Why would you have a hard time learning and understanding physiology if you did not also understand anatomy?
2. Kidney function, bone growth, and beating of the heart are all topics of anatomy. True or false?

For answers, see Appendix A.

### CONCEPTLINK

Throughout this text, Concept Links will highlight links between concepts and/or organ systems. Keep in mind that although discussions of the systems are separated into chapters for detailed study, the overall goal of this text is for you not only to gain an understanding of each individual system, but also to learn how the body systems interact to sustain life. ←

# Levels of Structural Organization

→ **Learning Objectives**

☐ Name the six levels of structural organization that make up the human body, and explain how they are related.

☐ Name the organ systems of the body, and briefly state the major functions of each system.

☐ Identify and classify by organ system all organs discussed.

## From Atoms to Organisms

The human body exhibits many levels of structural complexity **(Figure 1.1)**. The simplest level of the structural ladder is the *chemical level* (covered in Chapter 2). At this level, **atoms**, tiny building blocks of matter, combine to form *molecules* such as water, sugar, and proteins, like those that make up our muscles. Molecules, in turn, associate in specific ways to form microscopic **cells**, the smallest units of all living things. (We will examine the *cellular level* in Chapter 3.) All cells have some common structures and functions, but individual cells vary widely in size, shape, and their particular roles in the body.

The simplest living creatures are composed of single cells, but in complex organisms such as trees or human beings, the structural ladder continues on to the *tissue level*. **Tissues** consist of groups of similar cells that have a common function. There are four basic tissue types, and each plays a definite but different role in the body. (We discuss tissues in Chapter 3.)

An **organ** is a structure composed of two or more tissue types that performs a specific function for the body. At the *organ level* of organization, extremely complex functions become possible. For example, the small intestine, which digests and absorbs food, is composed of all four tissue types. An **organ system** is a group of organs that work together to accomplish a common purpose. For example, the heart and blood vessels of the cardiovascular system circulate blood continuously to carry nutrients and oxygen to all body cells.

In all, 11 organ systems make up the living human being, or the **organism**, which represents the highest level of structural organization, the *organismal level*. The organismal level is the sum total of all structural levels working together to keep us alive. The major organs of each system

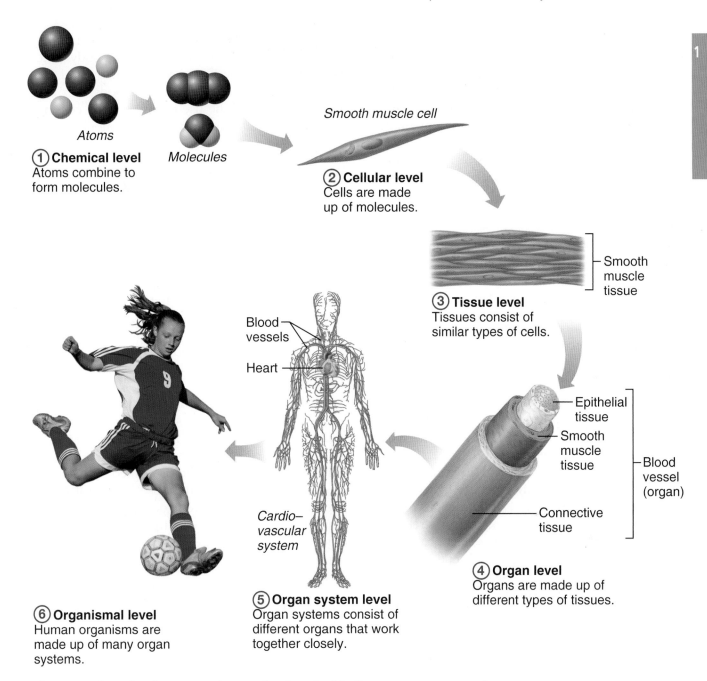

**Figure 1.1 Levels of structural organization.** In this diagram, components of the cardiovascular system are used to illustrate the levels of structural organization in a human being.

are shown in **Figure 1.2** on pp. 31–32. Refer to the figure as you read through the following descriptions of the organ systems.

## Organ System Overview

### Integumentary System

The **integumentary** (in-teg″u-men′tar-e) **system** is the external covering of the body, or the skin, including the hair and fingernails (Figure 1.2a). It waterproofs the body and cushions and protects the deeper tissues from injury. With the help of sunlight, it produces vitamin D. It also excretes salts in perspiration and helps regulate body temperature. Sensory receptors located in the skin alert us to what is happening at the body surface.

### Skeletal System

The **skeletal system** consists of bones, cartilages, and joints (Figure 1.2b). It supports the body and provides a framework that the skeletal muscles use to cause movement. It also has protective functions (for example, the skull encloses and protects the brain), and the cavities of the skeleton are the sites where blood cells are formed. The hard substance of bones acts as a storehouse for minerals.

## Muscular System

The muscles of the body have only one function—to *contract*, or shorten. When this happens, movement occurs. The mobility of the body as a whole reflects the activity of *skeletal muscles*, the large, fleshy muscles attached to bones (Figure 1.2c). When these contract, you are able to stand erect, walk, jump, grasp, throw a ball, or smile. The skeletal muscles form the **muscular system**. These muscles are distinct from the muscles of the heart and of other hollow organs, which move fluids (such as blood or urine) or other substances (such as food) along definite pathways within the body.

## Nervous System

The **nervous system** is the body's fast-acting control system. It consists of the brain, spinal cord, nerves, and sensory receptors (Figure 1.2d). The body must be able to respond to stimuli coming from outside the body (such as light, sound, or changes in temperature) and from inside the body (such as decreases in oxygen or stretching of tissue). The *sensory receptors* detect changes in temperature, pressure, or light, and send messages (via electrical signals called *nerve impulses*) to the central nervous system (brain and spinal cord) so that it is constantly informed about what is going on. The central nervous system then assesses this information and responds by activating the appropriate body *effectors* (muscles or glands, which are organs that produce secretions).

## Endocrine System

Like the nervous system, the **endocrine** (en'do-krin) **system** controls body activities, but it acts much more slowly. *Endocrine glands* produce chemical molecules called *hormones* and release them into the blood to travel to relatively distant target organs.

The endocrine glands include the pituitary, thyroid, parathyroids, adrenals, thymus, pancreas, pineal, ovaries (in the female), and testes (in the male) (Figure 1.2e). The endocrine glands are not connected anatomically in the same way that the parts of other organ systems are. What they have

in common is that they all secrete hormones, which regulate other structures. The body functions controlled by hormones are many and varied, involving every cell in the body. Growth, reproduction, and the use of nutrients by cells are all controlled (at least in part) by hormones.

## Cardiovascular System

The primary organs of the **cardiovascular system** are the heart and blood vessels (Figure 1.2f). Using blood as a carrier, the cardiovascular system delivers oxygen, nutrients, hormones, and other substances to, and picks up wastes such as carbon dioxide from, cells near sites of exchange. White blood cells and chemicals in the blood help to protect the body from such foreign invaders as bacteria, viruses, and tumor cells. The heart propels blood out of its chambers into blood vessels to be transported to all body tissues.

## Lymphatic System

The role of the **lymphatic system** complements that of the cardiovascular system. Its organs include lymphatic vessels, lymph nodes, and other lymphoid organs such as the spleen and tonsils (Figure 1.2g). When fluid is leaked into tissues from the blood, lymphatic vessels return it to the bloodstream so that there is enough blood to continuously circulate through the body. The lymph nodes and other lymphoid organs help to cleanse the blood and house white blood cells involved in immunity.

## Respiratory System

The job of the **respiratory system** is to keep the body supplied with oxygen and to remove carbon dioxide. The respiratory system consists of the nasal passages, pharynx, larynx, trachea, bronchi, and lungs (Figure 1.2h). Within the lungs are tiny air sacs. Gases are exchanged with the blood through the thin walls of these air sacs.

## Digestive System

The **digestive system** is basically a tube running through the body from mouth to anus. The organs of the digestive system include the oral cavity (mouth), esophagus, stomach, small and large intestines, and rectum plus a number of accessory organs (liver, salivary glands, pancreas, and others) (Figure 1.2i). Their role is to break down food and deliver the resulting nutrients to the blood for dispersal to body cells. The breakdown activities

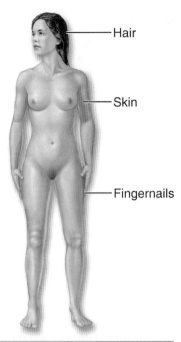

Hair

Skin

Fingernails

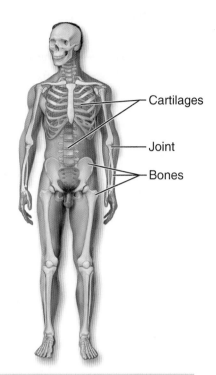

Cartilages

Joint

Bones

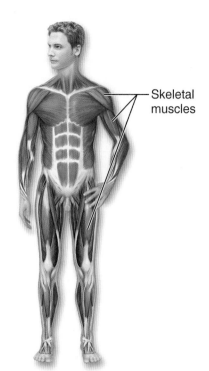

Skeletal muscles

**(a) Integumentary System**

Forms the external body covering; protects deeper tissue from injury; synthesizes vitamin D; location of sensory receptors (pain, pressure, etc.) and sweat and oil glands.

**(b) Skeletal System**

Protects and supports body organs; provides a framework the muscles use to cause movement; blood cells are formed within bones; stores minerals.

**(c) Muscular System**

Allows manipulation of the environment, locomotion, and facial expression; maintains posture; produces heat.

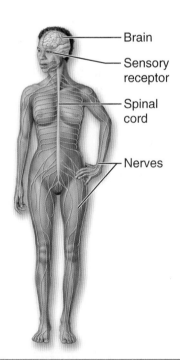

Brain

Sensory receptor

Spinal cord

Nerves

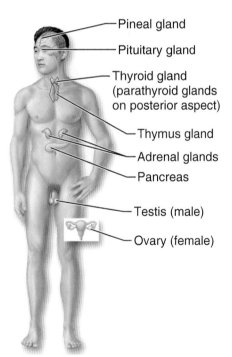

Pineal gland

Pituitary gland

Thyroid gland (parathyroid glands on posterior aspect)

Thymus gland

Adrenal glands

Pancreas

Testis (male)

Ovary (female)

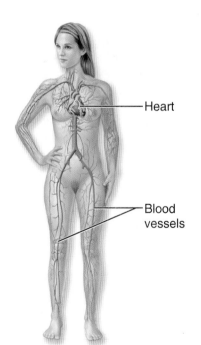

Heart

Blood vessels

**(d) Nervous System**

Fast-acting control system of the body; responds to internal and external changes by activating appropriate muscles and glands.

**(e) Endocrine System**

Glands secrete hormones that regulate processes such as growth, reproduction, and nutrient use by body cells.

**(f) Cardiovascular System**

Blood vessels transport blood, which carries oxygen, nutrients, hormones, carbon dioxide, wastes, etc.; the heart pumps blood.

**Figure 1.2 The body's organ systems.**

*(Figure continues on page 32.)*

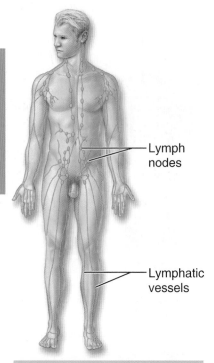

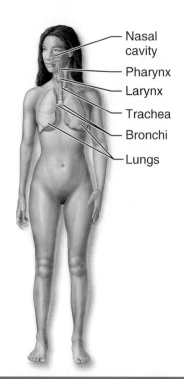

Nasal cavity

Pharynx

Larynx

Trachea

Bronchi

Lungs

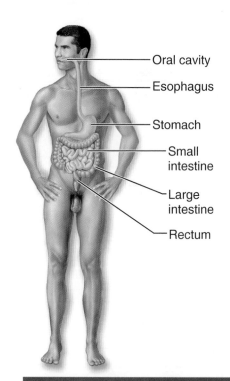

Oral cavity

Esophagus

Stomach

Small intestine

Large intestine

Rectum

Lymph nodes

Lymphatic vessels

**(g) Lymphatic System**

Picks up fluid leaked from blood vessels and returns it to blood; disposes of debris in the lymphatic stream; houses white blood cells involved in immunity.

**(h) Respiratory System**

Keeps blood constantly supplied with oxygen and removes carbon dioxide; the gaseous exchanges occur through the walls of the air sacs of the lungs.

**(i) Digestive System**

Breaks food down into absorbable nutrients that enter the blood for distribution to body cells; indigestible foodstuffs are eliminated as feces.

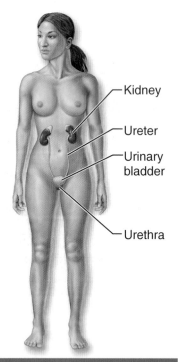

Kidney

Ureter

Urinary bladder

Urethra

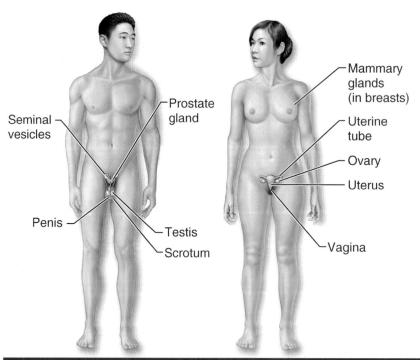

Seminal vesicles

Prostate gland

Penis

Testis

Scrotum

Mammary glands (in breasts)

Uterine tube

Ovary

Uterus

Vagina

**(j) Urinary System**

Eliminates nitrogen-containing wastes from the body; regulates water, electrolyte, and acid-base balance of the blood.

**(k) Male Reproductive System   (l) Female Reproductive System**

Overall function of the reproductive system is production of offspring. Testes produce sperm and male sex hormone; ducts and glands aid in delivery of viable sperm to the female reproductive tract. Ovaries produce eggs and female sex hormones; remaining structures serve as sites for fertilization and development of the fetus. Mammary glands of female breasts produce milk to nourish the newborn.

**Figure 1.2** *(continued)* **The body's organ systems.**

that begin in the mouth are completed in the small intestine. From that point on, the major function of the digestive system is to reabsorb water. The undigested food that remains in the tract leaves the body through the anus as feces. The liver is considered a digestive organ because the bile it produces helps to break down fats. The pancreas, which delivers digestive enzymes to the small intestine, has both endocrine and digestive functions.

## Urinary System

A normal part of healthy body function is the production of waste by-products, which must be disposed of. One type of waste contains nitrogen (examples are urea and uric acid), which results when the body cells break down proteins and nucleic acids, which are genetic information molecules. The **urinary system** removes the nitrogen-containing wastes from the blood and flushes them from the body in *urine*. This system, often called the *excretory system*, is composed of the kidneys, ureters, bladder, and urethra (Figure 1.2j). Other important functions of this system include maintaining the body's water and salt (electrolyte) balance, regulating the acid-base balance of the blood, and helping to regulate normal blood pressure.

## Reproductive System

The role of the **reproductive system** is to produce offspring. The male testes produce sperm. Other male reproductive system structures are the scrotum, penis, accessory glands, and the duct system, which carries sperm to the outside of the body (Figure 1.2k). The female ovaries produce eggs, or ova; the female duct system consists of the uterine tubes, uterus, and vagina (Figure 1.2l). The uterus provides the site for the development of the fetus (immature infant) once fertilization has occurred.

## Did You Get It?

3. At which level of structural organization is the stomach? At which level is a glucose molecule?
4. Which organ system includes the trachea, lungs, nasal cavity, and bronchi?
5. Which system functions to remove wastes and help regulate blood pressure?

**For answers, see Appendix A.**

# Maintaining Life

## → Learning Objectives

☐ List eight functions that humans must perform to maintain life.

☐ List the five survival needs of the human body.

## Necessary Life Functions

Now that we have introduced the structural levels composing the human body, a question naturally follows: What does this highly organized human body do? Like all complex animals, human beings maintain their boundaries, move, respond to environmental changes, take in and digest nutrients, carry out metabolism, dispose of wastes, reproduce themselves, and grow.

Organ systems do not work in isolation; instead, they work together to promote the well-being of the entire body **(Figure 1.3)**. Because this theme is emphasized throughout this text, it is worthwhile to identify the most important organ systems contributing to each of the necessary life functions. Also, as you study this figure, you may want to refer to the more detailed descriptions of the organ systems just provided (pp. 29–33 and in Figure 1.2).

### Maintaining Boundaries

Every living organism must be able to maintain its boundaries so that its "inside" remains distinct from its "outside." Every cell of the human body is surrounded by an external membrane that separates its contents from the outside interstitial fluid (fluid between cells) and allows entry of needed substances while generally preventing entry of potentially damaging or unnecessary substances. The body as a whole is also enclosed by the integumentary system, or skin. The integumentary system protects internal organs from drying out (which would be fatal), from pathogens, and from the damaging effects of heat, sunlight, and an unbelievable number of chemical substances in the external environment.

### Movement

**Movement** includes all the activities promoted by the muscular system, such as propelling ourselves from one place to another (by walking, swimming, and so forth) and manipulating the external environment with our fingers. The skeletal system provides the bones that the muscles pull on as they work. Movement also occurs when substances such as blood, foodstuffs, and urine

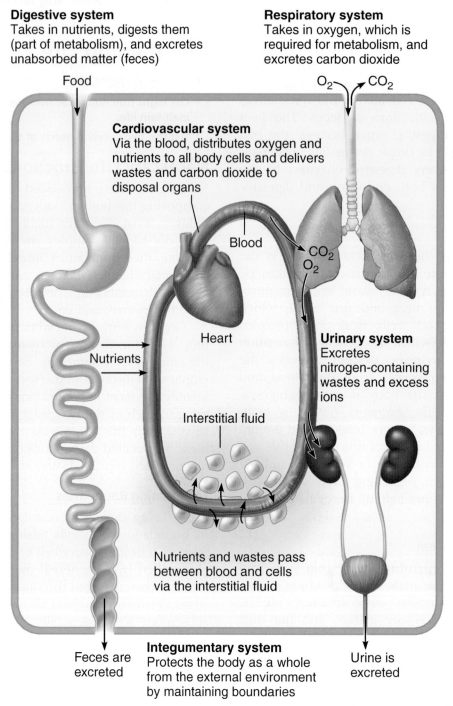

**Digestive system**
Takes in nutrients, digests them (part of metabolism), and excretes unabsorbed matter (feces)

**Respiratory system**
Takes in oxygen, which is required for metabolism, and excretes carbon dioxide

Food

$O_2$   $CO_2$

**Cardiovascular system**
Via the blood, distributes oxygen and nutrients to all body cells and delivers wastes and carbon dioxide to disposal organs

Blood

$CO_2$
$O_2$

Heart

Nutrients

**Urinary system**
Excretes nitrogen-containing wastes and excess ions

Interstitial fluid

Nutrients and wastes pass between blood and cells via the interstitial fluid

Feces are excreted

**Integumentary system**
Protects the body as a whole from the external environment by maintaining boundaries

Urine is excreted

**Figure 1.3 Examples of interrelationships among organ systems that illustrate life functions.**

are propelled through the internal organs of the cardiovascular, digestive, and urinary systems, respectively.

### Responsiveness

**Responsiveness**, or **irritability**, is the ability to sense changes (stimuli) in the environment and then to react to them. For example, if you accidentally touch a hot pan, you involuntarily pull your hand away from the painful stimulus (the pan). You do not need to think about it—it just happens! Likewise, when the amount of carbon dioxide in your blood rises to a dangerously high level, your breathing rate speeds up to blow off the excess carbon dioxide.

Because nerve cells are highly irritable and can communicate rapidly with each other via electrical impulses, the nervous system bears the major responsibility for responsiveness. However, all body cells are responsive to some extent.

## Digestion

**Digestion** is the process of breaking down ingested food into simple molecules that can then be absorbed into the blood. The nutrient-rich blood is then distributed to all body cells by the cardiovascular system, where body cells use these simple molecules for energy and raw materials.

## Metabolism

**Metabolism** is a broad term that refers to all chemical reactions that occur within the body and all of its cells. It includes breaking down complex substances into simpler building blocks (as in digestion), making larger structures from smaller ones, and using nutrients and oxygen to produce molecules of *adenosine triphosphate (ATP)*, the energy-rich molecules that power cellular activities. Metabolism depends on the digestive and respiratory systems to make nutrients and oxygen available to the blood and on the cardiovascular system to distribute these needed substances throughout the body. Metabolism is regulated chiefly by hormones secreted by the glands of the endocrine system.

## Excretion

**Excretion** is the process of removing *excreta* (ek-skre′tah), or wastes, from the body. Several organ systems participate in excretion. For example, the digestive system rids the body of indigestible food residues in feces, the urinary system disposes of nitrogen-containing metabolic wastes in urine, and the skin disposes of various waste products as components of sweat.

## Reproduction

**Reproduction**, the production of offspring, can occur on the cellular or organismal level. In cellular reproduction, the original cell divides, producing two identical daughter cells that may then be used for body growth or repair. Reproduction of the human organism is the task of the organs of the reproductive system, which produce sperm and eggs. When a sperm unites with an egg, a fertilized egg forms, which then develops into a baby within the mother's body. The function of the reproductive system is regulated very precisely by hormones of the endocrine system.

## Growth

*Growth* can be an increase in cell size or an increase in body size that is usually accomplished by an increase in the number of cells. For growth to occur, cell-constructing activities must occur at a faster rate than cell-destroying ones. Hormones released by the endocrine system play a major role in directing growth.

## Survival Needs

The goal of nearly all body systems is to maintain life. However, life is extraordinarily fragile and requires that several factors be available. These factors, which we will call *survival needs*, include nutrients (food), oxygen, water, and appropriate temperature and atmospheric pressure.

**Nutrients**, which the body takes in through food, contain the chemicals used for energy and cell building. *Carbohydrates* are the major energy-providing fuel for body cells. *Proteins* and, to a lesser extent, *fats* are essential for building cell structures. Fats also cushion body organs and provide reserve fuel. *Minerals* and *vitamins* are required for the chemical reactions that go on in cells and for oxygen transport in the blood.

All the nutrients in the world are useless unless **oxygen** is also available. Because the chemical reactions that release energy from foods require oxygen, human cells can survive for only a few minutes without it. It is made available to the blood and body cells by the cooperative efforts of the respiratory and cardiovascular systems.

*Water* accounts for 60 to 80 percent of body weight, depending on the age of the individual. It is the single most abundant chemical substance in the body and provides the fluid base for body secretions and excretions. We obtain water chiefly from ingested foods or liquids, and we lose it by evaporation from the lungs and skin and in body excretions.

If chemical reactions are to continue at life-sustaining levels, **normal body temperature** must be maintained. If body temperature drops below 37°C (98.6°F), metabolic reactions become slower and slower and finally stop. If body temperature is too high, chemical reactions proceed too rapidly, and body proteins begin to break down. At either extreme, death occurs. Most body heat is generated by the activity of the skeletal muscles and dissipated via blood circulating close to the skin surface or by the evaporation of sweat.

The force exerted on the surface of the body by the weight of air is referred to as **atmospheric pressure**. Breathing and the exchange of oxygen and carbon dioxide in the lungs depend on

*(Text continues on page 38.)*

# Medical Imaging: Illuminating the Body

Imaging procedures are important diagnostic tools that are either minimally invasive or not invasive at all. By bombarding the body with different forms of energy, medical imaging techniques can reveal the structure of internal organs, show blood flow in real time, and even determine the density of bone.

Until about 50 years ago, the magical but murky X ray was the only means of peering into a living body without performing surgery. The X-ray film is a shadowy negative image of internal structures produced by directing electromagnetic waves of very short wavelength at the body. X rays are best used to visualize hard, bony structures and locate abnormally dense structures (tumors, tuberculosis nodules) in the lungs and breasts.

Two examples of low-dose X-ray procedures are mammography and bone densitometry. A **mammogram** is used to identify changes in breast tissue, including dense masses or calcifications. A mammogram is performed by compressing the breast in a special X-ray machine because thinner tissue results in a better image; see photo (a). **Bone densitometry** detects the amount of calcium and minerals stored in bone and is the major diagnostic test for osteoporosis; see photo (b).

The 1950s saw the birth of **ultrasound imaging (ultrasonography)**, which has some distinct advantages over other imaging techniques. The equipment employs high-frequency sound waves (ultrasound) as its energy source. Unlike X rays, ultrasound has no known harmful effects on living tissues. The body is probed with pulses of sound waves, which cause echoes when reflected and scattered by body tissues. The echoes are analyzed by computer to construct visual images of body organs, much like sonar is used to map the ocean floor. Because of its safety, ultrasound is the imaging technique of choice for determining fetal age and position and locating the placenta; see photo (c). Because sound waves have very low penetrating power and are rapidly scattered in air, sonography does not visualize air-filled structures (the lungs) or those surrounded by bone (the brain and spinal cord) well.

Though nuclear medicine, which uses radioisotopes to scan the body, was developed in the 1950s, it was not until the 1970s that a powerful technique for visualizing metabolic activities, PET, was introduced. The 1970s also introduced CT, an X-ray–based scanning technique, and MRI scanning techniques, which use powerful magnets.

**Positron emission tomography (PET)** requires an injection of short-lived radioisotopes that have been tagged to biological molecules (such as glucose) in order to view metabolic processes. The patient is positioned in the PET scanner, and as the radioisotopes are absorbed by the most active brain cells, high-energy gamma rays are produced. A computer analyzes the gamma emissions and produces a picture of the brain's biochemical activity in vivid colors. PET's greatest clinical value has been its ability to provide insights into brain activity in people affected by mental illness, Alzheimer's disease, and epilepsy.

Currently PET can reveal signs of trouble in people with undiagnosed Alzheimer's disease (AD) because regions of beta-amyloid accumulation (a defining characteristic of AD) show up in brilliant red and yellow when that molecule is tagged. By tagging glucose, we can observe that brain tissue in areas of impairment or areas with Alzheimer's plaques use far less glucose compared to normal brain tissue as shown in photo (d).

Perhaps the best known of these newer imaging devices is **computed tomography (CT)**, a refined version of X ray that eliminates the confusion resulting from images of overlapping structures. A CT scanner takes "pictures" of a thin slice of the body, about as thick as a dime. Different tissues absorb the radiation in varying amounts. The device's computer translates this information into a detailed, cross-sectional picture of the body region scanned; see photo (e). CT scans are at the forefront in evaluating most problems that affect the brain and abdomen, and their clarity has all but eliminated exploratory surgery. Special ultrafast CT scanners have produced a technique called **dynamic spatial reconstruction (DSR)**, which provides three-dimensional images of body organs from any angle. It also allows organ movements and changes in internal volumes to be observed at normal speed, in slow motion, and at a specific moment in time. The greatest value of DSR has been to visualize the heart beating and blood flowing through blood vessels. This allows medical personnel to assess heart defects, constricted blood vessels, and the status of coronary bypass grafts.

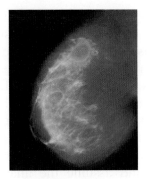

**(a) Mammogram showing breast cancer.**

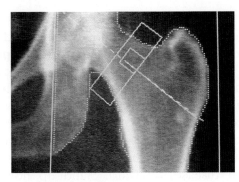

**(b) Bone densitometry scan showing osteoporosis in the small yellow box on the neck of the femur.**

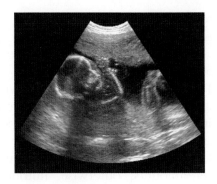

**(c) Sonogram of a fetus.**

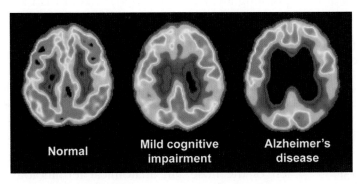

Normal

Mild cognitive impairment

Alzheimer's disease

**(d) A PET scan demonstrates glucose metabolism in a normal brain (*left*). Glucose metabolism is decreased in a brain with mild cognitive impairment (*middle*), and glucose is metabolized at an even lower level in a brain with Alzheimer's plaques (*right*).**

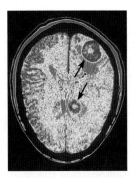

**(e) CT scan showing brain tumors (indicated by black arrows).**

A different technique is **magnetic resonance imaging (MRI)**, which uses magnetic fields up to 60,000 times stronger than Earth's to pry information from body tissues. The patient lies in a tubelike chamber within a huge magnet. Hydrogen molecules spin like tops in the magnetic field, and their energy is enhanced by the presence of radio waves. When the radio waves are turned off, energy is released and translated by a computer into a visual image (see Figure 1.5, p. 42). MRI is immensely popular because it can do many things a CT scan cannot. Dense structures do not show up in MRI, so bones of the skull and/or vertebral column do not impair the view of *soft tissues*, such as the brain or intervertebral discs, the cartilage pads between vertebrae. MRI is also particularly good at detecting signs of degenerative disease such as multiple sclerosis plaques, which do not show up well in CT scans but are dazzlingly clear in MRI scans.

A variation of MRI called **functional magnetic resonance imaging (fMRI)** allows tracking of blood flow into the brain in real time. Before 1992, PET was the only way to match brain activity to disease. With no need for injections of radioisotopes, fMRI provides a less invasive alternative. Despite its advantages, the powerful magnets of the MRI can pose risks. For example, metal objects, such as implanted pacemakers and loose tooth fillings, can be "sucked" through the body. Also, the long-term health risks of exposure to strong magnetic fields are not clear.

As you can see, modern medical science enlists remarkable diagnostic tools. CT and PET scans account for about 25 percent of all imaging. Ultrasonography, because of its safety and low cost, is the most widespread. However, conventional X rays remain the workhorse of diagnostic imaging and still account for more than half of all medical imaging.

appropriate atmospheric pressure. At high altitudes, where the air is thin and atmospheric pressure is lower, gas exchange may be too slow to support cellular metabolism.

The mere presence of these survival factors is not sufficient to maintain life. They must be present in appropriate amounts as well; excesses and deficits may be equally harmful. For example, the food ingested must be of high quality and in proper amounts; otherwise, nutritional disease, obesity, or starvation is likely.

### Did You Get It?

6. In addition to being able to metabolize, grow, digest food, and excrete wastes, what other functions must an organism perform if it is to survive?
7. Oxygen is a survival need. Why is it so important? In which life function does oxygen participate directly?

For answers, see Appendix A.

# The Language of Anatomy

→ **Learning Objectives**

☐ **Verbally describe or demonstrate the anatomical position.**

☐ **Use proper anatomical terminology to describe body directions, surfaces, and body planes.**

☐ **Locate the major body cavities, and list the chief organs in each cavity.**

Learning about the body is exciting, but it can get harder to maintain our interest when we are faced with all the new terminology of anatomy and physiology. Let's face it: You can't just pick up an anatomy and physiology book and read it as if it were a novel. Unfortunately, confusion is inevitable without specialized terminology. For example, if you are looking at a ball, "above" always means the area over the top of the ball. Other directional terms can also be used consistently because the ball is a sphere. All sides and surfaces are equal. The human body, of course, has many protrusions and bends. Thus, the question becomes: Above what? To prevent misunderstanding, anatomists use a set of terms that allow body structures to be located and identified clearly with just a few words. We present and explain this language of anatomy next.

## Anatomical Position

To accurately describe body parts and position, we must have an initial reference point and use directional terms. To avoid confusion, we always assume that the body is in a standard position called **anatomical position**. It is important to understand this position because most body terminology used in this text refers to this body positioning *regardless* of the position the body happens to be in. In the anatomical position, the body is erect with the feet parallel and the arms hanging at the sides with the palms facing forward. Stand up now, and assume the anatomical position. Notice that it is similar to "standing at attention" but is less comfortable because the palms are held unnaturally forward (with thumbs pointing away from the body) rather than hanging cupped toward the thighs.

## Directional Terms

**Directional terms** allow medical personnel and anatomists to explain exactly where one body structure is in relation to another. For example, we can describe the relationship between the ears and the nose informally by saying, "The ears are located on each side of the head to the right and left of the nose." In anatomical terminology, this condenses to, "The ears are lateral to the nose." Using anatomical terminology saves words and, once learned, is much clearer. Commonly used directional terms are defined and illustrated in **Table 1.1**. Although most of these terms are also used in everyday conversation, keep in mind that their anatomical meanings are very precise.

Before continuing, check your understanding of what you have read in the table. Give the relationship between the following body parts using the correct anatomical terms.

The wrist is _____ to the hand.

The breastbone is _____ to the spine.

The brain is _____ to the spinal cord.

The thumb is _____ to the fingers. (Be careful here. Remember the anatomical position.)

## Regional Terms

There are many visible landmarks on the surface of the body. Once you know their proper anatomical names, you can be specific in referring to different regions of the body.

### Anterior Body Landmarks

Look at **Figure 1.4a** to find the following body regions. Once you have identified all the anterior

## Table **1.1** Orientation and Directional Terms

| Term | Definition | Illustration | Example |
|---|---|---|---|
| Superior (cranial or cephalic) | Toward the head end or upper part of a structure or the body; above | | The forehead is superior to the nose. |
| Inferior (caudal)* | Away from the head end or toward the lower part of a structure or the body; below | | The navel is inferior to the breastbone. |
| Anterior (ventral)† | Toward or at the front of the body; in front of | | The breastbone is anterior to the spine. |
| Posterior (dorsal)† | Toward or at the backside of the body; behind | | The heart is posterior to the breastbone. |
| Medial | Toward or at the midline of the body; on the inner side of | | The heart is medial to the arm. |
| Lateral | Away from the midline of the body; on the outer side of | | The arms are lateral to the chest. |
| Intermediate | Between a more medial and a more lateral structure | | The collarbone is intermediate between the breastbone and the shoulder. |
| Proximal | Close to the origin of the body part or the point of attachment of a limb to the body trunk | | The elbow is proximal to the wrist (meaning that the elbow is closer to the shoulder or attachment point of the arm than the wrist is). |
| Distal | Farther from the origin of a body part or the point of attachment of a limb to the body trunk | | The knee is distal to the thigh. |
| Superficial (external) | Toward or at the body surface | | The skin is superficial to the skeleton. |
| Deep (internal) | Away from the body surface; more internal | | The lungs are deep to the rib cage. |

*The term caudal, literally "toward the tail," is synonymous with *inferior* only to the inferior end of the spine.

†*Anterior* and *ventral* are synonymous in humans, but not in four-legged animals. *Ventral* refers to an animal's "belly," making it the inferior surface. Likewise, posterior and dorsal surfaces are the same in humans, but *dorsal* refers to an animal's back, making it the superior surface.

**Q:** *Study this figure for a moment to answer these two questions. Where would you hurt if you (1) pulled a groin muscle or (2) cracked a bone in your olecranal area?*

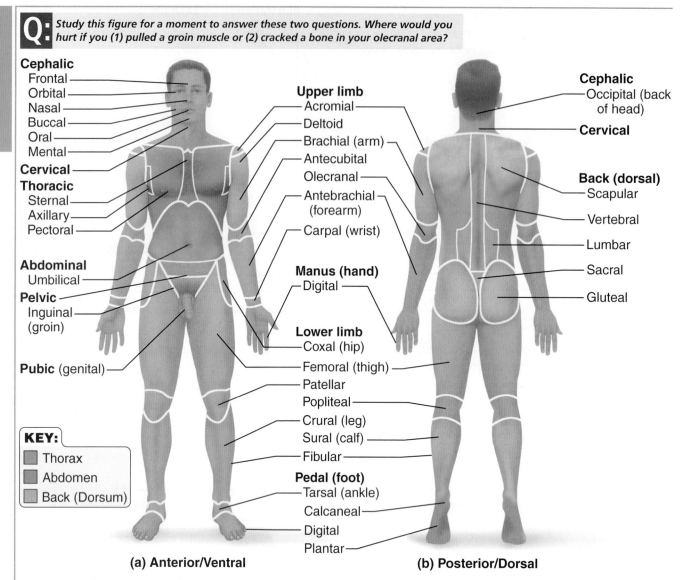

**Cephalic**
- Frontal
- Orbital
- Nasal
- Buccal
- Oral
- Mental

**Cervical**

**Thoracic**
- Sternal
- Axillary
- Pectoral

**Abdominal**
- Umbilical

**Pelvic**
- Inguinal (groin)

**Pubic** (genital)

**KEY:**
- ☐ Thorax
- ☐ Abdomen
- ☐ Back (Dorsum)

**Upper limb**
- Acromial
- Deltoid
- Brachial (arm)
- Antecubital
- Olecranal
- Antebrachial (forearm)
- Carpal (wrist)

**Manus (hand)**
- Digital

**Lower limb**
- Coxal (hip)
- Femoral (thigh)
- Patellar
- Popliteal
- Crural (leg)
- Sural (calf)
- Fibular

**Pedal (foot)**
- Tarsal (ankle)
- Calcaneal
- Digital
- Plantar

**(a) Anterior/Ventral**

**Cephalic**
- Occipital (back of head)

**Cervical**

**Back (dorsal)**
- Scapular
- Vertebral
- Lumbar
- Sacral
- Gluteal

**(b) Posterior/Dorsal**

**Figure 1.4 The anatomical position and regional terms.** This figure shows terms used to designate specific body areas. **(a)** Anterior view. **(b)** Posterior view. The heels are raised slightly to show the inferior plantar surface (sole) of the foot, which is actually on the inferior surface of the body.

body landmarks, cover the labels that describe what the structures are. Then go through the list again, pointing out these areas on your own body.

- **abdominal** (ab-dom′ĭ-nal): anterior body trunk inferior to ribs
- **acromial** (ah-kro′me-ul): point of shoulder
- **antebrachial** (an″te-bra′ke-al): forearm
- **antecubital** (an″te-ku′bĭ-tal): anterior surface of elbow

- **axillary** (ak′sĭ-lar″e): armpit
- **brachial** (bra′ke-al): arm
- **buccal** (buk′al): cheek area
- **carpal** (kar′pal): wrist
- **cervical** (ser′vĭ-kal): neck region
- **coxal** (kok′sal): hip
- **crural** (kroo′ral): anterior leg; the shin.
- **deltoid** (del′toyd): curve of shoulder formed by large deltoid muscle
- **digital** (dij′ĭ-tal): fingers, toes
- **femoral** (fem′or-al): thigh (applies to both anterior and posterior)

**A:** ʇuoɹʇɐℸ pʋ₃

(1) Your inguinal region. (2) Your posterior elbow region.

- **fibular** (fib′u-lar): lateral part of leg
- **frontal** (frun′tal): forehead
- **inguinal** (in′gwĭ-nal): area where thigh meets body trunk; groin
- **mental** (men′tul): chin
- **nasal** (na′zul): nose area
- **oral** (o′ral): mouth
- **orbital** (or′bĭ-tal): eye area
- **patellar** (pah-tel′er): anterior knee
- **pectoral** (pek′to-ral): relating to, or occurring in or on, the chest
- **pelvic** (pel′vik): area overlying the pelvis anteriorly
- **pubic** (pyu′bik): genital region
- **sternal** (ster′nul): breastbone area
- **tarsal** (tar′sal): ankle region
- **thoracic** (tho-ras′ik): area between the neck and abdomen, supported by the ribs, sternum and costal cartilages; chest
- **umbilical** (um-bil′ĭ-kal): navel

### Posterior Body Landmarks

Identify the following body regions in Figure 1.4b, and then locate them on yourself without referring to this text.

- **calcaneal** (kal-ka′ne-ul): heel of foot
- **cephalic** (seh-fă′lik): head
- **femoral** (fem′or-al): thigh
- **gluteal** (gloo′te-al): buttock
- **lumbar** (lum′bar): area of back between ribs and hips; the loin
- **occipital** (ok-sip′ĭ-tal): posterior surface of head or base of skull
- **olecranal** (ol-eh-kra′nel): posterior surface of elbow
- **popliteal** (pop-lit′e-al): posterior knee area
- **sacral** (sa′krul): area between hips at base of spine
- **scapular** (skap′u-lar): shoulder blade region
- **sural** (soo′ral): the posterior surface of leg; the calf
- **vertebral** (ver′tĕ-bral): area of spinal column

The **plantar** region, or the sole of the foot, actually on the inferior body surface, is illustrated with the posterior body landmarks (see Figure 1.4b).

### Did You Get It?

8. What is the anatomical position, and why is it important that an anatomy student understand it?
9. The axillary and the acromial areas are both in the general area of the shoulder. To what specific body area does each of these terms apply?
10. Use anatomical language to describe the location of a cut to the back of your left forearm.

**For answers, see Appendix A.**

## Body Planes and Sections

When preparing to look at the internal structures of the body, medical students make a **section**, or cut. When the section is made through the body wall or through an organ, it is made along an imaginary line called a **plane**. Because the body is three-dimensional, we can refer to three types of planes or sections that lie at right angles to one another **(Figure 1.5)**.

A **sagittal** (saj′ĭ-tal) **section** is a cut along the lengthwise, or longitudinal, plane of the body, dividing the body into right and left parts. If the cut is down the median plane of the body and the right and left parts are equal in size, it is called a **median (midsagittal) section**. All other sagittal sections are parasagittal sections (*para* = near).

A **frontal section** is a cut along a lengthwise plane that divides the body (or an organ) into anterior and posterior parts. It is also called a **coronal** (ko-ro′nal, "crown") **section**.

A **transverse section** is a cut along a horizontal plane, dividing the body or organ into superior and inferior parts. It is also called a **cross section**.

Sectioning a body or one of its organs along different planes often results in very different views. For example, a transverse section of the body trunk at the level of the kidneys would show kidney structure in cross section very nicely; a frontal section of the body trunk would show a different view of kidney anatomy; and a midsagittal section would miss the kidneys completely. Information on body organ positioning can be gained by taking magnetic resonance imaging (MRI) scans along different body planes (see Figure 1.5). MRI scans are described further in "A Closer Look" (pp. 36–37).

## Body Cavities

Anatomy and physiology textbooks typically describe two sets of internal body cavities, called the *dorsal* and *ventral body cavities*, that provide different degrees of protection to the organs

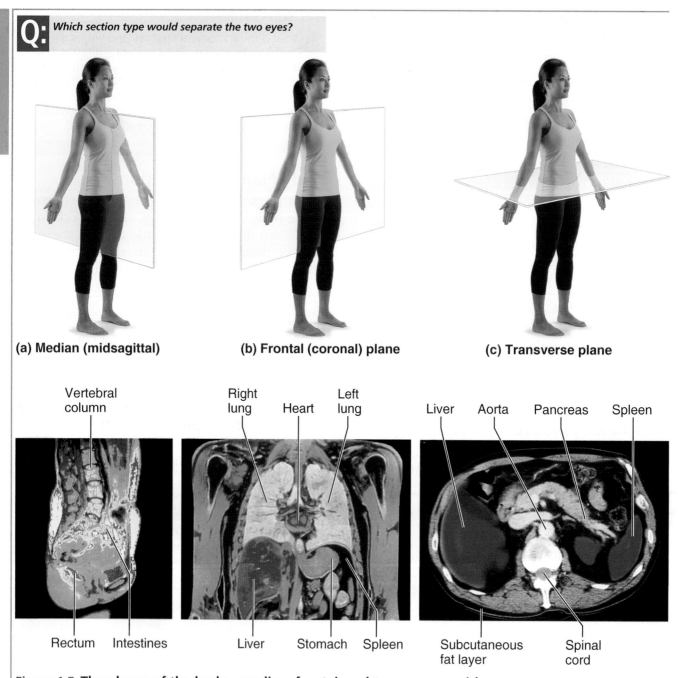

**(a) Median (midsagittal)**   **(b) Frontal (coronal) plane**   **(c) Transverse plane**

Vertebral column

Right lung   Heart   Left lung

Liver   Aorta   Pancreas   Spleen

Rectum   Intestines

Liver   Stomach   Spleen

Subcutaneous fat layer   Spinal cord

**Figure 1.5 The planes of the body—median, frontal, and transverse—with corresponding MRI scans.** Note that the planes are shown on a body in the anatomical position.

within them **(Figure 1.6)**. Because these cavities differ in their mode of embryological development and in their lining membranes, many anatomy reference books do not identify the dorsal, or neural, body cavity as an internal body cavity. However, the idea of two major sets of internal body cavities is a useful learning concept, so we will continue to use it here.

### Dorsal Body Cavity

The **dorsal body cavity** has two subdivisions, which are continuous with each other. The **cranial**

**cavity** is the space inside the bony skull. The brain is well protected because it occupies the cranial cavity. The **spinal cavity** extends from the cranial cavity to the end of the spinal cord. The spinal cord, which is a continuation of the brain, is protected by the bony vertebrae, which surround the spinal cavity and form the spine.

### Ventral Body Cavity

The **ventral body cavity** is much larger than the dorsal cavity. It contains all the structures within the chest and abdomen, that is, the visceral organs in those regions. Like the dorsal cavity, the ventral body cavity is subdivided. The superior **thoracic cavity** is separated from the rest of the ventral cavity by a dome-shaped muscle, the **diaphragm** (di′ah-fram). The organs in the thoracic cavity (lungs, heart, and others) are protected by the rib cage. A central region called the **mediastinum** (me″de-as-ti′num) separates the lungs into right and left cavities in the thoracic cavity. The mediastinum itself houses the heart, trachea, and several other visceral organs.

The cavity inferior to the diaphragm is the **abdominopelvic** (ab-dom″ĭ-no-pel′vik) **cavity**. Some prefer to subdivide it into a superior **abdominal cavity** containing the stomach, liver, intestines, and other organs, and an inferior **pelvic cavity** containing the reproductive organs, bladder, and rectum. However, there is no actual physical structure dividing the abdominopelvic cavity. The pelvic cavity is not immediately inferior to the abdominal cavity, but rather tips away from the abdominal cavity in the posterior direction (see Figure 1.6).

When the body is subjected to physical trauma (as often happens in an automobile accident, for example), the most vulnerable abdominopelvic organs are those within the abdominal cavity. The reason is that the abdominal cavity walls are formed only of trunk muscles and are not reinforced by bone. The pelvic organs receive some protection from the bony pelvis in which they reside.

Because the abdominopelvic cavity is quite large and contains many organs, it helps to divide it up into smaller areas for study. A scheme commonly used by medical personnel divides the abdomino-pelvic cavity into four more or less equal regions called *quadrants*. The quadrants are named according to their relative locations with respect to anatomical position—that is, right upper quadrant (RUQ), right lower quadrant (RLQ), left upper quadrant (LUQ), and left lower quadrant (LLQ) **(Figure 1.7)**.

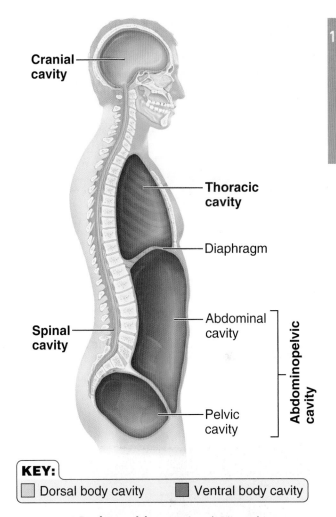

**Figure 1.6 Body cavities.** Notice the angular relationship between the abdominal and pelvic cavities.

Another system, used mainly by anatomists, divides the abdominopelvic cavity into nine separate *regions* by four planes **(Figure 1.8a)**. As you locate these regions, notice the organs they contain (Figure 1.8b).

- The **umbilical region** is the centermost region, deep to and surrounding the umbilicus (navel).
- The **epigastric** (ep″ĭ-gas′trik) **region** is located superior to the umbilical region (*epi* = upon, above; *gastric* = stomach).
- The **hypogastric (pubic) region** is inferior to the umbilical region (*hypo* = below).
- The **right iliac (inguinal) region** and **left iliac (inguinal) region** are lateral to the hypogastric region (*iliac* = superior part of the hip bone).

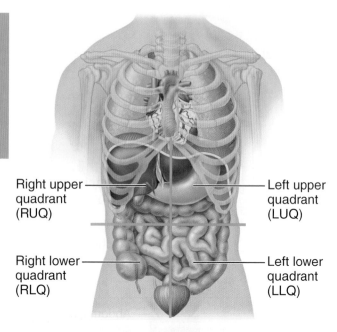

Right upper quadrant (RUQ)

Left upper quadrant (LUQ)

Right lower quadrant (RLQ)

Left lower quadrant (LLQ)

**Figure 1.7 The four abdominopelvic quadrants.** In this scheme used by medical personnel, the abdominopelvic cavity is divided into four quadrants by two planes.

- The **right lumbar region** and **left lumbar region** lie lateral to the umbilical region (*lumbus* = loins) and spinal column between the bottom ribs and the hip bones; see Figure 1.4b).

- The **right hypochondriac** (hi″po-kon′dre-ak) **region** and **left hypochondriac region** flank the epigastric region and contain the lower ribs (*chondro* = cartilage).

## Other Body Cavities

In addition to the large closed body cavities, there are several smaller body cavities. Most are in the head and open to the body exterior.

- **Oral cavity and digestive cavity.** The oral cavity, or the mouth, contains the teeth and tongue. This cavity is part of and continuous with the digestive organs, which open to the exterior at the anus.

- **Nasal cavity.** Located within and posterior to the nose, the nasal cavity is part of the respiratory system.

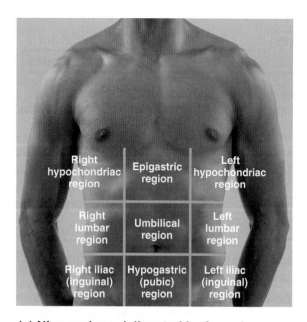

(a) Nine regions delineated by four planes

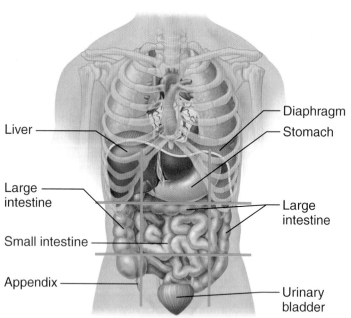

(b) Anterior view of the nine regions showing the superficial organs

**Figure 1.8 The nine abdominopelvic regions.** The abdominopelvic cavity is divided into nine regions by four planes in this scheme used mostly by anatomists. In **(a)**, the superior transverse plane is just superior to the ribs; the inferior transverse plane is just superior to the hip bones; and the parasagittal planes lie just medial to the nipples.

- **Orbital cavities.** The orbital cavities (orbits) in the skull house the eyes and present them in an anterior position.
- **Middle ear cavities.** The middle ear cavities carved into the skull lie just medial to the eardrums. These cavities contain tiny bones that transmit sound vibrations to the hearing receptors in the inner ears.

## Did You Get It?

11. If you were dissecting a cadaver and wanted to separate the thoracic cavity from the abdominal cavity, which type of section would you make?
12. Which of the following organs are in the abdominopelvic cavity: spinal cord, small intestine, uterus, heart?
13. Joe went to the emergency room, where he complained of severe pains in the lower right quadrant of his abdomen. What might be his problem?

**For answers, see Appendix A.**

# Homeostasis

→ **Learning Objectives**

☐ Define *homeostasis*, and explain its importance.

☐ Define *negative feedback*, and describe its role in maintaining homeostasis and normal body function.

When you really think about the fact that your body contains trillions of cells in nearly constant activity, and that remarkably little usually goes wrong with it, you begin to appreciate what a marvelous organism your body really is. The word **homeostasis** (ho″me-o-sta′sis) describes the body's ability to maintain relatively stable internal conditions even though the outside world is continuously changing. Although the literal translation of *homeostasis* is "unchanging" (*homeo* = the same; *stasis* = standing still), the term does not really mean an unchanging state. Instead, it indicates a *dynamic* state of equilibrium, or a balance in which internal conditions change and vary but always within relatively narrow limits.

In general, the body demonstrates homeostasis when its needs are being adequately met and it is functioning smoothly. Virtually every organ system plays a role in maintaining the constancy of the internal environment. Adequate blood levels of vital nutrients must be continuously present, and heart activity and blood pressure must be constantly monitored and adjusted so that the blood is propelled with adequate force to reach all body tissues. Additionally, wastes must not be allowed to accumulate, and body temperature must be precisely controlled.

Communication within the body is essential for homeostasis and is accomplished chiefly by the nervous and endocrine systems, which use electrical signals delivered by nerves or bloodborne hormones, respectively, as information carriers. The details of how these two regulating systems operate are the subjects of later chapters, but we explain the basic characteristics of the neural and hormonal control systems that promote homeostasis here.

## Components of Homeostatic Control Systems

Regardless of the factor or event being regulated (this is called the *variable*), all homeostatic control mechanisms have at least three components: a *receptor*, *control center*, and *effector* **(Figure 1.9)**. The **receptor** is a type of sensor that monitors and responds to changes in the environment. It responds to such changes, called *stimuli*, by sending information (input) to the second component, the control center. Information flows from the receptor to the control center along the *afferent pathway*. (It may help to remember that information traveling along the *afferent* pathway *approaches* the control center.)

The **control center** determines the level (set point) at which a variable is to be maintained. This component analyzes the information it receives and then determines the appropriate response or course of action.

The third component, the **effector**, provides the means for the control center's response (output) to the stimulus. Information flows from the control center to the effector along the *efferent pathway*. (*Efferent* information *exits* from the control center.) The results of the response then *feed back* to influence the stimulus (the original change), either by reducing the amount of change (negative feedback), so that the whole control mechanism is shut off; or by increasing the amount of change (positive feedback), so that the reaction continues at an even faster rate.

## Feedback Mechanisms

Most homeostatic control mechanisms are **negative feedback** mechanisms, as indicated in Figure 1.9. In such systems, the net effect of the response to the stimulus is to either shut off the original stimulus or reduce its intensity. A good example of a nonbiological negative feedback

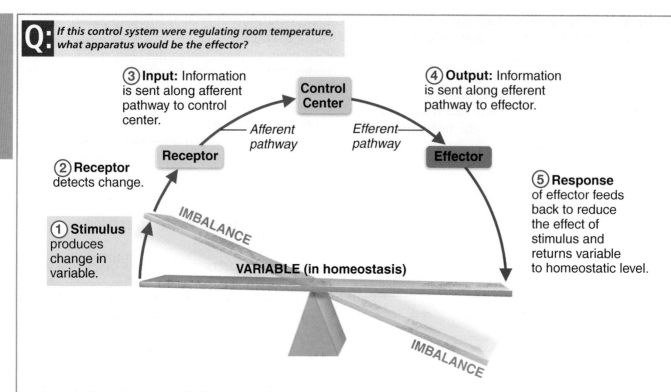

**Q:** *If this control system were regulating room temperature, what apparatus would be the effector?*

③ **Input:** Information is sent along afferent pathway to control center.

**Control Center**

④ **Output:** Information is sent along efferent pathway to effector.

*Afferent pathway*

*Efferent pathway*

**Receptor**

**Effector**

② **Receptor** detects change.

⑤ **Response** of effector feeds back to reduce the effect of stimulus and returns variable to homeostatic level.

① **Stimulus** produces change in variable.

IMBALANCE

**VARIABLE (in homeostasis)**

IMBALANCE

**Figure 1.9 The elements of a homeostatic control system.** Interaction between the receptor, control center, and effector is essential for normal operation of the system.

system is a home heating system connected to a thermostat. In this situation, the thermostat contains both the receptor and the control center. If the thermostat is set at 20°C (68°F), the heating system (effector) will be triggered ON when the house temperature drops below that setting. As the furnace produces heat, the air is warmed. When the temperature reaches 20°C or slightly higher, the thermostat sends a signal to shut off the furnace. Your body "thermostat" operates in a similar way to regulate body temperature. Other negative feedback mechanisms regulate heart rate, blood pressure, breathing rate, the release of hormones, and blood levels of glucose (blood sugar), oxygen, carbon dioxide, and minerals.

**Positive feedback** mechanisms are rare in the body because they tend to increase the original disturbance (stimulus) and to push the variable *farther* from its original value. Typically these mechanisms control infrequent events that occur explosively and do not require continuous adjustments. Blood clot-

ting and the birth of a baby are the most familiar examples of positive feedback mechanisms.

 **Homeostatic Imbalance 1.1**

Homeostasis is so important that most disease can be regarded as being the result of its disturbance, a condition called **homeostatic imbalance**. As we age, our body organs become less efficient, and our internal conditions become less and less stable. These events place us at an increasing risk for illness and produce the changes we associate with aging.

We provide examples of homeostatic imbalance throughout this text to enhance your understanding of normal physiological mechanisms. _____ ✚

### Did You Get It?

14. When we say that the body demonstrates homeostasis, do we mean that conditions in the body are unchanging? Explain your answer.
15. When we begin to become dehydrated, we usually get thirsty, which causes us to drink liquids. Is the thirst sensation part of a negative or a positive feedback control system? Defend your choice.

**A:** The heat-generating furnace or oil burner.

**For answers, see Appendix A.**

# Summary

## An Overview of Anatomy and Physiology (pp. 27–28)

1. Anatomy is the study of structure. Observation is used to see the sizes, shapes, and relationships of body parts.

2. Physiology is the study of how a structure (which may be a cell, an organ, or an organ system) functions or works.

3. Structure determines what functions can occur; therefore, if the structure changes, the function must also change.

## Levels of Structural Organization (pp. 28–33)

1. There are six levels of structural organization. Atoms (at the chemical level) combine, forming the unit of life, the cell. Cells are grouped into tissues, which in turn are arranged in specific ways to form organs. An organ system is a group of organs that performs a specific function for the body (which no other organ system can do). Together, all of the organ systems form the organism, or living body.

2. Eleven organ systems make up the human body: the integumentary, skeletal, muscular, nervous, endocrine, cardiovascular, lymphatic, respiratory, digestive, urinary, and reproductive systems. (For a description of the organ systems naming the major organs and functions, see pp. 29–33).

## Maintaining Life (pp. 33–36)

1. To sustain life, an organism must be able to maintain its boundaries, move, respond to stimuli, digest nutrients and excrete wastes, carry on metabolism, reproduce itself, and grow.

2. Survival needs include food, oxygen, water, appropriate temperature, and normal atmospheric pressure. Extremes of any of these factors can be harmful.

## The Language of Anatomy (pp. 36–45)

1. Anatomical terminology is relative and assumes that the body is in the anatomical position (standing erect, with palms facing forward).

2. Directional terms

   a. Superior (cranial, cephalic): above something else, toward the head.

   b. Inferior (caudal): below something else, toward the tail.

   c. Anterior (ventral): toward the front of the body or structure.

   d. Posterior (dorsal): toward the rear or back of the body or structure.

   e. Medial: toward the midline of the body.

   f. Lateral: away from the midline of the body.

   g. Proximal: closer to the point of attachment.

   h. Distal: farther from the point of attachment.

   i. Superficial (external): at or close to the body surface.

   j. Deep (internal): below or away from the body surface.

3. Regional terms. Visible landmarks on the body surface may be used to specifically refer to a body part or area. (See pp. 38–39 for terms referring to anterior and posterior surface anatomy.)

4. Body planes and sections

   a. Sagittal section: separates the body longitudinally into right and left parts.

   b. Frontal (coronal) section: separates the body on a longitudinal plane into anterior and posterior parts.

   c. Transverse (cross) section: separates the body on a horizontal plane into superior and inferior parts.

5. Body cavities

   a. Dorsal: well protected by bone; has two subdivisions.

      (1) Cranial: contains the brain.

      (2) Spinal: contains the spinal cord.

   b. Ventral: less protected than dorsal cavity; has two subdivisions.

      (1) Thoracic: The superior cavity that extends inferiorly to the diaphragm; contains heart and lungs, which are protected by the rib cage.

      (2) Abdominopelvic: The cavity inferior to the diaphragm that contains the digestive, urinary, and reproductive organs. The abdominal portion is vulnerable because it is protected only by the trunk muscles. The pelvic portion is somewhat protected by the bony pelvis. The abdominopelvic cavity is often divided into four quadrants or nine regions (see Figures 1.8 and 1.9).

   c. Smaller body cavities include the oral, nasal, orbital, and middle ear cavities. All are open to the outside of the body except the middle ear cavity.

## Homeostasis (pp. 45–46)

1. Body functions interact to maintain homeostasis, or a relatively stable internal environment within the body. Homeostasis is necessary for survival and good health; its loss results in illness.

2. All homeostatic control mechanisms have three components: (1) a receptor that responds to environmental changes (stimuli) and (2) a control center that assesses those changes and produces a response by activating (3) the effector.

3. Most homeostatic control systems are negative feedback systems, which act to reduce or stop the initial stimulus. Some are positive feedback systems, which act to increase the initial stimulus, as in the case of blood clotting.

## Review Questions

 Access additional practice questions using your smartphone, tablet, or computer:
Mastering A&P® > Study Area > Practice Tests & Quizzes

### Multiple Choice

*More than one choice may apply.*

1. Consider the following levels: (1) chemical; (2) tissue; (3) organ; (4) cellular; (5) organismal; (6) systemic. Which of the following choices has the levels listed in order of increasing complexity?

   a. 1, 2, 3, 4, 5, 6

   b. 1, 4, 2, 5, 3, 6

   c. 3, 1, 2, 4, 6, 5

   d. 1, 4, 2, 3, 6, 5

   e. 4, 1, 3, 2, 6, 5

2. Which of the following is (are) essential for the maintenance of life?

   a. Reproduction

   b. Effectors

   c. Growth

   d. Irritability

   e. Excretion

3. Using the terms listed below, fill in the blank with the proper term.

   anterior  superior  medial  proximal  superficial
   posterior  inferior  lateral  distal  deep

   The heart is located _____ to the diaphragm.

   The muscles are _____ to the bone.

   The shoulder is _____ to the elbow.

   In anatomical position, the thumb is _____ to the index finger.

   The vertebral region is _____ to the scapular region.

   The gluteal region is located on the _____ surface of the body.

4. Match the proper anatomical term (column B) with the common name (column A) for the body regions listed below.

   | Column A | Column B |
   |---|---|
   | ____ 1. buttocks | a. inguinal |
   | ____ 2. back | b. frontal |
   | ____ 3. heel | c. dorsal |
   | ____ 4. front of elbow | d. sural |
   | ____ 5. toes | e. gluteal |
   | ____ 6. groin | f. antecubital |
   | ____ 7. forehead | g. plantar |
   | ____ 8. calf | h. digital |
   | ____ 9. sole of foot | i. calcaneal |

5. Anatomical terms that apply to the backside of the body in the anatomical position include

   a. ventral and anterior.        c. posterior and dorsal.

   b. back and rear.        d. head and lateral.

6. A neurosurgeon orders a spinal tap for a patient. Into what body cavity will the needle be inserted?

   a. Ventral        c. Dorsal        e. Pelvic

   b. Thoracic        d. Cranial

7. Which of the following body regions are subdivisions of the abdominopelvic cavity?

   a. Thoracic        d. Right hypochondriac

   b. Epigastric        e. Spinal

   c. Hypogastric

8. Which of the following is (are) regulated by a negative feedback mechanism?

   a. Blood pressure

   b. Blood clotting

   c. Levels of blood sugar

   d. Release of hormones

   e. Heart rate

## Short Answer Essay

9. Define *anatomy* and *physiology*.

10. List the 11 organ systems of the body, briefly describe the function of each, and then name two organs in each system.

11. Many body structures are symmetrical. Are the kidneys symmetrical? What about the stomach?

12. What regions of the body do the following body landmarks refer to—acromial, crural, inguinal, popliteal, and sural?

13. Which of the following organ systems—digestive, respiratory, reproductive, circulatory, urinary, or muscular—are found in *both* subdivisions of the ventral body cavity? Which are found in the thoracic cavity only? In the abdominopelvic cavity only?

14. Explain the meaning of *homeostasis* as applied to the living organism.

15. What is the consequence of loss of homeostasis, or homeostatic imbalance?

 ## Critical Thinking and Clinical Application Questions

16. A diagnostic radiographer told John that she was going to take an image of his calcaneal region. What part of his body was she referring to? Later, a nurse told him that she was going to give him an injection containing pain killers in his gluteal region. Did he take off his shirt or his pants to get the injection? When John next saw a physiotherapist, the therapist noticed John had tenderness and swelling in the tarsal region. What part of his body was tender and swollen?

17. Jennifer fell off her motorcycle and tore a nerve in her axillary region. She also tore ligaments in her cervical and scapular regions and broke the only bone of her right brachial region. Explain where each of her injuries is located.

18. Students of an anatomy class are preparing to dissect a body to reach the heart. In which cavity is the heart located, and how can they gain access to it?

19. Parathyroid hormone (PTH) is secreted in response to a drop in the calcium level in the blood. The secretion of PTH is regulated by a negative feedback mechanism. What can you expect to happen to the calcium blood level as increased amounts of PTH are secreted, and why?

20. Mr. Harvey, a computer programmer, has been complaining about numbness and pain in his right hand. His physician diagnosed carpal tunnel syndrome and prescribed a splint. Where will Mr. Harvey apply the splint? Be sure to use correct anatomical terminology in your answer.

21. Dan "threw out his back" after lifting heavy furniture and complained of pain radiating down the back of his right leg. Use anatomical terminology to describe the body regions where the injury occurred and where Dan experienced pain. Which imaging technique would you prescribe to diagnose a spinal problem? Explain your choice.

22. Tom likes to discuss medical matters with his doctor. Since Tom has fever quite often, he wants to know what might happen if a person's body temperature deviates grossly from the normal. What would be the doctor's response?

23. How is the concept of homeostasis (or its loss) related to disease and aging? Provide examples to support your reasoning.

# 2 Basic Chemistry

**WHAT**
Chemistry is the basis for how the body transforms and uses energy and for how our cells use crucial molecules such as carbohydrates, lipids, proteins, and nucleic acids.

**HOW**
Everything that happens in the body, from cells responding to their surroundings to breaking down the food we eat, requires the movement of chemicals such as ions, carbohydrates, and lipids as they participate in chemical reactions.

**WHY**
Chemistry helps us understand the interactions of different molecules and why some interactions store energy, such as in fat, and other interactions release energy, such as when enzymes break down our food.

**INSTRUCTORS**
New **Building Vocabulary Coaching** Activities for this chapter are assignable in Mastering **A&P®**

Many short courses in anatomy and physiology lack the time to consider chemistry as a topic. So why include it here? The answer is simple: Your entire body is made up of chemicals—thousands of them—continuously interacting with one another at an incredible pace.

Chemical reactions underlie all body processes—movement, digestion, the pumping of your heart, and even your thoughts. In this chapter we present the basics of chemistry and biochemistry (the chemistry of living material), providing the background you will need to understand body functions.

## Concepts of Matter and Energy

→ **Learning Objectives**

☐ Differentiate matter from energy.

☐ List four major energy forms, and provide one example of how each is used in the body.

### Matter

**Matter** is the "stuff" of the universe. With some exceptions, it can be seen, smelled, and felt. More precisely, matter is anything that occupies space and has mass. *Weight* is a measure of gravity pulling on mass. Chemistry studies the nature of matter—how its building blocks are put together and how they interact.

Matter exists in solid, liquid, and gaseous states, all of which are found in the human body. *Solids*, such as bones and teeth, have a definite shape and volume. *Liquids* have a definite volume, but they conform to the shape of their container. Examples of body liquids are blood plasma and the interstitial fluid that bathes all body cells. *Gases* have neither a definite shape nor a definite volume. The air we breathe is a mixture of gases.

Matter may be changed both physically and chemically. *Physical changes* do not alter the basic nature of a substance. Examples include changes in state, such as ice melting to water and food being cut into smaller pieces. *Chemical changes* do alter the composition of the substance—often substantially. Fermenting grapes to make wine and the digestion of food in the body are examples of chemical changes.

## Energy

In contrast to matter, **energy** has no mass and does not take up space. It can be measured only by its effects on matter. We commonly define energy as the ability to do work or to put matter into motion. When energy is actually doing work (moving objects), we refer to it as **kinetic** (kĭ-neh′tik) **energy**. Kinetic energy is displayed in the constant movement of the tiniest particles of matter (atoms) as well as in larger objects, such as a bouncing ball. When energy is inactive or stored (as in the batteries of an unused toy), we call it **potential energy**. All forms of energy exhibit both kinetic and potential work capacities.

All living things are built of matter, and to grow and function they require a continuous supply of energy. Thus, matter is the substance, and energy is the mover of the substance. Because this is so, let's take a brief detour to introduce the forms of energy the body uses as it does its work.

### Forms of Energy

- **Chemical energy** is stored in the bonds of chemical substances. When the bonds are broken, the (potential) stored energy is unleashed and becomes kinetic energy (energy in action). For example, when gasoline molecules are broken apart in your automobile engine, the energy released powers your car. Similarly, the chemical energy harvested from the foods we eat fuels all body activities.

- **Electrical energy** results from the movement of charged particles. In your house, electrical energy is the flow of subatomic particles called electrons along the wiring. In your body, an electrical current is generated when charged particles (called *ions*) move across cell membranes. The nervous system uses electrical currents called *nerve impulses* to transmit messages from one part of the body to another.

- **Mechanical energy** is energy *directly* involved in moving matter. When you pedal a bicycle, your legs provide the mechanical energy that turns the wheels. We can take this example one step further back: As the muscles in your legs shorten (contract), they pull on your bones, causing your limbs to move (so that you can pedal the bike).

- **Radiant energy** travels in waves; that is, it is the energy of the electromagnetic spectrum, which includes X rays, infrared radiation (heat energy), visible light, radio, and ultraviolet (UV) waves. Light energy, which stimulates the retinas of your eyes, is important in vision. UV waves cause sunburn, but they also stimulate our bodies to make vitamin D.

### Energy Form Conversions

With a few exceptions, energy is easily converted from one form to another. For example, chemical energy (from gasoline) that powers the motor of a speedboat is converted into the mechanical energy of the whirling propeller that allows the boat to skim across the water. In the body, chemical energy from food is trapped in the bonds of a high-energy chemical called **ATP (adenosine triphosphate)**, and ATP's energy may ultimately be transformed into the electrical energy of a nerve impulse or the mechanical energy of contracting muscles.

Energy conversions are not very efficient, and some of the initial energy supply is always "lost" to the environment as heat (thermal energy). It is not really lost, because energy cannot be created or destroyed, but the part given off as heat is *unusable*. You can easily demonstrate this principle by putting your finger close to a lightbulb that has been lit for an hour or so. Notice that some of the electrical energy reaching the bulb is producing heat instead of light. Likewise, all energy conversions in the body liberate heat. This heat makes us warm-blooded animals and contributes to our relatively high body temperature, which has an important influence on body function. For example, when matter is heated, its particles begin to move more quickly; that is, their kinetic energy (energy of motion) increases. This is important to the chemical reactions that occur in the body because, up to a point, the higher the temperature, the faster those reactions occur.

### Did You Get It?

1. Matter and energy—how are they interrelated?
2. What form of energy is used to transmit messages from one part of the body to another? What form of energy fuels cellular processes?
3. What type of energy is available when we are still? When we are exercising?
4. What does it mean when we say that some energy is "lost" every time energy changes from one form to another in the body?

For answers, see Appendix A.

# Composition of Matter

## Elements and Atoms

→ **Learning Objectives**

☐ **Define** *element*, **and list the four elements that form the bulk of body matter.**

☐ **Explain how elements and atoms are related.**

All matter is composed of a limited number of substances called **elements**, unique substances that cannot be broken down into simpler substances by ordinary chemical methods. Examples of elements include many commonly known substances, such as oxygen, silver, gold, copper, and iron.

So far, 118 elements have been identified with certainty. Ninety-two of these occur in nature; the rest are made artificially in accelerator devices. A complete listing of the elements appears in the **periodic table**, an odd-shaped checkerboard (see Appendix C) that appears in chemistry classrooms the world over. Four of these elements—carbon, oxygen, hydrogen, and nitrogen—make up about 96 percent of the weight of the human body. Besides these four elements, several others are present in small or trace amounts. The most abundant elements found in the body and their major roles are listed in **Table 2.1**.

Each element is composed of very similar particles, or building blocks, called **atoms**. Because all elements are unique, the atoms of each element differ from those of all other elements. We designate each element by a one- or two-letter chemical shorthand called an **atomic symbol**. In most cases, the atomic symbol is simply the first or first two letters of the element's name. For example, C stands for carbon, O for oxygen, and Ca for calcium. In a few cases, the atomic symbol is taken from the Latin name for the element. For instance, Na (from the Latin word *natrium*) indicates sodium, and K (from *kalium*) potassium.

## Atomic Structure

→ **Learning Objective**

☐ **List the subatomic particles, and describe their relative masses, charges, and positions in the atom.**

### The Basic Atomic Subparticles

The word *atom* comes from the Greek word meaning "incapable of being divided," and historically this idea of an atom was accepted as a scientific truth. According to this notion, you could theoretically divide a pure element, such as a block of gold, into smaller and smaller particles until you got down to the individual atoms, at which point you could subdivide no further. We now know that atoms, although indescribably small, are clusters of even smaller components called **subatomic particles**, which include *protons*, *neutrons*, and *electrons*. Even so, the old idea of atomic indivisibility is still very useful because an atom loses the unique properties of its element when it is split into its component particles.

An atom's subatomic particles differ in their mass, electrical charge, and position within the atom (**Table 2.2**, p. 54). **Protons ($p^+$)** have a positive charge, whereas **neutrons ($n^0$)** are uncharged, or neutral. Protons and neutrons are heavy particles and have approximately the same mass (1 atomic mass unit, or amu). The tiny **electrons ($e^-$)** bear a negative charge equal in strength to the positive charge of the protons, but their mass is so small that it is usually designated as 0 amu.

The electrical charge of a particle is a measure of its ability to attract or repel other charged particles. Particles with the same type of charge (+ to + or − to −) repel each other, but particles with opposite charges (+ to −) attract each other.

| Table **2.1**   Common Elements Making Up the Human Body | | | |
|---|---|---|---|
| **Element** | **Atomic symbol** | **Percentage of body mass** | **Role** |
| **_Major (96.1%)_** | | | |
| Oxygen | O | 65.0 | A major component of both organic and inorganic molecules; as a gas, essential to the oxidation of glucose and other food fuels, during which cellular energy (ATP) is produced. |
| Carbon | C | 18.5 | The primary element in all organic molecules, including carbohydrates, lipids, proteins, and nucleic acids. |
| Hydrogen | H | 9.5 | A component of most organic molecules; as an ion (a charged atom), it influences the pH of body fluids. |
| Nitrogen | N | 3.2 | A component of proteins and nucleic acids (genetic material). |
| **_Lesser (3.9%)_** | | | |
| Calcium | Ca | 1.5 | Found as a salt in bones and teeth; in ionic form, required for muscle contraction, neural transmission, and blood clotting. |
| Phosphorus | P | 1.0 | Present as a salt, in combination with calcium, in bones and teeth; also present in nucleic acids and many proteins; forms part of the high-energy compound ATP. |
| Potassium | K | 0.4 | In its ionic form, the major intracellular cation; necessary for the conduction of nerve impulses and for muscle contraction. |
| Sulfur | S | 0.3 | A component of proteins (particularly contractile proteins of muscle). |
| Sodium | Na | 0.2 | As an ion, the major extracellular cation; important for water balance, conduction of nerve impulses, and muscle contraction. |
| Chlorine | Cl | 0.2 | In ionic (chloride) form, the most abundant extracellular anion. |
| Magnesium | Mg | 0.1 | Present in bone; also an important cofactor for enzyme activity in a number of metabolic reactions. |
| Iodine | I | 0.1 | Needed to make functional thyroid hormones. |
| Iron | Fe | 0.1 | A component of the functional hemoglobin molecule (which transports oxygen within red blood cells) and some enzymes. |
| **_Trace (less than 0.01%)*_** | | | |
| Chromium (Cr), Cobalt (Co), Copper (Cu), Fluorine (F), Manganese (Mn), Molybdenum (Mo), Selenium (Se), Silicon (Si), Tin (Sn), Vanadium (V), Zinc (Zn) | | | |

*Referred to as the _trace elements_ because they are required in very small amounts; many are found as part of enzymes or are required for enzyme activation.

Neutral particles are neither attracted to nor repelled by charged particles.

Because all atoms are electrically neutral, the number of protons an atom has must be balanced by its number of electrons (the + and − charges will then cancel the effect of each other). Thus, helium has two protons and two electrons, and iron has 26 protons and 26 electrons. For any atom, the number of protons and electrons is always equal. Atoms that have gained or lost electrons are called _ions_.

| Table **2.2** Subatomic Particles | | | |
|---|---|---|---|
| **Particle** | **Position in atom** | **Mass (amu)** | **Charge** |
| Proton (p$^+$) | Nucleus | 1 | + |
| Neutron (n$^0$) | Nucleus | 1 | 0 |
| Electron (e$^-$) | Orbits around the nucleus | 1/2000* | – |

*The mass of an electron is so small, that we will ignore it and assume a mass of 0 amu.

## Planetary and Orbital Models of an Atom

The **planetary model** of an atom portrays the atom as a miniature solar system **(Figure 2.1a)** in which the protons and neutrons are clustered at the center of the atom in the **atomic nucleus**. Because the nucleus contains all the heavy particles, it is fantastically dense and positively charged. The tiny electrons orbit around the nucleus in fixed, generally circular orbits, like planets around the sun. But we can never determine the exact location of electrons at a particular time because they jump around following unknown paths. So, instead of speaking of specific orbits, chemists talk about *orbitals*—regions around the nucleus in which electrons are

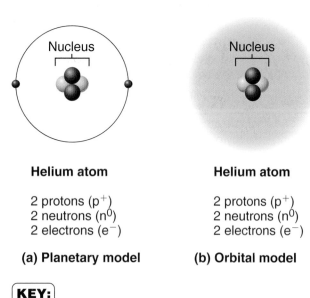

Helium atom

2 protons (p$^+$)
2 neutrons (n$^0$)
2 electrons (e$^-$)

**(a) Planetary model**

Helium atom

2 protons (p$^+$)
2 neutrons (n$^0$)
2 electrons (e$^-$)

**(b) Orbital model**

**KEY:**
- ● Proton
- ● Electron
- ○ Neutron
- ▦ Electron cloud

**Figure 2.1 The structure of an atom.** The dense central nucleus contains the protons and neutrons. **(a)** In the planetary model of an atom, the electrons move around the nucleus in fixed orbits. **(b)** In the orbital model, electrons are shown as a cloud of negative charge.

*likely* to be found. The **orbital model** depicts the general location of electrons outside the nucleus as a haze of negative charge referred to as the *electron cloud* (Figure 2.1b). Regions where electrons are most likely to be found are shown by denser shading rather than by orbit lines.

Notice that in both models, the electrons have the run of nearly the entire volume of the atom. Electrons also determine an atom's chemical behavior (that is, its ability to bond with other atoms). Though now considered outdated, the planetary model is simple and easy to understand and use. Most of the descriptions of atomic structure in this text use that model.

Hydrogen is the simplest atom, with just one proton and one electron. You can visualize the hydrogen atom by imagining it as a sphere with its diameter equal to the length of a football field. The nucleus could then be represented by a lead ball the size of a gumdrop in the exact center of the sphere and the lone electron pictured as a fly buzzing about unpredictably within the sphere. This mental picture should remind you that most of the volume of an atom is empty space, and most of the mass is in the central nucleus.

## Did You Get It?

5. Which four elements make up the bulk of living matter?
6. How is an atom related to an element?

For answers, see Appendix A.

## Identifying Elements

→ **Learning Objective**

☐ Define *radioisotope*, and describe briefly how radioisotopes are used in diagnosing and treating disease.

All protons are alike, regardless of the atom being considered. The same is true of all neutrons and all electrons. So what determines the unique properties

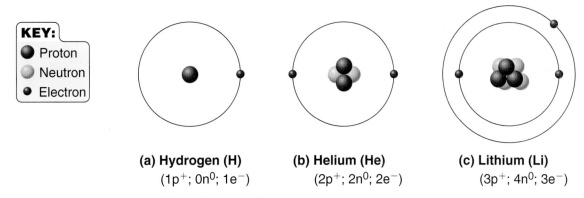

**Figure 2.2   Atomic structure of the three smallest atoms.**

of each element? The answer is that atoms of different elements are composed of *different numbers* of protons, neutrons, and electrons.

The simplest and smallest atom, hydrogen, has one proton, one electron, and no neutrons **(Figure 2.2)**. Next is the helium atom, with two protons, two neutrons, and two orbiting electrons. Lithium follows with three protons, four neutrons, and three electrons. If we continue this step-by-step listing of subatomic particles, we could describe all known atoms by adding one proton and one electron at each step. The number of neutrons is not as easy to pin down, but light atoms tend to have equal numbers of protons and neutrons, whereas in larger atoms neutrons outnumber protons. However, all we really need to know to identify a particular element is its *atomic number*, *atomic mass number*, and *atomic weight*. Taken together, these indicators provide a fairly complete picture of each element.

### Atomic Number

Each element is given a number, called its **atomic number**, that is equal to the number of protons its atoms contain. Atoms of each element contain a different number of protons from the atoms of any other element; hence, its atomic number is unique. Because the number of protons is always equal to the number of electrons, the atomic number *indirectly* also tells us the number of electrons that atom contains.

### Atomic Mass Number

The **atomic mass number** (or just **mass number**) of any atom is the sum of the masses of all the protons and neutrons contained in its nucleus.

(Remember, the mass of an electron is so small that we ignore it.) Because hydrogen has one proton and no neutrons in its nucleus, its atomic number and mass number are the same: 1. Helium, with two protons and two neutrons, has a mass number of 4. The mass number is written as a superscript to the left of the atomic symbol (see the examples in **Figure 2.3** on p. 56).

### Atomic Weight and Isotopes

At first glance, it seems that the **atomic weight** of an atom should be equal to its atomic mass. This would be so if there were only one type of atom representing each element. However, the atoms of almost all elements exhibit two or more structural variations; these varieties are called **isotopes** (ī′ sĭ-tōps). Isotopes have the same number of protons and electrons but vary in the number of *neutrons* they contain. Thus, the isotopes of an element have the same atomic number but different atomic masses. Because all of an element's isotopes have the same number of electrons (and electrons determine bonding properties), their chemical properties are *exactly* the same. As a general rule, the atomic weight of any element is approximately equal to the mass number of its most abundant isotope. For example, as we said before, hydrogen has an atomic number of 1, but it also has isotopes with atomic masses of 1, 2 (called *deuterium*), and 3 (called *tritium*) (see Figure 2.3). Its atomic weight is 1.008, which reveals that its lightest isotope is present in much greater amounts in our world than its $^2$H or $^3$H forms. (The atomic numbers, mass numbers, and atomic weights for elements commonly found in the body are provided in **Table 2.3**).

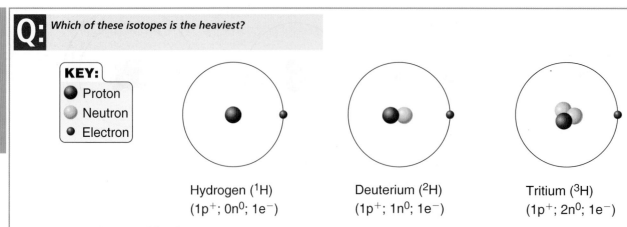

**Q:** *Which of these isotopes is the heaviest?*

**KEY:**
● Proton
○ Neutron
• Electron

Hydrogen ($^1$H)
(1p$^+$; 0n$^0$; 1e$^-$)

Deuterium ($^2$H)
(1p$^+$; 1n$^0$; 1e$^-$)

Tritium ($^3$H)
(1p$^+$; 2n$^0$; 1e$^-$)

**Figure 2.3 Isotopes of hydrogen.** Isotopes differ in their numbers of neutrons.

The heavier isotopes of certain atoms are unstable and tend to release energy in order to become more stable; such isotopes are called **radioisotopes**. Why this process occurs is very complex, but apparently the "glue" that holds the atomic nuclei together is weaker in the heavier isotopes. This process of spontaneous atomic decay, which is called **radioactivity**, can be compared to a tiny sustained explosion. All types of radioactive decay involve the ejection of particles (*alpha* or *beta particles*) or electromagnetic energy (*gamma rays*) from the atom's nucleus and are damaging to living cells. Emission of alpha particles has the least penetrating power; gamma

Table **2.3**   **Atomic Structures of the Most Abundant Elements in the Body**

| Element | Symbol | Atomic number (# of p) | Mass number (# of p + n) | Atomic weight | Electrons in valence shell |
|---------|--------|------------------------|--------------------------|---------------|----------------------------|
| Calcium | Ca | 20 | 40 | 40.078 | 2 |
| Carbon | C | 6 | 12 | 12.011 | 4 |
| Chlorine | Cl | 17 | 35 | 35.453 | 7 |
| Hydrogen | H | 1 | 1 | 1.008 | 1 |
| Iodine | I | 53 | 127 | 126.905 | 7 |
| Iron | Fe | 26 | 56 | 55.847 | 2 |
| Magnesium | Mg | 12 | 24 | 24.305 | 2 |
| Nitrogen | N | 7 | 14 | 14.007 | 5 |
| Oxygen | O | 8 | 16 | 15.999 | 6 |
| Phosphorus | P | 15 | 31 | 30.974 | 5 |
| Sodium | Na | 11 | 23 | 22.989 | 1 |
| Sulfur | S | 16 | 32 | 32.064 | 6 |

**A:**  Tritium.

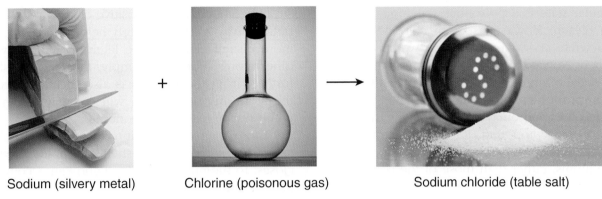

Sodium (silvery metal)    Chlorine (poisonous gas)    Sodium chloride (table salt)

**Figure 2.4 Properties of a compound differ from those of its atoms.**

radiation has the most. Contrary to what some believe, ionizing radiation does not damage the atoms in its path directly. Instead, it sends electrons flying, like a bowling ball through pins, all along its path. It is these electrons that do the damage.

Radioisotopes are used in minute amounts to tag biological molecules so that they can be followed, or traced, through the body and are valuable tools for medical diagnosis and treatment. For example, PET scans, which use radioisotopes, are discussed in "A Closer Look" (pp. 36–37). A radioisotope of iodine can be used to scan the thyroid gland of a patient suspected of having a thyroid tumor. Additionally, radium, cobalt, and certain other radioisotopes are used to destroy localized cancers.

### Did You Get It?

7. An atom has five neutrons, four protons, and four electrons. What is its atomic number? What is its atomic mass number?
8. What name is given to an unstable atom that has either more or fewer neutrons than its typical number?

For answers, see Appendix A.

## Molecules and Compounds

### → Learning Objective

☐ Define *molecule*, and explain how molecules are related to compounds.

When two or more atoms combine chemically, **molecules** are formed. If two or more atoms of the same element bond, or become chemically linked together, a molecule of that element is produced.

For example, when two hydrogen atoms bond, a molecule of hydrogen gas is formed:

$$H \text{ (atom)} + H \text{ (atom)} \rightarrow H_2 \text{ (molecule)}^*$$

In this example of a chemical reaction, the *reactants* (the atoms taking part in the reaction) are indicated by their atomic symbols, and the *product* (the molecule formed) is indicated by a *molecular formula* that shows its atomic makeup. The chemical reaction is shown as a *chemical equation*.

When two or more *different* atoms bind together to form a molecule, the molecule is more specifically referred to as a molecule of a **compound**. For example, four hydrogen atoms and one carbon atom can interact chemically to form methane:

$$4H + C = CH_4 \text{ (methane)}$$

Thus, a molecule of methane is a compound, but a molecule of hydrogen gas is not—it is instead called molecular hydrogen.

Compounds always have properties quite different from those of the atoms making them up, and it would be next to impossible to determine the atoms making up a compound without analyzing it chemically. Sodium chloride is an excellent example of the difference in properties between a compound and its constituent atoms **(Figure 2.4)**. Sodium is a silvery white metal, and chlorine in its

---

*Notice that when the number of atoms is written as a subscript, it indicates that the atoms are joined by a chemical bond. Thus, 2H represents two separate atoms, but $H_2$ indicates that the two hydrogen atoms are bound together to form a molecule. The atomic symbol by itself represents one atom.

molecular state is a poisonous green gas used to make bleach. However, sodium chloride is table salt, a white crystalline solid that we sprinkle on our food. Notice that just as an atom is the smallest particle of an element that still retains that element's properties, a molecule is the smallest particle of a compound that still retains the properties of that compound. If you break the bonds between the atoms of the compound, properties of the atoms, rather than those of the compound, will be exhibited.

## Did You Get It?

9. What is the meaning of the term molecule?
10. How does a molecular substance differ from a molecule of a compound?

**For answers, see Appendix A.**

# Chemical Bonds and Chemical Reactions

## → Learning Objectives

☐ **Recognize that chemical reactions involve the interaction of electrons to make and break chemical bonds.**

☐ **Differentiate between ionic, polar covalent, and nonpolar covalent bonds, and describe the importance of hydrogen bonds.**

☐ **Contrast synthesis, decomposition, and exchange reactions.**

**Chemical reactions** occur whenever atoms combine with or dissociate from other atoms. When atoms unite chemically, chemical bonds are formed.

## Bond Formation

A **chemical bond** is not an actual physical structure, like a pair of handcuffs linking two people together. Instead, it is an energy relationship that involves interactions between the electrons of the reacting atoms. Let's consider the role electrons play in forming bonds.

### Role of Electrons

The orbits, or generally fixed regions of space that electrons occupy around the nucleus (see Figure 2.2), are called **electron shells**, or **energy levels**. The maximum number of electron shells in any atom known so far is seven, and these are numbered 1 to 7 from the nucleus outward. The electrons closest to the nucleus are those most strongly attracted to its positive charge, and those farther away are less securely held. As a result, the more distant electrons are likely to interact with other atoms. Put more simply, the electron shell furthest from the nucleus is also the first part of the atom any other atom will come into contact with prior to reacting or bonding.

There is an upper limit to the number of electrons that each electron shell can hold. Shell 1, closest to the nucleus, is small and can accommodate only 2 electrons. Shell 2 holds a maximum of 8. Shell 3 can accommodate up to 18 electrons. Subsequent shells hold larger and larger numbers of electrons. In most (but not all) cases, the shells tend to be filled consecutively.

The only electrons that are important to bonding behavior are those in the atom's outermost shell. This shell is called the **valence shell**, and its *valence electrons* determine the chemical behavior of the atom. As a general rule, the electrons of inner shells do not take part in bonding.

The key to chemical reactivity is referred to as the *rule of eights*: In the absence of a full valence shell, atoms interact in such a way that they will have eight electrons in their valence shell. The first electron shell represents an exception to this rule, because it is "full" when it has two electrons. When the valence shell of an atom contains eight electrons, the atom is completely stable and is chemically inactive (inert). When the valence shell contains fewer than eight electrons, an atom will tend to gain, lose, or share electrons with other atoms to reach a stable state. When any of these events occurs, chemical bonds are formed. (Examples of chemically inert and reactive elements are shown in **Figure 2.5**). As you might guess, atoms must approach each other very closely for their electrons to interact in a chemical bond—in fact, their outermost electron shells must overlap.

### Types of Chemical Bonds

*Ionic Bonds*   **Ionic** (i-on′ik) **bonds** form when electrons are completely transferred from one atom to another. Atoms are electrically neutral, but when they gain or lose electrons during bonding, their positive and negative charges are no longer balanced, and charged particles, called **ions**, result. When an atom gains an electron, it acquires a net negative charge because it now has more electrons than protons. Negatively charged ions are more specifically called *anions*, and they are

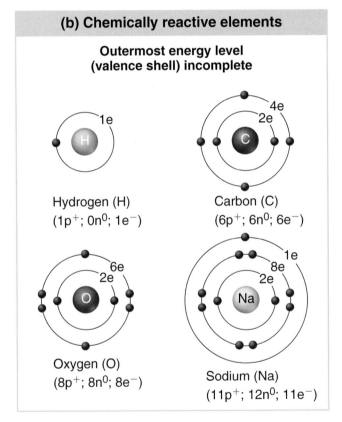

### (a) Chemically inert elements

**Outermost energy level (valence shell) complete**

Helium (He)
($2p^+$; $2n^0$; $2e^-$)

Neon (Ne)
($10p^+$; $10n^0$; $10e^-$)

### (b) Chemically reactive elements

**Outermost energy level (valence shell) incomplete**

Hydrogen (H)
($1p^+$; $0n^0$; $1e^-$)

Carbon (C)
($6p^+$; $6n^0$; $6e^-$)

Oxygen (O)
($8p^+$; $8n^0$; $8e^-$)

Sodium (Na)
($11p^+$; $12n^0$; $11e^-$)

**Figure 2.5 Chemically inert and reactive elements.** To simplify the diagrams, each atomic nucleus is shown as a circle with the atom's symbol in it; protons and neutrons are not shown.

indicated by a minus sign after the atomic symbol, such as $Cl^-$ for the chloride ion. When atoms lose an electron, they become positively charged ions, *cations*, because they now possess more protons than electrons. Cations are represented by their atomic symbol with a plus sign, such as $H^+$ for the hydrogen ion (it may help you to remember that a cation is positively charged by thinking of its "t" as

a plus [+] sign). Both anions and cations result when an ionic bond is formed. Because opposite charges attract, the newly created ions tend to stay close together.

The formation of sodium chloride (NaCl), common table salt, provides a good example of ionic bonding. Sodium's valence shell contains only one electron and so is incomplete (**Figure 2.6**, p. 60). However, if this single electron is "lost" to another atom, shell 2, which contains eight electrons, becomes the valence shell; thus sodium becomes a cation ($Na^+$) and achieves stability. Chlorine needs only one electron to fill its valence shell, and it is much easier to gain one electron (forming $Cl^-$) than it is to "give away" seven. Thus, the ideal situation is for sodium to donate its valence electron to chlorine, which is exactly what happens. Sodium chloride and most other compounds formed by ionic bonding fall into the general category of inorganic chemicals called *salts*.

***Covalent Bonds*** Electrons do not have to be completely lost or gained for atoms to become stable. Instead, they can be shared in such a way that each atom is able to fill its valence shell at least part of the time.

Molecules in which atoms share valence electrons are called *covalent molecules*, and their bonds are **covalent bonds** (*co* = with; *valent* = having power). For example, hydrogen, with its single electron, can become stable if it fills its valence shell (energy level 1) by sharing a pair of electrons—its own and one from another atom. A hydrogen atom can share an electron pair with another hydrogen atom to form a molecule of hydrogen gas (**Figure 2.7a**). The shared electron pair orbits the whole molecule and satisfies the stability needs of both hydrogen atoms. Likewise, two oxygen atoms, each with six valence electrons, can share two pairs of electrons (form double bonds) with each other (Figure 2.7b) to form a molecule of oxygen gas ($O_2$).

A hydrogen atom may also share its electron with an atom of a different element. Carbon has four valence electrons but needs eight to achieve stability. When methane gas ($CH_4$) is formed, carbon shares four electrons with four hydrogen atoms (each bond includes one electron from each atom to form a pair of electrons; Figure 2.7c). Because the shared electrons orbit and "belong to"

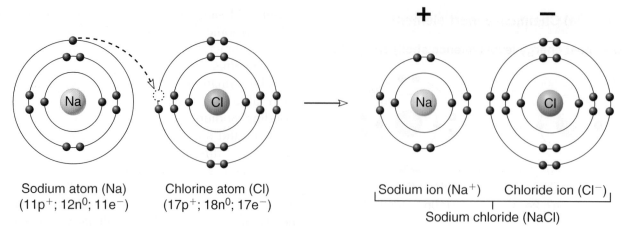

Sodium atom (Na)
($11p^+$; $12n^0$; $11e^-$)

Chlorine atom (Cl)
($17p^+$; $18n^0$; $17e^-$)

Sodium ion ($Na^+$)    Chloride ion ($Cl^-$)

Sodium chloride (NaCl)

**Figure 2.6  Formation of an ionic bond.** Both sodium and chlorine atoms are chemically reactive because their valence shells are incompletely filled. Sodium gains stability by losing one electron, whereas chlorine becomes stable by gaining one electron. After electron transfer, sodium becomes a sodium ion ($Na^+$), and chlorine becomes a chloride ion ($Cl^-$). The oppositely charged ions attract each other.

the whole molecule, each atom has a full valence shell enough of the time to satisfy its stability needs.

In the covalent molecules described thus far, electrons have been shared *equally* between the atoms of the molecule. Such molecules are called *nonpolar covalent molecules.* However, electrons are not in all cases shared equally. When covalent bonds are made, the molecule formed always has a definite three-dimensional shape. A molecule's shape plays a major role in determining just what other molecules (or atoms) it can interact with; the shape may also result in unequal electron-pair sharing. The following two examples illustrate this principle.

Carbon dioxide ($CO_2$) is formed when a carbon atom shares its four valence electrons with two oxygen atoms. Oxygen is a very "electron-hungry" atom, so it attracts the shared electrons much more strongly than does carbon. However, because the carbon dioxide molecule is linear (O=C=O), the electron-pulling power of one oxygen atom is offset by that of the other, like a tug-of-war at a standoff between equally strong teams (**Figure 2.8a**, p. 62). As a result, the electron pairs are shared equally and orbit the entire molecule, and carbon dioxide is a *nonpolar molecule.*

A water molecule ($H_2O$) is formed when two hydrogen atoms bind covalently to a single oxygen atom. Each hydrogen atom shares an electron pair with the oxygen atom, and again the

oxygen has the stronger electron-attracting ability. But in this case, the molecule formed is V-shaped (H    H). The two hydrogen atoms are located at
  O
one end of the molecule, and the oxygen atom is at the other (Figure 2.8b); consequently, the electron pairs are not shared equally. This arrangement allows them to spend more time in the vicinity of the oxygen atom. Because electrons are negatively charged, that end of the molecule becomes slightly more negative (indicated by $\delta^-$, the Greek letter delta with a minus sign) and the hydrogen end becomes slightly more positive (indicated by $\delta^+$). In other words, a *polar molecule,* a molecule with two charged *poles,* is formed.

Polar molecules orient themselves toward other polar molecules or charged particles (ions, proteins, and others), and they play an important role in chemical reactions that occur in body cells. Because body tissues are 60 to 80 percent water, the fact that water is a polar molecule is particularly significant.

**Hydrogen Bonds**  Hydrogen bonds are extremely weak bonds formed when a hydrogen atom bound to one "electron-hungry" nitrogen or oxygen atom is attracted by another such atom, and the hydrogen atom forms a "bridge" between them. Electrons are not involved in hydrogen bonds, as they are in ionic and covalent bonds. Hydrogen bonding is common between water

| **Reacting atoms** | **Resulting molecules** |
|---|---|

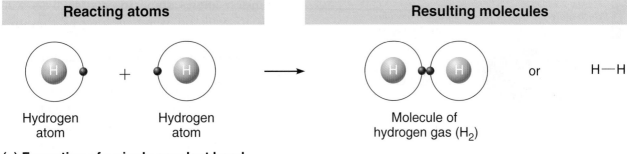

Hydrogen atom  +  Hydrogen atom  →  Molecule of hydrogen gas (H₂)    or    H—H

**(a) Formation of a single covalent bond**

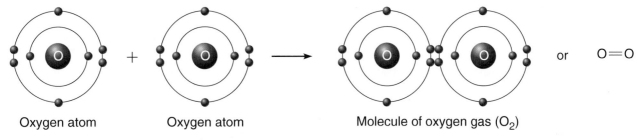

Oxygen atom  +  Oxygen atom  →  Molecule of oxygen gas (O₂)    or    O=O

**(b) Formation of a double covalent bond**

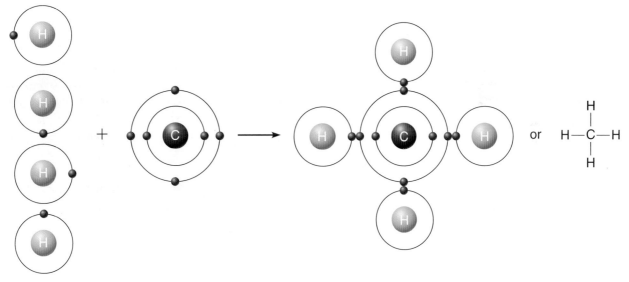

Hydrogen atoms    Carbon atom    Molecule of methane gas (CH₄)

**(c) Formation of four single covalent bonds**

**Figure 2.7  Formation of covalent bonds. (a)** Formation of a single covalent bond between two hydrogen atoms forms a molecule of hydrogen gas. **(b)** Formation of a molecule of oxygen gas. Each oxygen atom shares two electron pairs with its partner; thus, a double covalent bond is formed. **(c)** Formation of a molecule of methane. A carbon atom shares four electron pairs with four hydrogen atoms. In the structural formulas of the molecules shown at the far right, each pair of shared electrons is indicated by a single line connecting the sharing atoms.

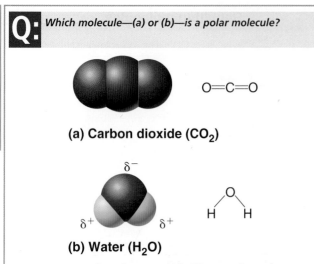

**Q:** *Which molecule—(a) or (b)—is a polar molecule?*

$$O=C=O$$

**(a) Carbon dioxide (CO$_2$)**

**(b) Water (H$_2$O)**

**Figure 2.8 Molecular models illustrating the three-dimensional structure of carbon dioxide and water molecules.**

molecules **(Figure 2.9a)** and is reflected in water's surface tension. The surface tension of water causes it to "ball up," or form spheres at the surface.

This tension allows some insects, such as water striders, to walk on water as long as they tread lightly (Figure 2.9b).

Hydrogen bonds are also important *intramolecular bonds*; that is, they help to bind different parts of the *same* molecule together into a special three-dimensional shape. These rather fragile bonds are very important in helping to maintain the structure of both protein molecules, which are essential functional molecules and body-building materials, and DNA, the genetic information molecule.

## Patterns of Chemical Reactions

Chemical reactions involve the making or breaking of bonds between atoms. The total number of atoms remains the same, but the atoms appear in new combinations. Most chemical reactions have one of the three recognizable patterns we describe next.

### Synthesis Reactions

**Synthesis reactions** occur when two or more atoms or molecules combine to form a larger,

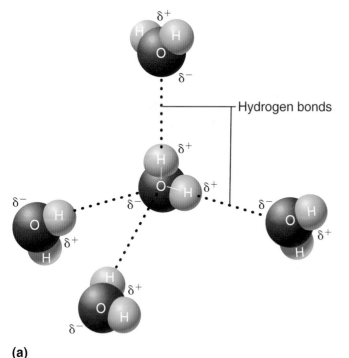

**(a)**    **(b)**

**Figure 2.9  Hydrogen bonding between polar water molecules.** **(a)** The slightly positive ends of the water molecules (indicated by $\delta^+$) become aligned with the slightly negative ends (indicated by $\delta^-$) of other water molecules. **(b)** Water's high surface tension, a result of the combined strength of its hydrogen bonds, allows a water strider to walk on a pond without breaking the surface.

**A:** Water is a polar molecule.

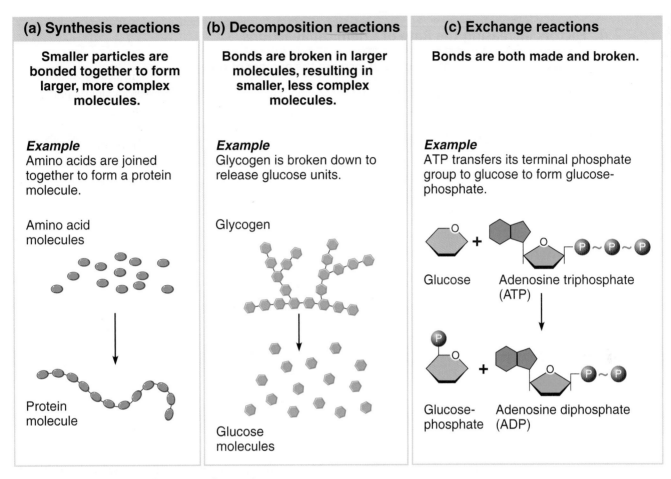

**(a) Synthesis reactions**

**Smaller particles are bonded together to form larger, more complex molecules.**

*Example*
Amino acids are joined together to form a protein molecule.

Amino acid molecules

Protein molecule

**(b) Decomposition reactions**

**Bonds are broken in larger molecules, resulting in smaller, less complex molecules.**

*Example*
Glycogen is broken down to release glucose units.

Glycogen

Glucose molecules

**(c) Exchange reactions**

**Bonds are both made and broken.**

*Example*
ATP transfers its terminal phosphate group to glucose to form glucose-phosphate.

Glucose       Adenosine triphosphate (ATP)

Glucose-phosphate       Adenosine diphosphate (ADP)

**Figure 2.10  Patterns of chemical reactions.**

more complex molecule, which can be simply represented as

$$A + B \rightarrow AB$$

Synthesis reactions always involve bond formation. Because energy must be absorbed to make bonds, synthesis reactions are energy-storing reactions.

Synthesis reactions underlie all anabolic (building) activities that occur in body cells. They are particularly important for growth and for repair of worn-out or damaged tissues. The formation of a protein molecule by the joining of *amino acids* (protein building blocks) into long chains is a synthesis reaction **(Figure 2.10a)**.

### Decomposition Reactions

**Decomposition reactions** occur when a molecule is broken down into smaller molecules, atoms, or ions and can be indicated by

$$AB \rightarrow A + B$$

Essentially, decomposition reactions are synthesis reactions in reverse. Bonds are always broken, and the products of these reactions are smaller and simpler than the original molecules. As bonds are broken, chemical energy is released.

Decomposition reactions underlie all catabolic (destructive) processes that occur in body cells; that is, they are reactions that break down molecules. Examples include the digestion of foods into their building blocks and the breakdown of glycogen (a large carbohydrate molecule stored in the liver) to release glucose (Figure 2.10b) when the blood sugar level starts to decline.

### Exchange Reactions

**Exchange reactions** involve simultaneous synthesis and decomposition reactions; in other words, bonds are both made and broken. During exchange reactions, a switch is made between molecule parts (changing partners, so to speak),

| Table **2.4** **Factors Increasing the Rate of Chemical Reactions** | |
|---|---|
| **Factor** | **Mechanism to increase the number of collisions** |
| ↑ temperature | ↑ the kinetic energy of the molecules, which in turn move more rapidly and collide more forcefully. |
| ↑ concentration of reacting particles | ↑ the number of collisions because of increased numbers of reacting particles. |
| ↓ particle size | Smaller particles have more kinetic energy and move faster than larger ones, hence they take part in more collisions. |
| Presence of catalysts | ↓ the amount of energy the molecules need to interact by holding the reactants in the proper positions to interact (see p. 77). |

and different molecules are made. Thus, an exchange reaction can be generally indicated as

$$AB + C \rightarrow AC + B \quad \text{and} \quad AB + CD \rightarrow AD + CB$$

An exchange reaction occurs, for example, when ATP reacts with glucose and transfers its end phosphate group (its full name is *adenosine triphosphate*; *tri-* = three) to glucose, forming glucose-phosphate (Figure 2.10c). At the same time, the ATP becomes ADP, adenosine diphosphate (*di-* = two). This important reaction, which occurs whenever glucose enters a body cell, traps glucose inside the cell.

Most chemical reactions are reversible. If chemical bonds can be made, they can be broken, and vice versa. Reversibility is indicated by a double arrow in a chemical equation. When the arrows differ in length, the longer arrow indicates the more rapid reaction or the major direction in which the reaction is proceeding. For example, in the reaction

$$A + B \rightleftharpoons AB$$

the reaction going to the right is occurring more rapidly, so over time AB will accumulate while A and B will decrease in amount. If the arrows are of equal length, the reaction is at chemical equilibrium. Thus, in

$$A + B \rightleftharpoons AB$$

for each molecule of AB made, a molecule of AB is breaking down to release A and B.

### Factors Influencing the Rate of Chemical Reactions

As mentioned earlier, for the atoms in molecules to react chemically, their outermost electron shells must overlap. In fact, for them to get close enough for this to happen, the particles must collide forcefully. Remember also that atoms move constantly because of their kinetic energy—this is what drives collisions between particles. Several factors, including temperature, concentration of the particles, and size of the particles, influence the kinetic energy and, hence, the speed of the particles and the force of collisions (Table 2.4).

### Did You Get It?

11. How do ionic bonds differ from covalent bonds?
12. What kind of bond forms between water molecules?
13. Which reaction type (see Figure 2.10) occurs when fats are digested in your small intestine?
14. How can you indicate that a chemical reaction is reversible?

For answers, see Appendix A.

# Biochemistry: The Chemical Composition of Living Matter

### → Learning Objective

☐ **Distinguish organic from inorganic compounds.**

All chemicals found in the body fall into one of two major classes of molecules: either inorganic or organic compounds. The class of the compound is determined solely by the presence or absence of carbon. Except for a few so far unexplainable exceptions (such as carbon dioxide gas [$CO_2$] and carbon monoxide [CO]), **inorganic compounds** lack carbon and tend to be small, simple molecules. Examples of inorganic compounds found in the body are *water*, *salts*, and many (but not all) *acids* and *bases*. **Organic compounds** contain carbon. The important organic compounds in the body are *carbohydrates*, *lipids*, *proteins*, and *nucleic acids*. All organic compounds are fairly (or very) large covalent molecules.

Inorganic and organic compounds are equally essential for life. Trying to determine which is more valuable can be compared to trying to decide whether the ignition system or the engine is more essential to the operation of a car.

## Inorganic Compounds

### → Learning Objectives

- ☐ Explain the importance of water to body homeostasis, and provide several examples of the roles of water.
- ☐ List several salts (or their ions) vitally important to body functioning.
- ☐ Differentiate a salt, an acid, and a base.
- ☐ Explain the concept of pH, and state the pH of blood.

### Water

Water is the most abundant inorganic compound in the body. It accounts for about two-thirds of body weight. Among the properties that make water so vital are the following:

- **High heat capacity.** Water has a *high heat capacity*; that is, it absorbs and releases large amounts of heat before its temperature changes. Thus, it prevents the sudden changes in body temperature that might otherwise result from intense sun exposure, chilling winter winds, or internal events (such as vigorous muscle activity) that liberate large amounts of heat.

- **Polarity/solvent properties.** Water is an excellent solvent because of its polarity; indeed, it is often called the "universal solvent." A *solvent* is a liquid or gas in which smaller amounts of other substances, called *solutes* (which may be gases, liquids, or solids), can be dissolved or suspended. The resulting mixture is called a *solution* when the solute particles are exceedingly tiny, and a *suspension* when the solute particles are fairly large. Translucent mixtures with solute particles of intermediate size are called *colloids*.

  Small reactive chemicals—such as salts, acids, and bases—dissolve easily in water and become evenly distributed. Molecules cannot react chemically unless they are in solution, so virtually all chemical reactions that occur in the body depend on water's solvent properties.

  Because nutrients, respiratory gases (oxygen and carbon dioxide), and wastes can dissolve in water, water can act as a transport and exchange medium in the body. For example, all these substances are carried around the body in blood plasma (the liquid part of blood that is mostly made up of water) and are exchanged between the blood and body cells by passing through the water-based interstitial fluid that bathes cells. Specialized molecules that lubricate the body, such as mucus and synovial fluid, also use water as their solvent. Synovial fluid "oils" the ends of bones as they move within joint cavities.

- **Chemical reactivity.** Water is an important *reactant* in some types of chemical reactions. For example, to digest foods or break down biological molecules, water molecules are added to the bonds of the larger molecules in order to break them. Such reactions are called *hydrolysis reactions*, a term that specifically recognizes this role of water (*hydro* = water; *lys* = splitting).

- **Cushioning.** Water also serves a protective function. In cerebrospinal fluid, water forms a cushion around the brain that helps to protect it from physical trauma. Amniotic fluid, which surrounds a developing fetus within the mother's body, plays a similar role in protecting the fetus.

### Salts

A **salt** is an ionic compound containing cations other than the *hydrogen ion* ($H^+$) and anions other than the *hydroxide ion* ($OH^-$). Salts of many metal elements are commonly found in the body, but the most plentiful salts are those containing calcium and phosphorus, found chiefly in bones and teeth. When dissolved in body fluids, salts easily separate into their ions. This process, called *dissociation*, occurs rather easily because the ions have already been formed. All that remains is for the ions to "spread out." This is accomplished by the polar water molecules, which orient themselves with their slightly negative ends toward the cations and their slightly positive ends toward the anions, thereby overcoming the attraction between them **(Figure 2.11)**.

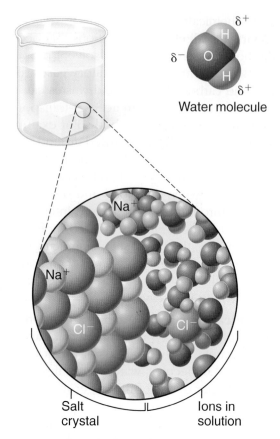

Water molecule

**Figure 2.11 Dissociation of salt in water.** The slightly negative ends of the water molecules ($\delta^-$) are attracted to $Na^+$, whereas the slightly positive ends of water molecules ($\delta^+$) orient toward $Cl^-$, causing the ions of the salt crystal to be pulled apart.

Salts, both in their ionic forms and in combination with other elements, are vital to body functioning. For example, sodium and potassium ions are essential for nerve impulses, and iron forms part of the hemoglobin molecule that transports oxygen within red blood cells.

Because ions are charged particles, all salts are **electrolytes**—substances that conduct an electrical current in solution. When electrolyte balance is severely disturbed, virtually nothing in the body works. (The functions of the elements found in body salts are summarized in Table 2.1 on p. 53.)

## Acids and Bases

Like salts, acids and bases are electrolytes. That is, they ionize, dissociate in water, and can then conduct an electrical current.

### Characteristics of Acids    Acids have a sour taste and can dissolve many metals or "burn" a hole in

your rug. Acids can have a devastating effect; for example, consider the damage to sea life, trees, and famous historical monuments caused by the vinegar-like acid rain. But the most useful definition of an acid is that it is a substance that can release *hydrogen ions (H⁺)* in detectable amounts. Because a hydrogen ion is a hydrogen nucleus (a "naked proton"), acids are also defined as **proton (H⁺) donors**. You may find it useful to think of acids as putting protons "in the game." As free protons, hydrogen ions can influence the acidity of body fluids.

When acids are dissolved in water, they release hydrogen ions and some anions. The anions are unimportant; it is the release of protons that determines an acid's effects on the environment. The ionization of hydrochloric acid (an acid produced by stomach cells that aids digestion) is shown in the following equation:

$$HCl \rightarrow H^+ + Cl^-$$
(hydrochloric    (proton)    (anion)
acid)

Other acids found or produced in the body include acetic acid (the acidic component of vinegar) and carbonic acid.

Acids that ionize completely and liberate all their protons are called *strong acids*; an example is hydrochloric acid. Acids that ionize incompletely, as do acetic and carbonic acid, are called *weak acids*. For example, when carbonic acid dissolves in water, only some of its molecules ionize to liberate $H^+$.

$$2H_2CO_3 \rightarrow H^+ + HCO_3^- + H_2CO_3$$
(carbonic    (proton)    (anion)    (carbonic
acid)                                 acid)

### Characteristics of Bases    Bases have a bitter taste, feel slippery, and are **proton (H⁺) acceptors**. (You can also think of them as taking protons "out of the game"; when protons are bound to a molecule, they are unable to affect the acidity of body fluids.) Hydroxides are common inorganic bases. Like acids, the hydroxides ionize and dissociate in water; but in this case, the *hydroxide ion (OH⁻)* and some cations are released. The ionization of sodium hydroxide (NaOH), commonly known as lye, is shown as

$$NaOH \rightarrow Na^+ + OH^-$$
(sodium    (cation)    (hydroxide
hydroxide)                      ion)

The hydroxide ion is an avid proton ($H^+$) seeker, and any base containing this ion is considered a strong base. By contrast, *bicarbonate ion ($HCO_3^-$)*, an important base in blood, is a fairly weak base.

***Acids, Bases, and Neutralization***   When acids and bases are mixed, they react with each other (in an exchange reaction) to form water and a salt:

$$HCl \quad + \quad NaOH \quad \rightarrow \quad H_2O \quad + \quad NaCl$$

(acid)    (base)    (water)    (salt)

This type of exchange reaction, in which an acid and a base interact, is more specifically called a **neutralization** reaction.

***pH: Acid-Base Concentrations***   The relative concentration of hydrogen (and hydroxide) ions in various body fluids is measured in concentration units called **pH** (pe-āch) units. The **pH scale**, which was devised in 1909 by a Danish biochemist (and part-time beer brewer) named Sørensen, is based on the number of protons in solution.

The pH scale runs from 0 to 14 **(Figure 2.12)**, and each successive change of 1 pH unit represents a tenfold change in hydrogen ion concentration. At a pH of 7, the number of hydrogen ions exactly equals the number of hydroxide ions, and the solution is neutral; that is, neither acidic nor basic. Solutions with a pH lower than 7 are acidic: The hydrogen ions outnumber the hydroxide ions. A solution with a pH of 6 has 10 times as many hydrogen ions as a solution with a pH of 7, and a pH of 3 indicates a 10,000-fold ($10 \times 10 \times 10 \times 10$) increase in hydrogen ion concentration from pH 7. Solutions with a pH number higher than 7 are basic, or alkaline, and solutions with a pH of 8 and 12 have 1/10 and 1/100,000 (respectively) the number of hydrogen ions present in a solution with a pH of 7.

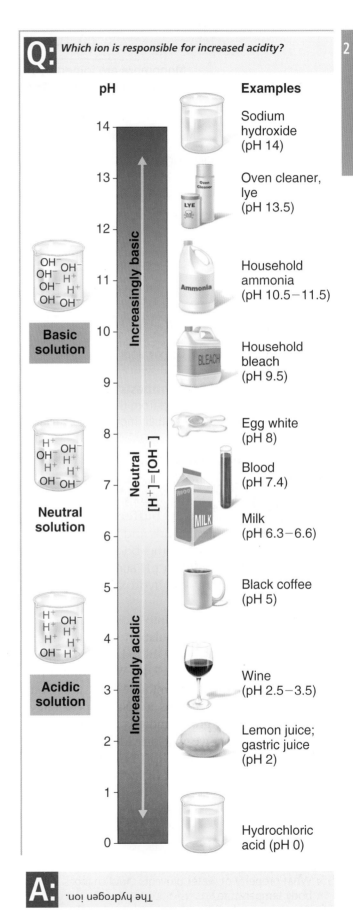

**Q:** *Which ion is responsible for increased acidity?*

**Figure 2.12 The pH scale and pH values of representative substances.** The pH scale is based on the number of hydrogen ions in solution. At a pH of 7, the number of $H^+$ = the number of $OH^-$, and the solution is neutral. A solution with a pH below 7 is acidic (more $H^+$ than $OH^-$); above 7, basic, or alkaline (less $H^+$ than $OH^-$).

**A:** The hydrogen ion.

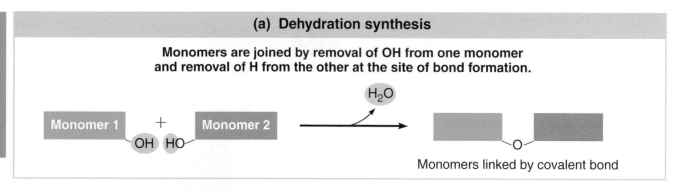

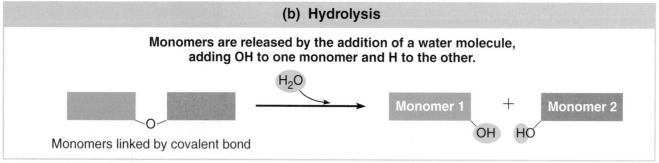

**Figure 2.13 Dehydration synthesis and hydrolysis of biological molecules.**
Biological molecules are formed from their monomers (units) by dehydration synthesis and broken down to their monomers by hydrolysis.

Living cells are extraordinarily sensitive to even slight changes in pH. Acid-base balance is carefully regulated by the kidneys, lungs, and a number of chemicals called **buffers** that are present in body fluids. Weak acids and weak bases are important components of the body's buffer systems, which act to maintain pH stability by taking up excess hydrogen or hydroxide ions (see Chapter 15).

Because blood comes into close contact with nearly every body cell, regulation of blood pH is especially critical. Normally, blood pH varies in a narrow range, from 7.35 to 7.45. When blood pH changes more than a few tenths of a pH unit from these limits, death becomes a distinct possibility. Although we could give hundreds of examples to illustrate this point, we will provide just one very important one: When blood pH begins to dip into the acid range, the amount of life-sustaining oxygen that the hemoglobin in blood can carry to body cells decreases rapidly to dangerously low levels.

### Did You Get It?

15. What property of water prevents rapid changes in body temperature?
16. Which is a proton donor—an acid or a base?
17. Is a pH of 11 acidic or basic? What is the difference in pH between a solution at pH 11 and a solution at pH 5?
18. Biochemistry is "wet" chemistry. What does this statement mean?
19. Salts are electrolytes. What does that mean?

For answers, see Appendix A.

## Organic Compounds

### → Learning Objectives

☐ **Explain the role of dehydration synthesis and hydrolysis in formation and breakdown of organic molecules.**

☐ **Compare and contrast carbohydrates and lipids in terms of their building blocks, structures, and functions in the body.**

Most organic compounds are very large molecules, but their interactions with other molecules typically involve only small, reactive parts of their structure called *functional groups* (acid groups, amines, and others).

Many organic compounds (carbohydrates and proteins, for example) are polymers. **Polymers** are chainlike molecules made of many similar or repeating units (**monomers**), which are joined together by **dehydration synthesis (Figure 2.13a)**. During dehydration synthesis, a hydrogen atom is

**(a) Simple sugar (monosaccharide)**    **(b) Double sugar (disaccharide)**

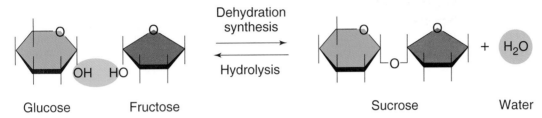

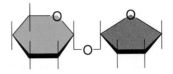

Glucose    Fructose    Sucrose    Water

**(c) Dehydration synthesis and hydrolysis of a molecule of sucrose**

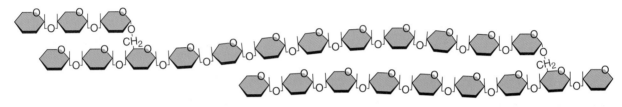

**(d) Starch (polysaccharide)**

**Figure 2.14 Carbohydrates. (a)** The generalized structure of a monosaccharide.
**(b)** and **(d)** The basic structures of a disaccharide and a polysaccharide, respectively.
**(c)** Formation and breakdown of the disaccharide sucrose by dehydration synthesis
and hydrolysis, respectively.

removed from one monomer and a hydroxyl group (OH) is removed from the monomer it is to be joined with. A covalent bond then forms, uniting the monomers, and a water molecule is released. This removal of a water molecule (dehydration) at the bond site occurs each time a monomer is added to the growing polymer chain.

When polymers must be broken down or digested to their monomers, the reverse process, called **hydrolysis**, occurs (Figure 2.13b). As a water molecule is added to each bond, the bond is broken, releasing the monomers. All organic molecules covered in this chapter—carbohydrates, lipids, proteins, and nucleic acids—are formed by dehydration synthesis and broken down by hydrolysis.

## Carbohydrates

**Carbohydrates**, which include sugars and starches, contain carbon, hydrogen, and oxygen. With slight variations, the hydrogen and oxygen atoms appear in the same ratio as in water; that is, two hydrogen atoms to one oxygen atom. This is reflected in the word *carbohydrate*, which means "hydrated carbon," and in the molecular formulas of sugars. For example, glucose is $C_6H_{12}O_6$, and ribose is $C_5H_{10}O_5$.

Carbohydrates are classified according to size and solubility in water as monosaccharides, disaccharides, or polysaccharides. Because monosaccharides are joined to form the molecules of the other two groups, they are the structural units, or building blocks, of carbohydrates.

***Monosaccharides*** **Monosaccharide** means one (*mono*) sugar (*saccharide*), and thus monosaccharides are also referred to as *simple sugars*. They are single-chain or single-ring structures (meaning the carbon backbone forms either a line or a circle), containing from three to seven carbon atoms **(Figure 2.14a)**.

The most important monosaccharides in the body are glucose, fructose, galactose, ribose, and deoxyribose. **Glucose**, also called *blood sugar*, is the universal cellular fuel. *Fructose* and *galactose* are converted to glucose for use by body cells.

*Ribose* and *deoxyribose* form part of the structure of nucleic acids, another group of organic molecules responsible for genetic information.

**Disaccharides** **Disaccharides**, or *double sugars* (Figure 2.14b), are formed when two simple sugars are joined by dehydration synthesis. In this reaction, as noted earlier, a water molecule is lost as the bond forms (Figure 2.14c).

Some of the important disaccharides in the diet are *sucrose* (glucose-fructose), which is cane sugar; *lactose* (glucose-galactose), found in milk; and *maltose* (glucose-glucose), or malt sugar. Because the double sugars are too large to pass through cell membranes, they must be broken down (digested) to their monosaccharide units to be absorbed from the digestive tract into the blood; this is accomplished by hydrolysis (see Figures 2.14b and 2.13b).

**Polysaccharides** Long, branching chains of linked simple sugars are called **polysaccharides** (literally, "many sugars") (Figure 2.14d). Because they are large, insoluble molecules, they are ideal storage products. Another consequence of their large size is that they lack the sweetness of simple and double sugars.

Only two polysaccharides, starch and glycogen, are of major importance to the body. *Starch* is the storage polysaccharide formed by plants. We ingest it in the form of "starchy" foods, such as grain products (corn, rice) and root vegetables (potatoes and carrots, for example). *Glycogen* is a slightly smaller, but similar, polysaccharide found in animal tissues (largely in the muscles and the liver). Like starch, it is a polymer of linked glucose units.

Carbohydrates provide a ready, easily used source of fuel for cells, and glucose is at the top of the "cellular menu." When glucose is oxidized ("burned" by combining with oxygen) in a complex set of chemical reactions, it is broken down into carbon dioxide and water. Some of the energy released as the glucose bonds are broken is trapped in the bonds of high-energy ATP molecules, the energy "currency" of all body cells. If not immediately needed for ATP synthesis, dietary carbohydrates are converted to glycogen or fat and stored. Those of us who have gained weight from eating too many carbohydrate-rich snacks have firsthand experience of this conversion process!

Small amounts of carbohydrates are used for structural purposes and represent 1 to 2 percent of cell mass. Some sugars are found in our genes, and others are attached to outer surfaces of cell membranes (boundaries), where they act as road signs to guide cellular interactions.

### Lipids

**Lipids**, or fats, are a large and diverse group of organic compounds (Table 2.5, p. 73). They enter the body in the form of meats, egg yolks, dairy products, and oils. The most abundant lipids in the body are *triglycerides*, *phospholipids*, and *steroids*. Like carbohydrates, all lipids contain carbon, hydrogen, and oxygen atoms, but in lipids, carbon and hydrogen atoms far outnumber oxygen atoms, as illustrated by the formula for a typical fat named tristearin, which is $C_{57}H_{110}O_6$. Lipids are insoluble in water but readily dissolve in other lipids and in organic solvents such as alcohol and acetone.

**Triglycerides** The **triglycerides** (tri-glis′er-īdz), or **neutral fats**, are composed of two types of building blocks, **fatty acids** and **glycerol**. Their synthesis involves the attachment of three fatty acids to a single glycerol molecule. The result is an E-shaped molecule that resembles the tines of a fork **(Figure 2.15a)**. Although the glycerol backbone is the same in all neutral fats, the fatty acid chains vary; this variation results in different kinds of neutral fats.

The length of a triglyceride's fatty acid chains and their type of C—C bonds determine how solid the molecule is at a given temperature. Fatty acid chains with only single covalent bonds between carbon atoms are referred to as **saturated fats**. Their fatty acid chains are straight **(Figure 2.16a, p. 72)**, and, at room temperature, the molecules of a saturated fat pack closely together, forming a solid; butter is an example. Fatty acids that contain one or more double bonds between carbon atoms are said to be **unsaturated fats** (**monounsaturated** and **polyunsaturated**, respectively). The double and triple bonds cause the fatty acid chains to kink (Figure 2.16b) so that they cannot pack closely enough to solidify. Hence, triglycerides with short fatty acid chains or unsaturated fatty acids are oils (liquid at room temperature). Plant lipids are typically oils. Examples include olive oil (rich in monounsaturated fats) and soybean and safflower oils, which contain a high percentage of

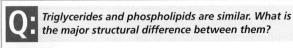

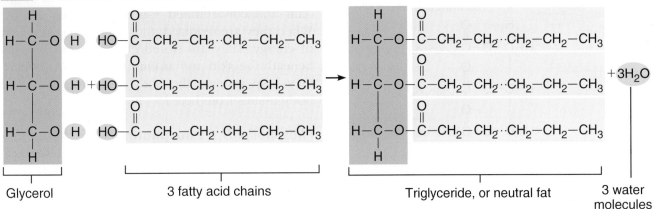

Glycerol      3 fatty acid chains      Triglyceride, or neutral fat      3 water molecules

**(a) Formation of a triglyceride. Fatty acid chains are bound to glycerol by dehydration synthesis.**

Polar "head"

Nonpolar "tail"
(schematic
phospholipid)

Phosphorus-containing group (polar head)     Glycerol backbone     2 fatty acid chains (nonpolar tail)

**(b) Typical structure of a phospholipid molecule (phosphatidylcholine). Two fatty acid chains and a phosphorous-containing group are attached to a glycerol backbone.**

**(c) Cholesterol. Simplified structure of cholesterol, a steroid, formed by four interlocking carbon rings.**

**Figure 2.15 Lipids. (a)** Triglycerides, or neutral fats, are synthesized by dehydration synthesis, and a water molecule is lost at each bond site. **(b)** Structure of a typical phospholipid molecule. Notice that one fatty acid chain in (b) is unsaturated (has one or more C=C [double] bonds). **(c)** The generalized structure of cholesterol (the basis for all steroids made in the body).

**Q:** *Triglycerides and phospholipids are similar. What is the major structural difference between them?*

**A:** They both have a glycerol backbone and fatty acid chains. However, triglycerides have three attached fatty acid chains, whereas phospholipids have only two; the third is replaced by a negatively charged phosphate-containing group.

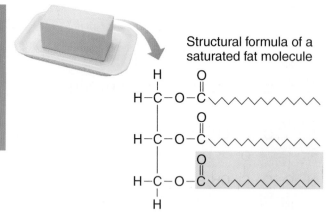

**(a) Saturated fat.** At room temperature, the molecules of a saturated fat such as this butter are packed closely together, forming a solid.

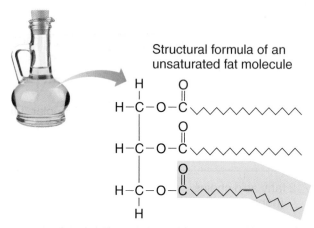

**(b) Unsaturated fat.** At room temperature, the molecules of an unsaturated fat such as this olive oil cannot pack together closely enough to solidify because of the kinks in some of their fatty acid chains.

**Figure 2.16 Examples of saturated and unsaturated fats and fatty acids.**

polyunsaturated fatty acids. Longer fatty acid chains and more saturated fatty acids are common in animal fats, such as butterfat and the fat of meats, which are solid at room temperature. Of the two types of fatty acids, the unsaturated variety, especially olive oil, is more "heart healthy."

**Trans fats**, common in many margarines and baked products, are oils that have been solidified by the addition of hydrogen atoms at sites of double carbon bonds, which reduces those bonds to single carbon bonds. They increase the risk of heart disease by raising "bad" cholesterol and decreasing "good" cholesterol. In contrast, the **omega-3 fatty acids**, found naturally in cold-water

fish, appear to decrease the risk of heart disease and some inflammatory diseases.

Triglycerides represent the body's most abundant and concentrated source of usable energy. When they are oxidized, they yield large amounts of energy. They are stored chiefly in fat deposits beneath the skin and around body organs, where they help insulate the body and protect deeper body tissues from heat loss and injury.

***Phospholipids.* Phospholipids** (fos'fo-lip"idz) are similar to triglycerides. The major difference is that a phosphorus-containing group is always part of the molecule and takes the place of one of the fatty acid chains. Thus, phospholipids have two instead of three attached fatty acids (see Figure 2.15b).

Because the phosphorus-containing portion (the "head") bears a negative charge, phospholipids have special chemical properties and polarity. For example, the charged region is **hydrophilic** ("water loving"), meaning it attracts and interacts with water and ions. The fatty acid chains (the nonpolar "tail") are **hydrophobic** ("water fearing") and do not interact with polar or charged molecules. The presence of phospholipids in *cell membranes* allows cells to be selective about what may enter or leave.

***Steroids* Steroids** are basically flat molecules formed of four interlocking carbon rings (see Figure 2.15c); thus, their structure differs quite a bit from that of other lipids. However, like fats, steroids are made largely of hydrogen and carbon atoms and are fat-soluble. The single most important steroid molecule is **cholesterol**. We ingest cholesterol in animal products such as meat, eggs, and cheese, and some is made by the liver, regardless of dietary intake.

Cholesterol has earned bad press because of its role in arteriosclerosis, but it is essential for human life. Cholesterol is found in cell membranes and is the raw material used to make vitamin D, steroid hormones (chemical signaling molecules in the body), and bile salts (which help break down fats during digestion). Although steroid hormones are present in the body in only small quantities, they are vital to homeostasis. Without sex hormones, reproduction would be impossible; and without the corticosteroids produced by the adrenal glands, the body would not survive.

Table **2.5**    **Representative Lipids Found in the Body**

| Lipid type | Location/function |
|---|---|
| **Triglycerides (neutral fats)** | Found in fat deposits (subcutaneous tissue and around organs); protect and insulate the body organs; the major source of stored energy in the body. |
| **Phospholipids** | Found in cell membranes; participate in the transport of lipids in plasma; abundant in the brain and the nervous tissue in general, where they help to form insulating white matter. |
| **Steroids** Cholesterol | The basis of all body steroids. |
| Bile salts | A breakdown product of cholesterol; released by the liver into the digestive tract, where they aid in fat digestion and absorption. |
| Vitamin D | A fat-soluble vitamin produced in the skin on exposure to UV (ultraviolet) radiation (sunshine); necessary for normal bone growth and function. |
| Sex hormones | Estrogen and progesterone (female hormones) and testosterone (a male hormone) produced from cholesterol; necessary for normal reproductive function; deficits result in sterility. |
| Corticosteroids (adrenal cortical hormones) | Cortisol, a glucocorticoid, is a long-term antistress hormone that is necessary for life; aldosterone helps regulate salt and water balance in body fluids by targeting the kidneys. |
| **Other lipid-based substances** Fat-soluble vitamins A | Found in orange-pigmented vegetables (carrots) and fruits (tomatoes); part of the photoreceptor pigment involved in vision. |
| E | Taken in via plant products such as wheat germ and green leafy vegetables; may promote wound healing and contribute to fertility, but not proven in humans; an antioxidant; may help to neutralize free radicals (highly reactive particles believed to be involved in triggering some types of cancers). |
| K | Made available largely by the action of intestinal bacteria; also prevalent in a wide variety of foods; necessary for proper clotting of blood. |
| Prostaglandins | Derivatives of fatty acids found in cell membranes; various functions depending on the specific class, including stimulation of uterine contractions (thus inducing labor and miscarriages), regulation of blood pressure, and control of motility of the gastrointestinal tract; involved in inflammation. |
| Lipoproteins | Lipoid and protein-based substances that transport fatty acids and cholesterol in the bloodstream; major varieties are high-density lipoproteins (HDLs) and low-density lipoproteins (LDLs). |
| Glycolipids | Component of cell membranes. Lipids associated with carbohydrate molecules that determine blood type, play a role in cell recognition or in recognition of foreign substances by immune cells. |

**Q:** *What is the importance of the R-groups (green here)?*

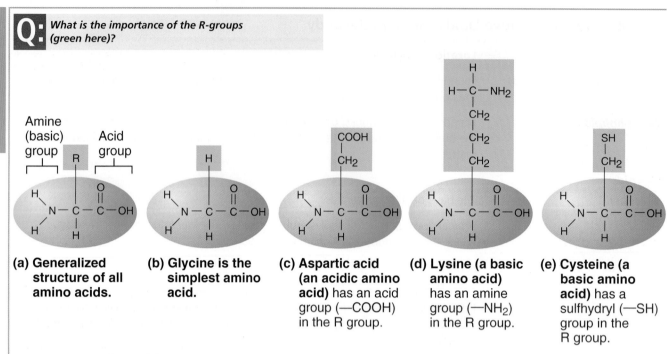

(a) **Generalized structure of all amino acids.**

(b) **Glycine is the simplest amino acid.**

(c) **Aspartic acid (an acidic amino acid)** has an acid group (—COOH) in the R group.

(d) **Lysine (a basic amino acid)** has an amine group (—NH₂) in the R group.

(e) **Cysteine (a basic amino acid)** has a sulfhydryl (—SH) group in the R group.

**Figure 2.17 Amino acid structures. (a)** Generalized structure of amino acids. All amino acids have both an amine ($NH_2$) group and an acid (COOH) group; they differ only in the atomic makeup of their R-groups (green). **(b–e)** Specific structures of four amino acids.

## Did You Get It?

20. What are the structural units, or building blocks, of carbohydrates? Of lipids?
21. Which type(s) of lipid(s) is/are found in cellular membranes?
22. Is the formation of glycogen from glucose an example of hydrolysis or dehydration synthesis?

For answers, see Appendix A.

## Proteins

→ **Learning Objectives**

☐ **Differentiate fibrous proteins from globular proteins.**

☐ **Define *enzyme*, and explain the role of enzymes.**

**Proteins** account for over 50 percent of the organic matter in the body, and they have the most varied functions of all organic molecules. Some are construction materials; others play vital roles in cell function. Like carbohydrates and lipids, all proteins contain carbon, oxygen, and hydrogen. In

**A:** The R-groups make each type of amino acid functionally unique.

addition, they contain nitrogen and sometimes sulfur atoms as well.

The building blocks of proteins are small molecules called **amino** (ah-me′no) **acids**. Twenty varieties of amino acids are found in human proteins. All amino acids have an *amine group* ($NH_2$), which gives them basic properties, and an *acid group* (COOH), which allows them to act as acids. In fact, all amino acids are identical except for a single group of atoms called their *R-group* **(Figure 2.17)**. Differences in the R-groups make each amino acid chemically unique. For example, an extra acid group (COOH) in the R-group (such as in aspartic acid) makes the amino acid more acidic, and a sulfhydryl group (—SH) in the R-group (such as in cysteine) is often involved in stabilizing protein structure by binding to another amino acid within the same protein.

Amino acids are joined together in chains to form *polypeptides* (fewer than 50 amino acids); proteins (more than 50 amino acids); and large, complex proteins (50 to thousands of amino acids). Because each type of amino acid has distinct properties, the sequence in which they are

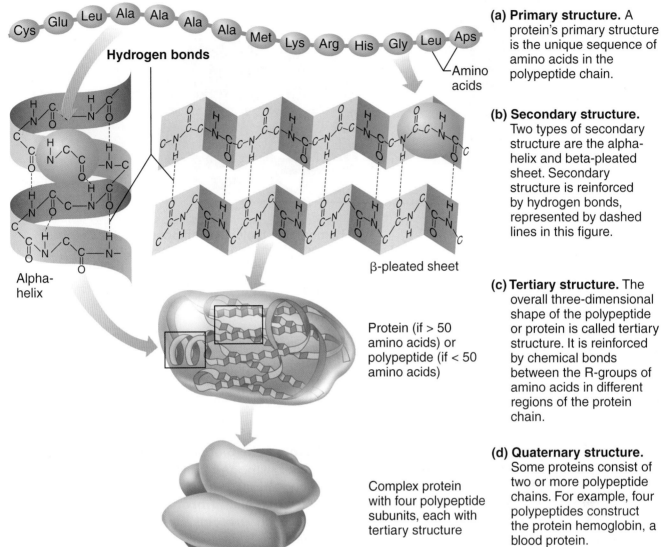

(a) **Primary structure.** A protein's primary structure is the unique sequence of amino acids in the polypeptide chain.

(b) **Secondary structure.** Two types of secondary structure are the alpha-helix and beta-pleated sheet. Secondary structure is reinforced by hydrogen bonds, represented by dashed lines in this figure.

(c) **Tertiary structure.** The overall three-dimensional shape of the polypeptide or protein is called tertiary structure. It is reinforced by chemical bonds between the R-groups of amino acids in different regions of the protein chain.

(d) **Quaternary structure.** Some proteins consist of two or more polypeptide chains. For example, four polypeptides construct the protein hemoglobin, a blood protein.

Protein (if > 50 amino acids) or polypeptide (if < 50 amino acids)

Complex protein with four polypeptide subunits, each with tertiary structure

**Figure 2.18 The four levels of protein structure.**

bound together produces proteins that vary widely both in structure and function. To understand this more easily, think of the 20 amino acids as a 20-letter alphabet. The letters (amino acids) are used in specific combinations to form words (a protein). Just as a change in one letter of any word can produce a word with an entirely different meaning (flour → floor) or one that is nonsensical (flour → fluur), changes in amino acids (letters) or in their positions in the protein allow literally thousands of different protein molecules to be made.

***Structural Levels of Proteins*** Proteins can be described in terms of four structural levels. The sequence of amino acids composing each amino acid chain is called the *primary structure*. This structure, which resembles a strand of amino acid "beads," is the backbone of a protein molecule in which the chemical properties of each amino acid will affect how the protein folds **(Figure 2.18a)**.

Most proteins do not function as simple, linear chains of amino acids. Instead, they twist or bend upon themselves to form a more complex *secondary structure*. The most common secondary structure is the **alpha (α)-helix**, which resembles a metal spring (Figure 2.18b). The α-helix is formed by coiling of the primary chain and is stabilized by hydrogen bonds. Hydrogen bonds in α-helices

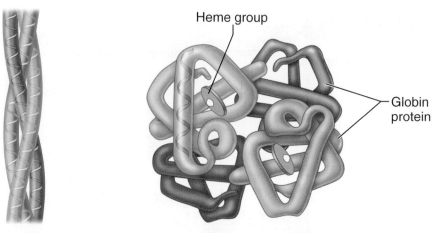

**(a) Triple helix of collagen (a fibrous or structural protein).**

**(b) Hemoglobin molecule composed of the protein globin and attached heme groups. (Globin is a globular, or functional, protein.)**

**Figure 2.19  General structure of (a) a fibrous protein and (b) a globular protein.**

always link different parts of the *same* chain together.

In another type of secondary structure, the **beta (β)-pleated sheet**, the primary polypeptide chains do not coil, but are linked side by side by hydrogen bonds to form a pleated, ribbonlike structure that resembles the pleats of a skirt or a sheet of paper folded into a fan (see Figure 2.18b). In this type of secondary structure, the hydrogen bonds may link together *different polypeptide chains* as well as *different parts* of the same chain that has folded back on itself.

Many proteins have *tertiary structure* (ter′she-a″re), the next higher level of complexity. Tertiary structure is achieved when α-helical or β-pleated regions of the amino acid chain fold upon one another to produce a compact ball-like, or *globular*, protein (Figure 2.18c). The unique structure is maintained by covalent and hydrogen bonds between amino acids that are often far apart in the primary chain. Finally, when two or more amino acid chains (polypeptide chains) combine in a regular manner to form a complex protein, the protein has *quaternary* (kwah′ter-na″re) *structure* (Figure 2.18d).

Although a protein with tertiary or quaternary structure looks a bit like a crumpled ball of tin foil, the final structure of any protein is very specific and is dictated by its primary structure. In other words, the types and positions of amino acids in the protein backbone determine where hydrogen bonds can form to keep water-loving (hydrophilic) amino acids near the surface and water-fearing (hydrophobic) amino acids buried in the protein's core so the protein remains water-soluble.

***Fibrous and Globular Proteins***   Based on their overall shape and structure, proteins are classed as either fibrous or globular proteins **(Figure 2.19)**. The strandlike **fibrous proteins**, also called **structural proteins**, appear most often in body structures. Some exhibit only secondary structure, but most have tertiary or even quaternary structure. They are very important in binding structures together and providing strength in certain body tissues. For example, *collagen* (kol′ah-jen) is found in bones, cartilage, and tendons and is the most abundant protein in the body (Figure 2.19a). *Keratin* (ker′ah-tin) is the structural protein of hair and nails and the material that makes skin tough.

**Globular proteins** are mobile, generally compact, spherical molecules that have at least tertiary structure. These water-soluble proteins play crucial roles in virtually all biological processes. Because they *do things* rather than just form structures, they are also called **functional proteins**; the scope of their activities is remarkable **(Table 2.6)**. For example, some proteins called antibodies help to provide immunity; others (hormones) help to regulate growth and development. Still others, called

Table **2.6**  **Representative Classes of Functional Proteins**

| Functional class | Role(s) in the body |
|---|---|
| Antibodies (immunoglobulins) | Highly specialized proteins that recognize, bind with, and inactivate bacteria, toxins, and some viruses; function in the immune response, which helps protect the body from "invading" foreign substances. |
| Hormones | Help to regulate growth and development. Examples include<br>• Growth hormone—an anabolic hormone necessary for optimal growth.<br>• Insulin—helps regulate blood sugar levels.<br>• Nerve growth factor—guides the growth of neurons in the development of the nervous system. |
| Transport proteins | Hemoglobin transports oxygen in the blood; other transport proteins in the blood carry iron, cholesterol, or other substances. |
| Enzymes (catalysts) | Essential to virtually every biochemical reaction in the body; increase the rates of chemical reactions by at least a millionfold; in their absence (or destruction), biochemical reactions cease. |

*enzymes* (en′zīmz), regulate essentially every chemical reaction that goes on within the body. The oxygen-carrying protein hemoglobin is an example of a globular protein with quaternary structure (Figure 2.19b).

The fibrous structural proteins are exceptionally stable, but the globular functional proteins are quite the opposite. Hydrogen bonds are critically important in maintaining their structure, but hydrogen bonds are fragile and are easily broken by heat and excesses of pH. When their three-dimensional structures are destroyed, the proteins are said to be *denatured* and can no longer perform their physiological roles. Why? Their function depends on their specific three-dimensional shapes. Hemoglobin becomes totally unable to bind and transport oxygen when blood pH becomes too acidic, as we stated earlier. Pepsin, a protein-digesting enzyme that acts in the stomach, is inactivated by alkaline pH. In each case, the improper pH has destroyed the structure required for function.

Because enzymes are important in the functioning of all body cells, we consider these incredibly complex molecules here.

### Enzymes and Enzyme Activity  Enzymes are functional proteins that act as biological catalysts. A **catalyst** is a substance that increases the rate of a chemical reaction without becoming part of the product or being changed itself. Enzymes can accomplish this feat because they have unique regions called **active sites** on their surfaces. These sites "fit" and interact chemically with other molecules of complementary shape and charge called *substrates* **(Figure 2.20)**. While the substrates are bound to the enzyme's active site, producing a structure called an *enzyme-substrate complex*, they undergo structural changes that result in a new product. Whereas some enzymes build larger molecules, others break things into smaller pieces or simply modify a substrate. Once the reaction has occurred, the enzyme releases the product. Because enzymes are not changed during the reaction, they are reusable, and the cells need only small amounts of each enzyme. Think of scissors cutting paper. The scissors (the enzyme) are unchanged while the paper (the substrate) is "cut" to become the product. The scissors are reusable and go on to "cut" other paper.

Enzymes are capable of catalyzing millions of reactions each minute. However, they do more than just increase the speed of chemical reactions; they also determine just which reactions are possible at a particular time. No enzyme, no reaction! Enzymes can be compared to a bellows used to fan a sluggish fire into flaming activity. Without enzymes, biochemical reactions would occur far too slowly to sustain life.

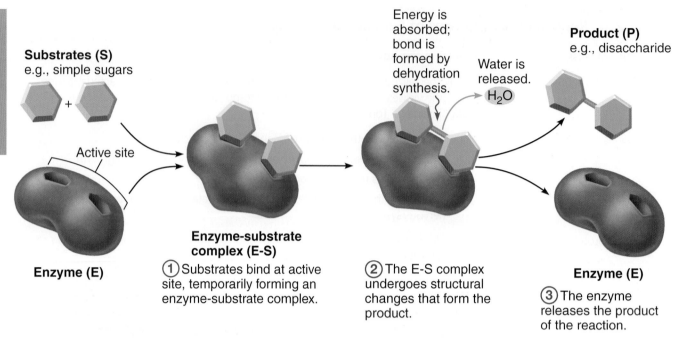

**Figure 2.20 A simplified view of enzyme action.**

Although there are hundreds of different kinds of enzymes in body cells, they are very specific in their activities, each controlling only one (or a small group of) chemical reaction(s) and acting only on specific molecules. Most enzymes are named according to the specific type of reaction they catalyze. For example, *hydrolases* add water, and *oxidases* cause oxidation. (In most cases, you can recognize an enzyme by the suffix **-ase** in its name.)

The activity of different enzymes is controlled in different ways. Many enzymes are produced in an inactive form and must be activated in some way before they can function. In other cases, enzymes are inactivated immediately after they have performed their catalytic function. Both events are true of enzymes that promote blood clotting when a blood vessel has been damaged. If this were not so, large numbers of unneeded and potentially lethal blood clots would be formed.

## Did You Get It?

23. What is the primary structure of proteins?
24. Which is more important for building body structures, fibrous or globular proteins?
25. How does an enzyme recognize its substrate(s)?

For answers, see Appendix A.

## Nucleic Acids

→ **Learning** Objectives

☐ **Compare and contrast the structures and functions of DNA and RNA.**

☐ **Explain the importance of ATP in the body.**

The role of **nucleic** (nu-kle′ik) **acids** is fundamental: They make up your genes, which provide the basic blueprint of life. They not only determined what type of organism you would be, but also directed your growth and development—and they did this largely by dictating protein structure. (Remember that enzymes, which catalyze all the chemical reactions that occur in the body, are proteins.)

Nucleic acids, composed of carbon, oxygen, hydrogen, nitrogen, and phosphorus atoms, are the largest biological molecules in the body. Their building blocks, **nucleotides** (nu′kle-o-tīdz), are quite complex. Each consists of three basic parts: (1) a nitrogen-containing base, (2) a pentose (5-carbon) sugar, and (3) a phosphate group (**Figure 2.21a** and **b**).

The bases come in five varieties: *adenine* (A), *guanine* (G), *cytosine* (C), *thymine* (T), and *uracil* (U). A and G are large, nitrogen-containing bases made up of two carbon rings, whereas C, T and U are smaller, single-ring structures. The nucleotides

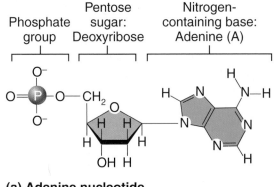

Phosphate group    Pentose sugar: Deoxyribose    Nitrogen-containing base: Adenine (A)

**(a) Adenine nucleotide**
(Chemical structure)

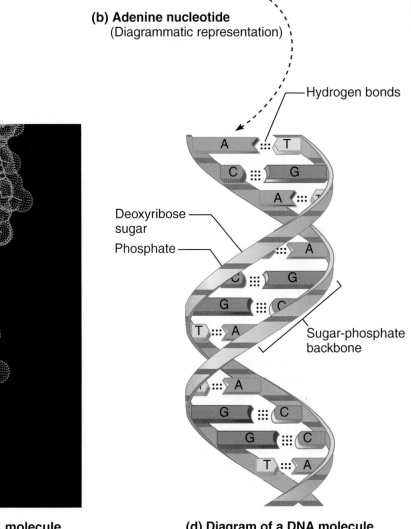

**(b) Adenine nucleotide**
(Diagrammatic representation)

Hydrogen bonds

Deoxyribose sugar

Phosphate

Sugar-phosphate backbone

**(d) Diagram of a DNA molecule**

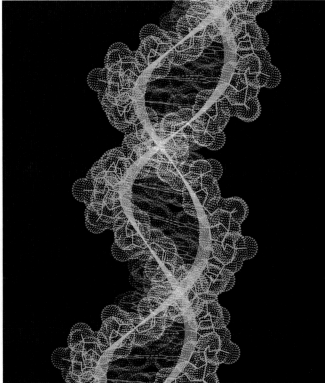

**(c) Computer-generated image of a DNA molecule**

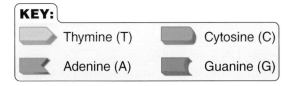

**KEY:**

Thymine (T)    Cytosine (C)

Adenine (A)    Guanine (G)

**Figure 2.21 Structure of DNA.**
**(a)** The unit of DNA, the nucleotide, is composed of a deoxyribose sugar molecule linked to a phosphate group. A nitrogen-containing base is attached to the sugar. The nucleotide illustrated, both in its **(a)** chemical and **(b)** diagrammatic structures, contains the base adenine. **(c)** Computer-generated image of DNA. **(d)** Diagrammatic structure of a DNA molecule—two nucleotide chains coiled into a double helix. The "backbones" of DNA are formed by alternating sugar and phosphate molecules. The "rungs" are formed by complementary bases (A to T, G to C) bound by two or three hydrogen bonds, respectively.

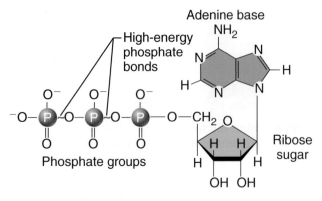

**(a) Adenosine triphosphate (ATP)**

**Figure 2.22 ATP—structure and hydrolysis.**
**(a)** The structure of ATP (adenosine triphosphate).
**(b)** Hydrolysis of ATP to yield ADP (adenosine diphosphate) and inorganic phosphate ($P_i$). High-energy bonds are indicated by a red ~.

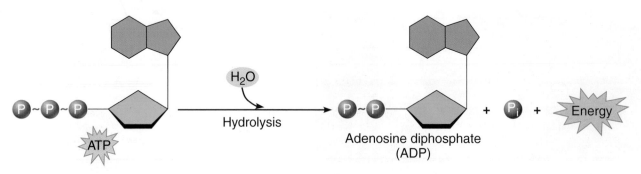

**(b) Hydrolysis of ATP**

are named according to the base they contain: A-containing bases are adenine nucleotides, C-containing bases are cytosine nucleotides, and so on.

The two major kinds of nucleic acids are **deoxyribonucleic** (de-ok″sĭ-ri″bo-nu-kle′ik) **acid (DNA)** and **ribonucleic acid (RNA)**. DNA and RNA differ in many respects. DNA is the genetic material found within the cell nucleus (the control center of the cell). It has two fundamental roles: (1) It replicates itself exactly before a cell can divide, thus ensuring that every body cell gets an identical copy of the genetic information; and (2) it provides the instructions for building every protein in the body. For the most part, RNA functions outside the nucleus and can be considered the "molecular assistant" of DNA; that is, RNA carries out the orders for protein synthesis issued by DNA.

Although both DNA and RNA are formed when nucleotides join together, their final structures are different. DNA is a long double chain of nucleotides (Figure 2.21c and d). Its bases are A, G, T, and C, and its sugar is *deoxyribose*. Its two nucleotide chains are held together by hydrogen bonds between the bases, forming a ladderlike molecule. Alternating sugar and phosphate molecules form the "uprights" of the ladder, called the *sugar-phosphate backbone*, and each "rung" is formed by two joined bases (one *base pair*). Binding between the bases is very specific: A always binds to T, and G always binds to C. Thus, A and T are said to be *complementary bases*, as are C and G. Therefore, a sequence of ACTGA on one nucleotide chain would be bound to the complementary sequence TGACT on the other nucleotide strand. The whole molecule is then coiled into a spiral-staircase-like structure called a *double helix*.

Whereas DNA is double-stranded, RNA molecules are single nucleotide strands. The RNA bases are A, G, C, and U (U replaces the T found in DNA), and its sugar is *ribose* instead of deoxyribose. Three major varieties of RNA exist—*messenger, transfer,* and *ribosomal RNA*—and each has a specific role to play in carrying out DNA's instructions for building proteins. Messenger RNA (mRNA) carries the information for building the protein from the DNA to the ribosomes, the protein-synthesizing sites. Transfer RNA (tRNA) ferries amino acids to the ribosomes. Ribosomal RNA (rRNA) forms part of the ribosomes, where it oversees the translation of the message and the binding together of amino acids to form the proteins. (We describe protein synthesis in greater detail in Chapter 3.)

### Adenosine Triphosphate (ATP)

The synthesis of **adenosine triphosphate** (ah-den′o-sēn tri-fos′fāt), or **ATP**, is all-important because it provides a form of chemical energy that all body cells can use. Without ATP, molecules cannot be made or broken down, cells cannot maintain their membrane boundaries, and all life processes grind to a halt.

Although glucose is the most important fuel for body cells, none of the chemical energy contained in its bonds can be used directly to power cellular work. Instead, energy is released as glucose is oxidized, and it is captured and stored in the bonds of ATP molecules as small packets of energy. Compare glucose and ATP to crude oil and gasoline: While oil stores lots of energy, it cannot be used directly to power a car without first being refined into gasoline.

Structurally, ATP is a modified RNA nucleotide; it consists of an adenine base, ribose sugar, and three phosphate groups instead of one **(Figure 2.22a)**. The phosphate groups are attached by unique chemical bonds called *high-energy phosphate bonds*. These bonds are high energy because the phosphate groups are negatively charged and the like charges repel one another, generating tension at these sites. As ATP is used to provide cellular energy, **adenosine diphosphate (ADP)** accumulates (Figure 2.22b), and ATP supplies are replenished by oxidation of food fuels. Essentially, the same amount of energy must be captured and used to reattach a phosphate group to ADP (that is, to reverse the reaction) as is liberated when the terminal phosphate is removed from ATP.

When the high-energy bonds of ATP are broken by hydrolysis, energy is liberated and can be used immediately by the cell to do work or power a particular activity—such as driving chemical reactions, transporting solutes across the membrane, or, in the case of muscle cells, contracting **(Figure 2.23)**. ATP can be compared to a tightly coiled spring that is ready to uncoil with tremendous

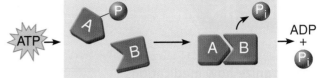

**(a) Chemical work.** ATP provides the energy needed to drive energy-absorbing chemical reactions.

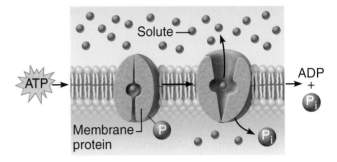

**(b) Transport work.** ATP drives the transport of certain solutes (amino acids, for example) across cell membranes.

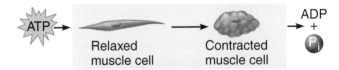

**(c) Mechanical work.** ATP activates contractile proteins in muscle cells so that the cells can shorten and perform mechanical work.

**Figure 2.23 Three examples of how ATP drives cellular work.** The high-energy bonds of ATP release energy for use by the cell when they are broken. ATP is regenerated (phosphate is again bound to ADP) as energy is released by the oxidation of food fuels and captured in the third high-energy bond.

energy when the "catch" is released. The consequence of breaking its terminal phosphate bond can be represented as follows:

$$\text{ATP} \rightleftharpoons \text{ADP} + \text{P}_i + \text{E}$$
$$\text{(adenosine} \quad \text{(adenosine} \quad \text{(inorganic} \quad \text{(energy)}$$
$$\text{triphosphate)} \quad \text{diphosphate)} \quad \text{phosphate)}$$

### Did You Get It?

**26.** How do DNA and RNA differ from each other in the kinds of bases and sugars they contain?

**27.** What is the vital importance of ATP to body cells?

**For answers, see Appendix A.**

# Pharmacy Technician

**To recognize how medications affect patients, pharmacy technicians need a thorough understanding of anatomy and physiology.**

When most people get a new medication, they open up the package and toss out the little pamphlet that goes into detail about how the medication works. Not Chris Green. "I love reading the package inserts," says Green, the lead pharmacy technician at a CVS drugstore in Birmingham, Alabama. Green's enthusiasm for those details is a lifesaver for his customers. Pharmacy technicians are a vital link in the chain between doctor and patient.

Although pharmacy technicians are legally prohibited from talking with patients about their symptoms, they can translate medical jargon and discuss a medication's side effects and other precautions the patient may need to take. For example, doctors may recommend that patients who are on certain medications for a long time have regular tests such as eye exams, bloodwork, or tests for liver function. A pharmacy technician can convey that information to the patient—and check on subsequent visits to make sure he or she is following up.

A busy retail pharmacy has various stations: data entry, where the patient's record is updated with a new prescription; production, where the prescription is filled; and verification, where the pharmacist reviews the prescription and makes sure it is filled and labeled correctly. Green's job is to make sure the process flows smoothly from station to station.

> Pharmacy technicians must have a good grasp of anatomy and physiology to understand each drug's chemical properties.

Green started working as a cashier at a drugstore when he was in high school and gradually became interested in the pharmacy itself.

"I was interested in how drugs work, how they can help people and improve their health," he says.

Having earned a bachelor's degree in biology, Green emphasizes that pharmacy technicians must have a good grasp of the sciences, especially basic chemistry and anatomy and physiology, to help them understand each drug's chemical makeup and properties.

"When all the data is entered, we see what potential side effects there are," he says. "It's important to know how the medications work, how they interact with each other, how they interact with the body. I might see something and bring it to the pharmacist's attention." In addition, communication skills and the ability to work with people are important. Good communication can be the difference between life and death for a patient, particularly when a doctor prescribes a

medication that could react badly with another medication the patient is already taking. Drug interactions happen commonly when you have multiple doctors. "Sometimes, we'll get two ACE inhibitors in the same category from two different doctors [prescribed for the same patient], and that could be lethal," Green says.

Pharmacy technicians work in retail and mail-order pharmacies, hospitals, nursing homes, assisted living facilities, and anywhere else patients have high needs for medication. As the Baby Boom generation ages and the number of senior citizens grows, so does the demand for pharmacists and pharmacy technicians.

Requirements to be a pharmacy technician vary from state to state, and many aspiring technicians simply receive on-the-job training. However, some pharmacies seek out technicians with specific training requiring classroom and laboratory work in a hospital, community college, or vocational program or sometimes through the military. Some of these programs also include internships in pharmacies.

For additional information on this career and others, click the Focus on Careers link at Mastering A&P®.

# Summary

## Concepts of Matter and Energy (pp. 50–52)

1. Matter
   a. Matter is anything that occupies space and has mass.
   b. Matter exists in three states: gas, liquid, and solid.

2. Energy
   a. Energy is the capacity to do work or to move matter. Energy has kinetic (active) and potential (stored) work capacities.
   b. Energy forms that are important in body function include chemical, electrical, mechanical, and radiant.
   c. Energy forms can be converted from one form to another, but some energy is always unusable (lost as heat) in such transformations.

## Composition of Matter (pp. 52–57)

1. Elements and atoms
   a. Each element is a unique substance that cannot be decomposed into simpler substances by ordinary chemical methods. A total of 118 elements exists; they differ from one another in their chemical and physical properties.
   b. Four elements (carbon, hydrogen, oxygen, and nitrogen) make up 96 percent of living matter. Several other elements are present in small or trace amounts.
   c. The building blocks of elements are atoms. Each atom is designated by an atomic symbol consisting of one or two letters.

2. Atomic structure
   a. Atoms are composed of three subatomic particles: protons, electrons, and neutrons. Protons are positively charged, electrons are negatively charged, and neutrons are neutral.
   b. The planetary model of the atom portrays all the mass of the atom (protons and neutrons) concentrated in a central nucleus. Electrons orbit the nucleus along specific orbits. The orbital model also locates protons and neutrons in a central nucleus, but it depicts electrons as occupying areas of space called orbitals and forming an electron cloud of negative charge around the nucleus.
   c. Each atom can be identified by an atomic number, which is equal to the number of protons contained in the atom's nucleus. Because all atoms are electrically neutral, the number of protons in any atom is equal to its number of electrons.
   d. The atomic mass number is equal to the sum of the protons and neutrons in the atom's nucleus.
   e. Isotopes are different atomic forms of the same element; they differ only in the number of neutrons. Many of the heavier isotopes are unstable and decompose to a more stable form by ejecting particles or energy from the nucleus, a phenomenon called radioactivity. Such radioisotopes are useful in medical diagnosis and treatment and in biochemical research.
   f. The atomic weight is approximately equal to the mass number of the most abundant isotope of any element.

## Molecules and Compounds (pp. 57–58)

1. A molecule is the smallest unit resulting from the bonding of two or more atoms. If the atoms are different, a molecule of a compound is formed.

2. Compounds exhibit properties different from those of the atoms that comprise them.

## Chemical Bonds and Chemical Reactions (pp. 58–64)

1. Bond formation
   a. Chemical bonds are energy relationships. Electrons in the outermost energy level (valence shell) of the reacting atoms are active in the bonding.
   b. Atoms with a stable valence shell (two electrons in shell 1, or eight in the subsequent shells) are chemically inactive. Those with an incomplete valence shell interact by losing, gaining, or sharing electrons to achieve stability (that is, to either fill the valence shell or meet the rule of eight).
   c. Ions are formed when valence electrons are completely transferred from one atom to another. The oppositely charged ions thus formed attract each other, forming an ionic bond. Ionic bonds are common in salts.
   d. Covalent bonds involve the sharing of electron pairs between atoms. If the electrons are shared equally, the molecule is a nonpolar covalent molecule. If the electrons are not shared equally, the molecule is a polar covalent molecule. Polar molecules orient themselves toward charged particles and other molecules.

e. Hydrogen bonds are fragile bonds that bind together water molecules or different parts of the same molecule (intramolecular bonds) but do not involve electrons. They are common in large, complex organic molecules, such as proteins and nucleic acids.

2. Patterns of chemical reactions

a. Chemical reactions involve the formation or breaking of chemical bonds. They are indicated by a chemical equation, which provides information about the atomic composition (formula) of the reactant(s) and product(s).

b. Chemical reactions that result in larger, more complex molecules are synthesis reactions; they involve storing energy in the bonds formed.

c. In decomposition reactions, larger molecules are broken down into simpler molecules or atoms. Bonds are broken, releasing energy.

d. Exchange reactions involve both the making and breaking of bonds. Atoms are replaced by other atoms.

e. Regardless of the type of reaction, most chemical reactions are reversible. Reversibility is indicated by a double arrow.

3. Factors increasing the rate of chemical reactions

a. For atoms to interact chemically, they must collide forcefully.

b. Factors that affect the number or force of collisions include the temperature, concentration of the reactants, particle size, and catalysts (enzymes).

## Biochemistry: The Chemical Composition of Living Matter (pp. 64–81)

1. Inorganic compounds

a. Inorganic compounds making up living matter do not contain carbon (exceptions include $CO_2$ and $CO$). They include water, salts, and some acids and bases.

b. Water is the single most abundant compound in the body. It acts as a universal solvent in which electrolytes (salts, acids, and bases) ionize and in which chemical reactions occur, and it is the basis of transport and lubricating fluids. It slowly absorbs and releases heat, thus helping to maintain homeostatic body temperature, and it protects certain body structures (such as the brain) by forming a watery cushion. Water is also a reactant in hydrolysis reactions.

c. Salts in ionic form (electrolytes) are involved in nerve transmission, muscle contraction, blood

clotting, transport of oxygen by hemoglobin, metabolism, and many other reactions. Additionally, calcium salts (as bone salts) contribute to bone hardness.

d. Acids are proton donors. When dissolved in water, they release hydrogen ions. Strong acids dissociate completely; weak acids dissociate incompletely.

e. Bases are proton acceptors. The most important inorganic bases are hydroxides. Bicarbonate ions are important bases in the body that act as buffers. When bases and acids interact, neutralization occurs—that is, a salt and water are formed.

f. The relative concentrations of hydrogen and hydroxide ions in various body fluids are measured using a pH scale. Each change of one pH unit represents a tenfold change in hydrogen ion concentration. A pH of 7 is neutral (that is, the concentrations of hydrogen and hydroxide ions are equal). A pH below 7 is acidic; a pH above 7 is alkaline (basic).

g. Normal blood pH ranges from 7.35 to 7.45. Slight deviations outside this range can be fatal.

2. Organic compounds

a. Organic compounds are the carbon-containing compounds that comprise living matter. Carbohydrates, lipids, proteins, and nucleic acids all contain carbon, oxygen, and hydrogen. Proteins and nucleic acids also contain substantial amounts of nitrogen, and nucleic acids also contain phosphorus.

b. Carbohydrates contain carbon, hydrogen, and oxygen in the general relationship of two hydrogen atoms to one oxygen atom and one carbon atom. Their building blocks are monosaccharides. Monosaccharides include glucose, fructose, galactose, deoxyribose, and ribose. Disaccharides include sucrose, maltose, and lactose; and polysaccharides include starch and glycogen. Carbohydrates are ingested as sugars and starches. Carbohydrates, and in particular glucose, are the major energy source for the formation of ATP.

c. Lipids include triglycerides (glycerol plus three fatty acid chains), phospholipids, and steroids (the most important of which is cholesterol). Triglycerides (neutral fats) are found primarily in fatty tissue, where they provide insulation and reserve body fuel. Phospholipids and cholesterol are found in all cell membranes. Cholesterol also forms the basis of certain hormones, bile salts, and vitamin D. Like carbohydrates, the lipids are degraded by hydrolysis and synthesized by dehydration synthesis.

d. Proteins are constructed from building blocks called amino acids; 20 amino acids are found in body proteins.

e. Levels of protein structure include the amino acid sequence (primary); the alpha helix and beta-pleated sheet (secondary); a three-dimensional structure superimposed on secondary structure(s) (tertiary); and a globular structure formed by two or more polypeptide chains (quaternary). Different amino acid sequences result in the construction of different proteins.

f. Fibrous, or structural, proteins are the basic structural materials of the body. Globular proteins are also called functional proteins; examples of these include enzymes, some hormones, and hemoglobin. Functional proteins become denatured and inactivated when their hydrogen bonds are disrupted.

g. Enzymes increase the rates of chemical reactions by binding temporarily and specifically with the reactants and holding them in the proper position to interact. Enzymes do not become part of the product. Many enzymes are produced in an inactive form or are inactivated immediately after use.

h. Nucleic acids include deoxyribonucleic acid (DNA) and ribonucleic acid (RNA). The monomer of nucleic acids is the nucleotide; each nucleotide consists of a nitrogen-containing base, a sugar (ribose or deoxyribose), and a phosphate group. DNA (the "stuff" of the genes) maintains genetic heritage by replicating itself before cell division and specifying protein structure. RNA executes the instructions of the DNA during protein synthesis.

i. ATP (adenosine triphosphate) is the universal energy compound used by all body cells. Some of the energy liberated by the oxidation of glucose is captured in the high-energy phosphate bonds of ATP molecules and stored for later use. Some liberated energy is lost as heat.

## Review Questions

 Access additional practice questions using your smartphone, tablet, or computer:
Mastering A&P® > Study Area > Practice Tests & Quizzes

### Multiple Choice

*More than one choice may apply.*

1. Which of the following is (are) true concerning energy?
   a. Energy has mass and takes up space.
   b. Energy can be kinetic and move matter.
   c. Chemical energy is potential energy stored in chemical bonds.
   d. Mechanical energy is indirectly involved in moving matter.
   e. Radiant energy directly involves electromagnetic waves.

2. Pick out the correct match(es) of element and number of valence electrons. Draw a planetary model of each atom to help you choose the best answer.
   a. Carbon—6         d. Oxygen—6
   b. Calcium—2        e. Sodium—2
   c. Iodine—7

3. Important functions of water include which of the following?
   a. Provides cushioning
   b. Acts as a transport medium
   c. Participates in chemical reactions
   d. Acts as a solvent for sugars, salts, and other solutes
   e. Reduces temperature fluctuations

4. Acidic substances include which of the following?
   a. Blood
   b. Water
   c. Gastric juice
   d. Egg white
   e. Lye

5. Glucose is to starch as
   a. a steroid is to a lipid.
   b. a nucleotide is to a nucleic acid.
   c. an amino acid is to a protein.
   d. a polypeptide is to an amino acid.

6. Which of the following is an acidic amino acid?
   a. Glycine
   b. Aspartic acid
   c. Cysteine
   d. Valine
   e. Lysine

7. Which of the following molecules is (are) present in RNA but not in DNA?
   a. Ribose
   b. Deoxyribose
   c. Phosphate group
   d. Adenine
   e. Uracil

8. ATP is *not* associated with

   a. a basic nucleotide structure.

   b. high-energy phosphate bonds.

   c. deoxyribose.

   d. inorganic phosphate.

   e. reversible reactions.

9. The element essential for normal thyroid function is

   a. iodine.                    d. selenium.

   b. iron.                      e. zinc.

   c. copper.

10. Which of the following is (are) true concerning enzymes?

    a. They are very specific in their activities.

    b. They cannot be reused.

    c. They contain active sites.

    d. They are needed in large amounts in cells.

## Short Answer Essay

11. Why is a study of basic chemistry essential to understanding human physiology?

12. Matter occupies space and has mass. Explain how energy *must be* described in terms of these two factors. Then define *energy*.

13. Explain the relationship between kinetic and potential energy.

14. Identify the energy form involved in each of the following examples:

    a. Knee movements when kicking a ball

    b. Digesting food

    c. Ions moving across cell membranes

    d. UV waves causing sunburn

15. According to Greek history, a Greek scientist went running through the streets announcing that he had transformed lead into gold. Both lead and gold are elements. On the basis of what you know about the nature of elements, explain why his rejoicing was short-lived.

16. Name and provide the atomic symbols of the four elements that make up the bulk of all living matter. Which of these is found primarily in proteins and nucleic acids?

17. All atoms are neutral. Explain the basis of this fact.

18. Fill in the following table to fully describe an atom's subatomic particles.

| Particle | Position in the atom | Charge | Mass |
|---|---|---|---|
| Proton |  |  |  |
| Neutron |  |  |  |
| Electron |  |  |  |

19. Define *radioactivity*. If an element has three isotopes, which of them (the lightest, the one with an intermediate mass, or the heaviest) is most likely to be a radioisotope, and why?

20. Define *molecule* as it relates to molecular substances and compounds.

21. Explain the basis of ionic bonding.

22. What are hydrogen bonds, and how are they important in the body?

23. Being polar, water molecules play a very important role in chemical reactions. Explain the polar nature of water molecules.

24. The formation of sodium chloride, common table salt, is an example of ionic bonding. Explain why.

25. Identify each of the following reactions as a synthesis, decomposition, or exchange reaction:

$$2H_2 + O_2 \rightarrow 2H_2O$$
$$MgCl_2 \rightarrow Mg^{2+} + Cl_2$$
$$Zn^{2+} + CuCl_2 \rightarrow ZnCl_2 + Cu^{2+}$$
$$Li_2CO_3 \rightarrow Li_2O + CO_2$$

26. Distinguish inorganic from organic compounds, and list the major categories of each in the body.

27. Salts, acids, and bases are electrolytes. What is an electrolyte?

28. Define *pH*. The pH range of blood is from 7.35 to 7.45. Circle the correct answer to complete the sentence: This is slightly (acidic / basic).

29. Explain why a pH value of 6.3 is 1000 times less acidic than a pH of 3.3.

30. Define *monosaccharide*, *disaccharide*, and *polysaccharide*. Give at least two examples of each. What is the primary function of carbohydrates in the body?

31. What are the general structures of triglycerides, phospholipids, and steroids? Give one or two important uses of each of these lipid types in the body.

32. The building block of proteins is the amino acid. Draw a diagram of the structure of a generalized amino acid. What is the importance of the R-group?

33. Name the two protein classes based on structure and function in the body, and give two examples of each.

34. Define *enzyme*, and describe enzyme action.

35. Virtually no chemical reaction can occur in the body in the absence of enzymes. How might excessively high body temperature interfere with enzyme activity?

36. How is the secondary structure of a protein molecule established? How does this differ from the way the tertiary structure is established?

37. What is ATP's central role in the body?

38. Explain why you can "stack" water slightly above the rim of a glass if you pour the water in very carefully.

39. Water is a precious natural resource all over the world, and it is said that supplies are dwindling. Desalinization (salt removal) of ocean water has been recommended as a solution to the problem. Why shouldn't we drink salt water?

40. Explain the unique chemical makeup and characteristics of phospholipids and what they have to do with the terms *hydrophilic* and *hydrophobic*.

## Critical Thinking and Clinical Application Questions

41. Several antibiotics act by binding to certain essential enzymes in the target bacteria. How might these antibiotics influence the chemical reaction controlled by the enzyme? What might be the effect on the bacteria? On the person taking the antibiotic prescription?

42. Mrs. Roberts, who is in a diabetic coma, has just been admitted to Noble Hospital. Her blood pH indicates that she is in severe acidosis (blood pH in the acid range), and the medical staff quickly institute measures to bring her blood pH back within normal limits. Note the normal pH of blood, and discuss why severe acidosis is a problem.

43. Stephen wants to bring some changes in his diet in order to make it more heart healthy. His friend advises him to choose the fats in his diet correctly. How can he improve his intake of fats? Why are some fats good for the heart and some are not?

44. Pediatricians become concerned about the potential for brain damage when an infant's temperature approaches 105°F. Which class of organic molecules is most likely to be damaged by high temperature? Explain why.

45. The genetic error that causes sickle-cell anemia begins at the DNA level and results in the synthesis of hemoglobin with one amino acid difference compared to normal hemoglobin. Explain how changing one amino acid could affect hemoglobin function.

# 3 Cells and Tissues

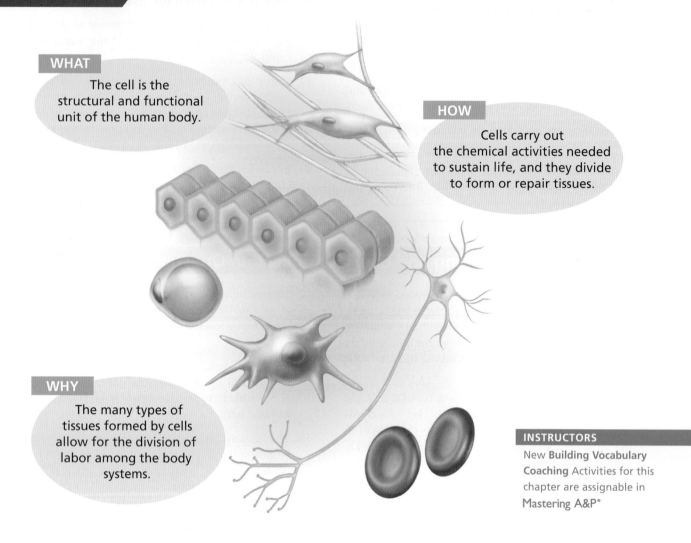

**WHAT**
The cell is the structural and functional unit of the human body.

**HOW**
Cells carry out the chemical activities needed to sustain life, and they divide to form or repair tissues.

**WHY**
The many types of tissues formed by cells allow for the division of labor among the body systems.

**INSTRUCTORS**
New **Building Vocabulary Coaching** Activities for this chapter are assignable in Mastering A&P°

## PART I: CELLS

Just as bricks and lumber are used to build a house, **cells** are the structural units of all living things, from one-celled organisms such as amoebas to complex multicellular organisms such as humans, dogs, and trees. The human body contains 50 to 100 trillion of these tiny building blocks.

In this chapter, we focus on structures and functions shared by all cells. We consider specialized cells and their unique functions in later chapters.

## Overview of the Cellular Basis of Life

→ **Learning Objectives**

☐ **Name and describe the four concepts of the cell theory.**

☐ **List four elements that make up the bulk of living matter.**

In the late 1600s, Robert Hooke was looking through a crude microscope at some plant tissue—cork. He saw some cubelike structures that reminded him of the long rows of monk's rooms (or cells) at the monastery, so he named these structures "cells." The living cells that had formed the cork were long since dead; only the plant cell walls remained. However, the name stuck and is still used to describe the smallest unit of all living things.

Since the late 1800s, cell research has been exceptionally fruitful and has provided us with the following four concepts collectively known as the **cell theory**:

- A cell is the basic structural and functional unit of living organisms. So, when you define cell properties, you are in fact defining the properties of life.

- The activity of an organism depends on the collective activities of its cells.

- According to the *principle of complementarity*, the activities of cells are dictated by their structure (anatomy), which determines function (physiology).

- Continuity of life has a cellular basis.

We will expand on all of these concepts as we go along. Let's begin with the idea that the cell is the smallest living unit. Whatever its form, however it behaves, the cell contains all the parts necessary to survive in a changing world. It follows, then, that loss of cell homeostasis underlies virtually every disease.

Perhaps the most striking thing about a cell is its organization. Yet if we chemically analyze cells, we find that they are made up primarily of the same four elements—carbon, oxygen, hydrogen, and nitrogen—plus much smaller amounts of several other elements. (A detailed account of body chemistry appears in Chapter 2.)

Strange as it may seem, especially when we feel our firm muscles, living cells are about 60 percent water, which is one of the reasons water is essential for life.

## Did You Get It?

1. Define *cell*.
2. According to the cell theory, what the organism can do depends on _____. (Fill in the blank.)

For answers, see Appendix A.

# Anatomy of a Generalized Cell

## → Learning Objectives

☐ Define *generalized cell*.

☐ Identify on a cell model or diagram the three major cell regions (nucleus, cytoplasm, and plasma membrane).

☐ List the structures of the nucleus, and explain the function of chromatin and nucleoli.

Although no one cell type is exactly like all others, cells *do* have the same basic parts, and there are certain functions common to all cells. Here we will talk about the **generalized cell**, which demonstrates most of these typical features.

In general, all cells have three main regions or parts—a *nucleus* (nu'kle-us), a *plasma membrane*, and the *cytoplasm* (si'to-plazm") **(Figure 3.1a)**. The nucleus is usually located near the center of the cell. It is surrounded by the semifluid cytoplasm, which in turn is enclosed by the plasma membrane, which forms the outer cell boundary. (If you look ahead to Figure 3.4 on p. 94, you can see a more detailed illustration of generalized cell structure.)

## The Nucleus

Anything that works, works best when it is controlled. For cells, "headquarters," or the control center, is the **nucleus** (*nucle* = kernel). The genetic material, or *deoxyribonucleic acid (DNA)*, is a blueprint that contains all the instructions needed for building the whole body; so, as you might expect, human DNA differs from frog DNA. More specifically, DNA has *genes*, which carry the instructions for building *proteins*. DNA is also absolutely necessary for cell reproduction. A cell that has lost or ejected its nucleus (for whatever reason) is destined to "self-destruct."

Although the nucleus is most often oval or spherical, its shape usually conforms to the shape of the cell. For example, if the cell is elongated, the nucleus is usually elongated as well. The nucleus has three recognizable regions or structures: the *nuclear envelope*, *nucleolus*, and *chromatin*.

### Nuclear Envelope

The nuclear boundary is a double membrane barrier called the **nuclear envelope**, or **nuclear membrane** (Figure 3.1b). Between the two membranes is a fluid-filled "moat," or space. At various points, the two layers of the nuclear envelope fuse, generating openings called **nuclear pores**. Like other cellular membranes, the nuclear envelope allows some but not all substances to pass through it, but substances pass through it much more freely than elsewhere because of its relatively large pores. The nuclear membrane encloses a jellylike fluid called *nucleoplasm* (nu'kle-o-plazm") in which other nuclear elements are suspended.

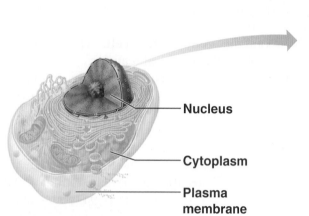

**(a) Generalized animal cell**

**Figure 3.1 Anatomy of the generalized animal cell nucleus. (a)** Orientation diagram: the three main regions of the generalized cell. **(b)** Structure of the nucleus surrounded by rough ER.

**(b) Nucleus**

## Nucleolus

The nucleus contains one or more small, dark-staining, essentially round bodies called **nucleoli** (nu-kle′o-li; "little nuclei"). Nucleoli (plural for nucleolus) are sites where cell structures called *ribosomes* are assembled. Most ribosomes eventually migrate into the cytoplasm, where they serve as the actual sites of protein synthesis.

## Chromatin

When a cell is not dividing, its DNA is carefully wound around proteins called histones to form a loose network of "beads on a string" called **chromatin** (kro′mah-tin) that is scattered throughout the nucleus. When a cell is dividing to form two daughter cells, the chromatin threads coil and condense to form dense, rodlike bodies called **chromosomes** (*chromo* = colored, *soma* = body)—much the way a stretched spring becomes shorter and thicker when allowed to relax. We discuss the functions of DNA and the events of cell division in the Cell Physiology section (beginning on p. 102).

## Did You Get It?

3. Name the three basic parts of a cell, and give the location of each.
4. How would you explain the meaning of a "generalized cell" to a classmate?
5. What is the general function of nucleoli?

**For answers, see Appendix A.**

# The Plasma Membrane

→ **Learning Objectives**

☐ Describe the chemical composition of the plasma membrane, and relate it to membrane functions.

☐ Compare the structure and function of tight junctions, desmosomes, and gap junctions.

The flexible **plasma membrane** is a fragile, transparent barrier that contains the cell contents and separates them from the surrounding environment. (The terms *cell membrane* or *cytoplasmic membrane* are sometimes used instead, but because nearly all cellular organelles are composed of membranes, in this text we will always refer to the cell's surface or outer limiting membrane as the plasma membrane.) Although the plasma membrane is important in defining the limits of the cell, it is much more than a passive envelope, or "baggie." As you will see, its unique structure allows it to play a dynamic role in many cellular activities.

## The Fluid Mosaic Model

The structure of the plasma membrane consists of two *phospholipid* (fat) layers arranged "tail to tail," with *cholesterol* and floating proteins scattered among them **(Figure 3.2)**. Some phospholipids may also have sugar groups attached, forming *glycolipids*. The proteins, some of which are free to move

**Q:** *Some proteins float freely in the lipid bilayer of the membrane, whereas others are anchored in specific locations. What structure(s) could anchor the stationary proteins?*

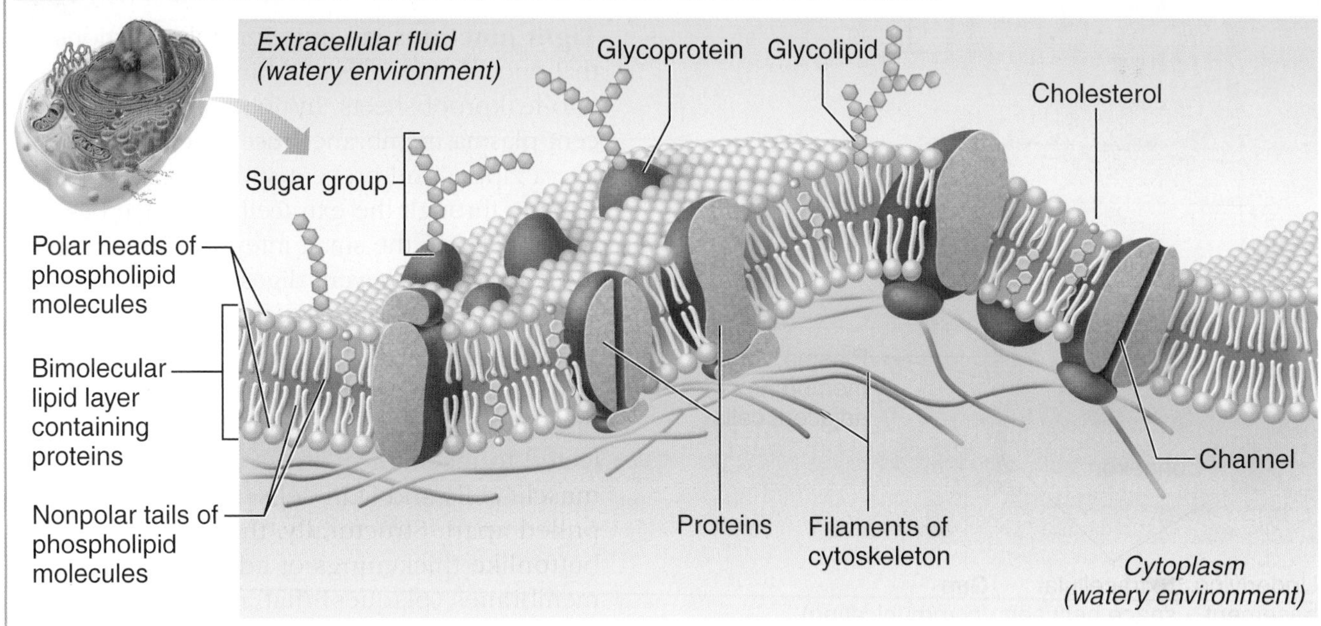

**Figure 3.2 Structure of the plasma membrane.**

and bob in the lipid layer, form a constantly changing pattern or *mosaic*, hence the name of the model that describes the plasma membrane.

### CONCEPTLINK

Remember, phospholipids are polar molecules: The charged end interacts with water, and the fatty acid chains do not (see Chapter 2, p. 72). It is this property of polarity that makes phospholipids a good foundation for cell membranes. ←

The olive oil–like phospholipid bilayer forms the basic "fabric" of the membrane. The polar "heads" of the lollipop-shaped phospholipid molecules are **hydrophilic** ("water loving") and are attracted to water, the main component of both the intracellular and extracellular fluids, and so they lie on both the inner and outer surfaces of the membrane. Their nonpolar fatty acid "tails," being **hydrophobic** ("water fearing"), avoid water and line up in the center (interior) of the membrane. The self-orienting property of the phospholipids allows biological membranes to

reseal themselves quickly when torn. The hydrophobic makeup of the membrane interior makes the plasma membrane relatively impermeable to most water-soluble molecules. The cholesterol helps to both stabilize the membrane and keep it flexible.

The proteins scattered in the lipid bilayer are responsible for most of the specialized functions of the membrane. Some proteins are enzymes. Many of the proteins protruding from the cell exterior are receptors for hormones or other chemical messengers or are binding sites for anchoring the cell to fibers or to other structures inside or outside the cell. Most proteins that span the membrane are involved in transport. For example, some cluster together to form protein channels (tiny pores) through which water and small water-soluble molecules or ions can move; others act as *carriers* that bind to a substance and move it through the membrane. Branching sugar groups are attached to most of the proteins abutting the extracellular space. Such "sugar-proteins" are called *glycoproteins*, and because of their presence, the cell surface is a fuzzy, sticky, sugar-rich area called the *glycocalyx* (gli-ko-ka′liks). (You can think of your cells as being sugar coated.) Among other things, these glycoproteins determine

**A:** Filaments of the cytoskeleton attached to membrane proteins.

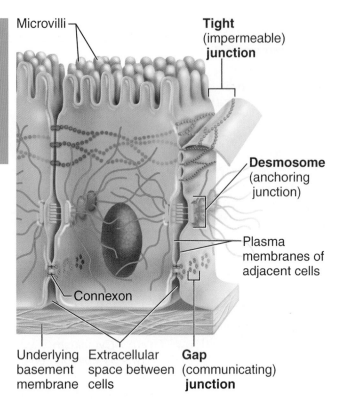

Microvilli

**Tight** (impermeable) **junction**

**Desmosome** (anchoring junction)

Plasma membranes of adjacent cells

Connexon

Underlying basement membrane

Extracellular space between cells

**Gap** (communicating) **junction**

**Figure 3.3 Cell junctions.** An epithelial cell is shown joined to adjacent cells by the three common types of cell junctions: tight junctions, desmosomes, and gap junctions. Also illustrated are microvilli, which increase the cell's surface area for absorption (seen projecting from the free cell surface).

your blood type, act as receptors that certain bacteria, viruses, or toxins can bind to, and play a role in cell-to-cell recognition and interactions. Definite changes in glycoproteins occur in cells that are being transformed into cancer cells. (We discuss cancer in "A Closer Look" on pp. 130–131.)

### Cell Membrane Junctions

Although certain cell types—blood cells, sperm cells, and some phagocytic cells (which ingest bacteria and foreign debris)—are "footloose" in the body, many other types, particularly epithelial cells, are knit into tight communities. Typically, cells are bound together in three ways:

- Glycoproteins in the glycocalyx act as an adhesive or cellular glue.

- Wavy contours of the membranes of adjacent cells fit together in a tongue-and-groove fashion.

- Special **cell membrane junctions** are formed **(Figure 3.3)**. These junctions vary structurally depending on their roles.

Because this last factor is the most important, let us look more closely at the main types of junctions: *tight junctions, desmosomes,* and *gap junctions.*

- **Tight junctions** are impermeable junctions that encircle the cells and bind them together into leakproof sheets. In tight junctions, adjacent plasma membranes fuse together tightly like a zipper and prevent substances from passing through the extracellular space between cells. In the small intestine, for example, these junctions prevent digestive enzymes from seeping into the bloodstream.

- **Desmosomes** (dez'mo-sōmz) are anchoring junctions scattered like rivets along the sides of adjacent cells. They prevent cells subjected to mechanical stress (such as heart muscle cells and skin cells) from being pulled apart. Structurally, these junctions are buttonlike thickenings of adjacent plasma membranes (plaques) that are connected by fine protein filaments. Thicker protein filaments extend from the plaques inside the cells to the plaques on the cells' opposite sides, thus forming an internal system of strong "guy wires."

- **Gap junctions**, or *communicating junctions,* function mainly to allow communication. These junctions are commonly found in the heart and between embryonic cells. In gap junctions, the neighboring cells are connected by hollow cylinders composed of proteins (called **connexons**) that span the entire width of the abutting membranes (which are therefore called *transmembrane proteins*). Chemical molecules, such as nutrients or ions, can pass directly through the water-filled connexon channels from one cell to another.

### Did You Get It?

6. Why do phospholipids (which form the bulk of plasma membranes) organize into a bilayer, tail to tail, in a watery environment?

7. The external faces of some membrane proteins have sugar groups attached to them. What are three roles these sugar-coated proteins play in the life of a cell?

8. What is the special function of gap junctions? Of tight junctions?

For answers, see Appendix A.

# The Cytoplasm

## → Learning Objective

☐ **Identify the organelles on a cell model or describe them, and indicate the major function of each.**

The **cytoplasm** is the cellular material outside the nucleus and inside the plasma membrane. It is the site of most cellular activities, so you might think of the cytoplasm as the "factory floor" of the cell. Although early scientists believed that the cytoplasm was a structureless gel, the electron microscope has revealed that it has three major components: the *cytosol, inclusions,* and *organelles.* Let's take a look at each of these.

## Cytosol and Inclusions

The **cytosol** is semitransparent fluid that suspends the other elements. Dissolved in the cytosol, which is largely water, are nutrients and a variety of other *solutes* (sol′yūtz; dissolved substances).

**Inclusions** are chemical substances that may or may not be present, depending on the specific cell type. Most inclusions are stored nutrients or cell products floating in the cytosol. They include the lipid droplets common in fat cells, glycogen granules abundant in liver and muscle cells, pigments such as melanin in skin and hair cells, mucus and other secretory products, and various kinds of crystals. It may help to think of an inclusion as a cellular "pantry" where items are kept on hand until needed.

## Organelles

The **organelles** (or″gah-nelz′; "little organs") are specialized cellular compartments (**Figure 3.4,** p. 94) that are the metabolic machinery of the cell. Each type of organelle is specialized to carry out a specific function for the cell as a whole, much like the organs carry out specialized functions for the whole body. Some synthesize proteins, others package those proteins, and so on.

Many organelles are bounded by a membrane similar to the plasma membrane. These membrane boundaries allow organelles to maintain an internal environment quite different from that of the surrounding cytosol. This compartmentalization is crucial to their ability to perform their specialized functions for the cell. Let's consider what goes on in each of these workshops of our cellular factory.

**Mitochondria** **Mitochondria** (mi″to-kon′dre-ah; singular: *mitochondrion*) are usually depicted as tiny, lozenge-like or sausage-shaped organelles (see Figure 3.4), but in living cells they lengthen and change shape almost continuously. The mitochondrial wall consists of a double membrane, equal to *two* plasma membranes placed side by side. The outer membrane is smooth and featureless, but the inner membrane has shelflike protrusions called *cristae* (kris′te; "crests"). Enzymes dissolved in the fluid within the mitochondria, as well as enzymes that form part of the cristae membranes, carry out the reactions in which oxygen is used to break down foods. As the foods are broken down, energy is released. Much of this energy escapes as heat, but some is captured and used to form *ATP molecules.*

ATP provides the energy for all cellular work, and every living cell requires a constant supply of ATP for its many activities. Because the mitochondria supply most of this ATP, they are the "powerhouses" of the cell. Metabolically "busy" cells, such as liver and muscle cells, use huge amounts of ATP and have hundreds of mitochondria, which replicate themselves by pinching in half. By contrast, cells that are relatively inactive (an unfertilized egg, for instance) have fewer.

**Ribosomes** **Ribosomes** (ri′bo-sōmz) are tiny, bilobed, dark bodies made of proteins and one variety of RNA called *ribosomal RNA.* Ribosomes are the actual sites of protein synthesis in the cell. Ribosomes that float freely in the cytoplasm manufacture proteins that function inside the cell, while others attach to membranes such as the rough ER, which produces proteins that function outside the cell.

**Endoplasmic Reticulum** The **endoplasmic reticulum** (en″do-plas′mik rĕ-tik′u-lum; "network within the cytoplasm"), or **ER**, is a system of fluid-filled tunnels (or canals) that coil and twist through the cytoplasm. It is continuous with the nuclear envelope and accounts for about half of a cell's membranes. It serves as a mini circulatory system for the cell because it provides a network of channels for carrying substances (primarily proteins) from one part of the cell to another. There are two forms of ER, *rough* and *smooth* (see Figure 3.4); a particular cell may have both forms or only one, depending on its specific functions.

The **rough endoplasmic reticulum** is so called because it is studded with ribosomes. Because essentially all of the building materials of cellular membranes are formed either in it or on it,

**Q:** *Which nuclear component contains your genes? Which produces protein-building "factories"?*

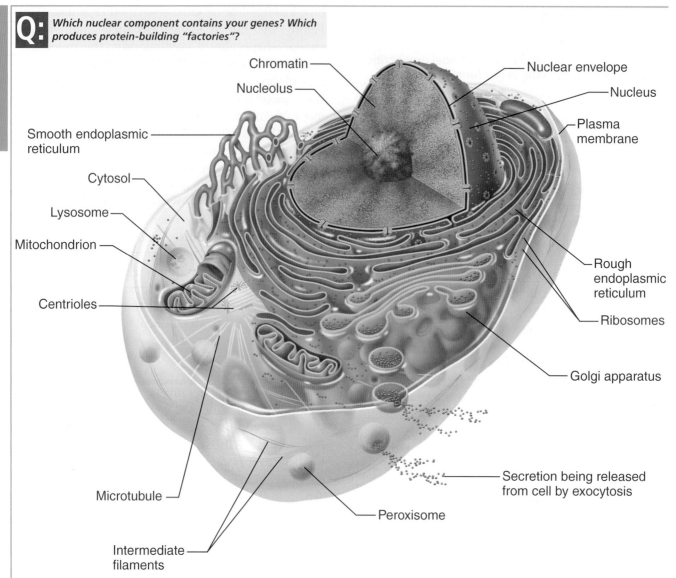

Chromatin

Nucleolus

Nuclear envelope

Nucleus

Plasma membrane

Smooth endoplasmic reticulum

Cytosol

Lysosome

Mitochondrion

Centrioles

Microtubule

Intermediate filaments

Peroxisome

Secretion being released from cell by exocytosis

Golgi apparatus

Ribosomes

Rough endoplasmic reticulum

**Figure 3.4 Structure of the generalized cell.** No cell is exactly like this one, but this generalized cell drawing illustrates features common to many human cells.

you can think of the rough ER as the cell's membrane factory. The proteins made on its ribosomes migrate into the rough ER tunnels, where they fold into their functional three-dimensional shapes. These proteins are then dispatched to other areas of the cell in small "sacs" of membrane called **transport vesicles (Figure 3.5)** that carry substances around the cell. Rough ER is especially abundant in cells that make (synthesize) and

export (secrete) proteins—for example, pancreatic cells, which produce digestive enzymes to be delivered to the small intestine. The enzymes that catalyze the synthesis of membrane lipids reside on the external (cytoplasmic) face of the rough ER, where the needed building blocks are readily available.

Although the **smooth endoplasmic reticulum** communicates with the rough variety, it plays no role in protein synthesis, because it lacks ribosomes. Instead it functions in lipid metabolism (cholesterol and fat synthesis and breakdown) and detoxification of drugs and pesticides. Hence it is

**A:**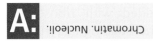
Chromatin. Nucleoli.

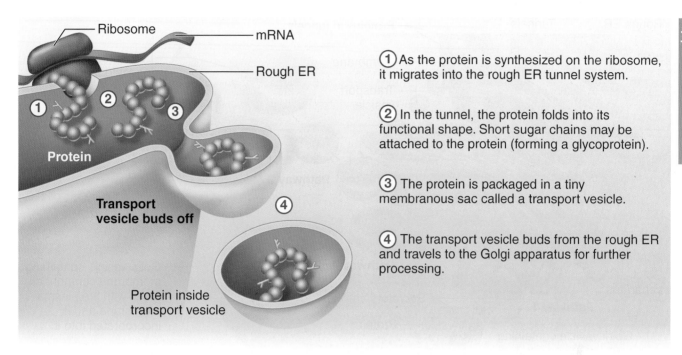

**Figure 3.5  Synthesis and export of a protein by the rough ER.**

① As the protein is synthesized on the ribosome, it migrates into the rough ER tunnel system.

② In the tunnel, the protein folds into its functional shape. Short sugar chains may be attached to the protein (forming a glycoprotein).

③ The protein is packaged in a tiny membranous sac called a transport vesicle.

④ The transport vesicle buds from the rough ER and travels to the Golgi apparatus for further processing.

not surprising that the liver cells are chock-full of smooth ER. So too are body cells that produce steroid-based hormones—for instance, cells of the male testes that manufacture testosterone.

***Golgi Apparatus***  The **Golgi** (gol′je) **apparatus** appears as a stack of flattened membranous sacs that are associated with swarms of tiny vesicles. It is generally found close to the ER and is the principal "traffic director" for cellular proteins. Its major function is to modify, package, and ship proteins (sent to it by the rough ER via transport vesicles) in specific ways, depending on their final destination **(Figure 3.6)**. Initially, all proteins leaving the Golgi apparatus accumulate in sacs called Golgi vesicles.

As proteins "tagged" for export accumulate in the Golgi apparatus, the sacs swell. Then their swollen ends, filled with protein, pinch off and form **secretory vesicles** (ves′ĭ-kulz), which travel to the plasma membrane. When the vesicles reach the plasma membrane, they fuse with it, the membrane ruptures, and the contents of the sac are ejected to the outside of the cell (pathway 1 in Figure 3.6). Mucus is packaged this way, as are digestive enzymes made by pancreatic cells.

In addition to its packaging-for-release functions, the Golgi apparatus pinches off *sacs* containing proteins and phospholipids destined for a "home" in the plasma membrane (pathway 2 in Figure 3.6) or other cellular membranes. It also packages hydrolytic enzymes into membrane-bound organelles called *lysosomes* that remain in the cell (pathway 3 in Figure 3.6 and discussed next).

***Lysosomes***  **Lysosomes** (li′so-sōmz; "breakdown bodies"), which appear in different sizes, are membranous "bags" containing powerful digestive enzymes. Because lysosomal enzymes are capable of digesting worn-out or nonusable cell structures and most foreign substances that enter the cell, lysosomes function as cellular "stomachs." Lysosomes are especially abundant in white blood cells called phagocytes, the cells that dispose of bacteria and cell debris. As we mentioned, the enzymes they contain are formed by ribosomes on the rough ER and packaged by the Golgi apparatus.

### Homeostatic Imbalance 3.1

The lysosomal membrane is ordinarily quite stable, but it becomes fragile when the cell is injured or deprived of oxygen and when excessive amounts of vitamin A are present. When lysosomes rupture, the cell self-digests. _____ ✚

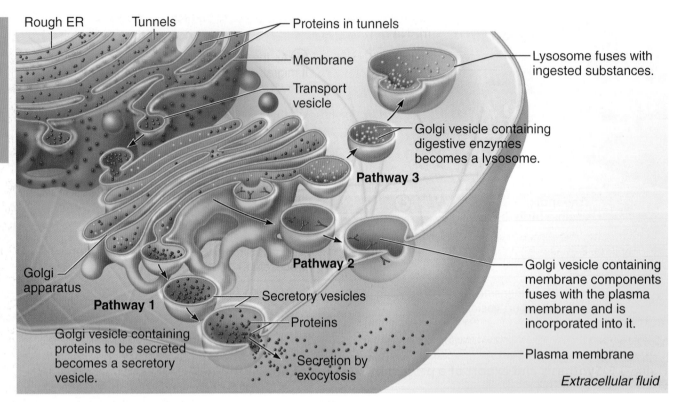

Rough ER    Tunnels    Proteins in tunnels

Membrane

Transport vesicle

Lysosome fuses with ingested substances.

Golgi vesicle containing digestive enzymes becomes a lysosome.

**Pathway 3**

**Pathway 2**

Golgi apparatus

**Pathway 1**

Secretory vesicles

Proteins

Golgi vesicle containing proteins to be secreted becomes a secretory vesicle.

Secretion by exocytosis

Golgi vesicle containing membrane components fuses with the plasma membrane and is incorporated into it.

Plasma membrane

*Extracellular fluid*

**Figure 3.6 Role of the Golgi apparatus in packaging the products of the rough ER.** Protein-containing transport vesicles pinch off the rough ER and migrate to fuse with the Golgi apparatus. As the proteins travel through the Golgi apparatus, they are sorted (and slightly modified if necessary), then packaged within vesicles. These vesicles then leave the Golgi apparatus and head for various destinations (pathways 1–3), as shown.

*Peroxisomes* Peroxisomes (per-ok′sih-sōmz) are membranous sacs containing powerful oxidase (ok′sĭ-dāz) enzymes that use molecular oxygen ($O_2$) to detoxify a number of harmful or poisonous substances, including alcohol and formaldehyde. However, their most important function is to "disarm" dangerous free radicals. **Free radicals** are highly reactive chemicals with unpaired electrons that can damage the structure of proteins and nucleic acids. Free radicals are normal by-products of cellular metabolism, but if allowed to accumulate, they can have devastating effects on cells. Peroxisomes convert free radicals to hydrogen peroxide ($H_2O_2$), a function indicated in their naming (*peroxisomes* = peroxide bodies). The enzyme *catalase* (kat′ah-lās) then converts excess hydrogen peroxide to water. Peroxisomes are especially numerous in liver and kidney cells, which are very active in detoxification.

Although peroxisomes look like small lysosomes (see Figure 3.4), they do not arise by budding from the Golgi apparatus. Instead, one way they replicate themselves is by simply pinching in half, like mitochondria, but most peroxisomes appear to bud directly from the ER.

### Did You Get It?

9. How do the cytosol and the cytoplasm differ?
10. Which two organelles are sacs of enzymes, and what is the function of each of these organelles?
11. Which organelle is the major site of ATP synthesis? Which packages proteins?

For answers, see Appendix A.

*Cytoskeleton* An elaborate network of protein structures extends throughout the cytoplasm. This network, or **cytoskeleton**, acts as a cell's "bones and muscles" by furnishing an internal framework that determines cell shape, supports other organelles, and provides the machinery for intracellular transport and various types of cellular movements. From its smallest to its largest elements, the cytoskeleton is made up of microfilaments, intermediate filaments, and microtubules **(Figure 3.7)**. Although there is some overlap in roles, generally speaking **microfilaments** (such as *actin* and

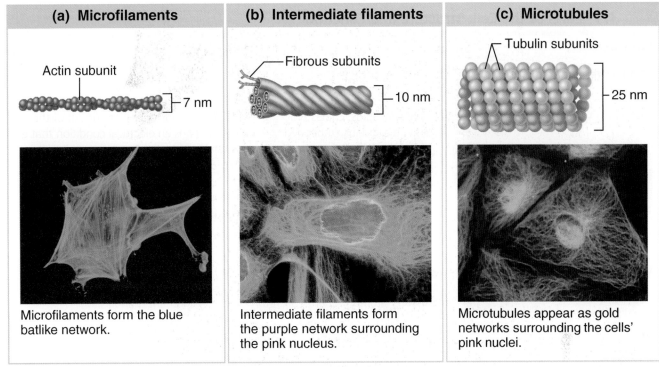

| **(a) Microfilaments** | **(b) Intermediate filaments** | **(c) Microtubules** |
| --- | --- | --- |
| Actin subunit — 7 nm | Fibrous subunits — 10 nm | Tubulin subunits — 25 nm |
| Microfilaments form the blue batlike network. | Intermediate filaments form the purple network surrounding the pink nucleus. | Microtubules appear as gold networks surrounding the cells' pink nuclei. |

**Figure 3.7 Cytoskeletal elements support the cell and help to generate movement.** Diagrammatic views (above) and photos (below). The photos are of cells with the structure of interest "tagged" by fluorescent molecules.

*myosin*) are most involved in cell motility and in producing changes in cell shape. (You could say that cells move when they get their act(in) together.) The strong, stable, ropelike **intermediate filaments** are made up of fibrous subunits. They help form desmosomes (see Figure 3.3) and provide internal guy wires to resist pulling forces on the cell. The tubelike **microtubules** are made up of repeating subunits of the protein tubulin. They determine the overall shape of a cell and the distribution of organelles. They are very important during cell division (see pp. 108–111).

*Centrioles*   The paired **centrioles** (sen'tre-ōlz), collectively called the *centrosome*, lie close to the nucleus (see Figure 3.4). They are rod-shaped bodies that lie at right angles to each other; internally they are made up of a pinwheel array of nine triplets of fine microtubules. Centrioles are best known for their role in generating microtubules and also for directing the formation of the *mitotic spindle* during cell division (look ahead to Figure 3.15, p. 110).

Cell parts and their structures and functions are summarized in **Table 3.1** (pp. 98–100).

## Cell Extensions

In addition to the cell structures described previously, some cells have obvious surface extensions. These come in two major "flavors," or varieties, depending on whether they have a core of microtubules or actin filaments.

### Cilia and Flagella

**Cilia** (sil'e-ah; "eyelashes") are whiplike cellular extensions that move substances along the cell surface. For example, mucus is carried up and away from the lungs by "crowd surfing" on the ciliated cells lining the respiratory system. Where cilia appear, there are usually many of them projecting from the exposed cell surface.

When a cell is about to make cilia, its centrioles multiply and then line up beneath the plasma membrane at the free cell surface. Microtubules then begin to "sprout" from the centrioles and put pressure on the membrane, forming the projections.

If the projections formed by the centrioles are substantially longer, they are called **flagella** (flah-jel'ah). The only example of a flagellated cell in the human body is the sperm, which has a single

*(Text continues on page 100.)*

## Table 3.1   Parts of the Cell: Structure and Function

| Cell part* | Structure | Functions |
|---|---|---|
| **Plasma Membrane (Figure 3.2)** | | |
| Sugar group / Protein / Cholesterol | Membrane made of a double layer of lipids (phospholipids, cholesterol, and so on) with proteins embedded within. Most externally facing proteins and some lipids have attached sugar groups. | Serves as an external cell barrier and acts in transport of substances into or out of the cell. Maintains an electrical condition that is essential for functioning of excitable cells. Externally facing proteins act as receptors (for hormones, neurotransmitters, and so on), transport proteins, and in cell-to-cell recognition. |
| **Cytoplasm** | Cellular region between the nuclear and plasma membranes. Consists of fluid **cytosol** containing dissolved solutes, **organelles** (the metabolic machinery of the cytoplasm), and **inclusions** (stored nutrients, secretory products, pigment granules). | |

**Organelles**

| Cell part* | Structure | Functions |
|---|---|---|
| • Mitochondria (Figure 3.4) | Rodlike, double-membrane structures; inner membrane folded into projections called cristae. | Site of aerobic respiration (the "burning" of glucose) and ATP synthesis; powerhouse of the cell. |
| • Ribosomes (Figures 3.4, 3.5) | Dense particles consisting of two subunits, each composed of ribosomal RNA and protein. Free or attached to rough endoplasmic reticulum. | The sites of protein synthesis. |
| • Rough endoplasmic reticulum (Figures 3.4, 3.5) | Membranous system enclosing a cavity, the tunnel, and coiling through the cytoplasm. Externally studded with ribosomes. | Sugar groups are attached to proteins within the tunnels. Proteins are bound in vesicles for transport to the Golgi apparatus and other sites. External face synthesizes phospholipids. |
| • Smooth endoplasmic reticulum (Figure 3.4) | Membranous system of tunnels and sacs; free of ribosomes. | Site of lipid and steroid (cholesterol) synthesis, lipid metabolism, and drug detoxification. |
| • Golgi apparatus (Figures 3.4, 3.6) | A stack of flattened membranes and associated vesicles close to the ER. | Packages, modifies, and segregates proteins for secretion from the cell, inclusion in lysosomes, or incorporation into the plasma membrane. |

*Individual cellular structures are not drawn to scale.

Table **3.1** *(continued)*

| Cell part* | Structure | Functions |
|---|---|---|
| • Peroxisomes (not shown) | Membranous sacs of oxidase and catalase enzymes. | The enzymes detoxify a number of toxic substances such as free radicals. The most important enzyme, catalase, breaks down hydrogen peroxide. |
| • Lysosomes (Figures 3.4, 3.6) | Membranous sacs containing acid hydrolases (powerful digestive enzymes). | Sites of intracellular digestion. The "stomach" of the cell. |
| • Microtubules (Figure 3.7) | Cylindrical structures made of tubulin proteins. | Support the cell and give it shape. Involved in intracellular and cellular movements. Form centrioles and cilia and flagella, if present. |
| • Microfilaments (Figure 3.7) | Fine filaments composed of the protein actin. | Involved in muscle contraction and other types of intracellular movement; help form the cell cytoskeleton and microvilli, if present. |
| • Intermediate filaments (Figure 3.7) | Protein fibers; composition varies. | The stable cytoskeletal elements; resist mechanical forces acting on the cell. |
| • Centrioles (Figure 3.4) | Paired cylindrical bodies, each composed of nine triplets of microtubules. | Organize a microtubule network during mitosis (cell division) to form the spindle and asters. Form the bases of cilia and flagella. |
| **Inclusions** | Varied; includes stored nutrients such as lipid droplets and glycogen granules, protein crystals, pigment granules. | Storage for nutrients, wastes, and cell products. |

### Nucleus (Figures 3.1, 3.4)

| | | |
|---|---|---|
| | Largest organelle. Surrounded by the nuclear envelope; contains fluid nucleoplasm, nucleoli, and chromatin. | Control center of the cell; responsible for transmitting genetic information and providing the instructions for protein synthesis. |
| • Nuclear envelope (Figure 3.1) | Double-membrane structure pierced by large pores. Outer membrane continuous with the endoplasmic reticulum. | Separates the nucleoplasm from the cytoplasm and regulates passage of substances to and from the nucleus. |

Table **3.1**  **Parts of the Cell: Structure and Function** *(continued)*

| Cell part* | Structure | Functions |
|---|---|---|
| • Nucleolus (Figure 3.1) | Dense spherical (non-membrane-bounded) bodies, composed of ribosomal RNA and proteins. | Site of ribosome subunit manufacture. |
| • Chromatin (Figure 3.1) | Granular, threadlike material composed of DNA and histone proteins. "Beads on a string." | DNA constitutes the genes, which carry instructions for building proteins. |

propulsive flagellum called its *tail* (look ahead to Figure 3.8g). Notice that cilia propel other substances across a cell's surface, whereas a flagellum propels the cell itself.

### Microvilli

**Microvilli** (mi″kro-vil′i; "little shaggy hairs") are tiny, fingerlike extensions of the plasma membrane that project from an exposed cell surface (see Figure 3.3). They increase the cell's surface area tremendously and so are usually found on the surface of cells active in absorption such as intestinal and kidney tubule cells. Microvilli have a core of actin filaments that extend into the internal cytoskeleton of the cell and stiffen the microvillus. Note that microvilli are "alcoves" projecting off one cell surface and do not involve microtubules.

### Did You Get It?

12. Which two types of cytoskeletal elements are involved in cell mobility?
13. Which of the cytoskeletal elements is the basis of centrioles? Of microvilli?
14. The major function of cilia is to move substances across the free cell surface. What is the major role of microvilli?

For answers, see Appendix A.

## Cell Diversity

→ **Learning Objective**

☐ Apply the principle of complementarity to different cell types by comparing overall shapes, internal structures, and special functions.

So far in this chapter, we have focused on a generalized human cell. However, the trillions of cells in the human body include over 200 different cell types that vary greatly in size, shape, and function. They include sphere-shaped fat cells, disc-shaped red blood cells, branching nerve cells, and cube-shaped cells of kidney tubules.

Depending on type, cells also vary greatly in length—ranging from 1/12,000 of an inch in the smallest cells to over a yard in the nerve cells that cause you to wiggle your toes. A cell's shape reflects its function. For example, the flat, tilelike epithelial cells that line the inside of your cheek fit closely together, forming a living barrier that protects underlying tissues from bacterial invasion.

The shapes of cells and the relative numbers of the various organelles they contain relate to specialized cell functions **(Figure 3.8)**. Let's take a look at some examples of specialized cells.

- **Cells that connect body parts** (Figure 3.8a)
  - *Fibroblast.* This cell has an elongated shape, like the cable-like fibers that it secretes. It has an abundant rough ER and a large Golgi apparatus to make and secrete the protein building blocks of these fibers.
  - *Erythrocyte (red blood cell).* This cell carries oxygen in the blood. Its biconcave disc shape provides extra surface area for the uptake of oxygen and streamlines the cell so it flows easily through the bloodstream. So much oxygen-carrying pigment is packed in erythrocytes that all other organelles have been shed to make room.

- **Cells that cover and line body organs** (Figure 3.8b)

  - *Epithelial cell.* The hexagonal shape of this cell is exactly like a "cell" in a honeycomb of a beehive. This shape allows epithelial cells to pack together in sheets. An epithelial cell has abundant intermediate filaments and desmosomes that resist tearing when the epithelium is rubbed or pulled.

- **Cells that move organs and body parts** (Figure 3.8c)

  - *Skeletal, cardiac, and smooth muscle cells.* These cells are elongated and filled with abundant contractile filaments, so they can shorten forcefully and move the bones, pump blood, or change the size of internal organs to move substances around the body.

- **Cell that stores nutrients** (Figure 3.8d)

  - *Fat cell.* The huge spherical shape of a fat cell is produced by a large lipid droplet in its cytoplasm.

- **Cell that fights disease** (Figure 3.8e)

  - *White blood cells such as the macrophage (a phagocytic cell).* This cell extends long *pseudopods* ("false feet") to crawl through tissue to reach infection sites. The many lysosomes within the cell digest the infectious microorganisms (such as bacteria) that it "eats."

- **Cell that gathers information and controls body functions** (Figure 3.8f)

  - *Nerve cell (neuron).* This cell has long processes (extensions) for receiving messages and transmitting them to other structures in the body. The processes are covered with an extensive plasma membrane, and a plentiful rough ER synthesizes membrane components and signaling molecules called *neurotransmitters.*

- **Cells of reproduction** (Figure 3.8g)

  - *Oocyte (female).* The largest cell in the body, this egg cell contains several copies of all organelles, for distribution to the

**Figure 3.8  Cell diversity.** The shape of human cells and the relative abundances of their various organelles determine their function in the body. (Note that these cells are not drawn to the same scale.)

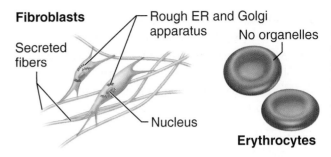

**(a) Cells that connect body parts**

**(b) Cells that cover and line body organs**

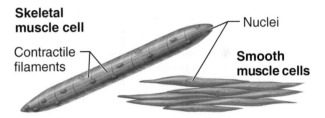

**(c) Cells that move organs and body parts**

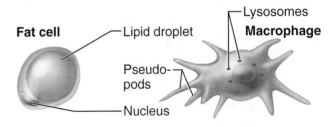

**(d) Cell that stores nutrients**          **(e) Cell that fights disease**

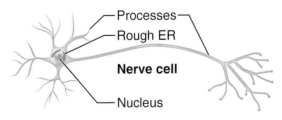

**(f) Cell that gathers information and controls body functions**

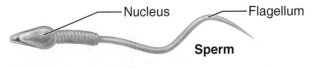

**(g) Cell of reproduction**

daughter cells that arise when the fertilized egg divides to become an embryo.

- *Sperm (male).* This cell is long and streamlined, built for swimming to the egg for fertilization. Its flagellum acts as a motile whip to propel the sperm.

## Did You Get It?

15. Name the two cell types involved in connecting body parts or regions.
16. What is the main function of a neuron?

For answers, see Appendix A.

# Cell Physiology

Each of the cell's internal parts is designed to perform a specific function for the cell. As mentioned earlier, most cells have the ability to *metabolize* (use nutrients to build new cell material, break down substances, and make ATP), *digest foods, dispose of wastes, reproduce, grow, move,* and *respond to a stimulus* (irritability). We consider most of these functions in detail in later chapters. (For example, we cover metabolism in Chapter 14, and the ability to react to a stimulus in Chapter 7). In this chapter, we consider only the functions of membrane transport (the means by which substances get through plasma membranes), protein synthesis, and cell reproduction (cell division).

## Membrane Transport

### → Learning Objectives

☐ Define *selective permeability, diffusion* (including *simple* and *facilitated diffusion* and *osmosis*), *active transport, passive transport, solute pumping, exocytosis, endocytosis, phagocytosis, pinocytosis, hypertonic, hypotonic,* and *isotonic.*

☐ Describe plasma membrane structure, and explain how the various transport processes account for the directional movements of specific substances across the plasma membrane.

The fluid environment on both sides of the plasma membrane is an example of a *solution.* It is important that you really understand solutions before we dive into an explanation of membrane transport. In the most basic sense, a **solution** is a homogeneous mixture of two or more components. Examples include the air we breathe (a mixture of gases), seawater (a mixture of water and salts), and rubbing alcohol (a mixture of water and alcohol). The substance present in the largest amount in a solution is called the **solvent** (or dissolving medium). Water is the body's chief solvent. Components or substances present in smaller amounts are called **solutes**. The solutes in a solution are so tiny that the molecules cannot be seen with the naked eye and do not settle out.

**Intracellular fluid** (collectively, the nucleoplasm and the cytosol) is a solution containing small amounts of gases (oxygen and carbon dioxide), nutrients, and salts, dissolved in water. So too is **extracellular fluid**, or **interstitial fluid**, the fluid that continuously bathes the exterior of our cells. You can think of interstitial fluid as a rich, nutritious, and rather unusual "soup." It contains thousands of ingredients, including nutrients (amino acids, sugars, fatty acids, vitamins), regulatory substances such as hormones and neurotransmitters, salts, and waste products. To remain healthy, each cell must extract from this soup the exact amounts of the substances it needs at specific times and reject the rest.

The plasma membrane is a selectively permeable barrier. **Selective permeability** means that a barrier allows some substances to pass through it while excluding others. Thus, it allows nutrients to enter the cell but keeps many undesirable or unnecessary substances out. At the same time, valuable cell proteins and other substances are kept within the cell, and wastes are allowed to pass out of it.

## ⚖ Homeostatic Imbalance 3.2

The property of selective permeability is typical only of healthy, unharmed cells. When a cell dies or is badly damaged, its plasma membrane can no longer be selective and becomes permeable to nearly everything. We see this problem when someone has been severely burned. Precious fluids, proteins, and ions "weep" (leak out) from the dead and damaged cells at the burn site. _____ ✚

Substances move through the plasma membrane in basically two ways—passively or actively. In passive processes, substances are transported across the membrane without any energy input from the cell. In active processes, the cell provides the metabolic energy (ATP) that drives the transport process.

### Passive Processes: Diffusion and Filtration

*Diffusion* (dĭ-fu′zhun) is an important means of passive membrane transport for every cell of the body. The other passive transport process, *filtration*, generally occurs only across capillary walls. Let's examine how these two types of passive transport differ.

***Diffusion*** **Diffusion** is the process by which molecules (and ions) move away from areas where they are more concentrated (more numerous) to areas where they are less concentrated (with fewer of them). All molecules possess *kinetic energy*, or energy of motion (as described in Chapter 2), and as the molecules move about randomly at high speeds, they collide and change direction with each collision. The overall effect of this erratic movement is that molecules move *down* their **concentration gradient** (spread out). The greater the difference in concentration between the two areas, the faster diffusion occurs. Because the driving force (source of energy) is the kinetic energy of the molecules themselves, the speed of diffusion is affected by the size of the molecules (the smaller the faster) and temperature (the warmer the faster).

An example should help you understand diffusion. Picture yourself dropping a tea bag into a cup of boiling water, but not stirring the cup. The bag itself represents the semipermeable cell membrane, which lets only some molecules leave the tea bag. As the tea molecules dissolve in the hot water and collide repeatedly, they begin to "spread out" from the tea bag even though it was never stirred. Eventually, as a result of their activity, the entire cup will have the same concentration of tea molecules and will appear uniform in color. (A laboratory example that might be familiar to some students is illustrated **Figure 3.9**).

The hydrophobic core of the plasma membrane is a physical barrier to diffusion. However, molecules will diffuse through the plasma membrane if any of the following are true:

- The molecules are small enough to pass through the membrane's pores (channels formed by membrane proteins).
- The molecules are lipid-soluble.
- The molecules are assisted by a membrane carrier.

The unassisted diffusion of solutes through the plasma membrane (or any selectively permeable membrane) is called **simple diffusion (Figure 3.10a)**. Solutes transported this way are lipid-soluble (such as fats, fat-soluble vitamins, oxygen, carbon dioxide).

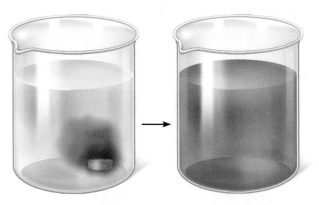

**Figure 3.9 Diffusion.** Particles in solution move continuously and collide constantly with other particles. As a result, particles tend to move away from areas where they are most highly concentrated and to become evenly distributed, as illustrated by the diffusion of dye molecules in a beaker of water.

Diffusion of water through a selectively permeable membrane such as the plasma membrane is specifically called **osmosis** (oz-mo′sis). Because water is highly polar, it is repelled by the (nonpolar) lipid core of the plasma membrane, but it can and does pass easily through special pores called *aquaporins* ("water pores") created by proteins in the membrane (Figure 3.10b). Osmosis into and out of cells is occurring all the time as water moves down its concentration gradient. The movement of water across the membrane occurs quickly. Anyone administering an IV (intravenous, into the vein) solution must use the correct solution to protect the patient's cells from life-threatening dehydration or rupture (see "A Closer Look" on p. 105).

Still another example of diffusion is **facilitated diffusion**. Facilitated diffusion provides passage for certain needed substances (notably glucose) that are both lipid-insoluble and too large to pass through the membrane pores, or charged, as in the case of chloride ions passing through a membrane protein channel. Although facilitated diffusion follows the laws of diffusion—that is, the substances move down their own concentration gradients—a protein membrane protein channel is used (Figure 3.10c), or a membrane protein that acts as a carrier (Figure 3.10d) is needed to move glucose and certain other solutes passively across the membrane into the cell.

Substances that pass into and out of cells by diffusion save the cell a great deal of energy. When you consider how vitally important water, glucose, and oxygen are to cells, you can understand just how necessary these passive transport

**Q:** *What "facilitates" facilitated diffusion?*

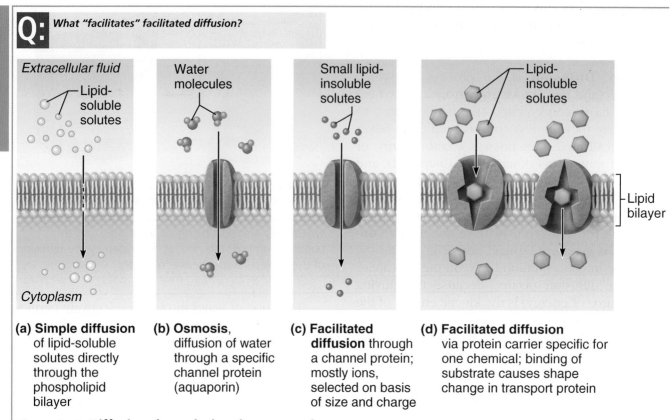

**(a) Simple diffusion** of lipid-soluble solutes directly through the phospholipid bilayer

**(b) Osmosis,** diffusion of water through a specific channel protein (aquaporin)

**(c) Facilitated diffusion** through a channel protein; mostly ions, selected on basis of size and charge

**(d) Facilitated diffusion** via protein carrier specific for one chemical; binding of substrate causes shape change in transport protein

**Figure 3.10 Diffusion through the plasma membrane.**

processes really are. Glucose and oxygen continually move into the cells (where they are in lower concentration because the cells keep using them up), and carbon dioxide (a waste product of cellular activity) continually moves out of the cells into the blood (where it is in lower concentration).

***Filtration*** **Filtration** is the process by which water and solutes are forced through a membrane (or capillary wall) by *fluid,* or *hydrostatic, pressure.* In the body, hydrostatic pressure is usually exerted by the blood. Like diffusion, filtration is a passive process, and a gradient is involved. In filtration, however, the gradient is a **pressure gradient** that actually pushes solute-containing fluid (*filtrate*) from the higher-pressure area through the filter to the lower-pressure area. In the kidneys, water and small solutes filter out of the capillaries into the kidney tubules because the blood pressure in the capillaries is greater than the fluid pressure in the tubules. Part of the filtrate formed

in this way eventually becomes urine. Filtration is not very selective. For the most part, only blood cells and protein molecules too large to pass through the membrane pores are held back.

### Did You Get It?

17. What is the energy source for all types of diffusion?
18. What determines the net direction of any diffusion process?
19. What are the two types of facilitated diffusion, and how do they differ?
20. Explain the phrase "move down the concentration gradient."

For answers, see Appendix A.

### Active Processes

Whenever a cell uses ATP to move substances across the membrane, the process is *active.* Substances moved actively are usually unable to pass in the desired direction by diffusion. They may be too large to pass through membrane channels, the membrane may lack special protein carriers for their transport, they may not be able to dissolve in the fat core, or they may have to move "uphill" *against* their concentration gradients. The

**A:** Carrier proteins or protein channels.

# IV Therapy and Cellular "Tonics"

**W**hy is it essential that medical personnel give only the proper *intravenous (IV)*, or into-the-vein, *solutions* to patients?

The tendency of a solution to hold water or "pull" water into it is called *osmotic pressure*. Osmotic pressure is directly related to the concentration of solutes in the solution. The higher the solute concentration, the greater the osmotic pressure and the greater the tendency of water to move into the solution. Many molecules, particularly proteins and some ions, are prevented from diffusing through the plasma membrane. Consequently, any change in their concentration on one side of the membrane forces water to move from one side of the membrane to the other, causing cells to lose or gain water. The ability of a solution to change the size and shape of cells by altering the amount of water they contain is called *tonicity* (ton-is′ĭ-te; *ton* = strength).

**Isotonic** (i″so-ton′ik; "same tonicity") solutions (such as 5 percent glucose and 0.9 percent saline) have the same solute and water concentrations as cells do. Isotonic solutions cause no visible changes in cells, and when such solutions are infused into the bloodstream, red blood cells (RBCs) retain their normal size and disclike shape (photo a). As you might guess, interstitial fluid and most intravenous solutions are isotonic solutions.

If red blood cells are exposed to a **hypertonic** (hi″per-ton′ik) solution—a solution that contains more solutes, or dissolved substances, than there are inside the cells—the cells begin to shrink. This is because water is in higher concentration inside the cell than outside, so it travels down its concentration gradient and leaves the cell (photo b). Hypertonic solutions are sometimes given to patients who have *edema* (swelling of the feet and hands due to fluid retention). Such solutions draw water out of the tissue spaces into the bloodstream so that the kidneys can eliminate excess fluid.

When a solution contains fewer solutes (and therefore more water) than the cell does, it is **hypotonic** (hi″po-ton′ik) to the cell. Cells placed in hypotonic solutions plump up rapidly as water rushes into them (photo c). Distilled water represents the most extreme example of a hypotonic fluid. Because it contains no solutes at all, water will enter cells until they finally burst, or *lyse*. Hypotonic solutions are sometimes infused intravenously (slowly and with care) to rehydrate extremely dehydrated patients. In less extreme cases, drinking hypotonic fluids usually does the trick.

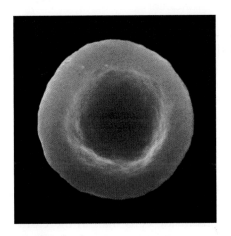

**(a) RBC in isotonic solution**

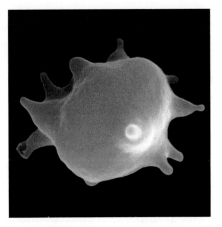

**(b) RBC in hypertonic solution**

**(c) RBC in hypotonic solution**

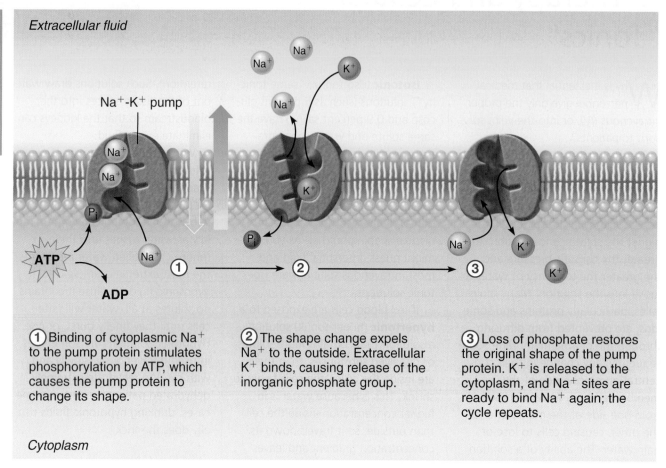

*Extracellular fluid*

Na$^+$-K$^+$ pump

ATP

ADP

Na$^+$

**P$_i$**

Na$^+$

**P$_i$**

Na$^+$

K$^+$

K$^+$

Na$^+$

K$^+$

K$^+$

① Binding of cytoplasmic Na$^+$ to the pump protein stimulates phosphorylation by ATP, which causes the pump protein to change its shape.

② The shape change expels Na$^+$ to the outside. Extracellular K$^+$ binds, causing release of the inorganic phosphate group.

③ Loss of phosphate restores the original shape of the pump protein. K$^+$ is released to the cytoplasm, and Na$^+$ sites are ready to bind Na$^+$ again; the cycle repeats.

*Cytoplasm*

**Figure 3.11 Operation of the sodium-potassium pump, a solute pump.** ATP provides the energy for a "pump" protein to move three sodium ions out of the cell and two potassium ions into the cell. The pump moves both ions against their concentration gradients. The numbered steps indicate the sequence of ion and phosphate binding a pump protein goes through as it expels Na$^+$ and imports K$^+$.

two most important active processes are active transport and vesicular transport.

***Active Transport*** Sometimes called *solute pumping*, **active transport** is similar to facilitated diffusion in that both processes require protein carriers that interact specifically and reversibly with the substances to be transported across the membrane. However, facilitated diffusion is driven by the kinetic energy of the diffusing molecules, whereas active transport uses ATP to energize its protein carriers, which are called **solute pumps**. Amino acids, some sugars, and most ions are transported by solute pumps, and in most cases these substances move *against* concentration (or electrical) gradients. This is opposite to the direction in which substances would naturally flow by diffusion, which explains the need for energy in the form of ATP.

The **sodium-potassium (Na$^+$-K$^+$) pump** alternately carries sodium ions (Na$^+$) out of and potassium ions (K$^+$) into the cell **(Figure 3.11)**. This process is absolutely necessary for normal transmission of nerve impulses. There are more sodium ions outside the cells than inside, so those inside tend to remain in the cell unless the cell uses ATP to force, or "pump," them out. ATP is split into ADP and P$_i$ (inorganic phosphate), and the phosphate is then attached to the sodium-potassium pump in a process called *phosphorylation*. Likewise, there are more potassium ions inside cells than in the extracellular fluid, and potassium ions that leak out of cells must be actively pumped back inside. Because each of the pumps in the plasma membrane transports only specific substances, active transport provides a way for the cell to be very selective in cases where substances cannot pass by diffusion. (No pump—no transport.)

***Vesicular Transport***  Some substances cannot get through the plasma membrane by active or passive transport. **Vesicular transport**, which involves help from ATP to fuse or separate membrane vesicles and the cell membrane, moves substances into or out of cells "in bulk" without their actually crossing the plasma membrane directly. The two types of vesicular transport are *exocytosis* and *endocytosis*.

**Exocytosis** (ek″so-si-to′sis; "out of the cell") **(Figure 3.12)** is the mechanism that cells use to actively secrete hormones, mucus, and other cell products or to eject certain cellular wastes. The product to be released is first "packaged" (typically by the Golgi apparatus) into a secretory **vesicle**. The vesicle migrates to the plasma membrane, fuses with it, and then ruptures, spilling its contents out of the cell (also look back at pathway 1 of Figure 3.6). Exocytosis involves a "docking" process in which docking proteins on the vesicles recognize plasma membrane docking proteins and bind with them. This binding causes the membranes to "corkscrew" together and fuse (see Figure 3.12).

**Endocytosis** (en″do-si-to′sis; "into the cell") includes those ATP-requiring processes that take up, or engulf, extracellular substances by enclosing them in a vesicle (**Figure 3.13a**, p. 108). Once the vesicle is formed, it detaches from the plasma membrane and moves into the cytoplasm, where it typically fuses with a lysosome and its contents are digested (by lysosomal enzymes). However, in some cases, the vesicle travels to the opposite side of the cell and releases its contents by exocytosis there.

If the engulfed substances are relatively large particles, such as bacteria or dead body cells, and the cell separates them from the external environment by pseudopods, the endocytosis process is more specifically called **phagocytosis** (fag″o-si-to′sis), a term that means "cell eating" (Figure 3.13b). Certain white blood cells, such as the macrophage, and other "professional" phagocytes of the body act as scavenger cells that police and protect the body by ingesting bacteria and other foreign debris. Hence, phagocytosis is a protective mechanism—a way to "clean house"—not a means of getting nutrients.

Cells eat by phagocytosis and drink by a form of endocytosis called **pinocytosis** (pĭ″no-si-to′sis; "cell drinking"), during which the cell "gulps" droplets of extracellular fluid. (It may help you to remember the name of this process to think of your mother drinking a glass of pinot noir wine.) The plasma membrane indents to form a tiny pit, or "cup," and then

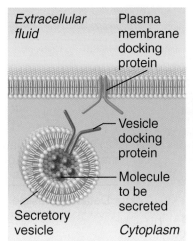

① The membrane-bound vesicle migrates to the plasma membrane.

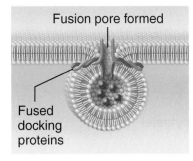

② There, docking proteins on the vesicle and plasma membrane bind, the vesicle and membrane fuse, and a pore opens up.

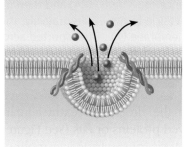

③ Vesicle contents are released to the cell exterior.

**(a) The process of exocytosis**

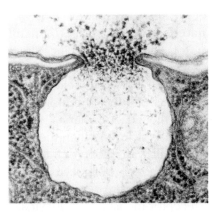

**(b) Electron micrograph of a secretory vesicle in exocytosis (190,000×)**

**Figure 3.12  Exocytosis.**

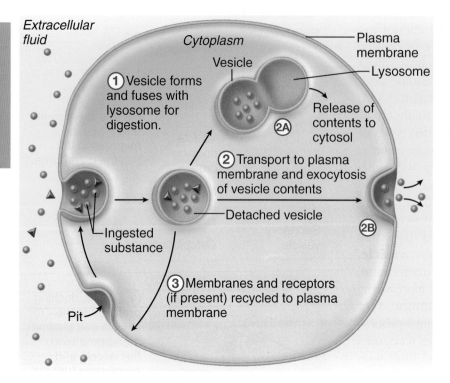

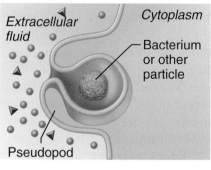

**(b) Phagocytosis**

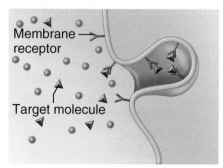

**(a) Endocytosis (pinocytosis)**

**(c) Receptor-mediated endocytosis**

**Figure 3.13 Events and types of endocytosis. (a)** Sequence of events in endocytosis. A vesicle forms by forming a pit in the membrane ①. Once the vesicle detaches from the plasma membrane, its contents may be digested within a lysosome ②A and then released to the cytosol. Alternatively, the vesicle may be transported across the cell intact and then released to the cell exterior by exocytosis ②B. If present, its membrane components and receptors are recycled to the plasma membrane ③. The type illustrated is pinocytosis. **(b)** Phagocytosis. **(c)** Receptor-mediated endocytosis.

its edges fuse around the droplet of extracellular fluid containing dissolved proteins or fats (see Figure 3.13a). Unlike phagocytosis, pinocytosis is a routine activity of most cells. It is especially important in cells that function in absorption (for example, cells forming the lining of the small intestine).

**Receptor-mediated endocytosis** is the main cellular mechanism for taking up specific target molecules (Figure 3.13c). In this process, receptor proteins on the plasma membrane bind exclusively with certain substances. Both the receptors and high concentrations of the attached target molecules are internalized in a vesicle, and then the contents of the vesicle are dealt with in one of the ways shown in Figure 3.13a. Although phagocytosis and pinocytosis are important, they are not very selective compared to receptor-mediated endocytosis. Specific substances taken in by receptor-mediated endocytosis include enzymes, some hormones, cholesterol, and iron. Unfortunately, flu viruses exploit this route to enter and attack our cells.

## Did You Get It?

21. What happens when the Na$^+$-K$^+$ pump is phosphorylated? When K$^+$ binds to the pump protein?
22. Which vesicular transport process moves large particles into the cell?
23. Which process is more selective—pinocytosis or receptor-mediated endocytosis?

For answers, see Appendix A.

## Cell Division

→ **Learning Objectives**

☐ **Briefly describe the process of DNA replication and of mitosis. Explain the importance of mitotic cell division.**

☐ **Describe the roles of DNA and of the three major varieties of RNA in protein synthesis.**

The **cell life cycle** is the series of changes a cell goes through from the time it is formed until it divides. The cycle has two major periods: **interphase**, in which the cell grows and carries on its usual metabolic activities, and **cell division**, during which it reproduces itself. Although the term *interphase* might lead you to believe that it is merely a resting time between the phases of cell division, this is not the case. During interphase, which is by far the longer phase of the cell cycle, the cell is very active and is preparing for cell division. A more accurate name for interphase would be *metabolic phase*.

### Preparations: DNA Replication

The function of cell division is to produce more cells for growth and repair processes. Because it is essential that all body cells have the same genetic material, an important event *always precedes* cell division: The DNA molecule (the genetic material) is duplicated exactly in a process called **DNA replication**. This occurs toward the end of interphase.

### CONCEPTLINK

Recall that DNA is a very complex molecule (see Chapter 2, p. 79). It is composed of building blocks called *nucleotides*, each consisting of a deoxyribose sugar, a phosphate group, and a nitrogen-containing base. Essentially, DNA is a *double helix*, a ladderlike molecule that is coiled into a spiral staircase shape. The upright parts of the DNA "ladder," or backbone, are alternating phosphate and sugar units, and the rungs of the ladder are made of pairs of nitrogen-containing bases. ←

The precise trigger for DNA synthesis is unknown, but once it starts, it continues until all the DNA has been replicated. The process begins as the DNA helix "unzips," gradually separating into its two nucleotide chains **(Figure 3.14)**. Each nucleotide strand then serves as a *template*, or set of instructions, for building a new nucleotide strand.

Remember that nucleotides join in a *complementary* way: Adenine (A) always bonds to thymine (T), and guanine (G) always bonds to cytosine (C). Hence, the order of the nucleotides on the template strand also determines the order on the new strand. For example, a TACTGC sequence on a template strand would generate a new strand with the order ATGACG. The end result is two DNA molecules that are identical to the original DNA helix, each consisting of one old and one newly assembled nucleotide strand.

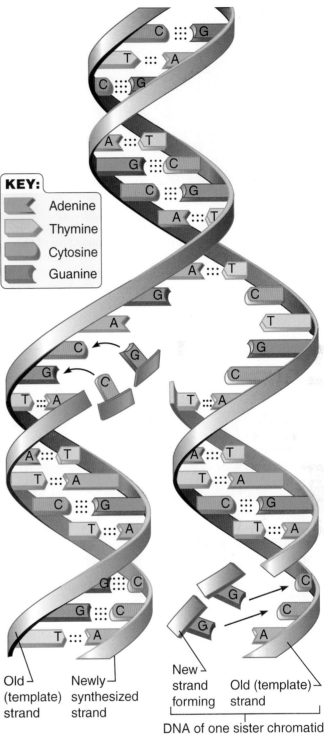

**KEY:**
Adenine
Thymine
Cytosine
Guanine

Old (template) strand

Newly synthesized strand

New strand forming

Old (template) strand

DNA of one sister chromatid

**Figure 3.14 Replication of the DNA molecule at the end of interphase.** In a process controlled by enzymes, the DNA helix unwinds (center), and its nucleotide strands are separated. Each strand then acts as a template for building a new complementary strand. As a result, two helices, each identical to the original DNA helix, are formed.

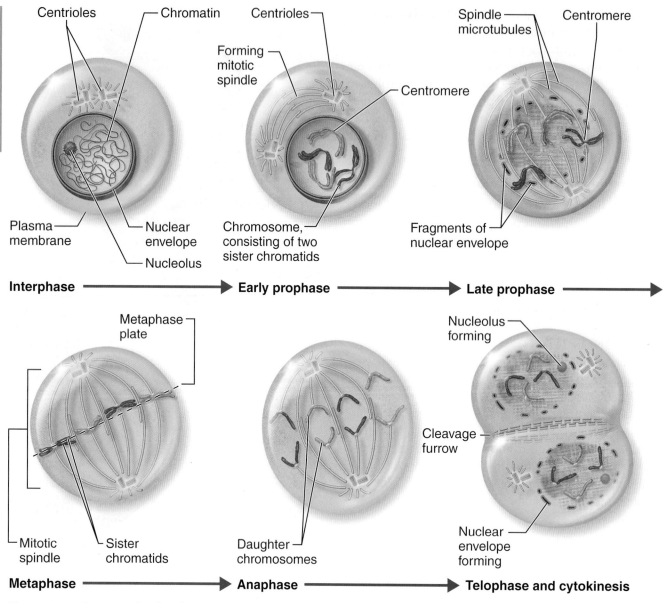

**Figure 3.15 Stages of mitosis.**

## Events of Cell Division

In all cells other than bacteria and some cells of the reproductive system, cell division consists of two events. *Mitosis* (mi-to′sis), or division of the nucleus, occurs first. The second event is division of the cytoplasm, *cytokinesis* (si″to-kĭ-ne′sis), which begins when mitosis is nearly completed.

***Mitosis*** **Mitosis** is the process of dividing a nucleus into two *daughter nuclei* with exactly the same genes as the "mother" nucleus. As explained previously, DNA replication precedes mitosis, so that for a short time the cell nucleus contains a double dose of genes. When the nucleus divides,

each daughter nucleus ends up with *exactly* the same genetic information as the original mother cell.

The stages of mitosis include the following events **(Figure 3.15)**:

- **Prophase** (pro′fāz). As cell division begins, the chromatin threads coil and shorten so that the barlike chromosomes become visible under a microscope. Because DNA has already been replicated, each chromosome is actually made up of two identical strands called sister **chromatids** (kro′mah-tidz), held together by a small buttonlike body called a **centromere** (sen′tro-mēr) (see Figure 3.15). The centrioles separate from each other and begin to move toward opposite sides of the cell, directing the assembly of a **mitotic spindle** (composed of microtubules) between them as they move. The spindle provides scaffolding for the attachment and movement of the chromosomes during the later mitotic stages. By the end of prophase, the nuclear envelope and the nucleoli have broken down and temporarily disappeared, and the chromosomes have attached randomly to the spindle fibers by their centromeres.

- **Metaphase** (met′ah-fāz). In this short stage, the chromosomes line up at the *metaphase plate* (the center of the spindle midway between the centrioles) so that a straight line of chromosomes is seen.

- **Anaphase** (an′ah-fāz). During anaphase, the centromeres that have held the chromatids together split. The chromatids (now called chromosomes again) begin to move slowly apart, drawn toward opposite ends of the cell. The chromosomes seem to be pulled by their half-centromeres, with their "arms" dangling behind them. This careful division of sister chromatids ensures that each daughter cell gets one copy of every chromosome. Anaphase is over when the chromosomes stop moving.

- **Telophase** (tel′o-fāz). Telophase is essentially prophase in reverse. The chromosomes at opposite ends of the cell uncoil to become threadlike chromatin again. The spindle breaks down and disappears, a nuclear envelope forms around each chromatin mass, and nucleoli appear in each of the daughter nuclei.

   Mitosis is basically the same in all animal cells. Depending on the type of tissue, it takes from 5 minutes to several hours to complete, but typically it lasts about 2 hours. Centriole replication is deferred until late interphase of the next cell cycle, when DNA replication begins before the onset of mitosis.

***Cytokinesis*** **Cytokinesis**, or the division of the cytoplasm, usually begins during late anaphase and completes during telophase. A contractile ring made of microfilaments forms a **cleavage furrow** over the midline of the spindle, and it eventually squeezes, or pinches, the original cytoplasmic mass into two parts. Thus, at the end of cell division, two daughter cells exist. Each is smaller with less cytoplasm than the mother cell had but is genetically identical to the mother cell. The daughter cells grow and carry out normal cell activities (interphase) until it is their turn to divide.

   Mitosis and cytokinesis usually go hand in hand, but in some cases the cytoplasm is not divided. This condition leads to the formation of *binucleate* (two nuclei) or *multinucleate* cells. This is fairly common in the liver and in the formation of skeletal muscle.

## Protein Synthesis

Proteins are key substances for all aspects of cell life. *Fibrous (structural) proteins* are the major building materials for cells (see Chapter 2). Other proteins, the *globular (functional) proteins*, perform functional roles in the body. For example, all **enzymes**, biological catalysts that speed up every chemical reaction that occurs in cells, are functional proteins. It follows, then, that every cell needs to produce proteins, a process called *protein synthesis*. This is accomplished with the DNA blueprints known as *genes* and with the help of the nucleic acid *RNA*. Let's look at this process and its components more closely.

### Genes: The Blueprint for Protein Structure

In addition to replicating itself for cell division, DNA serves as the master blueprint for protein synthesis. Traditionally, a **gene** is defined as a DNA segment that carries the information for building one protein.

   DNA's information is encoded in the sequence of bases. Each sequence of *three* bases (a *triplet*) calls for a particular *amino acid*. (Amino acids are the building blocks of proteins and are joined during protein synthesis; see Chapter 2.) For example, a DNA base sequence of AAA specifies an amino acid called phenylalanine, and CCT calls for glycine. Just as different arrangements of notes on sheet music are played as different chords, variations in the arrangements of A, C, T, and G in each gene allow cells to make all the different kinds of proteins needed. A single gene contains an estimated 300 to 3,000 base pairs in sequence.

### The Role of RNA

By itself, DNA is rather like a coded message; its information is not useful until it is decoded. Furthermore, most ribosomes—the manufacturing sites for proteins—are in the cytoplasm, but DNA never leaves the nucleus during interphase. Thus,

DNA requires not only a decoder but also a trusted messenger to carry the instructions for building proteins to the ribosomes. These messenger and decoder functions are carried out by a second type of nucleic acid, called **ribonucleic** (ri″bo-nu-kle′ik) **acid**, or **RNA**.

RNA differs from DNA in being single-stranded, in having ribose sugar instead of deoxyribose, and in having a uracil (U) base instead of thymine (T) (recall what you learned in Chapter 2). Three varieties of RNA play a special role in protein synthesis. **Ribosomal RNA (rRNA)** helps form the ribosomes, where proteins are built. **Messenger RNA (mRNA)** molecules are long, single nucleotide strands that resemble half of a DNA molecule. They carry the "message" containing instructions for protein synthesis from the DNA (gene) in the nucleus to the ribosomes in the cytoplasm. **Transfer RNA (tRNA)** molecules are small, cloverleaf-shaped molecules that escort amino acids to the ribosome.

### The Process of Protein Synthesis

Protein synthesis involves two major phases: *transcription*, when complementary mRNA (the messenger) is made using the information in the DNA gene, and *translation*, when the information carried in mRNA molecules is "decoded" and translated from nucleic acids into proteins **(Figure 3.16)**.

*Transcription* The word *transcription* often refers to one of the jobs done by a secretary—converting notes from one form (shorthand notes or an audio recording) into another form (a letter, for example). In other words, the same information is transformed from one form or format to another. In cells, **transcription** involves the transfer of information from the sequence of bases in a DNA gene into the *complementary* sequence of mRNA by an enzyme (see Figure 3.16, step ①). DNA is the template for transcription, and mRNA is the product. Each three-base sequence specifying a particular amino acid on the DNA gene is called a **triplet**, and the corresponding three-base sequences on mRNA are called **codons**. The form is different, but the same information is being conveyed. Thus, if the (partial) sequence of DNA triplets is AAT-CGT-TCG, RNA base-pairing rules (A:U, G:C) tell us that the corresponding codons on mRNA would be UUA-GCA-AGC.

*Translation* A translator takes words in one language and restates them in another language. In the **translation** phase of protein synthesis, the language of nucleic acids (base sequence) is "translated" into the language of proteins (amino acid sequence). Translation occurs in the cytoplasm and involves three major varieties of RNA. Translation consists of the following series of events (see Figure 3.16, steps 2–5). Once the mRNA attaches to the ribosome (②), tRNA transfers, or delivers, amino acids to the ribosome, where they are linked together by peptide bonds (formed by dehydration synthesis; see Chapter 2) in the exact sequence specified by the gene (and its mRNA). There are about 45 common types of tRNAs, each capable of carrying one of the 20 types of amino acids. But that is not the only job of the tRNAs. They also have to recognize the mRNA codons to "double check" that the amino acid they are toting will be added in the correct order. They can do this because they have a special three-base sequence called an **anticodon** on their "head" that can temporarily bind to the complementary codons (③).

Once the first tRNA has maneuvered itself into the correct position at the beginning of the mRNA message, the ribosome moves the mRNA strand along, bringing the next codon into position to be read by another tRNA. As amino acids are brought to their proper positions along the length of mRNA, they are joined together with peptide bonds catalyzed by the large ribosomal subunit (④).

### CONCEPTLINK

Recall that the joining of amino acids by the ribosome into peptide bonds is the result of dehydration synthesis reactions (Chapter 2, p. 68). To make room for the new peptide bond, water ($H_2O$) must be removed. A hydrogen atom is removed from one amino acid, and a hydroxyl group (OH) is removed from the other. ←

As each amino acid is added to the chain, its tRNA is released and moves away from the ribosome to pick up another amino acid (⑤). When the last codon (the termination, or "stop," codon) is read, the protein is released.

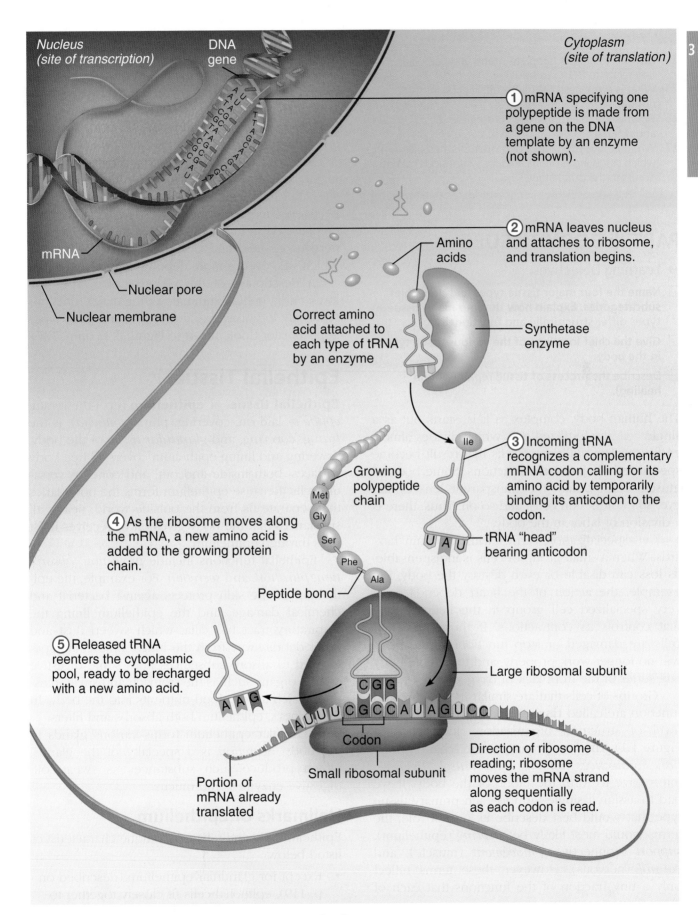

**Figure 3.16 Protein synthesis.** ① Transcription. ②–⑤ Translation.

## Did You Get It?

24. How do the terms template strand and complementary relate to DNA synthesis?
25. What is cytokinesis? What results if cytokinesis does not happen?
26. What is the role of mRNA in protein synthesis? What about tRNA?
27. What are the two stages of protein synthesis, and in which stage are proteins actually synthesized? What occurs in the other stage?

For answers, see Appendix A.

# PART II: BODY TISSUES

### → Learning Objectives

☐ **Name the four major tissue types and their chief subcategories. Explain how the four major tissue types differ structurally and functionally.**

☐ **Give the chief locations of the various tissue types in the body.**

☐ **Describe the process of tissue repair (wound healing).**

The human body, complex as it is, starts out as a single cell, the fertilized egg, which divides almost endlessly. The millions of cells that result become specialized for particular functions. Some become muscle cells, others the transparent lens of the eye, still others skin cells, and so on. Thus, there is a division of labor in the body.

Cell specialization carries with it certain hazards. When a small group of cells is indispensable, its loss can disable or even destroy the body. For example, the action of the heart depends on a very specialized cell group in the heart muscle that controls its contractions. If those particular cells are damaged or stop functioning, the heart will no longer work properly, and the whole body will suffer or die from lack of oxygen.

Groups of cells that are similar in structure and function are called **tissues** and represent the next level of structural organization (look back at Figure 1.1). The four primary tissue types—*epithelial, connective, nervous,* and *muscle tissues*—interweave to form the fabric of the body. If we had to assign a single term to each primary tissue type that would best describe its overall role, the terms would most likely be *covering* (epithelium), *support* (connective), *movement* (muscle), and *control* (nervous). However, these terms reflect only a tiny fraction of the functions that each of these tissues performs.

Tissues are organized into *organs* such as the heart, kidneys, and lungs. Most organs contain several tissue types, and the arrangement of the tissues determines each organ's structure and what it is able to do. Thus, a study of tissues will be helpful in your later study of the body's organs and how they work.

For now, we want to become familiar with the major similarities and differences in the primary tissues. Because epithelium and some types of connective tissue will not be considered again, they are emphasized more in this section than are muscle, nervous tissues, and bone (a connective tissue), which we cover in more depth in later chapters.

A description of the various tissue types follows, and a helpful summary of the major functions and body locations of the four primary tissue types appears later (look ahead to Figure 3.22 on p. 127).

# Epithelial Tissue

**Epithelial tissue**, or **epithelium** (ep″ĭ-the′le-um; *epithe* = laid on, covering; plural *epithelia*), is the *lining, covering,* and *glandular tissue* of the body. Covering and lining epithelium covers all free body surfaces, both inside and out, and contains versatile cells. Because epithelium forms the boundaries that separate us from the outside world, nearly all substances that the body gives off or receives must pass through epithelium.

Epithelial functions include *protection, absorption, filtration,* and *secretion*. For example, the epithelium of the skin protects against bacterial and chemical damage, and the epithelium lining the respiratory tract has cilia, which sweep dust and other debris away from the lungs. Epithelium specialized to absorb substances lines some digestive system organs such as the stomach and small intestine, which absorb food nutrients into the body. In the kidneys, epithelium both absorbs and filters.

Glandular epithelium forms various glands in the body. *Secretion* is a specialty of the glands, which produce such substances as sweat, oil, digestive enzymes, and mucus.

## Hallmarks of Epithelium

Epithelium generally has the unique characteristics listed below:

- Except for glandular epithelium (described on p. 119), epithelial cells fit closely together to form continuous sheets. Neighboring cells are bound together at many points by specialized cell junctions, including desmosomes and tight junctions (see p. 92).

- The membranes always have one free (unattached) surface or edge. This **apical surface** is exposed to the body's exterior or to the cavity of an internal organ. The exposed surfaces of some epithelia are slick and smooth, but others exhibit cell surface modifications, such as microvilli or cilia.

- The anchored (basal) surface of epithelium rests on a **basement membrane**, a structureless material secreted by both the epithelial cells and the connective tissue cells deep to the epithelium. Think of the basement membrane as the "glue" holding the epithelium in place.

- Epithelial tissues have no blood supply of their own (that is, they are *avascular*) and depend on diffusion from the capillaries in the underlying connective tissue for food and oxygen.

- If well nourished, epithelial cells regenerate themselves easily.

## Classification of Epithelia

Each epithelium is given two names. The first indicates the number of cell layers it has **(Figure 3.17a)**. The classifications by cell arrangement (layers) are **simple epithelium** (one layer of cells) and **stratified epithelium** (more than one cell layer). The second describes the shape of its cells (Figure 3.17b). There are *squamous* (skwa′mus) *cells*, flattened like fish scales (*squam* = scale), *cuboidal* (ku-boi′dal) *cells*, which are cube-shaped like dice, and *columnar cells*, shaped like columns. The terms describing the shape and arrangement are then combined to describe the epithelium fully. Stratified epithelia are named for the cells at the *free surface* of the epithelial membrane, not those resting on the basement membrane, which can appear "squished" and variable in shape. Figure 3.17c shows the relationship between the structure of these tissues and their functions, a topic we will cover more as we look at each type.

### Simple Epithelia

The simple epithelia are most concerned with absorption, secretion, and filtration. Because simple epithelia are usually very thin, protection is not one of their specialties.

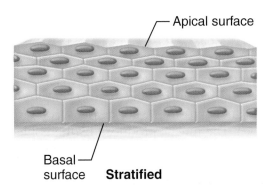

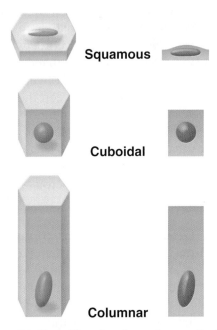

**(a) Classification based on number of cell layers**

**(b) Classification based on cell shape**

**Figure 3.17 Classification and functions of epithelia. (a)** Classification on the basis of arrangement (layers). **(b)** Classification on the basis of cell shape; for each category, a whole cell is shown on the left, and a longitudinal section is shown on the right.

*(Figure continues on page 116.)*

| Cell shape | Number of layers | |
| --- | --- | --- |
| | **One layer: simple epithelial tissues** | **More than one layer: stratified epithelial tissues** |
| **Squamous** | Diffusion and filtration<br>Secretion in serous membranes | Protection |
| **Cuboidal** | Secretion and absorption; ciliated types propel mucus or reproductive cells | Protection; these tissue types are rare in humans |
| **Columnar** | Secretion and absorption; ciliated types propel mucus or reproductive cells | |
| **Transitional** | No simple transitional epithelium exists | Protection; stretching to accommodate distension of urinary structures |

**(c) Function of epithelial tissue related to tissue type**

**Figure 3.17** *(continued)* **Classification and functions of epithelia. (c)** Function of epithelial tissue related to tissue type.

**Simple Squamous Epithelium**   **Simple squamous epithelium** is a single layer of thin squamous cells resting on a basement membrane. The cells fit closely together, much like floor tiles. This type of epithelium usually forms membranes where filtration or exchange of substances by rapid diffusion occurs. Simple squamous epithelium is in the air sacs of the lungs (called alveoli), where oxygen and carbon dioxide are exchanged (**Figure 3.18a**, p. 117), and it forms the walls of capillaries, where nutrients and gases pass between the blood in the capillaries and the interstitial fluid. Simple squamous epithelium also forms **serous membranes**, or **serosae** (se-ro′se), the slick membranes that line the ventral body cavity and cover the organs in that cavity. (We describe the serous membranes in more detail in Chapter 4.)

**Simple Cuboidal Epithelium**   **Simple cuboidal epithelium**, which is one layer of cuboidal cells resting on a basement membrane, is common in glands and their associated small tubes called *ducts* (for example, the salivary glands and pancreas). It also forms the walls of the kidney tubules (Figure 3.18b) and covers the surface of the ovaries.

**Simple Columnar Epithelium**   **Simple columnar epithelium** is made up of a single layer of tall cells that fit closely together. **Goblet cells**, which produce a lubricating mucus, are often seen in this type of epithelium. Simple columnar epithelium lines the entire length of the digestive tract from the stomach to the anus (Figure 3.18c). Epithelial membranes that line body cavities open to the body exterior are called **mucosae** (mu-ko′se), or **mucous membranes**.

**Pseudostratified Columnar Epithelium**   All of the cells of **pseudostratified** (soo″do-stra′ti-fid) **columnar epithelium** rest on a basement membrane. However, some of its cells are shorter than others, and their nuclei appear at different heights above the basement membrane. As a result, this epithelium gives the false (*pseudo*) impression that it is stratified; hence its name. Like simple columnar epithelium, this variety mainly functions in absorption and secretion. A ciliated variety (more precisely called *pseudostratified ciliated columnar epithelium*) lines most of the respiratory tract (Figure 3.18d). The mucus produced by the goblet cells in this epithelium acts as a "sticky trap" to catch dust and other debris, and the cilia propel the mucus upward and away from the lungs.

### Stratified Epithelia

Stratified epithelia consist of two or more cell layers. Being considerably more durable than the simple epithelia, these epithelia function primarily in protection.

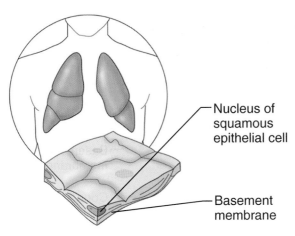

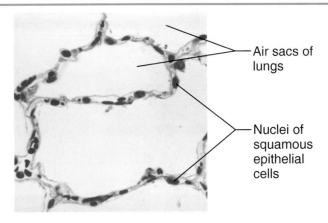

Nucleus of
squamous
epithelial cell

Basement
membrane

Air sacs of
lungs

Nuclei of
squamous
epithelial
cells

**(a) Diagram:** Simple squamous

**Photomicrograph:** Simple squamous
epithelium forming part of the alveolar
(air sac) walls (275×).

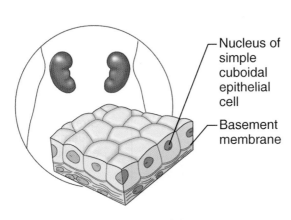

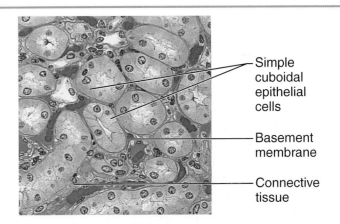

Nucleus of
simple
cuboidal
epithelial
cell

Basement
membrane

Simple
cuboidal
epithelial
cells

Basement
membrane

Connective
tissue

**(b) Diagram:** Simple cuboidal

**Photomicrograph:** Simple cuboidal
epithelium in kidney tubules (250×).

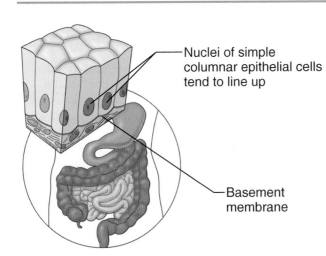

Nuclei of simple
columnar epithelial cells
tend to line up

Basement
membrane

Mucus of a
goblet cell

Simple columnar
epithelial cell

Basement
membrane

**(c) Diagram:** Simple columnar

**Photomicrograph:** Simple columnar
epithelium of the small intestine (575×).

**Figure 3.18 Types of epithelia and examples of common
locations in the body.**

View PAL Histology
Mastering A&P®

*(Figure continues on page 118.)*

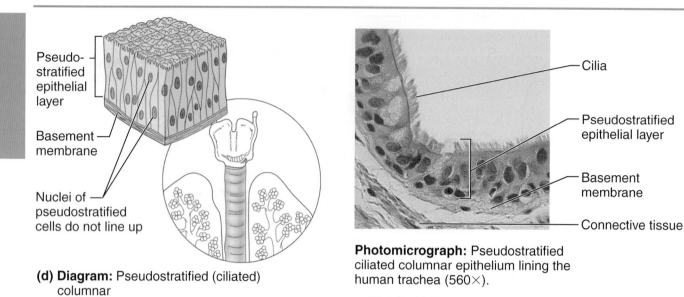

Pseudo-stratified epithelial layer

Basement membrane

Nuclei of pseudostratified cells do not line up

**(d) Diagram:** Pseudostratified (ciliated) columnar

Cilia

Pseudostratified epithelial layer

Basement membrane

Connective tissue

**Photomicrograph:** Pseudostratified ciliated columnar epithelium lining the human trachea (560×).

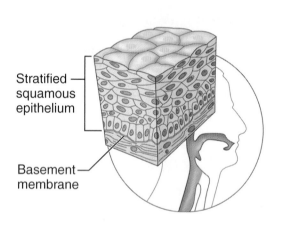

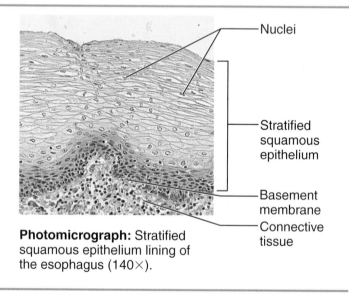

Stratified squamous epithelium

Basement membrane

**(e) Diagram:** Stratified squamous

Nuclei

Stratified squamous epithelium

Basement membrane

Connective tissue

**Photomicrograph:** Stratified squamous epithelium lining of the esophagus (140×).

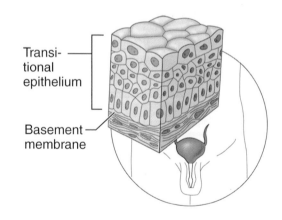

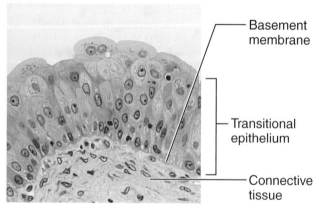

Transi-tional epithelium

Basement membrane

**(f) Diagram:** Transitional

Basement membrane

Transitional epithelium

Connective tissue

**Photomicrograph:** Transitional epithelium lining of the bladder, relaxed state (270×); surface rounded cells flatten and elongate when the bladder fills with urine.

**Figure 3.18** *(continued)* **Types of epithelia and examples of common locations in the body.**

***Stratified Squamous Epithelium*** Stratified **squamous epithelium** is the most common stratified epithelium in the body. It usually consists of many cell layers. The cells at the free edge are squamous cells, whereas those close to the basement membrane are cuboidal or columnar. Stratified squamous epithelium is found in sites that receive a good deal of abuse or friction, such as the surface of the skin, the mouth, and the esophagus (Figure 3.18e).

***Stratified Cuboidal and Stratified Columnar Epithelia*** **Stratified cuboidal epithelium** typically has just two cell layers with (at least) the surface cells being cuboidal in shape. The surface cells of **stratified columnar epithelium** are columnar cells, but its basal cells vary in size and shape. Both of these epithelia are fairly rare in the body, found mainly in the ducts of large glands. (Because the distribution of these two epithelia is extremely limited, they are not illustrated in Figure 3.18. We mention them here only to provide a complete listing of the epithelial tissues.)

***Transitional Epithelium*** **Transitional epithelium** is a highly modified, stratified squamous epithelium that forms the lining of only a few organs—the urinary bladder, the ureters, and part of the urethra. As part of the urinary system, *all* of these organs are subject to considerable stretching (Figure 3.18f). Cells of the basal layer are cuboidal or columnar; those at the free surface vary in appearance. When the organ is not stretched, the membrane is many-layered, and the superficial cells are rounded and domelike. When the organ is distended with urine, the epithelium thins like a rubber band being stretched, and the surface cells flatten and become squamouslike. This ability of transitional cells to slide past one another and change their shape (undergo "transitions") allows the ureter wall to stretch as a greater volume of urine flows through that tubelike organ. In the bladder, it allows more urine to be stored.

### Glandular Epithelium

A **gland** consists of one or more cells that make and secrete a particular product. This product, a **secretion**, typically contains protein molecules in an aqueous (water-based) fluid. The term *secretion* also indicates an active process in which the glandular cells obtain needed materials from the

blood and use them to make their products, which they then discharge by exocytosis.

Two major types of glands develop from epithelial sheets: endocrine and exocrine. **Endocrine** (en′do-krin) **glands** lose their ducts; thus they are often called *ductless* glands. Their secretions (all hormones) diffuse directly into the blood vessels that weave through the glands. Examples of endocrine glands include the thyroid, adrenals, and pituitary.

**Exocrine** (ek′so-krin) **glands** retain their ducts, and their secretions exit through the ducts to the epithelial surface. Exocrine glands, which include the sweat and oil glands, liver, and pancreas, are both internal and external. We discuss them with the organ systems to which their products are related.

### Did You Get It?

28. What two criteria are used to classify epithelial tissues?
29. How do endocrine and exocrine glands differ in structure and function?
30. Which of the following properties apply to epithelial tissues: has blood vessels; can repair itself; cells have specialized cell junctions?
31. Which two characteristics are hallmarks (distinguishing characteristics) of epithelial tissue?

**For answers, see Appendix A.**

# Connective Tissue

**Connective tissue**, as its name suggests, connects body parts. It is found everywhere in the body. It is the most abundant and widely distributed of the tissue types. Connective tissues perform many functions, but they are primarily involved in *protecting*, *supporting*, and *binding together* other body tissues.

## Hallmarks of Connective Tissue

The distinguishing characteristics of connective tissue include the following:

- **Variations in blood supply.** Most connective tissues are well *vascularized* (that is, they have a good blood supply), but there are exceptions. Tendons and ligaments, for example, have a poor blood supply, and cartilages are avascular. Consequently, all these structures heal very slowly when injured. (This is why some people say that, given a choice, they would rather have a broken bone than a torn ligament.)

- **Extracellular matrix.** Connective tissues are made up of many different types of cells plus varying amounts of a nonliving substance found outside the cells, called the *extracellular matrix.*

## Extracellular Matrix

The **extracellular matrix** deserves a bit more explanation because it is what makes connective tissue so different from the other tissue types. The matrix, which is produced by the connective tissue cells and then secreted to their exterior, has two main elements, a structureless ground substance and fibers. The *ground substance* of the matrix is composed largely of water plus some cell adhesion proteins and large, charged polysaccharide molecules. The cell adhesion proteins serve as a glue that allows the connective tissue cells to attach themselves to the matrix fibers embedded in the ground substance. The charged polysaccharide molecules trap water as they intertwine. As these polysaccharides become more abundant, they cause the matrix to vary from fluid to gel-like to firm in its consistency.

Depending on the connective tissue type, various types and amounts of fibers contribute to the matrix. These include collagen (white) fibers distinguished by their high tensile strength; elastic (yellow) fibers, which have the ability to stretch and recoil; and reticular fibers, which are fine collagen fibers that form the internal "skeleton" of soft organs such as the spleen. The building blocks, or *monomers*, of these fibers are made by the connective tissue cells and secreted into the ground substance in the extracellular space, where they join together to form the various fiber types.

Because of the variations in its extracellular matrix, connective tissue is able to form a soft packing tissue around other organs, to bear weight, and to withstand stretching and other abuses, such as abrasion, that no other tissue could endure.

## Types of Connective Tissue

As noted previously, all connective tissues consist of living cells surrounded by extracellular matrix. Their major differences reflect specific cell types, fiber types, and the number of fibers in the matrix.

At one extreme, bone and cartilage have very few cells and large amounts of hard matrix, which makes them extremely strong. At the opposite extreme, fat tissue is composed mostly of cells, and the matrix is soft. From most rigid to softest or most fluid, the major connective tissue classes are *bone*, *cartilage*, *dense connective tissue*, *loose connective tissue*, and *blood*. Find the various types of connective tissues in **Figure 3.19** as you read the following descriptions.

### Bone

**Bone**, sometimes called *osseous* (os′e-us) *tissue*, is composed of *osteocytes* (os′te-o-sītz; bone cells) sitting in cavities called *lacunae* (lah-ku′ne; "pits"). These pits are surrounded by layers of a very hard matrix that contains calcium salts in addition to large numbers of collagen fibers (Figure 3.19a). Because of its rocklike hardness, bone has an exceptional ability to protect and support other body organs (for example, the skull protects the brain).

### Cartilage

**Cartilage** is less hard and more flexible than bone. Its major cell type is *chondrocytes* (cartilage cells). It is found in only a few places in the body. Most widespread is **hyaline** (hi′ah-lin) **cartilage**, which has abundant collagen fibers hidden by a rubbery matrix with a glassy (*hyalin* = glass), blue-white appearance (Figure 3.19b). It forms the trachea, or windpipe, attaches the ribs to the breastbone, and covers bone ends at joints. The skeleton of a fetus is made largely of hyaline cartilage; but by the time the baby is born, most of that cartilage has been replaced by bone. The exceptions include the *epiphyseal*, or growth, *plates* in long bones, which allow the bones to grow in length.

Although hyaline cartilage is the most abundant type of cartilage in the body, there are others. Highly compressible **fibrocartilage** forms the cushionlike disks between the vertebrae of the spinal column (Figure 3.19c). **Elastic cartilage** is found in structures with elasticity, such as the external ear. (Elastic cartilage is not illustrated in Figure 3.19.)

### Dense Connective Tissue

In **dense connective tissue**, also called **dense fibrous tissue**, collagen fibers are the main matrix element (Figure 3.19d). Crowded between the collagen fibers are rows of *fibroblasts* (fiber-forming cells)

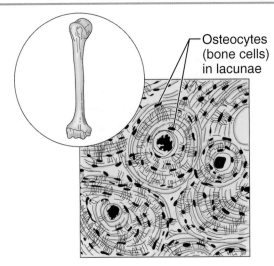

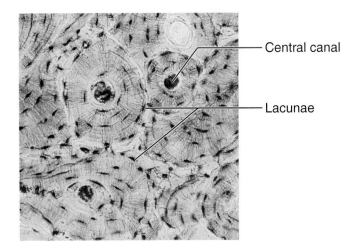

Osteocytes (bone cells) in lacunae

Central canal

Lacunae

**(a) Diagram:** Bone

**Photomicrograph:** Cross-sectional view of bone (165×).

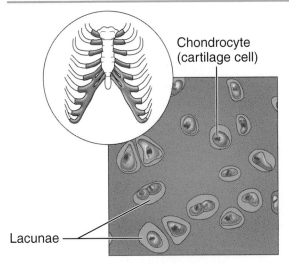

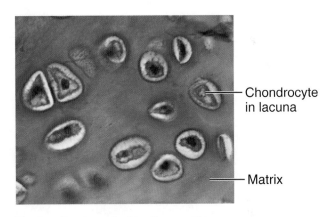

Chondrocyte (cartilage cell)

Chondrocyte in lacuna

Matrix

Lacunae

**(b) Diagram:** Hyaline cartilage

**Photomicrograph:** Hyaline cartilage from the trachea (400×).

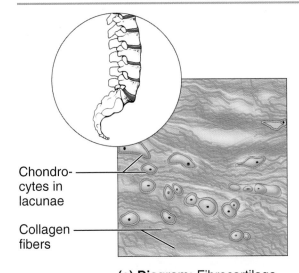

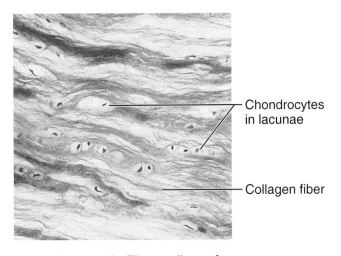

Chondro-cytes in lacunae

Collagen fibers

Chondrocytes in lacunae

Collagen fiber

**(c) Diagram:** Fibrocartilage

**Photomicrograph:** Fibrocartilage of an intervertebral disc (150×).

**Figure 3.19 Connective tissues and their common body locations.**

View PAL Histology
Mastering A&P®

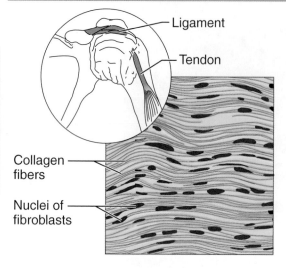

**(d) Diagram:** Dense fibrous

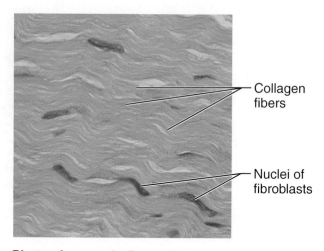

**Photomicrograph:** Dense fibrous connective tissue from a tendon (475×).

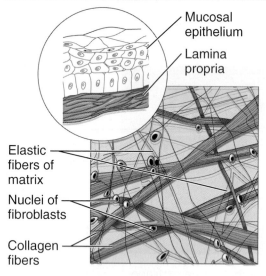

**(e) Diagram:** Areolar

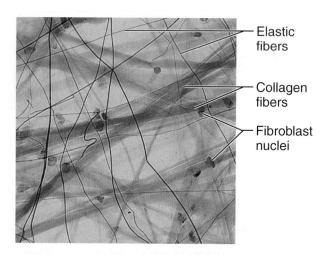

**Photomicrograph:** Areolar connective tissue, a soft packaging tissue of the body (270×).

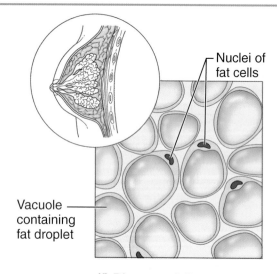

**(f) Diagram:** Adipose

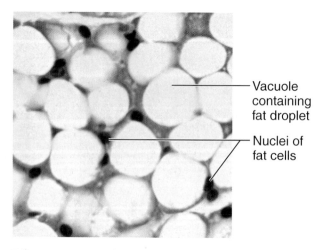

**Photomicrograph:** Adipose tissue from the subcutaneous layer beneath the skin (570×).

**Figure 3.19** *(continued)* **Connective tissues and their common body locations.**
**(e)** and **(f)** are subclasses of loose connective tissues.

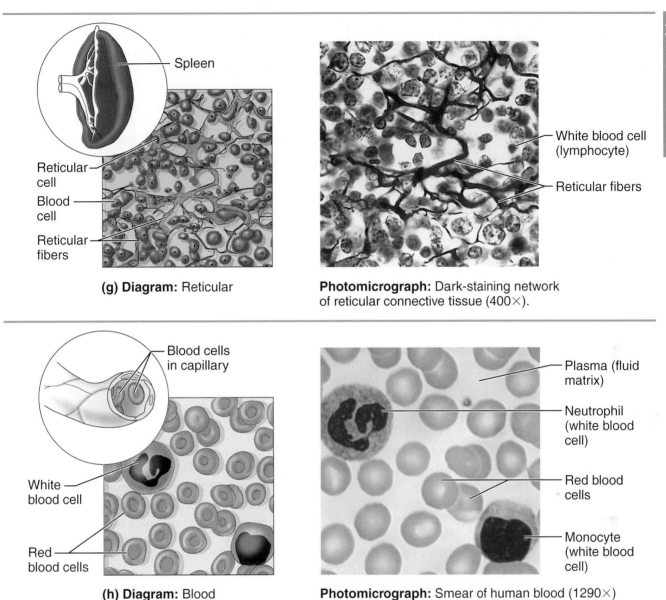

**(g) Diagram:** Reticular

**Photomicrograph:** Dark-staining network of reticular connective tissue (400×).

**(h) Diagram:** Blood

**Photomicrograph:** Smear of human blood (1290×)

**Figure 3.19** *(continued)* **(g)** is a subclass of loose connective tissue.

that manufacture the building blocks of the fibers. Dense connective tissue forms strong, ropelike structures such as tendons and ligaments. **Tendons** attach skeletal muscles to bones; **ligaments** connect bones to bones at joints. Ligaments are more stretchy and contain more elastic fibers than do tendons. Dense connective tissue also makes up the lower layers of the skin (dermis), where it is arranged in sheets.

## Loose Connective Tissue

Relatively speaking, **loose connective tissues** are softer and have more cells and fewer fibers than

any other connective tissue type except blood. There are three main types of loose connective tissue: *areolar*, *adipose*, and *reticular*.

***Areolar Connective Tissue*** **Areolar** (ah-re′o-lar) **connective tissue**, the most widely distributed connective tissue variety in the body, is a soft, pliable, "cobwebby" tissue that cushions and protects the body organs it wraps (Figure 3.19e). It functions as a universal packing tissue and connective tissue "glue" because it helps to hold the internal organs together and in their proper positions. A soft layer of areolar connective tissue called the *lamina propria* (lă′mĭ-nah pro′pre-ah) underlies all mucous membranes. Its fluid matrix contains all types of fibers, which form a loose network. In fact, when viewed through a microscope, most of the matrix appears to be empty space, which explains the name of this tissue type (*areola* = small open space). Because of its loose and fluid nature, areolar connective tissue provides a reservoir of water and salts for the surrounding tissues, and essentially all body cells obtain their nutrients from and release their wastes into this "tissue fluid." When a body region is inflamed, the local areolar tissue soaks up the excess fluid like a sponge, and the area swells and becomes puffy, a condition called **edema** (eh-de′mah). Many types of phagocytes wander through this tissue, scavenging for bacteria, dead cells, and other debris, which they destroy.

***Adipose Connective Tissue*** **Adipose** (ad′ĭ-pōs) **tissue** is commonly called *fat*. Basically, it is an areolar tissue in which *adipose* (fat) cells predominate (Figure 3.19f). A glistening droplet of oil occupies most of a fat cell's volume and compresses the nucleus, displacing it to one side.

Adipose tissue forms the subcutaneous tissue beneath the skin, where it insulates the body and protects it from bumps and extremes of both heat and cold. Adipose tissue also protects some organs individually—the kidneys are surrounded by a capsule of fat, and adipose tissue cushions the eyeballs in their sockets. There are also fat "depots" in the body, such as the hips, breasts, and belly, where fat is stored and available for fuel if needed.

***Reticular Connective Tissue*** **Reticular connective tissue** consists of a delicate network of interwoven reticular fibers associated with *reticular cells*, which resemble fibroblasts (Figure 3.19g). Reticular tissue is limited to certain sites: It forms the **stroma** (literally, "bed" or "mattress"), or internal framework of an organ. The stroma can support many free blood cells (largely white blood cells called lymphocytes) in lymphoid organs such as lymph nodes, the spleen, and bone marrow. It may help to think of reticular tissue as "cellular bleachers" where other cells can observe their surroundings.

### Blood

**Blood**, or *vascular tissue*, is considered a connective tissue because it consists of *blood cells* surrounded by a nonliving, fluid matrix called *blood plasma* (Figure 3.19h). The "fibers" of blood are soluble proteins that become visible only during blood clotting. Still, blood is quite atypical as connective tissues go. Blood is the transport vehicle for the cardiovascular system, carrying nutrients, wastes, respiratory gases, white blood cells, and many other substances throughout the body. (We consider blood in detail in Chapter 10.)

### Did You Get It?

32. What are the two hallmarks of connective tissue?
33. Which fiber type contributes strength to connective tissues?

For answers, see Appendix A.

# Muscle Tissue

**Muscle tissues** are highly specialized to *contract*, or *shorten*, which generates the force required to produce movement. There are three types of muscle tissue: *skeletal*, *cardiac*, and *smooth* **(Figure 3.20)**. Notice their similarities and differences as you read through the following descriptions.

## Skeletal Muscle

**Skeletal muscle** tissue is packaged by connective tissue sheets into organs called *skeletal muscles*, which are attached to the skeleton. These muscles, which can be controlled *voluntarily* (or consciously), form the flesh of the body, the so-called muscular system (see Chapter 6). When the skeletal muscles contract, they pull on bones or skin. As a result, gross body movements or changes in our facial expressions occur.

The cells of skeletal muscle are long, cylindrical, and multinucleate, and they have obvious *striations* (stripes). Because skeletal muscle cells are elongated to provide a long axis for contraction, they are often called *muscle fibers*.

## Cardiac Muscle

**Cardiac muscle** (covered in more detail in Chapter 11) is found only in the heart wall. As it contracts, the heart acts as a pump to propel blood through the blood vessels. Like skeletal muscle, cardiac muscle has striations, but cardiac cells have only a single nucleus and are relatively short, branching cells that fit tightly together (like clasped fingers) at junctions called **intercalated discs**.

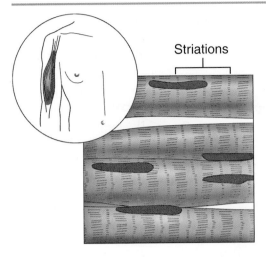
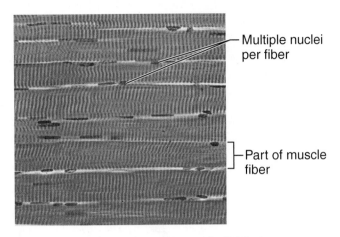

Striations

Multiple nuclei per fiber

Part of muscle fiber

**(a) Diagram:** Skeletal muscle

**Photomicrograph:** Skeletal muscle (195×).

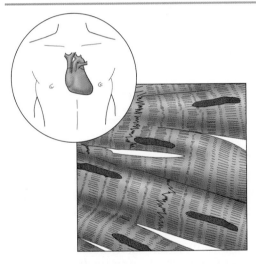
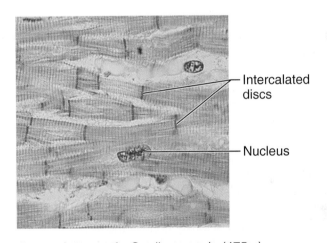

Intercalated discs

Nucleus

**(b) Diagram:** Cardiac muscle

**Photomicrograph:** Cardiac muscle (475×).

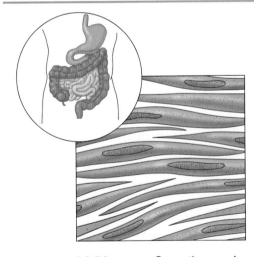
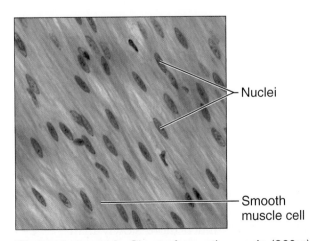

Nuclei

Smooth muscle cell

**(c) Diagram:** Smooth muscle

**Photomicrograph:** Sheet of smooth muscle (360×).

**Figure 3.20 Types of muscle tissue and their common locations in the body.**

View PAL Histology
Mastering A&P®

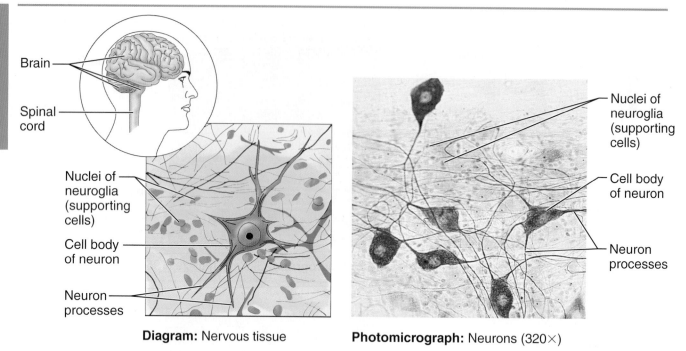

Diagram: Nervous tissue      **Photomicrograph:** Neurons (320×)

**Figure 3.21 Nervous tissue.** Neurons and supporting cells form the brain, spinal cord, and nerves.

View **PAL** Histology
Mastering A&P®

These intercalated discs contain gap junctions that allow ions to pass freely from cell to cell. This ties the cardiac cells into a functional syncytium (*syn =* together, *cyt* = cell), resulting in rapid conduction of the electrical signal to contract across the heart. Cardiac muscle is under *involuntary control,* which means that we cannot consciously control the activity of the heart. Can you imagine having to "tell" your heart to beat like you "tell" your legs to walk?

## Smooth Muscle

**Smooth (visceral) muscle** is so called because no striations are visible. The individual cells have a single nucleus and are tapered at both ends. Smooth muscle is found in the walls of hollow organs such as the stomach, uterus, and blood vessels. As smooth muscle in its walls contracts, the cavity of an organ alternately becomes smaller (constricts when smooth muscle contracts) or enlarges (dilates when smooth muscle relaxes) so that substances are mixed and/or propelled through the organ along a specific pathway. Smooth muscle contracts much more slowly than the other two muscle types, and these contractions tend to last longer. *Peristalsis* (per″ĭ-stal′sis), a wavelike motion that keeps food moving through the small intestine, is typical of its activity.

## Nervous Tissue

When we think of **nervous tissue**, we think of cells called **neurons**. All neurons receive and conduct electrochemical impulses from one part of the body to another; thus, *irritability* and *conductivity* are their two major functional characteristics. The structure of neurons is unique **(Figure 3.21)**. Their cytoplasm is drawn out into long processes (extensions), as long as 3 feet or more in the leg, which allows a single neuron to conduct an impulse to distant body locations.

However, the nervous system is more than just neurons. A special group of supporting cells called **neuroglia** insulate, support, and protect the delicate neurons in the structures of the nervous system—the brain, spinal cord, and nerves.

**Figure 3.22** summarizes the tissue types, including their hallmarks and functions in the body.

## Tissue Repair (Wound Healing)

The body has many techniques for protecting itself from uninvited guests or injury. Intact physical barriers such as the skin and mucous membranes, cilia, and the strong acid produced by

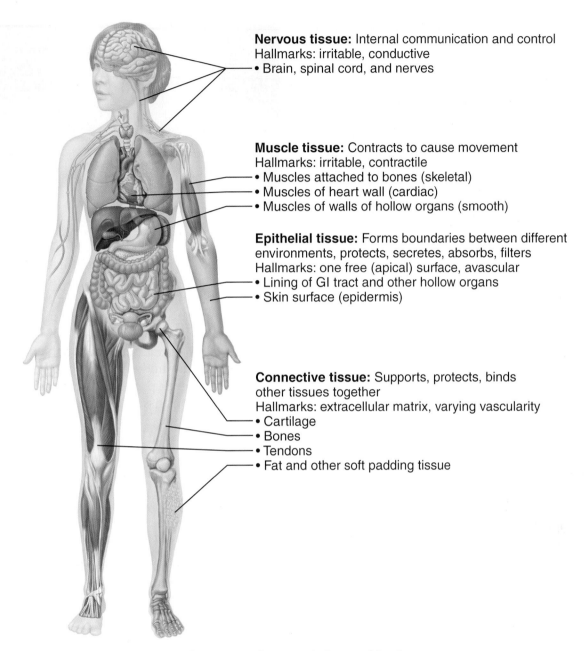

**Nervous tissue:** Internal communication and control
Hallmarks: irritable, conductive
• Brain, spinal cord, and nerves

**Muscle tissue:** Contracts to cause movement
Hallmarks: irritable, contractile
• Muscles attached to bones (skeletal)
• Muscles of heart wall (cardiac)
• Muscles of walls of hollow organs (smooth)

**Epithelial tissue:** Forms boundaries between different
environments, protects, secretes, absorbs, filters
Hallmarks: one free (apical) surface, avascular
• Lining of GI tract and other hollow organs
• Skin surface (epidermis)

**Connective tissue:** Supports, protects, binds
other tissues together
Hallmarks: extracellular matrix, varying vascularity
• Cartilage
• Bones
• Tendons
• Fat and other soft padding tissue

**Figure 3.22  Summary of the major functions, characteristics, and body
locations of the four tissue types: epithelial, connective, muscle, and
nervous tissues.**

stomach glands are just three examples of body defenses exerted at the tissue level. When tissue injury does occur, it stimulates the body's inflammatory and immune responses, and the healing process begins almost immediately. *Inflammation* is a general (nonspecific) body response that attempts to prevent further injury. The *immune response*, in contrast, is extremely specific and mounts a vigorous attack against recognized invaders, including bacteria, viruses, and toxins. (We consider these protective responses in detail

in Chapter 12.) Here we will concentrate on the process of tissue repair itself.

Tissue repair, or wound healing, occurs in two major ways: by regeneration and by fibrosis. **Regeneration** is the replacement of destroyed tissue by the same kind of cells, whereas **fibrosis** involves repair by dense (fibrous) connective tissue, that is, by the formation of *scar tissue*. Which occurs depends on (1) the type of tissue damaged and (2) the severity of the injury. Generally speaking, clean cuts (incisions) heal much more successfully than ragged tears of the tissue.

Tissue injury sets the following series of events into motion:

- **Inflammation sets the stage.** Injured tissue cells and others release inflammatory chemicals that make the capillaries very permeable. This allows fluid rich in clotting proteins and other substances to seep into the injured area from the bloodstream. The leaked clotting proteins construct a clot, which "plugs the hole" to stop blood loss and hold the edges of the wound together. The injured area becomes walled off, preventing bacteria or other harmful substances from spreading to surrounding tissues. Where the clot is exposed to air, it quickly dries and hardens, forming a scab.

- **Granulation tissue forms.** *Granulation tissue* is delicate pink tissue composed largely of new capillaries that grow into the damaged area from undamaged blood vessels nearby. These capillaries are fragile and bleed freely, as when a scab is picked away from a skin wound. Granulation tissue also contains phagocytes, which eventually dispose of the blood clot, and connective tissue cells (fibroblasts), which produce the building blocks of collagen fibers (scar tissue) to permanently bridge the gap.

- **Regeneration and fibrosis effect permanent repair.** As the surface epithelium begins to regenerate, it makes its way between the granulation tissue and the scab. The scab soon detaches, and the final result is a fully regenerated surface epithelium that covers an underlying area of fibrosis (the scar). The scar is either invisible or visible as a thin white line, depending on the severity of the wound.

The ability of the different tissue types to regenerate varies widely. Epithelial tissues such as the skin epidermis and mucous membranes regenerate beautifully. So, too, do most of the fibrous connective tissues and bone. Skeletal muscle regenerates poorly, and cardiac muscle and nervous tissue within the brain and spinal cord are replaced largely by scar tissue.

### Homeostatic Imbalance 3.3

Scar tissue is strong, but it lacks the flexibility of most normal tissues. Perhaps even more important

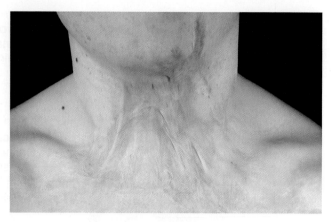

Photo showing post-burn contracture scars on the neck.

is its inability to perform the normal functions of the tissue it replaces. Thus, if scar tissue forms in the wall of the bladder, heart, or another muscular organ, it may severely hamper the functioning of that organ.

A *contracture* is a permanent tightening of the skin affecting the underlying tendons or muscles. Contractures develop during the healing process as inelastic fibrous tissue replaces the normal elastic connective tissues. Because fibrous tissue resists stretching, movement of the affected area may be limited. _____ **+**

### Did You Get It?

34. Which muscle type(s) is(are) injured when you pull a muscle while exercising?
35. How does the extended length of a neuron's processes aid its function in the body?
36. Nervous tissue is irritable. What does this mean?
37. Which tissue type is the basis for scar tissue?

For answers, see Appendix A.

# PART III: DEVELOPMENTAL ASPECTS OF CELLS AND TISSUES

### → Learning Objectives

☐ Define *neoplasm,* and distinguish benign and malignant neoplasms.

☐ Explain the significance of the fact that some tissue types (muscle and nerve) are largely amitotic after the growth stages are over.

Very early in embryonic development, cells begin to specialize to form the primary tissues, and by birth, most organs are well formed and functioning. The body continues to grow and enlarge by forming new tissue throughout childhood and adolescence.

Cell division is extremely important during the body's growth period. Most cells (except neurons and mature red blood cells) undergo mitosis until the end of puberty, when adult body size is reached and overall body growth ends. After this time, only certain cells routinely divide (are mitotic)—for example, cells exposed to abrasion that continually wear away, such as skin and intestinal cells. Liver cells stop dividing, but they retain this ability should some of them die or become damaged and need to be replaced. Still other cell groups (for example, heart muscle and nervous tissue) almost completely lose their ability to divide when they are fully mature; that is, they become *amitotic* (am″ĭ-tot′ik). Amitotic tissues are severely handicapped by injury because the lost cells cannot be replaced by the same type of cells. This is why the heart of an individual who has had several severe heart attacks becomes weaker and weaker. Damaged cardiac muscle does not regenerate and is replaced by scar tissue that cannot contract, so the heart becomes less and less capable of acting as an efficient blood pump.

The aging process begins once maturity has been reached. (Some think it begins at birth.) No one has been able to explain just *what* causes aging, but there have been many suggestions. Some think it is a result of little "chemical insults" that occur continually through life—for example, the presence of toxic chemicals (such as alcohol, certain drugs, or carbon monoxide) in the blood, or the temporary absence of needed substances such as glucose or oxygen. Perhaps the effect of these chemical insults is cumulative and finally succeeds in upsetting the delicate chemical balance of the body cells. Others think that external physical factors such as radiation (X rays or ultraviolet waves) contribute to the aging process. Still another theory is that the aging "clock" is built into our genes. We all know of cases like the radiant woman of 50 who looks about 35 or the young man of 24 who looks 40. It appears that such traits can run in families.

There is no question that certain events are part of the aging process. For example, with age, epithelial membranes thin and are more easily damaged, and the skin loses its elasticity and begins to sag. The exocrine glands of the body (epithelial tissue) become less active, and we begin to "dry out" as less oil, mucus, and sweat are produced. Some endocrine glands produce decreasing amounts of hormones, and the body processes that they control (such as metabolism and reproduction) slow down or stop altogether.

Connective tissue structures also show changes with age. Bones become porous and weaken, and tissue repair slows. Muscles begin to waste away. Although a poor diet may contribute to some of these changes, there is little doubt that decreased efficiency of the circulatory system, which reduces nutrient and oxygen delivery to body tissues, is a major factor.

Besides the tissue changes associated with aging, which accelerate in the later years of life, other modifications of cells and tissues may occur at any time. For example, when cells fail to honor normal controls on cell division and multiply wildly, an abnormal mass of proliferating cells, known as a **neoplasm** (ne′o-plazm″; "new growth"), results. Neoplasms may be benign or malignant (cancerous). (See "A Closer Look" on pp. 130–131.)

However, not all increases in cell number involve neoplasms. Certain body tissues (or organs) may enlarge because there is some local irritant or condition that stimulates the cells. This response is called **hyperplasia** (hi″per-pla′ze-ah). For example, a woman's breasts enlarge during pregnancy in response to increased hormones; this is a normal but temporary situation that doesn't have to be treated. In contrast, **atrophy** (at′ro-fe), or decrease in size, can occur in an organ or body area that loses its normal stimulation. For example, the muscles of a broken leg atrophy while in a cast during the healing period.

## Did You Get It?

38. Which of the four types of tissue is most likely to remain mitotic throughout life?
39. What cellular error allows a neoplasm to form?
40. How does the activity of endocrine glands change as the body ages?

**For answers, see Appendix A.**

# Cancer—An Intimate Enemy

When controls of the cell cycle and cell division malfunction, an abnormal cell mass called a **neoplasm** ("new growth") or **tumor** develops. Not all neoplasms are cancerous. **Benign** (be-nīn'; "kindly") neoplasms are usually surrounded by a capsule, grow slowly, and seldom kill their hosts if they are removed before they compress vital organs. In contrast, **malignant** ("bad") neoplasms (cancers) are not encapsulated. They grow more relentlessly and may become killers. Their cells resemble immature cells, and they invade their surroundings rather than push them aside, as reflected in the name **cancer**, from the Latin word for "crab." Normal cells die when they lose contact with the surrounding matrix, but malignant cells tend to break away from the parent mass and spread via the blood to distant parts of the body. The formation of new masses at other body sites is called **metastasis**

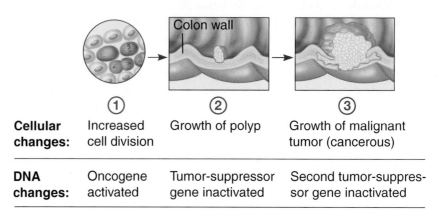

| Cellular changes: | ① Increased cell division | ② Growth of polyp | ③ Growth of malignant tumor (cancerous) |
|---|---|---|---|
| DNA changes: | Oncogene activated | Tumor-suppressor gene inactivated | Second tumor-suppressor gene inactivated |

**Stepwise development of a typical colon cancer**

(mě-tas'tă-sis). Additionally, cancer cells consume an exceptional amount of the body's nutrients, leading to weight loss and tissue wasting.

What causes *transformation*—the changes that convert a normal cell into a cancerous one? Radiation, mechanical trauma, certain viral infections, and many chemicals (tobacco tars, organic solvents such as mineral spirits) can all act as **carcinogens** (cancer-causers).

What all of these factors have in common is that they cause *mutations*—changes in DNA that alter the expression of certain genes. However, the immune system or peroxisomal and lysosomal enzymes eliminate most carcinogens. Furthermore, transformation requires several genetic changes.

The discovery of *oncogenes* (cancer-causing [*onco* = tumor] genes), and later *proto-oncogenes*, provided

# Summary

## PART I: CELLS (pp. 88–114)

### Overview of the Cellular Basis of Life (pp. 88–89)

1. A cell is composed primarily of four elements—carbon, hydrogen, oxygen, and nitrogen—plus many trace elements. Living matter is over 60 percent water. The major building material of the cell is protein.

2. Cells vary in size from microscopic to over a meter in length. Shape reflects function. For example, muscle cells have a long axis to allow shortening.

### Anatomy of a Generalized Cell (pp. 89–102)

1. Cells have three major regions—nucleus, cytoplasm, and plasma membrane.

a. The nucleus, or control center, directs cell activity and is necessary for reproduction. The nucleus contains genetic material (DNA), which carries instructions for synthesis of proteins.

b. The plasma membrane encloses the cytoplasm and acts as a semipermeable barrier to the movement of substances into and out of the cell. It is composed of a phospholipid bilayer containing proteins, sugars, and cholesterol. The water-impermeable lipid portion forms the basic membrane structure. The proteins (many of which are glycoproteins) act as enzymes or carriers in membrane transport, form membrane channels, provide receptor sites for hormones and other chemicals, or play a role in cellular recognition and interactions during development and immune reactions.

clues to the role of genes in cancer. Proto-oncogenes code for proteins that are needed for normal cell division and growth. However, many have fragile sites that break when they are exposed to carcinogens, which converts them into oncogenes. This may "switch on" dormant genes, allowing cells to become invasive and metastasize. There are more than 100 known oncogenes. Another group of genes, the tumor-suppressor genes (such as *p53* and *p16*) put the "brakes" on cell division, so inactivation of these genes leads to uncontrolled growth, allowing oncogenes to "do their thing." Tumor-suppressor genes also help inactivate carcinogens, aid DNA repair, and enhance the ability of the immune system to destroy cancer cells.

Consider colorectal cancer (see the figure). One of its first signs is a polyp (a small, benign growth of apparently normal mucosa cells. As cell division continues, the polyp enlarges and may become a malignant neoplasm. In most cases, these changes parallel cellular changes at the DNA level and include the activation of oncogenes and the inactivation of tumor-suppressor genes. The seeds of cancer appear to be in our own genes—an intimate enemy indeed.

Almost half of all Americans develop cancer in their lifetime, and a fifth of us will die of it. The most common cancers originate in the skin, lung, colon, breast, and prostate.

Screening procedures, such as self-examination of the breasts or testicles for lumps and checking fecal samples for blood, aid early detection. Unfortunately, most cancers are diagnosed after symptoms have already appeared. A *biopsy* involves removing a sample of the tumor surgically and examining it microscopically for changes typical of malignant cells. Increasingly, diagnosis includes genetic or chemical analysis of tissue samples to determine which genes are switched on or off, and which drugs may be most effective. MRI and CT scans can detect large cancers or the degree of metastasis.

Most neoplasms are removed surgically if possible. Radiation and drugs (*chemotherapy*) may also be used. Anticancer drugs have unpleasant side effects because most target *all* rapidly dividing cells, including normal ones. Side effects include nausea, vomiting, and hair loss. X rays also have side effects because they kill healthy cells in their path as well as cancer cells.

Current cancer treatments—"cut, burn, and poison"—are recognized as crude and painful. Promising new treatments include:

- Delivering radiation or anticancer drugs precisely to the cancer (via monoclonal antibodies that bind specifically to a protein "target" on a cancer cell or magnetic delivery systems for chemotherapy drugs).

- Strengthening the immune system by using "handpicked" immune cells from the patient.

- Starving tumors by cutting off their blood supply.

- Destroying cancer cells with viruses.

- Testing the tumor for genetic markers to customize treatment plans.

Today, about half of cancer cases are cured, and patients' quality of life has improved.

---

Specializations of the plasma membrane include microvilli (which increase the absorptive surface area) and cell junctions (desmosomes, tight junctions, and gap junctions).

c. The cytoplasm is where most cellular activities occur. Its fluid substance, the cytosol, contains inclusions, stored or inactive materials in the cytoplasm (fat globules, water vacuoles, crystals, and the like) and specialized bodies called organelles, each with a specific function. For example, mitochondria are sites of ATP synthesis, ribosomes are sites of protein synthesis, and the Golgi apparatus packages proteins for export from the cell. Lysosomes carry out intracellular digestion, and peroxisomes disarm dangerous chemicals in the cells. Cytoskeletal elements function in cellular support and motion. The centrioles play a role in cell division and form the bases of cilia and flagella.

## Cell Physiology (pp. 102–114)

1. All cells exhibit irritability (the ability to respond to stimuli), digest foods, excrete wastes, and are able to reproduce, grow, move, and metabolize.

2. Transport of substances through the cell membrane:

   a. Passive processes include diffusion and filtration.

      (1) Diffusion is the movement of a substance from an area of its higher concentration to an area of its lower concentration. It occurs because of kinetic energy of the molecules themselves; no ATP is required. The diffusion of dissolved solutes through the plasma membrane is simple diffusion. The diffusion of water across the plasma membrane is osmosis. Diffusion that requires a protein channel or carrier is facilitated diffusion.

(2) Filtration is the movement of substances through a membrane from an area of high hydrostatic pressure to an area of lower fluid pressure. In the body, the driving force of filtration is blood pressure.

b. Active processes (active transport and vesicular transport) use energy (ATP) provided by the cell.

   (1) In active transport, substances are moved across the membrane against an electrical or a concentration gradient by proteins called solute pumps. This accounts for the transport of amino acids, some sugars, and most ions.

   (2) The two types of ATP-activated vesicular transport are exocytosis and endocytosis. Exocytosis moves secretions and other substances out of cells; a membrane-bounded vesicle fuses with the plasma membrane, ruptures, and ejects its contents to the cell exterior. Endocytosis, in which particles are taken up by enclosure in a plasma membrane sac, includes phagocytosis (uptake of solid particles), pinocytosis (uptake of fluids), and the highly selective receptor-mediated endocytosis. In the latter, membrane receptors bind with and internalize only selected target molecules.

3. Osmotic pressure, which reflects the solute concentration of a solution, determines whether cells gain or lose water. (Discussed in "A Closer Look" on p. 105.)

   a. Hypertonic solutions contain more solutes (and less water) than do cells. In these solutions, cells lose water by osmosis and crenate.

   b. Hypotonic solutions contain fewer solutes (and more water) than do the cells. In these solutions, cells swell and may rupture (lyse) as water rushes in by osmosis.

   c. Isotonic solutions, which have the same solute-to-solvent ratio as cells, cause no changes in cell size or shape.

4. Cell division has two phases, mitosis (nuclear division) and cytokinesis (division of the cytoplasm).

   a. Mitosis begins after DNA has been replicated (during interphase); it consists of four stages—prophase, metaphase, anaphase, and telophase. The result is two daughter nuclei, each identical to the mother nucleus.

   b. Cytokinesis usually begins during anaphase and progressively pinches the cytoplasm in half. Cytokinesis does not always occur; in such cases, binucleate or multinucleate cells result.

   c. Mitotic cell division provides an increased number of cells for growth and repair.

5. Protein synthesis involves both DNA (the genes) and RNA.

   a. A gene is a segment of DNA that carries the instructions for building one protein. The information is in the sequence of bases in the nucleotide strands. Each three-base sequence (triplet) specifies one amino acid in the protein.

   b. Messenger RNA carries the instructions for protein synthesis from the DNA (gene) to the ribosomes. Transfer RNA transports amino acids to the ribosomes. Ribosomal RNA forms part of the ribosomal structure and helps coordinate the protein building process.

## PART II: BODY TISSUES (pp. 114–128)

1. Epithelium is covering, lining, and glandular tissue. Its functions include protection, absorption, and secretion. Epithelia are named according to number of layers (simple, stratified) and cell shape (squamous, cuboidal, columnar).

2. Connective tissue is the supportive, protective, and binding tissue. It is characterized by the presence of a nonliving, extracellular matrix (ground substance plus fibers) produced and secreted by the cells; it varies in amount and consistency. Fat, ligaments and tendons, bones, and cartilage are all connective tissues or connective tissue structures.

3. Muscle tissue is specialized to contract, or shorten, which causes movement. There are three types—skeletal (attached to the skeleton), cardiac (in the heart wall), and smooth (in the walls of hollow organs).

4. Nervous tissue is composed of irritable cells called neurons, which are highly specialized to receive and transmit nerve impulses, and supporting cells called neuroglia. Neurons are important in control of body processes. Nervous tissue is located in nervous system structures—brain, spinal cord, and nerves.

5. Tissue repair (wound healing) may involve regeneration, fibrosis, or both. In regeneration, the injured tissue is replaced by the same type of cells. In fibrosis, the wound is repaired with scar tissue. Epithelia and connective tissues regenerate well. Mature cardiac muscle and nervous tissue are repaired by fibrosis.

## PART III: DEVELOPMENTAL ASPECTS OF CELLS AND TISSUES (pp. 128–131)

1. Growth by cell division continues through puberty. Cell populations exposed to friction (such as epithelium) replace lost cells throughout life.

Connective tissue remains mitotic and forms repair (scar) tissue. With some exceptions, muscle tissue becomes amitotic by the end of puberty, and nervous tissue becomes amitotic shortly after birth. Injury can severely handicap amitotic tissues.

2. The cause of aging is unknown, but chemical and physical insults, as well as genetic programming, have been proposed as possible causes.

3. Neoplasms, both benign and cancerous, represent abnormal cell masses in which normal controls on cell division are not working. Hyperplasia (increase in size) of a tissue or organ may occur when tissue is strongly stimulated or irritated. Atrophy (decrease in size) of a tissue or organ occurs when the organ is no longer stimulated normally.

## Review Questions

Access additional practice questions using your smartphone, tablet, or computer:
Mastering A&P® > Study Area > Practice Tests & Quizzes

### Multiple Choice

*More than one choice may apply.*

1. Which of the following would you expect to find in or on cells whose main function is absorption?
   a. Microvilli
   b. Cilia
   c. Gap junctions
   d. Secretory vesicles

2. Which of the following cell types would not have tight junctions, desmosomes, or connexons?
   a. Phagocytes
   b. Epithelial cells
   c. Muscle cells
   d. Red blood cells

3. Which element of the plasma membrane contributes to its hydrophobic characteristics?
   a. Polar heads of phospholipid molecules
   b. Nonpolar tails of phospholipid molecules
   c. Floating proteins
   d. Cholesterol

4. A cell with abundant peroxisomes would most likely be involved in
   a. secretion.
   b. storage of glycogen.
   c. ATP manufacture.
   d. movement.
   e. detoxification activities.

5. A cell stimulated to increase its steroid production will have abundant
   a. ribosomes.
   b. rough ER.
   c. smooth ER.
   d. Golgi apparatus.
   e. secretory vesicles.

6. For osmosis to occur, there must be
   a. a selectively permeable membrane.
   b. isotonic conditions on both sides of the membrane.
   c. a difference in solute concentrations.
   d. a sodium/potassium ATPase.
   e. a hydrostatic pressure gradient.

7. In which of the following tissue types might you expect to find goblet cells?
   a. Simple cuboidal
   b. Simple columnar
   c. Simple squamous
   d. Stratified squamous
   e. Transitional

8. Which epithelium is "built" to withstand friction?
   a. Simple squamous
   b. Stratified squamous
   c. Simple cuboidal
   d. Simple columnar
   e. Pseudostratified

9. What kind of connective tissue acts as a sponge, soaking up fluid when edema occurs?
   a. Areolar connective
   b. Adipose connective
   c. Dense irregular connective
   d. Reticular connective
   e. Vascular

10. What type of connective tissue prevents muscles from pulling away from bones during contraction?
    a. Dense connective
    b. Areolar
    c. Elastic connective
    d. Hyaline cartilage

11. Which of the following is (are) *not* true of skeletal muscles?
    a. They pull on bones or skin when they contract.
    b. They have intercalated discs.
    c. They are controlled voluntarily.
    d. They have striations.
    e. They contain gap junctions.

12. Cancer is the same as
    a. all tumors.
    b. all neoplasms.
    c. all malignant neoplasms.
    d. benign tumors.
    e. AIDS.

## Short Answer Essay

13. Define *cell* and *organelle*.

14. Although cells have differences that reflect their special functions in the body, what functional abilities do *all* cells exhibit?

15. Describe the special function of DNA found in the nucleus. Which nuclear structures contain DNA? Which help to form ribosomes?

16. Describe the structures and functions of plasma membrane proteins in tight junctions, desmosomes, and gap junctions.

17. Name the cellular organelles, and explain the function of each.

18. What two structural characteristics of cell membranes determine whether substances can pass through them passively? What determines whether or not a substance can be actively transported through the membrane?

19. Explain the effect of the following solutions on living cells: hypertonic, hypotonic, and isotonic.

20. Briefly describe the process of DNA replication.

21. Define *mitosis*. Why is mitosis important?

22. What is the role of the spindle in mitosis?

23. Explain what the terms *triplet*, *codon*, and *anticodon* mean.

24. Why can an organ be permanently damaged if its cells are amitotic?

25. Describe the relative roles of DNA and RNA in protein synthesis.

26. Define *tissue*. List the four major types of tissue. Which of these types is most widely distributed in the body?

27. Describe the general characteristics of epithelial tissue. List the most important functions of epithelial tissues, and give examples of each.

28. What is the function of transitional epithelium, and how does it carry out this function?

29. What are the general structural characteristics of connective tissues? What are the functions of connective tissues? How are their functions reflected in their structures?

30. Why is blood considered to be a connective tissue?

31. What is the function of muscle tissue?

32. Tell where each of the three types of muscle tissue would be found in the body. What is meant by the statement, "Smooth muscles are involuntary in action"?

33. How does tissue healing by fibrosis differ from healing by regeneration? Which is more desirable and why?

34. Define *contracture*.

35. What is a neoplasm?

## Critical Thinking and Clinical Application Questions

36. Two examples of chemotherapeutic drugs (used to treat cancer) and their cellular actions are given below. Explain why each drug could be fatal to a cell.

    • Vincristine: damages the mitotic spindle.

    • Doxorubicin: binds to DNA and blocks messenger RNA synthesis.

37. Hydrocortisone is an anti-inflammatory drug that stabilizes lysosomal membranes. Explain how this effect reduces cell damage and inflammation.

38. John has severely injured his knee during football practice. He is told that he has a torn knee cartilage and to expect that recovery and repair will take a long time. Why will it take a long time?

39. Three patients in an intensive care unit are examined by the resident doctor. One patient has brain damage from a stroke, another had a heart attack that severely damaged his heart muscle, and the third has a severely damaged liver (a gland) from a crushing injury in a car accident. All three patients have stabilized and will survive, but only one will have full functional recovery through regeneration. Which one, and why?

40. Michael had a nervous habit of chewing on the inner lining of his lip with his front teeth. The lip grew thicker and thicker from years of continual irritation. Michael's dentist noticed his greatly thickened lip, then told him to have it checked to see if the thickening was a tumor. A biopsy revealed hyperplasia, but no evidence of neoplasia. What do these terms mean? Did Michael have cancer of the mouth?

41. Think carefully about the chemistry of the plasma membrane, and then answer this question: Why is minor damage to the membrane usually *not* a problem?

42. Lisa, an elderly lady, is overweight and has a sedentary lifestyle. She has recently developed puffy ankles. What is this condition called, and how do you think this has developed?

# Skin and Body Membranes

**4**

**WHAT**

The integument includes body membranes, skin, hair, fingernails, and sweat and oil glands.

**HOW**

The skin has multiple layers that protect the body, help regulate body temperature, and help excrete wastes via sweat.

**WHY**

The skin is the first barrier to keep good things such as water in, and bad things such as harmful bacteria out.

**INSTRUCTORS**

New **Building Vocabulary Coaching** Activities for this chapter are assignable in Mastering A&P®

*Body membranes* cover surfaces, line body cavities, and form protective (and often lubricating) sheets around organs. They fall into two major groups: (1) *epithelial membranes*, which include the cutaneous, mucous, and serous membranes; and (2) *connective tissue membranes*, represented by synovial membranes. The *cutaneous membrane*, generally called the *skin* or *integumentary system*, is the outer covering that we all rely on for protection. In this chapter, the skin will receive most of our attention, but first we will consider the other body membranes.

## Classification of Body Membranes

→ **Learning Objectives**

☐ **List the general functions of each membrane type—cutaneous, mucous, serous, and synovial—and give its location in the body.**

☐ **Compare the structure (tissue makeup) of the major membrane types.**

The two major categories of body membranes—epithelial and connective tissue—are classified in part according to their tissue makeup.

## Epithelial Membranes

The **epithelial membranes**, also called *covering* and *lining membranes*, include the cutaneous membrane (skin), the mucous membranes, and the serous membranes **(Figure 4.1)**. However, calling these membranes "epithelial" is misleading because it is not the whole story. Although they all *do* contain an epithelial layer, it is always combined with an underlying layer of connective tissue. Hence these membranes are actually simple organs. Because we will discuss the skin in detail shortly, we will list it here solely as a subcategory of the epithelial membranes.

### Cutaneous Membrane

The **cutaneous** (ku-ta′ne-us) **membrane** is composed of two layers, the superficial *epidermis* and the underlying *dermis*. The epidermis is composed of stratified squamous epithelium, whereas the dermis is mostly dense (fibrous) connective tissue. Unlike other epithelial membranes, the cutaneous membrane is exposed to air and is a dry membrane (Figure 4.1a).

### Mucous Membranes

A **mucous** (myu′kus) **membrane (mucosa)** is composed of epithelium (the type varies with the site) resting on a loose connective tissue membrane called a *lamina propria* (Figure 4.1b). This membrane type lines all body cavities that open to the exterior, such as those of the hollow organs of the respiratory, digestive, urinary, and reproductive tracts. Notice that the term *mucosa* refers only to the location of the epithelial membranes, *not* their cellular makeup, which varies. However, most mucosae contain either stratified squamous epithelium (as in the mouth and esophagus) or simple columnar epithelium (as in the rest of the digestive tract). In all cases, they are moist membranes that are almost continuously bathed in secretions or, in the case of the urinary mucosae, urine.

The epithelium of mucosae is often adapted for absorption or secretion. Although many mucosae secrete mucus, not all do. The mucosae of the respiratory and digestive tracts secrete large amounts of protective, lubricating mucus; that of the urinary tract does not.

## Serous Membranes

A **serous membrane**, or **serosa**, is composed of a layer of simple squamous epithelium resting on a thin layer of areolar connective tissue. In contrast to mucous membranes, which line open body cavities, serous membranes line body cavities that are closed to the exterior (except for the dorsal body cavity and joint cavities).

Serous membranes occur in pairs (Figure 4.1c). The *parietal* (pah-ri′ĕ-tal: *parie* = wall) *layer* lines a specific portion of the wall of the ventral body cavity. It folds in on itself to form the *visceral* (vis′er-al) *layer*, which covers the outside of the organ(s) in that cavity.

You can visualize the relationship between the serosal layers by pushing your fist into a limp balloon only partially filled with air (Figure 4.1d). The part of the balloon that clings to your fist can be compared to the visceral serosa clinging to the organ's external surface. The outer wall of the balloon represents the parietal serosa that lines the walls of the cavity and that, unlike the balloon, is never exposed but is always fused to the cavity wall.

In the body, the serous layers are separated not by air but by a scanty amount of thin, clear fluid, called **serous fluid**, which is secreted by both membranes. Although there is a potential space between the two membranes, they tend to lie very close to each other. The lubricating serous fluid allows the organs to slide easily across the cavity walls and one another without friction as they carry out their routine functions. This is extremely important when mobile organs such as the pumping heart and expanding lungs are involved.

The specific names of the serous membranes depend on their locations. The serosa lining the abdominal cavity and covering its organs is the **peritoneum** (per″ĭ-to-ne′um). In the thorax, serous membranes isolate the lungs and heart from one another. The membranes surrounding the lungs are the **pleurae** (ploo′re); those around the heart are the **pericardia** (per″ĭ-kar′de-ah).

## Connective Tissue Membranes

**Synovial** (sĭ-no′ve-al) **membranes** are composed of loose areolar connective tissue and contain no epithelial cells at all. These membranes line the

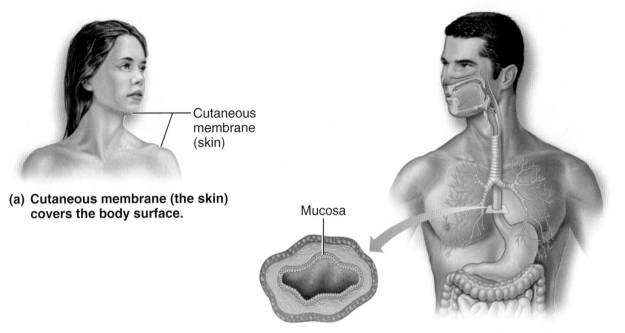

**(a) Cutaneous membrane (the skin) covers the body surface.**

Cutaneous membrane (skin)

Mucosa

**(b) Mucous membranes line body cavities open to the exterior.**

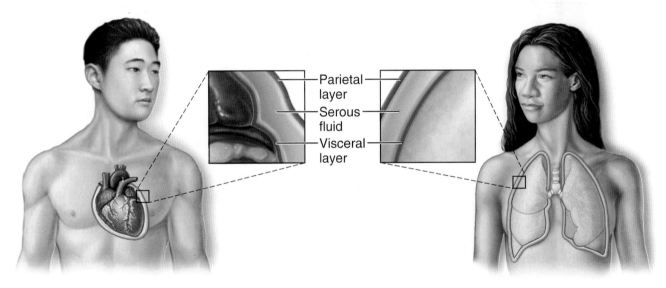

Parietal layer
Serous fluid
Visceral layer

**(c) Serous membranes line body cavities closed to exterior.**

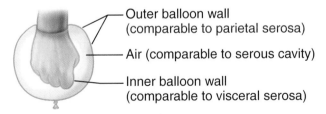

Outer balloon wall (comparable to parietal serosa)

Air (comparable to serous cavity)

Inner balloon wall (comparable to visceral serosa)

**(d) A fist thrust into a limp balloon demonstrates the relationship between the parietal and visceral serous membrane layers.**

Figure 4.1  **Classes of epithelial membranes.**

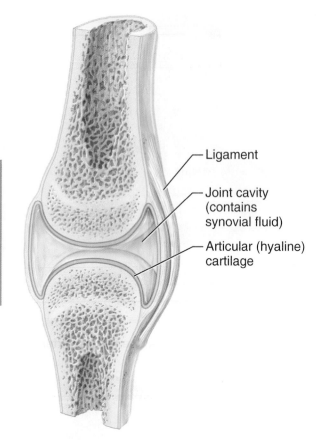

Ligament

Joint cavity (contains synovial fluid)

Articular (hyaline) cartilage

**Figure 4.2 A typical synovial joint.**

fibrous capsules surrounding joints **(Figure 4.2)**, where they provide a smooth surface and secrete a lubricating fluid. They also line small sacs of connective tissue called *bursae* (ber'se) and the tubelike *tendon sheaths*. Both of these structures cushion organs moving against each other during muscle activity—such as the movement of a tendon across a bone's surface.

### Did You Get It?

1. How do the body locations of serous and mucous membranes differ?
2. A scalpel penetrates the left lung and enters the heart. Name the six serous membrane layers the blade passes through as it moves from the body surface into the heart.
3. Where would you find a synovial membrane?

For answers, see Appendix A.

# The Integumentary System (Skin)

→ **Learning Objectives**

☐ **List several important functions of the integumentary system, and explain how these functions are accomplished.**

☐ **When provided with a model or diagram of the skin, recognize and name the following skin structures: epidermis, dermis (papillary and reticular layers), hair and hair follicle, sebaceous gland, and sweat gland.**

☐ **Name the layers of the epidermis, and describe the characteristics of each.**

Would you be enticed by an advertisement for a coat that is waterproof, stretchable, and washable; that invisibly repairs small cuts, rips, and burns; and that is guaranteed to last a lifetime with reasonable care? Sounds too good to be true, but you already have such a coat—your cutaneous membrane, or **skin**.

Skin is absolutely essential because it keeps water and other precious molecules in the body. It also keeps excess water (and other things) out. (This is why you can swim for hours without becoming waterlogged.) Structurally, the skin is a marvel. It is pliable yet tough, which allows it to take constant punishment from external agents. Without our skin, we would quickly fall prey to bacteria and perish from water and heat loss.

The skin and its appendages (sweat and oil glands, hair, and nails) are collectively called the **integumentary** (in-teg"u-men'tuh-re) **system**. Let's take a look at the structure and function of this often underappreciated body system.

## Functions of the Integumentary System

The integumentary system, also called the **integument** (in-teg'u-ment), which simply means "covering," performs a variety of functions; most, but not all, of which are protective **(Table 4.1)**. It insulates and cushions the deeper body organs and protects the entire body from mechanical damage (bumps and cuts), chemical damage (such as from acids and bases), thermal damage (heat and cold), ultraviolet (UV) radiation (in sunlight), and microbes. The uppermost layer of the skin is hardened, to help prevent water loss from the body surface.

The skin's rich capillary network and sweat glands (both controlled by the nervous system)

Table **4.1**    **Functions of the Integumentary System**

| Functions | How accomplished |
|---|---|
| Protects deeper tissues from | |
| • Mechanical damage (bumps) | Physical barrier contains keratin, which toughens cells; fat cells to cushion blows; and both pressure and pain receptors, which alert the nervous system to possible damage. |
| • Chemical damage (acids and bases) | Has relatively impermeable keratinized cells; contains pain receptors, which alert the nervous system to possible damage. |
| • Microbe damage | Has an unbroken surface and "acid mantle" (skin secretions are acidic and thus inhibit microbes, such as bacteria). Phagocytes ingest foreign substances and pathogens, preventing them from penetrating into deeper body tissues. |
| • Ultraviolet (UV) radiation (damaging effects of sunlight or tanning beds) | Melanin produced by melanocytes offers protection from UV damage. |
| • Thermal (heat or cold) damage | Contains heat/cold/pain receptors. |
| • Desiccation (drying out) | Contains a water-resistant glycolipid and keratin. |
| Aids in body heat loss or heat retention (controlled by the nervous system) | *Heat loss:* By activating sweat glands and by allowing blood to flush into skin capillary beds so that heat can radiate from the skin surface. *Heat retention:* By not allowing blood to flush into skin capillary beds. |
| Aids in excretion of urea and uric acid | Contained in perspiration produced by sweat glands. |
| Synthesizes vitamin D | Modified cholesterol molecules in skin converted to vitamin D in the presence of sunlight. |

play an important role in regulating heat loss from the body surface. The skin acts as a mini-excretory system; urea, salts, and water are lost when we sweat. The skin is also a chemical plant; it manufactures several proteins important to immunity and synthesizes vitamin D. (Modified cholesterol molecules in the skin are converted to vitamin D by sunlight.) The skin also produces acidic secretions, called the *acid mantle*, that protect against bacterial invasion.

## Did You Get It?

4. Explain the relationships between the words *skin*, *cutaneous membrane*, *integument*, and *integumentary system*.
5. What are three important functions of the integumentary system?

For answers, see Appendix A.

## Structure of the Skin

As we mentioned, the skin is composed of two kinds of tissue **(Figure 4.3)**. The outer **epidermis**

(ep″ĭ-der′mis) is made up of stratified squamous epithelium that is capable of becoming hard and tough. The underlying **dermis** is made up mostly of dense connective tissue. The epidermis and dermis are firmly connected and the dermis is fairly tear resistant. However, a burn or friction (such as the rubbing of a poorly fitting shoe) may cause them to separate, allowing interstitial fluid to accumulate in the cavity between the layers, which results in a *blister*.

Deep to the dermis is the **subcutaneous** tissue, or **hypodermis**, which essentially is adipose (fat) tissue. It is not considered part of the skin, but it does anchor the skin to underlying organs and provides a site for nutrient storage. Subcutaneous tissue serves as a shock absorber and insulates the deeper tissues from extreme temperature changes occurring outside the body. It is also responsible for the curves that are more a part of a woman's anatomy than a man's.

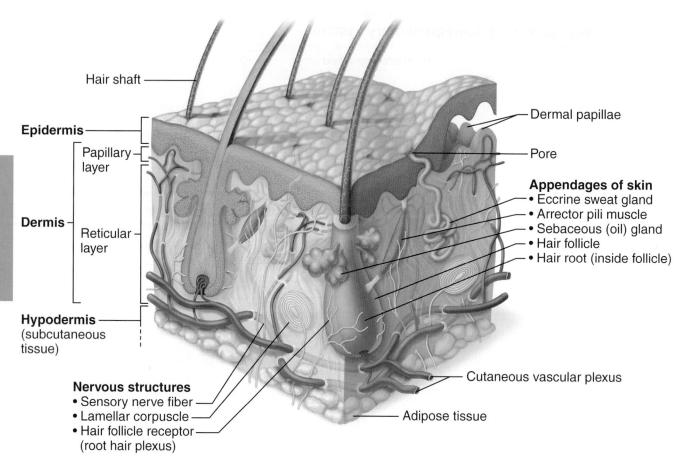

Hair shaft

**Epidermis**

Papillary layer

**Dermis**

Reticular layer

**Hypodermis** (subcutaneous tissue)

**Nervous structures**
- Sensory nerve fiber
- Lamellar corpuscle
- Hair follicle receptor (root hair plexus)

Dermal papillae

Pore

**Appendages of skin**
- Eccrine sweat gland
- Arrector pili muscle
- Sebaceous (oil) gland
- Hair follicle
- Hair root (inside follicle)

Cutaneous vascular plexus

Adipose tissue

**Figure 4.3 Skin structure.** Microscopic view of the skin and underlying subcutaneous tissue.

## Epidermis

Most cells of the epidermis are **keratinocytes** (keratin cells), which produce **keratin**, the fibrous protein that makes the epidermis a tough protective layer in a process called *keratinization*. These keratinocytes are connected by desmosomes throughout the epidermis (refer to Figure 3.3, p. 92).

Like all other epithelial tissues, the epidermis is *avascular*; that is, it has no blood supply of its own. This explains why a man can shave daily and not bleed even though he cuts off many cell layers each time he shaves.

The epidermis is composed of up to five layers, or *strata* (strah′tah). From the inside out these are the *stratum basale, spinosum, granulosum, lucidum*, and *corneum*, illustrated in **Figure 4.4** (except the stratum lucidum, which is found only in thick skin).

The deepest cell layer of the epidermis, the **stratum basale** (stra′tum ba-sah′le), lies closest

to the dermis and is connected to it along a wavy border that resembles corrugated cardboard. This basal layer contains the most adequately nourished of the epidermal cells because nutrients diffusing from the dermis reach them first. Stem cells in this layer are constantly dividing, and millions of new cells are produced daily; hence its alternate name, *stratum germinativum* (jer″min-ah-tiv′um; "germinating layer"). Of the new cells produced, some become epidermal cells, and others maintain the population of stem cells by continuing to divide. The daughter cells destined to become epidermal cells are pushed upward, away from the source of nutrition, to become part of the epidermal layers closer to the skin surface. As they move away from the dermis and become part of the more superficial layers, the **stratum spinosum** and then the **stratum granulosum**, they become flatter and increasingly keratinized.

As these cells leave the stratum granulosum, they die, forming the clear **stratum lucidum**

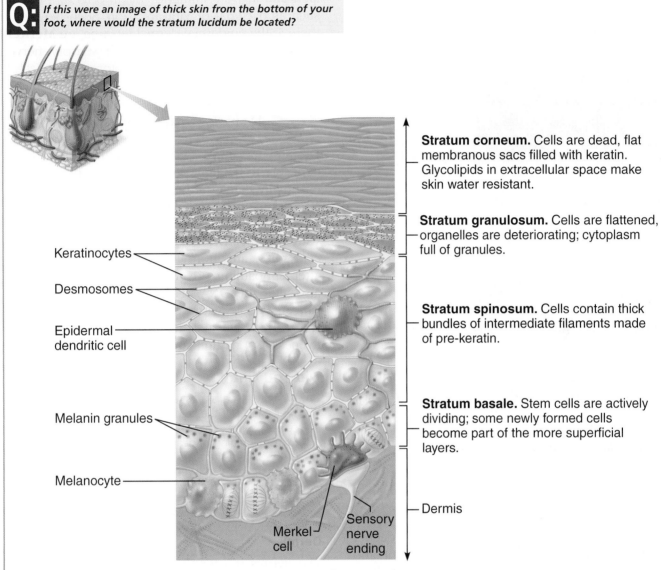

**Q:** *If this were an image of thick skin from the bottom of your foot, where would the stratum lucidum be located?*

Keratinocytes

Desmosomes

Epidermal dendritic cell

Melanin granules

Melanocyte

Merkel cell

Sensory nerve ending

**Stratum corneum.** Cells are dead, flat membranous sacs filled with keratin. Glycolipids in extracellular space make skin water resistant.

**Stratum granulosum.** Cells are flattened, organelles are deteriorating; cytoplasm full of granules.

**Stratum spinosum.** Cells contain thick bundles of intermediate filaments made of pre-keratin.

**Stratum basale.** Stem cells are actively dividing; some newly formed cells become part of the more superficial layers.

Dermis

4

**Figure 4.4  The main structural features of the epidermis.** Keratinocytes (tan), connected by many desmosomes, form most of the epidermis. Melanocytes (gray) make the pigment melanin, and star-shaped epidermal dendritic cells (blue) are protective immune cells. Occasional Merkel cells (purple), each associated with a nerve ending, act as touch receptors. (Only part of the stratum corneum is illustrated.)

(lu′sid-um). This latter epidermal layer is not present in all skin regions. It occurs only where the skin is hairless and extra thick, that is, on the palms of the hands and soles of the feet. The combination of accumulating keratin inside them, secreting a water-repellent glycolipid into the extracellular space, and their increasing distance from the blood supply (in the dermis) effectively dooms the stratum lucidum cells and the more superficial epidermal cells because they are unable to get adequate nutrients and oxygen.

The outermost layer, the **stratum corneum** (kor′ne-um), is 20 to 30 cell layers thick, but it accounts for about three-quarters of the epidermal thickness. The shinglelike dead cell remnants, completely filled with keratin, are referred to as *corni-fied*, or *horny, cells* (*cornu* = horn). The common saying "Beauty is only skin deep" is especially interesting in light of the fact that nearly everything we see when we look at someone is dead! The abundance of the tough keratin protein in the stratum corneum allows that layer to provide a durable

**A:** Stratum lucidum is located between stratum granulosum and stratum corneum.

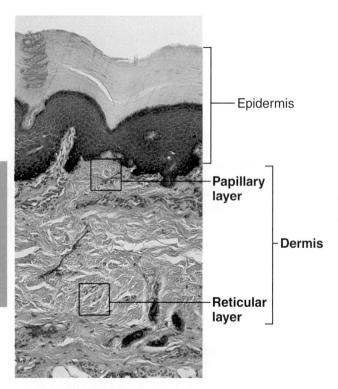

Figure 4.5 **Light micrograph of the two layers of the dermis (100×).** The dermal papillae are extensions of the superficial papillary layer, which consists of areolar connective tissue. The deeper reticular layer is dense irregular fibrous connective tissue.

View **PAL** Histology
Mastering A&P®

"overcoat" for the body, which protects deeper cells from the hostile external environment and from water loss, and helps the body resist biological, chemical, and physical assaults. The stratum corneum rubs and flakes off slowly and steadily as the *dandruff* familiar to everyone. The average person sheds about 18 kg (40 lb) of these flakes in a lifetime, providing a food source for the dust mites that inhabit our homes and bed linens. This layer is replaced by cells produced by the division of the deeper stratum basale cells. Indeed, we have a totally "new" epidermis every 25 to 45 days.

**Melanin** (mel′ah-nin), a pigment that ranges in color from yellow to brown to black, is produced by special spider-shaped cells called **melanocytes** (mel′ah-no-sītz), found chiefly in the stratum basale. *Freckles* and *moles* are seen where melanin is concentrated in one spot.

Scattered in the epidermis are **epidermal dendritic cells**, which are important "sentries" that

alert and activate immune system cells to a threat such as bacterial or viral invasion. Seen here and there at the epidermal-dermal junction are **Merkel cells**, which are associated with sensory nerve endings and serve as touch receptors called *Merkel discs*.

### Homeostatic Imbalance 4.1

Despite melanin's protective effects, excessive exposure to UV light (via sunlight or tanning beds) eventually damages the skin, leading to a leathery appearance. It also depresses the immune system. This may help to explain why people infected with the ***human herpesvirus 1***, which causes *cold sores*, are more likely to have an eruption after sunbathing. Overexposure to the sun can also alter the DNA of skin cells, leading to skin cancer. People with very dark skin seldom have skin cancer, attesting to melanin's amazing effectiveness as a natural sunscreen. _____ **+**

### Did You Get It?

6. What cell type is most abundant in the epidermis?
7. Which layer of the epidermis produces new epidermal cells?
8. Excess shedding of dead cells from the superficial layer of the skin of the scalp causes dandruff. What is the name of that skin layer?
9. In which epidermal layer do cells begin to die?

For answers, see Appendix A.

### Dermis

The dermis is your "hide." It is a strong, stretchy envelope that helps to bind the body together. When you purchase leather goods (bags, belts, shoes, and the like), you are buying the treated dermis of animals.

The connective tissue making up the dermis consists of two major regions—the *papillary* and the *reticular* areas **(Figure 4.5)**, which are composed of areolar and dense irregular connective tissue, respectively. Like the epidermis, the dermis varies in thickness. For example, it is particularly thick on the palms of the hands and soles of the feet but is quite thin on the eyelids.

The **papillary layer** is the superficial dermal region. It is uneven and has peglike projections from its superior surface, called **dermal papillae** (pah-pil′e; *papill* = nipple), which indent the epidermis above. Many of the dermal papillae contain capillary loops, which furnish nutrients to the

# A Wrinkle Out of Time

When it comes to preventing wrinkles, it helps to have good genes, to not smoke, to use a good sunscreen, and to think pleasant thoughts. Smoking ages the skin by increasing production of an enzyme that destroys collagen. Collagen supports the skin and keeps the skin hydrated, so with less of it, wrinkles appear. Too much unprotected exposure to the sun or tanning beds leads to UV radiation damage, which causes elastic fibers to clump, resulting in leathery skin. For those wrinkled by years of smoking and sun damage, a surgical face-lift that removes the excess and sagging skin followed by laser resurfacing or microdermabrasion seems to be the only way to banish the wrinkles.

However, for those who sport frown lines, furrowed brows, or crow's feet due to frequent and repetitive facial expressions, cosmetic injections of Botox may be the answer to regaining younger-looking skin. Botulinum toxin type A, more familiarly called Botox Cosmetic, is a neurotoxin produced by the bacterium that causes botulism, a deadly form of food poisoning. Injected in extremely small doses, the purified toxin blocks acetylcholine (ACh) release by nerve cells (ACh is a chemical signal from nerves telling muscles to contract), thus inhibiting the muscle's ability to contract. This "induced paralysis" smooths lines in about a week.

Botox was approved in 1989 to treat two eye muscle disorders—uncontrollable blinking (blepharospasm) and misaligned eyes (strabismus). The discovery that Botox could be used cosmetically was pure luck—physicians using the toxin to counter abnormal eye contractions noticed that frown lines between the eyes had softened.

The process has some risks. If too much toxin is injected, a person can

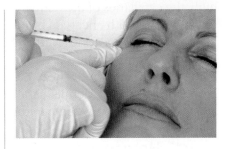

end up with droopy eyelid muscles or temporary muscle weakness for weeks (the effects of Botox Cosmetic last 3 to 6 months), and the toxin can spread to nearby tissues. Still, battling the signs of age in a noninvasive way is appealing to many people, and the fact that there is little or no recovery time allows treatment during a lunch hour. The attraction of Botox to physicians is both professional (a new tool to fight wrinkles) and monetary (truly dedicated patients are back for injections every 3 to 6 months).

---

epidermis. Others house pain receptors (*free nerve endings*) and touch receptors. On the palms of the hands and soles of the feet, the papillae are arranged in definite patterns that form looped and whorled ridges on the epidermal surface that increase friction and enhance the gripping ability of the fingers and feet. Papillary patterns are genetically determined. The ridges of the fingertips are well provided with sweat pores and leave unique, identifying films of sweat called *fingerprints* on almost anything they touch.

The **reticular layer** is the deepest skin layer. It contains dense irregular connective tissue, as well as blood vessels, sweat and oil glands, and deep pressure receptors called *lamellar corpuscles.*

Other *cutaneous sensory receptors*, which are actually part of the nervous system, are also located in the skin. These tiny sensors, which include touch, pressure, temperature, and pain receptors, provide us with a great deal of information about our external environment. They alert us to sources of heat or cold, or the tickle of a bug exploring our skin. In addition to detecting stimuli from our environment, phagocytes found here (and, in fact, throughout the dermis) act to prevent microbes that have managed to get through the epidermis from penetrating any deeper into the body.

Both *collagen* and *elastic fibers* are found throughout the dermis. Collagen fibers are responsible for the toughness of the dermis; they also attract and bind water and thus help to keep the skin hydrated. Elastic fibers give the skin its elasticity when we are young. As we age, the number of collagen and elastic fibers decreases, and the subcutaneous tissue loses fat. As a result, the skin loses its elasticity and begins to sag and wrinkle.

The dermis is abundantly supplied with blood vessels that play a role in maintaining body temperature homeostasis. When body temperature is high, the capillaries of the dermis become engorged, or swollen, with heated blood, and the skin becomes reddened and warm. This allows body heat to radiate from the skin surface. If the environment is cool and body heat must be conserved, blood bypasses the dermis capillaries temporarily, allowing internal body temperature to remain high.

## Homeostatic Imbalance 4.2

Any restriction of the normal blood supply to the skin results in cell death and, if severe or prolonged enough, skin ulcers. **Decubitus** (de-ku'bĭ-tus) **ulcers** (bedsores) occur in bedridden patients who are not turned regularly or who are dragged or pulled across the bed repeatedly. The weight of the body puts pressure on the skin, especially over bony projections. Because this pressure restricts the blood supply, the skin becomes pale or blanched at pressure points. At first, the skin reddens when pressure is released, but if the situation is not corrected, the cells begin to die, and small cracks or breaks in the skin appear at compressed sites. Permanent damage to the superficial blood vessels and tissue eventually results in degeneration and ulceration of the skin (see the photos).

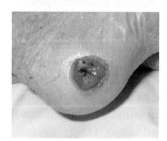

**(a) Stage II decubitus ulcer**

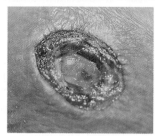

**(b) Stage III decubitus ulcer**

The dermis also has a rich nerve supply. As mentioned earlier, many of the nerve endings are designed to detect different types of stimuli (pressure, temperature, pain, etc.), then send messages to the central nervous system for interpretation. (We discuss these cutaneous receptors in more detail in Chapter 7.)

## Skin Color

### → Learning Objective

☐ Name the factors that determine skin color, and describe the function of melanin.

Three pigments contribute to skin color: melanin, carotene, and hemoglobin.

- The amount and kind (yellow, reddish brown, or black) of *melanin* in the epidermis. Skin exposure to sunlight stimulates melanocytes to produce more melanin pigment, resulting in tanning of the skin. As the melanocytes produce melanin, it accumulates in their cytoplasm in membrane-bound granules called *melanosomes*. These granules then move to the ends of the melanocytes' spidery arms, where they are taken up by nearby keratinocytes. Inside the keratinocytes, the melanin forms a pigment umbrella over the superficial, or "sunny," side of their nuclei and shields their genetic material (DNA) from the damaging effects of UV radiation in sunlight (see Figure 4.4). People who produce a lot of melanin have brown-toned skin, whereas people with less melanin are light skinned.

- The amount of *carotene* deposited in the stratum corneum and subcutaneous tissue. (Carotene is an orange-yellow pigment plentiful in carrots and other orange, deep yellow, or leafy green vegetables.) In people who eat large amounts of carotene-rich foods, the skin tends to take on a yellow-orange cast.

- The amount of oxygen-rich *hemoglobin* (pigment in red blood cells) in the dermal blood vessels. In light-skinned people, the crimson color of oxygen-rich hemoglobin in the dermal blood supply flushes through the transparent cell layers above and gives the skin a rosy glow.

## Homeostatic Imbalance 4.3

When hemoglobin is poorly oxygenated, both the blood and the skin of light-skinned people appear blue, a condition called **cyanosis** (si"ah-no'sis). Cyanosis is common during heart failure and severe breathing disorders. In dark-skinned people, the skin does not appear cyanotic in the same situations because of the masking effects of melanin, but cyanosis is apparent in their mucous membranes and nail beds. _____+

Emotions also influence skin color, and many alterations in skin color signal certain disease states:

- *Redness, or erythema (er"ĭ-the'mah).* Reddened skin may indicate embarrassment (blushing), fever, hypertension, inflammation, or allergy.

- *Pallor, or blanching.* Under certain types of emotional stress (fear, anger, and others), some people become pale. Pale skin may also signify anemia, low blood pressure, or impaired blood flow into the area.

- *Jaundice (jon'dis), or a yellow cast.* An abnormal yellow skin tone usually signifies a liver disorder in which excess bile pigments accumulate in the blood, circulate throughout the body, and become deposited in body tissues.

- *Bruises.* The black-and-blue marks of bruising reveal sites where blood has escaped from the circulation and has clotted in the tissue spaces. Such clotted blood masses are called *hematomas.* An unusual tendency to bruise may signify a deficiency of vitamin C in the diet or hemophilia (bleeder's disease).

## Did You Get It?

10. Which structure(s) are responsible for fingerprints?
11. You have just gotten a paper cut, but it doesn't bleed. Which layer(s) of the skin has/have the cut penetrated?
12. What pigments determine skin color?

For answers, see Appendix A.

## Appendages of the Skin

→ **Learning Objective**

☐ **Describe the distribution and function of the epidermal appendages—sebaceous glands, sweat glands, hair, and nails.**

The **skin appendages** include cutaneous glands, hair and hair follicles, and nails (see Figure 4.3). Each of these appendages arises from the epidermis and plays a unique role in maintaining body homeostasis.

### Cutaneous Glands

The cutaneous glands are all **exocrine glands** that release their secretions to the skin surface via ducts. They fall into two groups: *sebaceous glands* and *sweat glands.* As these glands are formed by the cells of the stratum basale, they push into the deeper skin regions and ultimately reside almost entirely in the dermis.

**Sebaceous (Oil) Glands** The **sebaceous** (se-ba'shus) **glands**, or oil glands, are found all over the skin, except on the palms of the hands and the soles of the feet. Their ducts usually empty into a hair follicle (**Figure 4.6a**, p. 146), but some open directly onto the skin surface.

The product of the sebaceous glands, **sebum** (se'bum; *seb* = grease), is a mixture of oily substances and fragmented cells. Sebum is a lubricant that keeps the skin soft and moist and prevents the hair from becoming brittle. Sebum also contains chemicals that kill bacteria, so it is important in preventing bacterial infection of the skin. The sebaceous glands become very active when *androgens* (male sex hormones) are produced in increased amounts (in both sexes) during adolescence. Thus, the skin tends to become oilier during this period of life.

### ⚖ Homeostatic Imbalance 4.4

When sebaceous gland ducts are blocked by sebum, **acne** appears on the skin surface. Acne is an active infection of the sebaceous glands. If the accumulated material oxidizes and dries, it darkens, forming a **blackhead**. If the material does not dry or darken, a **whitehead** forms. Acne can be mild or extremely severe, leading to permanent scarring.

**Seborrhea** (seb"o-re'ah; "fast-flowing sebum"), known as "cradle cap" in infants, is caused by overactivity of the sebaceous glands. It begins on the scalp as pink, raised lesions that gradually

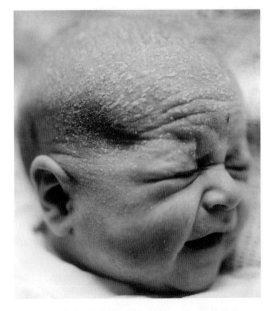

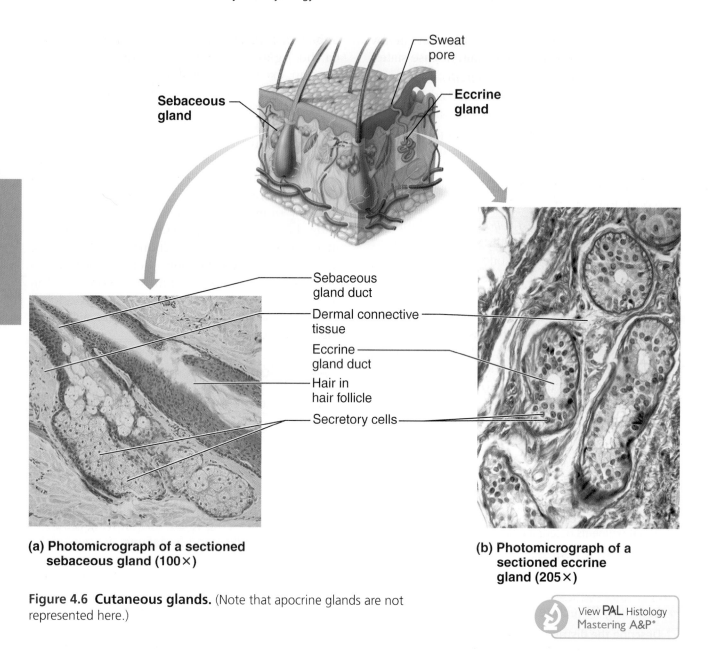

(a) **Photomicrograph of a sectioned sebaceous gland (100×)**

(b) **Photomicrograph of a sectioned eccrine gland (205×)**

**Figure 4.6  Cutaneous glands.** (Note that apocrine glands are not represented here.)

View **PAL** Histology
Mastering A&P®

form a yellow-to-brown crust that sloughs off oily scales and dandruff. Careful washing to remove the excessive oil often helps cradle cap in a newborn baby._____+

***Sweat Glands***  **Sweat glands**, also called **sudoriferous** (su"do-rif′er-us; *sudor* = sweat) **glands**, are widely distributed in the skin. Their number is staggering—more than 2.5 million per person. There are two types of sweat glands, *eccrine* and *apocrine*.

The **eccrine** (ek′rin) **glands** are far more numerous and are found all over the body. They produce **sweat**, a clear secretion that is primarily water plus some salts (sodium chloride), vitamin C,

traces of metabolic wastes (ammonia, urea, uric acid), and lactic acid (a chemical that accumulates during vigorous muscle activity). Sweat is acidic (pH from 4 to 6), a characteristic that inhibits the growth of certain bacteria, which are always present on the skin surface. Typically, sweat reaches the skin surface via a duct that opens externally as a funnel-shaped *sweat pore* (Figure 4.6b). Notice, however, that the facial "pores" commonly referred to when we talk about our complexion are the external outlets of hair follicles where sebaceous ducts empty, *not* these sweat pores.

The eccrine sweat glands are an important and highly efficient part of the body's heat-regulating equipment. They are supplied with nerve endings that cause them to secrete sweat when the external temperature or body temperature is too high. When sweat evaporates off the skin surface, it carries large amounts of body heat with it. On a hot day, it is possible to lose up to 7 liters of body water in this way. The heat-regulating functions of the body are important—if internal temperature changes more than a few degrees from the normal 37°C (98.6°F), life-threatening changes occur in the body. (We discuss body temperature regulation in more detail in Chapter 14.)

**Apocrine** (ap′o-krin) **glands** are largely confined to the axillary (armpit) and genital areas of the body. They are usually larger than eccrine glands, and their ducts empty into hair follicles. Their secretion contains fatty acids and proteins, as well as all the substances present in eccrine sweat; consequently, it may have a milky or yellowish color. The secretion is odorless, but when bacteria that live on the skin use its proteins and fats as a source of nutrients for their growth, it can take on a musky, sometimes unpleasant odor.

Apocrine glands begin to function during puberty under the influence of androgens. Although they produce their secretions almost continuously, apocrine glands play a minimal role in thermoregulation. Their precise function is not yet known, but they are activated by nerve fibers during pain and stress and during sexual arousal.

### Hair and Hair Follicles

Hair is an important part of our body image. Consider, for example, the spiky hair style of punk rockers and the flowing locks of high-fashion models. Millions of *hairs*, produced by *hair follicles*, are found all over the body surface except on the palms of the hands, soles of the feet, nipples, and lips. Humans are born with as many hair follicles as they will ever have, and hairs are among the fastest growing tissues in the body. Hair serves a few minor protective functions, such as guarding the head against bumps, shielding the eyes (via eyelashes), and helping to keep foreign particles out of the respiratory tract (via nose hairs). Hair may also help to attract sexual partners, as evidenced by its place in our body image. Hormones account for the development of hairy regions—the scalp and, in the adult, the pubic and axillary areas. Despite these functions, however, our body hair has lost much of its usefulness. Hair served early humans (and still serves hairy animals) by providing insulation in cold weather, but now we have other means of keeping warm.

*Hairs*    A **hair** is a flexible epithelial structure. The part of the hair enclosed in the hair follicle is called the *root*, and the part projecting from the surface of the scalp or skin is called the *shaft* (**Figure 4.7a**, p. 148). A hair forms by division of the well-nourished stratum basale epithelial cells in the **matrix** (growth zone) of the hair bulb at the deep end of the follicle. As the daughter cells are pushed farther away from the growing region, they become keratinized and die. Thus the bulk of the hair shaft, like the bulk of the epidermis, is dead material and almost entirely protein.

Each hair is made up of a central core called the *medulla* (mě-dul′ah), consisting of large cells and air spaces, surrounded by a bulky *cortex* layer composed of several layers of flattened cells (Figure 4.7b). The cortex is, in turn, enclosed by an outermost *cuticle* formed by a single layer of cells that overlap one another like shingles on a roof. This arrangement of the cuticle cells helps to keep the hairs apart and keeps them from matting (Figure 4.7c). The cuticle is the most heavily keratinized region; it provides strength and helps keep the inner hair layers tightly compacted. Because it is most subject to abrasion, the cuticle tends to wear away at the tip of the shaft, allowing the keratin fibrils in the inner hair regions to frizz out, a phenomenon called "split ends." Hair pigment is made by melanocytes in the hair bulb, and varying amounts of different types of melanin (yellow, rust, brown, and black) combine to produce *all* varieties of hair color from pale blond to red to pitch black.

Hairs come in a variety of sizes and shapes. They are short and stiff in the eyebrows, long and flexible on the head, and usually nearly invisible almost everywhere else. When the hair shaft is oval, hair is smooth, silky, and wavy. When the shaft is flat and ribbonlike, the hair is curly or kinky. If it is perfectly round, the hair is straight and tends to be coarse.

*Hair Follicles*    **Hair follicles** (*folli* = bag) are actually compound structures. The inner *epithelial root sheath* is composed of epithelial tissue and forms the hair. The outer *fibrous sheath* is actually dermal connective tissue (Figure 4.7d). This dermal region supplies blood vessels to the epidermal portion and reinforces it. Its nipplelike *hair papilla*

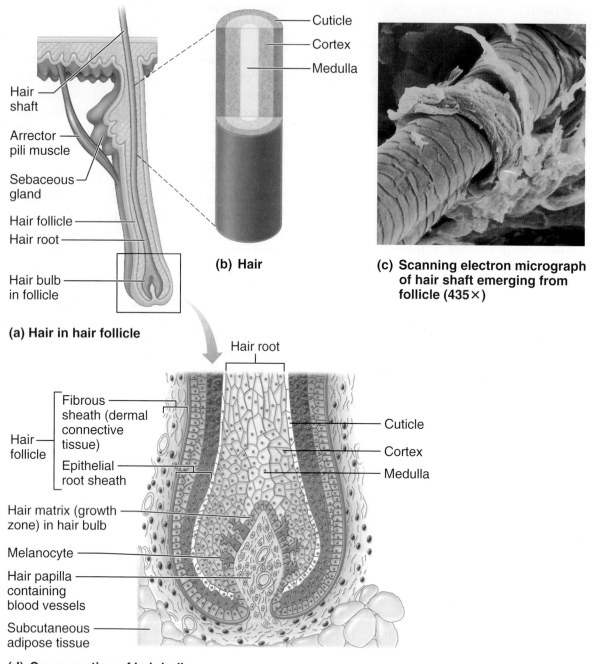

(a) Hair in hair follicle

Hair shaft
Arrector pili muscle
Sebaceous gland
Hair follicle
Hair root
Hair bulb in follicle

(b) Hair

Cuticle
Cortex
Medulla

(c) Scanning electron micrograph of hair shaft emerging from follicle (435×)

(d) Cross section of hair bulb

Hair root

Hair follicle
- Fibrous sheath (dermal connective tissue)
- Epithelial root sheath

Cuticle
Cortex
Medulla

Hair matrix (growth zone) in hair bulb
Melanocyte
Hair papilla containing blood vessels
Subcutaneous adipose tissue

**Figure 4.7 Structure of a hair root and follicle. (a)** Longitudinal section of a hair within its follicle. **(b)** Enlarged longitudinal section of a hair. **(c)** Scanning electron micrograph showing a hair shaft emerging from a follicle at the skin surface. **(d)** Enlarged longitudinal view of the expanded hair bulb in the follicle showing the root and matrix, the region of actively dividing epithelial cells that produces the hair.

provides the blood supply to the matrix in the **hair bulb** (the deepest part of the follicle).

If you look carefully at the structure of the hair follicle, you will notice that it is slightly slanted (see Figure 4.7a). Small bands of smooth muscle cells—**arrector pili** (ah-rek′tor pi′li; "raiser of hair")—connect each side of the hair follicle to the dermal tissue. When these muscles contract (as when we are cold or frightened), the hair is pulled upright, dimpling the skin surface with "goose bumps."

## Nails

A **nail** is a scalelike modification of the epidermis that corresponds to the hoof or claw of other animals. Each nail has a *free edge*, a *body* (visible attached portion), and a *root* (embedded in the skin) **(Figure 4.8)**. The borders of the nail are overlapped by folds of skin called *nail folds*. The edge of the thick proximal nail fold is commonly called the *cuticle*.

The stratum basale of the epidermis extends beneath the nail as the *nail bed*. Its thickened proximal area, called the *nail matrix*, is responsible for nail growth. As the matrix produces nail cells, they become heavily keratinized and die. Thus, nails, like hairs, are mostly nonliving material.

Nails are transparent and nearly colorless, but they look pink because of the rich blood supply in the underlying dermis. The exception to this is the region over the thickened nail matrix that appears as a white crescent and is called the *lunule* (loo'nyul; *lunul* = crescent). As noted earlier, when the supply of oxygen in the blood is low, the nail beds take on a cyanotic (blue) cast.

### Did You Get It?

13. Which of the cutaneous gland types can make your hair limp and oily?
14. How do secretions of apocrine glands differ from those of the eccrine sweat glands?
15. When a factory worker caught his finger in a machine, the entire nail, plus the nail matrix and bed, was torn off. Will his nail grow back? Why or why not?

For answers, see Appendix A.

## Homeostatic Imbalances of Skin

→ **Learning Objectives**

☐ **Differentiate among first-, second-, third, and fourth-degree burns.**

☐ **Explain the importance of the "rule of nines."**

☐ **Summarize the characteristics of basal cell carcinoma, squamous cell carcinoma, and malignant melanoma.**

Loss of homeostasis in body cells and organs reveals itself on the skin, sometimes in startling ways. The skin can develop more than 1,000 different ailments. The most common skin disorders are infections with *pathogens* (path'o-jenz) such as bacteria, viruses, or fungi. *Allergies*, which are caused by abnormally strong immune responses, are also commonly seen in the skin. Less common,

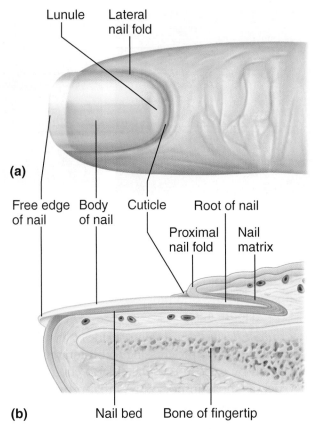

**(a)**

**(b)**

**Figure 4.8  Structure of a nail.**

but far more damaging to body well-being, are burns and skin cancers. We briefly summarize these homeostatic imbalances of the skin here.

### Infections and Allergies

Infections and allergies cause the following commonly occurring skin disorders:

- *Athlete's foot.* An itchy, red, peeling condition of the skin between the toes, resulting from an infection with the fungus *Tinea pedis*.

- *Boils* (furuncles) *and carbuncles* (kar'bun-kulz). Boils are caused by inflammation of hair follicles and surrounding tissues, commonly on the dorsal neck. Carbuncles are clusters of boils often caused by the bacterium *Staphylococcus aureus.*

- *Cold sores* (fever blisters). Small fluid-filled blisters that itch and sting, caused by *human herpesvirus 1* infection. The virus localizes in a cutaneous nerve, where it remains dormant until activated by emotional upset, fever, or UV

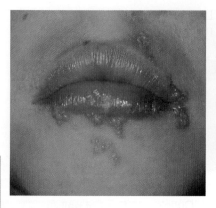

**(a) Cold sores**

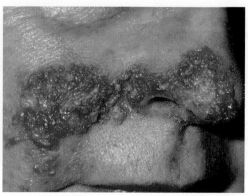

**(b) Impetigo**

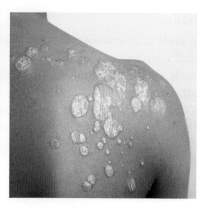

**(c) Psoriasis**

**Figure 4.9   Cutaneous lesions.**

radiation. Cold sores usually occur around the lips and in the oral mucosa of the mouth and nose **(Figure 4.9a)**.

- *Contact dermatitis.* Itching, redness, and swelling of the skin, progressing to blistering. It is caused by exposure of the skin to chemicals (such as those in poison ivy) that provoke allergic responses in sensitive individuals.

- *Impetigo* (im-peh-ti′go; *impet* = an attack). Pink, fluid-filled, raised lesions (commonly around the mouth and nose) that develop a yellow crust and eventually rupture (Figure 4.9b). Caused by highly contagious staphylococcus or streptococcus infections, impetigo is common in elementary school–aged children.

- *Psoriasis* (so-ri′ah-sis). Characterized by reddened epidermal lesions covered with dry, silvery scales that itch, burn, crack, and sometimes bleed (Figure 4.9c). A chronic condition, psoriasis is believed to be an autoimmune disorder in which the immune system attacks a person's own tissues, leading to the rapid overproduction of skin cells. Attacks are often triggered by trauma, infection, hormonal changes, or stress. When severe, psoriasis may be disfiguring.

### Burns

There are few threats to life more serious than burns. A **burn** is tissue damage and cell death caused by intense heat, electricity, UV radiation (sunburn), or certain chemicals (such as acids), which denature proteins and cause cell death in the affected areas.

When the skin is burned and its cells are destroyed, two life-threatening problems result. First, without an intact boundary, the body loses its precious supply of fluids containing proteins and electrolytes as these seep from the burned surfaces. Dehydration and electrolyte imbalance follow and can lead to a shutdown of the kidneys and *circulatory shock* (inadequate circulation of blood caused by low blood volume). To save the patient, lost fluids must be replaced immediately. The volume of fluid lost can be estimated indirectly by determining how much of the body surface is burned (extent of burns), using the **rule of nines**. This method divides the body into 11 areas, each accounting for 9 percent of the total body surface area, plus an additional area surrounding the genitals (the perineum) representing 1 percent of body surface area **(Figure 4.10a)**.

Later, infection becomes the most important threat and is the leading cause of death in burn victims. Burned skin is sterile for about 24 hours. But after that, pathogens easily invade areas where the skin has been destroyed and multiply rapidly in the nutrient-rich environment of dead tissues. To make matters worse, the patient's immune system becomes depressed within one to two days after severe burn injury.

Burns are classified according to their severity (depth) as first-degree (superficial), second-degree (superficial **partial-thickness burns**), third-degree (**full-thickness burns**), or fourth-degree (full-thickness burns with deep-tissue involvement) (Figure 4.10b). In **first-degree burns**, only the superficial epidermis is damaged. The area becomes red and swollen. Except for temporary discomfort, first-degree burns are not usually serious and generally heal in two to three days. Sunburn without blistering is a first-degree burn.

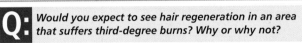

**Q:** *Would you expect to see hair regeneration in an area that suffers third-degree burns? Why or why not?*

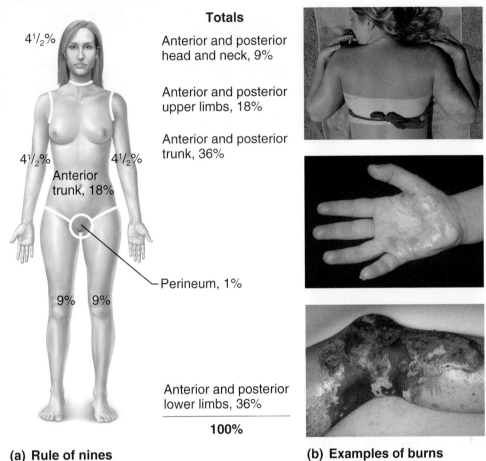

4½%

**Totals**

Anterior and posterior
head and neck, 9%

Anterior and posterior
upper limbs, 18%

Anterior and posterior
trunk, 36%

4½%    4½%

Anterior
trunk, 18%

Perineum, 1%

9%    9%

Anterior and posterior
lower limbs, 36%

**100%**

**(a) Rule of nines**    **(b) Examples of burns**

**Figure 4.10 Burns. (a)** Estimating the extent of burns using the rule of nines. The surface areas for the anterior body surface are indicated on the human figure. Total surface area (anterior and posterior body surfaces) is tabulated to the right of the figure. **(b)** Burns of increasing severity, from top to bottom: first-degree, second-degree, and third-degree.

**Second-degree burns** involve injury to the epidermis and the superficial part of the dermis. The skin is red, painful, and blistered. Because sufficient numbers of epithelial cells are still present, regrowth (regeneration) of the epithelium can occur. Ordinarily, no permanent scars result if care is taken to prevent infection. **Third-degree burns** destroy both the epidermis and the dermis and often extend into the subcutaneous tissue, reflecting their categorization as full thickness burns.

Blisters are usually present, and the burned area appears blanched (gray-white) or blackened. Because the nerve endings in the area are destroyed, the burned area is not painful. In third-degree burns, regeneration is not possible, and skin grafting must be done to cover the underlying exposed tissues.

**Fourth-degree burns** are also full-thickness burns, but they extend into deeper tissues such as bone, muscle, or tendons. These burns appear dry and leathery, and they require surgery and grafting to cover exposed tissue. In severe cases, amputation may be required to save the patient's life.

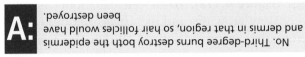

**A:** No. Third-degree burns destroy both the epidermis and dermis in that region, so hair follicles would have been destroyed.

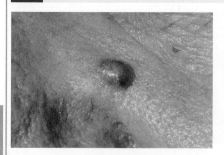 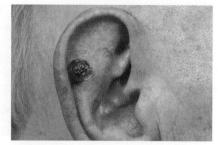 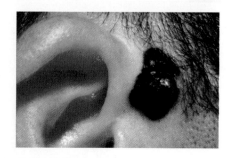

**(a) Basal cell carcinoma**    **(b) Squamous cell carcinoma**    **(c) Melanoma**

Figure 4.11   **Photographs of skin cancers.**

In general, burns are considered *critical* if any of the following conditions exists:

- Over 30 percent of the body has second-degree burns.
- Over 10 percent of the body has third- or fourth-degree burns.
- There are third- or fourth-degree burns of the face, hands, feet, or genitals.
- Burns affect the airway.
- Circumferential (around the body or limb) burns have occurred.

Facial burns are particularly dangerous because of the possibility of burns in respiratory passageways, which can swell and cause suffocation. Joint injuries are troublesome because the scar tissue that eventually forms can severely limit joint mobility. Circumferential burns can restrict movement, and depending on location, can interfere with normal breathing.

## Skin Cancer

Numerous types of *neoplasms* (tumors) arise in the skin. Most skin neoplasms are benign and do not spread (*metastasize*) to other body areas. For example, *warts* are caused by human papillomaviruses but are benign and do not spread. However, some skin neoplasms are *malignant*, or cancerous, and they tend to invade other body areas.

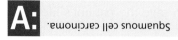

**A:** Squamous cell carcinoma.

### CONCEPTLINK

Recall that mitosis gone wild is the basis for cancer (Chapter 3, pp. 130–131). In malignant cancers, the stages of mitosis occur so quickly that errors are made. As a result, these cells lack normal control of cell division. Cells experiencing rapid, uncontrolled growth become cancerous. ←

Skin cancer is the single most common type of cancer in humans. One in five Americans now develops skin cancer at some point in his or her life. The most important risk factor is overexposure to UV radiation in sunlight and tanning beds. Frequent irritation of the skin by infections, chemicals, or physical trauma also seems to be a predisposing factor. Let's look in a bit more detail at the three most common types of skin cancer: basal cell carcinoma, squamous cell carcinoma, and malignant melanoma.

***Basal Cell Carcinoma*** Basal cell carcinoma (kar″sĭ-no′mah) is the least malignant and most common skin cancer. Cells of the stratum basale, altered so that they cannot form keratin, no longer honor the boundary between epidermis and dermis. They proliferate, invading the dermis and subcutaneous tissue. The cancerous lesions occur most often on sun-exposed areas of the face and often appear as shiny, dome-shaped nodules that later develop a central ulcer with a "pearly" beaded edge **(Figure 4.11a)**. Basal cell carcinoma is relatively slow-growing, and metastasis seldom occurs before the lesion is noticed. When the lesion is removed surgically, 99 percent of cases are completely cured.

### Squamous Cell Carcinoma

*Squamous cell carcinoma* arises from the cells of the stratum spinosum. The lesions appear as scaly, reddened *papules* (small, rounded swellings) that gradually form shallow ulcers with firm, raised borders (Figure 4.11b). This variety of skin cancer appears most often on the scalp, ears, back of the hands, and lower lip, but can appear anywhere on the skin. It grows rapidly and metastasizes to adjacent lymph nodes if not removed. This epidermal cancer is also believed to be induced by UV exposure. If it is caught early and removed surgically or by radiation therapy, the chance of complete cure is good.

### Malignant Melanoma

*Malignant melanoma* (mel″ah-no′mah) is a cancer of melanocytes. It accounts for only about 5 percent of skin cancers, but it is often deadly. Melanoma can begin wherever there is pigment; most such cancers appear spontaneously, but some develop from pigmented moles. It arises from accumulated DNA damage in a skin cell and usually appears as a spreading brown to black patch (Figure 4.11c) that metastasizes rapidly to surrounding lymph and blood vessels. The chance for survival is about 50 percent, and early detection helps. The American Cancer Society suggests that people who sunbathe frequently or attend tanning parlors examine their skin periodically for new moles or pigmented spots and apply the **ABCDE rule** for recognizing melanoma:

*(A) Asymmetry.* Any two sides of the pigmented spot or mole do not match.

*(B) Border irregularity.* The borders of the lesion are not smooth but exhibit indentations.

*(C) Color.* The pigmented spot contains areas of different colors (black, brown, tan, and sometimes blue or red).

*(D) Diameter.* The lesion is larger than 6 millimeters (mm) in diameter (the size of a pencil eraser).

*(E) Evolution.* One or more of these characteristics (ABCD) is *evolving*, or changing.

The usual therapy for malignant melanoma is wide surgical excision along with immunotherapy, a treatment that involves the patient's immune system. Large lesions may also require radiation or chemotherapy after surgical removal.

### Did You Get It?

16. What are the two life-threatening consequences of a severe burn?
17. What are the criteria for classifying burns as first-, second-, third-, or fourth-degree?
18. What name is given to the rule for recognizing the signs of melanoma?
19. What is the single most common risk factor for skin cancer?
20. Why do no skin cancers develop from stratum corneum cells?

**For answers, see Appendix A.**

# Developmental Aspects of Skin and Body Membranes

➔ **Learning Objective**

☐ **List several examples of integumentary system aging.**

During the fifth and sixth months of development, a fetus is covered with a downy type of hair called *lanugo* (lah-noo′go). By the time the infant is born, it has usually shed this hairy cloak, and instead its skin is covered with an oily coating called the *vernix caseosa* (ver′niks kah-se-o′sah). This white, cheesy-looking substance, produced by the sebaceous glands, protects the baby's skin while it is floating in its water-filled sac inside the mother. The newborn's skin is very thin, and blood vessels are easily seen through it. Commonly, there are accumulations in the sebaceous glands, which appear as small white spots called *milia* (mil′e-ah) on the baby's nose and forehead. These normally disappear by the third week after birth. As the baby grows, its skin becomes thicker, and more subcutaneous fat is deposited.

During adolescence, the skin and hair become oilier as sebaceous glands are activated, and acne may appear. Acne usually subsides in early adulthood, and the skin reaches its optimal appearance when we are in our twenties and thirties. Then visible changes in the skin begin to appear, as it is continually assaulted by abrasion, chemicals, wind, sun, and other irritants and as its pores become clogged with air pollutants and bacteria. As a result, pimples, scales, and various kinds of

*(Text continues on page 156.)*

# Medical Transcriptionist

**"If you have a basic understanding of anatomy and medical terminology, you will be much more accurate at interpreting and transcribing what you hear."**

Every time you consult a doctor or are hospitalized, your medical record gets longer. Medical transcriptionists play a key role in creating and maintaining these vital documents.

A medical transcriptionist is a medical language specialist who interprets and transcribes notes dictated by physicians and other healthcare professionals. These reports, which cover all aspects of a patient's assessment, diagnosis, treatment, and outcome, become part of the person's confidential medical record. Medical transcriptionists work in hospitals, clinics, doctors' offices, transcription services, insurance companies, and home healthcare agencies.

What does it take to be a transcriptionist? "Certainly, you need a good English background," says Pamela Shull, an experienced transcriptionist in San Jose, California. "Strong grammar, spelling, and punctuation skills are crucial. Physicians often dictate these records on the go, and a good transcriptionist must be able to edit the dictated material for grammar and clarity."

Knowledge of anatomy and physiology, however, is even more important. Notes Shull, "If you understand anatomy and medical terminology, you will be much more accurate at interpreting and transcribing what you

hear. A hospital transcriptionist deals with terms from a wide variety of medical specialties—one dictation might be from a gynecologist, the next from an orthopedic surgeon, and the next from a pediatrician." This is why anatomy and physiology, medical terminology, and the study of disease processes make up most of the curriculum in medical transcription training programs.

> ## Anatomy and physiology classes make up a large part of the curriculum in medical transcription training programs.

All health professionals who treat a patient rely on these typed documents, so accurate transcription is vital: "I see the transcriptionist as a partner with physicians. We work with them to create excellent medical records, so patients will always be assured of receiving the best and most appropriate care possible."

Shull enjoys the variety of medical transcription work. "It's fascinating

because you get to follow each patient's story, from the initial problem to diagnosis and treatment," she says. "You feel like you know these people." Of course, patient confidentiality is also a big responsibility.

Classes for medical transcription, which are offered through community colleges, proprietary schools, and home-study programs, vary in length from several months to two years. Accreditation procedures vary from state to state. The Association for Healthcare Documentation Integrity (AHDI) evaluates medical transcription programs and posts a list of recommended programs on its website.

For more information, contact:

Association for Healthcare Documentation Integrity

4120 Dale Rd, Suite J8-233
Modesto, CA 95356
(800) 982-2182 or (209) 527-9620
http://ahdionline.org/

For additional information on this career and others, click the Focus on Careers link at Mastering A&P®.

# Homeostatic Relationships between the **Integumentary System** and Other Body Systems

**Nervous System**
- Skin protects nervous system organs; cutaneous sensory receptors located in skin
- Nervous system regulates diameter of blood vessels in skin; activates sweat glands (contributing to thermoregulation); interprets cutaneous sensation; activates arrector pili muscles

**Endocrine System**
- Skin protects endocrine organs
- Androgens produced by the endocrine system activate sebaceous glands and help regulate hair growth; estrogen helps maintain skin hydration

**Respiratory System**
- Skin protects respiratory organs
- Respiratory system furnishes oxygen to skin cells and removes carbon dioxide via gas exchange with blood

**Lymphatic System/Immunity**
- Skin protects lymphoid organs; prevents pathogen invasion
- Lymphatic system prevents edema by picking up excessive leaked fluid; immune system protects skin cells

**Cardiovascular System**
- Skin protects cardiovascular organs; prevents fluid loss from body surface; serves as blood reservoir
- Cardiovascular system transports oxygen and nutrients to skin and removes wastes from skin; provides substances needed by skin glands to make their secretions

**Digestive System**
- Skin protects digestive organs; provides vitamin D needed for calcium absorption
- Digestive system provides nutrients needed by the skin

**Reproductive System**
- Skin protects reproductive organs; highly modified sweat glands (mammary glands) produce milk. During pregnancy, skin stretches to accommodate growing fetus; changes in skin pigmentation may occur

**Integumentary System (Skin)**

**Urinary System**
- Skin protects urinary organs; excretes salts and some nitrogen-containing wastes in sweat
- Urinary system activates vitamin D made by keratinocytes; disposes of nitrogen-containing wastes of skin metabolism

**Skeletal System**
- Skin protects bones; skin synthesizes vitamin D that bones need for normal calcium absorption and deposit of bone (calcium) salts, which make bones hard
- Skeletal system provides support for the skin

**Muscular System**
- Skin protects muscles
- Active muscles generate large amounts of heat, which increases blood flow to the skin and may promote activation of sweat glands of skin

*dermatitis* (der"mah-ti'tis), or skin inflammation, become more common.

As we get even older, the amount of subcutaneous tissue decreases, leading to sensitivity to cold. The skin also becomes drier (because of decreased oil production), and as a result, it may become itchy and bothersome. Thinning of the skin, another result of the aging process, makes it more susceptible to bruising and other types of injuries. The decreasing elasticity of the skin, along with the loss of subcutaneous fat, allows bags to form under our eyes, and our jowls begin to sag. Smoking and sunlight speed up this loss of elasticity, so three of the best things you can do for your skin are to avoid smoking, shield your skin from the sun, and do not go to tanning beds. In doing so, you will also be decreasing the chance of skin cancer. There is no way to avoid aging skin, but good nutrition, plenty of fluids, and cleanliness help delay the process.

Hair loses its luster as we age, and by age 50 the number of hair follicles has dropped by one-third and continues to decline, resulting in hair thinning and some degree of baldness, or *alopecia* (al"o-pe'she-ah), in most people. Many men become bald as they age, a phenomenon called *male pattern baldness*. A bald man is not really hairless—he does have hairs in the bald area. But, because those hair follicles have begun to degen-erate, the *vellus* (*vell* = wool) hairs are colorless and very tiny (and may not even emerge from the follicle). Another phenomenon of aging is graying hair. Like balding, this is usually genetically controlled by a "delayed-action" gene. Once the gene takes effect, the amount of melanin deposited in the hair decreases or becomes entirely absent, which results in gray-to-white hair.

## Homeostatic Imbalance 4.5

Certain events can cause hair to gray or fall out prematurely. For example, many people have claimed that they turned gray nearly overnight because of some emotional crisis in their life. In addition, we know that anxiety, protein-deficient diets, therapy with certain chemicals (chemotherapy), radiation, excessive vitamin A, and certain fungal diseases (ringworm) can cause both graying and hair loss. However, when the cause of these conditions is not genetic, hair loss is usually not permanent. _____ **+**

### Did You Get It?

21. What change in aging skin accounts for wrinkles and cold intolerance in older adults?
22. What is the source of the vernix caseosa that covers the skin of the newborn baby?

**For answers, see Appendix A.**

# Summary

### Classification of Body Membranes (pp. 135–138)

1. Epithelial: Simple organs, epithelium, and connective tissue components.

   a. Cutaneous (the skin): epidermis (stratified squamous epithelium) underlain by the dermis (dense connective tissue); protects body surface.

   b. Mucous: epithelial sheet underlain by a lamina propria (areolar connective tissue); lines body cavities open to the exterior.

   c. Serous: simple squamous epithelium resting on a scant connective tissue layer; lines the ventral body cavity.

2. Connective tissue: Synovial; lines joint cavities.

### The Integumentary System (Skin) (pp. 138–153)

1. Skin functions include protecting the deeper tissue from chemicals, bacteria, bumps, and drying; regu-lating body temperature through radiation and sweating; and synthesizing defensive proteins and vitamin D. Sensory receptors are located in the skin.

2. The epidermis, the superficial part of the skin, is stratified squamous keratinized epithelium and is avascular. Its layers from superficial to deep are the stratum corneum, stratum lucidum (in thick skin only), stratum granulosum, stratum spinosum, and stratum basale. Cells at its surface are dead and continually flake off. They are replaced by dividing cells in the basal cell layer. As the cells move away from the basal layer, they accumulate keratin and die. Melanin, a pigment produced by melanocytes, protects the nuclei of epithelial cells from the damaging rays of the sun.

3. The dermis is dense connective tissue. It contains blood vessels, nerves, and epidermal appendages.

It has two regions, the papillary and reticular layers. The papillary layer has ridges, which push outward on the epidermis to produce fingerprints.

4. Skin appendages are formed from the epidermis but reside in the dermis.

   a. Sebaceous glands produce an oily product (sebum), released at a hair follicle via a duct. Sebum keeps the skin and hair soft and contains chemicals that kill bacteria.

   b. Sweat (sudoriferous) glands, under the control of the nervous system, produce sweat, which is released at the epithelial surface. These glands participate in regulating body temperature. There are two types: eccrine (the most numerous) and apocrine (their product includes fatty acids and proteins).

   c. A hair is primarily dead keratinized cells and is produced by the matrix in the root of the hair bulb. The root is enclosed in a sheath, the hair follicle.

   d. Nails are hooflike derivatives of the epidermis. Like hair, nails are primarily dead keratinized cells.

5. Most minor afflictions of the skin result from infections or allergic responses; more serious are burns and skin cancer. Because they interfere with skin's protective functions, burns represent a major threat to the body.

   a. Burns result in loss of body fluids and invasion of bacteria. The extent of burns is assessed by the rule of nines. The severity (depth) of burns is described as first-degree (epidermal damage only), second-degree (epidermal and some dermal injury), third-degree (epidermis and dermis totally destroyed, may involve some subcutaneous tissue) and fourth-degree (all of skin and involves deeper tissue such as bone or muscle). Third- and fourth-degree burns require skin grafting.

   b. The most common cause of skin cancer is exposure to ultraviolet (UV) radiation. Basal cell carcinoma and squamous cell carcinoma can be completely cured if they are removed before metastasis. Malignant melanoma, a cancer of melanocytes, is still fairly rare but metastasizes quickly and is fatal in about half the cases.

## Developmental Aspects of Skin and Body Membranes (pp. 153–156)

1. The skin is thick, resilient, and well hydrated in our youth but loses elasticity and thins as we age. Skin cancer is a major threat to skin exposed to excessive sunlight.

2. Balding and/or graying occurs with aging. Both are genetically determined but can also be caused by other factors (drugs, emotional stress, and so on).

---

# Review Questions

Access additional practice questions using your smartphone, tablet, or computer: Mastering A&P® > Study Area > Practice Tests & Quizzes

## Multiple Choice

*More than one choice may apply.*

1. Select the one *false* statement about mucous and serous membranes.

   a. The epithelial type is the same in all serous membranes, but there are different epithelial types in different mucous membranes.

   b. Serous membranes line closed body cavities, whereas mucous membranes line body cavities open to the outside.

   c. Serous membranes always produce serous fluid, and mucous membranes always secrete mucus.

   d. Both membranes contain an epithelium plus a layer of loose connective tissue.

2. Serous membranes

   a. line the mouth.

   b. have parietal and visceral layers.

   c. consist of epidermis and dermis.

   d. have a connective tissue layer called the lamina propria.

   e. secrete a lubricating fluid.

3. Which of the following is (are) true of synovial membranes?

   a. They are composed of loose areolar connective tissue.

   b. They line bursae.

   c. They line tendon sheaths.

   d. They line pericardia.

   e. They line peritonea.

4. Which structure can be found in the hair bulb?

   a. Hair papilla

   b. Lunule

   c. Melanocyte

   d. Arrector pili muscle

   e. Blood vessels

5. In investigating the cause of thinning hair, which of the following questions needs to be asked?

   a. Is the diet deficient in proteins?

   b. Is the person taking megadoses of vitamin C?

c. Has the person been exposed to excessive radiation?

d. Has the person recently suffered severe emotional trauma?

6. Sebaceous glands

   a. are present on the soles of the feet.

   b. produce sebum.

   c. are present on the palms of the hands.

   d. are activated by androgens.

7. Which one of the following is *not* associated with the production of perspiration?

   a. Sweat glands              d. Eccrine gland

   b. Sweat pores               e. Apocrine gland

   c. Arrector pili

8. Which of the following is *not* a skin appendage?

   a. Eccrine sweat gland       c. Hair follicle

   b. Lamellar corpuscle        d. Sensory nerve fiber

9. Match the structures on the right with their function listed on the left.

   ____ 1. Protection from ultraviolet radiation

   ____ 2. Insulation, energy storage

   ____ 3. Waterproofing and preventing water loss

   ____ 4. Temperature regulation

   ____ 5. Excretion of water, urea, and salts

   ____ 6. Produces the patterns for fingerprints

   a. Reticular layer of the dermis

   b. Dermal blood supply

   c. Papillary layer of the dermis

   d. Hypodermis

   e. Melanocytes

   f. Stratum corneum

   g. Eccrine sweat glands

## Short Answer Essay

10. What are freckles?

11. From what types of damage does the skin protect the body?

12. Explain why we become tanned after sitting in the sun.

13. What is jaundice? Why does it occur?

14. Name two different categories of skin secretions and the glands that manufacture them.

15. How does the skin help to regulate body temperature?

16. How does smoking contribute to skin aging and formation of wrinkles?

17. What are arrector pili? What do they do?

18. Which skin cancer arises from the youngest epidermal cells?

19. Why does hair turn gray?

20. What kind of a disorder is psoriasis? What triggers it?

21. Is a bald man really hairless? Explain.

22. Contrast the secretions of eccrine and apocrine glands.

23. Explain the difference between a blackhead, whitehead, boil, and carbuncle. Which is most serious? Why?

## Critical Thinking and Clinical Application Questions

24. A nurse tells a doctor that a patient is cyanotic. What is cyanosis? What does its presence indicate?

25. Jane's newborn baby's scalp is covered with a brownish crust that is full of scales and dandruff. What is the likely cause?

26. A 40-year-old beachboy is complaining to you that his suntan made him popular when he was young—but now his face is all wrinkled, and he has several darkly pigmented moles that are growing rapidly and are as big as large coins. He shows you the moles, and immediately you think "ABCDE." What does that mean, and why should he be concerned?

27. Fred's new cycling shorts have got special silicone grip strips on the inside of the material to hold the shorts in place. After a day on the saddle, Fred notices that he has got more than just a tan line. Two bright red circles appeared on his thighs a few hours after he changed out of his gear. Explain what type of skin condition Fred might be experiencing.

28. The water of a swimming pool is hypotonic to our cells. Why do we not swell and pop when we go for a swim?

29. Mr. Rossi, a fisherman in his late sixties, comes to the clinic because he has small ulcers on both forearms as well as on his face and ears. Although he has had them for several years, he has not had any other problems. What is the likely diagnosis, and what is the likely cause?

30. Mr. Grayson is receiving a drug treatment transdermally (through the skin). Explain why drugs delivered by this route are fat-soluble rather than water-soluble.

31. Which type of injection would allow a drug to be absorbed more rapidly—intradermal or subcutaneous (a shallow injection just deep to the epidermis)? Why?

32. Think about the types and characteristics of cell junctions (Chapter 3). How does this help to explain why sunburned skin often peels in sheets?

33. A burn victim exhibits a red and swollen arm with blistering; a hand that has been charred black, exposing bone; and a cheek that appears blanched. How serious is each burn, and is this patient critical? Explain.

4

# 5 The Skeletal System

**WHAT**

The skeletal system is the internal frame of the body and includes bones, cartilages, and joints.

**HOW**

In addition to providing structure, bones articulate, or come together, at joints to allow body movement.

**WHY**

The skeleton is essential for protecting organs, producing blood cells, storing essential minerals, and anchoring skeletal muscles so that their contractions cause body movements.

**INSTRUCTORS**

New **Building Vocabulary Coaching** Activities for this chapter are assignable in Mastering A&P®

Although the word *skeleton* comes from the Greek word meaning "dried-up body," our internal framework is beautifully formed and proportioned. Strong yet light, it is perfectly adapted for its functions of protecting the body and allowing motion. No other animal has such long legs (compared to the arms or forelimbs) or such a strange foot, and few have grasping hands with opposable thumbs.

The bones of the skeleton are part of the **skeletal system**, which also includes *joints*, *cartilages*, and *ligaments* (fibrous cords that bind the bones together at joints). The skeleton is divided into two parts: the **axial skeleton**, the bones that form the longitudinal axis of the body, and the **appendicular skeleton**, the bones of the limbs and girdles that attach them to the axial skeleton. Joints give these parts of the skeleton flexibility and allow movement to occur.

## Bones: An Overview

→ Learning Objectives

☐ **Identify the subdivisions of the skeleton as axial or appendicular.**

☐ **List at least three functions of the skeletal system.**

☐ **Name the four main classifications of bones.**

At one time or another, all of us have heard the expressions "bone tired," "dry as a bone," or "bag of bones"—pretty unflattering and inaccurate images of some of our most phenomenal organs. Our brains, not our bones, convey feelings of fatigue, and living bones are far from dry. As for "bag of bones," they are indeed more obvious in some of us, but without bones to form our internal skeleton, we would creep along the ground like slugs, lacking any definite shape or form. Let's examine how our bones contribute to overall body homeostasis.

## Functions of the Bones

Besides contributing to body structure, our bones perform several important body functions:

- **Support.** Bones, the "steel girders" and "reinforced concrete" of the body, form the internal framework that supports the body and cradles its soft organs. The bones of the legs act as pillars to support the body trunk when we stand, and the rib cage supports the thoracic wall.

- **Protection.** Bones protect soft body organs. For example, the fused bones of the skull provide a snug enclosure for the brain, allowing someone to head a soccer ball without worrying about injuring the brain. The vertebrae surround the spinal cord, and the rib cage helps protect the vital organs of the thorax.

- **Allow movement.** Skeletal muscles, attached to bones by tendons, use the bones as levers to move the body and its parts. As a result, we can breathe, walk, swim, and throw a ball. Before continuing, take a moment to imagine that your bones have turned to putty. What if you were running when this change took place? Now imagine your bones forming a rigid metal framework inside your body (without joints). What problems could you envision with this arrangement? These images should help you understand how well our skeletal system provides support and protection while allowing movement.

- **Storage.** Fat is stored in the internal (marrow) cavities of bones. Bone itself serves as a storehouse for minerals, the most important of which are calcium and phosphorus. Most of the body's calcium is deposited in the bones as calcium salts, but a small amount of calcium in its ion form ($Ca^{2+}$) must be present in the blood at all times for the nervous system to transmit messages, for muscles to contract, and

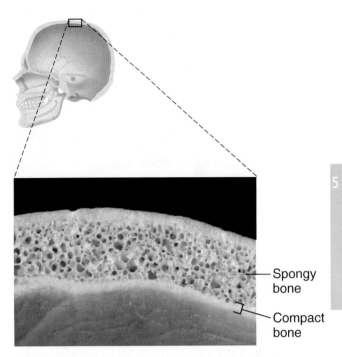

**Figure 5.1 Flat bones consist of a layer of spongy bone sandwiched between two thin layers of compact bone.**

Spongy bone

Compact bone

Explore PAL Cadaver
Mastering A&P®

for blood to clot. Problems occur not only when there is too little calcium in the blood, but also when there is too much. Hormones control the movement of calcium to and from the bones and blood according to the needs of the body. Indeed, "deposits" and "withdrawals" of calcium (and other minerals) to and from bones go on almost all the time.

- **Blood cell formation.** Blood cell formation, or *hematopoiesis* (hem″ah-to-poi-e′sis), occurs within the marrow cavities of certain bones.

## Classification of Bones

The adult skeleton is composed of 206 bones. There are two basic types of osseous, or bone, tissue: **Compact bone** is dense and looks smooth and homogeneous, whereas **spongy bone** has a spiky, open appearance like a sponge **(Figure 5.1)**.

Additionally, bones come in many sizes and shapes. For example, the tiny pisiform bone of the wrist is the size and shape of a pea, whereas the femur, or thigh bone, is nearly 2 feet long and has a large, ball-shaped head. The unique shape of

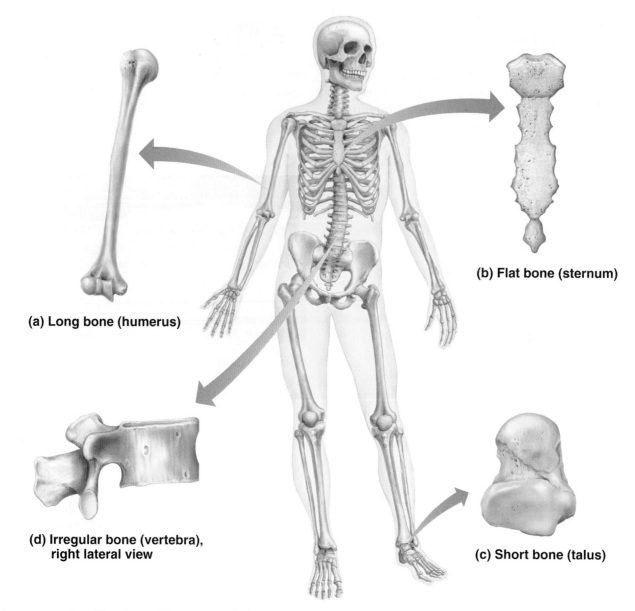

**(a) Long bone (humerus)**

**(b) Flat bone (sternum)**

**(d) Irregular bone (vertebra),
right lateral view**

**(c) Short bone (talus)**

**Figure 5.2 Classification of bones on the basis of shape.**

each bone fulfills a particular need. Bones are classified according to shape into four groups: long, short, flat, and irregular **(Figure 5.2)**.

As their name suggests, **long bones** are typically longer than they are wide. As a rule, they have a shaft with enlarged ends. Long bones are mostly compact bone but also contain spongy bone at the ends. All the bones of the limbs, except the patella (kneecap) and the wrist and ankle bones, are long bones.

**Flat bones** are thin, flattened, and usually curved. They have two thin layers of compact bone sandwiching a layer of spongy bone between

them (see Figure 5.1). Most bones of the skull, the ribs, and the sternum (breastbone) are flat bones.

**Short bones** are generally cube-shaped and contain mostly spongy bone with an outer layer of compact bone. The bones of the wrist and ankle are short bones. *Sesamoid* (ses'ah-moyd) *bones*, which form within tendons, are a special type of short bone. The best-known example is the patella.

Bones that do not fit one of the preceding categories are called **irregular bones**. The vertebrae, which make up the spinal column, fall into this group. Like short bones, they are mainly spongy bone with an outer layer of compact bone.

## Did You Get It?

1. What is the relationship between muscle function and bones?
2. What are two possible functions of a bone's marrow cavity?
3. Where are most long bones found in the body?

For answers, see Appendix A.

## Structure of Bone

### → Learning Objectives

☐ **Identify the major anatomical areas of a long bone.**

☐ **Describe the microscopic structure of compact bone.**

☐ **Explain the role of bone salts and the organic matrix in making bone both hard and flexible.**

### Gross Anatomy of a Long Bone

> **CONCEPTLINK**
>
> As we learn about the structure and organization of bones, remember the levels of structural organization (Figure 1.1, p. 29). Bones are organs, so they contain not only osseous tissue but also other connective tissues: fibrous tissue, cartilage, adipose tissue, and blood. ←

In a long bone, the **diaphysis** (di-af′ĭ-sis), or shaft, makes up most of the bone's length and is composed of compact bone (**Figure 5.3**, p. 164). The diaphysis is covered and protected by a fibrous connective tissue membrane, the **periosteum** (per-e-os′te-um). Hundreds of connective tissue fibers, called **perforating fibers**, or **Sharpey's fibers**, secure the periosteum to the underlying bone.

The **epiphyses** (ĕ-pif′ĭ-sēz) are the ends of long bones. Each epiphysis consists of a thin layer of compact bone enclosing an area filled with spongy bone. Instead of a periosteum, **articular cartilage** covers its external surface. Because the articular cartilage is glassy hyaline cartilage, it provides a smooth surface that decreases friction at the joint when covered by lubricating fluid. Imagine how slick a marble floor (the articular cartilage) is when wet; this is analogous to the lubrication of a joint.

In adult bones, there is a thin line of bony tissue spanning the epiphysis that looks a bit different from the rest of the bone in that area. This is the **epiphyseal line**. The epiphyseal line is a remnant of the **epiphyseal plate** (a flat plate of hyaline cartilage) seen in a young, growing bone. Epiphyseal plates cause the lengthwise growth of a long bone. By the end of puberty, when hormones inhibit long bone growth, epiphyseal plates have been completely replaced by bone, leaving only the epiphyseal lines to mark their previous location.

The inner bony surface of the shaft is covered by a delicate connective tissue called endosteum. In infants, the cavity of the shaft, called the **medullary cavity**, is a storage area for **red marrow**, which produces blood cells. Children's bones contain red marrow until the age of 6 or 7, when it is gradually replaced by **yellow marrow**, which stores adipose (fat) tissue. In adult bones, red marrow is confined to cavities in the spongy bone of the axial skeleton, the hip bones, and the epiphyses of long bones such as the humerus and femur.

Even when looking casually at bones, you can see that their surfaces are not smooth but scarred with bumps, holes, and ridges. These **bone markings** (described and illustrated in Table 5.1 on p. 166) reveal where muscles, tendons, and ligaments attach and where blood vessels and nerves pass. There are two categories of bone markings: (a) *projections*, or *processes*, which grow out from the bone surface, and (b) *depressions*, or *cavities*, which are indentations in the bone. There is a little trick for remembering some of the bone markings listed in the table: All the terms beginning with **T** are projections, and the terms beginning with **F** (except *facet*) are depressions.

### Microscopic Anatomy

The appearance of spongy bone and compact bone to the naked eye only hints at their underlying complexity. Under a microscope, you can see that spongy bone is composed of small needlelike pieces of bone called *trabeculae* and lots of "open" space filled by marrow, blood vessels and nerves (**Figure 5.4a**, p. 165).

In compact bone, the mature bone cells, **osteocytes** (os′te-o-sītz″), are found within the bone matrix in tiny cavities called **lacunae** (lah-ku′ne). The lacunae are arranged in concentric circles called **lamellae** (lah-mel′e) around **central canals** (also called **Haversian canals**). Each complex consisting of a central canal and matrix rings is called an **osteon**, or **Haversian system**, and is the structural and functional unit of compact bone.

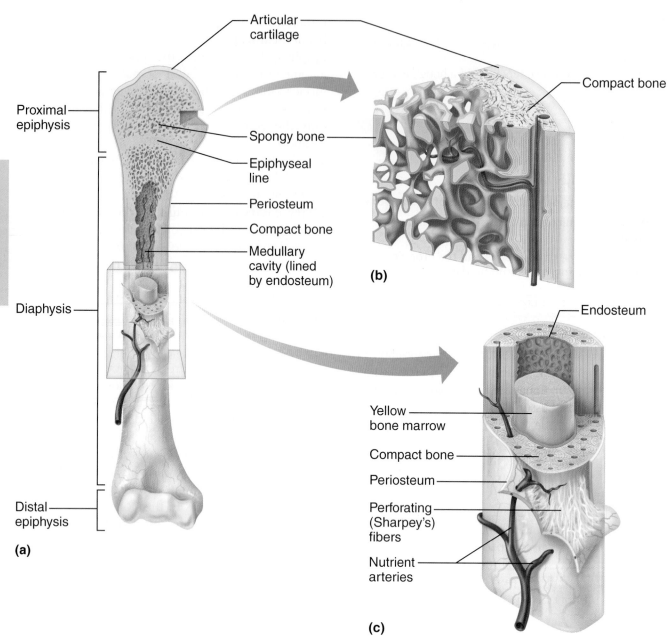

**Figure 5.3 The structure of a long bone (humerus of arm).** **(a)** Anterior view with longitudinal section cut away. **(b)** Pie-shaped, three-dimensional view of spongy bone and compact bone of the epiphysis. **(c)** Cross section of the shaft (diaphysis). Note that the external surface of the diaphysis is covered by a periosteum but that the articular surface of the epiphysis (see a and b) is covered with hyaline cartilage.

Central canals run lengthwise through the bony matrix, carrying blood vessels and nerves to all areas of the bone. Tiny canals, **canaliculi** (kan"ah-lik′u-li), radiate outward from the central canals to all lacunae. The canaliculi form a transportation system that connects all the bone cells to the nutrient supply and waste removal services through the hard bone matrix. Because of this elaborate network of canals, bone cells are well nourished in spite of the hardness of the matrix, and bone injuries heal quickly. The communication pathway from the outside of the bone to its interior (and

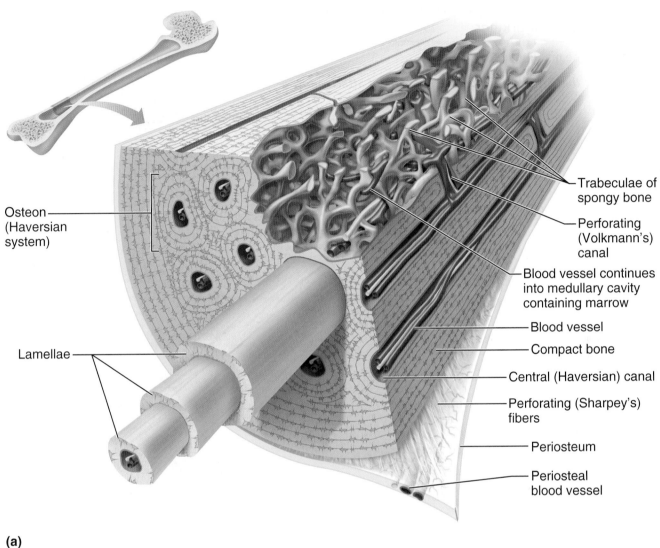

Osteon (Haversian system)

Lamellae

Trabeculae of spongy bone

Perforating (Volkmann's) canal

Blood vessel continues into medullary cavity containing marrow

Blood vessel

Compact bone

Central (Haversian) canal

Perforating (Sharpey's) fibers

Periosteum

Periosteal blood vessel

**(a)**

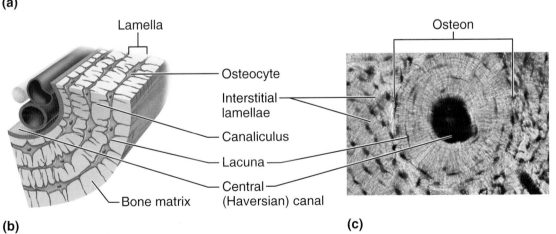

Lamella

Osteocyte

Interstitial lamellae

Canaliculus

Lacuna

Central (Haversian) canal

Bone matrix

Osteon

**(b)**

**(c)**

**Figure 5.4 Microscopic structure of bone. (a)** Diagram of a pie-shaped segment of compact and spongy bone illustrating their structural units (osteons) and trabeculae, respectively. **(b)** Higher magnification view of part of one osteon. Notice the position of osteocytes in lacunae (cavities in the matrix). **(c)** Photo of a cross-sectional view of an osteon.

View **PAL** Histology
Mastering A&P®

Table **5.1**  **Bone Markings**

| Name of bone marking | Description | Illustration |
|---|---|---|
| *Projections that are sites of muscle and ligament attachment* | | |
| Tuberosity | Large, rounded projection; may be roughened | |
| Crest | Narrow ridge of bone; usually prominent | |
| Trochanter (tro-kan'ter) | Very large, blunt, irregularly shaped process (the only examples are on the femur) | |
| Line | Narrow ridge of bone; less prominent than a crest | |
| Tubercle (too'ber-kl) | Small, rounded projection or process | |
| Epicondyle | Raised area on or above a condyle | |
| Spine | Sharp, slender, often pointed projection | |
| Process | Any bony prominence | |
| *Projections that help to form joints* | | |
| Head | Bony expansion carried on a narrow neck | |
| Facet | Smooth, nearly flat articular surface | |
| Condyle (kon'dīl) | Rounded articular projection | |
| Ramus (ra'mus) | Armlike bar of bone | |
| *Depressions and openings* | | |
| *For passage of blood vessels and nerves* | | |
| Groove | Furrow | |
| Fissure | Narrow, slitlike opening | |
| Foramen (fo-ra'men) | Round or oval opening through a bone | |
| Notch | Indentation at the edge of a structure | |
| *Others* | | |
| Meatus (me-a'tus) | Canal- or tunnel-like passageway | |
| Sinus | Cavity within a bone, filled with air and lined with mucous membrane | |
| Fossa (fos'ah) | Shallow, basinlike depression in a bone, often serving as an articular surface | |

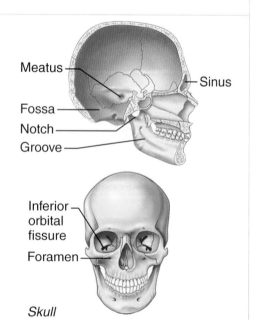

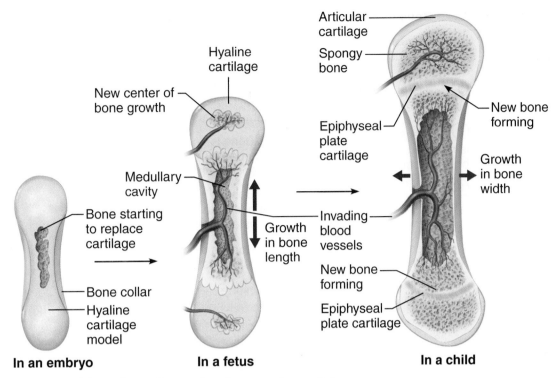

**Figure 5.5  Stages of long-bone formation in an embryo, fetus, and young child.**

the central canals) is completed by **perforating canals** (also called **Volkmann's canals**), which run in the compact bone at right angles to the shaft (diaphysis) and central canals.

Bone is one of the hardest materials in the body, and although relatively light in weight, it has a remarkable ability to resist tension and other forces acting on it. Nature has given us an extremely strong and exceptionally simple supporting system that also allows mobility. The calcium salts deposited in the matrix give bone its hardness, which resists compression. The organic parts (especially the collagen fibers) provide for bone's flexibility and great tensile strength (ability to be stretched without breaking).

### Did You Get It?

4. What is the anatomical name for the shaft of a long bone? For its ends?
5. How does the structure of compact bone differ from the structure of spongy bone?
6. What is the importance of canaliculi?

For answers, see Appendix A.

# Bone Formation, Growth, and Remodeling

→ **Learning Objective**

☐ **Describe briefly the process of bone formation in the fetus, and summarize the events of bone remodeling throughout life.**

### Bone Formation and Growth

The skeleton is formed from two of the strongest and most supportive tissues in the body—cartilage and bone. In embryos, the skeleton is primarily made of hyaline cartilage, but in young children, most of the cartilage has been replaced by bone. Cartilage remains only in isolated areas such as the bridge of the nose, parts of the ribs, and the joints.

Except for flat bones, which form on fibrous membranes, most bones develop using hyaline cartilage structures as their "models." This process of bone formation, or **ossification** (os"ĭ-fĭ-ka'shun), involves two major phases **(Figure 5.5)**. First, the hyaline cartilage model is completely covered with bone matrix (a bone "collar") by bone-building cells called **osteoblasts**. So, as the embryo develops into a fetus, for a short period it has cartilage "bones" enclosed by actual bone matrix. Then, in

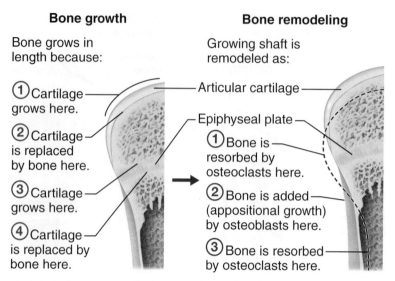

**Bone growth**

Bone grows in
length because:

① Cartilage
grows here.

② Cartilage
is replaced
by bone here.

③ Cartilage
grows here.

④ Cartilage
is replaced by
bone here.

**Bone remodeling**

Growing shaft is
remodeled as:

Articular cartilage

Epiphyseal plate

① Bone is
resorbed by
osteoclasts here.

② Bone is added
(appositional growth)
by osteoblasts here.

③ Bone is resorbed
by osteoclasts here.

**Figure 5.6 Growth and remodeling of long bones.** The events on the left depict the process of ossification that occurs at the articular cartilages and epiphyseal plates as the bone grows in length. The events on the right show bone remodeling and appositional growth during long-bone growth to maintain proper bone proportions.

the fetus, the enclosed hyaline cartilage model is replaced by bone, and the center is digested away, opening up a medullary cavity within the newly formed bone.

By birth or shortly after, most hyaline cartilage models have been converted to bone except for two regions—the articular cartilages (that cover the bone ends) and the epiphyseal plates. In order for bones to increase in length as the infant grows into a child, new cartilage is formed continuously on the external face (joint side) of the articular cartilage and on the epiphyseal plate surface that faces the bone end (is farther away from the medullary cavity). At the same time, the old cartilage abutting the internal face of the articular cartilage and the medullary cavity is broken down and replaced by bony matrix **(Figure 5.6)**.

Growing bones also widen as they lengthen to maintain proper proportion. How do they widen? Simply, osteoblasts in the periosteum add bone matrix to the outside of the diaphysis as cells called osteoclasts in the endosteum remove bone from the inner face of the diaphysis wall, enlarging the medullary cavity (see Figure 5.6). Because these two processes occur at about the same rate, the circumference of the long bone expands, and the bone widens. This process by which bones increase in diameter is called *appositional growth*, and like growth in length, is controlled by hormones. The most important hormones are *growth*

*hormone* and, during puberty, the *sex hormones*. It ends during adolescence, when the epiphyseal plates are completely converted to bone.

### Bone Remodeling

Many people mistakenly think that bones are lifeless structures that never change once long-bone growth has ended. Nothing could be further from the truth; bone is a dynamic and active tissue. Bones are remodeled continually in response to changes in two factors: (1) the calcium ion level in the blood and (2) the pull of gravity and muscles on the skeleton.

When the blood calcium ion level drops below its homeostatic level, the parathyroid glands (located in the throat) are stimulated to release parathyroid hormone (PTH) into the blood. PTH activates **osteoclasts**, giant bone-destroying cells in bones, to break down bone matrix and release calcium ions into the blood. When the blood calcium ion level is too high (*hypercalcemia* [hi″per-kal-se′me-ah]), calcium is deposited in bone matrix as hard calcium salts by osteoblasts.

**Bone remodeling** is essential if bones are to retain normal proportions and strength during long-bone growth as the body increases in size and weight. It also accounts for the fact that bones become thicker and form large projections to increase their strength in areas where bulky muscles are attached. At such sites, osteoblasts lay

# Radiologic Technologist

**Radiologic technologists supply critical information that allows doctors to make accurate diagnoses.**

"You never know what's going to walk in the door, really," says Maggie Regalado, a radiologic technologist at Dell Children's Hospital in Austin, Texas. "In an emergency room, you see kids who swallowed something, car accident victims, all kinds of things." Regalado and her coworkers operate X-ray equipment and must be ready to do everything from preparing patients for chest X-ray exams to MRIs.

Fortunately for Regalado, anatomy was her favorite class, because it's an important one for radiologic technologists. After getting her associate's degree in diagnostic imaging, she completed both state and national certification. To keep her certification current, she must complete 24 hours of continuing education every 2 years.

> You don't want to make errors, because one thing you do wrong could cost this patient his or her life.

"I didn't realize how big a field it was," she says. "With X-ray exams you're constantly moving from here to there, from surgery to the neonatal intensive care unit and so on." As you might guess, radiologic technologists, especially in hospitals, must be prepared to spend a lot of time on their feet and to think quickly. Regalado described one case when a two-car accident sent five children to the trauma unit. The radiologic technologists had to work quickly to help the doctors see what injuries the children suffered—and equally important, to make sure not to mix up anyone's X-ray exams.

"You don't want to make errors, because one thing you do wrong could cost this patient his or her life," she says. "Even though radiology can get emotional, you have to stay technical with your job."

"We can't see your bones with our bare eyes, so we have to make sure we position you correctly. Then also, if you say, 'It hurts here,' I'll call the doctor and see if he wants to do a different type of X-ray exam."

Regalado enjoys working with the patients at Dell. Getting children to remain perfectly still and positioned correctly is a challenge, but the imaging department has toys and televisions to distract them. For babies who cannot easily hold still or understand why they need to, there are various devices to position them appropriately.

"We have a lot of interaction with the patients and the patient's family; we try to joke around and make them happy," she says. "When we make the child happy, then the parents are happy."

In a hospital setting, radiologic technologists are needed 24 hours a day and often are required to be on-call in addition to their regular shifts. Technologists who work in clinics usually have a more traditional 9-to-5

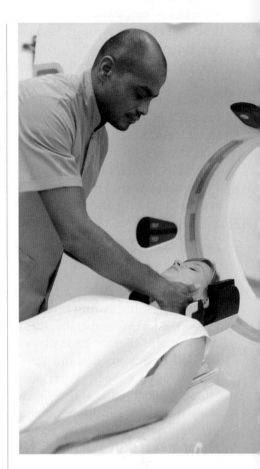

schedule. Depending on the clinic, these technologists may also specialize in areas such as ultrasound, mammography, magnetic resonance imaging (MRI), or computed tomography (CT).

For more information, contact:

American Society of Radiologic Technologists
15000 Central Ave. SE
Albuquerque, NM 87123-3909
(800) 444-2778
http://www.asrt.org

For additional information on this career and others, click the Focus on Careers link at Mastering A&P®.

down new matrix and become trapped within it. (Once they are trapped, they become *osteocytes*, or mature bone cells.) In contrast, the bones of bedridden or physically inactive people tend to lose mass and to atrophy because they are no longer subjected to stress.

These two controlling mechanisms—calcium uptake and release as well as bone remodeling—work together. PTH determines *when* bone is to be broken down or formed in response to the need for more or fewer calcium ions in the blood. The stresses of muscle pull and gravity acting on the skeleton determine *where* bone matrix is to be broken down or formed so that the skeleton can remain as strong and vital as possible.

### Homeostatic Imbalance 5.1

**Rickets** is a disease of children in which the bones fail to calcify. As a result, the bones soften, and the weight-bearing bones of the legs become bowed. Rickets is usually due to a lack of calcium in the diet or lack of vitamin D, which is needed to absorb calcium into the bloodstream. Rickets is not seen very often in the United States. Milk, bread, and other foods are fortified with vitamin D, and most children drink enough calcium-rich milk. However, it can happen in infants nursed by mothers who become vitamin D deficient, and it remains a problem in some other parts of the world.

This child suffering from rickets is a member of the el-Molo tribe in Kenya, whose diet consists primarily of fish.

＋

**Did You Get It?**

7. Bones don't begin as bones. What do they begin as?
8. Which stimulus—PTH (a hormone) or mechanical forces acting on the skeleton—is more important in maintaining the blood calcium ion level than in maintaining bone strength?
9. If osteoclasts in a long bone are more active than osteoblasts, what change in bone mass is likely to occur?

For answers, see Appendix A.

## Bone Fractures

→ **Learning Objective**

☐ **Name and describe the various types of fractures.**

### Homeostatic Imbalance 5.2

For their relatively low mass, bones are amazingly strong. Consider, for example, the forces endured in football and professional hockey. Despite their remarkable strength, bones are susceptible to **fractures**, or breaks, throughout life. During youth, most fractures result from exceptional trauma that twists or smashes the bones. Sports activities such as football, skating, and skiing jeopardize the bones, and automobile accidents certainly take their toll. In old age, bones thin and weaken, and fractures occur more often.

A fracture in which the bone breaks cleanly but does not penetrate the skin is a *closed* (or *simple*) *fracture*. When the broken bone ends penetrate through the skin, the fracture is *open* (or *compound*). Some of the many common types of fractures are illustrated and described in **Table 5.2**. _____ ＋

A fracture is treated by **reduction**, which is the realignment of the broken bone ends, followed by immobilization. In a *closed reduction*, the bone ends are coaxed back into their normal position by the physician's hands. In an *open reduction*, surgery is performed, and the bone ends are secured together with pins or wires. After the broken bone is reduced, it is immobilized by a cast or traction to allow the healing process to begin. The healing time for a simple fracture is 6 to 8 weeks but is much longer for large bones and for the bones of older people (because of their poorer circulation).

The repair of bone fractures involves four major events **(Figure 5.7)**:

① **A hematoma forms.** Blood vessels are ruptured when the bone breaks. As a result, a

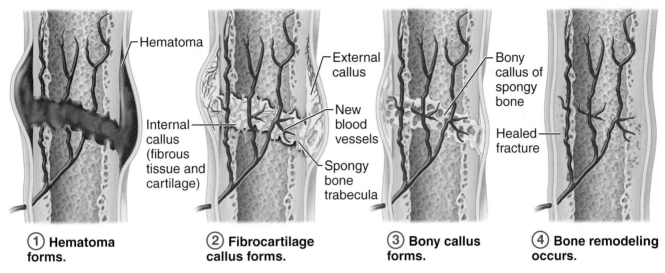

**Figure 5.7  Stages in the healing of a bone fracture.**

blood-filled swelling, or bruise, called a **hematoma** (he-mah-to'mah) forms. Bone cells deprived of nutrition die.

② **A fibrocartilage callus forms.** Two early events of tissue repair are the growth of new capillaries (*granulation tissue*) into the clotted blood at the site of the damage and disposal of

dead tissue by phagocytes. As this goes on, connective tissue cells of various types form internal and external masses of repair tissue, which collectively form the **fibrocartilage callus** (kal'us). The internal and external masses, called calluses, originate from cells of the endosteum and periosteum, respectively,

## Table **5.2**  Common Types of Fractures

| Fracture type | Illustration | Description | Comment |
|---|---|---|---|
| Comminuted | | Bone breaks into three or more fragments | Particularly common in older people, whose bones are more brittle |
| Compression | | Bone is crushed | Common in porous bones (i.e., osteoporotic bones of older people) |
| Depressed | | Broken bone portion is pressed inward | Typical of skull fracture |
| Impacted | | Broken bone ends are forced into each other | Commonly occurs when someone attempts to break a fall with outstretched arms |
| Spiral | | Ragged break occurs when excessive twisting forces are applied to a bone | Common sports fracture |
| Greenstick | | Bone breaks incompletely, much in the way a green twig breaks | Common in children, whose bones are more flexible than those of adults |

and contain several elements—cartilage matrix, bone matrix, and collagen fibers—which act to "splint" the broken bone, closing the gap.

③ **The bony callus forms.** As more osteoblasts and osteoclasts migrate into the area and multiply, the fibrocartilage callus is gradually replaced by the **bony callus** made of spongy bone.

④ **Bone remodeling occurs.** Over the next few weeks to months, depending on the bone's size and site of the break, the bony callus is remodeled in response to the mechanical stresses placed on it, so that it forms a strong, permanent "patch" at the fracture site.

### Did You Get It?

10. What is a fracture? What two fracture types are particularly common in older people?

For the answer, see Appendix A.

# Axial Skeleton

As noted earlier, the skeleton is divided into two parts, the axial and appendicular skeletons. The axial skeleton forms the longitudinal axis of the body (it is shown as the green portion of **Figure 5.8**). It can be divided into three parts—the *skull*, the *vertebral column*, and the *thoracic cage*.

### CONCEPT**LINK**

Recall the regional body terms you have already learned (look back at Figure 1.4, p. 40). Many of these terms can be associated with a bone name or group of bones. For example, the carpal region is the location of the carpals, or wrist bones. ←

## Skull

→ **Learning Objectives**

☐ **On a skull or diagram, identify and name the bones of the skull and the four major skull sutures.**

☐ **Describe how the skull of a newborn infant (or fetus) differs from that of an adult, and explain the function of fontanels.**

The **skull** is formed by two sets of bones. The **cranium** encloses and protects the fragile brain tissue. The **facial bones** form a cradle for the eyes that is open to the anterior and allow the facial muscles to show our feelings through smiles or frowns. All but one of the bones of the skull are

joined together by *sutures*, which are interlocking, immovable joints. Only the *mandible* (jawbone) is attached to the rest of the skull by a freely movable joint.

### Cranium

The boxlike cranium is composed of eight large flat bones. Except for two sets of paired bones (the *parietal* and *temporal*), they are all single bones.

**Frontal Bone**   The frontal bone forms the forehead, the bony projections under the eyebrows, and the superior part of each eye's *orbit* **(Figure 5.9)**.

**Parietal Bones**   The paired parietal bones form most of the superior and lateral walls of the cranium (see Figure 5.9). The **sagittal suture** is formed at the midline where the two parietal bones meet (shown in Figure 5.12, p. 176), and the **coronal suture** is formed where the paired parietal bones meet the frontal bone.

**Temporal Bones**   The temporal bones lie inferior to the parietal bones and join them at the **squamous sutures**. Several important bone markings appear on the temporal bones (see Figure 5.9):

- The **external acoustic meatus** is a canal that leads to the eardrum and the middle ear. It is the route by which sound enters the ear.

- The **styloid process**, a sharp, needlelike projection, is just inferior to the external auditory meatus. Many neck muscles use the styloid process as an attachment point.

- The **zygomatic** (zi"go-mat′ik) **process** is a thin bridge of bone that joins with the cheekbone (zygomatic bone) anteriorly.

- The **mastoid** (mas′toid) **process**, which is full of air cavities (the *mastoid sinuses*), is a rough projection posterior and inferior to the external acoustic meatus. It provides an attachment site for some muscles of the neck.

   The mastoid sinuses are so close to the middle ear—a high-risk spot for infections—that they may become infected too, a condition called *mastoiditis*. Also, this area is so close to the brain that mastoiditis may spread to the brain.

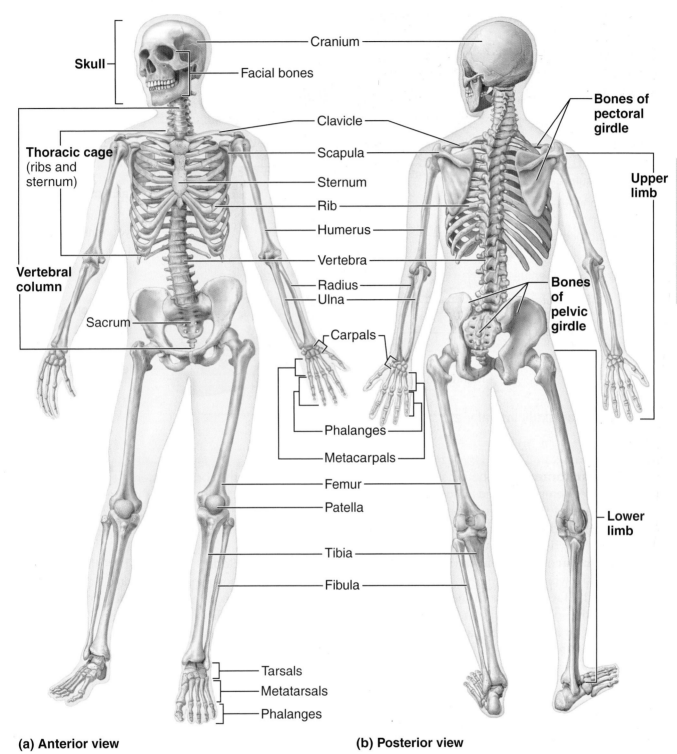

**(a) Anterior view**                    **(b) Posterior view**

**Figure 5.8  The human skeleton.** The bones of the axial skeleton are colored green. Bones of the appendicular skeleton are gold.

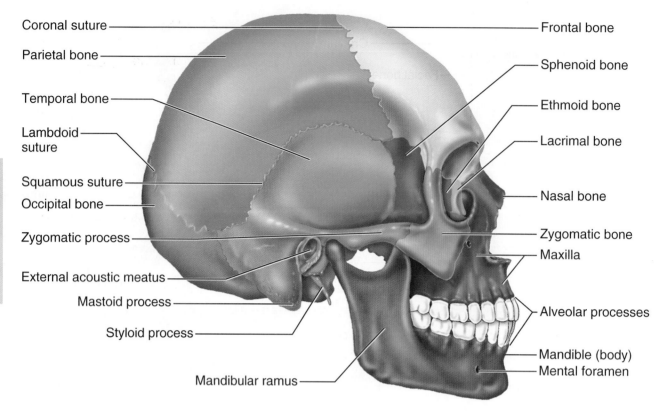

Coronal suture
Parietal bone
Temporal bone
Lambdoid suture
Squamous suture
Occipital bone
Zygomatic process
External acoustic meatus
Mastoid process
Styloid process
Mandibular ramus

Frontal bone
Sphenoid bone
Ethmoid bone
Lacrimal bone
Nasal bone
Zygomatic bone
Maxilla
Alveolar processes
Mandible (body)
Mental foramen

**Figure 5.9 Human skull, lateral view.**

- The **jugular foramen**, at the junction of the occipital and temporal bones **(Figure 5.10)**, allows passage of the *jugular vein*, the largest vein of the head, which drains blood from the brain. Just anterior to it in the cranial cavity is the **internal acoustic meatus** (see Figure 5.10), which transmits cranial nerves VII and VIII (the facial and vestibulocochlear nerves). Anterior to the jugular foramen on the skull's inferior aspect is the **carotid canal** (see Figure 5.11), through which the *internal carotid artery* runs, supplying blood to most of the brain.

**Occipital Bone**  The occipital (ok-sip'ĭ-tal) bone is the most posterior bone of the cranium **(Figure 5.11)**. It forms the base and back wall of the skull. The occipital bone joins the parietal bones anteriorly at the **lambdoid** (lam'doyd) **suture**. In the base of the occipital bone is a large opening, the **foramen magnum** (literally, "large hole"). The foramen magnum surrounds the lower part of the brain and allows the spinal cord to

connect with the brain. Lateral to the foramen magnum on each side are the rockerlike **occipital condyles** (see Figure 5.11), which rest on the first vertebra of the spinal column.

**Sphenoid Bone**  The butterfly-shaped sphenoid (sfe'noid) bone spans the width of the skull and forms part of the floor of the cranial cavity (see Figure 5.10). In the midline of the sphenoid is a small depression, the **sella turcica** (sel'ah tur'sĭ-kah), or *Turk's saddle*, which forms a snug enclosure for the pituitary gland. The **foramen ovale**, a large oval opening in line with the posterior end of the sella turcica (see Figure 5.10), allows fibers of cranial nerve V (the trigeminal nerve) to pass to the chewing muscles of the lower jaw (mandible). Parts of the sphenoid bone, seen exteriorly forming part of the eye orbits **(Figure 5.12**, p. 176), have two important openings, the **optic canal**, which allows the optic nerve to pass to the eye, and the slitlike **superior orbital fissure**, through which the cranial nerves controlling eye movements (III, IV, and VI) pass (see Figure 5.12). The central part

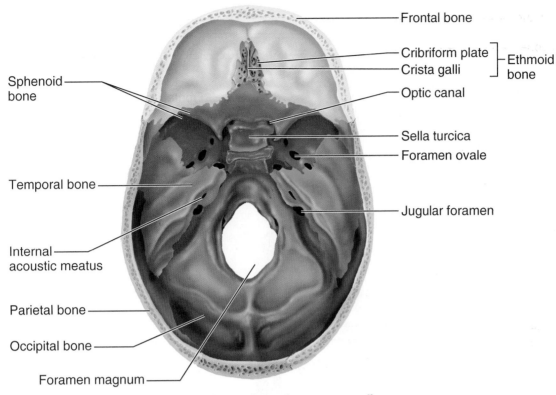

**Figure 5.10 Human skull, superior view (top of cranium removed).**

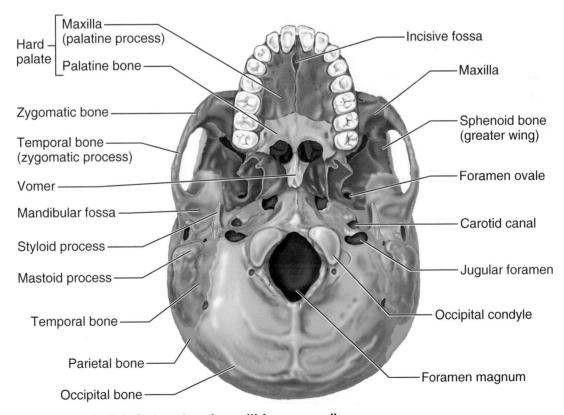

**Figure 5.11 Human skull, inferior view (mandible removed).**

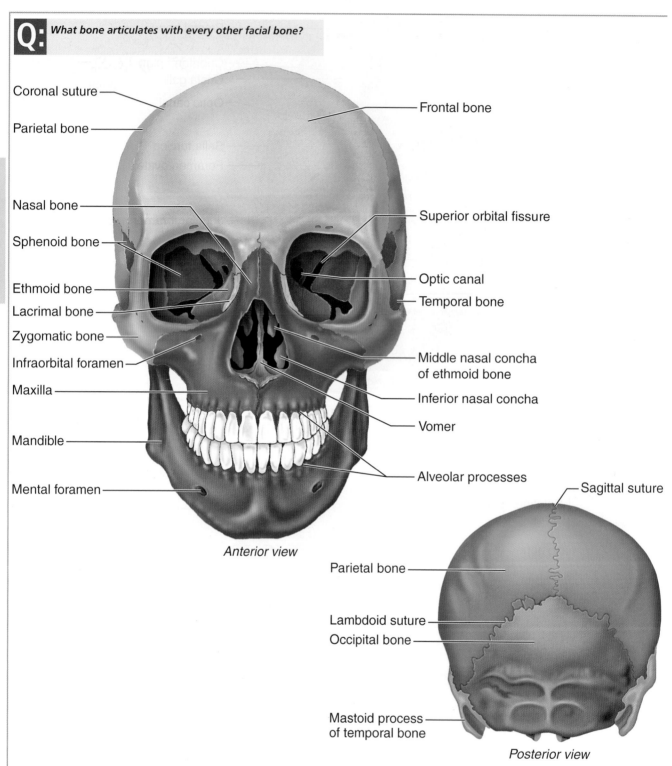

**Q:** *What bone articulates with every other facial bone?*

Coronal suture

Parietal bone

Frontal bone

Nasal bone

Sphenoid bone

Superior orbital fissure

Ethmoid bone

Optic canal

Lacrimal bone

Temporal bone

Zygomatic bone

Infraorbital foramen

Middle nasal concha of ethmoid bone

Maxilla

Inferior nasal concha

Mandible

Vomer

Mental foramen

Alveolar processes

*Anterior view*

Sagittal suture

Parietal bone

Lambdoid suture

Occipital bone

Mastoid process of temporal bone

*Posterior view*

**Figure 5.12 Human skull, anterior and posterior views.**

**A:** The maxilla.

of the sphenoid bone is riddled with air cavities, the **sphenoidal sinuses (Figure 5.13)**.

**■ *Ethmoid Bone*** The ethmoid (eth′moid) bone is very irregularly shaped and lies anterior to the sphenoid (see Figure 5.10). It forms the roof of the nasal cavity and part of the medial walls of the orbits. Projecting from its superior surface is the **crista galli** (kris′tah gah′le), literally "cock's comb" (see Figure 5.10). The outermost covering of the brain attaches to this projection. On each side of the crista galli are many small holes. These holey areas, the **cribriform** (krib′rĭ-form) **plates**, allow nerve fibers carrying impulses from the olfactory (smell) receptors of the nose to reach the brain. Extensions of the ethmoid bone, the **superior nasal conchae** (kong′ke) and **middle nasal conchae**, form part of the lateral walls of the nasal cavity (see Figure 5.12).

### Facial Bones

Fourteen bones make up the face. Twelve are paired; only the *mandible* and *vomer* are single. (Figures 5.9 and 5.12 show most of the facial bones.)

**■ *Maxillae*** The two maxillae (mak-sĭ′le), or **maxillary bones**, fuse to form the upper jaw. All facial bones except the mandible join the maxillae; thus they are the main, or "keystone," bones of the face. The maxillae carry the upper teeth in the **alveolar process**.

Extensions of the maxillae called the **palatine** (pal′ah-tīn) **processes** form the anterior part of the hard palate of the mouth (see Figure 5.11). Like many other facial bones, the maxillae contain **sinuses**, which drain into the nasal passages (see Figure 5.13). These **paranasal sinuses**, whose naming reveals their position surrounding the nasal cavity, lighten the skull bones and amplify the sounds we make as we speak.

### Homeostatic Imbalance 5.3

The paranasal sinuses cause many people a great deal of misery. Because the mucosa lining these sinuses is continuous with that in the nose and throat, infections in these areas tend to migrate into the sinuses, causing *sinusitis*. Depending on which sinuses are infected, a headache or upper jaw pain is the usual result. _____ ✚

**■ *Palatine Bones*** The paired palatine bones lie posterior to the palatine processes of the maxillae.

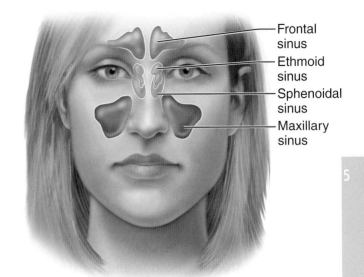

**(a) Anterior view**

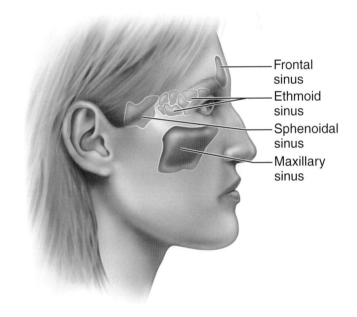

**(b) Medial view**

**Figure 5.13 Paranasal sinuses.**

They form the posterior part of the hard palate (see Figure 5.11). Failure of these or the palatine processes to fuse medially results in *cleft palate*.

**■ *Zygomatic Bones*** The zygomatic bones are commonly referred to as the cheekbones. They also form a good-sized portion of the lateral walls of the orbits.

**■ *Lacrimal Bones*** The lacrimal (lak′rĭ-mal) bones are fingernail-sized bones forming part of the medial

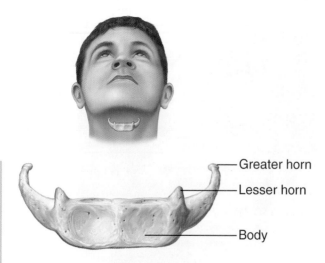

—Greater horn

—Lesser horn

—Body

**Figure 5.14 Anatomical location and structure of the hyoid bone.** Anterior view.

wall of each orbit. Each lacrimal bone has a groove that serves as a passageway for tears (*lacrima* = tear).

**Nasal Bones** The small rectangular bones forming the bridge of the nose are the nasal bones. (The lower part of the skeleton of the nose is made up of hyaline cartilage.)

**Vomer Bone** The single bone in the median line of the nasal cavity is the vomer. (*Vomer* means "plow," which refers to the bone's shape.) The vomer forms the inferior part of the bony nasal septum, which separates the two nostrils.

**Inferior Nasal Conchae** The inferior nasal conchae (kon'ke) are thin, curved bones projecting medially from the lateral walls of the nasal cavity. (As mentioned earlier, the superior and middle conchae are similar but are parts of the ethmoid bone.)

**Mandible** The mandible, or lower jaw, is the largest and strongest bone of the face. It joins the temporal bones on each side of the face, forming the only freely movable joints in the skull. You can find these joints on yourself by placing your fingers just anterior to your ears and opening and closing your mouth. The horizontal part of the mandible (the *body*) forms the chin. Two upright bars of bone (the *rami*) extend from the body to connect the mandible with the temporal bone. The lower teeth lie in *alveoli* (sockets) in the **alveolar process** at the superior edge of the mandibular body.

## The Hyoid Bone

Though not really part of the skull, the **hyoid** (hi'oid) **bone (Figure 5.14)** is closely related to the mandible and temporal bones. The hyoid bone is unique in that it is the only bone of the body that does not articulate (form a joint) with any other bone. Instead, it is suspended in the midneck region about 2 cm (1 inch) above the larynx (voicebox), where it is anchored by ligaments to the styloid processes of the temporal bones. Horseshoe-shaped, with a *body* and two pairs of *horns*, the hyoid bone serves as a movable base for the tongue and as an attachment point for neck muscles that raise and lower the larynx when we swallow and speak.

### Did You Get It?

11. What are the three main parts of the axial skeleton?
12. Johnny was vigorously exercising the only joints in the skull that are freely movable. What would you guess he was doing?
13. Which skull bone(s) form the "keystone of the face"?
14. Which bone has the cribriform plate and crista galli?
15. Which bones are connected by the coronal suture? By the sagittal suture?

**For answers, see Appendix A.**

## Vertebral Column (Spine)

### → Learning Objectives

☐ **Name the parts of a typical vertebra, and explain in general how the cervical, thoracic, and lumbar vertebrae differ from one another.**

☐ **Discuss the importance of the intervertebral discs and spinal curvatures.**

☐ **Explain how the abnormal spinal curvatures (scoliosis, lordosis, and kyphosis) differ from one another.**

Serving as the axial support of the body, the **vertebral column**, or **spine**, extends from the skull, which it supports, to the pelvis, where it transmits the weight of the body to the lower limbs. Some people think of the vertebral column as a rigid supporting rod, but that picture is inaccurate. Instead, the spine is formed from 26 irregular bones connected and reinforced by ligaments in such a way that a flexible, curved structure results **(Figure 5.15)**. Running through the central cavity of the vertebral column is the delicate spinal cord, which the vertebral column surrounds and protects.

Before birth, the spine consists of 33 separate bones called **vertebrae**, but 9 of these eventually fuse to form the two composite bones, the *sacrum* and the *coccyx*, that construct the inferior portion of

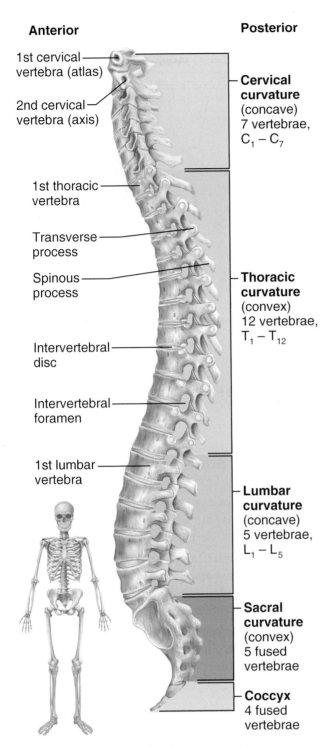

**Anterior**     **Posterior**

1st cervical vertebra (atlas)

2nd cervical vertebra (axis)

**Cervical curvature** (concave) 7 vertebrae, $C_1 – C_7$

1st thoracic vertebra

Transverse process

Spinous process

**Thoracic curvature** (convex) 12 vertebrae, $T_1 – T_{12}$

Intervertebral disc

Intervertebral foramen

1st lumbar vertebra

**Lumbar curvature** (concave) 5 vertebrae, $L_1 – L_5$

**Sacral curvature** (convex) 5 fused vertebrae

**Coccyx** 4 fused vertebrae

**Figure 5.15 The vertebral column.** Thin discs between the thoracic vertebrae allow great flexibility in the thoracic region; thick discs between the lumbar vertebrae reduce flexibility. Notice that the terms *convex* and *concave* refer to the curvature of the posterior aspect of the vertebral column.

the vertebral column. Of the 24 single bones, the 7 vertebrae of the neck are *cervical vertebrae*, the next 12 are the *thoracic vertebrae*, and the remaining 5 supporting the lower back are *lumbar vertebrae*.

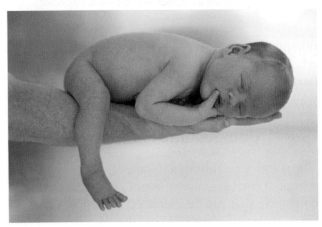

**Figure 5.16 The C-shaped spine typical of a newborn.**

Remembering common meal times, 7 a.m., 12 noon, and 5 p.m., may help you to recall the number of bones in these three regions of the vertebral column.

The individual vertebrae are separated by pads of flexible fibrocartilage—**intervertebral discs**—that cushion the vertebrae and absorb shock while allowing the spine flexibility. In a young person, the discs have a high water content (about 90 percent) and are spongy and compressible. But as a person ages, the water content of the discs decreases (as it does in other tissues throughout the body), and the discs become harder and less compressible.

### Homeostatic Imbalance 5.4

Drying of the discs, along with a weakening of the ligaments of the vertebral column, predisposes older people to **herniated** ("slipped") **discs**. However, herniation also may result when the vertebral column is subjected to exceptional twisting forces. If the protruding disc presses on the spinal cord or the spinal nerves exiting from the cord, numbness and excruciating pain can result. _____ ✚

The discs and the S-shaped structure of the vertebral column work together to prevent shock to the head when we walk or run. They also make the body trunk flexible. The spinal curvatures in the thoracic and sacral regions are referred to as **primary curvatures** because they are present when we are born. Together the two primary curvatures produce the C-shaped spine of the newborn baby **(Figure 5.16)**. The curvatures in the cervical and lumbar regions are referred to as **secondary curvatures** because they develop some time after birth. In adults, the secondary curvatures allow us to center our body weight on our lower limbs with minimum effort. The cervical curvature appears when a baby begins to raise its head, and the lumbar curvature develops when the baby begins to walk.

## Homeostatic Imbalance 5.5

Were you ever given a "spine check" in middle school? There are several types of abnormal spinal curvatures that can be identified by simple observation. Three of these are **scoliosis** (sko″le-o′sis), **kyphosis** (ki-fo′sis), and **lordosis** (lor-do′sis). These spinal abnormalities may be congenital (present at birth) or may result from disease, poor posture, unequal muscle pull on the spine, or injury. Generally, unless there is a congenital problem, young healthy people have no skeletal problems, assuming that their diet is nutritious and they stay reasonably active. As you look at these photos, try to pinpoint how each of these conditions differs from a normal healthy spine. The usual treatments for these abnormal curvatures are braces, casts, or surgery.

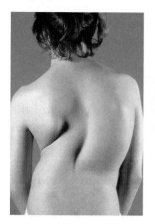  

**(a) Scoliosis**　　**(b) Kyphosis**　　**(c) Lordosis**

Abnormal spinal curvatures.

＋

All vertebrae have a similar structural pattern **(Figure 5.17)**. The common features of vertebrae include the following:

- **Body,** or **centrum:** disclike, weight-bearing part of the vertebra facing anteriorly in the vertebral column.
- **Vertebral arch:** arch formed from the joining of all posterior extensions, the **laminae** and **pedicles**, from the vertebral body.
- **Vertebral foramen:** canal through which the spinal cord passes.

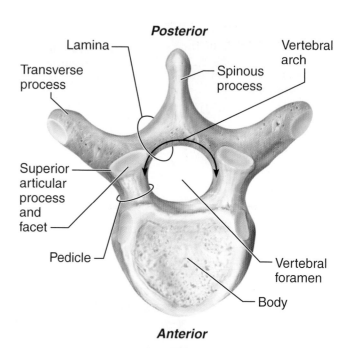

**Figure 5.17  A typical vertebra, superior view.** (Inferior articulating surfaces are not shown.)

- **Transverse processes:** two lateral projections from the vertebral arch.
- **Spinous process:** single projection arising from the posterior aspect of the vertebral arch (actually the fused laminae).
- **Superior articular process and inferior articular process:** paired projections lateral to the vertebral foramen, allowing a vertebra to form joints with adjacent vertebrae.

In addition to these common features, vertebrae in the different regions of the spine have very specific structural characteristics, which we describe next.

### Cervical Vertebrae

The seven **cervical vertebrae** (identified as $C_1$ to $C_7$) form the neck region of the spine. The first two vertebrae (*atlas* and *axis*) are different because they perform functions not shared by the other cervical vertebrae. As you can see in **Figure 5.18a**, the **atlas** ($C_1$) has no body. The superior surfaces of its transverse processes contain large depressions that receive the occipital condyles of the skull. This joint allows you to nod "yes." The **axis** ($C_2$) acts as a pivot for the rotation of the atlas (and skull) above. It has a large upright

**(a) ATLAS AND AXIS**

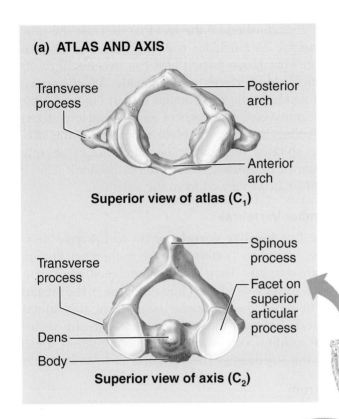

Transverse process

Posterior arch

Anterior arch

**Superior view of atlas (C$_1$)**

Transverse process

Spinous process

Facet on superior articular process

Dens

Body

**Superior view of axis (C$_2$)**

**(b) TYPICAL CERVICAL VERTEBRAE**

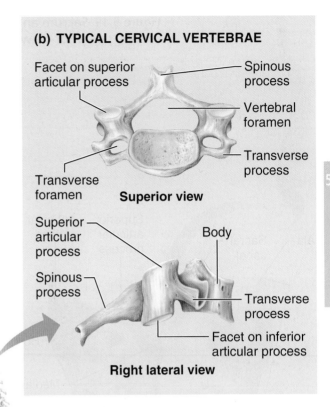

Facet on superior articular process

Spinous process

Vertebral foramen

Transverse process

Transverse foramen

**Superior view**

Superior articular process

Body

Spinous process

Transverse process

Facet on inferior articular process

**Right lateral view**

**(c) THORACIC VERTEBRAE**

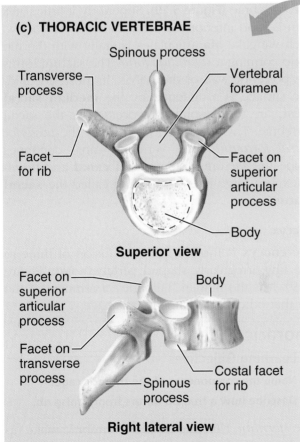

Spinous process

Transverse process

Vertebral foramen

Facet for rib

Facet on superior articular process

Body

**Superior view**

Facet on superior articular process

Body

Facet on transverse process

Spinous process

Costal facet for rib

**Right lateral view**

**(d) LUMBAR VERTEBRAE**

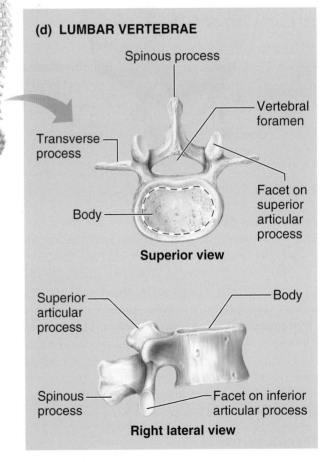

Spinous process

Vertebral foramen

Transverse process

Facet on superior articular process

Body

**Superior view**

Superior articular process

Body

Spinous process

Facet on inferior articular process

**Right lateral view**

**Figure 5.18 Regional characteristics of vertebrae.** (The types of vertebrae are not shown to scale with respect to one another.)

5

**Figure 5.19 Sacrum and coccyx, posterior view.**

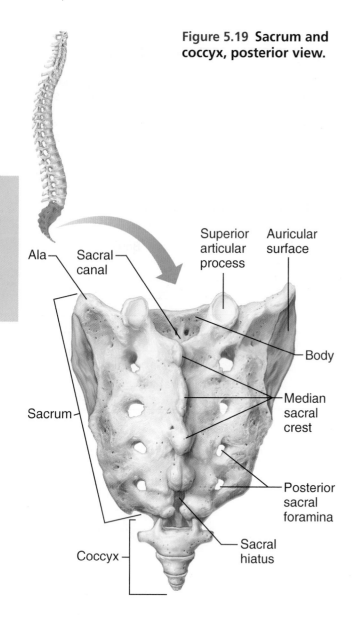

are distinguished by the fact that they are the only vertebrae to articulate with the ribs. The body is somewhat heart-shaped and has two *costal facets* (articulating surfaces) on each side, which receive the heads of the ribs (dotted line in Figure 5.18c). The transverse processes of each thoracic vertebra articulate with the knoblike tubercles of the ribs. The spinous process is long and hooks sharply downward, causing the vertebra to look like a giraffe's head viewed from the side.

### Lumbar Vertebrae

The five **lumbar vertebrae** ($L_1$ to $L_5$) have massive, blocklike bodies that are somewhat kidney bean–shaped. Their short, hatchet-shaped spinous processes (dotted line in Figure 5.18d) make them look like a moose head from the lateral aspect. Because most of the stress on the vertebral column occurs in the lumbar region, these are the sturdiest of the vertebrae.

### Sacrum

The **sacrum** (sa′krum) is formed by the fusion of five vertebrae **(Figure 5.19)**. Superiorly it articulates with $L_5$, and inferiorly it connects with the coccyx. Each winglike **ala** articulates laterally with the hip bone, forming a *sacroiliac joint*. The sacrum forms the posterior wall of the pelvis. Its posterior midline surface is roughened by the **median sacral crest**, the fused spinous processes of the sacral vertebrae. This is flanked laterally by the *posterior sacral foramina*. The vertebral canal continues inside the sacrum as the **sacral canal** and terminates in a large inferior opening called the **sacral hiatus**.

### Coccyx

The **coccyx** is formed from the fusion of three to five tiny, irregularly shaped vertebrae (see Figure 5.19). It is the human "tailbone," a remnant of the tail that other vertebrate animals have.

## Thoracic Cage

→ **Learning Objectives**

☐ **Name the components of the thoracic cage.**

☐ **Describe how a true rib differs from a false rib.**

The *sternum*, *ribs*, and thoracic vertebrae make up the **bony thorax (Figure 5.20)**. The bony thorax is routinely called the **thoracic cage** because it forms a protective cage of slender bones and

process, the **dens**, which acts as the pivot point. The joint between $C_1$ and $C_2$ allows you to rotate your head from side to side to indicate "no."

The "typical" cervical vertebrae are $C_3$ through $C_7$ (Figure 5.18b). They are the smallest, lightest vertebrae, and most often their spinous processes are short and divided into two branches. The transverse processes of the cervical vertebrae contain foramina (openings) through which the vertebral arteries pass on their way to the brain above. Any time you see these foramina in a vertebra, you should know immediately that it is a cervical vertebra.

### Thoracic Vertebrae

The 12 **thoracic vertebrae** ($T_1$ to $T_{12}$) are all typical. They are larger than the cervical vertebrae and

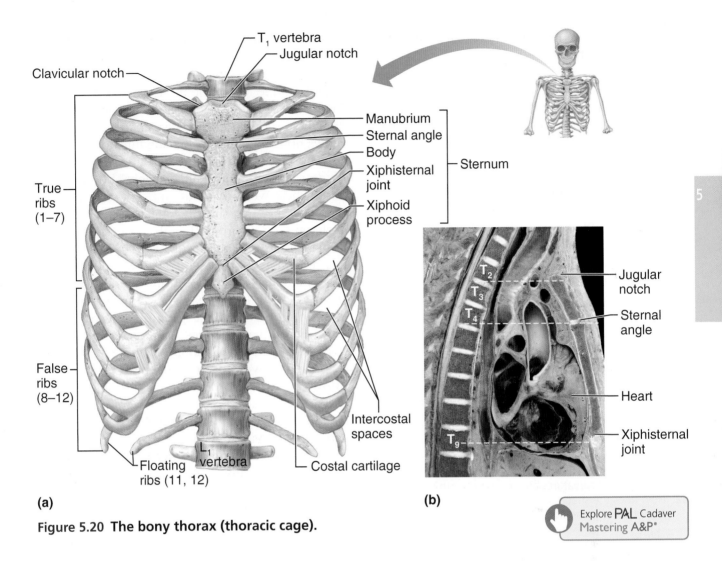

**Figure 5.20 The bony thorax (thoracic cage).**

cartilages around the organs of the thoracic cavity (heart, lungs, and major blood vessels).

### Sternum

The **sternum** (breastbone) is a typical flat bone and the result of the fusion of three bones—the **manubrium** (mah-nu′bre-um), **body**, and **xiphoid** (zif′oid) **process**. It is attached directly to the first seven pairs of ribs via costal cartilages.

The sternum has three important bony landmarks—the jugular notch, the sternal angle, and the xiphisternal joint.

- The **jugular notch** (concave upper border of the manubrium) can be palpated easily; generally it is at the level of the third thoracic vertebra.

- The **sternal angle** results where the manubrium and body meet at a slight angle to each other, so that a transverse ridge is

formed at the level of the second ribs. It provides a handy reference point for counting ribs to locate the second intercostal space for listening to certain heart valves. To clarify the location of these structures, you can palpate your own sternal angle and jugular notch.

- The **xiphisternal** (zi′fe-ster″nal) **joint**, the point where the sternal body and xiphoid process fuse, lies at the level of the ninth thoracic vertebra.

Because the sternum is so close to the body surface, it is easy to obtain samples from it of blood-forming (hematopoietic) tissue for the diagnosis of suspected blood diseases. A needle is inserted into the marrow of the sternum, and the sample is withdrawn; this procedure is called a *sternal puncture*. Because the heart lies immediately posterior to the sternum, the physician must take extreme care not to penetrate beyond the sternum during this procedure.

### Ribs

Twelve pairs of **ribs** form the walls of the bony thorax. All the ribs articulate with the vertebral column posteriorly and then curve downward and toward the anterior body surface. The **true ribs**, the first seven pairs, attach directly to the sternum by costal cartilages. **False ribs**, the next five pairs, either attach indirectly to the sternum or are not attached to the sternum at all. The last two pairs of false ribs lack the sternal attachments, so they are also called **floating ribs**.

The *intercostal spaces* (spaces between the ribs) are filled with the intercostal muscles, which aid in breathing.

### Did You Get It?

16. What are the five major regions of the vertebral column?
17. How can you distinguish a lumbar vertebra from a cervical vertebra?
18. What is a true rib? A false rib?
19. Besides the ribs and sternum, there is a third group of bones forming the thoracic cage. What is it?
20. What bone class do the ribs and skull bones fall into?
21. Which spinal curvatures are present at birth?
22. How does the shape of a newborn baby's spine differ from that of an adult?

**For answers, see Appendix A.**

# Appendicular Skeleton

### → Learning Objectives

☐ **Identify on a skeleton or diagram the bones of the shoulder and pelvic girdles and their attached limbs.**

☐ **Describe important differences between a male and a female pelvis.**

The appendicular skeleton (shaded gold in Figure 5.8) is composed of 126 bones of the *limbs* (*appendages*) and the *pectoral* and *pelvic girdles*, which attach the limbs to the axial skeleton.

## Bones of the Shoulder Girdle

Each **pectoral girdle**, or **shoulder girdle**, consists of two bones—a clavicle and a scapula **(Figure 5.21a)**.

The paired **clavicles** (klav′ĭ-kulz), or *collarbones*, are slender, doubly curved bones (Figure 5.21b). Each clavicle attaches to the manubrium of the sternum medially (at its sternal end) and to the scapula laterally, where it helps to form the shoulder joint. The clavicle acts as a brace to hold the arm away from the top of the thorax and helps prevent shoulder dislocation. When the clavicle is broken, the whole shoulder region caves in medially, which shows how important its bracing function is.

The paired **scapulae** (skap′u-le), or *shoulder blades*, are commonly called "wings" because they flare when we move our arms posteriorly. The scapulae are not directly attached to the axial skeleton (see Figure 5.21a); they are loosely held in place by trunk muscles. Each triangular scapula has a flattened body with three borders—superior, medial (vertebral), and lateral (axillary) (Figure 5.21c). It also has three angles—superior, inferior, and lateral. The **glenoid cavity**, a shallow socket that receives the head of the arm bone, is in the lateral angle (Figure 5.21d). The scapula has two important processes—the **acromion** (ah-kro′me-on), which is the enlarged lateral end of the spine of the scapula, and the beaklike **coracoid** (kor′ah-koid) **process**. The acromion connects with the clavicle laterally at the **acromioclavicular joint**. The coracoid process points laterally over the top of the shoulder and anchors some of the muscles of the arm. Just medial to the coracoid process is the large **suprascapular notch**, which is a nerve passageway.

The shoulder girdle is very light and allows the upper limb exceptionally free movement. This is due to the following factors:

- Each shoulder girdle attaches to the axial skeleton at only one point—the *sternoclavicular joint*.
- The loose attachment of the scapula allows it to slide back and forth against the thorax as muscles act.
- The glenoid cavity is shallow, and the shoulder joint is poorly reinforced by ligaments.

However, this exceptional flexibility also has a drawback; the shoulder girdle is very easily dislocated.

## Bones of the Upper Limbs

Thirty separate bones form the skeletal framework of each upper limb, including the arm, forearm, and hand.

### Arm

The arm is formed by a single bone, the **humerus** (hu′mer-us), which is a typical long

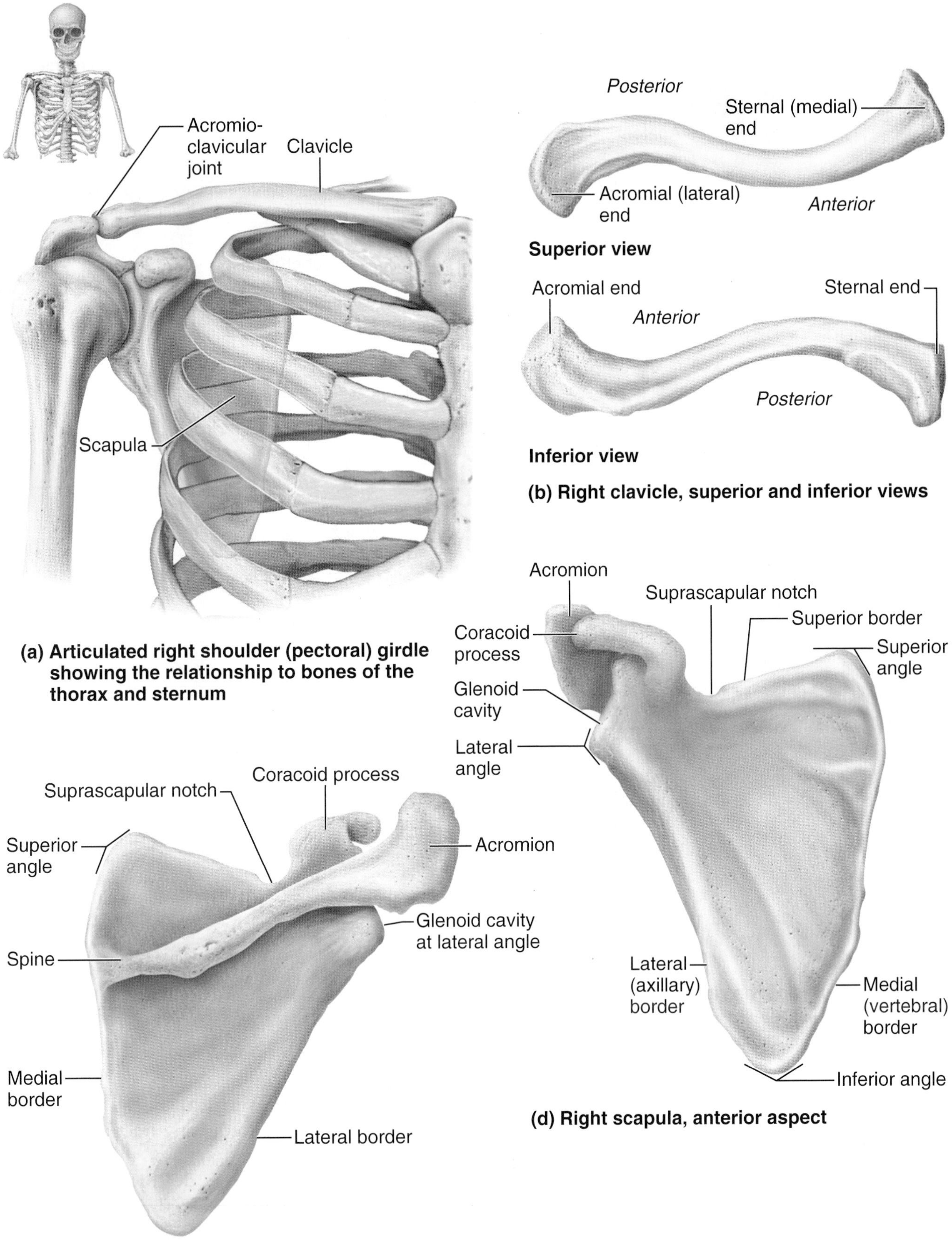

Acromio-clavicular joint

Clavicle

Scapula

**(a) Articulated right shoulder (pectoral) girdle showing the relationship to bones of the thorax and sternum**

Posterior

Sternal (medial) end

Acromial (lateral) end

Anterior

**Superior view**

Acromial end

Anterior

Sternal end

Posterior

**Inferior view**

**(b) Right clavicle, superior and inferior views**

Suprascapular notch

Coracoid process

Superior angle

Superior angle

Spine

Medial border

Glenoid cavity at lateral angle

Acromion

Lateral border

**(c) Right scapula, posterior aspect**

Acromion

Coracoid process

Glenoid cavity

Lateral angle

Suprascapular notch

Superior border

Superior angle

Lateral (axillary) border

Medial (vertebral) border

Inferior angle

**(d) Right scapula, anterior aspect**

Figure 5.21 **Bones of the shoulder girdle.**

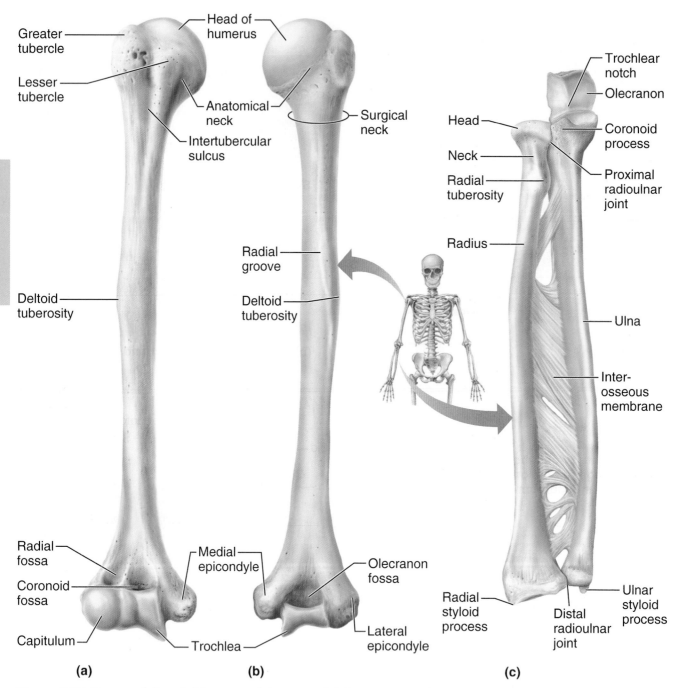

**Figure 5.22 Bones of the right arm and forearm. (a)** Humerus, anterior view. **(b)** Humerus, posterior view. **(c)** Anterior view of the bones of the forearm: the radius and the ulna.

bone (**Figure 5.22a** and **b**). At its proximal end is a rounded head that fits into the shallow glenoid cavity of the scapula. Immediately inferior to the head is a slight constriction called the **anatomical neck**. Anterolateral to the head are two bony projections separated by the **intertubercular sulcus**—the **greater tubercle** and **lesser tubercle**, which are sites of muscle attachment. Just distal to

the tubercles is the **surgical neck**, so named because it is the most frequently fractured part of the humerus. In the midpoint of the shaft is a roughened area called the **deltoid tuberosity**, where the large, fleshy deltoid muscle of the shoulder attaches. Nearby, the **radial groove** runs obliquely down the posterior aspect of the shaft. This groove marks the course of the radial nerve, an important nerve of the upper limb. At the distal end of the humerus is the medial **trochlea** (trok′le-ah), which looks somewhat like a spool, and the lateral ball-like **capitulum** (kah-pit′u-lum), which can be "outlined" with a "C" from the anterior view. Both of these processes articulate with bones of the forearm. Above the trochlea anteriorly is a depression, the **coronoid fossa**; on the posterior surface is the **olecranon** (o-lek′rah-non) **fossa**. These two depressions, which are flanked by the **medial epicondyle** and **lateral epicondyle**, allow the corresponding processes of the ulna to move freely when the elbow is bent and extended.

### Forearm

Two bones, the *radius* and the *ulna*, form the skeleton of the forearm (Figure 5.22c). When the body is in the anatomical position, the **radius** is the lateral bone; that is, it is on the thumb side of the forearm. When the hand is rotated so that the palm faces backward, the distal end of the radius crosses over and ends up medial to the ulna. Both proximally and distally the radius and ulna articulate at small **radioulnar joints**, and the two bones are connected along their entire length by the flexible **interosseous membrane**. Both the ulna and the radius have a **styloid process** at their distal end.

The disc-shaped head of the radius also forms a joint with the capitulum of the humerus. Just below the head is the **radial tuberosity**, where the tendon of the biceps muscle attaches.

When the upper limb is in the anatomical position, the **ulna** is the medial bone (on the little-finger side) of the forearm. On its proximal end are the anterior **coronoid process** and the posterior **olecranon**, which are separated by the **trochlear notch**. Together these two processes grip the trochlea of the humerus in a curved "tongue-in-groove"–like joint.

### Hand

The skeleton of the hand consists of the *carpals*, the *metacarpals*, and the *phalanges* (fah-lan′jēz) **(Figure 5.23)**. The eight **carpal bones**, arranged in two irregular rows of four bones each, form the part of the hand called the **carpus**, or the *wrist*. The carpals are bound together by ligaments that restrict movements between them.

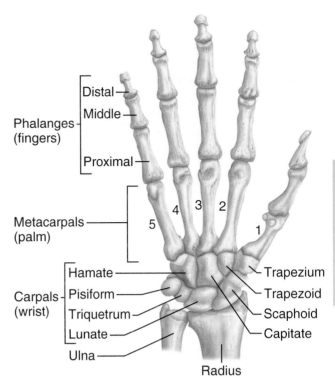

**Figure 5.23 Bones of the right hand, anterior view.**

The palm of the hand consists of the **metacarpals**. The metacarpals are numbered 1 to 5 from the thumb side of the hand toward the little finger. When the fist is clenched, the heads of the metacarpals become obvious as the "knuckles." The **phalanges** are the bones of the fingers. Each hand contains 14 phalanges. There are three in each finger (proximal, middle, and distal), except in the thumb, which has only two (proximal and distal).

### Did You Get It?

23. Contrast the general function of the axial skeleton to that of the appendicular skeleton.
24. What is the single point of attachment of the shoulder girdle to the axial skeleton?
25. What bone forms the skeleton of the arm?
26. Where are the carpals found, and what type (long, short, irregular, or flat) of bone are they?
27. Which bones of the upper limb have a styloid process?

**For answers, see Appendix A.**

## Bones of the Pelvic Girdle

The **pelvic girdle** is formed by two **coxal** (kok′sal) **bones**, commonly called **hip bones**, and the sacrum (described on p. 182). Together with the coccyx, the pelvic girdle forms the *pelvis* **(Figure 5.24a)**. Note that the terms *pelvic girdle* (the coxal bones and the sacrum) and *pelvis* (the coxal bones, sacrum, and coccyx) have slightly different meanings.

The bones of the pelvic girdle are large and heavy, and they are attached securely to the axial skeleton via the sacral attachment to the L₅ lumbar vertebra. The sockets, which receive the thigh bones, are deep and heavily reinforced by ligaments that attach the limbs firmly to the girdle. Bearing weight is the most important function of this girdle because the total weight of the upper body rests on the pelvis. The reproductive organs, urinary bladder, and part of the large intestine lie within and are protected by the pelvis.

Each hip bone is formed by the fusion of three bones: the *ilium, ischium,* and *pubis* (Figure 5.24b). The **ilium** (il′e-um), which connects posteriorly with the sacrum at the **sacroiliac** (sak″ro-il′e-ac) **joint**, is a large, flaring bone that forms most of the hip bone (imagine a peacock's fanned tail feathers). When you put your hands on your hips, they are resting over the *alae,* or winglike portions, of the ilia. The upper edge of an ala, the **iliac crest**, is an important anatomical landmark that is always kept in mind by those who give intramuscular injections. The iliac crest ends anteriorly in the **anterior superior iliac spine** and posteriorly in the **posterior superior iliac spine**. Small inferior spines are located below these.

The **ischium** (is′ke-um) is the "sit-down bone," so called because it forms the most inferior part of the coxal bone. The **ischial tuberosity** is a roughened area that receives body weight when you are sitting. The **ischial spine**, superior to the tuberosity, is another important anatomical landmark, particularly in the pregnant woman, because it narrows the outlet of the pelvis through which the baby must pass during birth. Another important structural feature of the ischium is the **greater sciatic notch**, which allows blood vessels and the large sciatic nerve to pass from the pelvis posteriorly into the thigh. Injections in the buttock should always be given well away from this area to avoid possible nerve damage.

The **pubis** (pu′bis) is the most anterior and inferior part of a coxal bone. Fusion of the *rami* of the pubis anteriorly and the ischium posteriorly forms a bar of bone enclosing the **obturator** (ob′tu-ra″tor) **foramen**, an opening that allows blood vessels and nerves to pass into the anterior part of the thigh. The pubic bones of each hip bone articulate anteriorly to form a cartilaginous joint, the **pubic symphysis** (pu′bik sim′fĭ-sis).

The ilium, ischium, and pubis fuse at the deep socket called the **acetabulum** (as″ĕ-tab′u-lum), which means "vinegar cup." The acetabulum receives the head of the thigh bone.

The bony pelvis is divided into two regions. The **false pelvis** is superior to the true pelvis; it is the area medial to the flaring portions of the ilia. The **true pelvis** is surrounded by bone and lies inferior to the flaring parts of the ilia and the pelvic brim (Figure 5.24c). The dimensions of the true pelvis of a woman are very important because they must be large enough to allow the infant's head (the largest part of the infant) to pass during childbirth. The dimensions of the cavity, particularly the **outlet** (the inferior opening of the pelvis measured between the ischial spines) and the **inlet** (superior opening between the right and left sides of the pelvic brim), are critical, and they are carefully measured by the obstetrician.

Of course, individual pelvic structures vary, but there are fairly consistent differences between a male and a female pelvis, including the following characteristics (see Figure 5.24c):

- The female inlet is larger and more circular.
- The female pelvis as a whole is shallower, and the bones are lighter and thinner.
- The female ilia flare more laterally, giving women curvy hips.
- The female sacrum is shorter and less curved.
- The female ischial spines are shorter and farther apart; thus, the outlet is larger.
- The female pubic arch is more rounded because the angle of the pubic arch is greater.

### Did You Get It?

28. What three bones form the hip bone? What bones form each pelvic girdle?
29. Describe three ways the bony pelvis of a woman differs from that of a man?

**For answers, see Appendix A.**

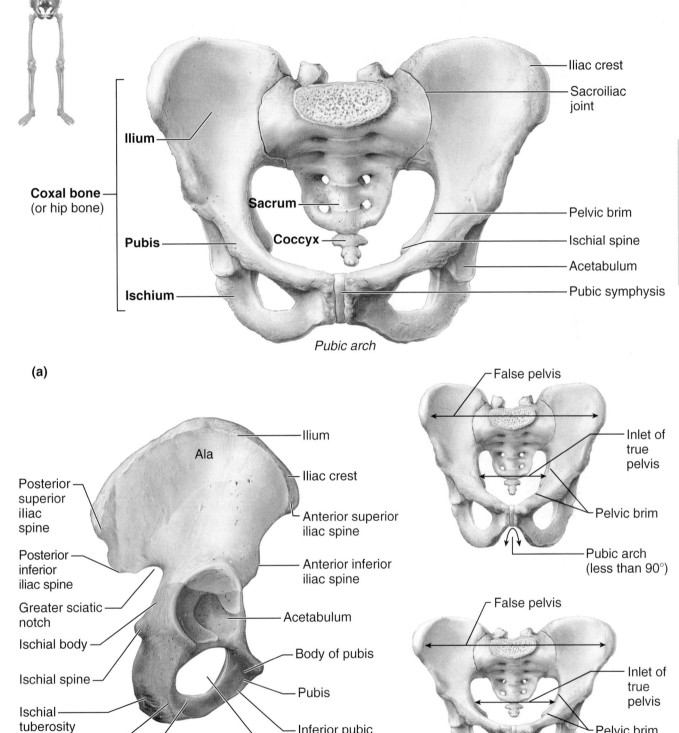

**(a)**

**(b)**                                                  **(c)**

**Figure 5.24 The bony pelvis. (a)** Articulated pelvis. **(b)** Coxal (hip) bone, showing the point of fusion of the ilium, ischium, and pubic bones at the acetabulum. **(c)** Comparison of the pelves of the male (above) and female (below).

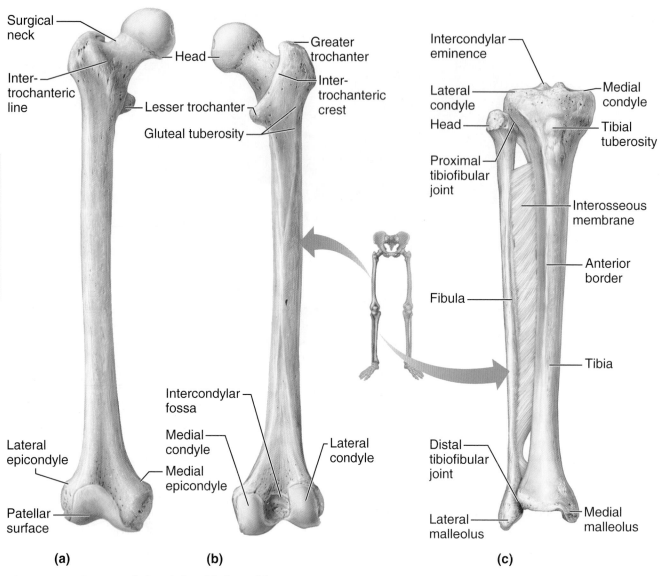

**Figure 5.25  Bones of the right thigh and leg.** Anterior view of femur **(a).**
Posterior view of femur **(b)**. Anterior view of leg **(c).**

## Bones of the Lower Limbs

The lower limbs carry our total body weight
when we are erect. Hence, it is not surprising that
the bones forming the three segments of the
lower limbs (thigh, leg, and foot) are much
thicker and stronger than the comparable bones
of the upper limb.

### Thigh

The **femur** (fe′mur), or *thigh bone*, is the only
bone in the thigh (**Figure 5.25a** and **b**). It is the
heaviest, strongest, and longest bone in the body.
Its proximal end has a ball-like head, a neck, and a
**greater trochanter** and **lesser trochanter** (sepa-

rated anteriorly by the **intertrochanteric line** and
posteriorly by the **intertrochanteric crest**). These
markings and the **gluteal tuberosity**, located on
the proximal end of the shaft, all serve as sites for
muscle attachment. The head of the femur articulates with the deep, secure socket of the acetabulum of the hip bone. However, the surgical neck of
the femur is still a common fracture site, especially
in old age.

The femur slants medially as it runs downward
to join with the leg bones; this brings the knees in
line with the body's center of gravity. The medial
course of the femur is more noticeable in women
because the female pelvis is typically wider than that

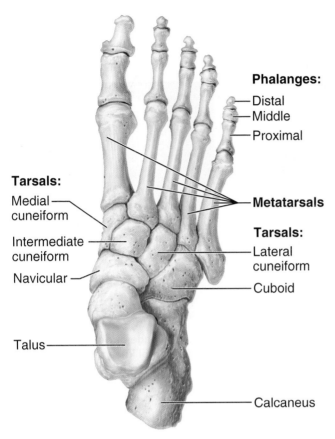

**Figure 5.26  Bones of the right foot, superior view.**

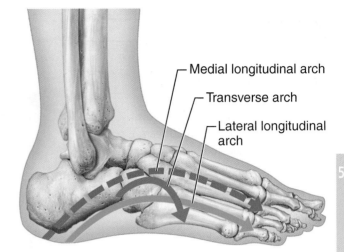

**Figure 5.27  Arches of the foot.**

of the male. Distally on the femur are the **lateral condyle** and **medial condyle**, which articulate with the tibia below. Posteriorly these condyles are separated by the deep **intercondylar fossa**. Anteriorly on the distal femur is the smooth **patellar surface**, which forms a joint with the patella, or kneecap.

## Leg

Connected along their length by an **interosseous membrane**, two bones, the *tibia* and *fibula*, form the skeleton of the leg (Figure 5.25c). The **tibia**, or *shinbone*, is larger and more medial. At the proximal end, the **medial condyle** and **lateral condyle** (separated by the **intercondylar eminence**) articulate with the distal end of the femur to form the knee joint. The patellar (kneecap) ligament, which encloses the **patella**, a sesamoid bone (look ahead to Figure 6.20c and d on p. 237), attaches to the **tibial tuberosity**, a roughened area on the anterior tibial surface. Distally, a process called the **medial malleolus** (mal-le′o-lus)

forms the inner bulge of the ankle. The anterior surface of the tibia is a sharp ridge, the **anterior border**, that is unprotected by muscles; thus, it is easily felt beneath the skin.

The **fibula**, which lies alongside the tibia laterally, forms joints with it both proximally and distally, and is thin and sticklike. The fibula has no part in forming the knee joint. Its distal end, the **lateral malleolus**, forms the outer part of the ankle.

## Foot

The foot, composed of the tarsals, metatarsals, and phalanges, has two important functions. It supports our body weight and serves as a lever that allows us to propel our bodies forward when we walk and run.

The **tarsus**, forming the posterior half of the foot, is composed of seven **tarsal bones (Figure 5.26)**. Body weight is carried mostly by the two largest tarsals, the **calcaneus** (kal-ka′ne-us), or heel bone, and the **talus** (ta′lus; "ankle"). The talus lies superior to the calcaneus, articulates with the tibia, and allows pointing of the toes. Five **metatarsals** form the sole, and 14 **phalanges** form the toes. Like the fingers of the hand, each toe has three phalanges, except the great toe, which has two.

The bones in the foot are arranged to form three strong arches: two longitudinal (medial and lateral) and one transverse **(Figure 5.27)**. *Ligaments*, which bind the foot bones together, and *tendons* of the foot muscles help to hold the bones firmly in the arched position but still allow a certain

# Joint Ventures

The technology for creating the prostheses (artificial joints) used in medicine today developed, in relative terms, in a flash—less than 60 years. Over 1 million knee and hip joint replacement surgeries are performed each year, mostly because of the destructive effects of osteoarthritis or rheumatoid arthritis.

The production of durable, mobile joints requires a substance that is strong, nontoxic, and resistant to the corrosive effects of organic acids in blood and that will not cause an immune response. In 1963, Sir John Charnley, an English orthopedic surgeon, performed the first total hip replacement using a modern prosthetic, revolutionizing the therapy of arthritic hips. His device consisted of a metal ball on a stem (the femoral head) and a cup-shaped polyethylene plastic socket (the acetabulum) anchored to the pelvis by polymethyl

**(a) A hip prosthesis.**

methacrylate cement (bone cement). This cement proved to be exceptionally strong and relatively problem free. Hip prostheses were followed by knee prostheses (see photos a and b), and replacements are now available for many other joints, including shoulders, fingers, and elbows.

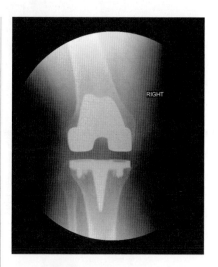

**(b) X-ray image of right knee showing total knee replacement prosthesis.**

Total hip and knee replacements last over 20 years in elderly patients who do not excessively stress the joint but may not last as long in younger (age 45 to 60), more active people. Most such operations are

---

amount of give, or springiness. Weak arches are referred to as "fallen arches," or "flat feet."

### Did You Get It?

30. What two bones form the skeleton of the leg?
31. Which bone allows us to "point our toes"?
32. Which bone of the lower limb has an intertrochanteric line and crest and an intercondylar fossa?

For answers, see Appendix A.

## Joints

→ **Learning Objective**

☐ Name the three major structural categories of joints, and compare the amount of movement allowed by each.

With one exception (the hyoid bone of the neck), every bone in the body forms a joint with at least one other bone. **Joints**, also called **articulations**, are the sites where two or more bones meet. They have two functions: They hold the bones together securely but also give the rigid skeleton mobility.

The graceful movements of a ballet dancer and the rough-and-tumble grapplings of a football player illustrate the great variety of motion that joints allow. With fewer joints, we would move like robots. Nevertheless, the bone-binding function of joints is just as important as their role in mobility. The immovable joints of the skull, for instance, form a snug enclosure for the vital brain.

Joints are classified in two ways—functionally and structurally. The functional classification focuses on the amount of movement the joint

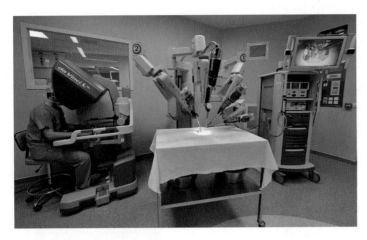

**(c) Physician with the ROBODOC machine used to perform hip joint replacement surgery.**

reduced the time and cost of creating individualized joints. The computer draws from a database of hundreds of healthy joints, generates possible designs, and produces a program to direct the machines that shape it.

Equally exciting are techniques that call on the patient's own tissues to regenerate, such as these three:

- Osteochondral grafting: Healthy bone and cartilage are removed from one part of the body and transplanted to the injured joint.

- Autologous chondrocyte implantation: Healthy chondrocytes are removed from the body, cultivated in the lab, and implanted at the damaged joint.

- Stem cell regeneration: Undifferentiated stem cells are removed from bone marrow and placed in a gel, which is packed into an area of eroded cartilage.

These techniques offer hope for younger patients because they could stave off the need for a joint prosthesis for several years.

done to reduce pain and restore joint function.

Replacement joints are not yet strong or durable enough for young, active people. The problem is that the prostheses work loose over time. One solution is to strengthen the cement that binds the implant to the bone. Another solution is robot-assisted surgery using ROBODOC (see photo c), which uses the information from a CT scan to precisely drill a better-fitting hole and so achieve better alignment for the femoral prosthesis in hip surgery. In uncemented prostheses, the bone will eventually grow so that it binds strongly to the implant, but this requires extended recovery time.

Dramatic changes are also occurring in the way artificial joints are made. CAD/CAM (computer-aided design and computer-aided manufacturing) techniques have significantly

allows. On this basis, there are **synarthroses** (sin″ar-thro′sēz), or immovable joints; **amphiarthroses** (am″fe-ar-thro′sēz), or slightly movable joints; and **diarthroses** (di″ar-thro′sēz), or freely movable joints. Immovable and slightly movable joints are restricted mainly to the axial skeleton, where firm attachments and protection of internal organs are priorities. Freely movable joints predominate in the limbs, where mobility is important.

Structurally, there are *fibrous, cartilaginous,* and *synovial joints.* These classifications are based on whether fibrous tissue, cartilage, or a joint cavity separates the bony regions at the joint. As a general rule, fibrous joints are immovable, and synovial joints are freely movable. Although cartilaginous joints have both immovable and slightly movable examples, most are amphiarthrotic.

Because the structural classification is more clear-cut, we will focus on that classification scheme here. The joint types are shown in **Figure 5.28** on p. 194 and described next.

### CONCEPTLINK

To understand the structural classes of joints more clearly, recall the properties of tissues that form the joints. Fibrous connective tissue contains many collagen fibers for strength. The three types of cartilage (hyaline, fibrocartilage, elastic) provide structure with some degree of flexibility, and fibrocartilage also has the ability to absorb compressive shock (Chapter 3, pp. 120, 121). Synovial membranes contain areolar connective tissue and line the joint cavities of synovial joints (Chapter 4, pp. 136, 138). ←

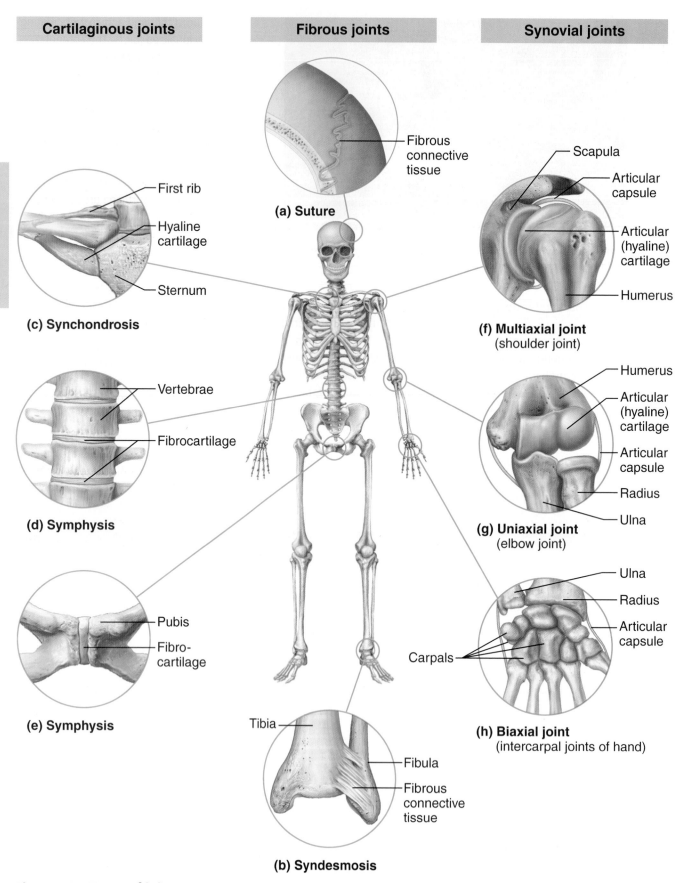

**Cartilaginous joints**

**Fibrous joints**

**Synovial joints**

Fibrous connective tissue

**(a) Suture**

First rib

Hyaline cartilage

Sternum

**(c) Synchondrosis**

Scapula

Articular capsule

Articular (hyaline) cartilage

Humerus

**(f) Multiaxial joint** (shoulder joint)

Vertebrae

Fibrocartilage

**(d) Symphysis**

Humerus

Articular (hyaline) cartilage

Articular capsule

Radius

Ulna

**(g) Uniaxial joint** (elbow joint)

Pubis

Fibro-cartilage

**(e) Symphysis**

Ulna

Radius

Articular capsule

Carpals

**(h) Biaxial joint** (intercarpal joints of hand)

Tibia

Fibula

Fibrous connective tissue

**(b) Syndesmosis**

**Figure 5.28 Types of joints.** Joints above and below the skeleton are fibrous joints **(a, b)**, joints to the left of the skeleton are cartilaginous joints **(c–e)**, and joints to the right of the skeleton are synovial joints **(f–h)**.

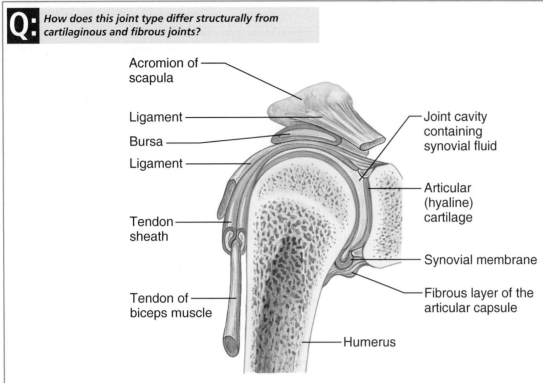

**Q:** *How does this joint type differ structurally from cartilaginous and fibrous joints?*

Acromion of scapula

Ligament

Bursa

Ligament

Tendon sheath

Tendon of biceps muscle

Joint cavity containing synovial fluid

Articular (hyaline) cartilage

Synovial membrane

Fibrous layer of the articular capsule

Humerus

**Figure 5.29  General structure of a synovial joint.**

## Fibrous Joints

In **fibrous joints**, the bones are united by fibrous tissue. The best examples of this type of joint are the sutures of the skull (Figure 5.28a). In sutures, the irregular edges of the bones interlock and are bound tightly together by connective tissue fibers, allowing no movement. *Gomphoses* (gom-fo′sēz), are "peg-in-socket" fibrous joints that are found where the teeth meet the facial bones (not shown in Figure 5.28). In another type of fibrous joint, *syndesmoses* (sin-dez-mo′sēz), the connecting fibers are longer than those of sutures; thus the joint has more "give." The joint connecting the distal ends of the tibia and fibula is a syndesmosis (Figure 5.28b).

## Cartilaginous Joints

**Cartilaginous joints** come in two varieties, which differ in the type of cartilage involved. *Synchondroses* are immoveable (synarthrotic) joints linked by hyaline cartilage. Examples include the epiphyseal plates of growing long bones and the joints between ribs 1–7 and the sternum (Figure 5.28c). *Symphyses* are amphiarthrotic (slightly moveable) joints linked by discs of fibrocartilage. Examples include the *intervertebral discs* of the spinal column (Figure 5.28d) and the *pubic symphysis* of the pelvis (Figure 5.28e).

## Synovial Joints

**Synovial joints** are joints in which the articulating bone ends are separated by a joint cavity containing synovial fluid (Figure 5.28f–h). All joints of the limbs are synovial joints.

All synovial joints have four distinguishing features **(Figure 5.29)**:

- **Articular cartilage.** Articular (hyaline) cartilage covers the ends of the bones forming the joint.

- **Articular capsule.** The joint surfaces are enclosed by a sleeve, or layer, of fibrous connective tissue, which is lined with a smooth **synovial membrane** (the reason these joints are called synovial joints).

**A:** It has a joint cavity filled with lubricating fluid instead of cartilage or fibrous tissue separating the articulating bones.

Table **5.3**  **Summary of Joint Classes**

| Structural class | Structural characteristics | Types | | Mobility |
|---|---|---|---|---|
| Fibrous | Bone ends/parts united by collagenic fibers | Suture (short fibers) | | Immobile (synarthrosis) |
| | | Syndesmosis (longer fibers) | | Slightly mobile (amphiarthrosis) and immobile |
| | | Gomphosis (periodontal ligament) | | Immobile |
| Cartilaginous | Bone ends/parts united by cartilage | Synchondrosis (hyaline cartilage) | | Immobile |
| | | Symphysis (fibrocartilage) | | Slightly movable |
| Synovial | Bone ends/parts covered with articular cartilage and enclosed within an articular capsule lined with synovial membrane | Plane Hinge Pivot | Condylar Saddle Ball and socket | Freely movable (diarthrosis; movements depend on design of joint) |

- **Joint cavity.** The articular capsule encloses a cavity, called the *joint cavity*, which contains lubricating **synovial fluid** secreted by the synovial membrane (look back at Chapter 4, pp. 136, 138).

- **Reinforcing ligaments.** The fibrous layer of the capsule is usually reinforced with ligaments.

   *Bursae* and *tendon sheaths* are not strictly part of synovial joints, but they are often found closely associated with them (see Figure 5.29). Essentially bags of lubricant, they help cushion and act like ball bearings to reduce friction between adjacent structures during joint activity. **Bursae** (ber′se; "purses") are flattened fibrous sacs lined with synovial membrane and containing a thin film of synovial fluid. They are common where ligaments, muscles, skin, tendons, or bones rub together. A **tendon sheath** is essentially an elongated bursa that wraps completely around a tendon subjected to friction, like a bun around a hot dog.

   The structural joint types are summarized in Table 5.3.

### Homeostatic Imbalance 5.6

A **dislocation** happens when a bone is forced out of its normal position in the joint cavity. The process of returning the bone to its proper position, called **reduction**, should be done only by a physician. Attempts by an untrained person to "snap the bone back into its socket" are usually more harmful than helpful. _____ ✚

## Types of Synovial Joints Based on Shape

The shapes of the articulating bone surfaces determine what movements are allowed at a joint. Based on such shapes, our synovial joints can be classified as *plane, hinge, pivot, condylar* (kon′di-ler), *saddle,* or *ball-and-socket joints.* Examples of these types of joints are shown in **Figure 5.30**; the arrows indicate movement around one or more axes in each type.

- In a **plane joint** (Figure 5.30a), the articular surfaces are essentially flat, and only short slipping or gliding movements are allowed. The movements of plane joints are *nonaxial*; that is, gliding back and forth does not involve rotation around any axis. The intercarpal joints of the wrist are the best examples of plane joints.

- In a **hinge joint** (Figure 5.30b), the cylindrical end of one bone fits into a trough-shaped surface on another bone. Angular movement is allowed in just one plane, like a door hinge. Examples are the elbow joint, ankle joint, and the joints between the phalanges of the fingers. Hinge joints are classified as *uniaxial* (u″ne-aks′e-al; "one axis"); they allow movement around one axis only.

- In a **pivot joint** (Figure 5.30c), the rounded end of one bone fits into a sleeve or ring of bone (and possibly ligaments). Because the rotating bone can turn only around its long axis,

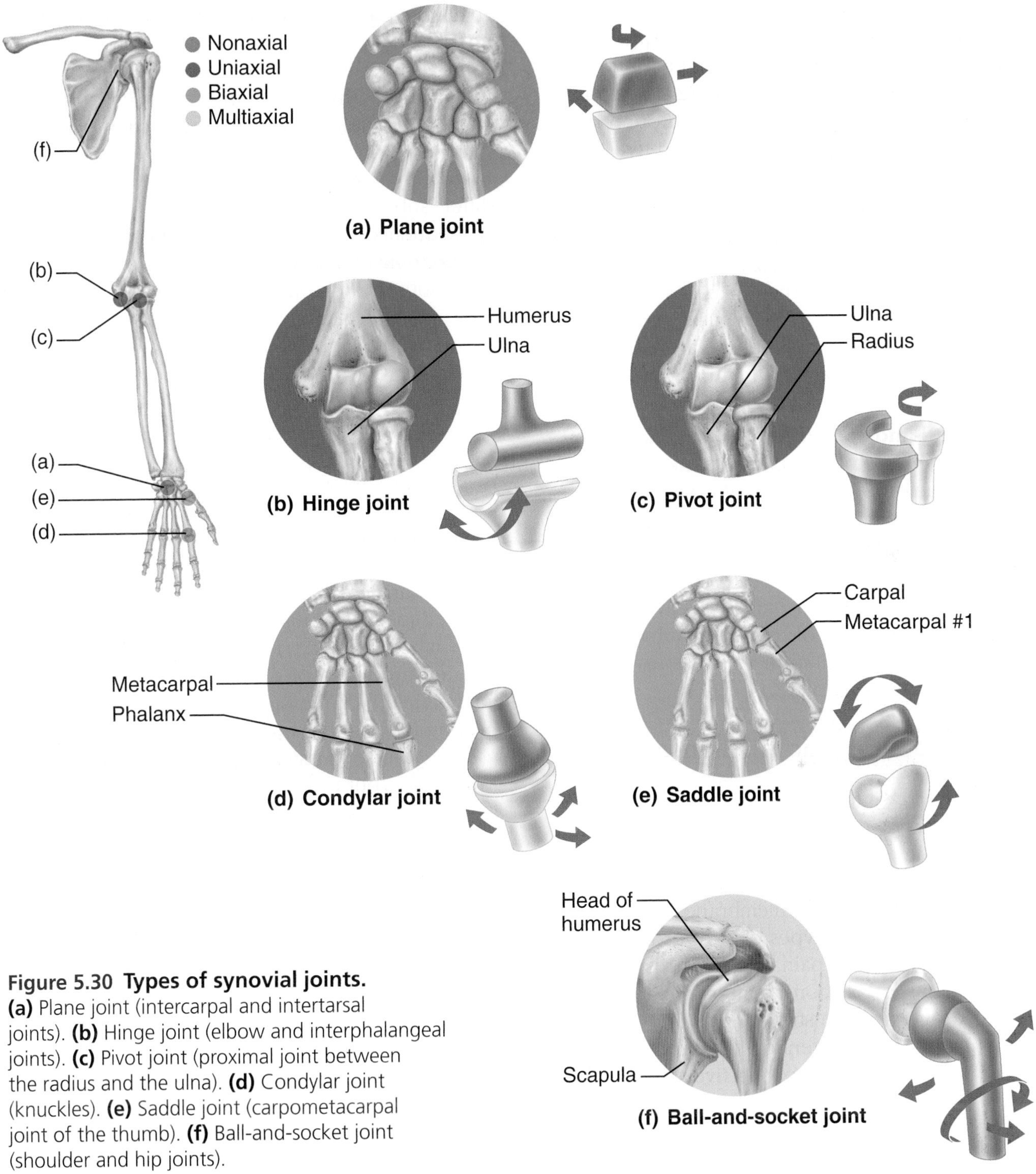

● Nonaxial
● Uniaxial
● Biaxial
● Multiaxial

(f)

(b)

(c)

(a)
(e)
(d)

**(a) Plane joint**

Humerus
Ulna

**(b) Hinge joint**

Ulna
Radius

**(c) Pivot joint**

Metacarpal
Phalanx

**(d) Condylar joint**

Carpal
Metacarpal #1

**(e) Saddle joint**

Head of
humerus

Scapula

**(f) Ball-and-socket joint**

**Figure 5.30 Types of synovial joints.**
**(a)** Plane joint (intercarpal and intertarsal
joints). **(b)** Hinge joint (elbow and interphalangeal
joints). **(c)** Pivot joint (proximal joint between
the radius and the ulna). **(d)** Condylar joint
(knuckles). **(e)** Saddle joint (carpometacarpal
joint of the thumb). **(f)** Ball-and-socket joint
(shoulder and hip joints).

pivot joints are also uniaxial joints. To understand the motion of this type of joint, think
about a basketball player pivoting around one
foot that stays planted in one spot. The proximal radioulnar joint and the joint between the
atlas and the dens of the axis are examples.

• In a **condylar joint** ("knucklelike"), the egg-
shaped articular surface of one bone fits into

an oval concavity in another (Figure 5.30d). Both of these articular surfaces are oval. Condylar joints allow the moving bone to travel (1) from side to side and (2) back and forth, but the bone cannot rotate around its long axis. Movement occurs around two axes; hence, these joints are *biaxial* (*bi* = two), as in knuckle (metacarpophalangeal) joints.

- In **saddle joints**, each articular surface has both convex and concave areas, like a saddle for a horse (Figure 5.30e). These biaxial joints allow essentially the same movements as condylar joints. The best examples of saddle joints are the carpometacarpal joints in the thumb, which are responsible for our opposable thumbs.

- In a **ball-and-socket joint** (Figure 5.30f), the spherical head of one bone fits into the round socket in another. These *multiaxial* joints allow movement in all axes, including rotation, and are the most freely moving synovial joints. The shoulder and hip are examples.

Because they relate to muscle activity, we discuss the various types of movements that occur at synovial joints in detail in the chapter covering the muscular system (Chapter 6).

## Homeostatic Imbalance 5.7

Few of us pay attention to our joints unless we experience joint pain or inflammation. For example, falling on one's knee can cause painful **bursitis**, called "water on the knee," due to inflammation of bursae or a synovial membrane. Sprains and dislocations are other types of joint problems that result in swelling and pain. A **sprain** results when the ligaments or tendons reinforcing a joint are damaged by excessive stretching or are torn away from the bone. Both tendons and ligaments are cords of dense fibrous connective tissue with a poor blood supply; thus, sprains heal slowly and can be extremely painful.

Few inflammatory joint disorders cause more pain and suffering than arthritis. The term **arthritis** (*arth* = joint, *itis* = inflammation) describes over 100 different inflammatory or degenerative diseases that damage joints. In all its forms, arthritis is the most widespread, crippling disease in the United States. All forms of arthritis have the same initial symptoms: pain, stiffness, and swelling of the joint. Then, depending on the specific form, certain changes in the joint structure occur.

Acute forms of arthritis usually result from bacterial invasion and are treated with antibiotic drugs. During such an infection, the synovial membrane thickens and fluid production decreases, leading to increased friction and pain. Chronic forms of arthritis include *osteoarthritis*, *rheumatoid arthritis*, and *gouty arthritis*, which differ substantially in their later symptoms and consequences. We will focus on these forms here.

**Osteoarthritis (OA)**, the most common form of arthritis, is a chronic degenerative condition that typically affects the aged: Eighty-five percent of people in the United States eventually develop this condition. OA, also called degenerative joint disease (DJD) or "wear-and-tear arthritis," affects the articular cartilages. Over the years, the cartilage softens, frays, and eventually breaks down. As the disease progresses, the exposed bone thickens, and extra bone tissue, called **bone spurs**, grows around the margins of the eroded cartilage and restricts joint movement. Patients complain of stiffness on arising that lessens with activity, and the affected joints may make a crunching noise (**crepitus**) when moved. The joints most commonly affected are those of the fingers, the cervical and lumbar joints of the spine, and the large, weight-bearing joints of the lower limbs (knees and hips).

The course of osteoarthritis is usually slow and irreversible, but it is rarely crippling. In most cases, its symptoms are controllable with a mild analgesic such as aspirin, moderate activity to maintain joint mobility, and rest when the joint becomes very painful. Some people with OA claim that rubbing capsaicin (a hot pepper extract) on the skin over painful joints provides relief. Others swear to the pain-reducing ability of glucosamine sulfate, a nutritional supplement.

**Rheumatoid** (roo′mah-toid) **arthritis (RA)** is a chronic inflammatory disorder. Its onset is insidious and usually occurs between the ages of 40 and 50, but it may occur at any age. It affects three times as many women as men. Many joints, particularly those of the fingers, wrists, ankles, and feet, are affected at the same time and usually in a symmetrical manner. For example, if the right elbow is affected, most likely the left elbow will be affected also. The course of RA varies and is marked by remissions and flare-ups (*rheumat* = susceptible to change or flux).

RA is an autoimmune disease—a disorder in which the body's immune system mistakenly attempts to destroy its own tissues. The initial trigger for this reaction is unknown but could involve genetic factors, certain bacterial or viral infections, hormones of the normal stress response, or a combination of these.

RA begins with inflammation of the synovial membranes. The membranes thicken and the joints swell as synovial fluid accumulates. Inflammatory cells (white blood cells and others) enter the joint cavity from the blood and inappropriately release a deluge of inflammatory chemicals that lead to the destruction of body tissues. In time, the inflamed synovial membrane thickens into a **pannus** ("rag"), an abnormal tissue that clings to and erodes articular cartilages. As the cartilage is destroyed, scar tissue forms and connects the bone ends. The scar tissue eventually ossifies, and the bone ends become firmly fused (**ankylosis**) and often deformed. Not all cases of RA progress to the severely crippling ankylosis stage, but all cases involve restricted joint movement and extreme pain.

Current therapy for RA involves many different kinds of drugs. Some, such as methotrexate, are immunosuppressants. Others, such as etanercept (Enbrel), neutralize the inflammatory chemicals in the joint space and (hopefully) prevent joint deformity. However, drug therapy often begins with aspirin, which in large doses is an effective anti-inflammatory agent. Exercise is recommended to maintain as much joint mobility as possible. Cold packs are used to relieve the swelling and pain, and heat helps to relieve morning stiffness. Replacement joints or bone removal are the last resort for severely crippled RA patients.

**Gouty** (gow'te) **arthritis**, or **gout**, is a disease in which uric acid (a normal waste product of nucleic acid metabolism) accumulates in the blood and may be deposited as needle-shaped crystals in the soft tissues of joints. This leads to an agonizingly painful attack that typically affects a single joint, often in the great toe. Gout is most common in men and rarely appears before the age of 30. It tends to run in families, so genetic factors are definitely implicated.

Untreated gout can be very destructive; the bone ends fuse, and the joint becomes immobilized. Fortunately, several drugs (colchicine, ibuprofen, and others) are successful in preventing acute gout attacks. Patients are advised to lose weight if obese; to avoid foods such as liver, kidneys, and sardines, which are high in nucleic acids; and to avoid alcohol, which inhibits excretion of uric acid by the kidneys.

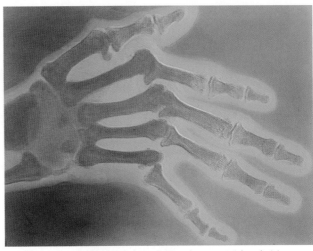

X-ray image of a hand deformed by rheumatoid arthritis.

## Did You Get It?

33. What are the functions of joints?
34. What is the major difference between a fibrous joint and a cartilaginous joint?
35. Where is synovial membrane found, and what is its role?
36. What two joints of the body are ball-and-socket joints? What is the best example of a saddle joint?

**For answers, see Appendix A.**

# Developmental Aspects of the Skeleton

### → Learning Objectives

☐ **Describe major changes in skeletal development at fetal, infant, youth, and adult stages of life.**

☐ **Identify some of the causes of bone and joint problems throughout life.**

As we have seen, the skeleton goes through many developmental changes in a person's lifetime. Let's look in more detail at the times when it changes the most.

## Birth to Adulthood

As we described earlier, the first "long bones" in the very young fetus are formed of hyaline cartilage, and the earliest "flat bones" of the skull are actually fibrous membranes. As the fetus develops and grows, all of the bone models are converted to bone **(Figure 5.31)**. At birth, the skull has fibrous regions that have yet to be converted to bone. These fibrous membranes connecting the cranial bones are called **fontanels** (fon"tah-nelz') **(Figure 5.32)**. The rhythm of the baby's pulse can be felt in these "soft spots,"

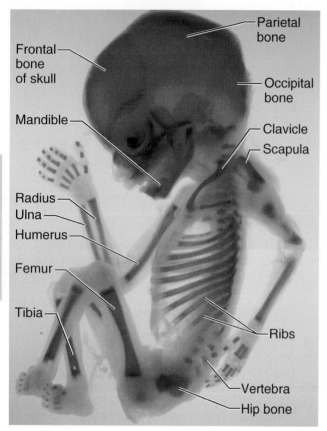

**Figure 5.31 Ossification centers in the skeleton of a 12-week-old fetus are indicated by the darker areas.** Lighter regions are still fibrous or cartilaginous.

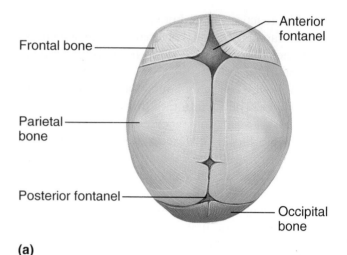

**(a)**

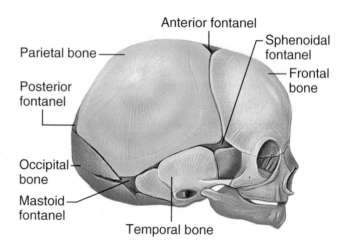

**(b)**

**Figure 5.32 The fetal skull.**

which explains their name (*fontanel* = little fountain). The largest fontanel is the diamond-shaped *anterior fontanel*. The fontanels allow the fetal skull to be compressed slightly during birth. In addition, because they are flexible, they allow the infant's brain to grow during the latter part of pregnancy and early infancy. This would not be possible if the cranial bones were fused by sutures, as in the adult skull. The fontanels usually fully ossify by 2 years of age.

The skeleton changes throughout life, but the changes in childhood are most dramatic. At birth and in infancy, the baby's cranium is huge relative to its face **(Figure 5.33a)**. This is why some young kids resemble "bobble head" dolls, because their heads are quite large compared to their body size. The rapid growth of the cranium before and after birth is related to the growth of the brain. By 2 years, the skull is three-quarters of its adult size; and, by 8 to 9 years, the skull is almost adult in size and proportion. Between the ages of 6 and

11, the head appears to enlarge substantially as the face literally grows out from the skull. The jaws increase in size, and the cheekbones and nose become more prominent as respiratory passages expand and the permanent teeth develop. During youth, growth of the skeleton not only increases overall body height and size but also changes body proportions (Figure 5.33b). At birth, the head and trunk are proportionately much longer than the lower limbs. The lower limbs grow more rapidly than the trunk from this time on, and by the age of 10, the head and trunk are approximately the same height as the lower limbs and change little thereafter. During puberty, the female pelvis broadens in preparation for childbearing, and the entire male skeleton becomes more

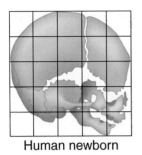
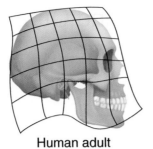

Human newborn          Human adult

**(a)**

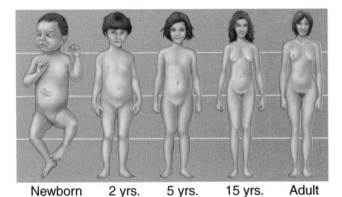

Newborn    2 yrs.    5 yrs.    15 yrs.    Adult

**(b)**

**Figure 5.33 Differences in the growth rates for some parts of the body compared to others determine body proportions. (a)** Differential growth transforms the rounded, foreshortened skull of a newborn to the sloping skull of an adult. **(b)** During growth of a human, the arms and legs grow faster than the head and trunk, as shown in this conceptualization of different-aged individuals all drawn at the same height.

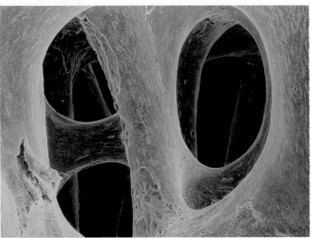

**Figure 5.34 Osteoporosis.** The architecture of osteoporotic bone, above, is contrasted with that of normal bone, below.

robust. By the end of adolescence, the epiphyseal plates of long bones that provide for longitudinal growth in childhood have become fully ossified, and long-bone growth ends. Once adult height is reached, a healthy skeleton changes very little until late middle age. In old age, losses in bone mass become obvious.

## Older Adults

Bones have to be physically stressed to remain healthy. When we remain active physically and muscles and gravity pull on the skeleton, the bones respond by becoming stronger. By contrast, if we are totally inactive, they become thin and fragile. **Osteoporosis** is a bone-thinning disease that afflicts half of women over 65 and some 20 percent of men over the age of 70. The ratio of bone formation to bone breakdown decreases as osteoblast

activity becomes sluggish. Osteoporosis can make the bones so fragile that even a hug or a sneeze can cause bones to fracture **(Figure 5.34)**. The bones of the spine and the neck of the femur are particularly susceptible. Vertebral collapse frequently results in a hunched-over posture (kyphosis) known as dowager's hump **(Figure 5.35**, p. 203).

Estrogen helps to maintain the health and normal density of a woman's skeleton, and the estrogen deficiency that occurs after a woman goes through *menopause* (when menstruation ceases) is strongly implicated as a cause of osteoporosis. Other factors that may contribute to osteoporosis are a diet poor in calcium and protein, lack of vitamin D, smoking, and insufficient weight-bearing exercise to stress the bones. Sadly, many older people feel that they are helping themselves by "saving their strength" and not doing anything too physical. Their reward for this

# Homeostatic Relationships between the Skeletal System and Other Body Systems

## Nervous System
- Skeletal system protects brain and spinal cord; provides a depot for calcium ions needed for neural function
- Nerves innervate bone and articular capsules, providing for pain and joint sense

## Endocrine System
- Skeletal system provides some bony protection
- Hormones regulate uptake and release of calcium from bone; hormones promote long-bone growth and maturation

## Respiratory System
- Skeletal system (rib cage) protects lungs by enclosure
- Respiratory system provides oxygen; disposes of carbon dioxide

## Lymphatic System/Immunity
- Skeletal system provides some protection to lymphoid organs; lymphocytes involved in immune response originate in bone marrow
- Lymphatic system drains leaked interstitial fluid; immune cells protect against pathogens

## Cardiovascular System
- Bone marrow cavities provide site for blood cell formation; matrix stores calcium needed for muscle, nerve, and blood-clotting activities
- Cardiovascular system delivers nutrients and oxygen to bones; carries away wastes

## Digestive System
- Skeletal system provides some bony protection to intestines, pelvic organs, and liver
- Digestive system provides nutrients needed for bone health and growth

## Reproductive System
- Skeletal system protects some reproductive organs by enclosure
- Gonads produce hormones that influence form of skeleton and epiphyseal closure

## Urinary System
- Skeletal system protects pelvic organs (bladder, etc.)
- Urinary system activates vitamin D; disposes of nitrogen-containing wastes

## Integumentary System
- Skeletal system provides support for body organs including the skin
- Skin provides vitamin D needed for proper calcium absorption and use

## Muscular System
- Skeletal system provides levers plus calcium for muscle activity
- Muscle pull on bones increases bone strength and viability; helps determine bone shape as bone matrix is deposited at points of stress

**Skeletal System**

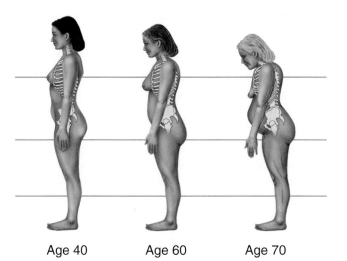

Age 40　　　Age 60　　　Age 70

**Figure 5.35 Vertebral collapse due to osteoporosis.** Women with postmenopausal osteoporosis are at risk for vertebral fractures. Eventually these vertebrae collapse, producing spinal curvature that causes loss of height, a tilted rib cage, a dowager's hump, and a protruding abdomen.

is *pathologic fractures* (breaks that occur spontaneously without apparent injury), which increase dramatically with age and are the single most common skeletal problem for this age group.

Advancing years also take their toll on joints. Weight-bearing joints in particular begin to degenerate, and *osteoarthritis* is common. Such degenerative joint changes lead to the complaint often heard from an aging person: "My joints are getting so stiff."

## Did You Get It?

37. Ninety-year-old Mrs. Pelky is groaning in pain. Her grandson has just given her a bear hug. What do you think might have happened to her spine, and what bone condition may she be suffering from?
38. Which two regions of the skeleton grow most rapidly during childhood?

For answers, see Appendix A.

## Summary

### Bones: An Overview (pp. 160–172)

1. Bones support and protect body organs; serve as levers for the muscles to pull on to cause movement at joints; and store calcium, fats, and other substances for the body. Some contain red marrow, the site of blood cell production.

2. Bones are classified into four groups—long, short, flat, and irregular—on the basis of their shape and the amount of compact or spongy bone they contain. Bone markings are important anatomical landmarks that reveal where muscles attach and where blood vessels and nerves pass.

3. A long bone is composed of a shaft (diaphysis) with two ends (epiphyses). The shaft is compact bone; its cavity contains yellow marrow. The epiphyses are covered with hyaline cartilage; they contain spongy bone (where red marrow is found).

4. The organic parts of the matrix make bone flexible; calcium salts deposited in the matrix make bone hard.

5. Bones form in the fetus on hyaline cartilage or fibrous membrane "models." Epiphyseal plates persist to provide for longitudinal growth of long bones during childhood and become inactive (by calcifying) when adolescence ends.

6. Bones change in shape throughout life. This remodeling occurs in response to hormones (for example, PTH, which regulates blood calcium levels) and mechanical stresses acting on the skeleton.

7. A fracture is a break in a bone. Common types of fractures include simple, compound, compression, comminuted, and greenstick. Bone fractures must be reduced to heal properly.

### Axial Skeleton (pp. 172–184)

1. The skull is formed by cranial and facial bones. Eight cranial bones protect the brain: frontal, occipital, ethmoid, and sphenoid bones, and the paired parietal and temporal bones. The 14 facial bones are all paired (maxillae, zygomatics, palatines, nasals, lacrimals, and inferior nasal conchae), except for the vomer and mandible. The hyoid bone, not really a skull bone, is supported in the neck by ligaments.

2. The vertebral column is formed from 24 vertebrae, the sacrum, and the coccyx. There are 7 cervical vertebrae, 12 thoracic vertebrae, and 5 lumbar vertebrae, which have common and unique features. The vertebrae are separated by fibrocartilage discs that allow the vertebral column to be flexible. The vertebral column is C-shaped at birth (thoracic and sacral curvatures are present); the secondary curvatures

(cervical and lumbar) form when the baby begins to lift its head and walk. After infancy, the vertebral column is S-shaped to allow for upright posture.

3. The bony thorax is formed from the sternum and 12 pairs of ribs. All ribs attach posteriorly to thoracic vertebrae. Anteriorly, the first 7 pairs attach directly to the sternum (true ribs); the last 5 pairs attach indirectly or not at all (false ribs). The bony thorax encloses the lungs, heart, and other organs of the thoracic cavity.

## Appendicular Skeleton (pp. 184–192)

1. The shoulder girdles are each composed of two bones—a scapula and a clavicle—and attach the upper limbs to the axial skeleton. They are light, poorly reinforced girdles that allow the upper limbs a great deal of freedom of movement.

2. The bones of the upper limb include the humerus of the arm, the radius and ulna of the forearm, and the carpals, metacarpals, and phalanges of the hand.

3. The pelvic girdle is formed by the two coxal bones, or hip bones, and the sacrum (part of the axial skeleton). Each hip bone is the result of fusion of the ilium, ischium, and pubis bones. The pelvic girdle is securely attached to the vertebral column, and the socket for the thigh bone is deep and heavily reinforced. This girdle receives the weight of the upper body and transfers it to the lower limbs. The female pelvis is lighter and broader than the male's; its inlet and outlet are larger, reflecting the childbearing function.

4. The bones of the lower limb include the femur of the thigh; the tibia and fibula of the leg; and the tarsals, metatarsals, and phalanges of the foot.

## Joints (pp. 192–199)

1. Joints hold bones together and allow movement of the skeleton.

2. Joints fall into three functional categories: synarthroses (immovable), amphiarthroses (slightly movable), and diarthroses (freely movable).

3. Joints also can be classified structurally as fibrous, cartilaginous, or synovial joints, depending on the substance separating the articulating bones.

4. Most fibrous joints are synarthrotic, and most cartilaginous joints are amphiarthrotic. Fibrous and cartilaginous joints occur mainly in the axial skeleton.

5. All joints of the limbs are synovial joints. In synovial joints, the articulating bone surfaces are covered with articular cartilage and enclosed within the joint cavity by a fibrous capsule lined with a synovial membrane that secretes lubricating synovial fluid. All synovial joints are diarthroses.

6. The most common joint problem is arthritis, or inflammation of the joints. Osteoarthritis, or degenerative arthritis, is a result of the "wear and tear" on joints over many years and is a common affliction of the aged. Rheumatoid arthritis occurs in both young and older adults; it is believed to be an autoimmune disease. Gouty arthritis, caused by the deposit of uric acid crystals in joints, typically affects a single joint.

## Developmental Aspects of the Skeleton (pp. 199–203)

1. Fontanels, which allow brain growth and ease birth passage, are present in the skull at birth. Growth of the cranium after birth is related to brain growth; the increase in size of the facial skeleton follows tooth development and enlargement of the respiratory passageways.

2. Long bones continue to grow in length until late adolescence. By the age of 10, the head and trunk are approximately the same height as the lower limbs and change little thereafter.

3. Fractures are the most common bone problem in elderly people. Osteoporosis, a condition of bone wasting that results mainly from hormone deficit or inactivity, is also common in older individuals.

## Review Questions

 Access additional practice questions using your smartphone, tablet, or computer: Mastering A&P® > Study Area > Practice Tests & Quizzes

### Multiple Choice

*More than one choice may apply.*

1. Which of the following are correctly matched?

   a. Short bone—wrist

   b. Long bone—leg

   c. Irregular bone—sternum

   d. Flat bone—cranium

2. A passageway connecting neighboring osteocytes in an osteon is a

   a. central canal.

   b. lamella.

   c. lacuna.

   d. canaliculus.

   e. perforating canal.

3. Which of the following are the elements of an osteon?

   a. Haversian canals    c. Lacunae

   b. Sharpey's fibers    d. Periostea

4. Bone pain behind the external acoustic meatus probably involves the

   a. maxilla.    d. temporal bone.

   b. ethmoid bone.    e. lacrimal bone.

   c. sphenoid bone.

5. Bones that articulate with the sphenoid include which of the following?

   a. Parietal    d. Zygomatic

   b. Frontal    e. Ethmoid

   c. Occipital

6. The clavicle articulates with the

   a. sternal manubrium.    d. coracoid process.

   b. sternal body.    e. head of humerus.

   c. acromion.

7. Every cervical vertebra has

   a. a body.

   b. a dens.

   c. costal facets.

   d. transverse processes with foramina for the vertebral arteries.

   e. a short spinous process divided into two branches.

8. Which of the following bones are found in the feet?

   a. Lunate    d. Metatarsal

   b. Talus    e. Calcaneus

   c. Metacarpal

9. The shoulder joint is an example of a

   a. plane joint.

   b. condylar joint.

   c. ball-and-socket joint.

   d. pivot joint.

   e. hinge joint.

10. At what stage of life do the lower limbs attain the same height as the head and trunk?

    a. At birth

    b. By 10 years of age

    c. At puberty

    d. When the epiphyseal plates fuse

    e. Never

11. Match the types of joints to the descriptions that apply to them. (More than one description might apply.)

    a. Fibrous joints

    b. Cartilaginous joints

    c. Synovial joints

    ____ 1. Have no joint cavity

    ____ 2. Types are sutures and syndesmoses

    ____ 3. Dense connective tissue fills the space between the bones

    ____ 4. Almost all joints of the skull

    ____ 5. Types are synchondroses and symphyses

    ____ 6. All are diarthroses

    ____ 7. The most common type of joint in the body

    ____ 8. Nearly all are synarthrotic

    ____ 9. Shoulder, hip, knee, and elbow joints

12. Match the bone markings listed on the right with their function listed on the left.

    1. Attachment site for muscle or ligament    ____ a. Trochanter

    2. Forms a joint surface    ____ b. Condyle

    ____ c. Foramen

    ____ d. Process

    3. Passageway for vessels or nerves    ____ e. Facet

    ____ f. Tuberosity

## Short Answer Essay

13. Name three functions of the skeletal system.

14. What is yellow marrow? How do spongy and compact bone look different?

15. Why do bone injuries heal much more rapidly than injuries to cartilage?

16. Compare and contrast the role of PTH (hormone) and mechanical forces acting on the skeleton in bone remodeling.

17. What type of tissue are epiphyseal plates composed of, and what is their role in growing children?

18. Identify the four phases of fracture repair.

19. Name the skull bones that harbor paranasal sinuses.

20. What facial bone forms the chin? The cheekbone? The upper jaw? The bony eyebrow ridges?

21. Where is the jugular foramen located, and which structure passes through it?

22. What is a human tailbone?

23. Diagram the normal spinal curvatures and then the curvatures seen in scoliosis and lordosis.

24. What is the function of the intervertebral discs? What is a herniated (slipped) disc?

25. What is the function of intercostal muscles?

26. What are bursae? Where are they commonly found?

27. Name the bones that constitute the tarsus.

28. Name all the bones with which the tibia articulates.

29. What bones make up each hip bone (coxal bone)? Which of these is the largest? Which has tuberosities that we sit on? Which is the most anterior?

30. Name the long bones of the upper limb from superior to inferior.

31. Compare the amount of movement possible in synarthrotic, amphiarthrotic, and diarthrotic joints. Relate these terms to the structural classification of joints—that is, to fibrous, cartilaginous, and synovial joints.

32. Describe the structure of a synovial joint. Use this description to briefly explain what occurs in a joint with osteoarthritis.

33. Professor Rogers pointed to the foramen magnum of the skull and said, "The food passes through this hole when you swallow." Some students believed him, but others said that this was a big mistake. What do you think? Support your answer.

34. Yolanda is asked to review a bone slide that has been set up under a microscope. She sees concentric layers surrounding a central cavity or canal. Is this bone section taken from the diaphysis or the epiphyseal plate of the bone specimen?

35. List two factors that keep bones healthy. List two factors that can cause bones to become soft or to atrophy.

36. Contrast the location and characteristics of foramen magnum and obturator foramen.

## Critical Thinking and Clinical Application Questions

37. A 75-year-old woman and her 9-year-old granddaughter were in a car accident in which both sustained trauma to the chest while seated next to each other. X-ray images showed that the grandmother had several fractured ribs, but her granddaughter had none. Explain these different findings.

38. The pediatrician explains to parents of a newborn that their son suffers from cleft palate. She tells them that the normal palate fuses in an anterior-to-posterior pattern. The child's palatine processes of the maxilla have not fused. Have his palatine bones fused normally?

39. After having a severe cold accompanied by nasal congestion, Nicole complained that she had a frontal headache and that the right side of her face ached. What bony structures probably became infected by the bacteria or viruses causing the cold?

40. Deborah, a 75-year-old woman, stumbled slightly while walking, then felt a terrible pain in her left hip. At the hospital, X-ray images revealed that the hip was broken. Also, the compact bone and spongy bone throughout her spine were very thin. What was her probable condition?

41. At work, a box fell from a shelf onto Ella's acromial region. In the emergency room, the physician felt that the head of her humerus had moved into the axilla. What had happened to Ella?

42. An X-ray image of the arm of an accident victim reveals a faint line curving around and down the shaft. What kind of fracture might this indicate?

43. Bone X-ray studies are sometimes used to determine whether a person has reached his or her final height. What are the clinicians checking out?

44. A patient complains of pain starting in the jaw and radiating down the neck. When he is questioned further, he states that when he is under stress he grinds his teeth. What joint is causing his pain?

45. 73-year-old Alice has developed an abnormal spinal curvature that has progressed into a hump. What could have led to this condition?

46. Mary, a 70-year-old woman, has been experiencing severe pain, swelling, and stiffness in her hip joint. Tests reveal Mary has osteoarthritis in her hip joint. Which treatment can help her?

# 6

# The Muscular System

**WHAT**

Muscles are responsible for body movements, stabilizing joints, and generating heat.

**HOW**

Muscles generate the force required to cause movement by contracting, a process in which proteins inside the muscle fibers overlap more than when they are at rest.

**WHY**

In addition to whole body movements, muscles are needed to move substances inside our bodies: for example, air into and out of our lungs, food through our digestive tracts, and blood through our heart and blood vessels.

**INSTRUCTORS**

New **Building Vocabulary Coaching** Activities for this chapter are assignable in Mastering **A&P°**

Because flexing muscles look like mice scurrying beneath the skin, a scientist long ago dubbed them *muscles*, from the Latin word *mus*, meaning "little mouse." Indeed, the rippling muscles of professional athletes often come to mind when we hear the word *muscle*. But muscle is also the dominant tissue in the heart and in the walls of other hollow organs of the body such as the intestines and blood vessels, and it makes up nearly half the body's mass.

The essential function of muscle is to *contract*, or *shorten*—a unique characteristic that sets it apart from other body tissues. As a result of this ability, muscles are responsible for all body movements and can be viewed as the "machines" of the body.

## Overview of Muscle Tissues

### → Learning Objectives

☐ **Describe similarities and differences in the structure and function of the three types of muscle tissue, and indicate where they are found in the body.**

☐ **Define** *muscular system.*

☐ **Define and explain the role of the following:** *endomysium, perimysium, epimysium, tendon,* **and** *aponeurosis.*

### Muscle Types

There are three types of muscle tissue—skeletal, smooth, and cardiac (**Table 6.1**, p. 208). These differ

Table **6.1**  **Comparison of Skeletal, Cardiac, and Smooth Muscles**

| Characteristic | Skeletal | Cardiac | Smooth |
|---|---|---|---|
| Body location | Attached to bones or, for some facial muscles, to skin | Walls of the heart | Mostly in walls of hollow visceral organs (other than the heart) |
| Cell shape and appearance | Single, very long, cylindrical, multinucleate cells with very obvious striations | Branching chains of cells; uninucleate, striations; intercalated discs | Single, fusiform, uninucleate; no striations |
| Connective tissue components | Epimysium, perimysium, and endomysium | Endomysium attached to the fibrous skeleton of the heart | Endomysium |
| Regulation of contraction | Voluntary; via nervous system controls | Involuntary; the heart has a pacemaker; also nervous system controls; hormones | Involuntary; nervous system controls; hormones, chemicals, stretch |
| Speed of contraction | Slow to fast | Slow | Very slow |
| Rhythmic contraction | No | Yes | Yes, in some |

in their cell structure, body location, and how they are stimulated to contract. But before we explore their differences, let's look at how they are similar.

First, skeletal and smooth muscle cells are elongated. For this reason, these types of muscle cells (but not cardiac muscle cells) are called **muscle fibers**. Second, the ability of muscle to shorten, or contract, depends on two types of *myofilaments*, the muscle cell equivalents of the microfilaments of the cytoskeleton (studied in Chapter 3). A third similarity has to do with terminology. Whenever you see the prefixes *myo-* or *mys-* ("muscle") or *sarco-* ("flesh"), you will know that muscle is being referred to. For example, in muscle cells, the cytoplasm is called *sarcoplasm* (sar′ko-plaz″um).

### Skeletal Muscle

**Skeletal muscle fibers** are packaged into organs called *skeletal muscles* that attach to the skeleton. As the skeletal muscles cover our bone and cartilage framework, they help form the smooth contours of the body. Skeletal muscle fibers are large, cigar-shaped, multinucleate cells. They are the largest muscle fibers—some ranging up to 30 cm (nearly 1 foot) in length. Indeed, the fibers of large, hardworking muscles, such as the antigravity muscles of the hip, are so big and coarse that they can be seen with the naked eye.

Skeletal muscle is also known as **striated muscle** (because its fibers have obvious stripes) and as **voluntary muscle** (because it is the only muscle type subject to conscious control). However, it is important to recognize that skeletal muscles can be activated by reflexes (without our "willed command") as well. Skeletal muscle tissue can contract rapidly and with great force, but it tires easily and must rest after short periods of activity. When you think of skeletal muscle tissue, the key words to remember are *skeletal*, *striated*, and *voluntary*.

Skeletal muscle fibers are soft and surprisingly fragile. Yet skeletal muscles can exert tremendous power—indeed, the force they generate while lifting a weight is often much greater than that required to lift the weight. The reason they are not ripped apart as they exert force is that connective tissue bundles thousands of their fibers together, which strengthens and supports the muscle as a whole **(Figure 6.1)**.

Each muscle fiber is enclosed in a delicate connective tissue sheath called **endomysium**

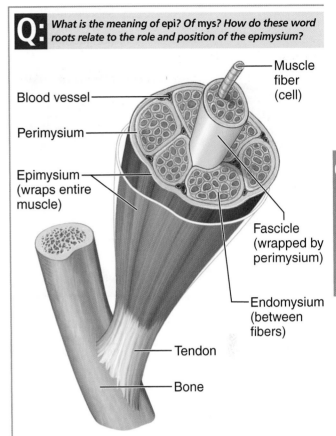

**Q:** *What is the meaning of epi? Of mys? How do these word roots relate to the role and position of the epimysium?*

Blood vessel
Perimysium
Epimysium (wraps entire muscle)
Muscle fiber (cell)
Fascicle (wrapped by perimysium)
Endomysium (between fibers)
Tendon
Bone

**Figure 6.1 Connective tissue wrappings of skeletal muscle.**

(en″do-mis′e-um). Several sheathed muscle fibers are then wrapped by a coarser fibrous membrane called **perimysium** to form a bundle of fibers called a **fascicle** (fas′ĭ-kul). Many fascicles are bound together by an even tougher "overcoat" of connective tissue called an **epimysium**, which covers the entire muscle. The ends of the epimysium that extend beyond the muscle (like the wrapper on a piece of candy) blend either into a strong, cordlike **tendon** or a sheetlike **aponeurosis** (ap″o-nu-ro′sis), which indirectly attaches the muscle to bone, cartilage, or another connective tissue covering.

In addition to anchoring muscles, tendons perform several other functions. The most important are providing durability and conserving space. Tendons are mostly tough collagen fibers, so they

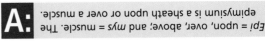

**A:** *Epi* = upon, over, above; and *mys* = muscle. The epimysium is a sheath upon or over a muscle.

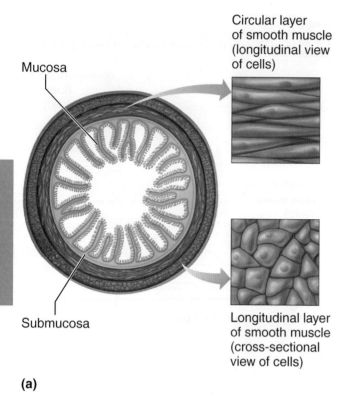

Mucosa

Circular layer
of smooth muscle
(longitudinal view
of cells)

Submucosa

Longitudinal layer
of smooth muscle
(cross-sectional
view of cells)

**(a)**

**(b)**

Cardiac
muscle
bundles

**Figure 6.2 Arrangement of smooth and cardiac muscle cells. (a)** Diagrammatic view of a cross section of the intestine. **(b)** Longitudinal view of the heart showing the spiral arrangement of the cardiac muscle cells in its walls.

can cross rough bony projections, which would tear the more delicate muscle tissues. Because of their relatively small size, more tendons than fleshy muscles can pass over a joint.

Many people think of muscles as always having an enlarged "belly" that tapers down to a tendon at each end. However, muscles vary considerably in the way their fibers are arranged. Many are spindle-shaped as just described, but in others, the fibers are arranged in a fan shape or a circle (as described on pp. 228–229).

### Smooth Muscle

**Smooth muscle** has no striations and is involuntary, which means that we cannot consciously control it. Found mainly in the walls of hollow (tubelike) visceral organs such as the stomach, urinary bladder, and respiratory passages, smooth muscle propels substances along a pathway. Think of smooth muscle as *visceral, nonstriated,* and *involuntary.*

Smooth muscle fibers are spindle-shaped, uninucleate, and surrounded by scant endomysium (see Table 6.1). They are arranged in layers, and most often there are two such layers, one running circularly and the other longitudinally **(Figure 6.2a)**. As the two layers alternately contract and relax, they change the size and shape of the organ. Moving food through the digestive tract and emptying the bowels and bladder are examples of "housekeeping" activities normally handled by smooth muscles. Smooth muscle contraction is slow and sustained. To use a running analogy, if skeletal muscle is like a sprinter, who runs fast but tires quickly, then smooth muscle is like a marathoner, who runs more slowly but keeps up the pace for many miles.

### Cardiac Muscle

**Cardiac muscle** is found in only one place in the body—the heart, where it forms the bulk of the heart walls. The heart serves as a pump, propelling blood through blood vessels to all body tissues. Like skeletal muscle, cardiac muscle is striated, and like smooth muscle, it is uninucleate and its control is involuntary. Important key words for this muscle type are *cardiac, striated,* and *involuntary.*

The cardiac cells are cushioned by small amounts of endomysium and are arranged in spiral or figure 8–shaped bundles (Figure 6.2b). When the heart contracts, its internal chambers become smaller, forcing blood into the large arteries leaving the heart. Cardiac muscle fibers are branching cells joined by special gap junctions called *intercalated discs* (see Figure 3.20 on p. 125 and Chapter 3, p. 124). These two structural features and the spiral arrangement of the muscle bundles in the heart allow heart activity to be closely coordinated.

Cardiac muscle usually contracts at a fairly steady rate set by the heart's "in-house" pacemaker. However, the nervous system can also stimulate the heart to shift into "high gear" for short periods, as when you run to catch a bus.

As you can see, each of the three muscle types has a structure and function well suited for its job in the body. But because the term *muscular system* applies specifically to skeletal muscle, we will concentrate on this muscle type in this chapter.

## Muscle Functions

All muscle types produce movement, but skeletal muscle plays three other important roles in the body as well: it *maintains posture and body position, stabilizes joints*, and *generates heat*. Let's take a look.

### Produce Movement

Skeletal muscles are responsible for our body's mobility, including all locomotion (walking, swimming, and cross-country skiing, for instance) and manipulating things with your agile upper limbs. They enable us to respond quickly to changes in the external environment. For example, their speed and power enable us to jump out of the way of a runaway car and then follow its flight with our eyes. They also allow us to express our emotions with the silent language of smiles and frowns.

They are distinct from the smooth muscle of blood vessel walls and cardiac muscle of the heart, which work together to circulate blood and maintain blood pressure, and the smooth muscle of other hollow organs, which forces fluids (urine, bile) and other substances (food, a baby) through internal body channels.

### Maintain Posture and Body Position

We are rarely aware of the workings of the skeletal muscles that maintain body posture. Yet they function almost continuously, making one tiny adjustment after another so that we maintain an erect or seated posture, even when we slouch, despite the never-ending downward pull of gravity.

### Stabilize Joints

As skeletal muscles pull on bones to cause movements, they also stabilize the joints of the skeleton. Muscles and tendons are extremely important in reinforcing and stabilizing joints that have poorly articulating surfaces, such as the shoulder and knee joints. In fact, physical therapy for knee injuries includes exercise to strengthen thigh muscles because they support the knee.

### Generate Heat

Muscle activity generates body heat as a by-product. As ATP is used to power muscle contraction, nearly three-quarters of its energy escapes as heat. This heat is vital in maintaining normal body temperature. Skeletal muscle accounts for at least 40 percent of body mass, so it is the muscle type most responsible for generating heat.

### Additional Functions

Muscles perform other important functions as well. Smooth muscles form valves that regulate the passage of substances through internal body openings, dilate and constrict the pupils of our eyes, and make up the arrector pili muscles that cause our hairs to stand on end. Skeletal muscles form valves that are under voluntary control, and they enclose and protect fragile internal organs.

### Did You Get It?

1. How do cells of the three types of muscle tissues differ from one another anatomically?
2. Which muscle type has the most elaborate connective tissue wrappings?
3. What does striated mean relative to muscle cells?
4. How do the movements promoted by skeletal muscle differ from those promoted by smooth or cardiac muscle?

**For answers, see Appendix A.**

## Microscopic Anatomy of Skeletal Muscle

→ **Learning Objective**

☐ Describe the microscopic structure of skeletal muscle, and explain the role of actin- and myosin-containing myofilaments.

As mentioned previously, skeletal muscle fibers (cells) are multinucleate (**Figure 6.3a**, p. 212). Many oval nuclei can be seen just beneath the plasma membrane, which is called the **sarcolemma** (sar″ko-lem′ah;

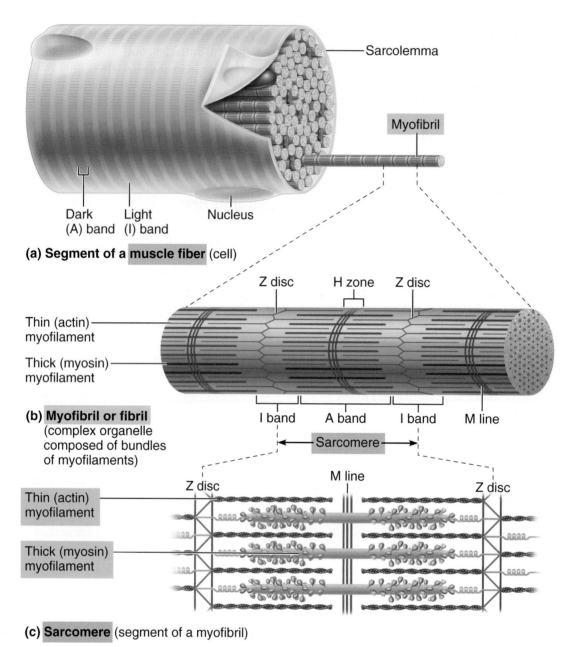

**(a) Segment of a muscle fiber** (cell)

**(b) Myofibril or fibril** (complex organelle composed of bundles of myofilaments)

**(c) Sarcomere** (segment of a myofibril)

**Figure 6.3 Anatomy of a skeletal muscle fiber (cell). (a)** A portion of a muscle fiber. One myofibril has been extended. **(b)** Enlarged view of a section of a myofibril showing its banding pattern. **(c)** Enlarged view of one sarcomere (contractile unit) of a myofibril.

"muscle husk") in muscle fibers. The nuclei are pushed aside by long ribbonlike organelles, the **myofibrils** (mi″o-fi′brilz), which nearly fill the cytoplasm. Alternating **light (I) bands** and **dark (A) bands** along the length of the perfectly aligned myofibrils give the muscle fiber its striated (banded) appearance. (Think of the second letter of *light*, I, and the second letter of *dark*, A, to help you remember which band is which.) A closer look at the banding pattern reveals that the light I band has

a midline interruption, a darker area called the *Z disc*, and the dark A band has a lighter central area called the *H zone* (Figure 6.3b). The *M line* in the center of the H zone contains tiny protein rods that hold adjacent thick filaments together.

So why are we bothering with all these terms— dark this and light that? Because the banding pattern reveals the working structure of the myofibrils. First, we find that the myofibrils are actually chains of tiny contractile units called **sarcomeres** (sar′ko-mērz), which are the structural and functional units of skeletal muscle. The sarcomeres are aligned end to end like boxcars in a train along the length of the myofibrils. Second, it is the precise arrangement of even smaller structures (myofilaments) *within* sarcomeres that produces the striations in skeletal muscle fibers.

Let's examine how the arrangement of the myofilaments leads to the banding pattern. There are two types of threadlike protein **myofilaments** within each sarcomere (Figure 6.3c). The **thick filaments** are made mostly of bundled molecules of the protein **myosin**, but they also contain ATPase enzymes, which split ATP to release the energy used for muscle contraction. Notice that the thick filaments extend the entire length of the dark A band. Also, notice that the midparts of the thick filaments are smooth, but their ends are studded with small projections (Figure 6.3c). These projections, or myosin *heads*, form **cross bridges** when they link the thick and thin filaments together during contraction. Myosin filaments are attached to the Z discs by *titin*, elastic filaments that run through the core of the thick filament.

The **thin filaments** are composed of the contractile protein called **actin**, plus some regulatory proteins that play a role in allowing (or preventing) binding of myosin heads to actin. The thin filaments are anchored to the Z disc (a disclike membrane). Notice that the light I band includes parts of two adjacent sarcomeres and contains *only* the thin filaments. Although they overlap the ends of the thick filaments, the thin filaments do not extend into the middle of a relaxed sarcomere, and thus the central region (the H zone) looks a bit lighter. When the actin-containing thin filaments slide toward each other during contraction, the H zones disappear because the actin and myosin filaments completely overlap.

Another very important muscle fiber organelle—the **sarcoplasmic reticulum (SR)**—is a specialized smooth endoplasmic reticulum (not shown in Figure 6.3). The interconnecting tubules and sacs of the SR surround every myofibril just as the sleeve of a loosely crocheted sweater surrounds your arm. The major role of this elaborate system is to store calcium and to release it on demand when the muscle fiber is stimulated to contract. As you will see, calcium provides the final "go" signal for contraction.

# Skeletal Muscle Activity

## Stimulation and Contraction of Single Skeletal Muscle Fibers

### → **Learning** Objective

☐ **Describe how an action potential is initiated in a muscle cell.**

Muscle fibers have several special functional properties that enable them to perform their duties. The first of these is *irritability*, also termed *responsiveness*, which is the ability to receive and respond to a stimulus. The second, *contractility*, is the ability to forcibly shorten when adequately stimulated. This property sets muscle apart from all other tissue types. *Extensibility* is the ability of muscle fibers to stretch, whereas *elasticity* is their ability to recoil and resume their resting length after being stretched.

### The Nerve Stimulus and the Action Potential

To contract, skeletal muscle fibers must be stimulated by nerve impulses. One motor neuron (nerve cell) may stimulate a few muscle fibers or hundreds of them, depending on the particular muscle and the work it does. A **motor unit** consists of one neuron and all the skeletal muscle fibers it stimulates (**Figure 6.4**, p. 214). When a long, threadlike extension of the neuron, called the **axon**, reaches the muscle, it branches into a number of **axon terminals**, each of which forms junctions with the sarcolemma of a different muscle cell (**Figure 6.5**, p. 215). These junctions, called **neuromuscular** (literally, "nerve-muscle") **junctions**, contain *synaptic vesicles* filled with a chemical referred to as a **neurotransmitter**. The specific neurotransmitter that stimulates skeletal muscle fibers is **acetylcholine** (as″e-til-ko′lēn), or **ACh**. Although the nerve endings and the muscle fiber membranes are very close, they never touch. The gap between them, the **synaptic cleft**, is filled with interstitial fluid.

### ⚖ Homeostatic Imbalance 6.1

In some cases, a motor nerve impulse is unable to reach the muscle. In **ALS**, or **amyotrophic lateral**

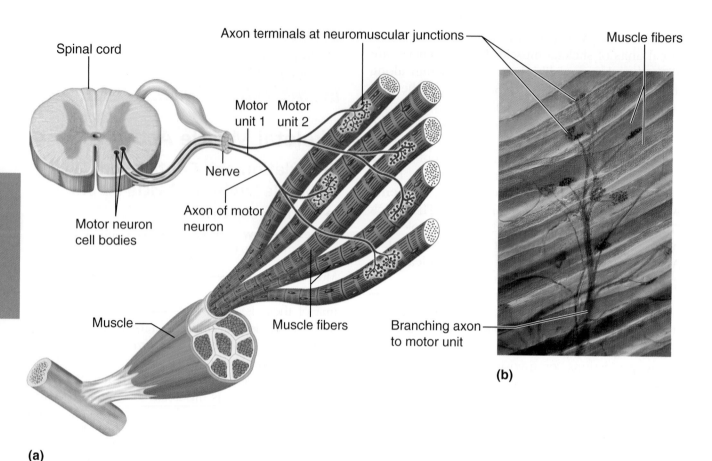

**Figure 6.4 Motor units.** Each motor unit consists of a motor neuron and all the muscle fibers it activates. **(a)** Portions of two motor units are shown. The motor neurons reside in the spinal cord, and their axons extend to the muscle. Within the muscle, each axon divides into a number of axon terminals distributed to muscle fibers scattered throughout the muscle. **(b)** Photo of a portion of a motor unit (1150×).

**sclerosis** (also called *Lou Gehrig's disease*), motor neurons degenerate over time, resulting in paralysis that gradually worsens. The cause of ALS is unknown, but common characteristics include malfunctioning mitochondria, inflammation, and the generation of free radicals that damage DNA and tissue much like intense UV light. The prognosis for patients with ALS is generally death within three to five years because the breathing muscles will eventually be affected, resulting in suffocation. _____ ✚

## Did You Get It?

6. What two structures are closely associated at a neuromuscular junction?

For the answer, see Appendix A.

Now that we have described the structure of the neuromuscular junction, refer to the numbered steps in Figure 6.5 as we examine what happens there. When a nerve impulse reaches the axon terminals ①, calcium channels open, and calcium

($Ca^{2+}$) enters the terminal ②. Calcium entry causes some of the synaptic vesicles in the axon terminal to fuse with the cell membrane and release acetylcholine ③, which then diffuses across the synaptic cleft and attaches to membrane receptors in highly folded regions of the sarcolemma ④. If enough acetylcholine is released, the sarcolemma at that point becomes *temporarily* even more permeable to sodium ions ($Na^+$), which rush into the muscle fiber, and to potassium ions ($K^+$), which diffuse out of the muscle fiber. However, more $Na^+$ enters than $K^+$ leaves. This imbalance gives the cell interior an excess of positive ions, which reverses the resting electrical conditions of the sarcolemma. This event, called *depolarization*, opens more channels that only allow $Na^+$ entry ⑤. This movement of ions generates an electrical current called an **action potential**. Once begun, the action potential is unstoppable; it travels over the entire surface of the sarcolemma, conducting the electrical

6

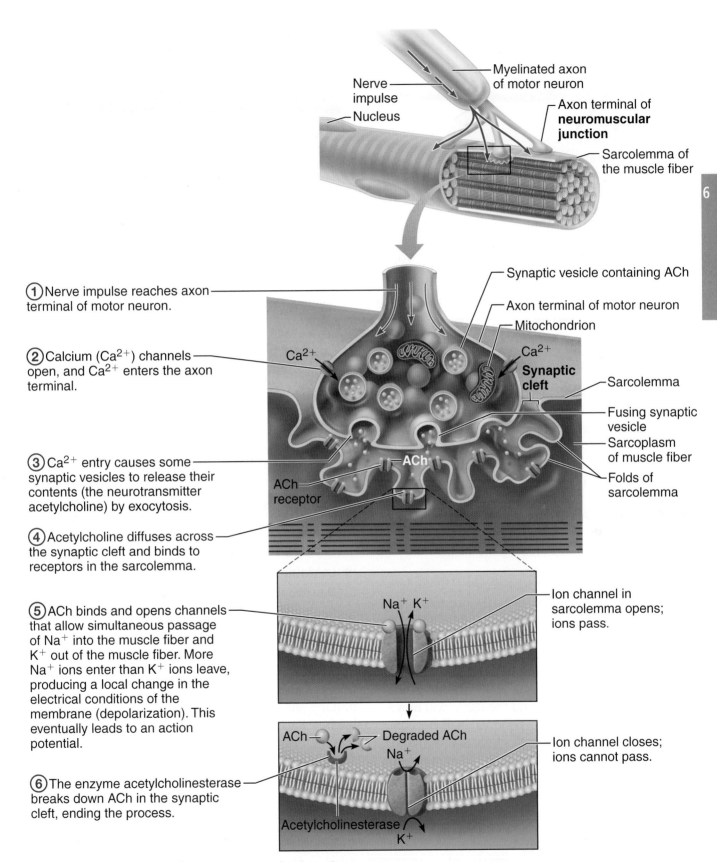

Nerve impulse

Nucleus

Myelinated axon of motor neuron

Axon terminal of **neuromuscular junction**

Sarcolemma of the muscle fiber

① Nerve impulse reaches axon terminal of motor neuron.

Synaptic vesicle containing ACh

Axon terminal of motor neuron

Mitochondrion

② Calcium ($Ca^{2+}$) channels open, and $Ca^{2+}$ enters the axon terminal.

$Ca^{2+}$

$Ca^{2+}$

**Synaptic cleft**

Sarcolemma

Fusing synaptic vesicle

Sarcoplasm of muscle fiber

Folds of sarcolemma

③ $Ca^{2+}$ entry causes some synaptic vesicles to release their contents (the neurotransmitter acetylcholine) by exocytosis.

ACh

ACh receptor

④ Acetylcholine diffuses across the synaptic cleft and binds to receptors in the sarcolemma.

⑤ ACh binds and opens channels that allow simultaneous passage of $Na^+$ into the muscle fiber and $K^+$ out of the muscle fiber. More $Na^+$ ions enter than $K^+$ ions leave, producing a local change in the electrical conditions of the membrane (depolarization). This eventually leads to an action potential.

$Na^+$  $K^+$

Ion channel in sarcolemma opens; ions pass.

ACh

Degraded ACh

$Na^+$

Ion channel closes; ions cannot pass.

⑥ The enzyme acetylcholinesterase breaks down ACh in the synaptic cleft, ending the process.

Acetylcholinesterase

$K^+$

**Figure 6.5 Events at the neuromuscular junction.**

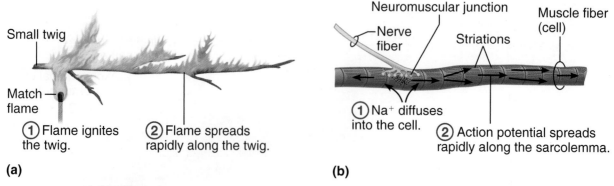

**(a)**

**(b)**

**Figure 6.6 Comparing the action potential to a flame consuming a dry twig. (a)** The first event in igniting a dry twig is holding the match flame under one area of the twig. The second event is the twig's bursting into flame when it has been heated enough and spreading of the flame to burn the entire twig. **(b)** The first event in exciting a muscle fiber is the rapid diffusion of sodium ions (Na⁺) into the cell when the permeability of the sarcolemma changes. The second event is the spreading of the action potential along the sarcolemma when enough sodium ions have entered to upset the electrical conditions in the cell.

impulse from one end of the cell to the other. The result is contraction of the muscle fiber.

Note that while the action potential is occurring, the enzyme acetylcholinesterase (AChE), present on the sarcolemma and in the synaptic cleft, breaks down acetylcholine to acetic acid and choline ⑥. For this reason, a single nerve impulse produces only one contraction. This prevents continued contraction of the muscle fiber in the absence of additional nerve impulses. The muscle fiber relaxes until stimulated by the next round of acetylcholine release.

Let's compare this series of events to lighting a match under a small dry twig **(Figure 6.6)**. The charring of the twig by the flame can be compared to the change in membrane permeability that allows sodium ions into the cell. When that part of the twig becomes hot enough (when enough sodium ions have entered the cell), the twig will suddenly burst into flame, and the flame will move along the twig (the action potential will be conducted along the entire length of the sarcolemma). We explain this series of events more fully in the discussion of nerve physiology (Chapter 7, pp. 260–265).

The events that return the cell to its resting state include (1) diffusion of potassium ions (K⁺) out of the cell and (2) operation of the sodium-potassium pump, the active transport mechanism that moves the sodium and potassium ions back to their initial positions.

## Did You Get It?

7. Which ions enter the muscle cell during the generation of an action potential?
8. What is the role of calcium ions in muscle contraction?

For answers, see Appendix A.

### Mechanism of Muscle Contraction: The Sliding Filament Theory

→ **Learning** Objective

☐ **Describe the events of muscle cell contraction.**

What causes the filaments to slide? This question brings us back to the myosin heads that protrude all around the ends of the thick filaments.

When the nervous system activates muscle fibers as just described, the myosin heads attach to binding sites on the thin filaments, and the sliding begins. Each cross bridge attaches and detaches several times during a contraction, generating tension that helps pull the thin filaments toward the center of the sarcomere. This "walking" of the myosin cross bridges, or heads, along the thin filaments during muscle shortening is much like a centipede's gait. Some myosin heads ("legs") are always in contact with actin ("the ground"), so that the thin filaments cannot slide backward, and this cycle repeats again and again during contraction. As this event occurs simultaneously in sarcomeres throughout the muscle fiber, the cell shortens **(Figure 6.7)**. Notice that the myofilaments themselves do not shorten during contraction; they simply slide past each other.

The formation of cross bridges—when the myosin heads attach to actin—requires calcium ions ($Ca^{2+}$) and ATP (to "energize" the myosin heads). So where does the calcium come from? Action potentials pass deep into the muscle fiber along membranous tubules that fold inward from the sarcolemma. Inside the cell, the action potentials stimulate the sarcoplasmic reticulum to release calcium ions into the cytoplasm. The calcium ions trigger the binding of myosin to actin, initiating filament sliding (**Figure 6.8**, p. 218). When the action potential ends, calcium ions are immediately returned to the SR storage areas, the regulatory proteins return to their resting shape and block myosin-binding sites, and the muscle fiber relaxes and settles back to its original length. This whole series of events takes a few thousandths of a second.

### Did You Get It?

9. Which chemical—ATP or $Ca^{2+}$—triggers sliding of the muscle filaments?
10. Which is a cross-bridge attachment more similar to: a synchronized rowing team or a person pulling a bucket on a rope out of a well?

For answers, see Appendix A.

## Contraction of a Skeletal Muscle as a Whole

→ **Learning Objective**

☐ Define *graded response, tetanus, isotonic* and *isometric contractions,* and *muscle tone* as these terms apply to a skeletal muscle.

### Graded Responses

In skeletal muscles, the "all-or-none" law of muscle physiology applies to the *muscle fiber*, not to the whole muscle. It states that a muscle fiber will contract to its fullest extent when it is stimulated adequately; it never partially contracts. However, the whole muscle reacts to stimuli with **graded responses**, or different degrees of shortening, which generate different amounts of force. In general, graded muscle contractions can be produced two ways: (1) by changing the *frequency* of muscle stimulation and (2) by changing the *number* of

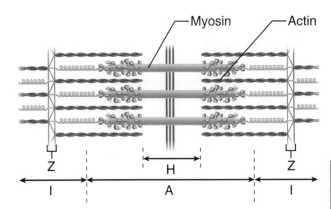

**(a) Relaxed sarcomere**

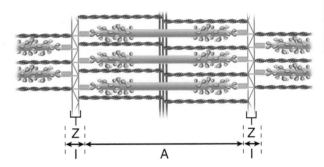

**(b) Fully contracted sarcomere**

**Figure 6.7 Diagrammatic views of a sarcomere.** Notice that in the contracted sarcomere **(b)**, the light H zone in the center of the A band has disappeared, the Z discs are closer to the thick filaments, and the I bands have nearly disappeared. The A bands of adjacent sarcomeres move closer together but do not change in length.

muscle fibers being stimulated at one time. Next, let's describe a muscle's response to each of these.

***Muscle Response to Increasingly Rapid Stimulation***
Although **muscle twitches** (single, brief, jerky contractions) sometimes result from certain nervous system problems, this is *not* the way our muscles normally operate. In most types of muscle activity, nerve impulses are delivered to the muscle at a very rapid rate—so rapid that the muscle does not get a chance to relax completely between stimuli. As a result, the effects of the successive contractions are "summed" (added) together, and the contractions of the muscle get stronger and smoother. The muscle exhibits **unfused tetanus** (tet′ah-nus), or **incomplete tetanus**. When the muscle is stimulated so rapidly that no evidence of relaxation is seen and the contractions are completely smooth and sustained, the muscle is in

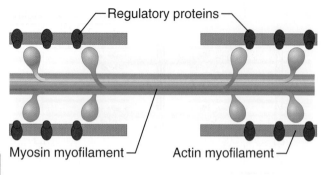

**(a)**

In a relaxed muscle fiber, the regulatory proteins forming part of the actin myofilaments prevent myosin binding (see **a**). When an action potential (AP) sweeps along its sarcolemma and a muscle fiber is excited, calcium ions ($Ca^{2+}$) are released from intracellular storage areas (the sacs of the sarcoplasmic reticulum).

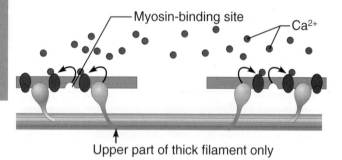

**(b)**

The flood of calcium acts as the final trigger for contraction, because as calcium binds to the regulatory proteins on the actin filaments, the proteins undergo a change in both their shape and their position on the thin filaments. This action exposes myosin-binding sites on the actin, to which the myosin heads can attach (see **b**), and the myosin heads immediately begin seeking out binding sites.

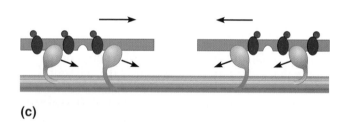

**(c)**

The free myosin heads are "cocked," much like an oar ready to be pulled on for rowing. Myosin attachment to actin causes the myosin heads to snap (pivot) toward the center of the sarcomere in a rowing motion. When this happens, the thin filaments are slightly pulled toward the center of the sarcomere (see **c**). ATP provides the energy needed to release and recock each myosin head so that it is ready to attach to a binding site farther along the thin filament.

**Figure 6.8 Schematic representation of contraction mechanism: the sliding filament theory.**

Play A&P Flix Animation
Mastering A&P®

**fused tetanus**, or **complete tetanus**, or in tetanic contraction* **(Figure 6.9)**.

***Muscle Response to Stronger Stimuli***   Tetanus produces stronger (more forceful) muscle contractions, but its primary role is to produce smooth and prolonged muscle contractions. How forcefully a muscle contracts depends to a large extent on how many of its cells are stimulated. When only a few cells are stimulated, the muscle as a whole contracts only slightly. When all the motor units are active and all the muscle fibers are stimulated, the muscle contraction is as strong as it can get. Thus, muscle contractions can range from slight to vigorous depending on the work to be done. The same hand that lifts a single sheet of paper can also lift a heavy backpack full of books!

---

* Tetanic contraction is normal and desirable. It is quite different from the pathological condition of tetanus (commonly called *lockjaw*), which is caused by a toxin made by bacteria. Lockjaw causes muscles to go into uncontrollable spasms, finally causing respiratory arrest.

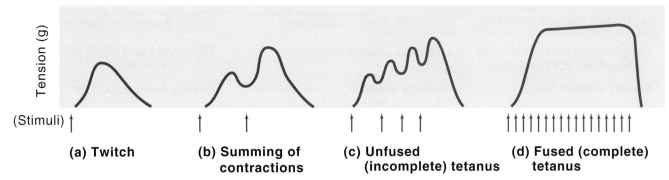

**(a) Twitch**    **(b) Summing of contractions**    **(c) Unfused (incomplete) tetanus**    **(d) Fused (complete) tetanus**

**Figure 6.9  A whole muscle's response to different stimulation rates.** A single stimulus **(a)** causes muscle contraction and relaxation (a twitch). As stimuli are delivered more frequently **(b)**, the muscle does not have time to completely relax before the next stimulus; contraction force increases because effects of the individual twitches are summed (added). Further fusion **(c)** of the twitches (unfused tetanus) occurs as stimuli are delivered at a still faster rate. Fused tetanus **(d)**, a smooth continuous contraction without any evidence of relaxation, results from a very rapid rate of stimulation. (Points at which stimuli are delivered are indicated by red arrows. Tension [measured in grams] on the vertical axis refers to the relative force of muscle contraction.)

## Providing Energy for Muscle Contraction

### → Learning Objective

☐ **Describe three pathways for ATP regeneration during muscle activity.**

As a muscle contracts, the bonds of ATP molecules are hydrolyzed to release the needed energy.

### CONCEPT**LINK**

Recall that ATP can be compared to a tightly coiled spring that is ready to uncoil with tremendous energy when the "catch" is released (Chapter 2, p. 81). Remember that all bonds store energy and that the "catch" in this example is one of the characteristic high-energy phosphate bonds in ATP. ←

Surprisingly, muscles store very limited supplies of ATP—only a few seconds' worth, just enough to get you going. Because ATP is the *only* energy source that can be used *directly* to power muscle activity, ATP must be regenerated continuously if contraction is to continue.

Working muscles use three pathways to regenerate ATP:

• **Direct phosphorylation of ADP by creatine phosphate** (**Figure 6.10a**, p. 220). The unique high-energy molecule **creatine phosphate (CP)** is found in muscle fibers but not other cell types. As ATP is depleted, interactions between CP and ADP result in transfers of a high-energy phosphate group from CP to ADP, thus regenerating more ATP in a fraction of a second. Although muscle fibers store perhaps five times as much CP as ATP, the CP supplies are also soon exhausted (in less than 15 seconds).

• **Aerobic pathway** (Figure 6.10b). At rest and during light to moderate exercise, some 95 percent of the ATP used for muscle activity comes from aerobic respiration. **Aerobic respiration** occurs in the mitochondria and involves a series of metabolic pathways that use oxygen. These pathways are collectively referred to as *oxidative phosphorylation*. During aerobic respiration, glucose is broken down completely to carbon dioxide and water, and some of the energy released as the bonds are broken is captured in the bonds of ATP molecules. Although aerobic respiration provides a rich ATP harvest (about 32 ATP per 1 glucose), it is fairly slow and requires continuous delivery of oxygen and nutrient fuels to the muscle to keep it going.

• **Anaerobic glycolysis and lactic acid formation** (Figure 6.10c). The initial steps of glucose breakdown occur via a pathway called *glycolysis*, which does not use oxygen and hence is *anaerobic* (literally "without oxygen"). During glycolysis, which occurs in the cytosol, glucose is broken down to pyruvic acid, and small amounts of energy are captured in ATP bonds (2 ATP per 1 glucose molecule). As long as enough oxygen is present, the pyruvic

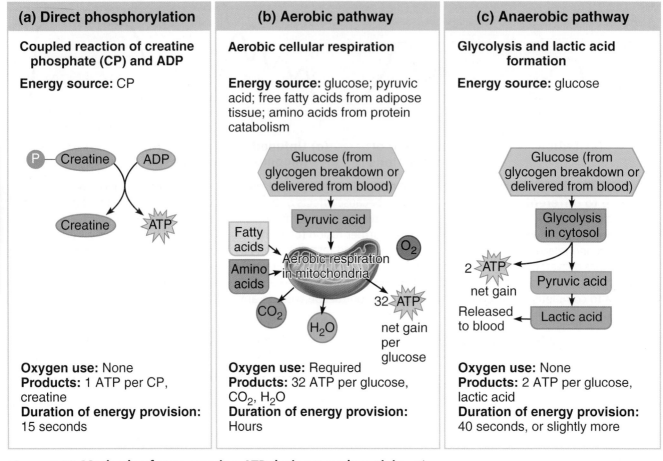

**Figure 6.10 Methods of regenerating ATP during muscle activity.** The fastest mechanism is **(a)** direct phosphorylation; the slowest is **(b)** aerobic respiration.

acid then enters the oxygen-requiring aerobic pathways that occur within the mitochondria to produce more ATP as described above. However, when muscle activity is intense, or oxygen and glucose delivery is temporarily inadequate to meet the needs of working muscles, the sluggish aerobic pathways cannot keep up with the demands for ATP. Under these conditions, the pyruvic acid generated during glycolysis is converted to **lactic acid**, and the overall process is referred to as **anaerobic glycolysis**.

Anaerobic glycolysis produces only about 5 percent as much ATP from each glucose molecule as aerobic respiration. However, it is some 2½ times faster, and it can provide most of the ATP needed for 30 to 40 seconds of strenuous muscle activity. Anaerobic glycolysis has two main shortcomings: it uses huge amounts of glucose for a small ATP harvest, and the accumulating lactic acid promotes muscle soreness.

**Did You Get It?**

11. What are the three processes used to generate energy for skeletal muscle contraction?
12. What is the direct source of energy used by muscle fibers for contraction?

For answers, see Appendix A.

### Muscle Fatigue and Oxygen Deficit

→ **Learning Objective**

☐ Define *oxygen deficit* and *muscle fatigue*, and list possible causes of muscle fatigue.

If we exercise our muscles strenuously for a long time, **muscle fatigue** occurs. A muscle is fatigued when it is unable to contract even though it is still being stimulated. Without rest, a working muscle begins to tire and contracts more weakly until it finally ceases reacting and stops contracting.

Factors that contribute to muscle fatigue are not fully known. Suspected causes are imbalances in ions ($Ca^{2+}$, $K^+$) and problems at the neuromuscular junction. However, many agree that the major factor is the **oxygen deficit** that occurs during prolonged muscle activity. Oxygen deficit is not a total lack of oxygen; rather, it happens when a person is not able to take in oxygen fast enough to keep the muscles supplied with all the oxygen they need when they are working vigorously. Obviously, then, the work that a muscle can do and how long it can work without becoming fatigued depend on how good its blood supply is. When muscles lack sufficient oxygen for aerobic respiration, lactic acid begins to accumulate in the muscle via the anaerobic pathway. We can recognize this event by the burning sensation we experience. In addition, the muscle's ATP supply starts to run low, and ionic imbalance tends to occur. Together these factors cause the muscle to contract less and less effectively and finally to stop contracting altogether.

True muscle fatigue, in which the muscle quits entirely, rarely occurs in most of us because we feel tired long before it happens and we simply slow down or stop our activity. It *does* happen in marathon runners. Many of them have literally collapsed when their muscles became fatigued and could no longer work.

Oxygen deficit, which always occurs to some extent during vigorous muscle activity, is like a loan that must be "paid back" whether fatigue occurs or not. During the recovery period after activity, the individual breathes rapidly and deeply. This continues until the muscles have received the amount of oxygen needed to get rid of the accumulated lactic acid and replenish ATP and creatine phosphate reserves.

### Types of Muscle Contractions—Isotonic and Isometric

Until now, we have been discussing contraction in terms of shortening, but muscles do not always shorten when they contract. (I can hear you saying, "What kind of double-talk is that?"—but pay attention.) The event that is common to all muscle contractions is that *tension* (force) develops in the muscle as the actin and myosin myofilaments interact and the myosin cross bridges attempt to slide the thin actin-containing filaments past the thick myosin filaments.

**Isotonic contractions** (literally, "same tone" or tension) are familiar to most of us. In isotonic contractions, the myofilaments are successful in their sliding movements, the muscle shortens, and

movement occurs. Bending the knee, lifting weights, and smiling are all examples of isotonic contractions.

Contractions in which the muscles do not shorten are called **isometric contractions** (literally, "same measurement" or length). In isometric contractions, the myosin filaments are "spinning their wheels," and the tension in the muscle keeps increasing. They are trying to slide, but the muscle is pitted against some more or less immovable object. For example, when you push the palms of your hands together in front of you, your arms and chest muscles are contracting isometrically.

### Muscle Tone

One aspect of skeletal muscle activity cannot be consciously controlled. Even when a muscle is voluntarily relaxed, some of its fibers are contracting—first one group and then another. These contractions are not visible, but thanks to them, the muscle remains firm, healthy, and constantly ready for action. This state of continuous partial contractions is called **muscle tone**. Muscle tone is the result of different motor units, which are scattered through the muscle, being stimulated by the nervous system in a systematic way. Think of these motor units as being "on duty" in case action is required.

 **Homeostatic Imbalance 6.2**

If the nerve supply to a muscle is destroyed (as in an accident), the muscle is no longer stimulated in this manner, and it loses tone. Soon after, it becomes **flaccid** (flă′sid), or soft and flabby, and begins to **atrophy** (waste away). This is called *flaccid paralysis*. Compare this with a condition that increases muscle tone until the muscle is no longer controllable—for example, the disease tetanus, which is caused by a bacterial toxin. This is called *spastic paralysis*. _____ ✚

### Effect of Exercise on Muscles

→ **Learning Objective**

☐ Describe the effects of aerobic and resistance exercise on skeletal muscles and other body organs.

The amount of work a muscle does changes the muscle. Muscle inactivity (due to a loss of nerve supply, immobilization, or whatever the cause) always leads to muscle weakness and wasting. Muscles

**(a)**　　　　**(b)**

**Figure 6.11 The effects of aerobic training versus strength training. (a)** A marathon runner. **(b)** A weight lifter.

are no exception to the saying "Use it or lose it!" Conversely, regular exercise increases muscle size, strength, and endurance. However, not all types of exercise produce these effects—in fact, there are important differences in the benefits of exercise.

**Aerobic exercise**, or **endurance exercise**, such as participating in an aerobics class, jogging, or biking **(Figure 6.11a)**, results in stronger, more flexible muscles with greater resistance to fatigue. These changes come about, at least partly, because the blood supply to the muscles increases, and the individual muscle fibers form more mitochondria and store more oxygen. Aerobic exercise helps us reach a steady rate of ATP production and improves the efficiency of aerobic respiration.

However, aerobic exercise benefits much more than the skeletal muscles. It makes overall body metabolism more efficient, improves digestion (and elimination), enhances neuromuscular coordination, and strengthens the skeleton. The heart enlarges (*hypertrophies*) and pumps out more blood with each beat, helping to clear more fat deposits from the blood vessel walls. The lungs become more efficient in gas exchange. These benefits may be permanent or temporary, depending on how often and how vigorously a person exercises.

Aerobic exercise does *not* cause the muscles to increase much in size, even though the exercise may go on for hours. The bulging muscles of a professional bodybuilder result mainly from **resistance**

**exercise**, or **isometric exercise** (Figure 6.11b), which pit the muscles against an immovable (or difficult to move) object.

Resistance exercises require very little time and little or no special equipment. A few minutes every other day is usually sufficient. You can push against a wall, and you can strongly contract buttock muscles even while standing in line at the grocery store. The key is forcing your muscles to contract with as much force as possible. The increased muscle size and strength that result are due mainly to enlargement of individual muscle fibers (they make more contractile myofilaments) rather than to an increase in their number. The amount of connective tissue that reinforces the muscle also increases.

Because endurance and resistance exercises produce different patterns of muscle response, it is important to know what your exercise goals are. Lifting weights will not improve your endurance for a marathon. By the same token, jogging will not make you stronger for lifting furniture. Obviously, the best exercise program for most people includes both types of exercise.

### Did You Get It?

13. Gary is trying with all his might to pull a tree stump out of the ground. It does not budge. Which type of contraction are his muscles performing?
14. What is meant by the term *oxygen deficit*?
15. To develop big, beautiful skeletal muscles, you should focus on which type of exercise: aerobic or resistance exercise?

**For answers, see Appendix A.**

# Muscle Movements, Roles, and Names

### → Learning Objectives

☐ **Define *origin, insertion, prime mover, antagonist, synergist*, and *fixator* as they relate to muscles.**

☐ **Demonstrate or identify the different types of body movements.**

There are five basic guidelines for understanding gross muscle activity. We refer to these as the *Five Golden Rules* of skeletal muscle activity because they make it easier to understand muscle movements and appreciate muscle interactions **(Table 6.2)**.

## Types of Body Movements

Every one of our 600-odd skeletal muscles is attached to bone, or to other connective tissue

| Table **6.2** | **The Five Golden Rules of Skeletal Muscle Activity** |
|---|---|
| 1. With a few exceptions, all skeletal muscles cross at least one joint. | |
| 2. Typically, the bulk of a skeletal muscle lies proximal to the joint crossed. | |
| 3. All skeletal muscles have at least two attachments: the origin and the insertion. | |
| 4. Skeletal muscles can only pull; they never push. | |
| 5. During contraction, a skeletal muscle insertion moves toward the origin. | |

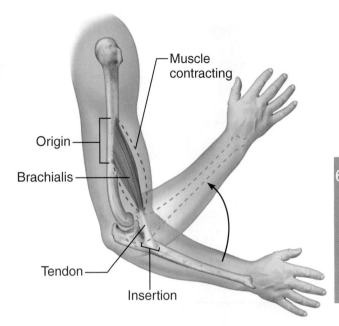

**Figure 6.12 Muscle attachments (origin and insertion).** When a skeletal muscle contracts, its insertion moves toward its origin.

structures, at no fewer than two points. One of these points, the **origin**, is attached to the immovable or less movable bone **(Figure 6.12)**. Think of the origin as the anchor, or leverage, point. Another point, the **insertion**, is attached to the movable bone. When the muscle contracts, the insertion moves toward the origin.

Some muscles have interchangeable origins and insertions, depending on the action being performed. For example, the rectus femoris muscle of the anterior thigh crosses both the hip and knee joints. Its most common action is to extend the knee, in which case the proximal pelvic attachment is the origin. However, when the knee bends (by other muscles), the rectus femoris can flex the hip, and then its distal attachment on the leg is considered the origin.

Generally speaking, body movement occurs when muscles contract across joints. The type of movement depends on the mobility of the joint and the location of the muscle in relation to the joint. The most obvious examples of the action of muscles on bones are the movements that occur at the joints of the limbs. However, less freely movable bones are also tugged into motion by the muscles, such as the vertebrae's movements when we bend to one side.

Next we describe the most common types of body movements (**Figure 6.13**, pp. 224–225). Try to act out each movement as you read the following descriptions:

- **Flexion.** Flexion is a movement, generally in the sagittal plane, that decreases the angle of the joint and brings two bones closer together (Figures 6.13a and 6.13b). Flexion is typical of hinge joints (bending the knee or elbow), but it is also common at ball-and-socket joints (for example, bending forward at the hip).

- **Extension.** Extension is the opposite of flexion, so it is a movement that increases the angle, or distance, between two bones or parts of the body (straightening the knee or elbow). Extension that is greater than 180° (as when you move your arm posteriorly beyond its normal anatomical position, or tip your head so that your chin points toward the ceiling) is called *hyperextension* (Figures 6.13a and 6.13b).

- **Rotation.** Rotation is movement of a bone around its longitudinal axis (Figure 6.13c). Rotation is a common movement of ball-and-socket joints and describes the movement of the atlas around the dens of the axis (as in shaking your head "no").

- **Abduction.** Abduction is moving a limb away (generally on the frontal plane) from the midline, or median plane, of the body (Figure 6.13d). The terminology also applies to the fanning movement of your fingers or toes when they are spread apart.

- **Adduction.** Adduction is the opposite of abduction, so it is the movement of a limb toward the body midline (Figure 6.13d). Think of adduction as "adding" a body part by bringing it closer to the trunk.

*(Text continues on page 226.)*

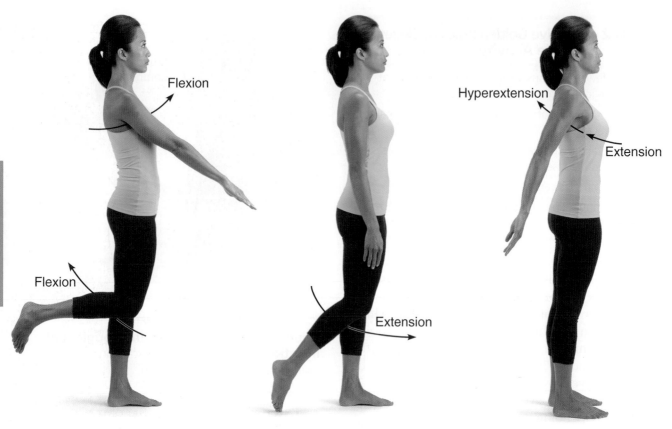

**(a) Flexion, extension, and hyperextension of the shoulder and knee**

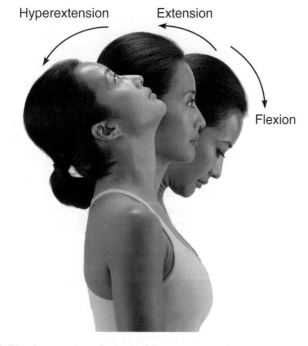

**(b) Flexion, extension, and hyperextension**

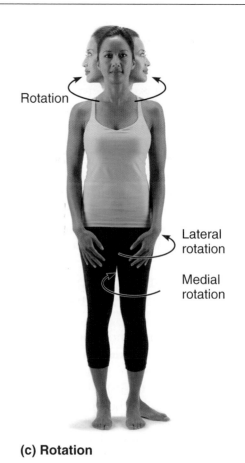

**Figure 6.13 Body movements.**

**(c) Rotation**

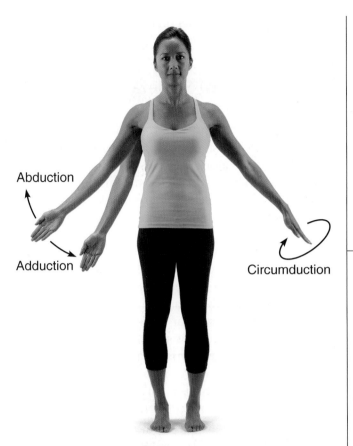

Abduction

Adduction

Circumduction

**(d) Abduction, adduction, and circumduction**

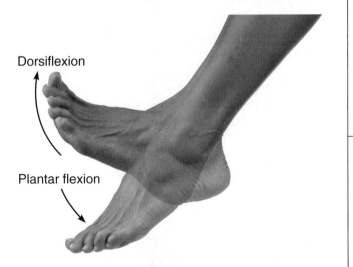

Dorsiflexion

Plantar flexion

**(e) Dorsiflexion and plantar flexion**

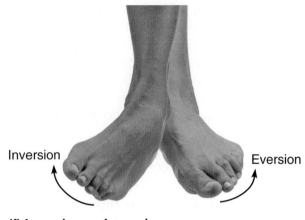

Inversion

Eversion

**(f) Inversion and eversion**

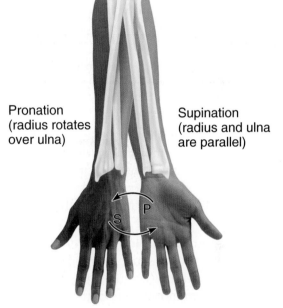

Pronation
(radius rotates
over ulna)

Supination
(radius and ulna
are parallel)

S    P

**(g) Supination (S) and pronation (P)**

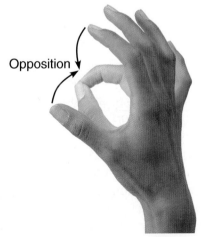

Opposition

**(h) Opposition**

**Figure 6.13** *(continued)*

- **Circumduction**. Circumduction is a combination of flexion, extension, abduction, and adduction commonly seen in ball-and-socket joints, such as the shoulder. The proximal end of the limb is stationary, and its distal end moves in a circle. The limb as a whole outlines a cone (Figure 6.13d), as when you do big arm circles.

### Special Movements

Certain movements do not fit into any of the previous categories and occur at only a few joints.

- **Dorsiflexion and plantar flexion.** Up-and-down movements of the foot at the ankle are given special names. Lifting the foot so that its superior surface approaches the shin (pointing your toe toward your head) is *dorsiflexion*, whereas pointing the toes away from your head is *plantar flexion* (Figure 6.13e). Dorsiflexion of the foot corresponds to extension and hyperextension of the hand at the wrist, whereas plantar flexion of the foot corresponds to flexion of the hand.

- **Inversion and eversion.** Inversion and eversion are also special movements of the foot (Figure 6.13f). To invert the foot, turn the sole medially, as if you were looking at the bottom of your foot. To evert the foot, turn the sole laterally.

- **Supination and pronation.** The terms *supination* (soo″pĭ-na′shun; "turning backward") and *pronation* (pro-na′shun; "turning forward") refer to movements of the radius around the ulna (Figure 6.13g). Supination occurs when the forearm rotates laterally so that the palm faces anteriorly (or up) and the radius and ulna are parallel, as in anatomical position. Pronation occurs when the forearm rotates medially so that the palm faces posteriorly (or down). Pronation brings the radius across the ulna so that the two bones form an X. A helpful memory trick: If you lift a cup of soup up to your mouth *on your palm*, you are supinating ("soup"-inating).

- **Opposition.** In the palm of the hand, the saddle joint between metacarpal 1 and the carpals allows opposition of the thumb (Figure 6.13h). This is the action by which you move your thumb to touch the tips of the other fingers on the same hand. This unique action makes the human hand a fine tool for grasping and manipulating objects.

## Interactions of Skeletal Muscles in the Body

Muscles can't push—they can only pull as they contract—so most often body movements result from two or more muscles acting together or against each other. Muscles are arranged so that whatever one muscle (or group of muscles) can do, other muscles can reverse. In general, groups of muscles that produce opposite movements lie on opposite sides of a joint **(Figure 6.14)**. Because of this arrangement, muscles are able to bring about an immense variety of movements.

The muscle that has the major responsibility for causing a particular movement is called the **prime mover**. Muscles that oppose or reverse a movement are **antagonists** (an-tag′o-nists). When a prime mover is active, its antagonist is stretched and relaxed. Antagonists can be prime movers in their own right. For example, the biceps brachii and brachialis muscles of the arm (prime movers of elbow flexion) are antagonized by the triceps brachii (a prime mover of elbow extension).

**Synergists** (sin′er-jists; *syn* = together, *erg* = work) help prime movers by producing the same movement or by reducing undesirable movements. When a muscle crosses two or more joints, its contraction will cause movement in all the joints crossed unless synergists are there to stabilize them. For example, the flexor muscles of the fingers cross both the wrist and the finger joints. You can make a fist without bending your wrist because synergist muscles stabilize the wrist joints and allow the prime mover to act on your finger joints.

**Fixators** are specialized synergists. They hold a bone still or stabilize the origin of a prime mover so all the tension can be used to move the insertion bone. The postural muscles that stabilize the vertebral column are fixators, as are the muscles that anchor the scapulae to the thorax.

In summary, although prime movers seem to get all the credit for causing certain movements, the actions of antagonistic and synergistic muscles are also important in producing smooth, coordinated, and precise movements.

**(a)  A muscle that crosses on the anterior side of a joint produces flexion***

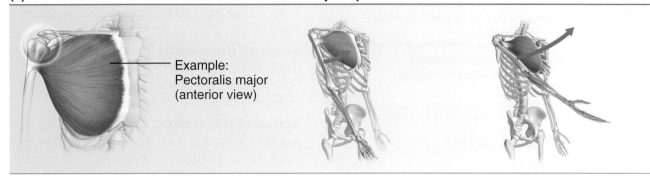

Example:
Pectoralis major
(anterior view)

**(b)  A muscle that crosses on the posterior side of a joint produces extension***

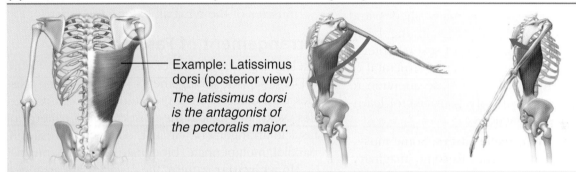

Example: Latissimus
dorsi (posterior view)

*The latissimus dorsi
is the antagonist of
the pectoralis major.*

**(c)  A muscle that crosses on the lateral side of a joint produces abduction**

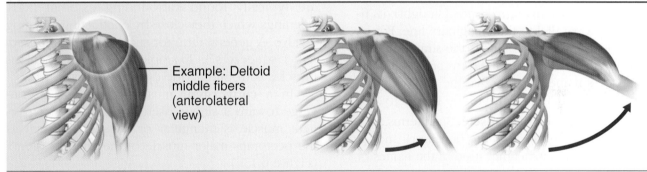

Example: Deltoid
middle fibers
(anterolateral
view)

**(d)  A muscle that crosses on the medial side of a joint produces adduction**

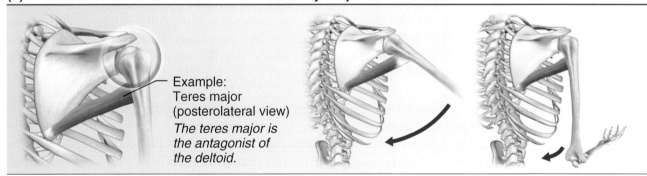

Example:
Teres major
(posterolateral view)

*The teres major is
the antagonist of
the deltoid.*

**\*** These generalities do not apply to the knee and ankle because the lower limb is rotated during development.
The muscles that cross these joints posteriorly produce flexion, and those that cross anteriorly produce extension.

**Figure 6.14  Muscle action.** The action of a muscle can be inferred
by the muscle's position as it crosses a joint.

Play A&P Flix Animation
Mastering A&P®

## Did You Get It?

16. What action is being performed by a person who sticks out his thumb to hitch a ride?
17. What actions take place at the neck when you nod your head up and down as if saying "yes"?
18. In what way are fixator and synergist muscles important?

For answers, see Appendix A.

## Naming Skeletal Muscles

→ **Learning** Objective

☐ **List seven criteria used in naming muscles.**

Like bones, muscles come in many shapes and sizes to suit their particular tasks in the body. Muscles are named on the basis of several criteria, each of which focuses on a particular structural or functional characteristic. Paying close attention to these cues can greatly simplify your task of learning muscle names and actions:

- **Direction of the muscle fibers.** Some muscles are named in reference to some imaginary line, usually the midline of the body or the long axis of a limb bone. When a muscle's name includes the term *rectus* (straight), its fibers run parallel to that imaginary line. For example, the rectus femoris is the straight muscle of the thigh. Similarly, the term *oblique* in a muscle's name tells you that the muscle fibers run obliquely (at a slant) to the imaginary line.

- **Relative size of the muscle.** Such terms as *maximus* (largest), *minimus* (smallest), and *longus* (long) are sometimes used in the names of muscles—for example, the gluteus maximus is the largest muscle of the gluteus muscle group.

- **Location of the muscle.** Some muscles are named for the bone with which they are associated. For example, the temporalis and frontalis muscles overlie the temporal and frontal bones of the skull, respectively.

- **Number of origins.** When the term *biceps*, *triceps*, or *quadriceps* forms part of a muscle name, you can assume that the muscle has two, three, or four origins, respectively. For example, the biceps muscle of the arm has two heads, or origins, and the triceps muscle has three.

- **Location of the muscle's origin and insertion.** Occasionally, muscles are named for their attachment sites. For example, the sternocleidomastoid muscle has its origin on the sternum (*sterno*) and clavicle (*cleido*) and inserts on the *mastoid* process of the temporal bone.

- **Shape of the muscle.** Some muscles have a distinctive shape that helps to identify them. For example, the deltoid muscle is roughly triangular (*deltoid* means "triangular").

- **Action of the muscle.** When muscles are named for their actions, terms such as *flexor*, *extensor*, and *adductor* appear in their names. For example, the adductor muscles of the thigh all bring about its adduction, and the extensor muscles of the wrist all extend the wrist.

## Arrangement of Fascicles

Skeletal muscles consist of fascicles (refer to Figure 6.1), but fascicle arrangements vary, producing muscles with different structures and functional properties. Next, let's look at the most common patterns of fascicle arrangement: circular, convergent, fusiform, parallel, multipennate, bipennate, and unipennate.

In a **circular** pattern, the fascicles are arranged in concentric rings **(Figure 6.15a)**. Circular muscles are typically found surrounding external body openings which they close by contracting, creating a valve. A general term for such muscles is *sphincters* ("squeezers"). Examples are the orbicularis muscles surrounding the eyes and mouth.

In a **convergent** muscle, the fascicles converge toward a single insertion tendon. A convergent muscle is triangular or fan-shaped, such as the pectoralis major muscle of the anterior thorax (Figure 6.15b).

In a **parallel** arrangement, the length of the fascicles run parallel to the long axis of the muscle, as in the sartorius of the anterior thigh. These muscles are straplike (Figure 6.15d). A modification of the parallel arrangement, called **fusiform**, results in a spindle-shaped muscle with an expanded belly (midsection); an example is the biceps brachii muscle of the arm (Figure 6.15c).

In a **pennate** (pen'āt; "feather") pattern, short fascicles attach obliquely to a central tendon. In the extensor digitorum muscle of the leg, the fascicles insert into only one side of the tendon, and the muscle is *unipennate* (Figure 6.15g). If the fascicles insert into opposite sides of the tendon, the muscle is *bipennate* (Figure 6.15f). If the fascicles insert from several different sides, the muscle is *multipennate* (Figure 6.15e).

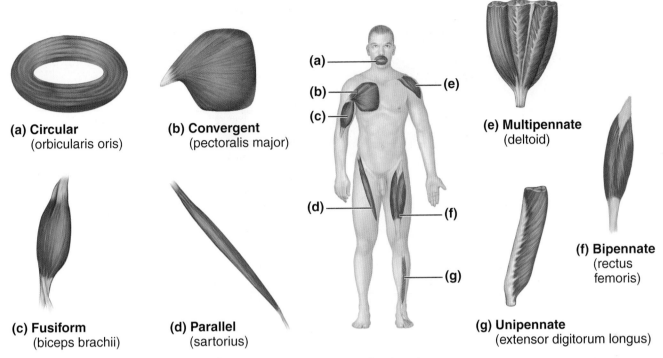

**(a) Circular**
(orbicularis oris)

**(b) Convergent**
(pectoralis major)

**(e) Multipennate**
(deltoid)

**(c) Fusiform**
(biceps brachii)

**(d) Parallel**
(sartorius)

**(f) Bipennate**
(rectus femoris)

**(g) Unipennate**
(extensor digitorum longus)

**Figure 6.15 Relationship of fascicle arrangement to muscle structure.**

A muscle's fascicle arrangement determines its range of motion and power. The longer and the more nearly parallel the fascicles are to a muscle's long axis, the more the muscle can shorten, but such muscles are not usually very powerful. Muscle power depends more on the total number of muscle fibers in the muscle. The stocky bipennate and multipennate muscles, which pack in the most fibers, shorten very little but are very powerful.

### Did You Get It?

19. Based on their names, deduce some characteristics of the following muscles: tibialis anterior, erector spinae, rectus abdominis, extensor carpi radialis longus.
20. What is the fascicle arrangement of the orbicularis oris muscle?

**For answers, see Appendix A.**

# Gross Anatomy of Skeletal Muscles

### → Learning Objective

☐ **Name and locate the major muscles of the human body (on a torso model, muscle chart, or diagram), and state the action of each.**

It is beyond the scope of this book to describe the hundreds of skeletal muscles of the human body.

We describe only the most important muscles here. All the superficial muscles we consider are summarized in Tables 6.3 and 6.4 and illustrated in the overall body views shown later in the chapter (Figures 6.22 and 6.23).

## Head and Neck Muscles

The head muscles (**Figure 6.16**, p. 230) are an interesting group. They have many specific functions but are usually grouped into two large categories—facial muscles and chewing muscles. Facial muscles are unique because they insert into soft tissues, such as other muscles or skin. When they pull on the skin of the face, they permit us to express ourselves by frowning, smiling, and so forth. The chewing muscles begin to break down food for the body. All head and neck muscles we describe are paired except for the platysma, orbicularis oris, frontalis, and occipitalis.

### Facial Muscles

***Frontalis*** The frontalis, which covers the frontal bone, runs from the cranial aponeurosis to the skin of the eyebrows, where it inserts. This muscle allows you to raise your eyebrows, as in surprise, and to wrinkle your forehead. At the posterior end of the cranial aponeurosis is the small occipitalis

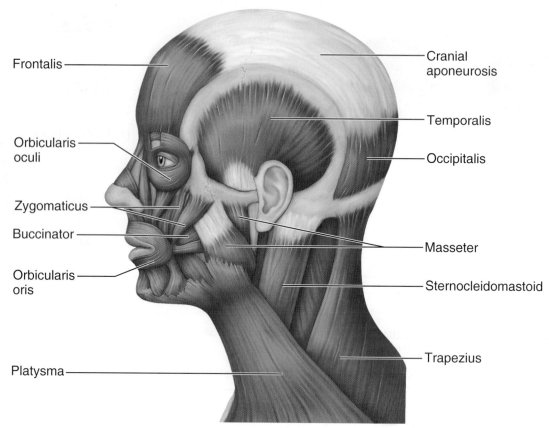

Frontalis

Orbicularis
oculi

Zygomaticus

Buccinator

Orbicularis
oris

Platysma

Cranial
aponeurosis

Temporalis

Occipitalis

Masseter

Sternocleidomastoid

Trapezius

**Figure 6.16 Superficial muscles of the head and neck.**

muscle, which covers the posterior aspect of the skull and pulls the scalp posteriorly.*

***Orbicularis Oculi*** The fibers of the orbicularis oculi (or-bik″u-la′ris ok′u-li) run in circles around the eyes. It allows you to close your eyes, squint, blink, and wink.

***Orbicularis Oris*** The orbicularis oris is the circular muscle of the lips. Often called the "kissing" muscle, it closes the mouth and protrudes the lips.

***Buccinator*** The fleshy buccinator (buck′sĭ-na″tor) muscle runs horizontally across the cheek and inserts into the orbicularis oris. It flattens the cheek (as in whistling or blowing a trumpet). It is also listed as a chewing muscle because it compresses

the cheek to hold food between the teeth during chewing.

***Zygomaticus*** The zygomaticus (zi″go-mat′i-kus) extends from the corner of the mouth to the cheekbone. It is often referred to as the "smiling" muscle because it raises the corners of the mouth.

***Masseter*** As it runs from the zygomatic process of the temporal bone to the mandible, the masseter (mă-se′ter) covers the angle of the lower jaw. This muscle closes the jaw by elevating the mandible.

***Temporalis*** The temporalis is a fan-shaped muscle overlying the temporal bone. It inserts into the mandible and acts as a synergist of the masseter in closing the jaw.

### Neck Muscles

For the most part, the neck muscles, which move the head and shoulder girdle, are small

*Although the current references on anatomic terminology refer to the frontalis and occipitalis as the *frontal* and *occipital bellies* of the *epicranius* ("over the cranium") muscle, we will continue to use the terms *frontalis* and *occipitalis* here.

# Anabolic Steroids: Dying to Win?

Many people feel the pressure to look and perform their best. It is not surprising that some will do anything to accomplish this—like taking performance-enhancing drugs such as anabolic steroids. These steroids are variants of testosterone, the hormone responsible for the changes during puberty that usher boys into manhood. The most notable effect is the increase in bone and muscle mass. Pharmaceutical companies introduced anabolic steroids in the 1950s to treat certain muscle-wasting diseases, anemia, and muscle atrophy in patients immobilized after surgery, and they became available to the general public during the 1980s.

Convinced that "doping" could enhance muscle mass and athletic endurance, many athletes (both men and women) have turned to anabolic steroids. Well-known athletes such as Mark McGwire (former St. Louis Cardinals first baseman), Marion Jones (2000 track and field Olympian), Lance Armstrong (seven-time Tour de France winner), and Alex Rodriguez (New York Yankees shortstop) admitted to steroid use and were stripped of medals and titles, and

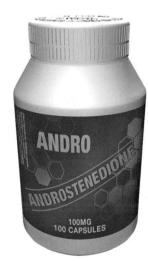

suspended from their respective sports.

Steroid use is not limited to the United States. In 2015, 11 Bulgarian weight lifters tested positive for anabolic steroid use prior to the European Championships; clearly this is a widespread problem that even affects adolescents. Studies by the National Institute on Drug Abuse (NIDA), National Institutes of Health (NIH), and the Centers for Disease Control (CDC) over a 20-year period estimated that 1.5–6% of adolescents 14–18 years old have used steroids.

Although most international athletic competitions and leagues have banned anabolic steroids, some athletes still claim they increase muscle mass, strength, and oxygen-carrying capacity of the blood. But do the drugs do all that is claimed? Research studies have reported increases in isometric strength and body weight in steroid users. Although these are results weight lifters dream about, there is hot dispute over whether the drugs also enhance fine muscle coordination and endurance (which are important to runners and other athletes).

Do the claimed slight advantages conferred by steroids outweigh their risks? Absolutely not! Side effects from steroid use vary but include increased risk of heart attack and stroke, high blood pressure, an enlarged heart, dangerous changes in cholesterol levels, fluid retention, and acne. Adolescents can experience stunted growth and early onset of puberty. Gender-specific side effects can also occur: Men may experience baldness, breast enlargement, and shrinking testicles, with an increased risk of prostate cancer. Women may grow facial hair but lose scalp hair, develop a deeper voice, become infertile, and have an increased risk of osteoporosis. It is also increasingly clear that steroids can be addictive and lead to serious psychological issues such as depression, delusions, paranoia, and aggressive or violent behavior called "'roid rage."

Some athletes say they are willing to do almost anything to win, short of killing themselves. Are those who use anabolic steroids unwittingly doing this as well?

and straplike. We consider only two neck muscles here (Figure 6.16).

***Platysma***    The platysma is a single sheetlike muscle that covers the anterolateral neck (see Figure 6.16). It originates from the connective tissue covering of the chest muscles and inserts into the area around the mouth. Its action is to pull the corners of the mouth inferiorly, producing a downward sag of the mouth (the "sad clown" face).

***Sternocleidomastoid***    The paired sternocleidomastoid (ster″no-kli″do-mas′toid) muscles are two-headed muscles, one found on each side of the neck. Of the two heads of each muscle, one arises from the sternum, and the other arises from the clavicle (see Figure 6.22, p. 240). The heads fuse before inserting into the mastoid process of the temporal bone. When both sternocleidomastoid muscles contract together, they flex your neck. (It is this action of bowing the head that has led some people to call these muscles the "prayer" muscles.) If just one muscle contracts, the face is rotated toward the shoulder on the opposite side and tilts the head to its own side.

### Homeostatic Imbalance 6.3

In some difficult births, one of the sternocleidomastoid muscles may be injured and develop spasms. A baby injured in this way has **torticollis** (tor″ti-kol′is), or wryneck. _____ ✚

#### Did You Get It?
21. Which muscle raises your eyebrow?
22. Which two muscles are synergists in closing your jaw?

For answers, see Appendix A.

## Trunk Muscles

The trunk muscles include (1) muscles that move the vertebral column (most of which are posterior antigravity muscles); (2) anterior thorax muscles, which move the ribs, head, and arms; and (3) muscles of the abdominal wall, which "hold your guts in" by forming a natural girdle and help to move the vertebral column.

### Anterior Muscles (Figure 6.17)
***Pectoralis Major***    The pectoralis (pek″to-ra′lis) major is a large fan-shaped muscle covering the upper part of the chest. Its origin is from the sternum, shoulder girdle, and the first six ribs. It inserts on the proximal end of the humerus. This muscle forms the anterior wall of the axilla (armpit) and acts to adduct and flex the arm.

***Intercostal Muscles***    The intercostal muscles are deep muscles found between the ribs. (Although they are not shown in Figure 6.17, which shows only superficial muscles, they are illustrated in Figure 6.22, p. 240.) The external intercostals are important in breathing because they help to raise the rib cage when you inhale. The internal intercostals, which lie deep to the external intercostals, depress the rib cage, helping to move air out of the lungs when you exhale forcibly.

***Muscles of the Abdominal Girdle***    The anterior abdominal muscles (rectus abdominis, external and internal obliques, and transversus abdominis) form a natural "girdle" that reinforces the body trunk. Taken together, they resemble the structure of plywood because the fibers of each muscle or muscle pair run in a different direction. Just as plywood is exceptionally strong for its thickness, the abdominal muscles form a muscular wall that is well suited for its job of containing and protecting the abdominal contents.

- **Rectus abdominis.** The paired straplike rectus abdominis muscles are the most superficial muscles of the abdomen. They run from the pubis to the rib cage, enclosed in an aponeurosis. Their main function is to flex the vertebral column. They also compress the abdominal contents during defecation and childbirth (they help you "push") and are involved in forced breathing.

- **External oblique.** The external oblique muscles are paired superficial muscles that make up the lateral walls of the abdomen. Their fibers run downward and medially from the last eight ribs and insert into the ilium. Like the rectus abdominis, they flex the vertebral column, but they also rotate the trunk and bend it laterally.

- **Internal oblique.** The internal oblique muscles are paired muscles deep to the external obliques. Their fibers run at right angles to those of the external obliques. They arise from

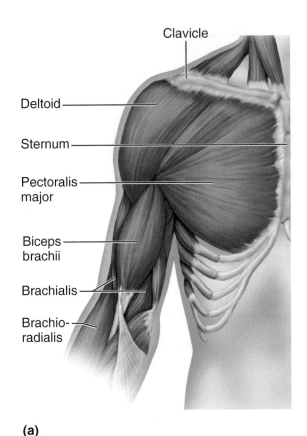

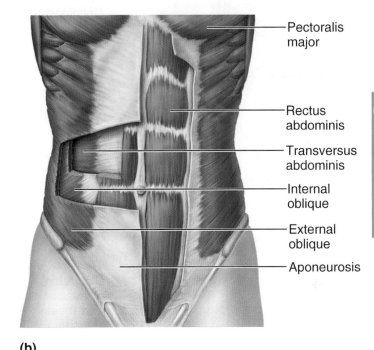

**(a)**                                                        **(b)**

**Figure 6.17 Muscles of the anterior trunk, shoulder, and arm. (a)** Muscles crossing the shoulder joint, causing movements of the arm. The platysma of the neck is removed. **(b)** Muscles of the abdominal wall. Portions of the superficial muscles of the right side of the abdomen are cut away to reveal the deeper muscles.

the iliac crest and insert into the last three ribs. Their functions are the same as those of the external obliques.

- **Transversus abdominis.** The deepest muscle of the abdominal wall, the transversus abdominis has fibers that run horizontally across the abdomen. It arises from the lower ribs and iliac crest and inserts into the pubis. This muscle compresses the abdominal contents.

**Posterior Muscles (Figure 6.18,** p. 234)

***Trapezius*** The trapezius (trah-pe′ze-us) muscles are the most superficial muscles of the posterior neck and upper trunk. When seen together, they form a diamond- or kite-shaped muscle mass. Their origin is very broad. Each muscle runs from the occipital bone of the skull down the vertebral column to the end of the thoracic vertebrae. They then flare laterally to insert on the scapular spine and clavicle. The trapezius muscles extend the head (thus they are antagonists of

the sternocleidomastoids). They also can elevate, depress, adduct, and stabilize the scapula.

***Latissimus Dorsi*** The latissimus (lah-tis′ĭ-mus) dorsi muscles are the two large, flat muscles that cover the lower back. They originate on the lower spine and ilium and then sweep superiorly to insert into the proximal end of the humerus. Each latissimus dorsi extends and adducts the humerus. These are very important muscles when the arm must be brought down in a power stroke, as when swimming or striking a blow.

***Erector Spinae*** The erector spinae (e-rek′tor spi′ne) group is the prime mover of back extension. These paired muscles are deep muscles of the back (Figure 6.18b). Each erector spinae is a composite muscle consisting of three muscle columns (longissimus, iliocostalis, and spinalis) that collectively span the entire length of the vertebral column. These muscles not only act as powerful back extensors ("erectors")

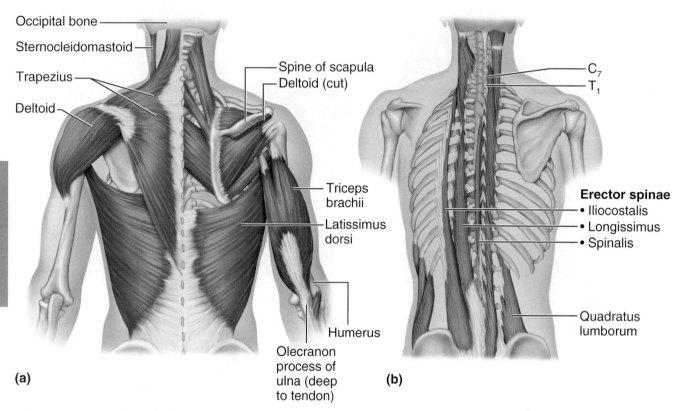

Occipital bone

Sternocleidomastoid

Trapezius

Deltoid

Spine of scapula

Deltoid (cut)

Triceps brachii

Latissimus dorsi

Humerus

Olecranon process of ulna (deep to tendon)

**(a)**

C₇

T₁

**Erector spinae**
• Iliocostalis
• Longissimus
• Spinalis

Quadratus lumborum

**(b)**

**Figure 6.18 Muscles of the posterior neck, trunk, and arm. (a)** Superficial muscles. **(b)** The erector spinae muscles (longissimus, iliocostalis, and spinalis), deep muscles of the back.

but also provide resistance that helps control the action of bending over at the waist. Following injury to back structures, these muscles go into spasms, a common source of lower back pain.

***Quadratus Lumborum***   The fleshy quadratus lumborum (qwad-ra′tus lum-bor′um) muscles form part of the posterior abdominal wall. Acting separately, each muscle of the pair flexes the spine laterally. Acting together, they extend the lumbar spine. These muscles arise from the iliac crests and insert into the upper lumbar vertebrae (Figure 6.18b).

***Deltoid***   The deltoids are fleshy, triangle-shaped muscles that form the rounded shape of your shoulders. Because they are so bulky, they are a favorite injection site **(Figure 6.19)** when relatively small amounts of medication (less than 5 ml) must be given intramuscularly (into muscle). The origin of each deltoid winds across the shoulder girdle from the spine of the scapula to the clavicle. It inserts into the proximal humerus. The deltoids are the prime movers of arm abduction.

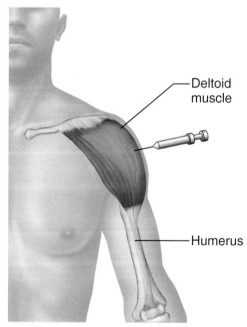

Deltoid muscle

Humerus

**Figure 6.19 The fleshy deltoid muscle is a favored site for administering intramuscular injections.**

## Did You Get It?

23. Which muscle group is the prime mover of back extension?
24. What structural feature makes the abdominal musculature especially strong for its thickness?
25. Which muscle of the posterior trunk is the synergist of the pectoralis major muscle in arm adduction?

For answers, see Appendix A.

## Muscles of the Upper Limb

The upper limb muscles fall into three groups. The first group includes muscles that arise from the shoulder girdle and cross the shoulder joint to insert into the humerus (see Figures 6.17 and 6.18a). We have already considered these muscles, which move the arm—they are the pectoralis major, latissimus dorsi, and deltoid.

The second group causes movement at the elbow joint. These muscles enclose the humerus and insert on the forearm bones. In this section we will focus on the muscles of this second group.

The third group of upper limb muscles includes the muscles of the forearm, which insert on the hand bones and cause their movement. The muscles of this last group are thin and spindle-shaped, and there are many of them. We will not consider them here except to mention their general naming and function. As a rule, the forearm muscles have names that reflect their activities. For example, the flexor carpi and flexor digitorum muscles, found on the anterior aspect of the forearm, flex the wrist and fingers, respectively. The extensor carpi and extensor digitorum muscles, found on the lateral and posterior aspect of the forearm, extend the same structures. (Table 6.4 and Figure 6.23 briefly describe some of these muscles; see pp. 241, 243).

### Muscles Causing Movement at the Elbow Joint

All *anterior* muscles of the humerus cause elbow flexion. In order of decreasing strength these are the brachialis, biceps brachii, and brachioradialis (Figures 6.17a and 6.22, p. 240).

***Biceps Brachii*** The biceps brachii (bra′ke-i) is the most familiar muscle of the arm because it bulges when you flex your elbow (see Figure 6.17a). It originates by two heads from the shoulder girdle and inserts into the radial tuberosity. This muscle is the powerful prime mover for flexion of the forearm and acts to supinate the forearm. The best way to remember its actions is to think of opening a bottle of wine. The biceps supinates the forearm to turn the corkscrew *and* then flexes the elbow to pull the cork.

***Brachialis*** The brachialis lies deep to the biceps brachii and, like the biceps, is a prime mover in elbow flexion. The brachialis lifts the ulna as the biceps lifts the radius.

***Brachioradialis*** The brachioradialis is a fairly weak muscle that arises on the humerus and inserts into the distal forearm (see Figure 6.22, p. 240). Hence, it resides mainly in the forearm.

***Triceps Brachii*** The triceps brachii is the only muscle fleshing out the posterior humerus (see Figure 6.18a). Its three heads arise from the shoulder girdle and proximal humerus, and it inserts into the olecranon process of the ulna. Being the powerful prime mover of elbow extension, it is the antagonist of the biceps brachii and brachialis. This muscle straightens the arm—for instance, to deliver a strong jab in boxing.

## Muscles of the Lower Limb

Muscles that act on the lower limb cause movement at the hip, knee, and foot joints. They are among the largest, strongest muscles in the body and are specialized for walking and balancing the body. Because the pelvic girdle is composed of heavy, fused bones that allow little movement, no special group of muscles is necessary to stabilize it. This is very different from the shoulder girdle, which requires several fixator muscles.

Many muscles of the lower limb span two joints and can cause movement at both of them. Therefore, in reference to these muscles, the terms *origin* and *insertion* are often interchangeable depending on the action being performed.

Muscles acting on the thigh are massive muscles that help hold the body upright against the pull of gravity and cause various movements at the hip joint. Muscles acting on the leg form the flesh of the thigh. (In common usage, the term *leg* refers to the whole lower limb, but anatomically the term refers only to that part between the knee and the ankle.) The thigh muscles cross the knee and cause its flexion or extension. Because many of the thigh muscles also have attachments on the pelvic girdle, they can cause movement at the hip joint as well.

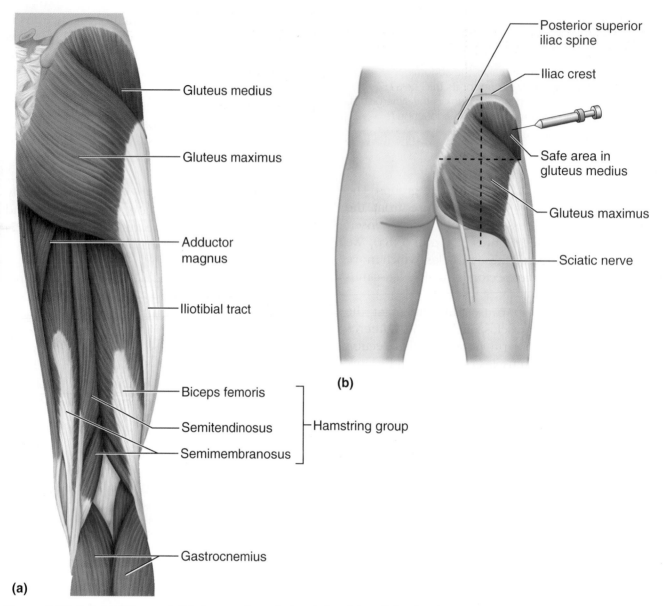

**Figure 6.20 Pelvic, hip, and thigh muscles of the right side of the body.**
**(a)** Posterior view of hip and thigh muscles. **(b)** Diagram showing deep structures of the gluteal region and the proper site for administering an injection into the gluteus medius muscle.

Muscles originating on the leg cause assorted movements of the ankle and foot. We will consider only three muscles of this group, but there are many others that extend and flex the ankle and toe joints.

### Muscles Causing Movement at the Hip Joint (Figure 6.20)

***Gluteus Maximus***    The gluteus maximus (gloo′te-us max′ĭ-mus) is a superficial muscle of the hip that forms most of the flesh of the buttock (Figure 6.20a). It is a powerful hip extensor that acts to bring the thigh in a straight line with the pelvis. Although it is not very important in walking, it is probably the most important muscle for extending the hip when power is needed, as when climbing stairs or jumping. It originates from the sacrum and iliac bones and inserts on the gluteal tuberosity of the femur and into the large tendinous *iliotibial* tract.

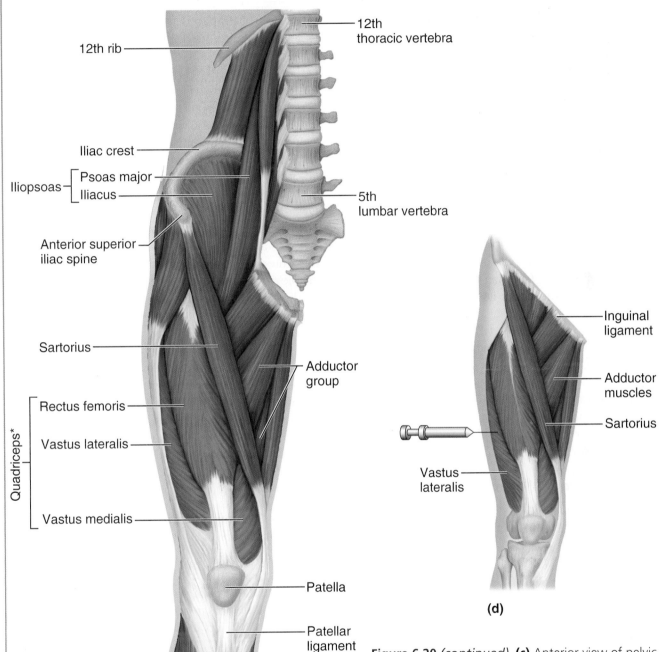

**(c)**

**Figure 6.20** *(continued)* **(c)** Anterior view of pelvic and thigh muscles. **(d)** Diagram showing the proper site for administering an injection into the lateral thigh (vastus lateralis muscle). *Note that the fourth quadriceps muscle, the vastus intermedius, is deep to the rectus femoris.

**Gluteus Medius**   The gluteus medius runs from the ilium to the femur, beneath the gluteus maximus for most of its length. The gluteus medius is a hip abductor and is important in steadying the pelvis during walking. It is also an important site for giving intramuscular injections, particularly when administering more than 5 ml (Figure 6.20b). Although it might appear that the large, fleshy gluteus maximus that forms the bulk of the buttock mass would be a better choice, notice that the medial part of each buttock overlies the large *sciatic nerve*; hence this area must be carefully avoided to prevent nerve damage. This can be accomplished by imagining the buttock is divided into four equal quadrants (shown by the division lines on Figure 6.20b). The superolateral quadrant then overlies the gluteus medius muscle, which is usually a very safe site for an intramuscular injection. Injections at or near the sciatic nerve can result in physical trauma from the needle or degeneration of the nerve itself.

**Iliopsoas**   The iliopsoas (il"e-o-so'as; the *p* is silent) is a fused muscle composed of two muscles, the *iliacus* and the *psoas major* (Figure 6.20c). It runs from the iliac bone and lower vertebrae deep inside the pelvis to insert on the lesser trochanter of the femur. It is a prime mover of hip flexion. It also acts to keep the upper body from falling backward when we are standing erect.

**Adductor Muscles**   The muscles of the adductor group form the muscle mass at the medial side of each thigh (Figure 6.20c). As their name indicates, they adduct, or press, the thighs together. However, because gravity does most of the work for them, they tend to become flabby very easily. Special exercises are usually needed to keep them toned. The adductors have their origin on the pelvis and insert on the proximal aspect of the femur.

## Muscles Causing Movement at the Knee Joint (Figure 6.20)

**Hamstring Group**   The muscles forming the muscle mass of the posterior thigh are the hamstrings (Figure 6.20a). The group consists of three muscles—the *biceps femoris*, *semimembranosus*, and *semitendinosus*—which originate on the ischial tuberosity and run down the thigh to insert on both sides of the proximal tibia. They are prime movers of thigh extension and knee flexion. Their name comes from the fact that butchers use their

tendons to hang hams (consisting of thigh and hip muscles) for smoking. You can feel these tendons at the back of your knee.

**Sartorius**   Compared with other thigh muscles described here, the thin, straplike sartorius (sarto're-us) muscle is not too important. However, it is the most superficial muscle of the thigh and so is rather hard to miss (Figure 6.20c). It runs obliquely across the thigh from the anterior iliac crest to the medial side of the tibia. It is a weak thigh flexor. The sartorius is commonly referred to as the "tailor's" muscle because it acts as a synergist to help tailors sit with both legs crossed in front of them.

**Quadriceps Group**   The quadriceps (kwod'riseps) group consists of four muscles—the *rectus femoris* and three *vastus muscles*—that flesh out the anterior thigh. (Only two vastus muscles are visible in Figure 6.20c. The third, the vastus intermedius, is obscured by the rectus femoris muscle, which lies over it.) The vastus muscles originate from the femur; the rectus femoris originates on the pelvis. All four muscles insert into the tibial tuberosity via the patellar ligament. The group as a whole acts to extend the knee powerfully, as when kicking a soccer ball. Because the rectus femoris crosses two joints, the hip and knee, it can also help to flex the hip. The vastus lateralis and rectus femoris are sometimes used as intramuscular injection sites (Figure 6.20d), particularly in infants, who have poorly developed gluteus muscles.

## Muscles Causing Movement at the Ankle and Foot (Figure 6.21)

**Tibialis Anterior**   The tibialis anterior is a superficial muscle on the anterior leg. It arises from the upper tibia and then parallels the anterior crest as it runs to the tarsal bones, where it inserts by a long tendon. It acts to dorsiflex and invert the foot.

**Extensor Digitorum Longus**   Lateral to the tibialis anterior, the extensor digitorum longus muscle arises from the lateral tibial condyle and proximal three-quarters of the fibula and inserts into the phalanges of toes 2 to 5. It is a prime mover of toe extension.

**Fibularis Muscles**   The three fibularis muscles—*longus*, *brevis*, and *tertius*—are found on the lateral part of the leg. They arise from the fibula and insert into the metatarsal bones of the foot. The

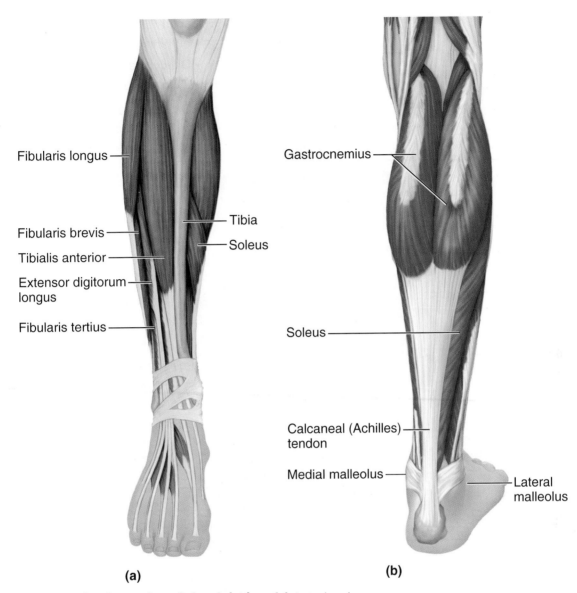

Fibularis longus

Fibularis brevis

Tibialis anterior

Extensor digitorum longus

Fibularis tertius

Tibia

Soleus

Gastrocnemius

Soleus

Calcaneal (Achilles) tendon

Medial malleolus

Lateral malleolus

**(a)**

**(b)**

**Figure 6.21 Superficial muscles of the right leg. (a)** Anterior view.
**(b)** Posterior view.

group as a whole plantar flexes and everts the foot, which is antagonistic to the tibialis anterior.

***Gastrocnemius*** The gastrocnemius (gas″trok-ne′me-us) muscle is a two-bellied muscle that forms the curved calf of the posterior leg. It arises by two heads, one from each side of the distal femur, and inserts through the large *calcaneal (Achilles) tendon* into the heel of the foot. It is a prime mover for plantar flexion of the foot; for this reason it is often called the "toe dancer's" muscle. If the calcaneal tendon is severely damaged or cut, walking is very difficult. The foot drags because it is not able to "push off" the toe (raise the heel).

***Soleus*** Deep to the gastrocnemius is the fleshy soleus muscle. Because it arises on the tibia and fibula (rather than the femur), it does not affect knee movement, but like the gastrocnemius, it inserts into the calcaneal tendon and is a strong plantar flexor of the foot.

**Figure 6.22** (p. 240) and **Figure 6.23** (p. 241) show anterior and posterior views of the whole body, including most of the superficial muscles we have described. **(Table 6.3)** (p. 242) and **(Table 6.4)** (p. 243) summarize these muscles. Take time to review these muscles again before continuing with this chapter.

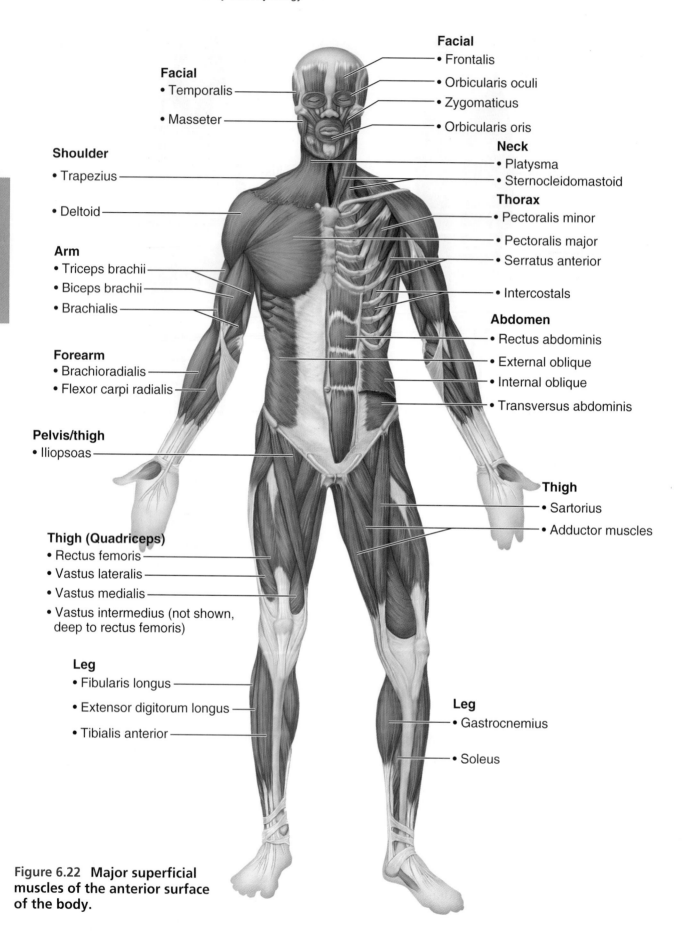

**Facial**
- Temporalis
- Masseter

**Facial**
- Frontalis
- Orbicularis oculi
- Zygomaticus
- Orbicularis oris

**Neck**
- Platysma
- Sternocleidomastoid

**Shoulder**
- Trapezius
- Deltoid

**Thorax**
- Pectoralis minor
- Pectoralis major
- Serratus anterior
- Intercostals

**Arm**
- Triceps brachii
- Biceps brachii
- Brachialis

**Abdomen**
- Rectus abdominis
- External oblique
- Internal oblique
- Transversus abdominis

**Forearm**
- Brachioradialis
- Flexor carpi radialis

**Pelvis/thigh**
- Iliopsoas

**Thigh**
- Sartorius
- Adductor muscles

**Thigh (Quadriceps)**
- Rectus femoris
- Vastus lateralis
- Vastus medialis
- Vastus intermedius (not shown, deep to rectus femoris)

**Leg**
- Fibularis longus
- Extensor digitorum longus
- Tibialis anterior

**Leg**
- Gastrocnemius
- Soleus

**Figure 6.22  Major superficial muscles of the anterior surface of the body.**

**Figure 6.23   Major superficial muscles of the posterior surface of the body.**

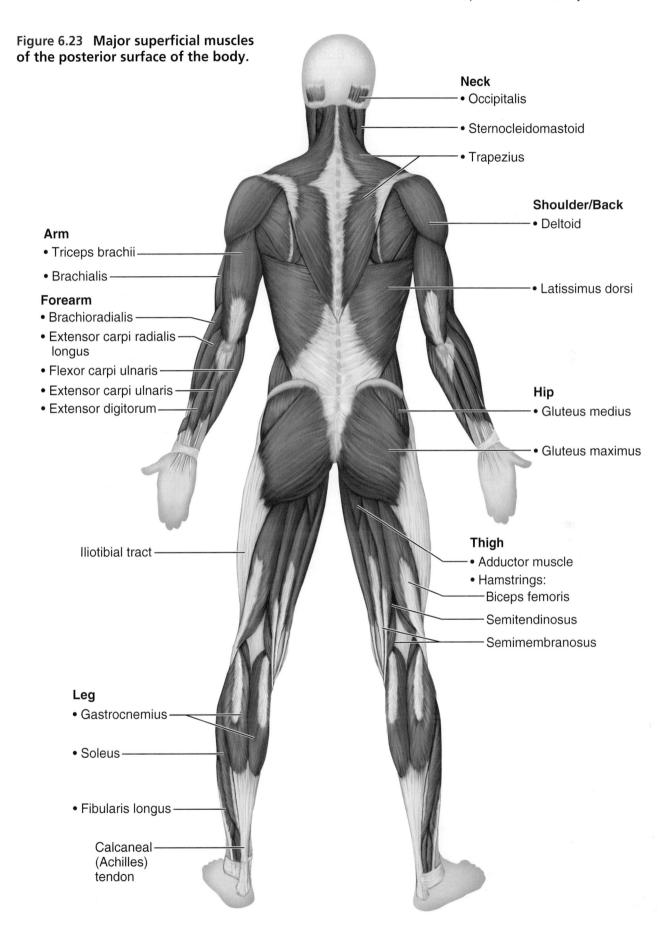

**Neck**
- Occipitalis
- Sternocleidomastoid
- Trapezius

**Shoulder/Back**
- Deltoid
- Latissimus dorsi

**Arm**
- Triceps brachii
- Brachialis

**Forearm**
- Brachioradialis
- Extensor carpi radialis longus
- Flexor carpi ulnaris
- Extensor carpi ulnaris
- Extensor digitorum

**Hip**
- Gluteus medius
- Gluteus maximus

Iliotibial tract

**Thigh**
- Adductor muscle
- Hamstrings:
  - Biceps femoris
  - Semitendinosus
  - Semimembranosus

**Leg**
- Gastrocnemius
- Soleus
- Fibularis longus

Calcaneal (Achilles) tendon

6

Table **6.3**    **Superficial Anterior Muscles of the Body (See Figure 6.22)**

| Name | Origin | Insertion | Primary action(s) |
|---|---|---|---|
| *Head/neck muscles* | | | |
| Frontalis | Cranial aponeurosis | Skin of eyebrows | Raises eyebrows |
| Orbicularis oculi | Frontal bone and maxilla | Tissue around eyes | Blinks and closes eye |
| Orbicularis oris | Mandible and maxilla | Skin and muscle around mouth | Closes and protrudes lips |
| Temporalis | Temporal bone | Mandible | Closes jaw |
| Zygomaticus | Zygomatic bone | Skin and muscle at corner of lips | Raises corner of mouth |
| Masseter | Temporal bone | Mandible | Closes jaw |
| Buccinator | Maxilla and mandible near molars | Orbicularis oris | Compresses cheek (as in sucking); holds food between teeth during chewing |
| Sternocleidomastoid | Sternum and clavicle | Temporal bone (mastoid process) | Flexes neck; laterally rotates head |
| Platysma | Connective tissue covering of superior chest muscles | Tissue around mouth | Tenses skin of neck (as in shaving) |
| *Trunk muscles* | | | |
| Pectoralis major | Sternum, clavicle, and first to sixth ribs | Proximal humerus | Adducts and flexes humerus |
| Rectus abdominis | Pubis | Sternum and fifth to seventh ribs | Flexes vertebral column |
| External oblique | Lower eight ribs | Iliac crest | Flexes and rotates vertebral column |
| *Arm/shoulder muscles* | | | |
| Biceps brachii | Scapula of shoulder girdle | Proximal radius | Flexes elbow and supinates forearm |
| Brachialis | Distal humerus | Proximal ulna | Flexes elbow |
| Deltoid | (See Table 6.4) | | Abducts arm |
| *Hip/thigh/leg muscles* | | | |
| Iliopsoas | Ilium and lumbar vertebrae | Femur (lesser trochanter) | Flexes hip |
| Adductor muscles | Pelvis | Proximal femur | Adduct and medially rotate thigh |
| Sartorius | Ilium | Proximal tibia | Flexes thigh on hip |
| Quadriceps group (vastus medialis, intermedius, and lateralis; and the rectus femoris) | Vasti: femur | Tibial tuberosity via patellar ligament | All extend knee; rectus femoris also flexes hip on thigh |
| | Rectus femoris: pelvis | Tibial tuberosity via patellar ligament | |
| Tibialis anterior | Proximal tibia | First cuneiform (tarsal) and first metatarsal of foot | Dorsiflexes and inverts foot |
| Extensor digitorum longus | Proximal tibia and fibula | Distal toes 2–5 | Extends toes |
| Fibularis muscles | Fibula | Metatarsals of foot | Plantar flex and evert foot |

Table **6.4**  **Superficial Posterior Muscles of the Body (Some Forearm Muscles Also Shown) (See Figure 6.23)**

| Name | Origin | Insertion | Primary action(s) |
|---|---|---|---|
| *Head/Neck/trunk/shoulder muscles* | | | |
| Occipitalis | Occipital and temporal bones | Cranial aponeurosis | Fixes aponeurosis and pulls scalp posteriorly |
| Trapezius | Occipital bone and all cervical and thoracic vertebrae | Scapular spine and clavicle | Raises, retracts, and rotates scapula |
| Latissimus dorsi | Lower spine and iliac crest | Proximal humerus | Extends and adducts humerus |
| Erector spinae* | Iliac crests, ribs 3–12, and vertebrae | Ribs, thoracic and cervical vertebrae | Extends and laterally flexes spine |
| Quadratus lumborum* | Iliac crest, lumbar fascia | Transverse processes of upper lumbar vertebrae | Flexes spine laterally; extends spine |
| Deltoid | Scapular spine and clavicle | Humerus (deltoid tuberosity) | Abducts humerus |
| *Arm/forearm muscles* | | | |
| Triceps brachii | Shoulder girdle and proximal humerus | Olecranon process of ulna | Extends elbow |
| Flexor carpi radialis | Distal humerus | Second and third metacarpals | Flexes wrist and abducts hand (see Figure 6.22) |
| Flexor carpi ulnaris | Distal humerus and posterior ulna | Carpals of wrist and fifth metacarpal | Flexes wrist and adducts hand |
| Flexor digitorum superficialis† | Distal humerus, ulna, and radius | Middle phalanges of second to fifth fingers | Flexes wrist and fingers |
| Extensor carpi radialis | Humerus | Base of second and third metacarpals | Extends wrist and abducts hand |
| Extensor digitorum | Distal humerus | Distal phalanges of second to fifth fingers | Extends fingers |
| *Hip/thigh/leg muscles* | | | |
| Gluteus maximus | Sacrum and ilium | Proximal femur (gluteal tuberosity) | Extends hip (when forceful extension is required) |
| Gluteus medius | Ilium | Proximal femur | Abducts thigh; steadies pelvis during walking |
| Hamstring muscles (biceps femoris, semitendinosus, semimembranosus) | Ischial tuberosity | Proximal tibia (head of fibula in the case of biceps femoris) | Flex knee and extend hip |
| Gastrocnemius | Distal femur | Calcaneus (heel via calcaneal tendon) | Plantar flexes foot and flexes knee |
| Soleus | Proximal tibia and fibula | Calcaneus | Plantar flexes foot |

*Erector spinae and quadratus lumborum are deep muscles (they are not shown in Figure 6.23; see Figure 6.18b).
†Although its name indicates that it is a superficial muscle, the flexor digitorum superficialis lies deep to the flexor carpi radialis and is not visible in a superficial view.

## Did You Get It?

26. Which muscle is the antagonist of the biceps brachii when the biceps flexes the elbow?
27. Which muscle group is the antagonist of the hamstring muscles?
28. What are two good sites for intramuscular injections in adults?
29. Which two muscles insert into the calcaneal tendon? What movement do they effect?

For answers, see Appendix A.

# Developmental Aspects of the Muscular System

→ **Learning Objectives**

☐ **Explain the importance of a nerve supply and exercise in keeping muscles healthy.**

☐ **Describe the changes that occur in aging muscles.**

In the developing embryo, the muscular system is laid down in segments, and then each segment is invaded by nerves. The muscles of the thoracic and lumbar regions become very extensive because they must cover and move the bones of the limbs. The muscles and their control by the nervous system develop rather early in pregnancy. The expectant mother is often astonished by the first movements (called the *quickening*) of the fetus, which usually occur by the 16th week of pregnancy.

## Homeostatic Imbalance 6.4

Very few congenital muscular problems occur. The exception to this is **muscular dystrophy**—a group of inherited muscle-destroying diseases that affect specific muscle groups. The muscles appear to enlarge because of fat and connective tissue deposits, but the muscle fibers degenerate and atrophy.

The most common and serious form is **Duchenne's muscular dystrophy**, which is expressed almost exclusively in boys. This tragic disease is usually diagnosed between the ages of 2 and 7 years. Active, normal-appearing children become clumsy and fall frequently as their muscles weaken. The disease progresses from the extremities upward, finally affecting the head and chest muscles. Children with this disease rarely live beyond their early twenties and generally die of respiratory failure. Although researchers have identified the cause of muscular dystrophy—the diseased muscle fibers lack a protein (called dystrophin) that helps maintain the sarcolemma—a cure is still elusive. _____ ✚

After birth, a baby's movements are all gross reflex types of movements. Because the nervous system must mature before the baby can control muscles, we can trace the increasing efficiency of the nervous system by observing a baby's development of muscle control. This development proceeds in a superior/inferior direction, and gross muscular movements precede fine ones. Babies can raise their heads before they can sit up and can sit up before they can walk. Muscular control also proceeds in a proximal/distal direction; that is, babies can perform the gross movements like waving "bye-bye" and pulling objects to themselves before they can use the pincer grasp to pick up a pin. All through childhood, the nervous system's control of the skeletal muscles becomes more and more precise. By midadolescence, we have reached the peak level of development of this natural control and can simply accept it or bring it to a fine edge by athletic training.

Because of its rich blood supply, skeletal muscle is amazingly resistant to infection throughout life, and given good nutrition, relatively few problems afflict skeletal muscles. We repeat, however, that muscles, like bones, *will* atrophy, even with normal tone, if they are not used continually. A lifelong program of regular exercise keeps the whole body operating at its best possible level.

## Homeostatic Imbalance 6.5

One rare autoimmune disease that can affect muscles during adulthood is **myasthenia gravis** (mi″as-the′ne-ah gră′vis; *asthen* = weakness; *gravi* = heavy), a disease characterized by drooping upper eyelids, difficulty in swallowing and talking, and generalized muscle weakness and fatigability. The disease involves a shortage of acetylcholine receptors at neuromuscular junctions caused by antibodies specific for acetylcholine receptors. Antibodies are immune molecules that will mark the receptors for destruction. Because the muscle fibers are not stimulated properly, they get progressively weaker. Death usually occurs when the respiratory muscles can no longer function, which leads to **respiratory failure**.

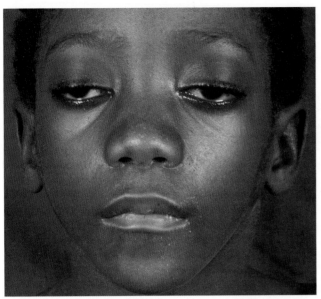

A female patient with myasthenia gravis. Notice the effect on the patient's eyelid muscles.

As we age, the amount of connective tissue in muscles increases, and the amount of muscle tissue decreases; thus the muscles become stringier, or more sinewy. Because skeletal muscles represent so much of our body mass, body weight begins to decline in older adults as this natural loss in muscle mass occurs. Muscle strength also decreases by about 50 percent by age 80. Regular exercise can help offset the effects of aging on the muscular system, and frail older people who begin to "pump iron" (use leg and hand weights) can rebuild muscle mass and dramatically increase their strength.

### Did You Get It?

30. What must happen before babies can control their muscles?
31. How does lifelong exercise affect our skeletal muscles and muscle mass in old age?

For answers, see Appendix A.

## Summary

### Overview of Muscle Tissues (pp. 207–211)

1. Skeletal muscle tissue forms the muscles attached to the skeleton, which move the limbs and other body parts. Its fibers (cells) are long, striated and multinucleate, and they are subject to voluntary control. Connective tissue coverings (endomysium, perimysium, and epimysium) enclose and protect the muscle fibers and increase the strength of skeletal muscles.

2. Smooth muscle fibers are uninucleate, spindle-shaped, and arranged in opposing layers in the walls of hollow organs. When they contract, substances (food, urine, a baby) are moved along internal pathways. Smooth muscle control is involuntary.

3. Cardiac muscle cells are striated, branching cells that fit closely together and are arranged in spiral bundles in the heart. Their contraction pumps blood through the blood vessels. Control is involuntary.

4. The sole function of muscle tissue is to contract, or shorten, to generate tension. As it contracts, it causes movement, maintains posture, stabilizes joints, and generates heat.

### Microscopic Anatomy of Skeletal Muscle (pp. 211–213)

1. The multinucleate cylindrical skeletal muscle fibers are packed with unique organelles called myofibrils. The banding pattern (striations) of the myofibrils and the fiber as a whole reflects the regular arrangement of thin (actin-containing) and thick (myosin) filaments within the sarcomeres, the contractile units (structurally and functionally) composing the myofibrils.

 Complete an interactive tutorial: Mastering A&P®> Study Area > Interactive Physiology > Muscular System > Anatomy Review: Skeletal Muscle Tissue

2. Each myofibril is loosely enclosed by a specialized ER, called the sarcoplasmic reticulum (SR), which plays an important role in storing and releasing calcium ions. Calcium ions are the final trigger for muscle fiber contraction.

### Skeletal Muscle Activity (pp. 213–222)

1. All skeletal muscle fibers are stimulated by motor neurons. When the neuron releases a neurotransmitter (acetylcholine), the permeability of the sarcolemma changes, allowing sodium ions to enter

# Homeostatic Relationships between the
# Muscular System and Other Body Systems

**Nervous System**
- Facial muscle activity allows emotions to be expressed
- Nervous system stimulates and regulates muscle activity

**Endocrine System**
- Growth hormone and androgens influence skeletal muscle strength and mass

**Respiratory System**
- Muscular exercise increases respiratory capacity
- Respiratory system provides oxygen and disposes of carbon dioxide

**Lymphatic System/Immunity**
- Physical exercise may enhance or depress immunity depending on its intensity
- Lymphatic vessels drain leaked interstitial fluids; immune system protects muscles from disease

**Cardiovascular System**
- Skeletal muscle activity increases efficiency of cardiovascular functioning; helps prevent atherosclerosis and causes cardiac hypertrophy
- Cardiovascular system delivers oxygen and nutrients to muscles; carries away wastes

**Digestive System**
- Physical activity increases gastrointestinal mobility when at rest
- Digestive system provides nutrients needed for muscle health; liver metabolizes lactic acid

**Reproductive System**
- Skeletal muscle helps support pelvic organs (e.g., uterus in females); assists erection of penis and clitoris
- Testicular androgen promotes increased skeletal muscle size

**Urinary System**
- Physical activity promotes normal voiding behavior; skeletal muscle forms the voluntary sphincter of the urethra
- Urinary system disposes of nitrogen-containing wastes

**Integumentary System**
- Muscular exercise enhances circulation to skin and improves skin health; exercise also increases body heat, which the skin helps dissipate
- Skin protects the muscles by external enclosure

## Muscular System

**Skeletal System**
- Skeletal muscle activity maintains bone health and strength
- Bones provide levers for muscle activity

the muscle fiber. This produces an electrical current (action potential), which flows across the entire sarcolemma, resulting in release of calcium ions from the SR.

2. Calcium binds to regulatory proteins on the thin filaments and exposes myosin-binding sites, allowing the myosin heads on the thick filaments to attach. The attached heads pivot, sliding the thin filaments toward the center of the sarcomere, and contraction occurs. ATP provides the energy for the sliding process, which continues as long as ionic calcium is present.

3. Although individual muscle fibers contract completely when adequately stimulated, a muscle (which is an organ) responds to stimuli to different degrees; that is, it exhibits graded responses.

4. Most skeletal muscle contractions are tetanic (smooth and sustained) because rapid nerve impulses are reaching the muscle, and the muscle cannot relax completely between contractions. The strength of muscle contraction reflects the relative number of muscle fibers contracting (more = stronger).

5. ATP, the immediate source of energy for muscle contraction, is stored in muscle fibers in small amounts that are quickly used up. ATP is regenerated via three routes. From the fastest to the slowest, these are via a coupled reaction of creatine phosphate with ADP, via anaerobic glycolysis and lactic acid formation, and via aerobic respiration. Only aerobic respiration requires oxygen.

6. If muscle activity is strenuous and prolonged, muscle fatigue occurs because ionic imbalances occur, lactic acid accumulates in the muscle, and the energy (ATP) supply decreases. After exercise, the oxygen deficit is repaid by rapid, deep breathing.

7. Muscle contractions are isotonic (the muscle shortens, and movement occurs) or isometric (the muscle does not shorten, but its tension increases).

8. Muscle tone keeps muscles healthy and ready to react. It is a result of a staggered series of nerve impulses delivered to different fibers within the muscle. If the nerve supply is destroyed, the muscle loses tone, becomes paralyzed, and atrophies.

9. Inactive muscles atrophy. Muscles challenged almost beyond their ability by resistance exercise will increase in size and strength. Muscles subjected to

regular aerobic exercise become more efficient and stronger and can work longer without tiring. Aerobic exercise also benefits other body organ systems.

## Muscle Movements, Roles, and Names (pp. 222–229)

1. All muscles are attached to bones at two points. The origin is the immovable (or less movable) attachment; the insertion is the movable bony attachment. When contraction occurs, the insertion moves toward the origin.

2. Body movements include flexion, extension, abduction, adduction, circumduction, rotation, pronation, supination, inversion, eversion, dorsiflexion, plantar flexion, and opposition.

3. On the basis of their general functions in the body, muscles are classified as prime movers, antagonists, synergists, and fixators.

4. Muscles are named according to several criteria, including muscle size, shape, number and location of origins, associated bones, and action of the muscle.

5. Muscles have several fascicle arrangements that influence their force and degree of shortening.

## Gross Anatomy of Skeletal Muscles (pp. 229–244)

1. Muscles of the head fall into two groups. The muscles of facial expression include the frontalis, orbicularis oris, orbicularis oculi, and zygomaticus. The chewing muscles are the masseter, temporalis, and buccinator (which is also a muscle of facial expression).

2. Muscles of the trunk and neck move the head, shoulder girdle, and trunk and form the abdominal girdle. Anterior neck and trunk muscles include the sternocleidomastoid, pectoralis major, external and internal intercostals, rectus abdominis, external and internal obliques, and transversus abdominis. Posterior trunk and neck muscles include the trapezius, latissimus dorsi, and deltoid. Deep muscles of the back are the erector spinae muscles.

3. Muscles of the upper limb include muscles that cause movement at the shoulder joint, elbow, and hand. Muscles causing movement at the elbow include the brachialis, biceps brachii, brachioradialis, and triceps brachii.

4. Muscles of the lower extremity cause movement at the hip, knee, and foot. They include the iliopsoas,

gluteus maximus, gluteus medius, adductors, quadriceps and hamstring groups, gastrocnemius, tibialis anterior, fibularis muscles, soleus, and extensor digitorum longus.

## Developmental Aspects of the Muscular System
(pp. 244–245)

1. Increasing muscular control reflects the maturation of the nervous system. Muscle control is achieved in a superior/inferior and proximal/distal direction.

2. To remain healthy, muscles must be regularly exercised. Without exercise, they atrophy; with extremely vigorous exercise, they hypertrophy.

3. As we age, muscle mass decreases, and the muscles become more sinewy. Exercise helps to retain muscle mass and strength.

## Review Questions

Access additional practice questions using your smartphone, tablet, or computer:
Mastering A&P®> Study Area > Practice Tests & Quizzes

### Multiple Choice

*More than one choice may apply.*

1. If you compare electron micrographs of a relaxed skeletal muscle fiber and a fully contracted muscle fiber, which would you see only in the *relaxed* fiber?
   a. Z discs
   b. Sarcomeres
   c. I bands
   d. A bands
   e. H zones

2. After ACh attaches to its receptors at the neuromuscular junction, what is the next step?
   a. Sodium channels open.
   b. Calcium binds to regulatory proteins on the thin filaments.
   c. Cross bridges attach.
   d. ATP is hydrolyzed.

3. Your ability to lift that heavy couch would be increased by which type of exercise?
   a. Aerobic
   b. Endurance
   c. Resistance
   d. Swimming

4. Which of the following can lead to muscle fatigue?
   a. Imbalances in ions
   b. Excessive amounts of oxygen
   c. Excessive amounts of ADP
   d. Oxygen deficit

5. Which are ways in which muscle names have been derived?
   a. Attachments
   b. Size
   c. Function
   d. Location

6. Which of the following muscles attach to the hip bones?
   a. Rectus abdominis
   b. Rectus femoris
   c. Vastus medialis
   d. Vastus lateralis

7. Which of these muscles is found on the posterior surface of the leg?
   a. Gastrocnemius
   b. Extensor digitorum longus
   c. Extensor carpi ulnaris
   d. Biceps femoris

8. Which of the following is (are) true concerning the pectoralis major?
   a. It raises the clavicle.
   b. It adducts the humerus.
   c. It abducts the arm.
   d. It flexes the humerus.

### Short Answer Essay

9. What is the major function of muscle?

10. Compare skeletal, smooth, and cardiac muscles in regard to their microscopic anatomy, location and arrangement in body organs, and function in the body.

11. What two types of muscle tissue are striated?

12. Why are the connective tissue wrappings of skeletal muscles important? Name these connective tissue coverings, beginning with the finest and ending with the coarsest.

13. What is the function of tendons?

14. Define *neuromuscular junction, motor unit, tetanus, graded response, aerobic respiration, anaerobic glycolysis, muscle fatigue,* and *neurotransmitter.*

15. Describe the events that occur from the time calcium ions enter the axon terminal at the neuromuscular junction until muscle cell contraction occurs.

16. How do isotonic and isometric contractions differ?

17. How does muscle activity contribute to the maintenance of body temperature? Explain how significant this contribution is.

18. A skeletal muscle is attached to bones at two points. Name each of these attachment points, and indicate which is movable and which is immovable.

19. Circumduction is a combination of four movements. List these four movements, and describe each.

20. How is a prime mover different from a synergist muscle? How can a prime mover also be considered an antagonist?

21. Which facial muscle allows you to whistle? Why is it also considered a chewing muscle?

22. Injury to which muscle can lead to torticollis?

23. Name two muscles that reverse the movement of the deltoid muscle.

24. Name the prime mover of elbow extension. Name its antagonist(s).

25. Other than acting to flex the spine and compress the abdominal contents, the abdominal muscles are extremely important in protecting and containing the abdominal viscera. What is it about the arrangement of these muscles that makes them so well suited for their job?

26. The hamstring and quadriceps muscle groups are antagonists of each other, and each group is a prime mover in its own right. What action does each muscle group perform?

27. Which muscles insert on the transverse processes of the upper lumbar vertebrae? What are their primary functions?

28. What happens to muscles when they are exercised regularly? Exercised vigorously as in weight lifting? Not used?

29. What is the effect of aging on skeletal muscles?

30. Should a triathlete engage in aerobic or resistance training? Explain.

31. Explain the general relationship between the muscles that flex and extend the wrist, including their location on the body.

32. Harry is pondering an exam question that asks, "What muscle type has elongated cells and is found in the walls of the urinary bladder?" How should he respond?

 ## Critical Thinking and Clinical Application Questions

33. Name three muscles or muscle groups used as sites for intramuscular injections. Which is most often used in babies?

34. While jogging, Mr. Ahmadi was forced to jump out of the way of a speeding car. He heard a snapping sound that was immediately followed by pain in his right lower calf. A gap was visible between his swollen calf and his heel, and he was unable to plantar flex that foot. What do you think happened?

35. Susan fell off her bicycle and fractured her right clavicle. Treatment prescribed by the emergency room physician included using a sling to immobilize the clavicle and speed its healing. What mus-

cles are temporarily "out of business" as a result of this injury?

36. When Eric returned from jogging, he was breathing heavily and sweating profusely, and he complained that his legs ached and felt weak. On the basis of what you have learned about muscle energy metabolism, respond to the following questions:

    a. Why is Eric breathing heavily?

    b. What ATP-harvesting pathway have his working muscles been using that leads to such a breathing pattern?

    c. What metabolic product(s) might account for his sore muscles and his feeling of muscle weakness?

37. Chemical A binds and blocks acetylcholine receptors of muscle cells. Chemical B floods the cytoplasm of muscle cells with calcium ions. Which chemical would make the best muscle relaxant, and why?

38. Kendra's broken leg was in a cast for 8 weeks. When the cast was removed, she was shocked that her leg was no longer muscular, but skinny. Explain what happened and why.

# 7 The Nervous System

**WHAT**

As the primary control system of the body, the nervous system provides for higher mental function and emotional expression, maintains homeostasis, and regulates the activities of muscles and glands.

**HOW**

Communication by the nervous system involves a combination of electrcial and chemical signals.

**WHY**

All body systems are under the control or regulation of the nervous system. If the nervous systems stops functioning, the body can stay alive only with the assistance of life-supporting machines.

You are driving down the freeway when a horn blares on your right, and you swerve to your left. You are with a group of friends but are not paying attention to any particular conversation, yet when you hear your name, you immediately focus in. What do these events have in common? They are everyday examples of nervous system function, which has your body cells humming with activity nearly all the time.

The **nervous system** is the master control and communication system of the body. Every thought, action, and emotion reflects its activity. It communicates with body cells using electrical impulses, which are rapid and specific and cause almost immediate responses.

The nervous system does not work alone to regulate and maintain body homeostasis; the endocrine system is a second important regulating system. Whereas the nervous system controls with rapid electrical nerve impulses, the endocrine system produces hormones that are released into the blood. Thus, the endocrine system acts in a more leisurely way. We will discuss the endocrine system in detail in Chapter 9.

To carry out its normal role, the nervous system has three overlapping functions (**Figure 7.1**, p. 252):

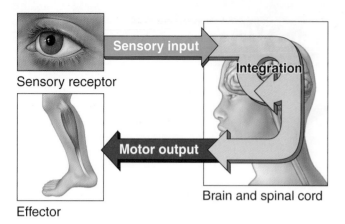

Sensory receptor

Effector

Brain and spinal cord

**Figure 7.1 The nervous system's functions.**

(1) It uses its millions of sensory receptors to *monitor changes* occurring both inside and outside the body. These changes are called *stimuli*, and the gathered information is called **sensory input**. (2) It *processes and interprets* the sensory input and decides what should be done at each moment—a process called **integration**. (3) It then *causes a response, or effect*, by activating muscles or glands (effectors) via **motor output**.

**CONCEPTLINK**

These three overlapping nervous system functions are very similar to a feedback loop (Chapter 1, p. 45). Recall that in a feedback loop, a receptor receives *sensory input*, which it sends to the brain (control center) for processing (*integration*); the brain then analyzes the information and determines the appropriate *output*, which leads to a motor response. ←

An example will illustrate how these functions work together. When you are driving and see a red light just ahead (sensory input), your nervous system integrates this information (red light means "stop") and sends motor output to the muscles of your right leg and foot. Your foot goes for the brake pedal (the response).

# Organization of the Nervous System

→ **Learning Objectives**

☐ **List the general functions of the nervous system.**

☐ **Explain the structural and functional classifications of the nervous system.**

☐ **Define *central nervous system* and *peripheral nervous system*, and list the major parts of each.**

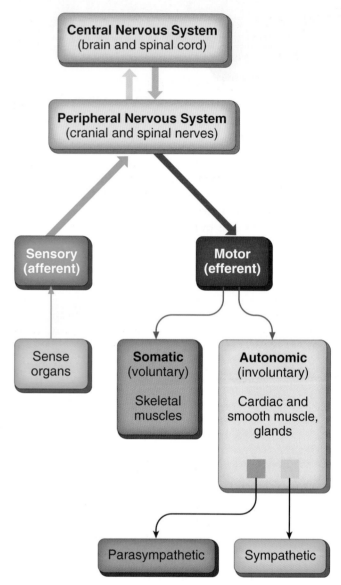

**Figure 7.2 Organization of the nervous system.** As the flowchart shows, the central nervous system receives input via sensory fibers and issues commands via motor fibers. The sensory and motor fibers together form the nerves that constitute the peripheral nervous system.

We have only one nervous system, but because of its complexity, it is difficult to consider all its parts at the same time. So, to simplify its study, we divide it in terms of its structures (structural classification) or in terms of its activities (functional classification). We briefly describe each of these classification schemes next, and **Figure 7.2** illustrates their relationships. You do not need to memorize this whole scheme now, but as you are reading the descriptions, try to get a "feel" for the major parts and how they fit together. This will make learning easier as you make your way through this chapter. Later you will see all these terms and concepts again in more detail.

## Structural Classification

The structural classification, which includes all nervous system organs, has two subdivisions—the central nervous system and the peripheral nervous system (see Figure 7.2).

The **central nervous system (CNS)** consists of the brain and spinal cord, which occupy the dorsal body cavity and act as the integrating and command centers of the nervous system. They interpret incoming sensory information and issue instructions based on past experience and current conditions.

The **peripheral** (pĕ-rif′er-al) **nervous system (PNS)** includes all parts of the nervous system outside the CNS. It consists mainly of the nerves that extend from the spinal cord and brain. *Spinal nerves* carry impulses to and from the spinal cord. *Cranial* (kra′ne-al) *nerves* carry impulses to and from the brain. These nerves serve as communication lines. They link all parts of the body by carrying impulses from the sensory receptors to the CNS and from the CNS to the appropriate glands or muscles.

## Functional Classification

The functional classification scheme is concerned only with PNS structures. It divides them into two principal subdivisions (see Figure 7.2).

The **sensory division**, or **afferent** (af′er-ent; literally "to go toward") **division**, consists of nerves (composed of many individual nerve fibers) that convey impulses *to* the central nervous system from sensory receptors located in various parts of the body. The sensory division keeps the CNS constantly informed of events going on both inside and outside the body. Sensory fibers delivering impulses from the skin, skeletal muscles, and joints are called *somatic* (*soma* = body) *sensory* (afferent) *fibers*, whereas those transmitting impulses from the visceral organs are called *visceral sensory* (afferent) *fibers*.

The **motor division**, or **efferent** (ef′er-rent) **division**, carries impulses *from* the CNS to effector organs, the muscles and glands. These impulses activate muscles and glands; that is, they *effect* (bring about or cause) a motor response.

The motor division in turn has two subdivisions (see Figure 7.2):

- The **somatic** (so-mat′ik) **nervous system** allows us to consciously, or voluntarily, control our skeletal muscles. Hence, we often refer to this subdivision as the **voluntary nervous system**. However, not all skeletal muscle activity controlled by this motor division is voluntary. Skeletal muscle reflexes, such as the stretch reflex (described later in the chapter), are initiated involuntarily by these same fibers.

- The **autonomic** (aw″to-nom′ik) **nervous system (ANS)** regulates events that are automatic, or involuntary, such as the activity of smooth muscle, cardiac muscle, and glands. This subdivision, commonly called the **involuntary nervous system**, itself has two parts, the *sympathetic* and *parasympathetic*, which typically bring about opposite effects. What one stimulates, the other inhibits. We will describe these later.

Although it is simpler to study the nervous system in terms of its subdivisions, remember that these subdivisions are made for the sake of convenience only. Remember that the nervous system acts as a coordinated unit, both structurally and functionally.

### Did You Get It?

1. Name the structures that make up the CNS and those that make up the PNS.

For the answer, see Appendix A.

# Nervous Tissue: Structure and Function

→ **Learning Objective**

☐ **Describe the structures and functions of neurons and neuroglia.**

Even though it is complex, nervous tissue is made up of just two principal types of cells—*supporting cells* and *neurons*.

## Supporting Cells

Supporting cells in the CNS are "lumped together" as **neuroglia** (nu-rog′le-ah), literally, "nerve glue," also called *glial cells* or **glia**. Neuroglia include many types of cells that support, insulate, and protect the delicate neurons **(Figure 7.3)**. In addition, each of the different types of neuroglia has special functions. CNS neuroglia include the following:

- **Astrocytes:** abundant star-shaped cells that account for nearly half of neural tissue

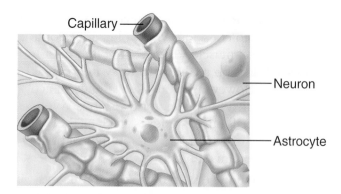

Capillary — Neuron — Astrocyte

**(a) Astrocytes are the most abundant and versatile neuroglia.**

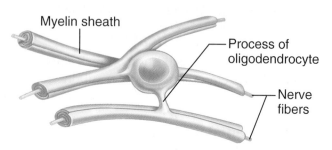

Myelin sheath — Process of oligodendrocyte — Nerve fibers

**(d) Oligodendrocytes have processes that form myelin sheaths around CNS nerve fibers.**

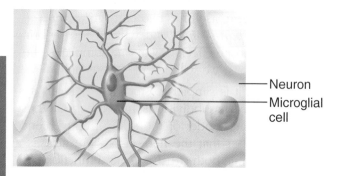

Neuron — Microglial cell

**(b) Microglial cells are phagocytes that defend CNS cells.**

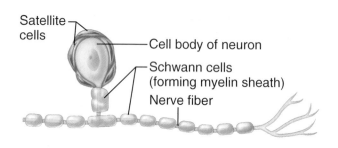

Satellite cells — Cell body of neuron — Schwann cells (forming myelin sheath) — Nerve fiber

**(e) Satellite cells and Schwann cells (which form myelin) surround neurons in the PNS.**

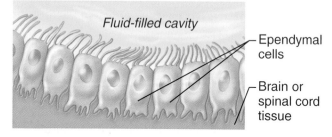

*Fluid-filled cavity* — Ependymal cells — Brain or spinal cord tissue

**(c) Ependymal cells line cerebrospinal fluid–filled cavities.**

**Figure 7.3 Supporting cells (neuroglia) of nervous tissue.**

(Figure 7.3a). Their numerous projections have swollen ends that cling to neurons, bracing them and anchoring them to their nutrient supply lines, the blood capillaries. Astrocytes form a living barrier between capillaries and neurons, help determine capillary permeability, and play a role in making exchanges between the two. In this way, they help protect the neurons from harmful substances that might be in the blood. Astrocytes also help control the chemical environment in the brain by "mopping up" leaked potassium ions, which are involved in generating a nerve impulse, and recapturing chemicals released for communication purposes.

• **Microglia** (mi-krog'le-ah): spiderlike phagocytes that monitor the health of nearby neurons and dispose of debris, such as dead brain cells and bacteria (Figure 7.3b).

• **Ependymal** (ĕ-pen'dĭ-mal) **cells:** neuroglia that line the central cavities of the brain and the spinal cord (Figure 7.3c). The beating of their cilia helps to circulate the cerebrospinal fluid that fills those cavities and forms a protective watery cushion around the CNS.

- **Oligodendrocytes** (ol″ĭ-go-den′dro-sītz): neuroglia that wrap their flat extensions (processes) tightly around the nerve fibers, producing fatty insulating coverings called *myelin sheaths* (Figure 7.3d).

Although neuroglia somewhat resemble neurons structurally (both cell types have cell extensions), they are not able to transmit nerve impulses, a function that is highly developed in neurons. Another important difference is that neuroglia never lose their ability to divide, whereas most neurons do. Consequently, most brain tumors are *gliomas*, or tumors formed by neuroglia.

Supporting cells in the PNS come in two major varieties—Schwann cells and satellite cells (Figure 7.3e). **Schwann cells** form the myelin sheaths around nerve fibers in the PNS. **Satellite cells** act as protective, cushioning cells for peripheral neuron cell bodies.

### Did You Get It?

2. Which neuroglia are most abundant in the body? Which produce the insulating material called myelin?
3. Why is a brain tumor more likely to be formed from neuroglia than from neurons?

**For answers, see Appendix A.**

## Neurons

### Anatomy

→ **Learning Objectives**

☐ Describe the general structure of a neuron, and name its important anatomical regions.

☐ Describe the composition of gray matter and white matter.

**Neurons**, also called **nerve cells**, are highly specialized to transmit messages (nerve impulses) from one part of the body to another. Although neurons differ structurally from one another, they have many common features (**Figure 7.4**, p. 256). All have a cell body, which contains the nucleus and one or more slender processes extending from the cell body.

*Cell Body*   The **cell body** is the metabolic center of the neuron. Its transparent nucleus contains a large nucleolus. The cytoplasm surrounding the nucleus contains the usual organelles, except that it lacks centrioles (which confirms the amitotic nature of most neurons). The rough ER, called **Nissl** (nis′l) **bodies**, and **neurofibrils** (intermediate

filaments that are important in maintaining cell shape) are particularly abundant in the cell body.

*Processes*   The armlike **processes**, or **fibers**, vary in length from microscopic to about 7 feet in the tallest humans. The longest ones in humans reach from the lumbar region of the spine to the great toe. Neuron processes that convey incoming messages (electrical signals) *toward* the cell body are **dendrites** (den′drītz), whereas those that generate nerve impulses and typically conduct them *away* from the cell body are **axons** (ak′sonz). Neurons may have hundreds of branching dendrites (*dendr* = tree), depending on the neuron type. However, each neuron has only one axon, which arises from a conelike region of the cell body called the **axon hillock**.

An occasional axon gives off a *collateral branch* along its length, but all axons branch profusely at their terminal end, forming hundreds to thousands of **axon terminals**. These terminals contain hundreds of tiny vesicles, or membranous sacs, that contain chemicals called **neurotransmitters** (review our discussion of events at the neuromuscular junction in Chapter 6). As we said, axons transmit nerve impulses away from the cell body. When these impulses reach the axon terminals, they stimulate the release of neurotransmitters into the extracellular space between neurons, or between a neuron and its target cell.

Each axon terminal is separated from the next neuron by a tiny gap called the **synaptic** (sĭ-nap′tik) **cleft**. Such a functional junction, where an impulse is transmitted from one neuron to another, is called a **synapse** (*syn* = to clasp or join). Although they are close, neurons never actually touch other neurons. You will learn more about synapses and the events that occur there a bit later.

*Myelin Sheaths*   Most long nerve fibers are covered with a whitish, fatty material called **myelin** (mi′ĕ-lin), which has a waxy appearance. Myelin protects and insulates the fibers and increases the transmission rate of nerve impulses. Compare myelin sheaths to the many layers of insulation that cover the wires in an electrical cord; the layers keep the electricity flowing along the desired path just as myelin does for nerve fibers. Axons outside the CNS are myelinated by Schwann cells, as just noted. Many of these cells wrap themselves around the axon in a jelly-roll fashion (**Figure 7.5**, p. 257). Initially, the membrane coil is loose, but the

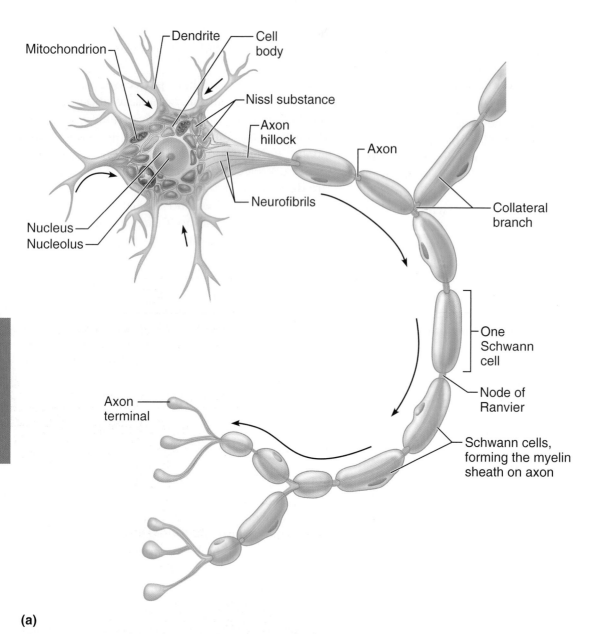

**(a)**

**(b)**

**Figure 7.4 Structure of a typical motor neuron. (a)** Diagrammatic view. **(b)** Scanning electron micrograph showing the cell body and dendrites (615×).

Schwann cell cytoplasm is gradually squeezed from between the membrane layers. When the wrapping process is done, a tight coil of wrapped membranes, the **myelin sheath**, encloses the axon. Most of the Schwann cell cytoplasm ends up just beneath the outermost part of its plasma membrane. This part of the Schwann cell, external to the myelin sheath, is called the **neurilemma** (nu″rĭ-lem′mah, "neuron husk"). Because the myelin sheath is formed by many individual Schwann cells, it has gaps, or indentations, called **nodes of Ranvier** (rahn-vēr), at regular intervals (see Figure 7.4).

As mentioned previously, myelinated fibers are also found in the central nervous system. Oligodendrocytes form CNS myelin sheaths (see Figure 7.3d). In the PNS, it takes many Schwann cells to make a single myelin sheath; but in the CNS, the oligodendrocytes with their many flat extensions can coil around as many as 60 different fibers at the same time. Thus, in the CNS, one oligodendrocyte can form many myelin sheaths. Although the myelin sheaths formed by oligodendrocytes and those formed by Schwann cells are similar, the CNS sheaths lack a neurilemma. Because the neurilemma remains intact (for the most part) when a peripheral nerve fiber is damaged, it plays an important role in fiber regeneration, an ability that is largely lacking in the central nervous system.

### Homeostatic Imbalance 7.1

The importance of myelin insulation is best illustrated by observing what happens when myelin is not there. The disease **multiple sclerosis (MS)** gradually destroys the myelin sheaths around CNS fibers by converting them to hardened sheaths called *scleroses*. As this happens, the electrical current is short-circuited and may "jump" to another demyelinated neuron. In other words, nerve signals do not always reach the intended target. The affected person may have visual and speech disturbances, lose the ability to control his or her muscles, and become increasingly disabled. Multiple sclerosis is an autoimmune disease in which the person's own immune system attacks a protein component of the sheath. As yet there is no cure, but injections of interferon (a hormonelike substance released by some immune cells) appear to hold the symptoms at bay and provide some relief. Other drugs aimed at slowing the autoimmune response are also being used, though further research is needed to determine their long-term effects. ------------------- +

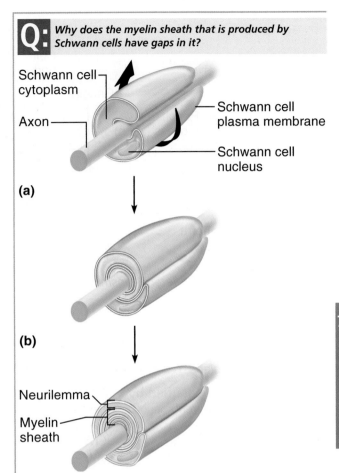

**Q:** *Why does the myelin sheath that is produced by Schwann cells have gaps in it?*

Schwann cell cytoplasm

Axon

Schwann cell plasma membrane

Schwann cell nucleus

(a)

(b)

Neurilemma

Myelin sheath

(c)

**Figure 7.5 Relationship of Schwann cells to axons in the peripheral nervous system. (a–c)** As illustrated (top to bottom), a Schwann cell envelops part of an axon in a trough and then rotates around the axon. Most of the Schwann cell cytoplasm comes to lie just beneath the exposed part of its plasma membrane. The tight coil of plasma membrane material surrounding the axon is the myelin sheath. The Schwann cell cytoplasm and exposed membrane are referred to as the *neurilemma*.

***Terminology*** Clusters of neuron cell bodies and collections of nerve fibers are named differently in the CNS and in the PNS. For the most part, cell bodies are found in the CNS in clusters called **nuclei**. This well-protected location within the bony skull or vertebral column is essential to the well-being of the nervous system—remember that neurons do not routinely undergo cell division

**A:** The sheath is produced by several Schwann cells that arrange themselves end to end along the nerve fiber, each Schwann cell forming only one tiny segment of the sheath.

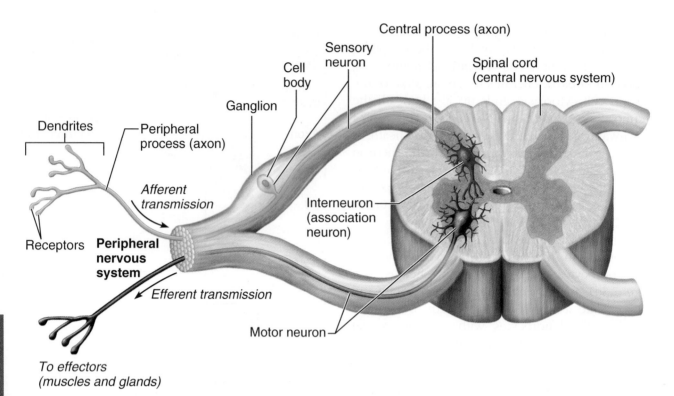

**Figure 7.6 Neurons classified by function.** Sensory (afferent) neurons conduct impulses from sensory receptors (in the skin, viscera, muscles) to the central nervous system; most cell bodies are in ganglia in the PNS. Motor (efferent) neurons transmit impulses from the CNS (brain or spinal cord) to effectors in the body periphery. Interneurons (association neurons) complete the communication pathway between sensory and motor neurons; their cell bodies reside in the CNS.

after birth. The cell body carries out most of the metabolic functions of a neuron, so if it is damaged, the cell dies and is not replaced. Small collections of cell bodies called **ganglia** (gang′le-ah; *ganglion*, singular) are found in a few sites outside the CNS in the PNS.

Bundles of nerve fibers (neuron processes) running through the CNS are called **tracts**, whereas in the PNS they are called **nerves**. The terms *white matter* and *gray matter* refer respectively to myelinated versus unmyelinated regions of the CNS. As a general rule, the **white matter** consists of dense collections of myelinated fibers (tracts), and **gray matter** contains mostly unmyelinated fibers and cell bodies.

## Classification

→ **Learning Objectives**

☐ **Classify neurons according to structure and function.**

☐ **List the types of general sensory receptors and describe their functions.**

Neurons may be classified based on their function or their structure.

*Functional Classification*    Functionally, neurons are grouped according to the direction the nerve impulse travels relative to the CNS. On this basis, there are sensory, motor, and association neurons (interneurons) **(Figure 7.6)**. Neurons carrying impulses from sensory receptors (in the internal organs or the skin) to the CNS are **sensory neurons**, or **afferent neurons**. (Recall that *afferent* means "to go toward.") The cell bodies of sensory neurons are always found in a ganglion outside the CNS. Sensory neurons keep us informed about what is happening both inside and outside the body.

The dendrite endings of the sensory neurons are usually associated with specialized **receptors** that are activated by specific changes occurring nearby. (We cover the very complex receptors of the special senses—vision, hearing, equilibrium, taste, and smell—separately in Chapter 8). The simpler types of sensory receptors in the skin are **cutaneous sense organs**, and those in the muscles and tendons are **proprioceptors** (pro″pre-o-sep′torz) (shown in Figure 4.3 and **Figure 7.7**). The pain receptors (actually bare nerve

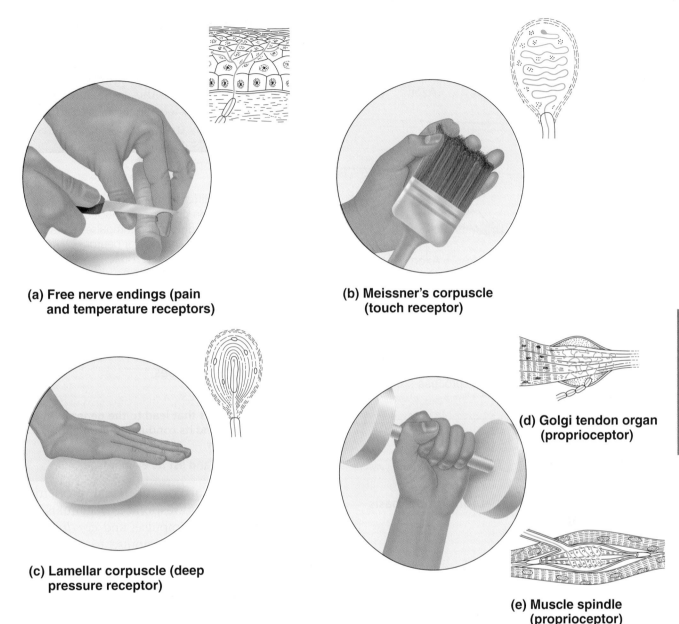

**(a) Free nerve endings (pain and temperature receptors)**

**(b) Meissner's corpuscle (touch receptor)**

**(c) Lamellar corpuscle (deep pressure receptor)**

**(d) Golgi tendon organ (proprioceptor)**

**(e) Muscle spindle (proprioceptor)**

**Figure 7.7  Types of sensory receptors.** Parts (d) and (e) represent two types of proprioceptors.

endings) are the least specialized of the cutaneous receptors. They are also the most numerous, because pain warns us that some type of body damage is occurring or is about to occur. However, strong stimulation of any of the cutaneous receptors (for example, by searing heat, extreme cold, or excessive pressure) is also interpreted as pain.

The proprioceptors detect the amount of stretch, or tension, in skeletal muscles, their tendons, and joints. They send this information to the brain so that the proper adjustments can be made to maintain balance and normal posture. *Propria*

comes from the Latin word meaning "one's own," and the proprioceptors constantly advise our brain of our own movements.

Neurons carrying impulses from the CNS to the viscera and/or muscles and glands are **motor neurons**, or **efferent neurons** (see Figure 7.6). The cell bodies of motor neurons are usually located in the CNS.

The third category of neurons consists of the **interneurons**, or **association neurons**. They connect the motor and sensory neurons in neural

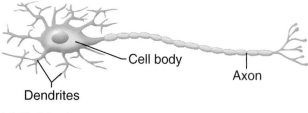

**(a) Multipolar neuron**

**(b) Bipolar neuron**

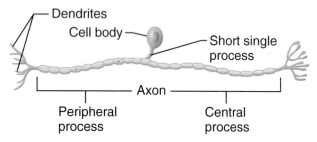

**(c) Unipolar neuron**

**Figure 7.8 Classification of neurons on the basis of structure.**

pathways. Their cell bodies are typically located in the CNS.

***Structural Classification***    Structural classification is based on the number of processes, including both dendrites and axons, extending from the cell body **(Figure 7.8)**. If there are several, the neuron is a **multipolar neuron**. Because all motor and association neurons are multipolar, this is the most common structural type. Neurons with two processes—one axon and one dendrite—are **bipolar neurons**. Bipolar neurons are rare in adults, found only in some special sense organs (eye, nose), where they act in sensory processing as receptor cells. **Unipolar neurons** have a single process emerging from the cell body as if the cell body were on a "cul-de-sac" off the "main road" that is the axon. However, the process is very short and divides almost immediately into proximal (central)

and distal (peripheral) processes. Unipolar neurons are unique in that only the small branches at the end of the peripheral process are dendrites. The remainder of the peripheral process and the central process function as the axon; thus, in this case, the axon actually conducts nerve impulses both toward *and* away from the cell body. Sensory neurons found in PNS ganglia are unipolar.

## Did You Get It?

4. How does a tract differ from a nerve?
5. How does a ganglion differ from a nucleus?
6. Which part of a neuron conducts impulses toward the cell body in multipolar and bipolar neurons? Which part releases neurotransmitters?
7. Your professor tells you that one neuron transmits a nerve impulse at the rate of 1 meter per second and another neuron conducts at the rate of 40 meters per second. Which neuron has the myelinated axon?

For answers, see Appendix A.

## Physiology: Nerve Impulses

→ **Learning Objectives**

☐ **Describe the events that lead to the generation of a nerve impulse and its conduction from one neuron to another.**

☐ **Define *reflex arc*, and list its elements.**

Neurons have two major functional properties: *irritability*, the ability to respond to a stimulus and convert it into a nerve impulse, and *conductivity*, the ability to transmit the impulse to other neurons, muscles, or glands.

***Electrical Conditions of a Resting Neuron's Membrane***    The plasma membrane of a resting, or inactive, neuron is **polarized**, which means that there are fewer positive ions sitting on the inner face of the neuron's plasma membrane than there are on its outer face (**Figure 7.9 ①**). The major positive ions inside the cell are potassium ($K^+$), whereas the major positive ions outside the cell are sodium ($Na^+$). As long as the inside remains more negative (fewer positive ions) than the outside, the neuron will stay inactive.

***Action Potential Initiation and Generation***    Many different types of stimuli excite neurons to become active and generate an impulse. For example, light excites the eye receptors, sound excites some of the ear receptors, and pressure excites some cutaneous

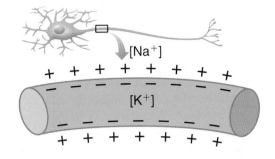

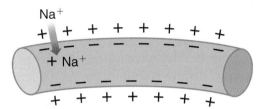

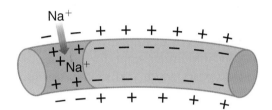

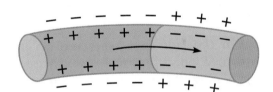

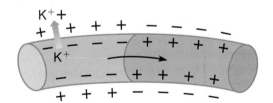

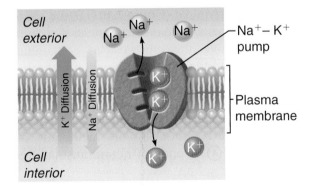

**Figure 7.9 The nerve impulse.**

① **Resting membrane is polarized.** In the resting state, the external face of the membrane is slightly positive; its internal face is slightly negative. The chief extracellular ion is sodium ($Na^+$), whereas the chief intracellular ion is potassium ($K^+$). The membrane is relatively impermeable to both ions.

② **Stimulus initiates local depolarization.** A stimulus changes the permeability of a local "patch" of the membrane, and sodium ions diffuse rapidly into the cell. This changes the polarity of the membrane (the inside becomes more positive; the outside becomes more negative) at that site.

③ **Depolarization and generation of an action potential.** If the stimulus is strong enough, depolarization causes membrane polarity to be completely reversed, and an action potential is initiated.

④ **Propagation of the action potential.** Depolarization of the first membrane patch causes permeability changes in the adjacent membrane, and the events described in step ② are repeated. Thus, the action potential propagates rapidly along the entire length of the membrane.

⑤ **Repolarization.** Potassium ions diffuse out of the cell as the membrane permeability changes again, restoring the negative charge on the inside of the membrane and the positive charge on the outside surface. Repolarization occurs in the same direction as depolarization.

⑥ **Initial ionic conditions restored.** The ionic conditions of the resting state are restored later by the activity of the sodium-potassium pump. Three sodium ions are ejected for every two potassium ions carried back into the cell.

receptors of the skin. However, *most* neurons in the body are excited by neurotransmitter chemicals released by other neurons, as we will describe shortly.

Regardless of the stimulus, the result is always the same—the permeability properties of the cell's plasma membrane change for a very brief period. *Normally*, sodium ions cannot diffuse through the plasma membrane to any great extent, but when the neuron is adequately stimulated, the "gates" of sodium channels in the membrane open. Because sodium is in much higher concentration outside the cell, it then diffuses quickly into the neuron. (Remember the laws of diffusion?) This inward rush of sodium ions changes the polarity of the neuron's membrane at that site, an event called **depolarization** (Figure 7.9 ②). Locally, the inside is now more positive, and the outside is less positive, a local electrical situation called a **graded potential**. However, if the stimulus is strong enough and the sodium influx is great enough, the *local depolarization* (graded potential) activates the neuron to initiate and transmit a long-distance signal called an **action potential**, also called a **nerve impulse** in neurons. ③ The nerve impulse is an *all-or-none response*, like starting a car. It is either propagated (conducted, or sent) over the entire axon ④, or it doesn't happen at all. The nerve impulse never goes partway along an axon's length, nor does it die out with distance, as do graded potentials.

Almost immediately after the sodium ions rush into the neuron, the membrane permeability changes again, becoming impermeable to sodium ions but permeable to potassium ions. So potassium ions are allowed to diffuse out of the neuron into the interstitial fluid, and they do so very rapidly. This outflow of positive ions from the cell restores the electrical conditions at the membrane to the polarized, or resting, state, an event called **repolarization** (Figure 7.9 ⑤). After repolarization of the electrical conditions, the sodium-potassium pump restores the initial concentrations of the sodium and potassium ions inside and outside the neuron ⑥. This pump uses ATP (cellular energy) to pump excess sodium ions out of the cell and to bring potassium ions back into it. *Until repolarization occurs, a neuron cannot conduct another impulse.* Once begun, these sequential events spread along the entire neuronal membrane.

The events just described explain propagation of a nerve impulse along unmyelinated fibers. Fibers that have myelin sheaths conduct impulses much faster because the nerve impulse literally jumps, or leaps, from node to node along the length of the fiber. This occurs because no electrical current can flow across the axon membrane where there is fatty myelin insulation. This faster type of electrical impulse propagation is called *saltatory* (sal′tah-to″re) *conduction* (*saltare* = to dance or leap).

### Homeostatic Imbalance 7.2

A number of factors can impair the conduction of impulses. For example, sedatives and anesthetics block nerve impulses by altering membrane permeability to ions, mainly sodium ions. As we have seen, no sodium entry = no action potential.

Cold and continuous pressure hinder impulse conduction because they interrupt blood circulation (and hence the delivery of oxygen and nutrients) to the neurons. For example, your fingers get numb when you hold an ice cube for more than a few seconds. Likewise, when you sit on your foot, it "goes to sleep." When you warm your fingers or remove the pressure from your foot, the impulses begin to be transmitted again, leading to an unpleasant prickly feeling. _____**✚**

***Transmission of the Signal at Synapses***   So far we have explained only the irritability aspect of neuronal functioning. What about conductivity—how does the electrical impulse traveling along one neuron get across the synapse to the next neuron (or effector cell) to influence its activity?

The answer is that the *impulse* doesn't! Instead, a neurotransmitter chemical crosses the synapse to transmit the signal from one neuron to the next, or to the target cell (as occurred at the neuromuscular junction; see Chapter 6). When the action potential reaches an axon terminal (**Figure 7.10 ①**), the electrical change opens calcium channels. Calcium ions, in turn, cause the tiny vesicles containing neurotransmitter to fuse with the axonal membrane ②, and porelike openings form, releasing the neurotransmitter into the synaptic cleft ③. The neurotransmitter molecules diffuse across the synaptic cleft[*] and bind to receptors on the membrane of the next neuron ④. If enough neurotransmitter is

---

[*]Although most neurons communicate via the *chemical* type of synapse described above, there are some examples of *electrical* synapses, in which the neurons are physically joined by gap junctions and electrical currents actually flow from one neuron to the next.

released, the whole series of events described above (sodium entry ①, depolarization, etc.) will occur, generating a graded potential and eventually a nerve impulse in the receiving neuron beyond the synapse. The electrical changes prompted by neurotransmitter binding are very brief because the neurotransmitter is quickly removed from the synaptic cleft ⑥, either by diffusing away, by reuptake into the axon terminal, or by enzymatic breakdown. This limits the effect of each nerve impulse to a period shorter than the blink of an eye.

Notice that the transmission of an impulse is an *electrochemical event*. Transmission down the length of the neuron's membrane is basically *electrical*, but the next neuron is stimulated by a neurotransmitter, which is a *chemical*. Because each neuron both receives signals from and sends signals to scores of other neurons, it carries on "conversations" with many different neurons at the same time.

## Physiology: Reflexes

Although there are many types of communication between neurons, much of what the body *must* do every day is programmed as reflexes. **Reflexes** are *rapid*, *predictable*, and *involuntary responses* to stimuli. They are much like one-way streets—once a reflex begins, it always goes in the same direction. Reflexes occur over neural pathways called **reflex arcs** and involve both CNS and PNS structures. Think of a reflex as a preprogrammed response to a given stimulus.

The types of reflexes that occur in the body are classed as either somatic or autonomic. **Somatic reflexes** include all reflexes that stimulate the skeletal muscles; these are still involuntary reflexes even though skeletal muscle normally is under voluntary control. When you quickly pull your hand away from a hot object, a somatic reflex is working. **Autonomic reflexes** regulate the activity of smooth muscles, the heart, and glands. Secretion of saliva (salivary reflex) and changes in the size of the eye pupils (pupillary reflex) are two such reflexes. Autonomic reflexes regulate such body functions as digestion, elimination, blood pressure, and sweating.

All reflex arcs have a minimum of five elements (**Figure 7.11a**, p. 264): a *receptor* (which reacts to a stimulus), an *effector* (the muscle or gland eventually stimulated), and *sensory* and *motor neurons* to connect the two. The synapse or interneurons between the sensory and motor neurons represents the fifth element—the CNS *integration center*.

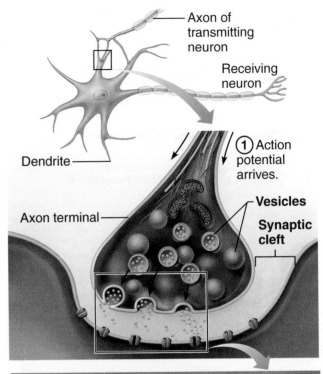

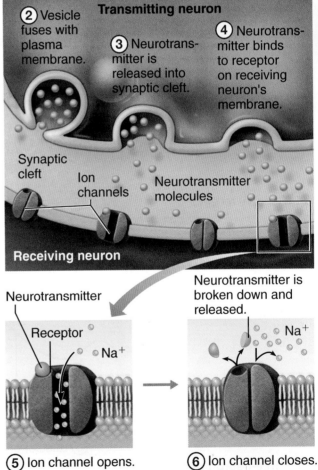

**Figure 7.10 How neurons communicate at chemical synapses.** The events occurring at the synapse are numbered in order.

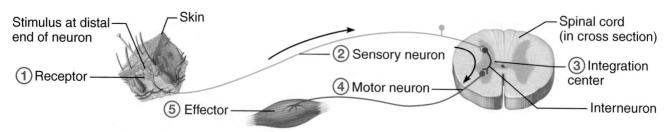

① Receptor — Stimulus at distal end of neuron — Skin — ② Sensory neuron — Spinal cord (in cross section) — ③ Integration center — ④ Motor neuron — Interneuron — ⑤ Effector

**(a) Five basic elements of reflex arc**

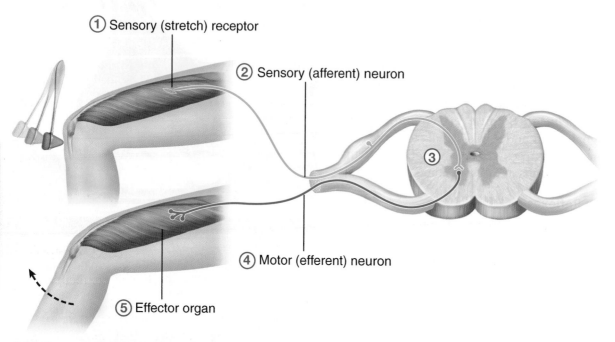

① Sensory (stretch) receptor — ② Sensory (afferent) neuron — ③ — ④ Motor (efferent) neuron — ⑤ Effector organ

**(b) Two-neuron reflex arc**

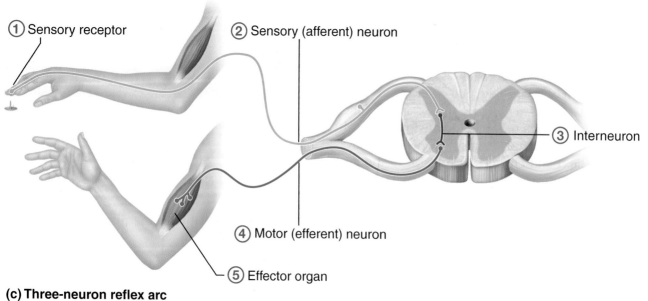

① Sensory receptor — ② Sensory (afferent) neuron — ③ Interneuron — ④ Motor (efferent) neuron — ⑤ Effector organ

**(c) Three-neuron reflex arc**

Figure 7.11 **Simple reflex arcs.**

The simple *patellar* (pah-tel'ar), or *knee-jerk, reflex* is an example of a two-neuron reflex arc, the simplest type in humans (Figure 7.11b). The patellar reflex (in which the quadriceps muscle attached to the hit tendon is stretched) is familiar to most of us. It is usually tested during a physical exam to determine the general health of the motor portion of our nervous system.

Most reflexes are much more complex than the two-neuron reflex, involving synapses between one or more interneurons in the CNS (integration center). The *flexor*, or *withdrawal, reflex* is a three-neuron reflex arc in which the limb is withdrawn from a painful stimulus (see Figure 7.11c). A three-neuron reflex arc also consists of five elements—receptor, sensory neuron, interneuron, motor neuron, and effector. Because there is always a delay at synapses (it takes time for neurotransmitter to diffuse through the synaptic cleft), the more synapses there are in a reflex pathway, the longer the reflex takes to happen.

Many spinal reflexes involve only spinal cord neurons and occur without brain involvement. As long as the spinal cord is functional, spinal reflexes, such as the flexor reflex, will work. By contrast, some reflexes require that the brain become involved because many different types of information have to be evaluated to arrive at the "right" response. The response of the pupils of the eyes to light is a reflex of this type.

As noted earlier, reflex testing is an important tool in evaluating the condition of the nervous system. Reflexes that are exaggerated, distorted, or absent indicate damage or disease in the nervous system. Reflex changes often occur before a pathological condition becomes obvious in other ways.

## Did You Get It?

8. What is the difference between a graded potential and an action potential?
9. Explain the difference between a synaptic cleft and a synapse. How is a stimulus transmitted across a synapse?
10. Which portion(s) of a neuron is (are) likely to be associated with a sensory receptor or a sensory organ?
11. What is a reflex?

For answers, see Appendix A.

# Central Nervous System

## Functional Anatomy of the Brain

### → Learning Objective

☐ Identify and indicate the functions of the major regions of the cerebral hemispheres, diencephalon, brain stem, and cerebellum on a human brain model or diagram.

The adult brain's unimpressive appearance gives few hints of its remarkable abilities. It is about two good fistfuls of pinkish gray tissue, wrinkled like a walnut and with the texture of cold oatmeal. It weighs a little over 3 pounds. Because the brain is the largest and most complex mass of nervous tissue in the body, we commonly discuss it in terms of its four major regions—*cerebral hemispheres, diencephalon* (di"en-sef'ah-lon), *brain stem*, and *cerebellum* (**Figure 7.12,** p. 266 and **Table 7.1**, p. 269).

### Cerebral Hemispheres

The paired **cerebral** (suh re'bral) **hemispheres**, collectively called the **cerebrum**, are the most superior part of the brain and together are a good deal larger than the other three brain regions combined. In fact, as the cerebral hemispheres develop and grow, they enclose and obscure most of the brain stem, so many brain stem structures cannot normally be seen unless a sagittal section is made. Picture how a mushroom cap covers the top of its stalk, and you have an idea of how the cerebral hemispheres cover the diencephalon and the superior part of the brain stem (see Figure 7.12).

The entire surface of the cerebrum exhibits elevated ridges of tissue called **gyri** (ji're; *gyrus,* singular; "twisters"), separated by shallow grooves called **sulci** (sul'ki; *sulcus,* singular; "furrows"). Less numerous are the deeper grooves called **fissures** (**Figure 7.13a**, p. 267), which separate large regions of the brain. Many of the fissures and gyri are important anatomical landmarks. The cerebral hemispheres are separated by a single deep fissure, the *longitudinal fissure.* Other fissures or sulci divide each cerebral hemisphere into a number of **lobes**, named for the cranial bones that lie over them (see Figure 7.13a and b).

Each cerebral hemisphere has three basic regions: a superficial *cortex* of gray matter, which looks gray in fresh brain tissue; an internal area of *white matter*; and the *basal nuclei*, islands of gray

7

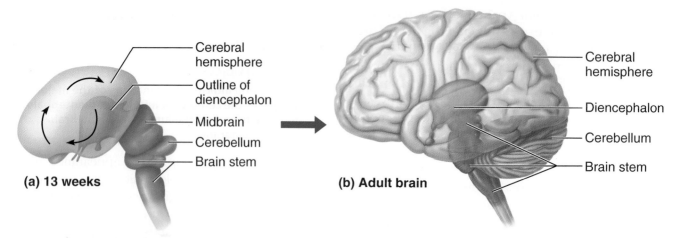

**Figure 7.12 Development and regions of the human brain.** The brain can be considered in terms of four main parts: cerebral hemispheres, diencephalon, brain stem, and cerebellum. **(a)** In the developing brain, the cerebral hemispheres, initially smooth, are forced to grow posteriorly and laterally over the other brain regions by the bones of the skull. **(b)** In the adult brain, the cerebral hemispheres, now highly convoluted, enclose the diencephalon and the superior part of the brain stem. The left cerebral hemisphere is drawn so that it looks transparent, to reveal the location of the deeply situated diencephalon and superior part of the brain stem.

matter situated deep within the white matter. We consider these regions next.

*Cerebral Cortex* Speech, memory, logical and emotional responses, consciousness, the interpretation of sensation, and voluntary movement are all functions of the **cerebral cortex**. Many of the functional areas of the cerebral hemispheres have been identified (Figure 7.13c). The **primary somatic sensory area** is located in the **parietal lobe** posterior to the **central sulcus**. Impulses traveling from the body's sensory receptors (except for the special senses) are localized and interpreted in this area of the brain. The primary somatic sensory area allows you to recognize pain, differences in temperature, or a light touch.

A spatial map, the **sensory homunculus** (ho-mung′ku-lus; "little man"), has been developed to show how much tissue in the primary somatic sensory area is devoted to various sensory functions. (**Figure 7.14**, p. 268; note that the body is represented in an upside-down manner). Body regions with the most sensory receptors—the lips and fingertips—send impulses to neurons that make up a large part of the sensory area. Furthermore, the sensory pathways are crossed pathways—meaning that the left side of the primary somatic sensory area receives impulses from the right side of the body, and vice versa.

Impulses from the special sense organs are interpreted in other cortical areas (see Figure 7.13b and c). For example, the visual area is located in the posterior part of the **occipital lobe**, the auditory area is in the **temporal lobe** bordering the *lateral sulcus*, and the olfactory area is deep inside the temporal lobe.

The **primary motor area**, which allows us to consciously move our skeletal muscles, is anterior to the central sulcus in the *frontal lobe*. The axons of these motor neurons form the major voluntary motor tract—the **pyramidal tract**, or **corticospinal** (kor″tĭ-ko-spi′nal) **tract**, which descends to the cord. As in the primary somatic sensory cortex, the body is represented upside-down, and the pathways are crossed. Most of the neurons in the primary motor area control body areas having the finest motor control; that is, the face, mouth, and hands (see Figure 7.14). The body map on the motor cortex, as you might guess, is called the **motor homunculus**.

A specialized cortical area that is very involved in our ability to speak, **Broca's** (bro′kahz) **area**, or *motor speech area* (see Figure 7.13c), is found at the base of the precentral gyrus (the gyrus anterior to the central sulcus). Damage to this area, which is located in only one cerebral hemisphere (usually the left), causes the inability to say words properly. You know what you want to say, but you can't vocalize the words.

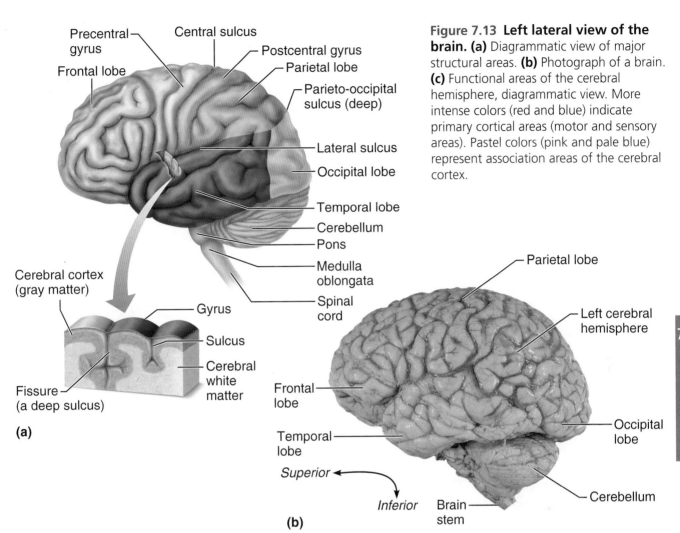

(a)

(b)

**Figure 7.13 Left lateral view of the brain. (a)** Diagrammatic view of major structural areas. **(b)** Photograph of a brain. **(c)** Functional areas of the cerebral hemisphere, diagrammatic view. More intense colors (red and blue) indicate primary cortical areas (motor and sensory areas). Pastel colors (pink and pale blue) represent association areas of the cerebral cortex.

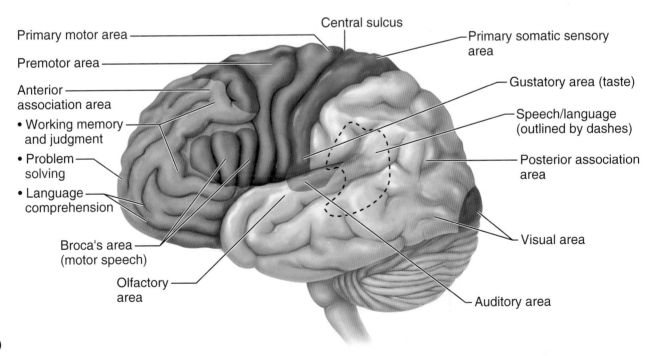

(c)

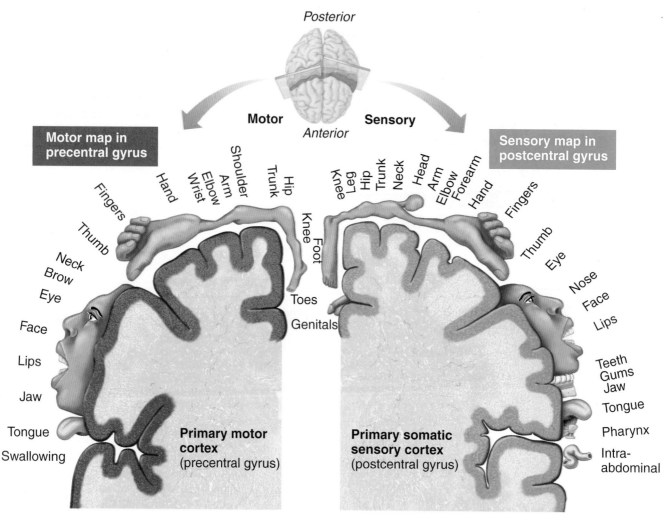

**Figure 7.14 Sensory and motor areas of the cerebral cortex.** The relative amount of cortical tissue devoted to each function is indicated by the amount of the gyrus occupied by the body area diagrams (homunculi). The primary motor cortex is shown on the left side of the diagram, and the somatic sensory cortex is on the right.

Areas involved in higher intellectual reasoning and socially acceptable behavior are believed to be in the anterior part of the frontal lobes, the **anterior association area**. The frontal lobes also house areas involved with language comprehension. Complex memories appear to be stored in the temporal and frontal lobes.

The **posterior association area** encompasses part of the posterior cortex. This area plays a role in recognizing patterns and faces, and blending several different inputs into an understanding of the whole situation. Within this area is the **speech area**, located at the junction of the temporal, parietal, and occipital lobes. The speech area allows you to sound out words. This area (like Broca's area) is usually in only one cerebral hemisphere.

***Cerebral White Matter*** Most of the remaining cerebral hemisphere tissue—the deeper **cerebral white matter** (see Figures 7.13a and 7.15)—is composed of fiber tracts carrying impulses to, from, or within the cortex. One very large fiber tract, the **corpus callosum** (kah-lo'sum), connects the cerebral hemispheres **(Figure 7.15)**. Such fiber tracts are called *commissures*. The corpus callosum arches above the structures of the brain stem and allows the cerebral hemispheres to communicate with one another. This is important because, as already noted, some of the cortical functional areas are in only one hemisphere. *Association fiber tracts* connect areas within a hemisphere, and *projection fiber tracts* connect the cerebrum with lower CNS centers, such as the brain stem.

| Table **7.1** | **Functions of Major Brain Regions** |
| --- | --- |

| Region | Function |
| --- | --- |
| ***Cerebral hemispheres***<br> | ▨ **Cortex: Gray matter:**<br>• Localizes and interprets sensory inputs<br>• Controls voluntary and skilled skeletal muscle activity<br>• Acts in intellectual and emotional processing<br>**Basal nuclei:**<br>• Subcortical motor centers help control skeletal muscle movements (see Figure 7.14) |
| ***Diencephalon***<br><br> | ▨ **Thalamus:**<br>• Relays sensory impulses to cerebral cortex<br>• Relays impulses between cerebral motor cortex and lower motor centers<br>• Involved in memory<br>■ **Hypothalamus:**<br>• Chief integration center of autonomic (involuntary) nervous system<br>• Regulates body temperature, food intake, water balance, and thirst<br>• Regulates hormonal output of anterior pituitary gland and acts as an endocrine organ (producing ADH and oxytocin)<br>■ **Limbic system—A functional system:**<br>• Includes cerebral and diencephalon structures (e.g., hypothalamus and anterior thalamic nuclei)<br>• Mediates emotional response; involved in memory processing |
| ***Brain stem***<br><br> | ▨ **Midbrain:**<br>• Contains visual and auditory reflex centers<br>• Contains subcortical motor centers<br>• Contains nuclei for cranial nerves III and IV; contains projection fibers (e.g., fibers of the pyramidal tracts)<br>▨ **Pons:**<br>• Relays information from the cerebrum to the cerebellum<br>• Cooperates with the medullary centers to control respiratory rate and depth<br>• Contains nuclei of cranial nerves V–VII; contains projection fibers<br>■ **Medulla oblongata:**<br>• Relays ascending sensory pathway impulses from skin and proprioceptors<br>• Contains nuclei controlling heart rate, blood vessel diameter, respiratory rate, vomiting, etc.<br>• Relays sensory information to the cerebellum<br>• Contains nuclei of cranial nerves VIII–XII; contains projection fibers<br>• Site of crossover of pyramids<br>▨ **Reticular formation—A functional system:**<br>• Maintains cerebral cortical alertness; filters out repetitive stimuli<br>• Helps regulate skeletal and visceral muscle activity |
| ***Cerebellum***<br> | ▨ **Cerebellum:**<br>• Processes information from cerebral motor cortex, proprioceptors, and visual and equilibrium pathways<br>• Provides "instructions" to cerebral motor cortex and subcortical motor centers, resulting in smooth, coordinated skeletal muscle movements<br>• Responsible for proper balance and posture |

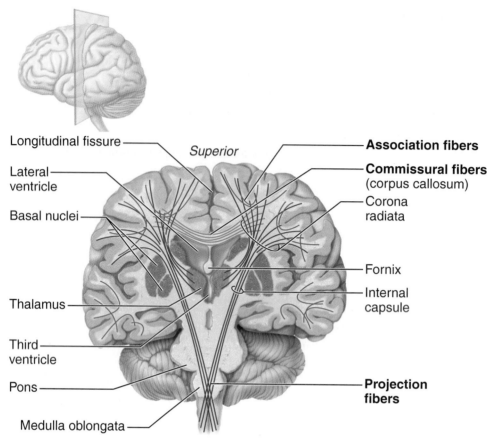

**Figure 7.15 Frontal section (facing posteriorly) of the brain showing commissural, association, and projection fibers running through the cerebrum and the lower CNS.** Notice the *internal capsule* that passes between the thalamus and the basal nuclei.

***Basal Nuclei***   Although most of the gray matter is in the cerebral cortex, there are several "islands" of gray matter, called the **basal nuclei**, buried deep within the white matter of the cerebral hemispheres (see Figure 7.15). The basal nuclei help regulate voluntary motor activities by modifying instructions (particularly in relation to starting or stopping movement) sent to the skeletal muscles by the primary motor cortex. A tight band of projection fibers, called the *internal capsule*, passes between the thalamus and the basal nuclei.

### Homeostatic Imbalance 7.3

Individuals who have problems with their basal nuclei are often unable to walk normally or carry out other voluntary movements in a normal way. *Huntington's disease* and *Parkinson's disease* are two examples of such syndromes. (See "A Closer Look" on pp. 278–279. ⎯⎯⎯⎯⎯⎯✛

### Did You Get It?

12. What are the three major regions of the cerebrum?
13. What is the composition of white matter of the brain?

For answers, see Appendix A.

### Diencephalon

The **diencephalon**, or **interbrain**, sits atop the brain stem and is enclosed by the cerebral hemispheres (see Figure 7.12). The major structures of the diencephalon are the *thalamus, hypothalamus*, and *epithalamus* **(Figure 7.16)**. The **thalamus**, which encloses the shallow *third ventricle* of the brain, is a relay station for sensory impulses passing upward to the sensory cortex. As impulses surge through the thalamus, we have a crude recognition of whether the sensation we are about to have is pleasant or unpleasant. It is the neurons of the sensory cortex that actually localize and interpret the sensation.

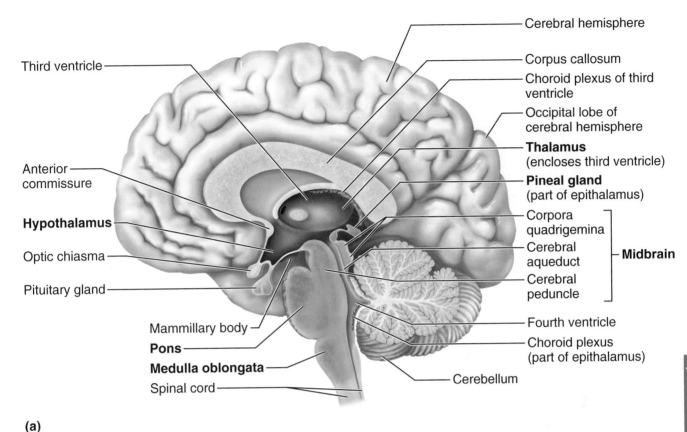

**(a)**

**Figure 7.16 Diencephalon and brain stem structures. (a)** A midsagittal section of the brain illustrating the diencephalon (purple) and brain stem (green). *(Figure continues on page 272.)*

The **hypothalamus** (literally, "under the thalamus") makes up the floor of the diencephalon. It is an important autonomic center because it plays a role in regulating body temperature, water balance, and metabolism. The hypothalamus is also the center for many drives and emotions, and as such it is an important part of the so-called **limbic system**, or "emotional-visceral brain." For example, thirst, appetite, sex, pain, and pleasure centers are in the hypothalamus. Additionally, the hypothalamus regulates the pituitary gland (an endocrine organ) and produces two hormones of its own. The **pituitary gland** hangs from the anterior floor of the hypothalamus by a slender stalk. (We discuss its function in Chapter 9.) The **mammillary bodies**, reflex centers involved in olfaction (the sense of smell), bulge from the floor of the hypothalamus posterior to the pituitary gland.

The **epithalamus** (ep″ĭ-thal′ah-mus) forms the roof of the third ventricle. Important parts of the epithalamus are the **pineal gland** (part of the endocrine system) and the **choroid** (ko′roid) **plexus** of the third ventricle. The choroid plexuses, which are knots of capillaries within each of the four ventricles, form the cerebrospinal fluid.

### Brain Stem

The **brain stem** is about the size of a thumb in diameter and approximately 3 inches (approximately 7.5 cm) long. Its structures are the *midbrain, pons,* and *medulla oblongata.* In addition to providing a pathway for ascending and descending tracts, the brain stem has many small gray matter areas. These nuclei produce the rigidly programmed autonomic behaviors necessary for survival. In addition, some are associated with the cranial nerves and control vital activities such as breathing and blood pressure. Identify the brain stem areas (see Figure 7.16) as you read the descriptions that follow.

*Midbrain* A relatively small part of the brain stem, the **midbrain** extends from the mammillary bodies to the pons inferiorly. The **cerebral aqueduct**, a tiny canal that travels through the midbrain, connects the third ventricle of the diencephalon to the fourth ventricle below. Anteriorly, the midbrain is composed primarily of two bulging fiber tracts, the **cerebral peduncles** (pe′dun klz) (literally, "little feet of the cerebrum"), which convey ascending and descending impulses. Dorsally

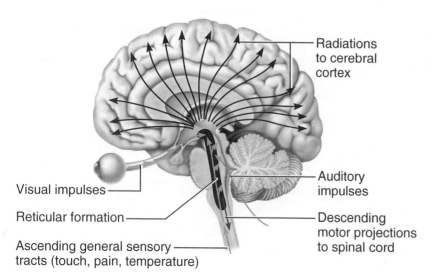

Radiations to cerebral cortex

Visual impulses

Reticular formation

Ascending general sensory tracts (touch, pain, temperature)

Auditory impulses

Descending motor projections to spinal cord

**(b)**

**Figure 7.16** *(continued)* **(b)** The reticular formation, which extends the length of the brain stem. Ascending arrows indicate sensory input to the cerebrum. Descending arrows indicate efferent output of reticular neurons.

located are four rounded protrusions called the **corpora quadrigemina** (kor′por-ah kwah″drĭ-jem′ĭ-nah) because they reminded some anatomist of two pairs of twins (*gemini*). These bulging nuclei are reflex centers involved with vision and hearing.

**Pons** The **pons** (ponz) is the rounded structure that protrudes just below the midbrain. *Pons* means "bridge," and this area of the brain stem is mostly fiber tracts (bundles of nerve fibers in the CNS). However, it does have important nuclei involved in the control of breathing.

**Medulla Oblongata** The **medulla oblongata** (mĕ-dul′ah ob″long-gă′tah) is the most inferior part of the brain stem. It merges into the spinal cord below without any obvious change in structure. Like the pons, the medulla is an important fiber tract area. Additionally, the medulla is the area where the important pyramidal tracts (motor fibers) cross over to the opposite side. The medulla also contains many nuclei that regulate vital visceral activities. It contains centers that control heart rate, blood pressure, breathing, swallowing, and vomiting, among others. The **fourth ventricle** lies posterior to the pons and medulla and anterior to the cerebellum.

**Reticular Formation** Extending the entire length of the brain stem is a diffuse mass of gray matter, the **reticular formation**. The neurons of the reticular formation are involved in motor control of the visceral organs—for example, controlling smooth

muscle in the digestive tract. A special group of reticular formation neurons, the **reticular activating system (RAS)**, plays a role in consciousness and the awake/sleep cycle (Figure 7.16b). The RAS also acts as a filter for the flood of sensory inputs that streams up the spinal cord and brain stem daily. Weak or repetitive signals are filtered out, but unusual or strong impulses do reach consciousness. Damage to this area can result in prolonged unconsciousness (coma).

### Cerebellum

The large, cauliflower-like **cerebellum** (ser″e-bel′um) projects dorsally from under the occipital lobe of the cerebrum. Like the cerebrum, the cerebellum has two hemispheres and a convoluted surface. The cerebellum also has an outer cortex made up of gray matter and an inner region of white matter.

The cerebellum provides the precise timing for skeletal muscle activity and controls our balance. Thanks to its activity, body movements are smooth and coordinated. It plays its role less well when it is sedated by alcohol. Fibers reach the cerebellum from the equilibrium apparatus of the inner ear, the eye, the proprioceptors of the skeletal muscles and tendons, and many other areas. The cerebellum can be compared to an automatic pilot, continuously comparing the brain's "intentions" with actual body performance by monitoring body position and the amount of tension in various body parts. When needed, the cerebellum sends messages to initiate the appropriate corrective measures.

## Homeostatic Imbalance 7.4

If the cerebellum is damaged (for example, by a blow to the head, a tumor, or a stroke), movements become clumsy and disorganized—a condition called **ataxia** (uh tax′e uh). Victims cannot keep their balance and may appear drunk because of the loss of muscle coordination. They are no longer able to touch their finger to their nose with eyes closed—a feat that healthy individuals accomplish easily. ————————————✚

### Did You Get It?

14. Which brain region controls such vital activities as breathing and blood pressure—cerebrum, brain stem, or cerebellum?
15. What is the function of the cerebellum?
16. In what major brain region are the thalamus, hypothalamus, and pineal gland found?

**For answers, see Appendix A.**

## Protection of the Central Nervous System

→ **Learning Objectives**

☐ **Name the three meningeal layers, and state their functions.**

☐ **Discuss the formation and function of cerebrospinal fluid and the blood-brain barrier.**

Nervous tissue is soft and delicate, and even slight pressure can injure the irreplaceable neurons. As we saw in Chapter 5, nature tries to protect the brain and spinal cord by enclosing them within bone (the skull and vertebral column). Now, let's focus on three additional protections for the CNS: the meninges, cerebrospinal fluid, and blood-brain barrier.

### Meninges

The three connective tissue membranes covering and protecting the CNS structures are **meninges** (mĕ-nin′jēz) (**Figure 7.17**, p. 274). The outermost layer, the leathery **dura mater** (du′rah ma′ter), meaning "tough or hard mother," is a double-layered membrane where it surrounds the brain. One of its layers is attached to the inner surface of the skull, forming the periosteum (*periosteal layer*). The other, called the *meningeal layer*, forms the outermost covering of the brain and continues as the dura mater of the spinal cord. The dural layers are fused together except in three areas where they separate to enclose *dural venous sinuses* that

collect venous blood, such as the superior sagittal sinus.

In several places, the inner dural membrane extends inward to form a fold that attaches the brain to the cranial cavity. Two of these folds, the **falx** (falks) **cerebri** and the **tentorium cerebelli**, separate the cerebellum from the cerebrum (shown in Figure 7.17).

The middle meningeal layer is the weblike **arachnoid** (ah-rak′noid) **mater** (see Figure 7.17). *Arachnida* means "spider," and some think the arachnoid membrane looks like a cobweb. Its threadlike extensions span the **subarachnoid space** to attach it to the innermost membrane, the **pia** (pi′ah) **mater** ("gentle mother"). The delicate pia mater clings tightly to the surface of the brain and spinal cord, following every fold.

The subarachnoid space is filled with cerebrospinal fluid. (Remember that the choroid plexuses produce CSF). Specialized projections of the arachnoid membrane, **arachnoid granulations**, protrude through the dura mater. The cerebrospinal fluid is absorbed into the venous blood in the dural sinuses through the arachnoid granulations. Next, we will discuss the production and flow of cerebrospinal fluid.

## Homeostatic Imbalance 7.5

**Meningitis**, an inflammation of the meninges, is a serious threat to the brain because bacterial or viral meningitis may spread into the nervous tissue of the CNS. This condition of brain inflammation is called **encephalitis** (en-sef-ah-li′tis). Meningitis is usually diagnosed by taking a sample of cerebrospinal fluid from the subarachnoid space surrounding the spinal cord. ————————✚

### Cerebrospinal Fluid

**Cerebrospinal** (ser″e-bro-spi′nal) **fluid (CSF)** is a watery "broth" with components similar to blood plasma, from which it forms. However, it contains less protein and more vitamin C, and its ion composition is different.

The choroid plexuses—clusters of capillaries hanging from the "roof" in each of the brain's **ventricles**, or enlarged chambers—continually form CSF from blood. The CSF in and around the brain and cord forms a watery cushion that protects the fragile nervous tissue from blows and other

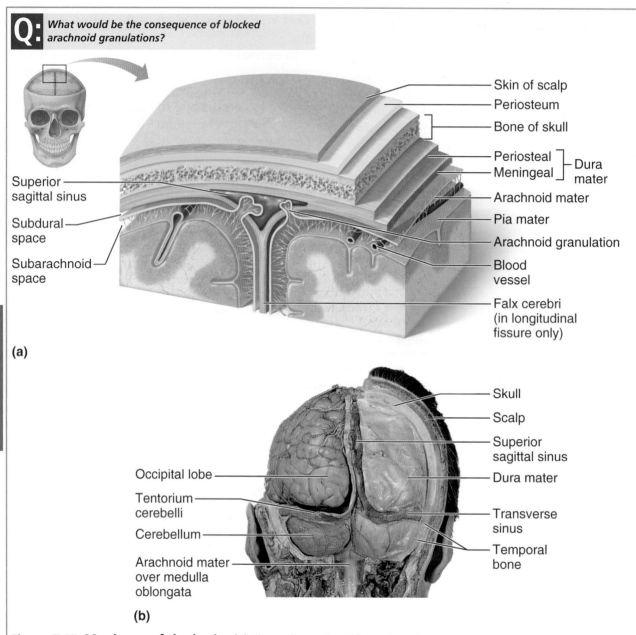

Skin of scalp
Periosteum
Bone of skull
Periosteal ⎫ Dura
Meningeal ⎭ mater
Arachnoid mater
Pia mater
Arachnoid granulation
Blood vessel
Falx cerebri (in longitudinal fissure only)

Superior sagittal sinus
Subdural space
Subarachnoid space

**(a)**

Skull
Scalp
Superior sagittal sinus
Dura mater
Transverse sinus
Temporal bone

Occipital lobe
Tentorium cerebelli
Cerebellum
Arachnoid mater over medulla oblongata

**(b)**

**Figure 7.17 Meninges of the brain. (a)** Three-dimensional frontal section showing the meninges—the dura mater, arachnoid mater, and pia mater—that surround and protect the brain. The relationship of the dura mater to the falx cerebri and the superior sagittal (dural) venous sinus is also shown. **(b)** Posterior view of the brain in place surrounded by the dura mater.

trauma, and helps the brain "float" so it is not damaged by the pressure of its own weight.

Inside the brain, CSF is continually moving (as indicated by the numbered pathway in **Figure 7.18c**).

It circulates from the two lateral ventricles (in the cerebral hemispheres) into the third ventricle (in the diencephalon) and then through the cerebral aqueduct of the midbrain into the fourth ventricle dorsal to the pons and medulla oblongata. Some of the fluid reaching the fourth ventricle simply continues down into the spinal cord, but most of it

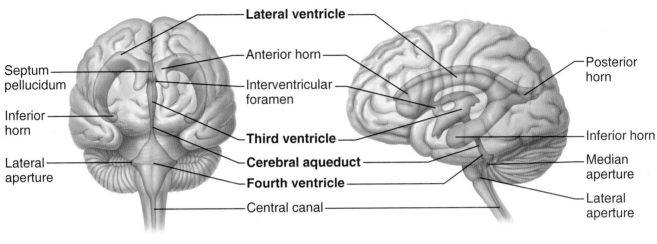

**(a) Anterior view**

**(b) Left lateral view**

Labels for (a) and (b):
- **Lateral ventricle**
- Septum pellucidum
- Anterior horn
- Interventricular foramen
- Inferior horn
- **Third ventricle**
- Lateral aperture
- **Cerebral aqueduct**
- **Fourth ventricle**
- Central canal
- Posterior horn
- Inferior horn
- Median aperture
- Lateral aperture

Labels for (c):
- Superior sagittal sinus
- Choroid plexuses of lateral and third ventricles
- Corpus callosum
- Interventricular foramen
- Third ventricle
- Cerebral aqueduct
- Lateral aperture
- Fourth ventricle
- Median aperture
- Central canal of spinal cord
- Arachnoid granulation
- Subarachnoid space
- Arachnoid mater
- Meningeal dura mater
- Periosteal dura mater
- Right lateral ventricle (deep to cut)
- Choroid plexus of fourth ventricle

**(c) CSF circulation**

① CSF is produced by the choroid plexus of each ventricle.

② CSF flows through the ventricles and into the subarachnoid space via the median and lateral apertures. Some CSF flows through the central canal of the spinal cord.

③ CSF flows through the subarachnoid space.

④ CSF is absorbed into the dural venous sinuses via the arachnoid granulations.

**Figure 7.18 Ventricles and location of the cerebrospinal fluid. (a, b)** Three-dimensional views of the ventricles of the brain.

**(c)** Circulatory pathway of the cerebrospinal fluid (indicated by arrows) within the central nervous system and the subarachnoid space.

(The relative position of the right lateral ventricle is indicated by the pale blue area deep to the corpus callosum.)

circulates into the subarachnoid space through three openings, the paired lateral apertures and the median aperture (*aper* = open), in the walls of the fourth ventricle. The CSF returns to the blood in the dural venous sinuses through the arachnoid granulations. In this way, CSF is continually replaced.

Ordinarily, CSF forms and drains at a constant rate so that its normal pressure and volume (150 ml—about half a cup) are maintained. Any significant changes in CSF composition (or the appearance of blood cells in it) could indicate meningitis or certain other brain pathologies (such as tumors or multiple sclerosis). A procedure called a *lumbar (spinal) puncture* can obtain a sample of CSF for testing. Because the withdrawal of fluid decreases CSF fluid pressure, the patient must remain horizontal (lying down) for 6 to 12 hours after the procedure to prevent an agonizingly painful "spinal headache."

## Homeostatic Imbalance 7.6

If something obstructs its drainage (for example, a tumor), CSF begins to accumulate and exert pressure on the brain. This condition is **hydrocephalus** (hi-dro-sef'ah-lus), literally, "water on the brain." Hydrocephalus in a newborn baby causes the head to enlarge as the brain increases

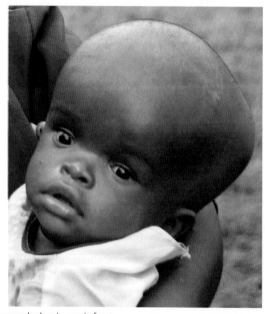

Hydrocephalus in an infant.

in size. This is possible in an infant because the skull bones have not yet fused. However, in an adult this condition is likely to result in brain damage because the skull is hard, and the accumulating fluid creates pressure that crushes soft nervous tissue and could restrict blood flow into the brain. Today hydrocephalus is treated surgically by inserting a shunt (a plastic tube) to drain the excess fluid into a vein in the neck or abdomen. ───────────────────── ✛

### The Blood-Brain Barrier

No other body organ is so absolutely dependent on a constant internal environment as is the brain. Other body tissues can withstand the rather small fluctuations in the concentrations of hormones, ions, and nutrients that continually occur, particularly after eating or exercising. If the brain were exposed to such chemical changes, uncontrolled neural activity might result—remember that certain ions (sodium and potassium) are involved in initiating nerve impulses and that some amino acids serve as neurotransmitters.

Consequently, neurons are kept separated from bloodborne substances by the **blood-brain barrier**, composed of the *least* permeable capillaries in the whole body. These capillaries are almost seamlessly bound together by tight junctions all around. Of water-soluble substances, only water, glucose, and essential amino acids pass easily through the walls of these capillaries. Metabolic wastes, such as urea, toxins, proteins, and most drugs, are prevented from entering brain tissue. Nonessential amino acids and potassium ions not only are prevented from entering the brain, but also are actively pumped from the brain into the blood across capillary walls. Although the bulbous "feet" of the astrocytes that cling to the capillaries may contribute to the barrier, the relative impermeability of the brain capillaries is most responsible for providing this protection.

The blood-brain barrier is virtually useless against fats, respiratory gases, and other fat-soluble molecules that diffuse easily through all plasma membranes. This explains why bloodborne alcohol, nicotine, and anesthetics can affect the brain.

### Did You Get It?

17. What name is given to the cerebrospinal fluid–filled cavities within the brain?
18. What name is given to the barrier that protects the brain from toxic chemicals?

**19.** Which meningeal layer provides the means for draining cerebrospinal fluid back into the blood—dura mater, arachnoid mater, or pia mater?

For answers, see Appendix A.

# Brain Dysfunctions

## → Learning Objectives

☐ Compare the signs of a CVA with those of Alzheimer's disease; of a contusion with those of a concussion.

☐ Define *EEG*, and explain how it evaluates neural functioning.

 **Homeostatic Imbalance 7.7**

Brain dysfunctions are unbelievably varied. We discuss some of them—the "terrible three"—in "A Closer Look" (pp. 278–279), and we will discuss developmental problems in the final section of this chapter. Here, we will focus on traumatic brain injuries and cerebrovascular accidents. We describe techniques used to diagnose many brain disorders in "A Closer Look" later in the chapter (pp. 296–297).

## Traumatic Brain Injuries

Head injuries are a leading cause of accidental death in the United States. Consider, for example, what happens if you crash into the rear end of another car. If you're not wearing a seatbelt, your head will jerk forward and then violently stop as it hits the windshield. Brain trauma is caused not only by injury at the site of the blow, but also by the effect of the ricocheting brain hitting the opposite end of the skull.

A **concussion** occurs when brain injury is slight. The victim may be dizzy, "see stars," or lose consciousness briefly, but typically little permanent brain damage occurs. A brain **contusion** results from marked tissue destruction. If the cerebral cortex is injured, the individual may remain conscious, but severe brain stem contusions always result in a coma lasting from hours to a lifetime due to injury to the reticular activating system.

After head blows, death may result from **intracranial hemorrhage** (bleeding from ruptured vessels) or **cerebral edema** (swelling of the brain due to inflammatory response to injury). Individuals who are initially alert and lucid following head trauma and then begin to deteriorate neurologically are most likely hemorrhaging or suffering the delayed consequences of edema, both of which compress vital brain tissue.

## Cerebrovascular Accidents

Commonly called *strokes*, **cerebrovascular (ser″e-bro-vas′ku-lar) accidents (CVAs)** are the fifth leading cause of death in the United States. CVAs occur when blood circulation to a brain area is blocked, as by a blood clot or a ruptured blood vessel, and vital brain tissue dies. After a CVA, it is often possible to determine the area of brain damage by observing the patient's symptoms. For example, if the patient has left-sided paralysis (a one-sided paralysis is called **hemiplegia**), the right motor cortex of the frontal lobe is most likely involved. **Aphasias** (ah-fa′ze-ahz) are a common result of damage to the left cerebral hemisphere, where the language areas are located. There are many types of aphasias, but the most common are *motor aphasia*, which involves damage to Broca's area and a loss of ability to speak, and *sensory aphasia*, in which a person loses the ability to understand written or spoken language. Aphasias are maddening to victims because, as a rule, their intellect is unimpaired. Brain lesions can also cause marked changes in a person's disposition (for example, a change from a sunny to a foul personality). In such cases, a tumor as well as a CVA might be suspected.

CVA mortality rates have declined over the past 15 years. More than 75 percent of patients who have a CVA survive 1 year, and more than 50 percent survive for 5 years. Some patients recover at least part of their lost faculties because undamaged neurons spread into areas where neurons have died and take over some lost functions. Indeed, most of the recovery seen after brain injury is due to this phenomenon.

Not all strokes are "completed." Temporary brain ischemia, or restricted blood flow, is called a **transient ischemic attack (TIA)**. TIAs last from 5 to 50 minutes and are characterized by symptoms such as numbness, temporary paralysis, and impaired speech. Although these defects are not permanent, they do constitute "red flags" that warn of impending, more serious CVAs. _____ ✚

# The "Terrible Three"

**A**lzheimer's (altz'hi-merz) **disease (AD)** is a progressive degenerative disease of the brain that ultimately results in dementia (mental deterioration). Alzheimer's patients represent between 5 and 15 percent of people over 65, and the disease is a major contributing cause in the deaths of as many as half of those over 85.

Its victims exhibit memory loss (particularly for recent events), a short attention span and disorientation, and eventual language loss. Over several years, formerly good-natured people may become moody, confused, delusional, and sometimes violent.

AD is associated with a shortage of acetylcholine (ACh). AD also exhibits structural changes in the brain, particularly in areas involved with thought and memory; the gyri shrink, and the brain atrophies. The precise cause of AD is unknown, but some cases appear to run in families.

Microscopic examinations of brain tissue reveal senile *plaques* (aggregations of *beta-amyloid peptide*) littering the brain like shrapnel between the neurons. It has been frustratingly difficult for researchers to uncover how beta-amyloid peptide acts as a neurotoxin, particularly because it is also present in healthy brain cells (but there is less of it). Just what tips things off balance to favor production of more beta-amyloid is not understood, but it is known that this tiny peptide does its damage by enhancing calcium entry into certain brain neurons. High calcium levels can play a role in initiating *apoptosis*, or programmed cell death.

Another line of research has implicated a protein called *tau*, which appears to bind microtubule "tracks" together, much like railroad ties. In the brains of AD victims, tau grabs onto other tau molecules, forming spaghetti-like neurofibrillary tangles within neuron cell bodies. These

degenerative changes develop over several years, during which time family members watch the person they love "disappear." It is a long and painful process. It is hoped that the lines of investigation, particularly stem cell research, will eventually merge and point to a treatment, but at present drugs that ease symptoms by inhibiting ACh breakdown are most useful.

**Parkinson's disease**, an example of a basal nuclei problem, typically strikes people in their fifties and sixties (actor Michael J. Fox is an unlucky exception who developed it around age 30). It results from a degeneration of specific neurons in the substantia nigra of the midbrain, which normally supply dopamine to the basal nuclei. The dopamine-deprived basal nuclei, which help regulate voluntary motor activity, become overactive, causing symptoms of the disease. Afflicted individuals have a persistent tremor at rest (exhibited by

## Did You Get It?

20. A person has been hit on the head with a baseball and soon becomes comatose. Has he suffered a concussion or a contusion?

**For the answer, see Appendix A.**

## Spinal Cord

→ **Learning Objectives**

☐ **List two important functions of the spinal cord.**

☐ **Describe spinal cord structure.**

The cylindrical **spinal cord**, which is approximately 17 inches (42 cm) long, is a glistening white continuation of the brain stem. The spinal cord provides a two-way conduction pathway to

and from the brain, and it is a major reflex center (spinal reflexes are completed at this level). Enclosed within the vertebral column, the spinal cord extends from the foramen magnum of the skull to the first or second lumbar vertebra, where it ends just below the ribs (**Figure 7.19**, p. 280).

Like the brain, the spinal cord is cushioned and protected by meninges. Meningeal coverings do not end at the second lumbar vertebra ($L_2$) but instead extend well beyond the end of the spinal cord in the vertebral canal. Because there is no possibility of damaging the cord beyond $L_3$, the meningeal sac inferior to that point provides a nearly ideal spot for a lumbar puncture to remove CSF for testing.

head nodding and a "pill-rolling" movement of the fingers), a forward-bent walking posture and shuffling gait, and a stiff facial expression. In addition, they have trouble initiating movement or getting their muscles going.

The cause of Parkinson's disease is still unknown. The drug L-*dopa*, which is converted to dopamine in the brain, helps alleviate some of the symptoms. However, L-dopa is not a cure, and as more and more neurons die, it becomes ineffective. It also has undesirable side effects: severe nausea, dizziness, and, in some, liver damage. In 2015, the FDA approved two new forms of L–dopa, RYTARY™ and Duopa™, which are extended-release versions of existing drugs. Both medications allow for more consistent dopamine levels and better control of symptoms.

Deep-brain (thalamic) stimulation via implanted electrodes alleviates tremors, but the procedure is risky and expensive. More promising are intrabrain transplants of embryonic substantia nigra tissue, genetically engineered adult nigral cells, or dopamine-producing cells from fetal pigs; all these have produced some regression of disease symptoms. However, the use of fetal tissue is controversial and riddled with ethical and legal roadblocks.

**Huntington's disease** is a genetic disease that strikes during middle age and leads to massive degeneration of the basal nuclei and later of the cerebral cortex. Its initial symptoms in many patients are wild, jerky, and almost continuous flapping movements called *chorea* (Greek for "dance"). Although the movements appear to be voluntary, they are not. Late in the disease, marked mental deterioration occurs, resulting in a lack of focus, fatigue, and irritability. Huntington's disease is progressive, and the prognosis varies from 10 to 30 years from the onset of symptoms.

The signs and symptoms of early Huntington's disease are essentially the opposite of Parkinson's disease (overstimulation rather than inhibition of the motor drive), and Huntington's is usually treated with drugs that block, rather than enhance, dopamine's effects. Late in the disease, however, dopamine levels fall below normal levels, as seen in Parkinson's patients. As with Parkinson's, fetal tissue or stem cell implants may help Huntington's patients in the future.

As you can see, neurotransmitters, which are the "vocabulary" of neurons, can cause garbled neural language when things go wrong.

---

In humans, 31 pairs of spinal nerves arise from the cord and exit from the vertebral column to serve the body area close by. The spinal cord is about the size of a thumb for most of its length, but it is enlarged in the cervical and lumbar regions where the nerves serving the upper and lower limbs arise and leave the cord. Because the vertebral column grows faster than the spinal cord, the spinal cord does not reach the end of the vertebral column, and the spinal nerves leaving its inferior end must travel through the vertebral canal for some distance before exiting. This collection of spinal nerves at the inferior end of the vertebral canal is called the **cauda equina** (kaw'da e-kwi'nah) because it looks so much like a horse's tail (the literal translation of *cauda equina*).

### Gray Matter of the Spinal Cord and Spinal Roots

The gray matter of the spinal cord looks like a butterfly or the letter H in cross section (**Figure 7.20**, p. 281). The two posterior projections are the **dorsal horns**, or **posterior horns**; the two anterior projections are the **ventral horns**, or **anterior horns**. The gray matter surrounds the **central canal** of the cord, which contains CSF.

Neurons with specific functions can be located in the gray matter. The dorsal horns contain interneurons. The cell bodies of the sensory neurons, whose fibers enter the cord by the **dorsal root**, are found in an enlarged area called the **dorsal root ganglion**. If the dorsal root or its ganglion is damaged, sensation from the body area served

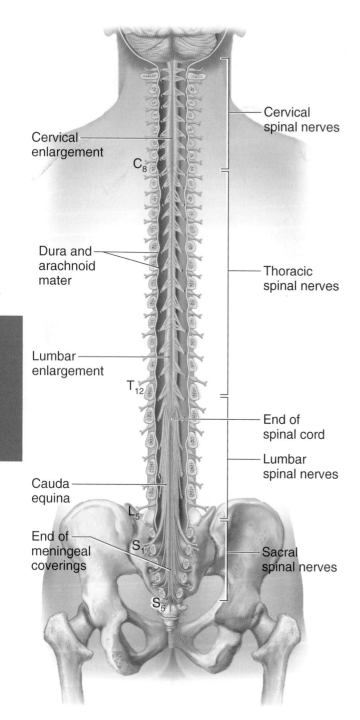

Cervical enlargement

Cervical spinal nerves

$C_8$

Dura and arachnoid mater

Thoracic spinal nerves

Lumbar enlargement

$T_{12}$

End of spinal cord

Lumbar spinal nerves

Cauda equina

$L_5$

End of meningeal coverings

$S_1$

Sacral spinal nerves

$S_5$

**Figure 7.19 Anatomy of the spinal cord, posterior view.**

will be lost. The ventral horns of the gray matter contain cell bodies of motor neurons of the somatic (voluntary) nervous system, which send their axons out the **ventral root** of the cord. The dorsal and ventral roots fuse to form the **spinal nerves**.

## Homeostatic Imbalance 7.8

Damage to the ventral root results in **flaccid paralysis** of the muscles served. In flaccid paralysis, as described in Chapter 6, nerve impulses do not reach the muscles affected; thus, no voluntary movement of those muscles is possible. The muscles begin to atrophy because they are no longer stimulated. ——————————————— ✚

### White Matter of the Spinal Cord

White matter of the spinal cord is composed of myelinated fiber tracts—some running to higher centers, some traveling from the brain to the cord, and some conducting impulses from one side of the spinal cord to the other (**Figure 7.21**, p. 282).

Because of the irregular shape of gray matter, the white matter on each side of the cord is divided into three regions—the **dorsal column**, **lateral column**, and **ventral column**. Each of the columns contains a number of fiber tracts made up of axons with the same destination and function. Tracts conducting sensory impulses to the brain are *sensory*, or *afferent*, *tracts*. Those carrying impulses from the brain to skeletal muscles are *motor*, or *efferent*, *tracts*. All tracts in the dorsal columns are ascending tracts that carry sensory input to the brain. The lateral and ventral columns contain both ascending and descending (motor) tracts.

## Homeostatic Imbalance 7.9

If the spinal cord is transected (cut crosswise) or crushed, **spastic paralysis** results. The affected muscles stay healthy because they are still stimulated by spinal reflex arcs, and movement of those muscles does occur. However, movements are involuntary and not controllable, as described in Chapter 6. This can be as much of a problem as complete lack of mobility. In addition, because the spinal cord carries both sensory and motor impulses, a loss of feeling or sensory input occurs in the body areas below the point of cord destruction. Physicians often use a pin to see whether a person can feel pain after spinal cord injury—to find out whether regeneration is occurring. Pain is a hopeful sign in such cases. If

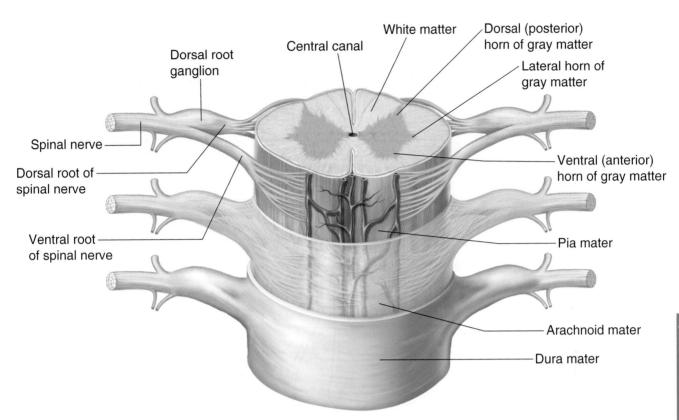

Central canal

White matter

Dorsal (posterior) horn of gray matter

Dorsal root ganglion

Lateral horn of gray matter

Spinal nerve

Dorsal root of spinal nerve

Ventral (anterior) horn of gray matter

Ventral root of spinal nerve

Pia mater

Arachnoid mater

Dura mater

**Figure 7.20 Spinal cord with meninges (three-dimensional, anterior view).**

the spinal cord injury occurs high in the cord, so that all four limbs are affected, the individual is a *quadriplegic* (kwod″rĭ-ple′jik). If only the legs are paralyzed, the individual is a *paraplegic* (par″ă-ple′jik). ⎯⎯⎯⎯⎯⎯⎯⎯⎯⎯✛

### Did You Get It?

21. What is found in the gray matter of the spinal cord?
22. Which spinal cord pathways are sensory pathways—ascending or descending?
23. Why is the leash of nerve fibers at the end of the spinal cord called the cauda equina?

**For answers, see Appendix A.**

# Peripheral Nervous System

The **peripheral nervous system (PNS)** consists of nerves and scattered ganglia (groups of neuronal cell bodies found outside the CNS). We have already considered one type of ganglion—the dorsal root ganglion of the spinal cord. We will cover others in the discussion of the autonomic nervous system. Here, we will concern ourselves only with nerves.

## Structure of a Nerve

→ **Learning Objective**

☐ **Describe the general structure of a nerve.**

As noted earlier in this chapter, a **nerve** is a bundle of neuron fibers found outside the CNS. Within a nerve, neuron fibers, or processes, are wrapped in protective connective tissue coverings. Each fiber is surrounded by a delicate connective tissue sheath, an **endoneurium** (en″do-nu′re-um). Groups of fibers are bound by a coarser connective tissue wrapping, the **perineurium** (per″ĭ-nu′re-um), to form fiber bundles, or **fascicles**. Finally, all the fascicles are bound together by a tough fibrous sheath, the **epineurium**, to form the cordlike nerve (**Figure 7.22**, p. 283).

7

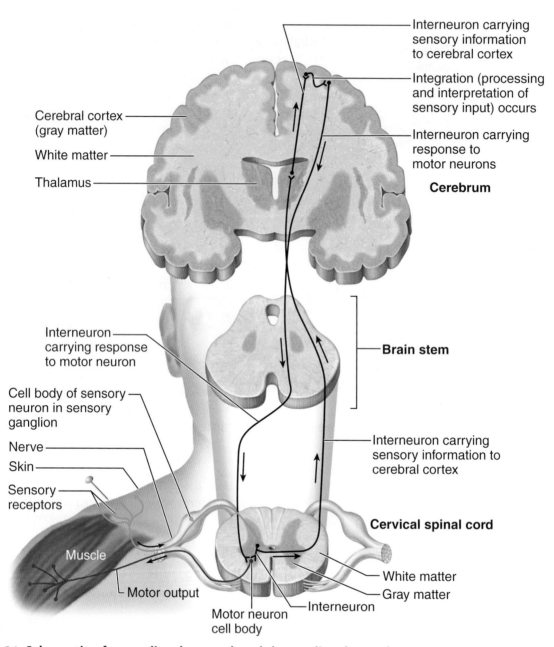

Cerebral cortex
(gray matter)

White matter

Thalamus

Interneuron carrying
sensory information
to cerebral cortex

Integration (processing
and interpretation of
sensory input) occurs

Interneuron carrying
response to
motor neurons

**Cerebrum**

Interneuron
carrying response
to motor neuron

Cell body of sensory
neuron in sensory
ganglion

Nerve

Skin

Sensory
receptors

Muscle

Motor output

Motor neuron
cell body

Interneuron

**Brain stem**

Interneuron carrying
sensory information to
cerebral cortex

**Cervical spinal cord**

White matter

Gray matter

**Figure 7.21  Schematic of ascending (sensory) and descending (motor)
pathways between the brain and the spinal cord.**

**CONCEPTLINK**

The terms for the connective tissue coverings of a
nerve should seem familiar: We discussed similar structures in the muscle chapter (Figure 6.1, p. 209). Names
of muscle structures include the root word *mys*,
whereas the root word *neuro* tells you that the structure relates to a nerve. For example, the endomysium
covers one individual muscle fiber, whereas the endoneurium covers one individual neuron fiber. ←

Like neurons, nerves are classified according
to the direction in which they transmit impulses.
Nerves that carry impulses only toward the CNS
are called **sensory (afferent) nerves**, whereas
those that carry only motor fibers are **motor
(efferent) nerves**. Nerves carrying both sensory
and motor fibers are called **mixed nerves**; all spinal nerves are mixed nerves.

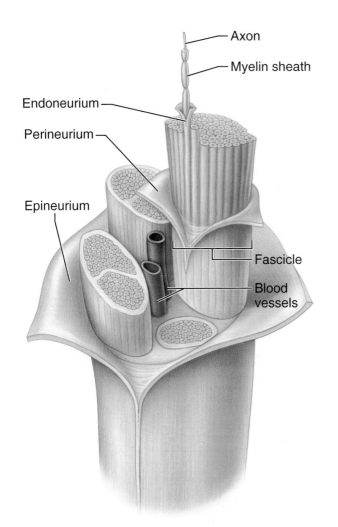

Axon

Myelin sheath

Endoneurium

Perineurium

Epineurium

Fascicle

Blood vessels

**Figure 7.22 Structure of a nerve.** Three-dimensional view of a portion of a nerve, showing its connective tissue wrappings.

## Cranial Nerves

→ **Learning Objective**

☐ Identify the cranial nerves by number and by name, and list the major functions of each.

The 12 pairs of **cranial nerves** primarily serve the head and neck. Only one pair (the vagus nerves) extends to the thoracic and abdominal cavities.

The cranial nerves are numbered in sequence, and in most cases their names reveal the most important structures they control. **Table 7.2** (pp. 286–287) describes cranial nerves by name, number, course, and major function. The last column of the table describes how cranial nerves

are tested, which is an important part of any neurological examination. You do not need to memorize these tests, but this information may help you understand cranial nerve function. As you read through the table, also look at **Figure 7.23**, p. 284, which shows the location of the cranial nerves on the brain's anterior surface.

Most cranial nerves are mixed nerves; however, three pairs—the optic, olfactory, and vestibulocochlear (ves-tib″u-lo-kok′le-ar) nerves—are purely sensory in function. (The older name for the vestibulocochlear nerve is *acoustic nerve*, a name that reveals its role in hearing but not in equilibrium.)

This study tool may help you remember the cranial nerves in order. The first letter of each word in the saying is the first letter of the cranial nerve to be remembered: "**O**h, **o**h, **o**h, **t**o **t**ouch **a**nd **f**eel **v**ery **g**ood **v**elvet **a**t **h**ome."

## Spinal Nerves and Nerve Plexuses

→ **Learning Objectives**

☐ Describe the origin and fiber composition of (1) ventral and dorsal roots, (2) the spinal nerve proper, and (3) ventral and dorsal rami.

☐ Name the four major nerve plexuses, give the major nerves of each, and describe their distribution.

The 31 pairs of human **spinal nerves** are formed by the combination of the ventral and dorsal roots of the spinal cord. Although each of the cranial nerves issuing from the brain is named specifically, the spinal nerves are named for the region of the cord from which they arise. (**Figure 7.24**, p. 285, shows how the nerves are named in this scheme.)

Almost immediately after being formed, each spinal nerve divides into the **dorsal ramus** and **ventral ramus** (plural *rami* [ra′mi]), making each spinal nerve only about ½ inch long. The rami, like the spinal nerves, contain both motor and sensory fibers. Thus, damage to a spinal nerve or either of its rami results both in loss of sensation and in flaccid paralysis of the area of the body served. The smaller dorsal rami serve the skin and muscles of the posterior body trunk. The ventral rami of spinal nerves $T_1$ through $T_{12}$ form the *intercostal nerves*, which supply the muscles

7

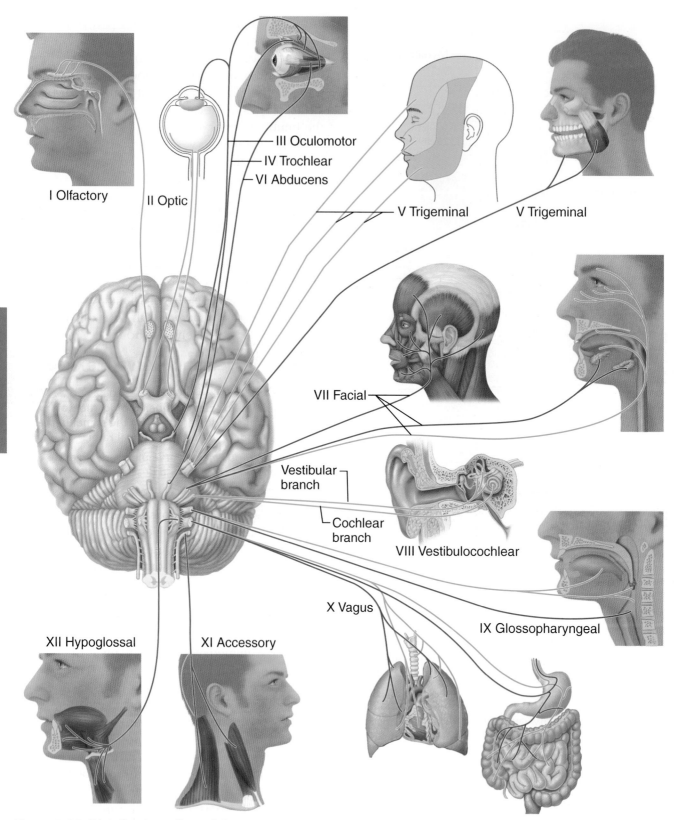

I Olfactory

II Optic

III Oculomotor
IV Trochlear
VI Abducens

V Trigeminal

V Trigeminal

VII Facial

Vestibular branch

Cochlear branch

VIII Vestibulocochlear

X Vagus

IX Glossopharyngeal

XII Hypoglossal

XI Accessory

**Figure 7.23 Distribution of cranial nerves.** Sensory nerves are shown in blue, motor nerves in red. Although cranial nerves III, IV, and VI have sensory fibers, these are not shown because the sensory fibers account for only minor parts of these nerves.

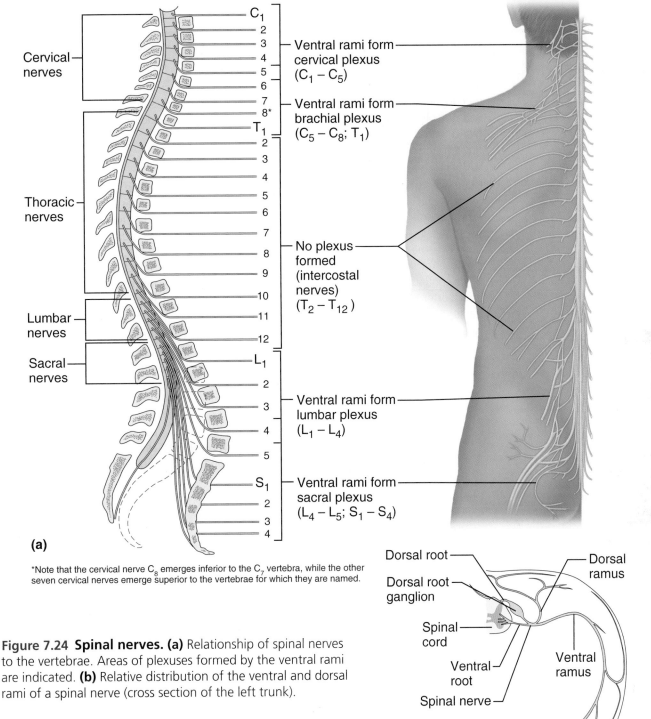

*Note that the cervical nerve $C_8$ emerges inferior to the $C_7$ vertebra, while the other seven cervical nerves emerge superior to the vertebrae for which they are named.

**Figure 7.24 Spinal nerves. (a)** Relationship of spinal nerves to the vertebrae. Areas of plexuses formed by the ventral rami are indicated. **(b)** Relative distribution of the ventral and dorsal rami of a spinal nerve (cross section of the left trunk).

between the ribs and the skin and muscles of the anterior and lateral trunk. The ventral rami of all other spinal nerves form complex networks of nerves called **plexuses**, which serve the motor and sensory needs of the limbs. (The four nerve plexuses are described in Table 7.3, p. 288; three of the four plexuses are shown in **Figure 7.25**, p. 289.)

*(Text continues on page 290.)*

Table **7.2**  **The Cranial Nerves**

| Name/number | Origin/course | Function | Test |
|---|---|---|---|
| I. Olfactory | Fibers arise from olfactory receptors in the nasal mucosa and synapse with the olfactory bulbs (which, in turn, send fibers to the olfactory cortex) | Purely sensory; carries impulses for the sense of smell | Subject is asked to sniff and identify aromatic substances, such as oil of cloves or vanilla |
| II. Optic | Fibers arise from the retina of the eye and form the optic nerve. The two optic nerves form the optic chiasma by partial crossover of fibers; the fibers continue to the optic cortex as the optic tracts | Purely sensory; carries impulses for vision | Vision and visual field are tested with an eye chart and by testing the point at which the subject first sees an object (finger) moving into the visual field; eye interior is viewed with an ophthalmoscope |
| III. Oculomotor | Fibers run from the midbrain to the eye | Supplies motor fibers to four of the six muscles (superior, inferior, and medial rectus, and inferior oblique) that direct the eyeball; to the eyelid; and to the internal eye muscles controlling lens shape and pupil size | Pupils are examined for size, shape, and size equality; pupillary reflex is tested with a penlight (pupils should constrict when illuminated); eye convergence is tested, as is the ability to follow moving objects |
| IV. Trochlear | Fibers run from the midbrain to the eye | Supplies motor fibers for one external eye muscle (superior oblique) | Tested in common with cranial nerve III for the ability to follow moving objects |
| V. Trigeminal | Fibers emerge from the pons and form three divisions that run to the face | Conducts sensory impulses from the skin of the face and mucosa of the nose and mouth; also contains motor fibers that activate the chewing muscles | Sensations of pain, touch, and temperature are tested with a safety pin and hot and cold objects; corneal reflex tested with a wisp of cotton; motor branch tested by asking the subject to open mouth against resistance and move jaw from side to side |
| VI. Abducens | Fibers leave the pons and run to the eye | Supplies motor fibers to the lateral rectus muscle, which rolls the eye laterally | Tested in common with cranial nerve III for the ability to move each eye laterally |

Table **7.2**  *(continued)*

| Name/number | Origin/course | Function | Test |
|---|---|---|---|
| VII. Facial | Fibers leave the pons and run to the face | Activates the muscles of facial expression and the lacrimal and salivary glands; carries sensory impulses from the taste buds of anterior tongue | Anterior two-thirds of tongue is tested for ability to taste sweet, salty, sour, and bitter substances; subject is asked to close eyes, smile, whistle, etc.; tearing of eyes is tested with ammonia fumes |
| VIII. Vestibulocochlear | Fibers run from the equilibrium and hearing receptors of the inner ear to the brain stem | Purely sensory; vestibular branch transmits impulses for the sense of balance, and cochlear branch transmits impulses for the sense of hearing | Hearing is checked by air and bone conduction, using a tuning fork |
| IX. Glossopharyngeal | Fibers emerge from the medulla and run to the throat | Supplies motor fibers to the pharynx (throat) that promote swallowing and saliva production; carries sensory impulses from taste buds of the posterior tongue and from pressure receptors of the carotid artery | Gag and swallowing reflexes are checked; subject is asked to speak and cough; posterior tongue may be tested for taste |
| X. Vagus | Fibers emerge from the medulla and descend into the thorax and abdominal cavity | Fibers carry sensory impulses from and motor impulses to the pharynx, larynx, and the abdominal and thoracic viscera; most motor fibers are parasympathetic fibers that promote digestive activity and help regulate heart activity | Tested in common with cranial nerve IX because they both serve muscles of the throat |
| XI. Accessory | Fibers arise from the superior spinal cord ($C_1$–$C_5$)* and travel to muscles of the neck and back | Mostly motor fibers that activate the sternocleidomastoid and trapezius muscles | Sternocleidomastoid and trapezius muscles are checked for strength by asking the subject to rotate head and shrug shoulders against resistance |
| XII. Hypoglossal | Fibers run from the medulla to the tongue | Motor fibers control tongue movements; sensory fibers carry impulses from the tongue | Subject is asked to stick out tongue, and any position abnormalities are noted |

*Until recently, it was thought that the accessory nerves also received a contribution from cranial rootlets, but it has now been determined that in almost all people, these cranial rootlets are instead part of the vagus nerves. This raises an interesting question: Should the accessory nerves still be considered cranial nerves? Some anatomists say yes because they pass through the cranium. Other anatomists say no because they don't arise from the brain. Stay tuned!

Table **7.3**  **Spinal Nerve Plexuses**

| Plexus | Origin (from ventral rami) | Important nerves | Body areas served | Result of damage to plexus or its nerves |
|---|---|---|---|---|
| Cervical | $C_1$–$C_5$ | Phrenic | Diaphragm; skin and muscles of shoulder and neck | Respiratory paralysis (and death if not treated promptly) |
| Brachial | $C_5$–$C_8$ and $T_1$ | Axillary | Deltoid muscle and skin of shoulder; muscles and skin of superior thorax | Paralysis and atrophy of deltoid muscle |
| | | Radial | Triceps and extensor muscles of the forearm; skin of posterior upper limb | Wristdrop—inability to extend hand at wrist |
| | | Median | Flexor muscles and skin of forearm and some muscles of hand | Decreased ability to flex and abduct hand and flex and abduct thumb and index finger—therefore, inability to pick up small objects |
| | | Musculocutaneous | Flexor muscles of arm; skin of lateral forearm | Decreased ability to flex forearm at elbow |
| | | Ulnar | Some flexor muscles of forearm; wrist and many hand muscles; skin of hand | Clawhand—inability to spread fingers apart |
| Lumbar | $L_1$–$L_4$ | Femoral (including lateral and anterior cutaneous branches) | Lower abdomen, anterior and medial thigh muscles (hip flexors and knee extensors), and skin of anteromedial leg and thigh | Inability to extend leg and flex hip; loss of cutaneous sensation |
| | | Obturator | Adductor muscles of medial thigh and small hip muscles; skin of medial thigh and hip joint | Inability to adduct thigh |
| Sacral | $L_4$–$L_5$ and $S_1$–$S_4$ | Sciatic (largest nerve in body; splits to common fibular and tibial nerves just above knee) | Lower trunk and posterior surface of thigh (hip extensors and knee flexors) | Inability to extend hip and flex knee; sciatica |
| | | • Common fibular (superficial and deep branches) | Lateral aspect of leg and foot | Footdrop—inability to dorsiflex foot |
| | | • Tibial (including sural and plantar branches) | Posterior aspect of leg and foot | Inability to plantar flex and invert foot; shuffling gait |
| | | Superior and inferior gluteal | Gluteus muscles of hip | Inability to extend hip (maximus) or abduct and medially rotate thigh (medius) |

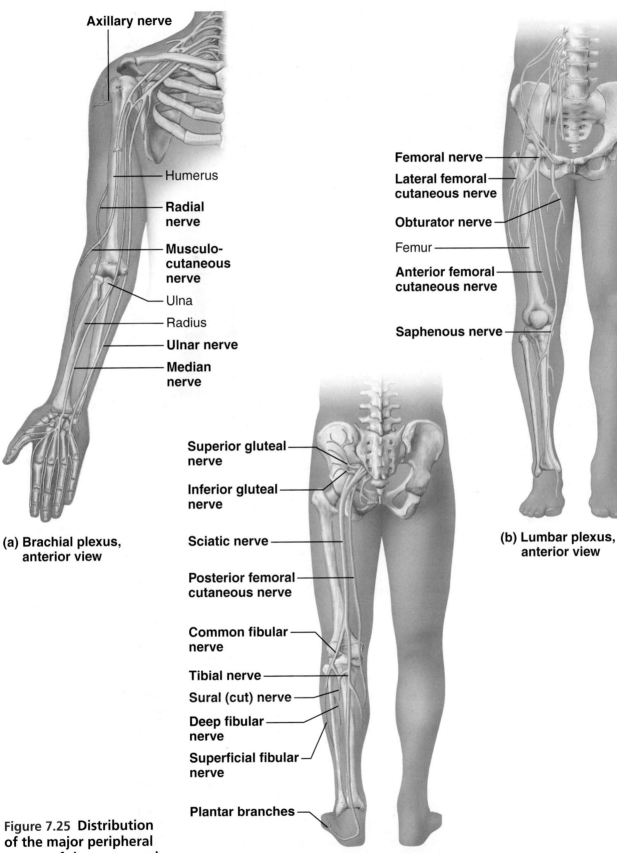

**Axillary nerve**

Humerus

**Radial nerve**

**Musculo-cutaneous nerve**

Ulna

Radius

**Ulnar nerve**

**Median nerve**

**(a) Brachial plexus, anterior view**

**Femoral nerve**

**Lateral femoral cutaneous nerve**

**Obturator nerve**

Femur

**Anterior femoral cutaneous nerve**

**Saphenous nerve**

**(b) Lumbar plexus, anterior view**

**Superior gluteal nerve**

**Inferior gluteal nerve**

**Sciatic nerve**

**Posterior femoral cutaneous nerve**

**Common fibular nerve**

**Tibial nerve**

**Sural (cut) nerve**

**Deep fibular nerve**

**Superficial fibular nerve**

**Plantar branches**

**(c) Sacral plexus, posterior view**

Figure 7.25 **Distribution of the major peripheral nerves of the upper and lower limbs.**

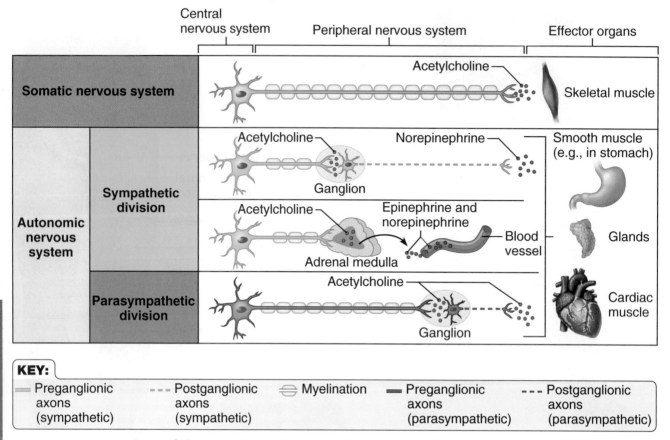

**Figure 7.26 Comparison of the somatic and autonomic nervous systems.**

## Did You Get It?

24. Where is the epineurium located?
25. Which cranial nerve pair is the only one to serve structures outside the head and neck?
26. What is a nerve plexus?
27. Ron has a horrible pain in his right buttock, thigh, and leg. He is told he has sciatica. Which spinal nerve is involved, and what plexus does it belong to?

For answers, see Appendix A.

## Autonomic Nervous System

→ **Learning** Objective

☐ Identify the site of origin, and explain the function of the sympathetic and parasympathetic divisions of the autonomic nervous system.

The **autonomic nervous system (ANS)** is the motor subdivision of the PNS that controls body activities automatically. It is composed of a specialized group of neurons that regulate cardiac muscle (the heart), smooth muscles (found in the walls of the visceral organs and blood vessels), and glands. Although all body systems contribute

to homeostasis, the relative stability of our internal environment depends largely on the workings of the ANS. At every moment, signals flood from the visceral organs into the CNS, and the autonomic nervous system makes adjustments as necessary to best support body activities. For example, blood flow may be shunted to more "needy" areas, heart and breathing rate may be sped up or slowed down, blood pressure may be adjusted, and stomach secretions may be increased or decreased. Most of this fine-tuning occurs without our awareness or attention—few of us realize when our pupils dilate or our arteries constrict— hence the ANS is also called the **involuntary nervous system**, as noted at the beginning of this chapter.

### Somatic and Autonomic Nervous Systems Compared

Our previous discussions of nerves in the efferent (motor) division of the PNS have focused on the somatic nervous system, the PNS subdivision that controls our skeletal muscles. So, before plunging

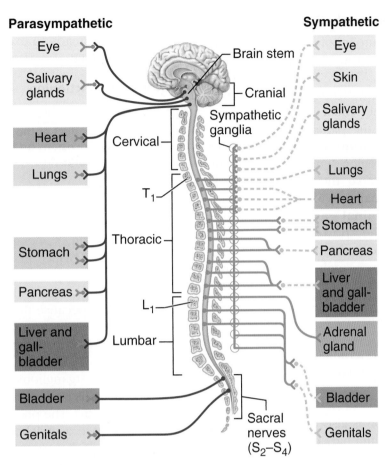

**Figure 7.27 Anatomy of the autonomic nervous system.** Parasympathetic fibers are shown in purple, sympathetic fibers in green. Solid lines represent preganglionic fibers; dashed lines indicate postganglionic fibers. Note the direct connection between the sympathetic nervous system and the adrenal gland.

into a description of autonomic nervous system anatomy, let's note some important differences between the somatic and autonomic subdivisions of the PNS.

Besides differences in their effector organs and in the neurotransmitters they release, the patterns of their motor pathways differ. In the somatic division, the cell bodies of the motor neurons are inside the CNS, and their axons (in spinal nerves) extend all the way to the skeletal muscles they serve. The autonomic nervous system, however, has a chain of *two* motor neurons. The first motor neuron of each pair, the **preganglionic neuron**, is in the brain or spinal cord. Its axon, the **preganglionic axon** (literally, the "axon before the ganglion"), leaves the CNS to form a synapse with the second motor neuron in a ganglion outside the

CNS. The axon of this ganglionic neuron, the **postganglionic axon**, then extends to the organ it serves. (**Figure 7.26** summarizes these differences.)

The autonomic nervous system has two arms, the sympathetic and the parasympathetic (**Figure 7.27**). Both serve the same organs but cause essentially opposite effects, counterbalancing each other's activities to keep body systems running smoothly. The **sympathetic division** mobilizes the body during extreme situations (such as fear, exercise, or rage), whereas the **parasympathetic division** allows us to "unwind" and conserve energy. We examine these differences in more detail shortly, but first let's consider the structural characteristics of the two arms of the ANS.

## Anatomy of the Parasympathetic Division

The preganglionic neurons of the parasympathetic division are located in brain nuclei of several cranial nerves—III, VII, IX, and X (the vagus being the most important of these) and in the $S_2$ through $S_4$ levels of the spinal cord (see Figure 7.27). For this reason, the parasympathetic division is also called the **craniosacral division**. The neurons of the cranial region send their axons out in cranial nerves to serve the head and neck organs. There they synapse with the ganglionic motor neuron in a **terminal ganglion**. From the terminal ganglion, the postganglionic axon extends a short distance to the organ it serves. In the sacral region, the preganglionic axons leave the spinal cord and form the *pelvic splanchnic* (splank′nik) *nerves*, also called the *pelvic nerves*, which travel to the pelvic cavity. In the pelvic cavity, the preganglionic axons synapse with the second motor neurons in terminal ganglia on, or close to, the organs they serve.

## Anatomy of the Sympathetic Division

The sympathetic division is also called the **thoracolumbar** (tho″rah-ko-lum′bar) **division** because its preganglionic neurons are in the gray matter of the spinal cord from $T_1$ through $L_2$ (see Figure 7.27). The preganglionic axons leave the cord in the ventral root, enter the spinal nerve, and then pass through a **ramus communicans**, or small communicating branch, to enter a **sympathetic trunk ganglion (Figure 7.28)**. The **sympathetic trunk**, or **sympathetic chain** lies alongside the vertebral column on each side. After it reaches the ganglion, the axon may synapse with the second (ganglionic) neuron in the sympathetic chain at the same level (Figure 7.28a) or a different level (Figure 7.28b), and the postganglionic axon then reenters the spinal nerve to travel to the skin. Or, the preganglionic axon may pass through the ganglion without synapsing and form part of the **splanchnic nerves** (Figure 7.28c). The splanchnic nerves travel to the viscera to synapse with the ganglionic neuron, found in a **collateral ganglion** anterior to the vertebral column. The major collateral ganglia—the celiac and the superior and inferior mesenteric ganglia—supply the abdominal and pelvic organs. The postganglionic axon then leaves the collateral ganglion and travels to serve a nearby visceral organ.

Now that we have described the anatomical details, we are ready to examine ANS functions in a little more detail.

## Autonomic Functioning

→ **Learning Objective**

☐ Contrast the effect of the parasympathetic and sympathetic divisions on the following organs: heart, lungs, digestive system, blood vessels.

Body organs served by the autonomic nervous system receive fibers from both divisions. Exceptions are most blood vessels and most structures of the skin, some glands, and the adrenal medulla, all of which receive only sympathetic fibers (Table 7.4, p. 294). When both divisions serve the same organ, they cause antagonistic effects, mainly because their postganglionic axons release different neurotransmitters (see Figure 7.26). The parasympathetic fibers, called *cholinergic* (ko″lin-er′jik) *fibers*, release acetylcholine. The sympathetic postganglionic fibers, called *adrenergic* (ad″ren-er′jik) *fibers*, release norepinephrine (nor″ep-ĭ-nef′rin). The preganglionic axons of *both* divisions release acetylcholine. To emphasize the *relative* roles of the two arms of the ANS, we will focus briefly on situations in which each division is "in control."

***Sympathetic Division*** The sympathetic division is often referred to as the "fight-or-flight" system. Its activity is evident when we are excited or find ourselves in emergency or threatening situations, such as being frightened by a stranger late at night. A pounding heart; rapid, deep breathing; cold, sweaty skin; a prickly scalp; and dilated eye pupils are sure signs of sympathetic nervous system activity. Under such conditions, the sympathetic nervous system increases heart rate, blood pressure, and blood glucose levels; dilates the bronchioles of the lungs; and brings about many other effects that help the individual cope with the stressor. Other examples are dilation of blood vessels in skeletal muscles (so that we can run faster or fight better) and withdrawal of blood from the digestive organs (so that the bulk of the blood can be used to serve the heart, brain, and skeletal muscles).

The sympathetic nervous system is working at full speed not only when you are emotionally upset but also when you are physically stressed.

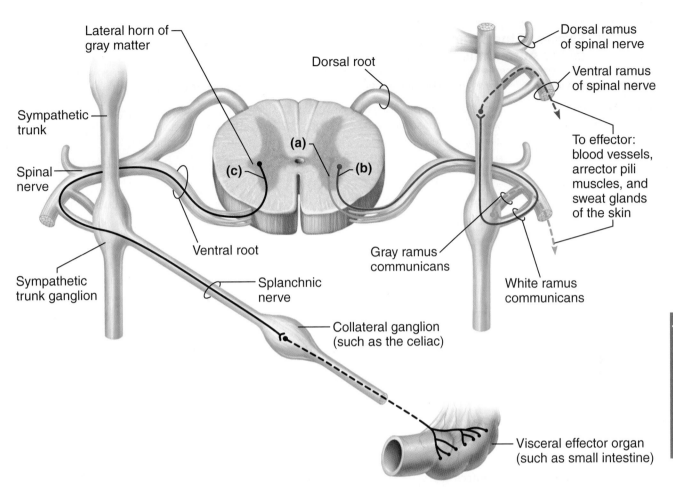

**Figure 7.28 Sympathetic pathways. (a)** Synapse in a sympathetic trunk ganglion at the same level. **(b)** Synapse in a sympathetic trunk ganglion at a different level. **(c)** Synapse in a collateral ganglion anterior to the vertebral column.

For example, if you have just had surgery or run a marathon, your adrenal glands (activated by the sympathetic nervous system) will be pumping out epinephrine and norepinephrine (see Figure 7.26). The effects of sympathetic nervous system activation continue for several minutes until its hormones are destroyed by the liver. Thus, although sympathetic nerve impulses themselves may act only briefly, the hormonal effects they provoke linger. The widespread and prolonged effects of sympathetic activation help explain why we need time to calm down after an extremely stressful situation.

The sympathetic division generates a head of steam that enables the body to cope rapidly and vigorously with situations that threaten homeostasis. Its function is to provide the best conditions for responding to some threat, whether the best response is to run, to see better, or to think more clearly.

### Homeostatic Imbalance 7.10

Some illnesses or diseases are at least aggravated, if not caused, by excessive sympathetic nervous system stimulation. Certain individuals, called type A people, always work at breakneck speed and push themselves continually. These are people who are likely to have heart disease, high blood pressure, and ulcers, all of which may be worsened by prolonged sympathetic nervous system activity or the rebound from it. ——————+

***Parasympathetic Division*** The parasympathetic division is most active when the body is at rest and not threatened in any way. This division, sometimes called the "rest-and-digest" system, is chiefly concerned with promoting normal digestion, with elimination of feces and urine, and with conserving body energy, particularly by decreasing demands

| Table **7.4** | **Effects of the Sympathetic and Parasympathetic Divisions of the Autonomic Nervous System** | |

| Target organ/system | Parasympathetic effects | Sympathetic effects |
|---|---|---|
| Digestive system | Increases smooth muscle mobility (peristalsis) and amount of secretion by digestive system glands; relaxes sphincters | Decreases activity of digestive system and constricts digestive system sphincters (for example, anal sphincter) |
| Liver | No effect | Causes glucose to be released to blood |
| Lungs | Constricts bronchioles | Dilates bronchioles |
| Urinary bladder/urethra | Relaxes sphincters (allows voiding) | Constricts sphincters (prevents voiding) |
| Kidneys | No effect | Decreases urine output |
| Heart | Decreases rate; slows and steadies | Increases rate and force of heartbeat |
| Blood vessels | No effect on most blood vessels | Constricts blood vessels in viscera and skin (dilates those in skeletal muscle and heart); increases blood pressure |
| Glands—salivary, lacrimal, gastric | Stimulates; increases production of saliva, tears, and gastric juice | Inhibits; result is dry mouth and dry eyes |
| Eye (iris) | Stimulates constrictor muscles; constricts pupils | Stimulates dilator muscles; dilates pupils |
| Eye (ciliary muscle) | Stimulates to increase bulging of lens for close vision | Inhibits; decreases bulging of lens; prepares for distant vision |
| Adrenal medulla | No effect | Stimulates medulla cells to secrete epinephrine and norepinephrine |
| Sweat glands of skin | No effect | Stimulates to produce perspiration |
| Arrector pili muscles attached to hair follicles | No effect | Stimulates; produces "goose bumps" |
| Penis | Causes erection due to vasodilation | Causes ejaculation (emission of semen) |
| Cellular metabolism | No effect | Increases metabolic rate; increases blood sugar levels; stimulates fat breakdown |
| Adipose tissue | No effect | Stimulates fat breakdown |

on the cardiovascular system. Its activity is best illustrated by a person who relaxes after a meal and reads the newspaper. Blood pressure and heart and respiratory rates are being regulated at low-normal levels, the digestive tract is actively digesting food, and the skin is warm (indicating that there is no need to divert blood to skeletal muscles or vital organs). The eye pupils are constricted to protect the retinas from excessive damaging light, and the lenses of the eyes are "set" for close vision. We might also consider the parasympathetic division as the "housekeeping" system of the body.

An easy way to remember the most important roles of the two ANS divisions is to think of the parasympathetic division as the **D** (digestion, def-

ecation, and diuresis [urination]) division and the sympathetic division as the **E** (exercise, excitement, emergency, and embarrassment) division. Remember, however, that although it is easiest to think of the sympathetic and parasympathetic divisions as working in an all-or-none fashion, this is rarely the case. A dynamic balance exists between the two divisions, and both are continuously making fine adjustments. Also, although we have described the parasympathetic division as the "at-rest" system, most blood vessels are controlled only by the sympathetic fibers regardless of whether the body is "on alert" or relaxing.

The major effects of each division are summarized in Table 7.4.

## Did You Get It?

28. Which regions or organs of the body are served by the autonomic nervous system? Which are served by the somatic nervous system?
29. How does the motor pathway of the autonomic nervous system differ from that of the somatic nervous system?
30. Which division of the autonomic nervous system is the "fight-or-flight" system?

**For answers, see Appendix A.**

# Developmental Aspects of the Nervous System

→ **Learning Objectives**

☐ **List several factors that may have harmful effects on brain development.**

☐ **Briefly describe the cause, signs, and consequences of the following congenital disorders: spina bifida, anencephaly, and cerebral palsy.**

☐ **Explain the decline in brain size and weight that occurs with age.**

☐ **Define senility, and list some possible causes.**

Because the nervous system forms during the first month of embryonic development, maternal infection early in pregnancy can have extremely harmful effects on the fetal nervous system. For example, German measles (*rubella*) in the mother often causes deafness and other types of CNS damage. Also, because nervous tissue has the highest metabolic rate in the body, lack of oxygen for even a few minutes kills neurons. Because smoking decreases the amount of oxygen that can be carried in the blood, a smoking mother may be sentencing her infant to possible brain damage. Radiation and various drugs (alcohol, opiates, cocaine, and others) can also be very damaging if taken during early fetal development.

 **Homeostatic Imbalance 7.11**

In difficult deliveries, temporary lack of oxygen may lead to **cerebral palsy** (pawl′ze), but this is only one of the suspected causes. Cerebral palsy is a neuromuscular disability in which the voluntary muscles are poorly controlled and spastic because of brain damage. About half of its victims have seizures, are intellectually disabled, and/or have impaired hearing or vision. Cerebral palsy is the largest single cause of physical disabilities in children. A number of other congenital

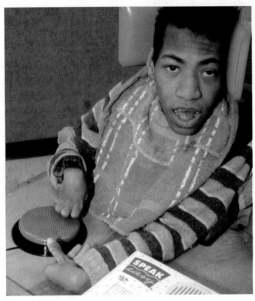

This adult patient with cerebral palsy presses a pad to communicate through a speaker.

malformations—triggered by genetic or environmental factors—also plague the CNS. Most serious are hydrocephalus (see p. 276), *anencephaly*, and spina bifida. **Anencephaly** is a birth defect in which the cerebrum fails to develop. Children with anencephaly cannot see, hear, or process sensory information; these babies typically die soon after birth. **Spina bifida** (spi′nah bi′fi-dah; "forked spine") results when the vertebrae form incompletely (typically in the lumbosacral region). There are several varieties of spina bifida. In the least serious, a dimple, and perhaps a tuft of hair, appears over the site of malformation, but no neurological problems occur. In the most serious, meninges, nerve roots, and even parts of the spinal cord protrude from the spine, rendering the lower part of the spinal cord functionless. The child is unable to control the bowels or bladder, and the lower limbs are paralyzed. ————✚

One of the last areas of the CNS to mature is the hypothalamus, which contains centers for regulating body temperature. For this reason, premature babies usually have problems controlling their loss of body heat and must be carefully monitored. The nervous system grows and matures all through childhood, largely as a result of myelination that goes on during this period. A good indication of the degree of myelination of particular neural pathways is the level of neuromuscular control in that body area. Neuromuscular coordination

# Tracking Down CNS Problems

When a doctor taps your patellar tendon with a reflex hammer, your thigh muscles contract, resulting in the knee-jerk response. This response shows that the spinal cord and upper brain centers are functioning normally. When reflex tests are abnormal, more sophisticated neurological tests may be ordered to try to localize and identify the problem.

An "oldie-but-goodie" procedure used to diagnose and localize many different types of brain lesions (such as epileptic lesions, tumors, and abscesses) is *electroencephalography* (e-lek"tro-en-sef-ah-lah'grah-fe). Normal brain function involves the continuous transmission of electrical impulses by neurons. A recording of that electrical activity, called an **electroencephalogram (EEG)**, can be made by placing electrodes at various points on the scalp and connecting these to a recording device, as in part (a) of the figure. The patterns of electrical activity are called *brain*

*waves*. Because people differ genetically, and because everything we have ever experienced has left its imprint in our brain, each of us has a brain wave pattern that is as unique as our fingerprints. The four most commonly seen brain waves are illustrated and described in part (b) of the figure.

As might be expected, brain-wave patterns while a person is wide awake differ from those that occur during relaxation or deep sleep. Brain waves that are too fast or too slow suggest that the function of the cerebral cortex has been compromised, and unconsciousness occurs at both extremes. Sleep and coma result in brain-wave patterns that are slower than normal, whereas epileptic seizures and some kinds of drug overdose cause abnormally fast brain waves. Because brain waves are seen even during coma, the absence of brain waves (a flat EEG) is taken as evidence of clinical death.

New imaging techniques (described in Chapter 1, pp. 43–44) have revolutionized the diagnosis of brain lesions. Together, **computed tomography (CT)** and **magnetic resonance imaging (MRI)** scans allow quick identification of most tumors, intracranial lesions, multiple sclerosis plaques, and areas of dead brain tissue (*infarcts*). **Positron emission tomography (PET)** scans can localize lesions that generate epileptic seizures.

Take, for example, a patient arriving in the ER with a stroke. The first step is to image the brain, most often with CT, to determine whether the stroke is due to a clot or a bleed. If the stroke is due to a clot, then the clot-busting drug tPA can be used, but only within the first hours. **Cerebral angiography**, which involves injecting a dye to make arteries in the brain stand out on X-ray imaging, might be used to visualize the clot or directly deliver tPA to the clot through a catheter. Less expensive and invasive

progresses in a superior to inferior direction and in a proximal to distal direction, and myelination occurs in the same sequence (as described in Chapter 6).

The brain reaches its maximum weight in the young adult. Over the next 60 years or so, neurons are damaged and die; and brain weight and volume steadily decline. However, a nearly unlimited number of neural pathways is always available and ready to be developed, allowing us to continue to learn throughout life.

As we grow older, the sympathetic nervous system gradually becomes less and less efficient, particularly in its ability to constrict blood vessels. When older people stand up quickly, they often become lightheaded or faint. The reason is that

the sympathetic nervous system is not able to react quickly enough to counteract the pull of gravity by activating the vasoconstrictor fibers, and blood pools in the feet. This condition, **orthostatic hypotension**, is a type of low blood pressure resulting from changes in body position as described. Orthostatic hypotension can be prevented to some degree if the person changes position slowly, giving the sympathetic nervous system time to adjust.

The usual cause of nervous system deterioration is circulatory system problems. For example, **arteriosclerosis** (ar-ter"e-o-sklĕ-ro'sis; decreased elasticity of the arteries) and high blood pressure reduce the supply of oxygen to brain neurons. A gradual decline of oxygen due to the aging process

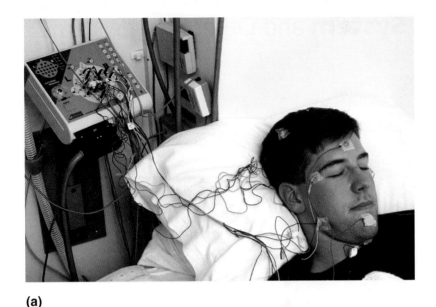

(a)

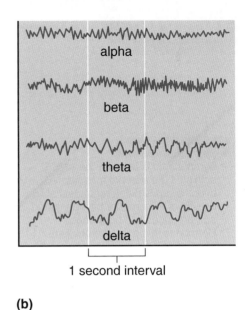

1 second interval

(b)

**Electroencephalography and brain waves. (a)** Electrodes are positioned on the scalp and attached to an electroencephalograph. **(b)** Typical EEGs. Alpha waves are typical of the awake, relaxed state; beta waves occur in the awake, alert state; theta waves are common in children but not in normal adults; and delta waves occur during deep sleep.

than angiography, ultrasound can be used to quickly examine the carotid arteries and even measure blood flow through them.

A new imaging procedure, DaTscan, can assist with diagnosing neurodegenerative conditions involving dopamine, such as Parkinson's disease. This imaging procedure uses a radioactive drug that binds dopamine transporters (DaT) paired with single-photon emission computed tomography (SPECT). The power of this technique is that it can image both the amount and distribution of dopamine in the brain. Although DaTscan cannot diagnose Parkinson's disease directly, it helps doctors distinguish Parkinson's from other neurological conditions characterized by abnormal dopamine levels.

can lead to **senility**, characterized by forgetfulness, irritability, difficulty in concentrating and thinking clearly, and confusion. A sudden loss of blood and oxygen delivery to the brain results in a CVA (stroke), as described earlier. However, many people continue to enjoy intellectual lives and mentally demanding tasks their entire lives. In fact, fewer than 5 percent of people over age 65 demonstrate true senility.

Sadly, many cases of "reversible senility," caused by certain drugs, low blood pressure, constipation, poor nutrition, depression, dehydration, and hormone imbalances, go undiagnosed. The best way to maintain your mental abilities in old age may be to seek regular medical checkups throughout life.

Although eventual shrinking of the brain is normal, some individuals (professional boxers and chronic alcoholics, for example) hasten the process. Whether a boxer wins the match or not, the likelihood of brain damage and atrophy increases with every blow. The expression "punch drunk" reflects the symptoms of slurred speech, tremors, abnormal gait, and dementia (mental illness) seen in many retired boxers.

Everyone recognizes that alcohol has a profound effect on the mind as well as the body. CT scans of chronic alcoholics reveal reduced brain size at a fairly early age. Like boxers, chronic alcoholics tend to exhibit signs of mental deterioration unrelated to the aging process.

# Homeostatic Relationships between the
# Nervous System and Other Body Systems

## Nervous System

### Endocrine System
- Sympathetic division of the ANS activates the adrenal medulla; hypothalamus helps regulate the activity of the anterior pituitary gland and produces two hormones
- Hormones influence metabolism of neurons

### Lymphatic System/Immunity
- Nerves innervate lymphoid organs; the brain plays a role in regulating immune function
- Lymphatic vessels carry away leaked interstitial fluids from tissues surrounding nervous system structures; immune elements protect all body organs from pathogens (CNS has additional mechanisms)

### Digestive System
- ANS (particularly the parasympathetic division) regulates digestive system activity
- Digestive system provides nutrients needed for healthy neurons

### Urinary System
- ANS regulates bladder emptying and renal blood pressure
- Kidneys help to dispose of metabolic wastes and maintain proper electrolyte composition and pH of blood for neural functioning

### Muscular System
- Somatic division of nervous system activates skeletal muscles; maintains muscle health
- Skeletal muscles are the effectors of the somatic division

### Respiratory System
- Nervous system regulates respiratory rhythm and depth
- Respiratory system provides life-sustaining oxygen; disposes of carbon dioxide

### Cardiovascular System
- ANS helps regulate heart rate and blood pressure
- Cardiovascular system delivers blood containing oxygen and nutrients to the nervous system; carries away wastes

### Reproductive System
- ANS regulates sexual erection and ejaculation in males; erection of the clitoris in females
- Testosterone masculinizes the brain and underlies sex drive and aggressive behavior

### Integumentary System
- Sympathetic division of the ANS regulates sweat glands and blood vessels of skin (therefore heat loss/retention)
- Skin serves as heat loss surface

### Skeletal System
- Nerves innervate bones
- Bones serve as depot for calcium needed for neural function; protect CNS structures

## Did You Get It?

**31.** Why must premature babies be placed in incubators until their hypothalamus matures?

**32.** What is orthostatic hypotension? Why do many older people suffer from this condition?

**For answers, see Appendix A.**

The human cerebral hemispheres—our "thinking caps"—are awesome in their complexity. No less amazing are the brain regions that oversee all our subconscious, autonomic body functions—the diencephalon and brain stem—particularly when you consider their relatively insignificant size. The spinal cord, which acts as a reflex center, and the peripheral nerves, which provide communication links between the CNS and body periphery, are equally important to body homeostasis.

We have introduced a good deal of new terminology in this chapter, and, as you will see, much of it will come up again in later chapters as we study the other organ systems of the body and examine how the nervous system helps to regulate their activity. The terms *are* essential, so try to learn them as you go along. (Use the Glossary at the back of the book or on Mastering A&P as often as you find it helpful.)

# Summary

## Organization of the Nervous System (pp. 252–253)

1. Structural: All nervous system structures are classified as part of the CNS (brain and spinal cord) or PNS (nerves and ganglia).

2. Functional: Motor nerves of the PNS are classified on the basis of whether they stimulate skeletal muscle (somatic division) or smooth/cardiac muscle and glands (autonomic division).

## Nervous Tissue: Structure and Function (pp. 253–265)

1. Supportive cells

   a. Neuroglia support and protect neurons in the CNS. Some neuroglia are phagocytes; others myelinate neuron processes in the CNS or line cavities; others brace neurons and filter nutrients.

   b. Schwann cells myelinate neuron processes in the PNS.

2. Neurons

   a. Anatomy: All neurons have a cell body, which contains the nucleus, and processes (fibers) of two types: (1) dendrites (one to many per cell) typically carry electrical currents toward the cell body, and (2) axons (one per cell) generate and conduct impulses away from the cell body and often release a neurotransmitter. Most large fibers are myelinated; myelin speeds nerve impulse transmission.

   b. Classification

      (1) On the basis of function (direction of impulse transmission) there are sensory (afferent) and motor (efferent) neurons and interneurons (association neurons). Dendritic endings of sensory neurons are bare (pain receptors) or are associated with sensory receptors.

      (2) On the basis of structure, there are unipolar, bipolar, and multipolar neurons; the terminology reveals the number of processes extending from the cell body. Motor and interneurons are multipolar; most sensory neurons are unipolar. The exceptions are sensory neurons in certain special sense organs (ear, eye), which are bipolar.

   c. Physiology

      (1) A nerve impulse is an electrochemical event (initiated by various stimuli) that causes a change in neuron plasma membrane permeability. This change allows sodium ions ($Na^+$) to enter the cell, causing depolarization. Once begun, the action potential, or nerve impulse, continues over the entire surface of the axon. Electrical conditions of the resting state are restored by the diffusion of potassium ions ($K^+$) out of the cell (repolarization). Ion concentrations of the resting state are restored by the sodium-potassium pump.

 Complete an interactive tutorial: Mastering A&P® > Study Area > Interactive Physiology > Nervous System > Resting Membrane Potential

 Complete an interactive tutorial: Mastering A&P® > Study Area > Interactive Physiology > Nervous System > Generation of an Action Potential

      (2) A neuron influences other neurons or effector cells by releasing neurotransmitters, chemicals that diffuse across the synaptic cleft and attach to membrane receptors on the postsynaptic cell. The result is opening of specific ion channels and activation or

inhibition, depending on the neurotransmitter released and the target cell.

(3) A reflex is a rapid, predictable response to a stimulus. There are two types—autonomic and somatic. A reflex arc has at least five components: receptor, sensory neurons, CNS integration center, motor neurons, and effector. Normal reflexes indicate normal nervous system function.

## Central Nervous System (pp. 265–281)

1. The brain is located within the cranial cavity of the skull and consists of the cerebral hemispheres, diencephalon, brain stem structures, and cerebellum.

a. The two cerebral hemispheres form the largest part of the brain. Their surface, the cortex, is gray matter, and their interior is white matter. The cortex is convoluted and has gyri, sulci, and fissures. The cerebral hemispheres are involved in logical reasoning, moral conduct, emotional responses, sensory interpretation, and the initiation of voluntary muscle activity. Several functional areas of the cerebral lobes have been identified (see p. 268). The basal nuclei, regions of gray matter deep within the white matter of the cerebral hemispheres, modify voluntary motor activity. Parkinson's disease and Huntington's disease are disorders of the basal nuclei.

b. The diencephalon is superior to the brain stem and is enclosed by the cerebral hemispheres. The major structures include the following:

(1) The thalamus, which encloses the third ventricle, is the relay station for sensory impulses passing to the sensory cortex for interpretation.

(2) The hypothalamus makes up the "floor" of the third ventricle and is the most important regulatory center of the autonomic nervous system (regulates water balance, metabolism, thirst, temperature, and the like).

(3) The epithalamus includes the pineal gland (an endocrine gland) and the choroid plexus of the third ventricle.

c. The brain stem is the short region inferior to the hypothalamus that merges with the spinal cord.

(1) The midbrain is most superior and is primarily fiber tracts (bundles of nerve fibers in the CNS connecting CNS structures).

(2) The pons is inferior to the midbrain and has fiber tracts and nuclei involved in respiration.

(3) The medulla oblongata is the most inferior part of the brain stem. In addition to fiber tracts, it contains autonomic nuclei involved in the regulation of vital life activities (breathing, heart rate, blood pressure, etc.).

d. The cerebellum is a large, cauliflower-like part of the brain posterior to the fourth ventricle. It coordinates muscle activity and body balance.

2. Protection of the CNS

a. Bones of the skull and vertebral column are the most external protective structures.

b. Meninges are three connective tissue membranes—dura mater (tough, outermost), arachnoid mater (middle, weblike), and pia mater (innermost, delicate). The meninges extend beyond the end of the spinal cord.

c. Cerebrospinal fluid (CSF) provides a watery cushion around the brain and cord. CSF is formed by the choroid plexuses of the brain. It is found in the subarachnoid space, ventricles, and central canal. CSF is continually formed and drained into dural venous sinuses.

d. The blood-brain barrier is composed of relatively impermeable capillaries.

3. Brain dysfunctions

a. Head trauma may cause concussions (reversible damage) or contusions (irreversible damage). When the brain stem is affected, unconsciousness (temporary or permanent) occurs. Trauma-induced brain injuries may be aggravated by intracranial hemorrhage or cerebral edema, both of which compress brain tissue.

b. Cerebrovascular accidents (CVAs, or strokes) result when blood circulation to brain neurons is blocked and brain tissue dies. The result may be visual impairment, paralysis, and aphasias (problems with speech and/or language).

c. Alzheimer's disease is a degenerative brain disease in which abnormal protein deposits and other structural changes appear. It results in slow, progressive loss of memory and motor control plus increasing dementia.

d. Techniques used to diagnose brain dysfunctions include the EEG, simple reflex tests, angiography, and DaTscan, CT, PET, and MRI scans.

4. The spinal cord is a reflex center and conduction pathway. Found within the vertebral canal, the cord extends from the foramen magnum to $L_1$ or $L_2$. The cord has a central butterfly-shaped area of gray matter surrounded by columns of white matter, which carry sensory and motor tracts to and from the brain.

## Peripheral Nervous System (pp. 281–295)

1. A nerve is a bundle of neuron processes wrapped in connective tissue coverings (endoneurium, perineurium, and epineurium).

2. Cranial nerves: 12 pairs of nerves that extend from the brain to serve the head and neck region. The exception is the paired vagus nerves, which extend into the thorax and abdomen.

3. Spinal nerves: 31 pairs of nerves are formed by the union of the dorsal and ventral roots of the spinal cord on each side. The spinal nerve proper is very short and splits into dorsal and ventral rami. Dorsal rami serve the posterior body trunk; ventral rami (except $T_2$ through $T_{12}$) form plexuses (cervical, brachial, lumbar, sacral) that serve the limbs.

4. Autonomic nervous system: Part of the PNS, composed of neurons that regulate the activity of smooth and cardiac muscle and glands. This system differs from the somatic nervous system in that there is a chain of two motor neurons from the CNS to the effector. Two subdivisions serve the same organs with different effects.

   a. The parasympathetic division is the "housekeeping" system and is in control most of the time. This division maintains homeostasis by seeing that normal digestion and elimination occur and that energy is conserved ("rest and digest"). The first motor neurons are in the brain or the sacral region of the cord. The second motor neurons are in the terminal ganglia close to the organ served. All parasympathetic axons secrete acetylcholine.

   b. The sympathetic division is the "fight-or-flight" subdivision, which prepares the body to cope with some threat. Its activation increases heart rate and blood pressure. The preganglionic neurons are in the gray matter of the cord. The ganglionic sympathetic neurons are in the sympathetic trunk or in collateral ganglia. Postganglionic axons secrete norepinephrine.

## Developmental Aspects of the Nervous System (pp. 295–299)

1. Maternal and environmental factors may impair embryonic brain development. Oxygen deprivation destroys brain cells. Severe congenital brain diseases include cerebral palsy, anencephaly, hydrocephalus, and spina bifida.

2. Premature babies have trouble regulating body temperature because the hypothalamus is one of the last brain areas to mature prenatally.

3. Development of motor control indicates the progressive myelination and maturation of a child's nervous system. Brain growth ends in young adulthood. Neurons die throughout life and are not replaced; thus, brain mass declines with age.

4. Healthy aged people maintain nearly optimal intellectual function. Disease—particularly cardiovascular disease—is the major cause of declining mental function with age.

---

## Review Questions

Access additional practice questions using your smartphone, tablet, or computer:
Mastering A&P® > Study Area > Practice Tests & Quizzes

### Multiple Choice

*More than one choice may apply.*

1. Which of the following is an example of integration by the nervous system?
   a. The feel of a cold breeze
   b. Shivering and goose bumps in response to cold
   c. Perceiving the sound of rain
   d. The decision to go back for an umbrella

2. Where might a gray matter nucleus be located?
   a. Alongside the vertebral column
   b. Within the brain
   c. Within the spinal cord
   d. In the sensory receptors

3. Which of the following does the blood-brain barrier prevent from entering the brain?
   a. Glucose          c. Nicotine
   b. Urea             d. Alcohol

4. Histological examination of a slice of neural tissue reveals a bundle of nerve fibers held together by cells whose multiple processes wrap around several fibers and form a myelin sheath. The specimen is likely to be
   a. a nucleus.        c. a nerve.
   b. a ganglion.       d. a tract.

5. The pineal gland is located in the
   a. hypothalamus.     c. epithalamus.
   b. mesencephalon.    d. corpus callosum.

6. Choose the proper term to respond to the statements describing various brain areas.
   a. cerebellum        e. medulla
   b. corpora           f. midbrain
      quadrigemina      g. pons
   c. corpus callosum   h. thalamus
   d. hypothalamus

_____ 1. Region where there is a gross crossover of fibers of descending pyramidal tracts

_____ 2. Control of temperature, autonomic nervous system reflexes, hunger, and water balance

_____ 3. Houses the substantia nigra and cerebral aqueduct

_____ 4. Relay stations for visual and auditory stimuli input; found in midbrain

_____ 5. Houses vital centers for control of the heart, respiration, and blood pressure

_____ 6. Brain area through which all the sensory input is relayed to get to the cerebral cortex

_____ 7. Brain area most concerned with equilibrium, body posture, and coordination of motor activity.

7. The nerves that arise from the spinal lumbar plexus include the
   a. sciatic.
   b. femoral.
   c. obturator.
   d. gluteal.

8. Which contains only motor fibers?
   a. Dorsal root
   b. Dorsal ramus
   c. Ventral root
   d. Ventral ramus

9. The cranial nerve responsible for activating the lacrimal glands is the
   a. trigeminal.
   b. facial.
   c. hypoglossal.
   d. vagus.

10. Motor functions of the extensor muscles of the arm, forearm, and fingers would be affected by damage to which one of these nerves?
    a. Radial
    b. Axillary
    c. Ulnar
    d. Median

11. Which cells are responsible for the formation of myelin in the CNS?
    a. Schwann cells
    b. Oligodendrocytes
    c. Satellite cells
    d. Astrocytes

12. Which of the following is (are) found in a three-neuron reflex arc?
    a. Sensory (efferent) neuron
    b. Motor (efferent) neuron
    c. Sensory receptor
    d. Effector organ

## Short Answer Essay

13. What distinguishes spinal nerves from cranial nerves?

14. Explain both the structural and functional classifications of the nervous system. In your explanation, include the subdivisions of each.

15. What is the basis for the functional classification of neurons?

16. Two major cell groups make up the nervous system—neurons and supporting cells such as astrocytes and Schwann cells. Which are "nervous" cells? Why? What are the major functions of the other cell group?

17. Briefly explain how nerve impulses are initiated and transmitted, and why conduction at synapses is always one-way.

18. Name four types of cutaneous sensory receptors. Which of the cutaneous receptor types is most numerous? Why?

19. What is the function of Broca's area? In which part of the brain is it found?

20. Make a rough drawing of the left cerebral hemisphere. On your drawing, locate at least five different functional areas, and then indicate their specific functions.

21. Other than serving as a conduction pathway, what is a major function of the pons? Why is the medulla the most vital part of the brain?

22. What are the basal nuclei made of, and what is their function?

23. Describe how the brain is protected by bone, membranes, fluid, and capillaries.

24. What is gray matter? White matter? How does the arrangement of gray and white matter differ in the cerebral hemispheres and the spinal cord?

25. What are two functions of the spinal cord?

26. How many pairs of cranial nerves are there? Which are purely sensory? Which activates the chewing muscles? Which helps regulate heart rate and activity of the digestive tract?

27. What general area of the body do the cranial nerves, other than the vagus nerves, serve? What do the vagus nerves serve?

28. Which glial cells promote the circulation of the cerebrospinal fluid?

29. What region of the body is served by the dorsal rami of the spinal nerves? By the ventral rami?

30. Name the four major nerve plexuses formed by the ventral rami and the body region served by each.

31. How does the autonomic nervous system differ from the somatic nervous system?

32. What is the difference in function of the sympathetic and parasympathetic divisions of the autonomic nervous system (a) in general and (b) as specifically relates to the operation of the cardiovascular and digestive systems?

33. The sympathetic and parasympathetic fibers serve the same organs. How can their opposing effects be explained?

34. How does a Schwann cell help insulate a nerve fiber?

35. Compare CVAs and TIAs in terms of causes, symptoms, and consequences.

36. Define *senility*. Name possible causes of permanent and reversible senility.

 ## Critical Thinking and Clinical Application Questions

37. Mrs. Jones has had a progressive decline in her mental capabilities in the past five or six years. At first, her family attributed her occasional memory lapses, confusion, and agitation to grief over her husband's death six years earlier. When examined, Mrs. Jones was aware of her cognitive problems and was shown to have an IQ score approximately 30 points lower than would be predicted by her work history. A CT scan showed diffuse cerebral atrophy. The physician prescribed a mild tranquilizer for Mrs. Jones and told her family that there was little else he could recommend. What is Mrs. Jones's problem?

38. Joseph, a man in his early 70s, was having problems chewing his food. He was asked to stick out his tongue. It deviated to the right, and its right side was quite wasted. What cranial nerve was impaired?

39. Andy is about to take the stage to play the guitar in front of the biggest audience he has ever played for. He can feel his heart pounding in his chest and his hands are sweating. Explain Andy's physiological signs in terms of the autonomic nervous system.

40. A semiconscious young woman is brought to the hospital by friends after falling from a roof. She did not lose consciousness immediately, and she was initially lucid. After a while, though, she became confused and then unresponsive. What is a likely explanation of her condition?

41. During John's checkup, one year after an accident severed his right accessory nerve, the physician noted severe muscle atrophy. What two prominent muscles have been affected?

42. Mrs. Chen, a new mother, brings her infant to the clinic because he has suffered repeated seizures. When questioned, she states that her labor was unusually long and difficult. What condition do you suspect? Will the infant's condition worsen?

43. Jake has been recently administered medication through an injection in his right buttock. Now he is not being able to dorsiflex his right foot. His doctor suspects that he has footdrop. What could have happened?

44. Mr. Harrison is an 82-year-old bedridden gentleman who has discovered a new interest in learning about his body. While being tended by the visiting nurse, he remarks that the supporting cells in nervous tissue (such as Schwann cells and oligodendrocytes) act like the rubber coating around household wiring. What does he mean by this analogy?

45. Maria had an automobile accident that caused injury to her head. As a result, she could not understand what was said to her, though she was able to pronounce words herself. What was her probable condition? What caused it?

46. Why does exposure to toxins have more devastating neural effects during early pregnancy than in late pregnancy?

47. Jason is the star of his hometown ice hockey team. During a game, he is hit with a hockey stick so hard that he hits the ice. When he tries to get up, he is unable to flex his left hip or extend his left knee, but he has no pain. Which nerve has been damaged?

48. Derek got hit in the back of the neck with a baseball, and now he can't shrug one shoulder. Which cranial nerve is involved?

49. When Taylor begins to feel drowsy while driving, she opens her window, turns up the volume of the car stereo, and sips her ice-cold water. How do these actions keep her awake?

50. As the aroma of freshly brewed coffee drifted by Joe's nose, his mouth began to water, and his stomach started to rumble. Explain these reactions in terms of ANS activity.

# 8 Special Senses

**WHAT**

The special senses respond to stimuli involved in vision, hearing, balance, smell, and taste.

**HOW**

A variety of receptors, housed in special sense organs such as the eye, ear, and nose, help detect stimuli in your surroundings.

**WHY**

Without your special senses, you could not smell or taste your favorite food, appreciate colors, or hear your favorite song.

**INSTRUCTORS**

New **Building Vocabulary Coaching** Activities for this chapter are assignable in Mastering A&P®

People are responsive creatures. Hold freshly baked bread before us, and our mouths water. Sudden thunder makes us jump. These stimuli (the bread and thunder) and many others continually greet us and are interpreted by our nervous system.

We are told that we have five senses that keep us in touch with what is going on in the external world: touch, taste, smell, sight, and hearing. Actually, touch is a mixture of the general senses that we considered in Chapter 7—the temperature, pressure, and pain receptors of the skin and the proprioceptors of muscles and joints. The other four "traditional" senses—*smell, taste, sight,* and *hearing*—are called **special senses**. Receptors for a fifth special sense, *equilibrium*, are housed in the ear, along with the organ of hearing. In contrast to the small and widely distributed general receptors, the **special sense receptors** are either large, complex sensory organs (eyes and ears) or localized clusters of receptors (taste buds and olfactory epithelium).

In this chapter we focus on the functional anatomy of each special sense organ, but keep in mind that sensory inputs overlap. What we experience—our "feel" of the world—is a blending of stimulus effects.

### ◄ CONCEPT**LINK**

Recall the three basic functions of the nervous system (Figure 7.1, p. 252). Each of the special senses gathers unique sensory information that, once integrated, will influence motor output. For example, if you saw a ball moving toward your head, this sensory input might result in a motor output that would move your body out of the path of the ball. Additionally, recall that each type of sensory information is processed in a specialized area of the cerebrum (Figure 7.13c, p. 268). ◄

## PART I: THE EYE AND VISION

Of all the senses, vision has been studied most. Nearly 70 percent of all sensory receptors in the body are in the eyes. The optic tracts that carry information from the eyes to the brain are massive bundles, containing over a million nerve fibers. We rely heavily on our sight and often have to "see it to believe it" **(Figure 8.1)**.

# Anatomy of the Eye

## External and Accessory Structures

→ **Learning Objective**

☐ **When provided with a model or diagram, identify the accessory eye structures, and list the functions of each.**

The adult eye is a sphere that measures about 1 inch (2.5 cm) in diameter. Only the anterior one-sixth of the eye's surface is normally seen. The rest of it is enclosed and protected by a cushion of fat and the walls of the bony orbit. The **accessory structures** of the eye include the extrinsic eye muscles, eyelids, conjunctiva, and lacrimal apparatus.

Anteriorly the eyes are protected by the **eyelids**, which meet at the medial and lateral corners of the eye, the **medial commissure (canthus)** and **lateral commissure (canthus)**, respectively (Figure 8.1). The space between the eyelids in an open eye is called the *palpebral fissure*. Projecting from the border of each eyelid are the **eyelashes**. Modified sebaceous glands associated with the eyelid edges are the **tarsal glands**. These glands produce an oily secretion that lubricates the eye (**Figure 8.2a**, p. 306). *Ciliary glands*, which are modified sweat glands, lie between the eyelashes (*cilium* = eyelash), and their ducts open at the eyelash follicles. On the medial aspect of each eye is the *lacrimal caruncle* (see Figure 8.1), a raised area containing sebaceous and sweat glands that produce an oily, whitish secretion that also lubricates the eye.

A delicate membrane, the **conjunctiva** (kon-junk″ti′vah), lines the eyelids and covers part of the outer surface of the eyeball (Figure 8.2a). It ends at the edge of the transparent cornea by fusing with the corneal epithelium. The conjunctiva secretes mucus, which helps to lubricate the eyeball and keep it moist.

8

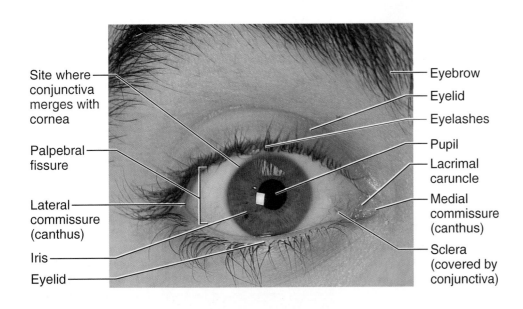

Site where conjunctiva merges with cornea

Palpebral fissure

Lateral commissure (canthus)

Iris

Eyelid

Eyebrow

Eyelid

Eyelashes

Pupil

Lacrimal caruncle

Medial commissure (canthus)

Sclera (covered by conjunctiva)

**Figure 8.1 Surface anatomy of the eye and accessory structures.**

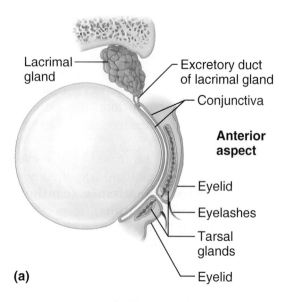

**(a)**

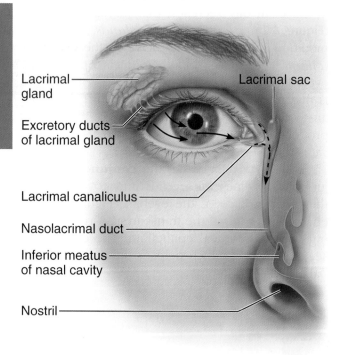

**(b)**

**Figure 8.2 Accessory structures of the eye.**
**(a)** Sagittal section of the accessory structures associated with the anterior part of the eye. **(b)** Anterior view of the lacrimal apparatus.

 **Homeostatic Imbalance 8.1**

Inflammation of the conjunctiva, called **conjunctivitis**, results in reddened, irritated eyes. **Pinkeye**, its infectious form caused by bacteria or viruses, is highly contagious. _____ ✚

The **lacrimal apparatus** (Figure 8.2b) consists of the lacrimal gland and a number of ducts that drain lacrimal secretions into the nasal cavity. The **lacrimal glands** are located above the lateral end of each eye. They continually release a dilute salt solution (_tears_) onto the anterior surface of the eyeball through several small ducts. The tears flush across the eyeball into the **lacrimal canaliculi** medially, then into the **lacrimal sac**, and finally into the **nasolacrimal duct**, which empties into the inferior meatus of the nasal cavity (see Figure 8.2b). Tears also contain mucus, antibodies, and **lysozyme** (li'so-zīm), an enzyme that destroys bacteria. Thus, they cleanse and protect the eye surface as they moisten and lubricate it. When lacrimal secretion increases substantially, tears spill over the eyelids and fill the nasal cavities, causing congestion and the "sniffles." This happens when the eyes are irritated by foreign objects or chemicals and when we are emotionally upset. In the case of irritation, the enhanced tearing acts to wash away or dilute the irritating substance. The importance of "emotional tears" is poorly understood, but some suspect that crying is important in reducing stress. Anyone who has had a good cry would probably agree, but this has been difficult to prove scientifically.

**Homeostatic Imbalance 8.2**

Because the nasal cavity mucosa is continuous with that of the lacrimal duct system, a cold or nasal inflammation often causes the lacrimal mucosa to become inflamed and swell. This impairs the drainage of tears from the eye surface, causing "watery" eyes. _____ ✚

Six **extrinsic eye muscles (external eye muscles)** are attached to the outer surface of each eye. These muscles produce gross eye movements and make it possible for the eyes to follow a moving object. **Figure 8.3** gives the names, locations, actions, and cranial nerve serving each of the extrinsic muscles.

**Did You Get It?**

1. What is the role of the eyelids?
2. Which four accessory glands or structures help lubricate the eye?
3. What is the role of lysozyme in tears?
4. What is the visual role of the external eye muscles?

For answers, see Appendix A.

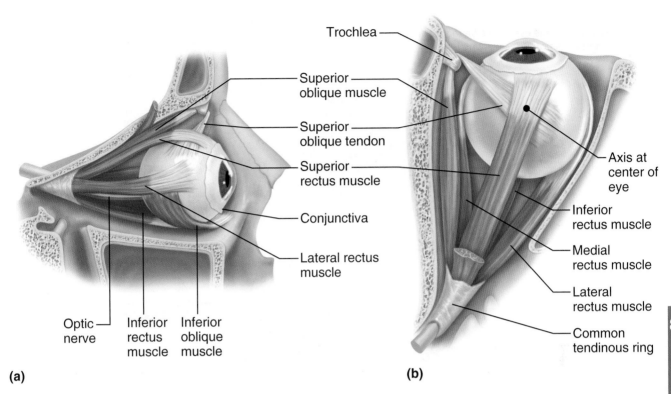

**Figure 8.3 Extrinsic muscles of the eye. (a)** Lateral view of the right eye. **(b)** Superior view of the right eye. The four rectus muscles originate from the common tendinous ring, a ringlike tendon at the back of the eye socket. **(c)** Summary of cranial nerve supply and actions of the extrinsic eye muscles.

| Name | Action | Controlling cranial nerve |
|------|--------|---------------------------|
| Lateral rectus | Moves eye laterally | VI (abducens) |
| Medial rectus | Moves eye medially | III (oculomotor) |
| Superior rectus | Elevates eye and turns it medially | III (oculomotor) |
| Inferior rectus | Depresses eye and turns it medially | III (oculomotor) |
| Inferior oblique | Elevates eye and turns it laterally | III (oculomotor) |
| Superior oblique | Depresses eye and turns it laterally | IV (trochlear) |

## Internal Structures: The Eyeball

### → Learning Objectives

☐ **Name the layers of the wall of the eye, and indicate the major function of each.**

☐ **Explain how the functions of rods and cones differ.**

☐ **Define** *blind spot*, *cataract*, **and** *glaucoma*.

☐ **Trace the pathway of light through the eye to the retina.**

☐ **Discuss the importance of an ophthalmoscopic examination.**

The eye itself, called the **eyeball**, is a hollow sphere (**Figure 8.4**, p. 308). Its wall is composed of three *tunics*, or layers, and its interior is filled with fluids called *humors* that help to maintain its shape. The lens, the main focusing apparatus of the eye, is supported upright within the eye cavity, dividing it into two chambers.

### Layers Forming the Wall of the Eyeball

Now that we have covered the general anatomy of the eyeball, we are ready to get specific.

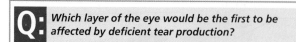

**Q:** *Which layer of the eye would be the first to be affected by deficient tear production?*

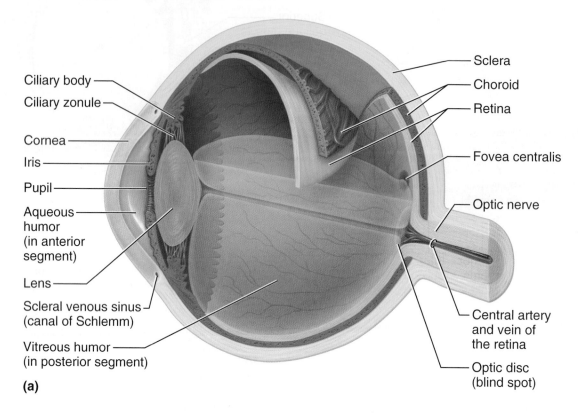

Ciliary body

Ciliary zonule

Cornea

Iris

Pupil

Aqueous humor (in anterior segment)

Lens

Scleral venous sinus (canal of Schlemm)

Vitreous humor (in posterior segment)

Sclera

Choroid

Retina

Fovea centralis

Optic nerve

Central artery and vein of the retina

Optic disc (blind spot)

**(a)**

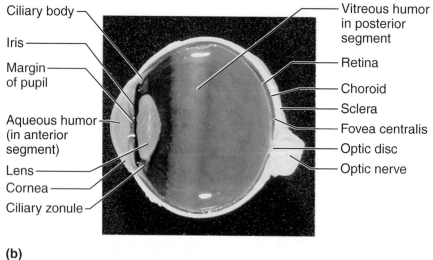

Ciliary body

Iris

Margin of pupil

Aqueous humor (in anterior segment)

Lens

Cornea

Ciliary zonule

Vitreous humor in posterior segment

Retina

Choroid

Sclera

Fovea centralis

Optic disc

Optic nerve

**(b)**

**Figure 8.4  Internal anatomy of the eye (sagittal section).**
**(a)** Diagrammatic view. **(b)** Photograph.

**A:** The outermost fibrous layer (the sclera and especially its cornea), which normally is continuously washed by tears.

*Fibrous Layer*  The outermost layer, called the **fibrous layer**, consists of the protective **sclera** (skle′rah) and the transparent **cornea** (kor′ne-ah). The sclera (thick white connective tissue) is seen anteriorly as the "white of the eye." The central anterior portion of the fibrous layer is crystal clear. This "window" is the cornea through which light enters the eye. The cornea is well supplied with nerve endings. Most are pain fibers, and when the cornea is touched, blinking and increased tear production occur. Even so, the cornea is the most exposed part of the eye, and it is very vulnerable to damage. Luckily, its ability to repair itself is extraordinary. Furthermore, the cornea is the only tissue in the body that is transplanted from one person to another without the worry of rejection. Because the cornea has no blood vessels, it is beyond the reach of the immune system.

*Vascular Layer*  The middle, or **vascular layer**, of the eyeball, has three distinguishable regions. Most posterior is the **choroid** (ko′roid), a blood-rich nutritive tunic that contains a dark pigment. The pigment prevents light from scattering inside the eye. Moving anteriorly, the choroid is modified to form two smooth muscle structures, the **ciliary** (sil′e-er-e) **body**, which is attached to the lens by a suspensory ligament called the **ciliary zonule**, and the **iris**. The pigmented iris has a rounded opening, the **pupil**, through which light passes. Circularly and radially arranged smooth muscle fibers form the iris, which acts like the diaphragm of a camera. That is, it regulates the amount of light entering the eye so that we can see as clearly as possible in the available light. In close vision and bright light, the circular muscles contract, and the pupil constricts, or gets smaller. In distant vision and dim light, the radial fibers contract to enlarge (dilate) the pupil, which allows more light to enter the eye. Cranial nerve III (oculomotor) controls the muscles of the iris.

*Sensory Layer*  The innermost **sensory layer** of the eye is the delicate two-layered **retina** (ret′ĭ-nah), which extends anteriorly only to the ciliary body. The outer **pigmented layer** of the retina is composed of pigmented cells that, like those of the choroid, absorb light and prevent light from scattering inside the eye. They also act as phago-cytes to remove dead or damaged receptor cells and store vitamin A needed for vision.

The transparent inner **neural layer** of the ret-ina contains millions of receptor cells, the **rods** and **cones**, which are called **photoreceptors** because they respond to light (**Figure 8.5**, p. 310). Electrical signals pass from the photoreceptors via a two-neuron chain—**bipolar cells** and then **ganglion cells**—before leaving the retina via the **optic nerve** and being transmitted to, and inter-preted by, the optic cortex. The result is vision.

The photoreceptor cells are distributed over the entire retina, except where the optic nerve leaves the eyeball; this site is called the **optic disc**. Since there are no photoreceptors at the optic disc, it results in a **blind spot** in our vision. When light from an object is focused on the optic disc, the object disappears from our view and we can-not see it.

The rods and cones are not evenly distributed in the retina. The rods are densest at the periph-ery, or edge, of the retina and decrease in number as the center of the retina is approached. The rods allow us to see in gray tones in dim light, and they provide our peripheral vision.

### ⚖ Homeostatic Imbalance 8.3

Anything that interferes with rod function hinders our ability to see at night. This condition, called **night blindness**, dangerously impairs our ability to drive safely at night. Its most common cause is prolonged vitamin A deficiency, which eventually causes the neural retina to deteriorate. Vitamin A is one of the building blocks of the pigments the photoreceptor cells need to respond to light (see "A Closer Look" on p. 311). Vitamin A supple-ments will restore function if taken before degen-erative changes in the neural retina occur._____ ✚

Cones are discriminatory receptors that allow us to see the details of our world in color under bright light conditions. They are densest in the cen-ter of the retina and decrease in number toward the retinal edge. Lateral to each blind spot is the **fovea centralis** (fo′ve-ah sen-trä′lis), a tiny pit that contains only cones (see Figure 8.4). Consequently, this is the area of greatest **visual acuity**, or point of sharpest vision, and anything we wish to view critically is focused on the fovea centralis.

There are three varieties of cones. Each type is most sensitive to particular wavelengths of visible light (**Figure 8.6**). One type responds most

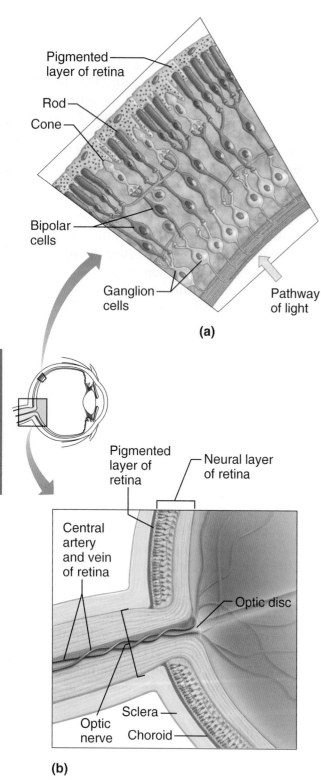

**(a)**

**(b)**

**Figure 8.5 The three major types of neurons composing the retina. (a)** Notice that light must pass through the thickness of the retina to excite the rods and cones. Electrical signals flow in the opposite direction: from the rods and cones to the bipolar cells and finally to the ganglion cells. The ganglion cells generate the nerve impulses that leave the eye via the optic nerve. **(b)** Schematic view of the posterior part of the eyeball illustrating how the axons of the ganglion cells form the optic nerve.

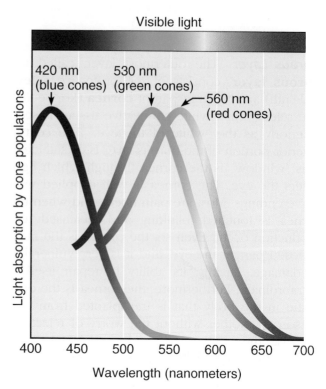

**Figure 8.6 Sensitivities of the three cone types to different wavelengths of visible light**.

vigorously to blue light, another to green light. The third cone variety responds to a range including both green and red wavelengths of light. However, this is the only cone population to respond to red light at all, so these are called the "red cones." Impulses received at the same time from more than one type of cone by the visual cortex are interpreted as *intermediate* colors, similar to what occurs when two colors of paint are mixed. For example, simultaneous impulses from blue and red color receptors are seen as purple or violet tones. When all three cone types are stimulated, we see white. If someone shines red light into one of your eyes and green into the other, you will see yellow, indicating that the "mixing" and interpretation of colors occur in the brain, not in the retina.

### Homeostatic Imbalance 8.4

Lack of all three cone types results in total **color blindness**, whereas lack of one cone type leads to partial color blindness. Most common is the lack of red or green receptors, which leads to two varieties of red-green color blindness. Red and green are seen as the same color—either red or

# Visual Pigments—The Actual Photoreceptors

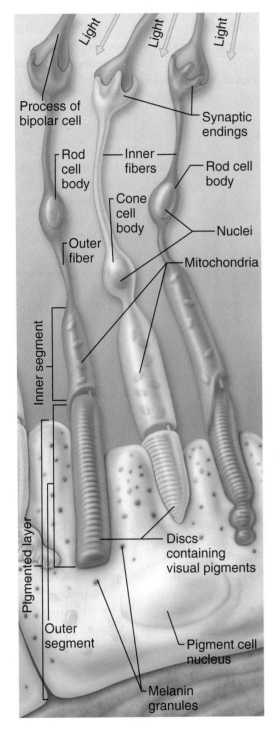

The names of the tiny photoreceptor cells of the retina reflect their shapes. As shown to the left, rods are slender, elongated neurons, whereas the fatter cones taper to pointed tips. In each type of photoreceptor, there is a region called an *outer segment*, attached to the cell body. The outer segment corresponds to a light-trapping dendrite, in which the discs containing the visual pigments are stacked like a row of pennies.

The behavior of the visual pigments is dramatic. When light strikes them, they lose their color, or are "bleached"; shortly afterward, they regenerate their pigment. Absorption of light and pigment bleaching cause electrical changes in the photoreceptor cells that ultimately cause nerve impulses to be transmitted to the brain for visual interpretation. Pigment regeneration ensures that you are not blinded and unable to see in bright sunlight.

A good deal is known about the structure and function of **rhodopsin**, the purple pigment found in rods (see figure below). It is formed from the union of a protein (**opsin**) and a modified vitamin A product (**retinal**). When combined in rhodopsin, retinal has a kinked shape that allows it to bind to opsin. But when light strikes rhodopsin, retinal straightens out and releases the protein. Once straightened out, the retinal continues its conversion until it is once again vitamin A. As these changes occur, the purple color of rhodopsin changes to the yellow of retinal and finally becomes colorless as the change to vitamin A occurs. Thus the term "bleaching of the pigment" accurately describes the color changes that occur when light hits the pigment. Rhodopsin is regenerated as vitamin A is again converted to the kinked form of retinal and recombined with opsin in an ATP-requiring process. The cone pigments, although similar to rhodopsin, differ in the specific kinds of proteins they contain.

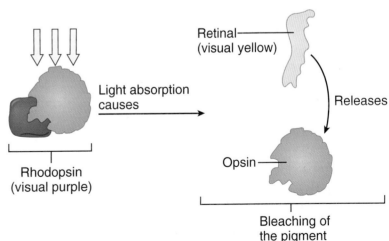

green, depending on the cone type *present*. Many color-blind people are unaware of their condition because they have learned to rely on other cues—such as differences in intensities of the same color—to distinguish green from red, for example, on traffic signals. Because the genes regulating color vision are on the X (female) sex chromosome, color blindness is a sex-linked condition. It occurs almost exclusively in males. _____ ✚

## Lens

Light entering the eye is focused on the retina by the lens, a flexible biconvex crystal-like structure. Recall the lens is held upright in the eye by the ciliary zonule and attached to the ciliary body (see Figure 8.4).

### ⚖ Homeostatic Imbalance 8.5

In youth, the lens is transparent and has the consistency of firm jelly, but as we age it becomes increasingly hard and cloudy. **Cataracts**, the loss of lens transparency, cause vision to become hazy and distorted and can eventually cause blindness. Other risk factors for forming cataracts include diabetes mellitus, frequent exposure to intense sunlight, and heavy smoking. Current treatment of cataracts is either special cataract glasses or surgical removal of the lens and replacement with a lens implant.

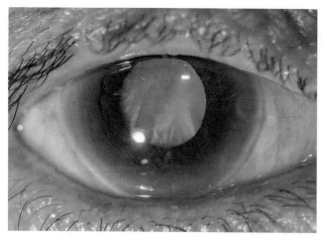

The cataract in this photo appears as a milky structure that seems to fill the pupil.

_____ ✚

The lens divides the eye into two segments, or chambers. The *anterior (aqueous) segment*, anterior to the lens, contains a clear watery fluid called **aqueous humor**. The *posterior (vitreous) segment*, posterior to the lens, is filled with a gel-like substance called **vitreous** (vit′re-us) **humor**, or the **vitreous body** (see Figure 8.4). Vitreous humor helps prevent the eyeball from collapsing inward by reinforcing it internally. Aqueous humor is similar to blood plasma and is continually secreted by a special area of the choroid. Like the vitreous humor, it helps maintain *intraocular* (in″trah-ok′u-lar) *pressure*, the pressure inside the eye. It also provides nutrients for the avascular lens and cornea. Aqueous humor is reabsorbed into the venous blood through the **scleral venous sinus**, or **canal of Schlemm** (shlĕm), which is located at the junction of the sclera and cornea.

### ⚖ Homeostatic Imbalance 8.6

If drainage of aqueous humor is blocked, fluid backs up like a clogged sink. Pressure within the eye may increase to dangerous levels and compress the delicate retina and optic nerve. The resulting condition, **glaucoma** (glaw-ko′mah; "vision going gray"), can lead to blindness unless detected early. Glaucoma is a common cause of blindness in the elderly. Unfortunately, many forms of glaucoma progress slowly and have almost no symptoms at first. Thus, sight deteriorates slowly and painlessly until the damage is done. Signs of advanced glaucoma include seeing halos around lights, headaches, and blurred vision. A simple instrument called a *tonometer* (to-nom′e-ter) is used to measure the intraocular pressure, which should be tested yearly in people over 40. Glaucoma is commonly treated with eyedrops that increase the rate of aqueous humor drainage. Laser or surgical enlargement of the drainage channel is another option. _____ ✚

The *ophthalmoscope* (of-thal′mo-skōp) is an instrument that illuminates the interior of the eyeball, allowing the retina, optic disc, and internal blood vessels at the **fundus**, or posterior wall of the eye, to be viewed and examined **(Figure 8.7)**. Such an examination can detect certain pathological conditions, such as diabetes, arteriosclerosis, and degeneration of the optic nerve and retina.

### Did You Get It?

5. What is the meaning of the term blind spot in relation to the eye?
6. What function does the choroid of the vascular layer have in common with the pigmented layer of the retina?
7. How do the rods and cones differ from each other?

For answers, see Appendix A.

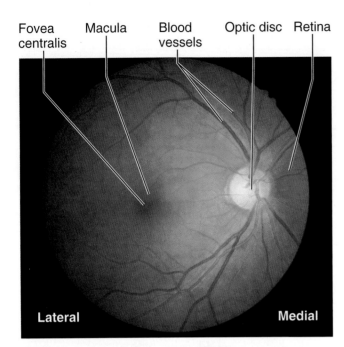

Fovea centralis    Macula    Blood vessels    Optic disc    Retina

Lateral                                          Medial

**Figure 8.7 The posterior wall (fundus) of the retina as seen with an ophthalmoscope.** Notice the optic disc, from which the blood vessels radiate.

# Physiology of Vision

## Pathway of Light through the Eye and Light Refraction

→ Learning Objectives

☐ Describe image formation on the retina.

☐ Define the following terms: *accommodation, astigmatism, emmetropia, hyperopia, myopia,* and *refraction.*

When light passes from one substance to another substance that has a different density, its speed changes and its rays are bent, or **refracted**. Light rays are bent in the eye as they encounter the cornea, aqueous humor, lens, and vitreous humor.

The refractive, or bending, power of the cornea and humors is constant. However, that of the lens can be changed by changing its shape—that is, by making it more or less convex, so that light can be properly focused on the retina. The greater the lens convexity, or bulge, the more it bends the light. The flatter the lens, the less it bends the light.

The resting eye is "set" for distant vision. In general, light from a distant source (over 20 feet away) approaches the eye as parallel rays **(Figure 8.8a)**, and the lens does not need to change shape to focus properly on the retina. However, light from a

**Q:** *As you look at this figure, are your lenses relatively thick or relatively thin?*

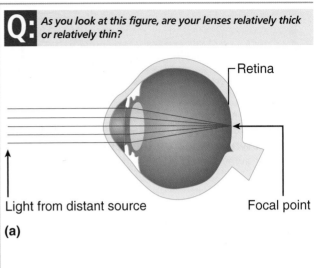

Retina

Light from distant source                     Focal point

**(a)**

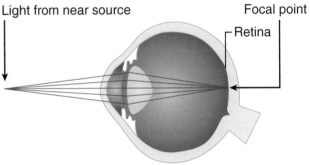

Light from near source                        Focal point

Retina

**(b)**

**Figure 8.8 Relative convexity of the lens during focusing for distant and close vision. (a)** Light rays from a distant object are nearly parallel as they reach the eye and can be focused without requiring changes in lens convexity. **(b)** Diverging light rays from close objects require that the lens bulge more to focus the image sharply on the retina.

close object tends to scatter and *diverge*, or spread out, and the lens must bulge more to make close vision possible (Figure 8.8b). To achieve this, the ciliary body contracts, allowing the lens to become more convex. This ability of the eye to focus specifically for close objects (those less than 20 feet away) is called **accommodation**. The image formed on the retina as a result of the light-bending activity of the lens is a **real image**—that is, it is reversed from left to right, upside down (inverted), and smaller than the object (**Figure 8.9**, p. 314).

The normal eye is able to accommodate properly. However, vision problems occur when a

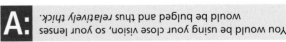

**A:** *You would be using your close vision, so your lenses would be bulged and thus relatively thick.*

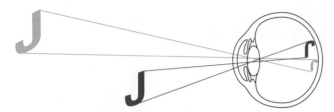

**Figure 8.9 Real image (reversed left to right, and upside down) formed on the retina.** Notice that the farther away the object, the smaller its image on the retina.

lens is too strong (overconverging) or too weak (underconverging) or when there are structural problems of the eyeball (as described in "A Closer Look" on near- and farsightedness on p. 315–316.

## Visual Fields and Visual Pathways to the Brain

→ **Learning Objective**

☐ **Trace the visual pathway to the visual cortex.**

Axons carrying impulses from the retina are bundled together at the posterior aspect of the eyeball and leave the back of the eye as the optic nerve. At the **optic chiasma** (ki-as'mah; *chiasm* = cross) the fibers from the medial side of each eye cross over to the opposite side of the brain. The fiber tracts that result are the **optic tracts**. Each optic tract contains fibers from the lateral side of the eye on the same side and the medial side of the opposite eye. The optic tract fibers synapse with neurons in the thalamus, whose axons form the **optic radiation**, which runs to the occipital lobe of the brain. There they synapse with the cortical cells, and visual interpretation, or seeing, occurs. (**Figure 8.10** shows the visual pathway from the eye to the brain).

Each side of the brain receives visual input from both eyes—from the lateral field of the eye on its own side and from the medial field of the other eye. Also notice that each eye "sees" a slightly different view but that their *visual fields* overlap quite a bit. As a result of these two phenomena, humans have *binocular vision*. Binocular vision, literally "two-eyed vision," provides for depth perception, also called "three-dimensional" vision, as our visual cortex fuses the two slightly different images delivered by the two eyes into one "picture."

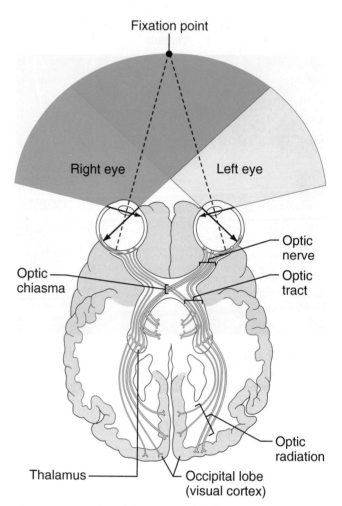

**Figure 8.10 Visual fields of the eyes and visual pathway to the brain (inferior view).** Notice that the visual fields overlap considerably (area of binocular vision). Notice also the retinal sites at which a real image would be focused when both eyes are fixed on a close, pointlike object.

## Homeostatic Imbalance 8.7

**Hemianopia** (hem″e-ah-no′pe-ah) is the loss of the same side of the visual field of both eyes, which results from damage to the visual cortex on one side only (as occurs in some strokes). Thus, the person would not be able to see things past the middle of the visual field on either the right or left side, depending on the site of the stroke. Such individuals should be carefully attended and warned of objects in the nonfunctional side of the visual field. Their food and personal objects should always be placed on their functional side, or they might miss them._____ ✛

# Bringing Things into Focus

The eye that focuses images correctly on the retina is said to have **emmetropia** (em"ĕ-tro'pe-ah), literally, "harmonious vision." Part (a) of the figure shows an emmetropic eye.

Nearsightedness, or **myopia** (mi"o'pe-ah; "short vision"), occurs when the parallel light rays from distant objects fail to reach the retina and instead are focused in front of it; see part (b) in the figure. Another way to think of this is the focal point of a myopic eye falls short of the retina. Therefore, *distant* objects appear blurry to myopic people. Nearby objects are in focus, however, because the lens "accommodates" (bulges) to focus the image properly on the retina. Myopia results from an eyeball that is too long, a lens that is too strong, or a cornea that is too curved. Correction requires *concave* corrective lenses that diverge the light rays before they enter the eye, so that they converge farther back on the retinal surface. In other words, *near*sighted people see *near* objects clearly and need corrective lenses to focus on distant objects.

Farsightedness, or **hyperopia** (hi"per-o'pe-ah; "far vision"), occurs when the parallel light rays from distant objects are focused *behind* the retina—at least in the resting eye, in which the lens is flat and the ciliary muscle is relaxed; see part (c) in the figure. Hyperopia usually results from an eyeball that is too short or from a "lazy" lens. People with hyperopia see distant objects clearly because their ciliary muscles contract continuously to increase the light-bending power of the lens, which moves the focal point forward onto the retina.

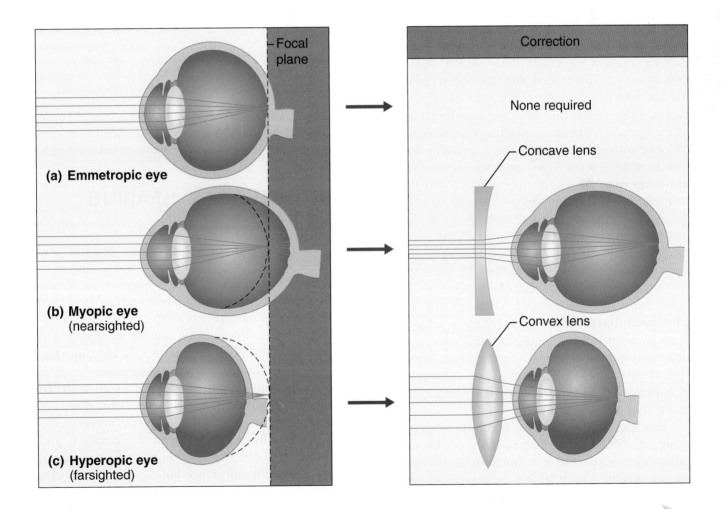

(a) Emmetropic eye

(b) Myopic eye (nearsighted)

(c) Hyperopic eye (farsighted)

Focal plane

Correction

None required

Concave lens

Convex lens

However, the diverging rays from *nearby* objects are focused so far behind the retina that even at full "bulge," the lens cannot focus the image on the retina. Therefore, nearby objects appear blurry, and hyperopic individuals are subject to eyestrain as their endlessly contracting ciliary muscles tire from overwork.

Correction of hyperopia requires *convex* corrective lenses that converge the light rays before they enter the eye. Thus, *far*sighted people can see *far*away objects clearly and require corrective lenses to focus on nearby objects.

Unequal curvatures in different parts of the cornea or lens cause

**astigmatism** (ah-stig'mah-tizm). In this condition, blurry images occur because points of light are focused not as points on the retina but as lines (*astigma* = not a point). Special cylindrically ground lenses or contacts are used to correct astigmatism.

## Eye Reflexes

→ **Learning Objective**

☐ **Discuss the importance of the convergence and pupillary reflexes.**

Both the internal and external (extrinsic) eye muscles are necessary for proper eye function. The autonomic nervous system controls the internal muscles. As mentioned earlier, these muscles include those of the ciliary body, which alters lens curvature, and the radial and circular muscles of the iris, which control pupil size. The external muscles are the rectus and oblique muscles attached to the eyeball exterior (see Figure 8.3), which control eye movements and make it possible to follow moving objects. They are also responsible for **convergence**, which is the reflexive movement of the eyes medially when we view close objects. When convergence occurs, both eyes are aimed toward the near object being viewed. The extrinsic muscles are controlled by somatic fibers of cranial nerves III, IV, and VI (see Figure 8.3).

When the eyes are suddenly exposed to bright light, the pupils immediately constrict; this is the **photopupillary reflex**. This protective reflex prevents excessively bright light from damaging the delicate photoreceptors. The pupils also constrict reflexively when we view close objects; this **accommodation pupillary reflex** provides more acute vision.

Reading requires almost continuous work by both sets of muscles. The muscles of the ciliary body bring about the lens bulge, and the circular (or *constrictor*) muscles of the iris produce the accommodation pupillary reflex. In addition, the extrinsic muscles must converge the eyes as well as move them to follow the printed lines. This is why long periods of reading tire the eyes and

often result in what is commonly called *eyestrain*. When you read for an extended time, look up from time to time and stare into the distance. This temporarily relaxes all the eye muscles.

**Did You Get It?**

8. What are the refractory media of the eye?
9. What name is given to the ability of the eye to focus on close objects?
10. What is the difference between the optic tract and the optic nerve?
11. In what way does the photopupillary reflex protect the eyes?
12. How is astigmatism different from myopia and hyperopia?

**For answers, see Appendix A**

## PART II: THE EAR: HEARING AND BALANCE

At first glance, the machinery for hearing and balance appears crude. Fluids must be stirred to stimulate the receptors of the ear: sound vibrations move fluid to stimulate hearing receptors, whereas gross movements of the head disturb fluids surrounding the balance organs. Receptors that respond to such physical forces are called **mechanoreceptors** (mek"ah-no-re-sep'terz).

Our hearing apparatus allows us to hear an extraordinary range of sound, and our highly sensitive equilibrium receptors keep our nervous system continually up to date on the position and movements of the head. Without this information, it would be difficult if not impossible to maintain our balance or to know which way is up. Although these two sense organs are housed together in the ear, their receptors respond to different stimuli and are activated independently of one another.

# Anatomy of the Ear

→ **Learning Objective**

☐ **Identify the structures of the external, middle, and internal ear, and list the functions of each.**

Anatomically, the ear is divided into three major areas: the external, or outer, ear; the middle ear; and the internal, or inner, ear **(Figure 8.11)**. The external and middle ear structures are involved with hearing *only*. The internal ear functions in both equilibrium and hearing.

## External (Outer) Ear

The **external ear**, or **outer ear**, is composed of the auricle and the external acoustic meatus. The **auricle** (aw′ri-kul), or **pinna** (pin′nah), is what most people call the "ear"—the shell-shaped structure surrounding the auditory canal opening. In many animals, the auricle collects and directs sound waves into the auditory canal, but in humans this function is largely lost.

The **external acoustic meatus** (or *auditory canal*) is a short, narrow chamber (about 1 inch long by ¼ inch wide) carved into the temporal bone of the skull. In its skin-lined walls are the **ceruminous** (sĕ-roo′mĭ-nus) **glands**, which secrete waxy yellow **cerumen**, or **earwax**, which provides a sticky trap for foreign bodies and repels insects.

Sound waves entering the auditory canal eventually hit the **tympanic** (tim-pan′ik; *tympanum* = drum) **membrane**, or **eardrum**, and cause it to vibrate. The canal ends at the eardrum, which separates the external from the middle ear.

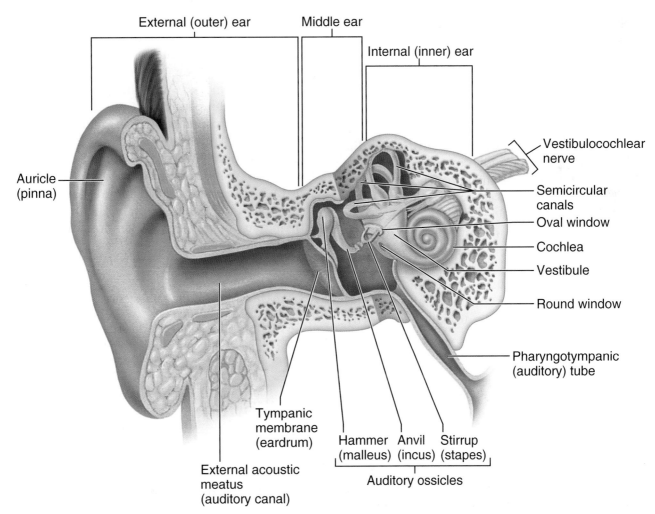

**Figure 8.11 Anatomy of the ear.** Note that the inner ear structures (vestibule, cochlea, semicircular canals) represent a cast of the cavity formed by the bony labyrinth.

## Middle Ear

The **middle ear cavity**, or **tympanic cavity**, is a small, air-filled, mucosa-lined cavity within the temporal bone. It is flanked laterally by the eardrum and medially by a bony wall with two openings, the **oval window** and the inferior, membrane-covered **round window**. The **pharyngotympanic** (think *throat-eardrum: pharynx-tympanic*) **tube**, or **auditory tube**, runs obliquely downward to link the middle ear cavity with the throat, and the mucosae lining the two regions are continuous. Normally, the pharyngotympanic tube is flattened and closed, but swallowing or yawning can open it briefly to equalize the pressure in the middle ear cavity with the external, or atmospheric, pressure. This is an important function because the eardrum does not vibrate freely unless the pressure on both of its surfaces is the same. When the pressures are unequal, the eardrum bulges inward or outward, causing hearing difficulty (voices may sound far away) and sometimes earaches. The ear-popping sensation of the pressures equalizing is familiar to anyone who has flown in an airplane.

 ### Homeostatic Imbalance 8.8

Inflammation of the middle ear, **otitis media** (o-ti′tis me′de-ah), is a fairly common result of a sore throat, especially in children, whose pharyngotympanic tubes run more horizontally. In otitis media, the eardrum bulges and often becomes inflamed. When large amounts of fluid or pus accumulate in the cavity, an emergency *myringotomy* (lancing of the eardrum) may be required to relieve the pressure. A tiny tube is implanted in the eardrum that allows pus to drain into the external ear canal. The tube usually falls out by itself within the year. _____ **✛**

The more horizontal course of the pharyngotympanic tube in infants also explains why it is never a good idea to "prop" a bottle or feed them when they are lying flat (a condition that favors the entry of the food into that tube).

The tympanic cavity is spanned by the three smallest bones in the body, the **ossicles** (os′sĭ-kulz), which transmit the vibratory motion of the eardrum to the fluids of the inner ear (see Figure 8.11). These bones, named for their shape, are the **hammer**, or **malleus** (mă′le-us); the **anvil**, or **incus** (in′kus); and the **stirrup**, or **stapes**

(sta′pēz). Like dominoes falling, when the eardrum moves, it moves the hammer and transfers the vibration to the anvil. The anvil, in turn, passes the vibration on to the stirrup, which presses on the oval window of the inner ear. The movement at the oval window sets the fluids of the inner ear into motion, eventually exciting the hearing receptors.

## Internal (Inner) Ear

The **internal ear** is a maze of bony chambers called the **bony labyrinth**, or **osseous labyrinth** (lab′ĭ-rinth; "maze"), located deep within the temporal bone behind the eye socket. The three subdivisions of the bony labyrinth are the spiraling, pea-sized **cochlea** (kok′le-ah, "snail"), the **vestibule** (ves′ti-būl), and the **semicircular canals**. The vestibule is situated between the semicircular canals and the cochlea. The views of the bony labyrinth typically seen in textbooks, including this one, are somewhat misleading because we are really talking about a cavity. Figure 8.11 can be compared to a *cast* of the bony labyrinth—that is, a labyrinth that was filled with plaster of paris and then had the bony walls removed after the plaster hardened. The shape of the plaster then reveals the shape of the *cavity* that worms through the temporal bone.

The bony labyrinth is filled with a plasmalike fluid called **perilymph** (per′i-limf). Suspended in the perilymph is a **membranous labyrinth**, a system of membrane sacs that more or less follows the shape of the bony labyrinth. The membranous labyrinth itself contains a thicker fluid called **endolymph** (en′do-limf).

### Did You Get It?

13. Which region(s) of the ear (external, middle, or internal) serve hearing only?
14. Which structures of the ear transmit sound vibrations from the eardrum to the oval window?

For answers, see Appendix A.

# Equilibrium

### → Learning Objectives

☐ **Distinguish between static and dynamic equilibrium.**

☐ **Describe how the equilibrium organs help maintain balance.**

The equilibrium sense is not easy to describe because it does not "see," "hear," or "feel." What it *does* is respond (frequently without our awareness) to various head movements. The equilibrium receptors of the inner ear, collectively called the **vestibular apparatus**, can be divided into two branches—one branch is responsible for monitoring *static equilibrium*, and the other monitors *dynamic equilibrium*.

## Static Equilibrium

Within the membrane sacs of the vestibule are receptors called **maculae** (mak'u-le; "spots") that are essential to our sense of **static equilibrium** (**Figure 8.12**). The maculae report on changes in the position of the head in space with respect to the pull of gravity when the body is not moving (*static* = at rest). Because they provide information on which way is up or down, they help us keep our head erect. The maculae are extremely important to divers swimming in the dark depths (where most other orienting cues are absent), enabling them to tell which way is up (to the surface). Each macula is a patch of receptor (hair) cells with their "hairs" embedded in the **otolithic membrane**, a gelatinous mass studded with **otoliths** (o'tŏ-liths), tiny stones made of calcium salts. As the head moves, the otoliths roll in response to changes in the pull of gravity. This movement creates a pull on the gel, which in turn bends the hairs of the hair cells. Bending activates the hair cells, which send impulses along the **vestibular nerve** (a division of cranial nerve VIII) to the cerebellum of the brain, informing it of the position of the head in space.

## Dynamic Equilibrium

The **dynamic equilibrium** receptors, found in the semicircular canals, respond to angular or rotational movements of the head rather than to straight-line movements. When you twirl on the dance floor or suffer through a spinning carnival ride, these receptors are working overtime. The semicircular canals (each about ½ inch, or 1.3 cm, around) are oriented in the three planes of space. Thus, regardless of which plane you move in, there will be receptors to detect the movement.

Within the *ampullae*, swollen regions at the base of each membranous semicircular canal (**Figure 8.13a**, p. 320), are multiple receptor regions, each called a **crista ampullaris** (kris'tah am"pu-lar'is), or simply *crista*, which consists of a tuft of hair cells covered with a gelatinous cap called the **cupula** (ku'pu-lah) (Figure 8.13b).

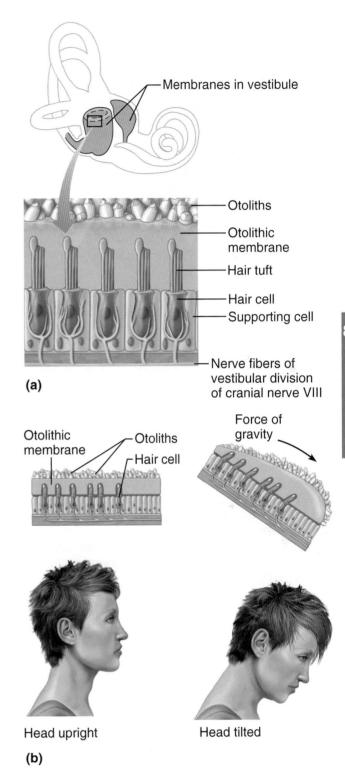

**(a)**

**(b)**

**Figure 8.12 Structure and function of maculae (static equilibrium receptors). (a)** Diagrammatic view of part of a macula. **(b)** When the head is tipped, the otoliths in the gelatinous otolithic membrane move in the direction of gravitational pull, stimulating the maculae. This creates a pull on the hair cells, causing them to bend.

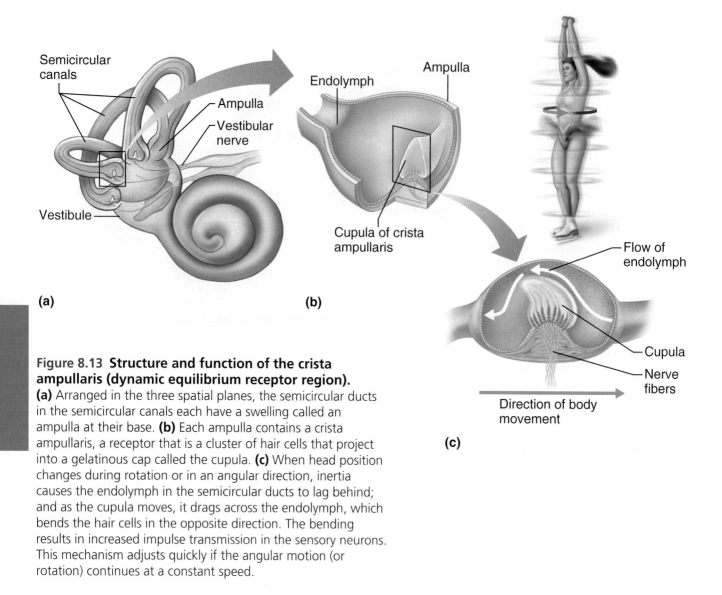

**Figure 8.13 Structure and function of the crista ampullaris (dynamic equilibrium receptor region).**
**(a)** Arranged in the three spatial planes, the semicircular ducts in the semicircular canals each have a swelling called an ampulla at their base. **(b)** Each ampulla contains a crista ampullaris, a receptor that is a cluster of hair cells that project into a gelatinous cap called the cupula. **(c)** When head position changes during rotation or in an angular direction, inertia causes the endolymph in the semicircular ducts to lag behind; and as the cupula moves, it drags across the endolymph, which bends the hair cells in the opposite direction. The bending results in increased impulse transmission in the sensory neurons. This mechanism adjusts quickly if the angular motion (or rotation) continues at a constant speed.

When your head moves in an arclike or angular direction, the endolymph in the canal lags behind the movement (Figure 8.13c). Then, as the cupula drags against the stationary endolymph, the cupula bends—like a diver's fins in water—with the body's motion. This stimulates the hair cells, and impulses are transmitted up the vestibular nerve to the cerebellum. Bending the cupula in the opposite direction reduces impulse generation. When you are moving at a constant rate, the receptors gradually stop sending impulses, and you no longer have the sensation of motion until your speed or direction of movement changes.

Although the receptors of the semicircular canals and vestibule are responsible for dynamic and static equilibrium, respectively, they usually act together. Besides these equilibrium senses, sight and the proprioceptors of the muscles and tendons are also important in providing the cerebellum with information used to control balance.

### Did You Get It?

15. What sense do the vestibule and semicircular canals serve?
16. Describe the different receptors for static and dynamic equilibrium and their locations.
17. What are otoliths, and what is their role in equilibrium?

**For answers, see Appendix A.**

# Physical Therapy Assistant

**Patients trying to regain mobility rely on physical therapy assistants.**

As the population ages, a growing number of people find themselves needing in-home medical care as they recover from injuries or surgical procedures. Many of these patients rely on physical therapy assistants like Leslie Burgess.

Burgess works for Amedisys Home Health Care, and 90 to 95 percent of her patients are senior citizens. Once a doctor prescribes physical therapy, a licensed physical therapist visits the patient and writes a treatment plan. Based on the nature of the problem, this regimen may incorporate strength, movement, and/or balance training, with the goal of improving mobility, reducing pain, and/or helping the patient function with a disability. The therapist also sets goals: for example, the patient will be able to walk 300 feet with a cane after 6 weeks.

Burgess's job is to help the patients carry out these treatment plans, visiting the patient two or three times a week, for 6 to 8 weeks or more, depending on the patient's progress. In some cases, she will use electrical stimulation or ultrasound to stimulate nerves or muscles. If the patient has a new piece of equipment, such as a cane or a walker, she helps him or her learn to use it. She reviews any prescribed medication to make sure the patient is taking it and discusses safety concerns with the patient and family, such as loose electric cords that the patient could trip on. Finally, she leaves instructions with the patient to exercise on his or her own.

Anatomy is an important part of physical therapy work, Burgess says. "Working with various deviations of movement, you need to know what bones and muscles are involved so that you know which bones and muscles to strengthen and show patients how to regain their mobility."

In some cases, part of her job is to help her patients and their families recognize that they will not be exactly the way they were before, particularly if they have suffered a stroke or other severe injury. The fact that patients may also be coping with hearing or vision loss complicates their therapy.

"As we start to age, we begin to lose our independence," she says. "So what can we do to change our lifestyle so that we can still be as independent as possible?"

> The fact that patients may also be coping with hearing or vision loss complicates their therapy.

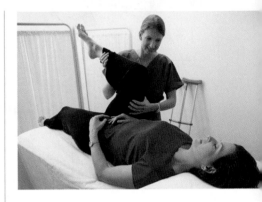

Physical therapy assistants work in hospitals, nursing homes, and clinics—anywhere physical therapists are found. They usually work directly with patients, putting them through exercises under the supervision of a physical therapist. Not all patients are geriatric—some are recovering from serious injuries or have conditions such as cerebral palsy.

Many states require that physical therapy assistants complete an associate's degree and pass a board exam, in addition to completing continuing education.

For more information, contact:

American Physical Therapy Association
1111 N. Fairfax St.
Alexandria, VA 22314-1488
(800) 999-2782
http://www.apta.org

For additional information on this career and others, click the Focus on Careers link at Mastering A&P®.

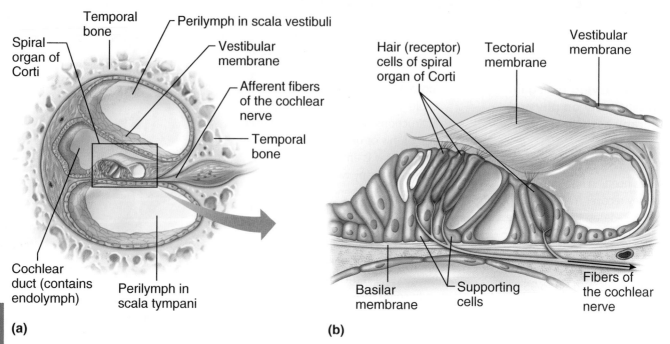

**(a)**

**(b)**

**Figure 8.14 Anatomy of the cochlea. (a)** A cross-sectional view of one turn of the cochlea, showing the position of the spiral organ of Corti in the cochlear duct. The cavities of the bony labyrinth contain perilymph. The cochlear duct contains endolymph. **(b)** Detailed structure of the spiral organ of Corti. The receptor cells (hair cells) rest on the basilar membrane.

# Hearing

→ **Learning Objectives**

☐ **Explain the function of the spiral organ of Corti in hearing.**

☐ **Define** *sensorineural deafness* **and** *conductive deafness*, **and list possible causes of each.**

☐ **Explain how a person is able to localize the source of a sound.**

Within the **cochlear duct**, the endolymph-containing membranous labyrinth of the cochlea is the **spiral organ of Corti** (kor′te), which contains the hearing receptors, or **hair cells (Figure 8.14a)**. The chambers (scalae) above and below the cochlear duct contain perilymph. Sound waves that reach the cochlea through vibrations of the eardrum, ossicles, and oval window set the cochlear fluids into motion **(Figure 8.15)**. As the sound waves are transmitted by the ossicles from the eardrum to the oval window, their force (amplitude) is increased by the lever activity of the ossicles. In this way, nearly the total force exerted on the much larger eardrum reaches the tiny oval window, which in turn sets the fluids of the inner ear into motion, and these pressure waves set up vibrations in the **basilar membrane**. The receptor cells, positioned on the basilar membrane in the spiral organ of Corti, are stimulated by the vibrating movement of the basilar membrane against the gel-like **tectorial** (tek-to′re-al) **membrane** that lies over them. The "hairs" of the receptor cells are embedded in the stationary tectorial membrane such that when the basilar membrane vibrates against it, the "hairs" bend (see Figure 8.14b). The length of the fibers spanning the basilar membrane "tunes" specific regions to vibrate at specific frequencies. In general, high-pitched sounds disturb the shorter, stiffer fibers of the basilar membrane and stimulate receptor cells close to the oval window, whereas low-pitched sounds affect longer, more floppy fibers and activate specific hair cells further along the cochlea **(Figure 8.16**, p. 324).

Once stimulated, the hair cells transmit impulses along the **cochlear nerve** (a division of cranial nerve VIII—the vestibulocochlear nerve) to the auditory cortex in the temporal lobe, where interpretation of the sound, or hearing, occurs. Because sound usually reaches the two ears at different times, we could say that we hear "in stereo." Functionally, this helps us to determine where sounds are coming from in our environment.

When the same sounds, or tones, keep reaching the ears, the auditory receptors tend to *adapt*,

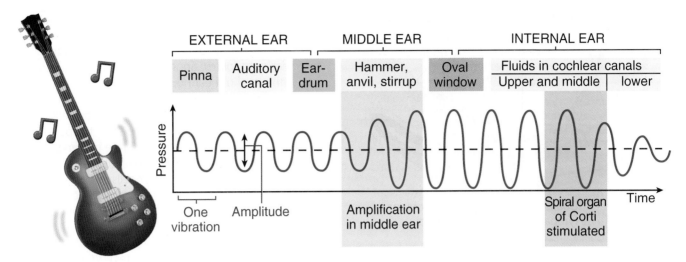

**Figure 8.15 Route of sound waves through the ear.** To excite the hair cells in the spiral organ of Corti in the inner ear, sound wave vibrations must pass through air, membranes, bone, and fluid.

or stop responding, to those sounds, and we are no longer aware of them. This is why the drone of a continuously running motor does not demand our attention after the first few seconds. However, hearing is the last sense to leave our awareness when we fall asleep or receive anesthesia and is the first to return as we awaken.

# Hearing and Equilibrium Deficits

 **Homeostatic Imbalance 8.9**

Children with ear problems or hearing deficits often pull on their ears or fail to respond when spoken to. Under such conditions, tuning fork or audiometry testing is done to try to diagnose the problem. **Deafness** is defined as *hearing loss of any degree*—from a slight loss to a total inability to hear sound. Generally speaking, there are two kinds of deafness, conduction and sensorineural. Temporary or permanent **conduction deafness** results when something interferes with the conduction of sound vibrations to the fluids of the inner ear. Something as simple as a buildup of earwax may be the cause. Other causes of conduction deafness include fusion of the ossicles (a problem called **otosclerosis** [o″to-sklĕ-ro′sis]), a ruptured eardrum, and *otitis media* (inflammation of the middle ear).

**Sensorineural deafness** occurs when there is degeneration or damage to the receptor cells in the spiral organ of Corti, to the cochlear nerve, or to neurons of the auditory cortex. This often results from extended listening to excessively loud sounds. Thus, whereas conduction deafness results from mechanical factors, sensorineural deafness is a problem with nervous system structures.

A person who has a hearing loss due to conduction deafness will still be able to hear by bone conduction, even though his or her ability to hear air-conducted sounds (the normal conduction route) is decreased or lost. In contrast, individuals with sensorineural deafness cannot hear better by *either* conduction route. Hearing aids, which use skull bones to conduct sound vibrations to the inner ear, are generally very successful in helping people with conduction deafness to hear. They are not helpful for sensorineural deafness.

Equilibrium problems are usually obvious. Nausea, dizziness, and problems in maintaining balance are common symptoms, particularly when impulses from the vestibular apparatus "disagree" with what we see (visual input). There also may be strange (jerky or rolling) eye movements.

A serious pathology of the inner ear is **Ménière's** (mān″e-airz′) **syndrome**. The exact cause of this condition is not fully known, but suspected causes are arteriosclerosis, degeneration of cranial nerve VIII, and increased pressure of the inner ear fluids. In Ménière's syndrome, progressive

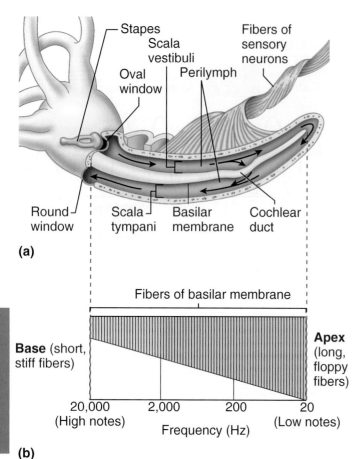

**(a)**

**(b)**

**Figure 8.16 Activation of the cochlear hair cells.**
**(a)** The cochlea is drawn as though it were uncoiled to make the events of sound transmission occurring there easier to follow. Sound waves of low frequency, below the level of hearing, travel entirely around the cochlear duct without exciting hair cells. But sounds of higher frequency penetrate through the cochlear duct and basilar membrane to reach the scala tympani. This causes the basilar membrane to vibrate maximally in certain areas in response to certain frequencies of sound, stimulating particular hair cells and sensory neurons. The differential stimulation of hair cells is perceived in the brain as sound of a certain pitch.
**(b)** The length and stiffness of the fibers spanning the basilar membrane tune specific regions to vibrate at specific frequencies. The higher notes—20,000 Hertz (Hz)—are detected by shorter, stiffer hair cells along the base of the basilar membrane.

deafness occurs. Affected individuals become nauseated and often have howling or ringing sounds in their ears and **vertigo** (a sensation of spinning) that is so severe that they cannot stand up without extreme discomfort. Anti–motion sickness drugs are often prescribed to decrease the discomfort. _____ **✛**

### Did You Get It?

18. From the air outside the body, through what substances do sound waves travel to excite the receptor cells of the cochlea?
19. Which nerve transmits impulses from the spiral organ of Corti to the brain?
20. Do high-pitched sounds peak close to or far from the oval window?
21. How do sensorineural deafness and conduction deafness differ from each other?

For answers, see Appendix A.

# PART III: CHEMICAL SENSES: SMELL AND TASTE

→ **Learning Objectives**

☐ Describe the location, structure, and function of the olfactory and taste receptors.

☐ Name the five basic taste sensations, and list factors that modify the sense of taste.

The receptors for taste and olfaction are classified as **chemoreceptors** (ke"mo-re-sep'terz) because they respond to chemicals in solution. Five types of taste receptors have been identified, but the olfactory receptors (for smell) are believed to be sensitive to a much wider range of chemicals. The receptors for smell and taste complement each other and respond to many of the same stimuli.

## Olfactory Receptors and the Sense of Smell

Even though our sense of smell is far less acute than that of many other animals, the human nose is still no slouch in picking up small differences

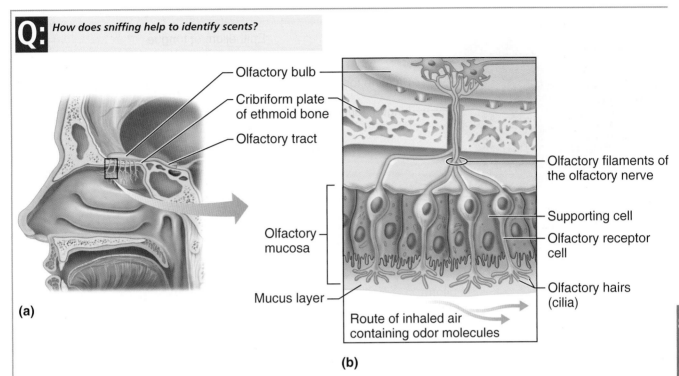

**Q:** *How does sniffing help to identify scents?*

Figure 8.17 **Location and cellular makeup of the olfactory epithelium.**

in odors. Some people capitalize on this ability by becoming tea and coffee blenders, perfumers, or wine tasters.

The thousands of **olfactory receptors**, receptors for the sense of smell, occupy a postage stamp–sized area in the roof of each nasal cavity **(Figure 8.17)**. Air entering the nasal cavities must make a 90° turn to enter the respiratory passageway below, so sniffing, which causes more air to flow superiorly across the olfactory receptors, intensifies the sense of smell.

The **olfactory receptor cells** are neurons equipped with **olfactory hairs**, long cilia that protrude from the nasal epithelium and are continuously bathed by a layer of mucus secreted by underlying glands. When the olfactory receptors located on the cilia are stimulated by chemicals dissolved in the mucus, they transmit impulses along the **olfactory filaments**, which are bundled axons of olfactory neurons that collectively make up the **olfactory nerve** (cranial nerve I).

The olfactory nerve conducts the impulses to the olfactory cortex of the brain. There the odor is interpreted, and an "odor snapshot" is made. The olfactory pathways are closely tied into the limbic system (emotional-visceral part of the brain). Thus, olfactory impressions are long-lasting and very much a part of our memories and emotions. For example, the smell of chocolate chip cookies may remind you of your grandmother, and the smell of a special cologne may make you think of your father. Our reactions to odors are rarely neutral. We tend to either like or dislike certain odors, and we change, avoid, or add odors according to our preferences.

The olfactory receptors are exquisitely sensitive—just a few molecules can activate them. Like the auditory receptors, the olfactory neurons tend to adapt rather quickly when they are exposed to an unchanging stimulus, in this case, an odor. This is why a woman stops smelling her own perfume after a while but will quickly pick up the scent of another perfume on someone else.

**A:** It brings more odor-containing air into contact with the olfactory receptors in the superior part of the nasal cavity.

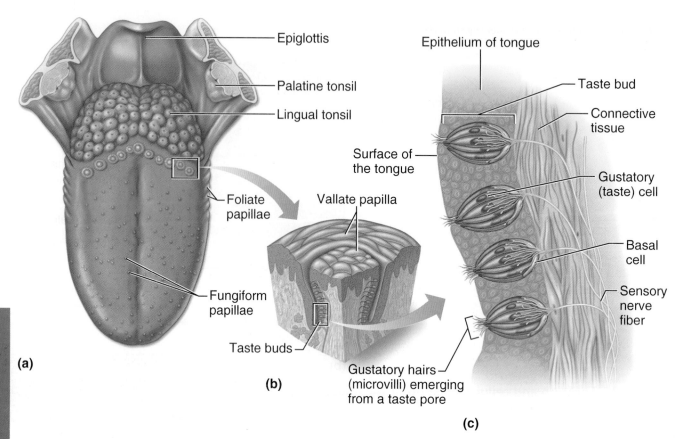

**Figure 8.18 Location and structure of taste buds. (a)** Taste buds on the tongue are associated with papillae, projections of the tongue mucosa. **(b)** A sectioned vallate papilla shows the position of the taste buds in its lateral walls. **(c)** An enlarged view of four taste buds.

### Homeostatic Imbalance 8.10

Although it is possible to have either taste or smell deficits, most people seeking medical help for loss of chemical senses have olfactory disorders, or **anosmias** (ă-noz′me-uz). Most anosmias result from head injuries, the aftereffects of nasal cavity inflammation (due to a cold, an allergy, or smoking), or aging. Some brain disorders can destroy the sense of smell or mimic it. For example, some epileptics experience **olfactory auras** (olfactory hallucinations) just before they go into seizures. _____ +

# Taste Buds and the Sense of Taste

The word *taste* comes from the Latin word *taxare*, which means "to touch, estimate, or judge." When we taste things, we are, in fact, testing or judging

our environment in an intimate way, and many of us consider the sense of taste to be the most pleasurable of our special senses.

The **taste buds**, or receptors for the sense of taste, are widely scattered in the oral cavity. Of the 10,000 or so taste buds that we have, most are on the tongue. A few are scattered on the soft palate, superior part of the pharynx, and inner surface of the cheeks.

The dorsal tongue surface is covered with small peglike projections, or **papillae** (pah-pil′e). The taste buds are found on the sides of the large round **vallate papillae**, or **circumvallate** (ser″kum-val′at) **papillae**, on the tops of the more numerous **fungiform** (fun′jĭ-form) **papillae** and in the **foliate papillae** on the sides of the tongue **(Figure 8.18)**. The specific receptor cells that respond to chemicals dissolved in the saliva are epithelial cells called **gustatory cells**. Their long microvilli—the **gustatory hairs**—protrude through the **taste pore**; and when they are stimulated, they depolarize, and impulses are transmitted to the brain. Three cranial nerves—VII, IX, and X—carry taste impulses from the various taste buds to the gustatory cortex. The **facial nerve** (VII) serves the anterior part of the tongue. The other two cranial nerves—the **glossopharyngeal nerve** and **vagus nerve**—serve the other taste bud–containing areas. Because of their location, taste bud cells are subjected to huge amounts of friction and are routinely burned by hot foods. Luckily, they are among the most dynamic cells in the body and are replaced every 7 to 10 days by **basal cells** (stem cells) found in the deeper regions of the taste buds.

There are five basic taste sensations, each corresponding to stimulation of one of the five major types of taste buds. The *sweet receptors* respond to substances such as sugars, saccharine, some amino acids, and some lead salts (such as those found in lead paint). *Sour receptors* respond to hydrogen ions ($H^+$), or the acidity of the solution; *bitter receptors* to alkaloids; and *salty receptors* to metal ions in solution. *Umami* (u-mah′me; "delicious"), a taste discovered by the Japanese, is elicited by the amino acid glutamate, which appears to be responsible for the "beef taste" of steak and the flavor of monosodium glutamate, a food additive.

Historically, the tip of the tongue was believed to be most sensitive to sweet and salty substances, its sides to sour, the back of the tongue to bitter, and the pharynx to umami. Actually there are only slight differences in the locations of the taste receptors in different regions of the tongue, but the bitter receptors do seem to be clustered more at the rear of the tongue. Most taste buds respond to two, three, four, or even all five taste modalities.

Taste likes and dislikes have homeostatic value. A liking for sugar and salt will satisfy the body's need for carbohydrates and minerals (as well as some amino acids). Many sour, naturally acidic foods (such as oranges, lemons, and tomatoes) are rich sources of vitamin C, an essential vitamin. Umami guides the intake of proteins, and because many natural poisons and spoiled foods are bitter, our dislike for bitterness is protective.

Many factors affect taste, and what is commonly referred to as our sense of taste depends heavily on stimulation of our olfactory receptors by aromas. Think of how bland food tastes when your nasal passages are congested by a cold.

Without the sense of smell, our morning coffee would simply taste bitter. In addition, the temperature and texture of food can enhance or spoil its taste for us. For example, some people will not eat foods that have a pasty texture (avocados) or that are gritty (pears), and almost everyone considers a cold greasy hamburger unfit to eat. "Hot" foods such as chili peppers actually excite pain receptors in the mouth.

### Did You Get It?

22. What general name is used to describe both taste and smell receptors? Why?
23. Where, relative to specific structures, are most taste buds located?
24. Why does it help to sniff substances that you want to smell?

**For answers, see Appendix A.**

# PART IV: Developmental Aspects of the Special Senses

→ **Learning Objective**

☐ **Describe changes that occur with age in the special sense organs.**

The special sense organs, essentially part of the nervous system, are formed very early in embryonic development. For example, the eyes, which are literally outgrowths of the brain, are developing by the fourth week. All of the special senses are functional, to a greater or lesser degree, at birth.

 **Homeostatic Imbalance 8.11**

Congenital eye problems are relatively uncommon, but we can give some examples. **Strabismus** (strah-biz′mus), which is commonly called "crossed eyes," results from unequal pulls by the external eye muscles that prevent the baby from coordinating movement of the two eyes. First, exercises are used to strengthen the weaker eye muscles, and/or the stronger eye may be covered with an eye patch to force the weaker muscles to become stronger. If these measures are not successful, surgery is always used to correct the condition because if it is allowed to persist, the brain may stop recognizing signals from the deviating eye, causing that eye to become functionally blind.

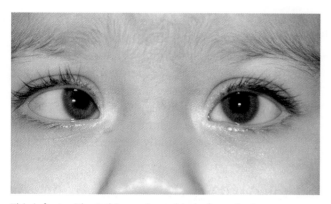

This infant with strabismus is unable to focus both eyes simultaneously on the same object.

Maternal infections that occur during early pregnancy, particularly *rubella* (German measles), may lead to congenital blindness or cataracts. If the mother has a type of sexually transmitted infection called *gonorrhea* (gon"o-re'ah), the bacteria will infect the baby's eyes during delivery. In the resulting *conjunctivitis*, specifically called **ophthalmia neonatorum** (of-thal'me-ah ne"o-na-to'rum), the baby's eyelids become red, swollen, and produce pus. All states have enacted laws requiring that all newborn babies' eyes be routinely treated with silver nitrate or antibiotics shortly after birth. _____ ✚

Generally speaking, vision is the only special sense that is not fully functional when the baby is born, and many years of "learning" are needed before the eyes are fully mature. The eyeballs continue to enlarge until the age of 8 or 9, but the lens grows throughout life. At birth, the eyeballs are foreshortened, and all babies are hyperopic (farsighted). As the eyes grow, this condition usually corrects itself. The newborn infant sees only in gray tones, makes uncoordinated eye movements, and often "sees" using only one eye at a time. Because the lacrimal glands are not fully developed until about 2 weeks after birth, the baby is tearless for this period, even though he or she may cry lustily.

By 5 months, the infant is able to focus on articles within easy reach and to follow moving objects, but visual acuity is still poor. For example, an object that someone with mature vision can see clearly 200 feet away has to be a mere 20 feet away before an infant can see it clearly. (Such vision is said to be 20/200.) By the time the child is 5 years old, color vision is well developed,

visual acuity has improved to about 20/30, and depth perception is present, providing a readiness to begin reading. By school age, the earlier hyperopia has usually been replaced by emmetropia (normal vision). This condition continues until about age 40, when **presbyopia** (pres"be-o'pe-ah) begins to set in. Presbyopia (literally, "old vision") results from decreasing lens elasticity that accompanies aging. This condition makes it difficult to focus for close vision; it is basically farsightedness. Your grandmother holding a magazine at arm's length to read it provides a familiar example of this developmental change in vision.

As aging occurs, the lacrimal glands become less active, and the eyes tend to become dry and more vulnerable to bacterial infection and irritation. The lens loses its crystal clarity and becomes discolored. As a result, it begins to scatter light, causing a distressing glare when the person drives at night. The dilator muscles of the iris become less efficient; thus, the pupils are always somewhat constricted. These last two conditions work together to decrease the amount of light reaching the retina, and visual acuity is dramatically lower by one's seventies. In addition to these changes, older people are susceptible to certain conditions that may result in blindness, such as glaucoma, cataracts, arteriosclerosis, and diabetes.

## ⚖ Homeostatic Imbalance 8.12

Congenital abnormalities of the ears are fairly common. Examples include partly or completely missing pinnas and closed or absent external acoustic meatuses. Maternal infections can have a devastating effect on ear development, and maternal rubella during the early weeks of pregnancy results in sensorineural deafness.

A newborn infant can hear after his or her first cry, but early responses to sound are mostly reflexive—for example, crying and clenching the eyelids in response to a loud noise. By the age of 3 or 4 months, the infant is able to localize sounds and will turn to the voices of family members. The toddler listens critically as he or she begins to imitate sounds, and good language skills are closely tied to an ability to hear well.

Except for ear inflammations (*otitis*) resulting from bacterial infections or allergies, few problems affect the ears during childhood and adult life. By the sixties, however, gradual deterioration and atrophy of the spiral organ of Corti begin and lead to a loss in the ability to hear high tones and speech sounds. This condition, **presbycusis** (pres″bĭ-kyu′sis), is a type of sensorineural deafness. In some cases, the ear ossicles fuse (**otosclerosis**), which compounds the hearing problem by interfering with sound conduction to the inner ear. Because many elderly people refuse to accept their hearing loss and resist using hearing aids, they begin to rely more and more on their vision for clues as to what is going on around them and may be accused of ignoring people. Although presbycusis was once considered a disability of old age, it is becoming much more common in younger people as our world grows noisier day by day. The damage caused by excessively loud sounds is progressive and cumulative. Music played and heard at deafening levels definitely contributes to the deterioration of hearing receptors.

The chemical senses, taste and smell, are sharp at birth, and infants relish some food that adults consider bland or tasteless. Some researchers claim the sense of smell is just as important as the sense of touch in guiding a newborn baby to its mother's breast. However, very young children seem indifferent to odors and can play happily with their own feces. As they get older, their emotional responses to specific odors increase.

There appear to be few problems with the chemical senses throughout childhood and young adulthood. Beginning in the midforties, our ability to taste and smell diminishes, which reflects the gradual decrease in the number of these receptor cells. Almost half of people over the age of 80 cannot smell at all, and their sense of taste is poor. This may explain why older adults are inattentive to formerly disagreeable odors and why they often prefer highly seasoned foods or lose their appetite entirely.

### Did You Get It?

25. Fifty-year-old Mrs. Bates is complaining that she can't read without holding the newspaper out at arm's length. What is her condition, and what is its cause?
26. Which of the special senses is least mature at birth?
27. What is presbycusis?

*For answers, see Appendix A.*

# Summary

## PART I: THE EYE AND VISION (pp. 305–316)

1. External/accessory structures of the eye:
   a. Extrinsic eye muscles aim the eyes for following moving objects and for convergence.
   b. The lacrimal apparatus includes a series of ducts and the lacrimal glands that produce tears, a saline solution that washes and lubricates the eyeball.
   c. Eyelids protect the eyes. Associated with the eyelashes are the ciliary glands (modified sweat glands) and the tarsal glands (which produce an oily secretion). Both help keep the eye lubricated.
   d. The conjunctiva is a mucous membrane that covers the anterior eyeball and lines the eyelids. It produces a lubricating mucus.

2. Three layers form the eyeball.
   a. The sclera forms most of the outer, tough, protective fibrous layer. The anterior portion is the cornea, which is transparent to allow light to enter the eye.
   b. The vascular layer, or middle coat, provides nutrition to the internal eye structures. Its posterior portion, the choroid, is pigmented and prevents light from scattering in the eye. Anterior modifications include two smooth muscle structures: the ciliary body (which changes lens shape) and the iris (which controls the size of the pupil).
   c. The sensory layer consists of the two-layered retina—a pigmented epithelium and the innermost (neural) coat, which contains the photoreceptors. Rods are dim light receptors. Cones are receptors that provide for color vision and high visual acuity. The fovea centralis, on which acute focusing occurs, contains only cones.

3. The optic disc is the point where the optic nerve leaves the back of the eyeball. Since no photoreceptors exist at the optic disc, it results in the blind spot in our vision.

4. The lens is the major light-bending (refractory) structure of the eye. Its convexity is increased by contraction of the ciliary body for close focus. Anterior to the lens is the aqueous humor; posterior

to the lens is the vitreous humor. Both humors reinforce the eye internally. The aqueous humor also provides nutrients to the avascular lens and cornea.

5. Errors of refraction include myopia, hyperopia, and astigmatism. All are correctable with specially ground lenses.

6. The pathway of light through the eye is cornea → aqueous humor → (through pupil) → aqueous humor → lens → vitreous humor → retina.

7. Overlap of the visual fields and inputs from both eyes to each optic cortex provide for depth perception.

8. The pathway of nerve impulses from the retina of the eye is optic nerve → optic chiasma → optic tract → thalamus → optic radiation → visual cortex in occipital lobe of brain.

9. Eye reflexes include the photopupillary, accommodation pupillary, and convergence reflexes.

## PART II: THE EAR: HEARING AND BALANCE (pp. 316–324)

1. The ear is divided into three major areas.

   a. External ear structures are the auricle (pinna), external acoustic meatus, and tympanic membrane (eardrum). Sound entering the external acoustic meatus causes the eardrum to vibrate. These structures are involved with sound transmission only.

   b. Middle ear structures are the ossicles and pharyngotympanic tube within the tympanic cavity. Auditory ossicles transmit the vibratory motion from the eardrum to the oval window. The pharyngotympanic tube allows pressure to be equalized on both sides of the eardrum. These structures are involved with sound transmission only.

   c. The internal ear, or bony labyrinth, consists of bony chambers (cochlea, vestibule, and semicircular canals) in the temporal bone. The bony labyrinth contains perilymph and membranous sacs filled with endolymph. Within the membranous sacs of the vestibule and semicircular canals are equilibrium receptors. Hearing receptors are found within the membranes of the cochlea.

2. Receptors of the semicircular canals (cristae ampullares) are dynamic equilibrium receptors, which respond to angular or rotational body movements. Receptors of the vestibule (maculae) are static equilibrium receptors, which respond to the pull of gravity and report on head position. Visual and proprioceptor input to the brain are also necessary for normal balance.

3. Symptoms of equilibrium apparatus problems include involuntary rolling of the eyes, nausea, vertigo, and an inability to stand erect.

4. Hair cells of the spiral organ of Corti (the receptor for hearing within the cochlea) are stimulated by sound vibrations transmitted through air, membranes, bone, and fluids.

5. Deafness is any degree of hearing loss. Conduction deafness results when the transmission of sound vibrations through the external and middle ears is hindered. Sensorineural deafness occurs when there is damage to the nervous system structures involved in hearing.

## PART III: CHEMICAL SENSES: SMELL AND TASTE (pp. 324–327)

1. Chemical substances must be dissolved in aqueous solution to excite the receptors for smell and taste.

2. The olfactory (smell) receptors are located in the superior aspect of each nasal cavity. Sniffing helps to bring more air (containing odors) over the olfactory mucosa.

3. Olfactory pathways are closely linked to the limbic system; odors stimulate the recall of memories and arouse emotional responses.

4. Gustatory (taste) cells are located in the taste buds, primarily on the tongue. The five major taste sensations are sweet, salt, sour, bitter, and umami.

5. Taste and appreciation of foods are influenced by the sense of smell and the temperature and texture of foods.

## PART IV: DEVELOPMENTAL ASPECTS OF THE SPECIAL SENSES (pp. 327–329)

1. Special sense organs are formed early in embryonic development. Maternal infections during the first 5 or 6 months of pregnancy may cause visual abnormalities as well as sensorineural deafness in the developing child. An important congenital eye problem is strabismus. The most important congenital ear problem is lack of the external acoustic meatus.

2. Vision requires the most learning. The infant has poor visual acuity (is farsighted) and lacks color vision and depth perception at birth. The eye continues to grow and mature until age 8 or 9.

3. Problems of aging associated with vision include presbyopia, glaucoma, cataracts, and arteriosclerosis of the eye's blood vessels.

4. The newborn infant can hear sounds, but initial responses are reflexive. By the toddler stage, the child is listening critically and beginning to imitate sounds as language development begins.

5. Sensorineural deafness (presbycusis) is a normal consequence of aging.

6. Taste and smell are most acute at birth and decrease in sensitivity after age 40 as the number of olfactory and gustatory receptors decreases.

## Review Questions

 Access additional practice questions using your smartphone, tablet, or computer: Mastering A&P® > Study Area > Practice Tests & Quizzes

### Multiple Choice

*More than one choice may apply.*

1. Gustatory cells are
   a. bipolar neurons.
   b. multipolar neurons.
   c. unipolar neurons.
   d. epithelial cells.

2. Alkaloids excite gustatory hairs mostly at the
   a. tip of the tongue.
   b. back of the tongue.
   c. vallate papillae.
   d. fungiform papillae.

3. Parts of the olfactory mucosa include the
   a. mucus layer.
   b. olfactory bulb.
   c. olfactory hairs.
   d. cribriform plate.

4. Which cranial nerve is *not* involved in controlling the action of the extrinsic eye muscles?
   a. Abducens
   b. Oculomotor
   c. Trochlear
   d. Trigeminal

5. The cornea is nourished by
   a. corneal blood vessels.
   b. aqueous humor.
   c. vitreous humor.
   d. scleral blood vessels.

6. When the eye focuses for far vision,
   a. the lens is at its thinnest.
   b. the ciliary muscles contract.
   c. the light rays are nearly parallel.
   d. suspensory fibers of the ciliary zonule are slack.

7. Convergence
   a. requires contraction of the medial rectus muscles of both eyes.
   b. is needed for near vision.
   c. involves transmission of impulses along the abducens nerves.
   d. can promote eyestrain.

8. Which of the following are paired *incorrectly*?
   a. Cochlear duct—cupula
   b. Vestibule—macula
   c. Ampulla—otoliths
   d. Semicircular duct—ampulla

9. Movement of the _____ membrane triggers bending of hairs of the hair cells in the spiral organ of Corti.
   a. tympanic
   b. tectorial
   c. basilar
   d. vestibular

10. Sounds entering the external acoustic meatus are eventually converted to nerve impulses via a chain of events including
    a. vibration of the eardrum.
    b. vibratory motion of the ossicles against the oval window.
    c. stimulation of hair cells in the spiral organ of Corti.
    d. resonance of the cupula.

### Short Answer Essay

11. Distinguish between tarsal and ciliary glands.

12. Explain crying in physiological terms.

13. Diagram and label the internal structures of the eye, and give the major function of each structure.

14. Name the extrinsic eye muscles that allow you to direct your eyes.

15. Locate and describe the functions of the two humors of the eye.

16. What is the blind spot, and why is it so called?

17. What name is given to the rounded opening in the pigmented iris through which light passes?

18. What is the fovea centralis, and why is it important?

19. Trace the pathway of light from the time it hits the cornea until it excites the rods and cones.

20. Trace the pathway of nerve impulses from the photoreceptors in the retina to the visual cortex of the brain.

21. Define *hyperopia*, *myopia*, and *emmetropia*.

22. Why do most people develop presbyopia as they age? Which of the conditions in question 21 does it most resemble?

23. There are only three types of cones. How can you explain the fact that we see many more colors?

24. A rise in intraocular pressure can be damaging for the retina and optic nerve. How is this pressure kept within the normal range?

25. Many students struggling through mountains of reading assignments are told that they need glasses for eyestrain. Why is it more of a strain on the extrinsic and intrinsic muscles to look at close objects than at far objects?

26. Name the structures of the outer, middle, and inner ears, and give the general function of each structure and each group of structures.

27. Sound waves hitting the eardrum set it into motion. Trace the pathway of vibrations from the eardrum to the spiral organ of Corti, where the hair cells are stimulated.

28. What is the difference between the perilymph and the endolymph?

29. Normal balance depends on information transmitted from a number of sensory receptor types. Name at least three of these receptors.

30. Which cranial nerves are involved in taste registration?

31. Name the five primary taste sensations.

32. Where are the olfactory receptors located, and why is that site poorly suited for their job?

33. What is Ménière's syndrome?

34. Which special sense requires the most learning?

35. For each of the following descriptions, indicate whether it applies to a macula or a crista ampullaris: inside a semicircular canal; contains otoliths; responds to linear acceleration and deceleration; has a cupula; responds to rotational acceleration and deceleration; inside the vestibule.

 ## Critical Thinking and Clinical Application Questions

36. An engineering student has been working in construction to earn money to pay for his education. After about eight months, he notices that he is having problems hearing high-pitched tones. What is the cause-and-effect relationship here?

37. Nine children attending the same day-care center developed red, inflamed eyes and eyelids. What are the most likely cause and name of this condition?

38. Dr. Nguyen used an instrument to press on Mr. Cruz's eye during his annual physical examination on his 60th birthday. The eye deformed very little, indicating that the intraocular pressure was too high. What was Mr. Cruz's probable condition?

39. Brandon suffered a ruptured artery in his middle cranial fossa, and a pool of blood compressed his left optic tract, destroying its axons. What part of the visual field was blinded?

40. Sylvia Marcus, age 70, recently had surgery for otosclerosis. The operation was a failure and did not improve her condition. What was the purpose of the surgery, and exactly what was it trying to accomplish?

41. Kathy visits her doctor as she has been struggling to read books, food labels, and her computer screen. An eye test reveals that Kathy is hyperopic. What sort of lenses would Kathy need? Explain.

42. Julie and her father love to find the constellations in the sky on starry nights. One evening, Julie came running into the house and whispered excitedly to her mother. "Mom, I've got power! When I look hard at a star it disappears!" What is happening?

43. While visiting her father's office on the 25th floor of the Harris Building, 5-year-old Emma wandered away into the hall. Fascinated by the buttons in the

elevator, she entered and pressed 1, and the high-speed elevator plummeted to the first floor. Later she told her father that she felt like she "kept on going" when the elevator stopped. Explain her sensation.

44. Mrs. Garson has an immune disorder that causes dry mouth, and she complains to her doctor that she's lost her sense of taste. How might her symptoms be explained?

45. Serge can see colors, but he confuses certain colors, especially red and green. What is the cause of his condition?

46. Explain the role of vitamin A in vision.

# 9

# The Endocrine System

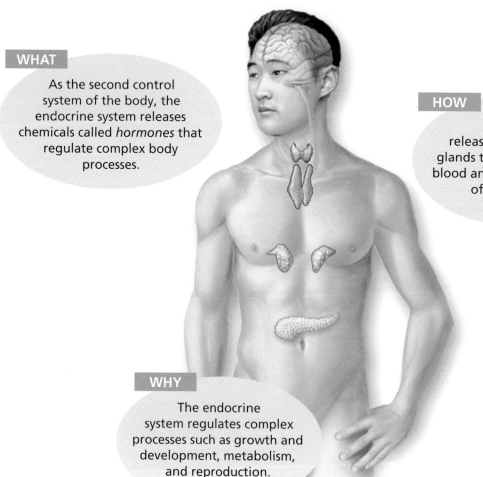

**WHAT**

As the second control system of the body, the endocrine system releases chemicals called *hormones* that regulate complex body processes.

**HOW**

Hormones released by endocrine glands travel through the blood and alter the activity of target cells.

**WHY**

The endocrine system regulates complex processes such as growth and development, metabolism, and reproduction.

**INSTRUCTORS**

New **Building Vocabulary Coaching** Activities for this chapter are assignable in Mastering A&P®

Your body cells have dynamic adventures on the microscopic level all the time. For instance, when insulin molecules leave the blood and bind tightly to protein receptors on nearby cells, the response is dramatic: bloodborne glucose molecules begin to enter the cells, and cellular activity accelerates. Such is the power of the second great controlling system of the body, the **endocrine system**. Along with the nervous system, it coordinates and directs the activity of the body's cells.

The speed of control in these two regulating systems is very different. The nervous system is "built for speed." It uses nerve impulses to prod the muscles and glands into immediate action so that rapid adjustments can be made in response to changes occurring both inside and outside the body. By contrast, the endocrine system acts more slowly by using chemical messengers called *hormones*, which are released into the blood to be transported throughout the body.

Although hormones have widespread effects, the major processes they control are reproduction; growth and development; mobilizing body defenses against stressors; maintaining electrolyte, water, and nutrient balance of the blood; and regulating cellular

**334**

metabolism and energy balance. As you can see, the endocrine system regulates processes that go on for relatively long periods and, in some cases, continuously. The scientific study of hormones and endocrine organs is called **endocrinology**.

# The Endocrine System and Hormone Function— An Overview

→ **Learning Objectives**

☐ Define *hormone* and *target organ*.

☐ Describe how hormones bring about their effects in the body.

Compared to other organs of the body, the organs of the endocrine system are small and unimpressive. The endocrine system also lacks the structural or anatomical continuity typical of most organ systems. Instead, bits and pieces of endocrine tissue are tucked away in separate regions of the body (see Figure 9.3, p. 338). However, functionally the *endocrine organs* are impressive, and when their role in maintaining body homeostasis is considered, they are true giants.

## The Chemistry of Hormones

**Hormones** are chemical substances secreted by endocrine cells into the extracellular fluids that regulate the metabolic activity of other cells in the body. Although the body produces many different hormones, nearly all of them can be classified chemically as either **amino acid–based molecules** (including proteins, peptides, and amines) or **steroids**. Steroid hormones (made from cholesterol) include the sex hormones made by the *gonads* (ovaries and testes) and the hormones produced by the adrenal cortex. All other hormones are nonsteroidal amino acid derivatives. If we also consider the hormones that act locally, called **prostaglandins** (pros"tah-glan'dinz), we must add a third chemical class. (Prostaglandins are described later in the chapter—see Table 9.2, p. 356). The prostaglandins are made from highly active lipids released from nearly all cell membranes.

## Hormone Action

Although hormones circulate to all the organs of the body via blood, a given hormone affects only certain tissue cells or organs, referred to as its **target cells** or **target organs**.

A hormone's relationship to its target cells resembles that of an enzyme to its substrate. Recall that enzymes interact very specifically with their substrates (Chapter 2, pp. 77–78). Hormone interactions with target cell receptors are also very specific and fit together like two pieces of a puzzle. ←

For a target cell to respond to a hormone, specific protein *receptors* to which that hormone can attach must be present on the cell's plasma membrane or in its interior. Only when this binding occurs can the hormone influence the workings of a cell.

The term *hormone* comes from a Greek word meaning "to arouse." In fact, the body's hormones do just that. They "arouse," or bring about their effects on, the body's cells primarily by *altering* cellular activity—that is, by increasing or decreasing the rate of a normal, or usual, metabolic process rather than by stimulating performance of a new one. The precise change(s) that follow hormone binding depend on the specific hormone and the target cell type. Hormones can:

- Change plasma membrane permeability or membrane potential (electrical state) by opening or closing ion channels
- Activate or inactive enzymes
- Stimulate or inhibit cell division
- Promote or inhibit secretion of a product
- Turn on or turn off transcription of certain genes (such as those encoding proteins or regulatory molecules)

### Direct Gene Activation

Despite the huge variety of hormones, there are really only two mechanisms by which hormones trigger changes in cells. Steroid hormones (and, strangely, thyroid hormone) can use the mechanism of *direct gene activation* (**Figure 9.1a**, p. 336). Because they are lipid-soluble molecules, the steroid hormones can diffuse through the plasma membranes of their target cells ①. Once inside, the steroid hormone enters the nucleus ② and binds to a specific hormone receptor ③. The hormone-receptor complex then binds to specific sites on the cell's DNA ④, activating certain genes to transcribe messenger RNA (mRNA) ⑤. The mRNA is translated in the cytoplasm ⑥, resulting

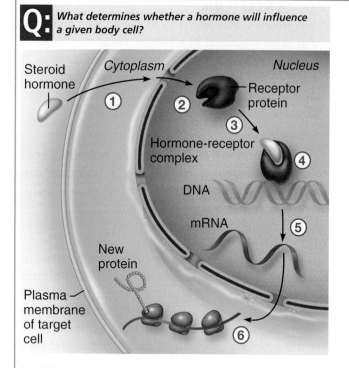

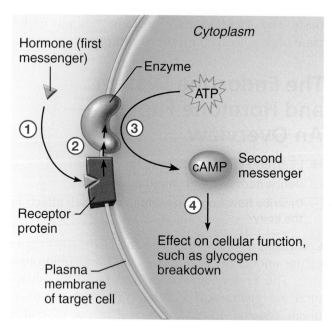

**Q:** *What determines whether a hormone will influence a given body cell?*

(a) **Direct gene activation**

(b) **Second-messenger system**

**Figure 9.1 Mechanisms of hormone action.** Steroid hormones can act through either **(a)** direct gene activation or **(b)** second-messenger systems. Amino acid–based hormones typically act through second-messenger systems.

in the synthesis of new proteins. Alternatively, the steroid hormone may bind to receptors in the cytoplasm, and then the complex moves into the nucleus to activate certain genes.

### Second-Messenger System

Steroid hormones can influence cell activity either by direct gene activation or by the indirect pathway of activating a second messenger. Protein and peptide hormones, however, are not water-soluble and are unable to enter target cells directly. Instead, they bind to hormone receptors situated on the target cell's plasma membrane and use a **second-messenger** system. In these cases (Figure 9.1b), the hormone (first messenger) binds to the receptor protein on the membrane ①, and the activated receptor sets off a series of reactions (a cascade) that activates an enzyme ②. The enzyme, in turn, catalyzes reactions that produce second-messenger molecules ③ (in this case,

cyclic AMP, also known as cAMP, or cyclic adenosine monophosphate) that oversee additional intracellular changes that promote the typical response of the target cell to the hormone ④. Think of a postal service employee (first messenger) delivering a letter to campus, and a college mailroom employee (second messenger) carrying the letter from the college mailbox (cell membrane) to your dorm mailbox (DNA). As you might guess, the same hormone may have a variety of possible second messengers (including *cyclic guanosine monophosphate*, or *cGMP*, and *calcium ions*) and many possible target cell responses, depending on the tissue type stimulated.

### Did You Get It?

1. Walking barefoot, you step on a piece of broken glass and immediately pull your foot back. Why is it important that the signal triggering this motion come from the nervous system and not from the endocrine system?
2. What is a hormone? What does *target organ* mean?
3. Why is cAMP called a *second messenger*?

**A:** A hormone can influence a body cell only if that cell has receptors for that hormone, either on its plasma membrane or internally.

For answers, see Appendix A.

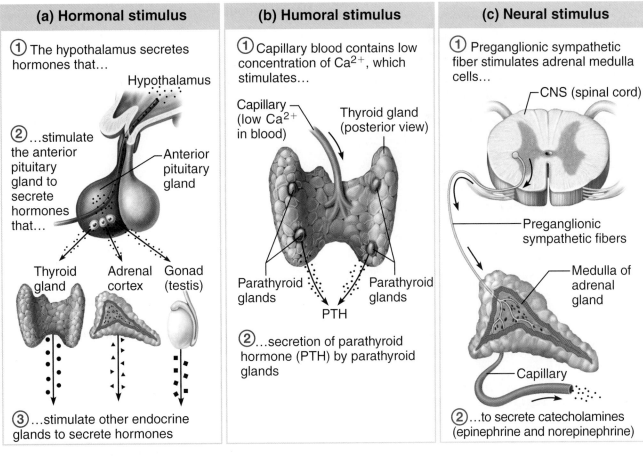

**(a) Hormonal stimulus**

① The hypothalamus secretes hormones that...

Hypothalamus

② ...stimulate the anterior pituitary gland to secrete hormones that...

Anterior pituitary gland

Thyroid gland    Adrenal cortex    Gonad (testis)

③ ...stimulate other endocrine glands to secrete hormones

**(b) Humoral stimulus**

① Capillary blood contains low concentration of $Ca^{2+}$, which stimulates...

Capillary (low $Ca^{2+}$ in blood)

Thyroid gland (posterior view)

Parathyroid glands    Parathyroid glands

PTH

② ...secretion of parathyroid hormone (PTH) by parathyroid glands

**(c) Neural stimulus**

① Preganglionic sympathetic fiber stimulates adrenal medulla cells...

CNS (spinal cord)

Preganglionic sympathetic fibers

Medulla of adrenal gland

Capillary

② ...to secrete catecholamines (epinephrine and norepinephrine)

**Figure 9.2  Endocrine gland stimuli.**

## Stimuli for Control of Hormone Release

→ **Learning Objectives**

☐ **Explain how various endocrine glands are stimulated to release their hormonal products.**

☐ **Define** *negative feedback*, **and describe its role in regulating blood levels of the various hormones.**

Now that we've discussed *how* hormones work, the next question is, "What prompts the endocrine glands to release or not release their hormones?" Let's take a look.

Recall that **negative feedback** mechanisms are the chief means of regulating blood levels of nearly all hormones (see Chapter 1, p. 45). In such systems, some internal or external stimulus triggers hormone secretion; then, a rising level of the hormone inhibits further hormone release (even while promoting a response in the target organ). As a result, blood levels of many hormones vary within a very narrow range. With hormones, "a little goes a long way"!

The stimuli that activate endocrine glands fall into three major categories—hormonal, humoral, and neural **(Figure 9.2)**. These three mechanisms represent the most common systems that control hormone release, but they by no means explain all of them. Some endocrine organs respond to many different stimuli.

### Hormonal Stimuli

The most common stimulus is a *hormonal stimulus*, in which endocrine organs are prodded into action by other hormones. For example, hormones of the hypothalamus stimulate the anterior pituitary gland to secrete its hormones, and many anterior pituitary hormones stimulate other endocrine organs to release their hormones into the blood (Figure 9.2a). As the hormones produced by the final target glands increase in the blood, they "feed back" to inhibit the release of anterior pituitary hormones and thus their own release. Hormone release promoted by this mechanism tends to be rhythmic, with hormone blood levels rising and falling again and again.

9

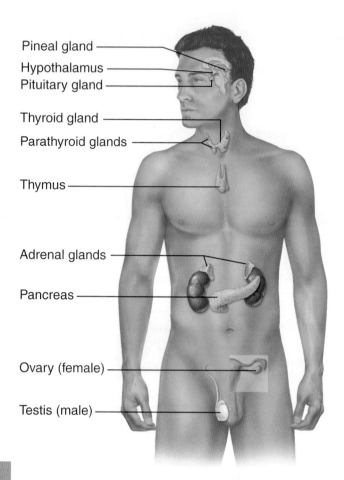

Pineal gland

Hypothalamus

Pituitary gland

Thyroid gland

Parathyroid glands

Thymus

Adrenal glands

Pancreas

Ovary (female)

Testis (male)

**Figure 9.3 Location of the major endocrine organs of the body.** (The parathyroid glands, which appear to be on the anterior surface of the thyroid gland in this illustration, are typically located on its posterior aspect.)

## Humoral Stimuli

Changing blood levels of certain ions and nutrients may also stimulate hormone release. Such stimuli are referred to as *humoral* (hyoo-mor'al) *stimuli* to distinguish them from hormonal stimuli, which are also bloodborne chemicals. The term *humoral* refers to the ancient use of the word *humor* to indicate the various body fluids (blood, bile, and others). For example, a decreasing blood calcium ion level in the capillaries serving the parathyroid glands prompts the release of parathyroid hormone (PTH). Because PTH acts by several routes to reverse that decline, the blood $Ca^{2+}$ level soon rises, ending the stimulus for PTH release (Figure 9.2b). Other hormones released in response to humoral stimuli include calcitonin, released by the thyroid gland, and insulin, produced by the pancreas.

## Neural Stimuli

In isolated cases, nerve fibers stimulate hormone release, and the endocrine cells are said to respond to *neural stimuli*. The classic example is sympathetic nervous system stimulation of the adrenal medulla to release the catecholamines norepinephrine and epinephrine during periods of stress (Figure 9.2c).

## Did You Get It?

4. What are three ways in which endocrine glands are stimulated to secrete their hormones?

For the answer, see Appendix A.

# The Major Endocrine Organs

### → Learning Objectives

☐ Describe the difference between endocrine and exocrine glands.

☐ On an appropriate diagram, identify the major endocrine glands and tissues.

☐ List hormones produced by the endocrine glands, and discuss their general functions.

☐ Discuss ways in which hormones promote body homeostasis by giving examples of hormonal actions.

☐ Describe major pathological consequences of hypersecretion and hyposecretion of the hormones considered in this chapter.

The major endocrine organs of the body include the *pituitary, pineal, thyroid, parathyroid, thymus* and *adrenal glands, pancreas,* and gonads (*ovaries* and *testes*) **(Figure 9.3)**. The **hypothalamus**, which is part of the nervous system, is also recognized as a major endocrine organ because it produces several hormones. Some hormone-producing glands (the anterior pituitary, thyroid, parathyroids, and adrenals) have purely endocrine functions, but others (pancreas and gonads) have both endocrine and exocrine functions and are thus mixed glands. Both types of glands are formed from epithelial tissue, but the *endocrine glands* are **ductless glands** that produce hormones that they release into the blood or lymph. (As you might expect, the endocrine glands have a rich blood supply.) Conversely, the *exocrine glands* release their products at the body's surface or into body cavities through ducts (they have an *exit*). We have already discussed the formation of and differences and similarities between these two types of glands (recall what you learned in Chapter 3). Here we will focus on the endocrine glands. A summary of the endocrine organs and their hormones' main actions and regulatory factors appears in Table 9.1, pp. 354–355.

### Did You Get It?

5. What are two important differences between endocrine and exocrine glands?

**For the answer, see Appendix A.**

## Pituitary Gland and Hypothalamus

The **pituitary** (pĭ-too′ĭ-tar″e) **gland** is approximately the size of a pea. It hangs by a stalk from the inferior surface of the hypothalamus of the brain, where it is snugly surrounded by the sella turcica ("Turk's saddle") of the sphenoid bone (look back at Figure 5.10, p. 175). It has two functional lobes—the *anterior pituitary* (glandular tissue) and the *posterior pituitary* (nervous tissue).

### Pituitary-Hypothalamus Relationships

→ **Learning Objective**

☐ **Describe the functional relationship between the hypothalamus and the pituitary gland.**

Despite its relatively small size, the anterior pituitary gland controls the activity of so many other endocrine glands that it has often been called the "master endocrine gland." Its removal or destruction has a dramatic effect on the body. The adrenal and thyroid glands and the gonads atrophy, and results of hyposecretion by those glands quickly become obvious. However, the anterior pituitary is not as all-powerful as it might appear, because the release of each of its hormones is controlled by **releasing hormones** and **inhibiting hormones** produced by the hypothalamus. The hypothalamus liberates these regulatory hormones into the blood of the portal circulation, which connects the blood supply of the hypothalamus with that of the anterior pituitary. (In a *portal circulation*, two capillary beds are connected by one or more veins; in this case, the capillaries of the hypothalamus are drained by veins that empty into the capillaries of the anterior pituitary.)

The hypothalamus also makes two additional hormones, **oxytocin** and **antidiuretic hormone**, which are transported along the axons of the hypothalamic **neurosecretory cells** to the posterior pituitary for storage **(Figure 9.4)**. They are later released into the blood in response to nerve impulses from the hypothalamus.

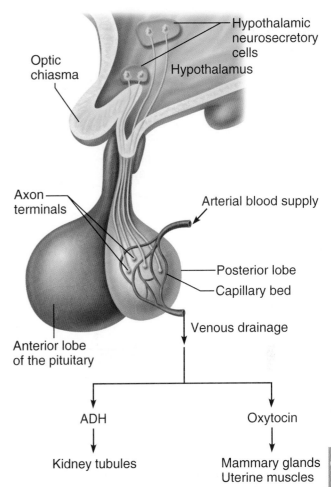

**Figure 9.4 Hormones released by the posterior pituitary and their target organs.** Neurosecretory cells in the hypothalamus synthesize oxytocin and antidiuretic hormone (ADH) and transport them down their axons to the posterior pituitary. There, the hormones are stored until nerve impulses from the hypothalamus trigger their release.

### *Posterior Pituitary and Hypothalamic Hormones*

The *posterior pituitary* is not an endocrine gland in the strict sense because it *does not make* the peptide hormones it releases. Instead, it acts as a storage area for hormones made by hypothalamic neurons.

**Oxytocin** (ok″se-to′sin) is releavsed in significant amounts only during childbirth and nursing. It stimulates powerful contractions of the uterine muscle during sexual relations, during labor, and during breastfeeding. It also causes milk ejection (the *let-down reflex*) in a nursing woman. Both

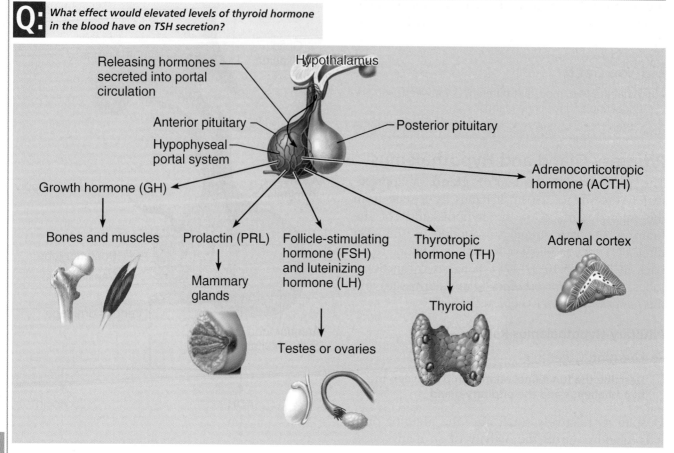

**Figure 9.5 Hormones of the anterior pituitary and their major target organs.** Releasing hormones secreted by hypothalamic neurons stimulate secretion of anterior pituitary hormones. The releasing hormones are secreted into a capillary network that connects via portal veins to a second capillary bed in the anterior lobe of the pituitary gland.

natural and synthetic oxytocic drugs (Pitocin and others) are used to induce labor or to hasten labor that is progressing at a slow pace. Less often, oxytocics are used to stop postpartum bleeding (by causing constriction of the ruptured blood vessels at the placental site) and to stimulate the let-down reflex.

The second hormone released by the posterior pituitary is **antidiuretic** (an″ti-di″u-ret′ik) **hormone (ADH)**. *Diuresis* is urine production. Thus, an *antidiuretic* is a chemical that inhibits or prevents urine production. ADH causes the kidneys to reabsorb more water from the forming urine; as a result, urine volume decreases, and blood volume increases. Water is a powerful inhibitor of ADH

release. In larger amounts, ADH also increases blood pressure by causing constriction of the arterioles (small arteries). For this reason, it is sometimes referred to as **vasopressin** (vās″o-pres′in).

Drinking alcoholic beverages inhibits ADH secretion and results in output of large amounts of urine. The dry mouth and intense thirst experienced "the morning after" reflect this dehydrating effect of alcohol. Certain drugs, classed together as *diuretics*, antagonize the effects of ADH, causing water to be flushed from the body. These drugs are used to manage the *edema* (water retention in tissues) typical of congestive heart failure.

### Homeostatic Imbalance 9.1

Hyposecretion of ADH leads to a condition of excessive urine output called **diabetes insipidus** (di″ah-be′tēz in-sip′ĭ-dus). People with this

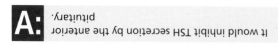

problem are continually thirsty and drink huge amounts of water. _____✚

## Did You Get It?

6. Both the anterior pituitary and the posterior pituitary release hormones, but the posterior pituitary is not an endocrine gland. What is it?
7. Barry is excreting huge amounts of urine. He has an endocrine system problem, but it is not diabetes mellitus, which has a similar sign. What is his possible problem?

**For answers, see Appendix A.**

***Anterior Pituitary Hormones*** The *anterior pituitary* produces several hormones that affect many body organs **(Figure 9.5)**. Two of the six anterior pituitary hormones in the figure—growth hormone and prolactin—exert their major effects on nonendocrine targets. The remaining four—follicle-stimulating hormone, luteinizing hormone, thyrotropic hormone, and adrenocorticotropic hormone—are all **tropic hormones**. Tropic (tro′pik = turn on) hormones stimulate their target organs, which are also endocrine glands, to secrete their hormones, which in turn exert their effects on other body organs and tissues. All anterior pituitary hormones (1) are proteins (or peptides), (2) act through second-messenger systems, and (3) are regulated by hormonal stimuli and, in most cases, negative feedback.

**Growth hormone (GH)** is a general metabolic hormone. However, its major effects are directed to the growth of skeletal muscles and long bones of the body, and thus it plays an important role in determining final body size. GH is a protein-sparing and anabolic hormone that causes the building of amino acids into proteins and stimulates most target cells to grow in size and divide. At the same time, it causes fats to be broken down and used for energy while it spares glucose, helping to maintain blood sugar homeostasis.

### ⚖ Homeostatic Imbalance 9.2

If untreated, both deficits and excesses of GH may result in structural abnormalities. Hyposecretion of GH during childhood leads to **pituitary dwarfism**. Body proportions are fairly normal, but the person as a whole is a living miniature (with a maximum adult height of 4 feet). Hypersecretion during childhood results in **gigantism**. The individual becomes

Average height of 5′10″

Disorders of pituitary growth hormone. This individual exhibiting gigantism (right) stands 8 feet, 1 inch tall. The pituitary dwarf (left) is 2 feet, 5.4 inches tall.

extremely tall; height of 8 to 9 feet is common. Again, body proportions are fairly normal.

If hypersecretion occurs after long-bone growth has ended, **acromegaly** (ak″ro-meg′ah-le) results. The facial bones, particularly the lower jaw and the bony ridges underlying the eyebrows, enlarge tremendously, as do the feet and hands. Thickening of soft tissues leads to coarse or malformed facial features. Most cases of hypersecretion by endocrine organs (the pituitary and the other endocrine organs) result from tumors of the affected gland. The tumor cells act in much the same way as the normal glandular cells do; that is, they produce the hormones normally made by that gland. Pharmacological doses of GH have been used to reverse some of the effects of aging. Read about this in "A Closer Look" (p. 342). _____✚

**Prolactin (PRL)** is a protein hormone structurally similar to growth hormone. Its only known target in humans is the breast (*pro* = for; *lact* = milk). After childbirth, it stimulates and maintains milk production by the mother's breasts. Its function in men is not known.

# Potential Uses for Growth Hormone

Growth hormone (GH) has been used for pharmaceutical purposes (that is, as a drug) since its discovery in the 1950s. Originally obtained from the pituitary glands of cadavers, it is now biosynthesized and administered by injection.

GH is administered legally to children who do not produce it naturally to allow these children to grow to near-normal heights. Unfortunately, some physicians succumb to parental pressures to prescribe GH to children who *do* produce it but are extremely short.

When GH is administered to *adults* with a growth hormone deficiency, body fat decreases, and bone density and muscle mass increase. GH also appears to increase the performance and muscle mass of the heart, decreases blood cholesterol, boosts the immune system, and perhaps improves a person's psychological outlook. Such effects (particularly those involving increased muscle mass and decreased body fat) have led to abuse of GH by bodybuilders and athletes. Based on these effects, GH is a banned performrance-enhancing drug, and its use remains restricted.

Because GH may also reverse some effects of aging, anti-aging clinics using GH injections to delay aging have sprung up. Many people naturally stop producing GH after age 60, and this may explain why their ratio of lean-to-fat mass declines and their skin thins. GH already is the drug treatment of choice for many aging Hollywood stars who dread the loss of their youth and vitality. Administration of GH to older patients reverses these declines. However, clinical studies reveal that the administered GH does not increase strength or exercise tolerance in older patients, and a careful study of very sick patients in intensive care units (where GH is routinely given to restore nitrogen balance) found that high doses of GH are associated with increased mortality. For these reasons, earlier media claims that GH is a "youth potion" have proven to be dangerously misleading.

GH may help AIDS patients. Because of improved antibiotics, fewer AIDS patients are dying from opportunistic infections. The other side of this picture is that more die from the weight loss called "wasting." Injections of GH can actually reverse wasting during AIDS, leading to weight gain and gain of lean muscle. In 1996, the U.S. Food and Drug Administration approved the use of GH to treat such wasting.

GH is not a wonder drug, even in cases where it is clearly beneficial. GH treatment is expensive and has undesirable side effects. It can lead to fluid retention and edema, joint and muscle pain, high blood sugar, glucose intolerance, and gynecomastia (breast enlargement in men). Hypertension, heart enlargement, diabetes, and colon cancer are other possible results of high doses of GH, and edema and headaches accompany even the lowest doses.

Intensive research into the potential benefits of GH should keep this hormone in the public eye for years to come. Let's hope its uncontrolled use does not become a public health problem.

**Can growth hormone help older patients?**

The **gonadotropic** (gō″nad-o-tro′pik) **hormones** regulate the hormonal activity of the **gonads** (ovaries and testes). In women, the gonadotropin **follicle-stimulating hormone (FSH)** stimulates follicle development in the ovaries. As the follicles mature, they produce estrogen, and eggs are readied for ovulation. In men, FSH stimulates sperm development by the testes. **Luteinizing** (lu′te-in-i″zing) **hormone (LH)** triggers ovulation of an egg from the ovary and causes the ruptured follicle to produce progesterone and some estrogen. In men, LH stimulates testosterone production by the interstitial cells of the testes.

 **Homeostatic Imbalance 9.3**

Hyposecretion of FSH or LH leads to **sterility**, the lack of ability to successfully reproduce, in both males and females. In general, hypersecretion does not appear to cause any problems. However, some drugs used to promote fertility stimulate the release of the gonadotropic hormones, and multiple births (indicating multiple ovulations at the same time rather than the usual single ovulation each month) are relatively common after their use. _____ ✚

**Thyrotropic hormone (TH)**, also called **thyroid-stimulating hormone (TSH)**, influences the growth and activity of the thyroid gland. **Adrenocorticotropic** (ad-re″no-kor″tĭ-ko-tro′pik) **hormone (ACTH)** regulates the endocrine activity of the cortex portion of the adrenal gland.

### Did You Get It?
8. What are tropic hormones?

**For the answer, see Appendix A.**

## Pineal Gland

The **pineal** (pin′e-al) **gland** is a small, cone-shaped gland that hangs from the roof of the third ventricle of the brain (see Figure 9.3). The endocrine function of this tiny gland is still somewhat of a mystery. Although many chemical substances have been identified in the pineal gland, only the hormone **melatonin** (mel″ah-to′nin) appears to be secreted in substantial amounts. The level of melatonin rises and falls during the course of the day and night. The peak level occurs at night and makes us drowsy; the lowest level occurs during daylight around noon. Melatonin is believed to be a "sleep trigger" that plays an important role in establishing the body's sleep-wake cycle. It is believed to coordinate the hormones of fertility and to inhibit the reproductive system (especially the ovaries of females) until the body matures.

## Thyroid Gland

The **thyroid gland** is located at the base of the throat, just inferior to the Adam's apple, where it is easily palpated during a physical examination. It is a fairly large gland consisting of two lobes joined by a central mass, or *isthmus* (**Figure 9.6a**, p. 344). The thyroid gland makes two hormones, one called *thyroid hormone*, the other called *calcitonin* (produced by the parafollicular cells). Internally, the thyroid gland is composed of hollow structures called **follicles** (Figure 9.6b), which store a sticky colloidal material. Thyroid hormone is derived from this colloid.

**Thyroid hormone**, often referred to as the body's major metabolic hormone, is actually two active iodine-containing hormones, **thyroxine** (thi-rok′sin), or $T_4$, and **triiodothyronine** (tri″i″o-do-thi′ro-nēn), or $T_3$. Thyroxine is the major hormone secreted by the thyroid follicles. Most triiodothyronine is formed at the target tissues by conversion of thyroxine to triiodothyronine. These two hormones are very much alike. Each is constructed from two tyrosine amino acids linked together, but thyroxine has four bound iodine atoms, whereas triiodothyronine has three (thus, $T_4$ and $T_3$, respectively).

Thyroid hormone controls the rate at which glucose is "burned," or oxidized, and converted to body heat and chemical energy (ATP). Because all body cells depend on a continuous supply of ATP to power their activities, every cell in the body is a target. Thyroid hormone is also important for normal tissue growth and development, especially in the reproductive and nervous systems.

 **Homeostatic Imbalance 9.4**

Without iodine, functional thyroid hormones cannot be made. The source of iodine is our diet, and foods richest in iodine are seafoods. Years ago many people who lived in the Midwest, in areas with iodine-deficient soil that were far from the seashore (and a supply of fresh seafood), developed **goiters** (goy′terz). So, that region of the country came to be known as the "goiter belt." A goiter is an enlargement of the thyroid gland (shown in the figure) that results when the diet is deficient in iodine.

9

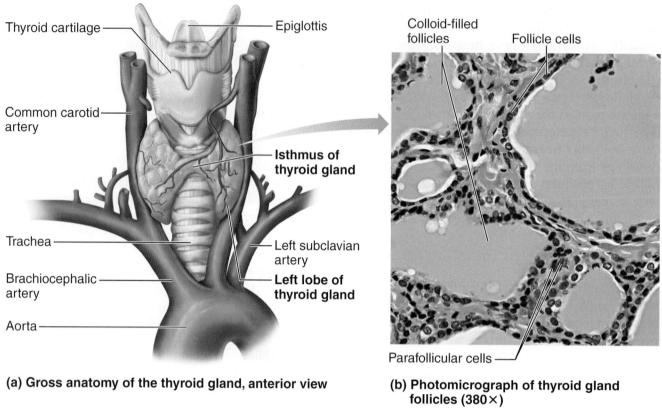

**(a) Gross anatomy of the thyroid gland, anterior view**

**(b) Photomicrograph of thyroid gland follicles (380×)**

**Figure 9.6 The thyroid gland.**

View **PAL** Histology
Mastering A&P®

Woman with an enlarged thyroid (goiter).

TSH "calls" for the release of thyroxine, but as the peptide part of the molecule is made, the thyroid gland enlarges because the protein is nonfunctional without iodine. This inactive form of thyroxine fails to provide negative feedback to inhibit TSH release, resulting in continued production of the peptide component, which makes the goiter larger. Simple goiter is uncommon in the United States today because most of our salt is iodized, but it is still a problem in some other areas of the world.

Hyposecretion of thyroxine may indicate problems other than iodine deficiency, such as lack of stimulation by TSH. If it occurs in early childhood, the result is **cretinism** (kre′tin-izm). Cretinism results in dwarfism in which adult body proportions remain childlike, with a proportionally longer torso and shorter legs compared to normal adults. Untreated individuals with cretinism are intellectually impaired. Their hair is scanty, and their skin is dry. If the hyposecretion is discovered early, hormone replacement will prevent mental impairment and other signs and symptoms of the deficiency. Hypothyroidism occurring in adults results in **myxedema** (mik″sah-de′mah), which is characterized by both physical and mental sluggishness (however, mental impairment does not occur). Other signs are puffiness of the face, fatigue, poor muscle tone, low body

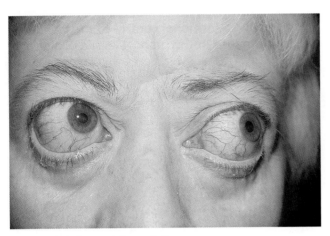

The exophthalmos of Graves' disease.

temperature (the person is always cold), obesity, and dry skin. Oral thyroxine is prescribed to treat this condition.

Hyperthyroidism generally results from a tumor of the thyroid gland. Extreme overproduction of thyroxine results in a high basal metabolic rate, intolerance of heat, rapid heartbeat, weight loss, nervous and agitated behavior, and a general inability to relax. **Graves' disease** is one form of hyperthyroidism. In addition to the symptoms of hyperthyroidism described earlier, the thyroid gland enlarges, and the eyes may bulge, or protrude anteriorly. This condition, called **exophthalmos** (ek″sof-thal′mus), is shown in the figure. Hyperthyroidism may be treated surgically by removal of part of the thyroid (and/or a tumor if present) or chemically with thyroid-blocking drugs or radioactive iodine, which destroys some of the thyroid cells. _____✚

The second important hormone product of the thyroid gland, **calcitonin**, decreases the blood calcium ion level by causing calcium to be deposited in the bones. It acts antagonistically to parathyroid hormone, the hormone produced by the parathyroid glands. Whereas thyroxine is made and stored in follicles before it is released to the blood, calcitonin is made by the **parafollicular cells** found in the connective tissue between the follicles (see Figure 9.6b). It is released directly to the blood in response to an increasing level of blood calcium ions. Few effects of hypo- or hypersecretion of calcitonin are known, and calcitonin production is meager or ceases entirely in adults. This may help to explain (at least in part) the progressive decalcification of bones that accompanies aging.

### Did You Get It?

9. Why is iodine important for proper thyroid gland function?
10. What hormone is called the "sleep hormone," and which endocrine organ produces it?

For answers, see Appendix A.

## Parathyroid Glands

The **parathyroid glands** are tiny masses of glandular tissue most often found on the posterior surface of the thyroid gland (look back at Figure 9.3). Typically, there are two parathyroid glands on each thyroid lobe, that is, a total of four parathyroids; but as many as eight have been reported, and some may be in other regions of the neck or even in the thorax. The parathyroids secrete **parathyroid hormone (PTH)**, which is the most important regulator of calcium ion ($Ca^{2+}$) homeostasis of the blood. When the blood calcium ion concentration drops below a certain level, the parathyroids release PTH, which stimulates bone destruction cells (osteoclasts) to break down bone matrix and release calcium ions into the blood. Thus, PTH is a *hypercalcemic* hormone (that is, it acts to increase the blood level of calcium ions), whereas calcitonin is a *hypocalcemic* hormone. (The negative feedback interaction between these two hormones as they control the blood calcium ion level during youth is illustrated in **Figure 9.7**, p. 346.) Although the skeleton is the major PTH target, PTH also stimulates the kidneys and intestine to absorb more calcium ions (from urinary filtrate and foodstuffs, respectively).

### Homeostatic Imbalance 9.5

If the blood calcium ion level falls too low, neurons become extremely irritable and overactive. They deliver impulses to the muscles so rapidly that the muscles go into uncontrollable spasms (**tetany**), which may be fatal. Before surgeons knew the importance of the tiny parathyroid glands on the backside of the thyroid, they would remove a hyperthyroid patient's thyroid gland entirely. Many times this resulted in death. Once it was revealed that the parathyroids are functionally very different from the thyroid gland, surgeons began to leave at least some parathyroid-containing tissue (if at all possible) to take care of blood calcium ion homeostasis.

Severe hyperparathyroidism causes massive bone destruction—an X-ray examination of the bones shows large punched-out holes in the bony

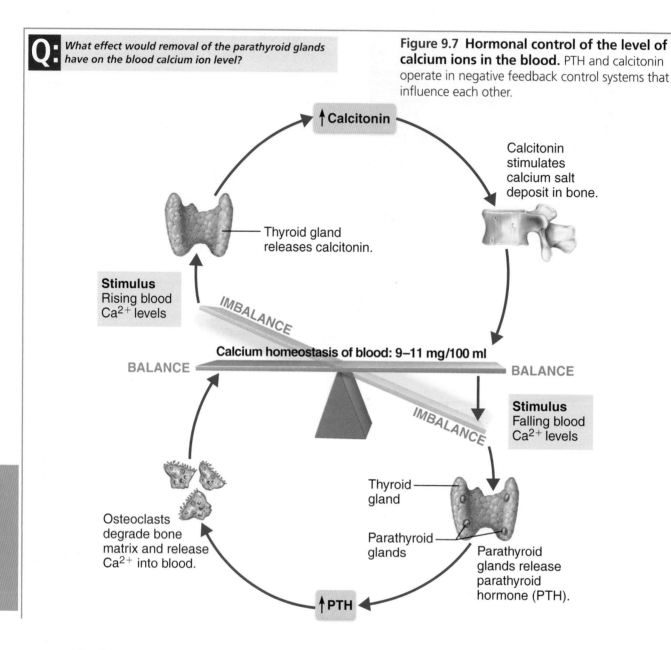

**Q:** *What effect would removal of the parathyroid glands have on the blood calcium ion level?*

**Figure 9.7 Hormonal control of the level of calcium ions in the blood.** PTH and calcitonin operate in negative feedback control systems that influence each other.

↑Calcitonin

Calcitonin stimulates calcium salt deposit in bone.

Thyroid gland releases calcitonin.

**Stimulus** Rising blood $Ca^{2+}$ levels

IMBALANCE

Calcium homeostasis of blood: 9–11 mg/100 ml

BALANCE                    BALANCE

IMBALANCE

**Stimulus** Falling blood $Ca^{2+}$ levels

Thyroid gland

Parathyroid glands

Parathyroid glands release parathyroid hormone (PTH).

Osteoclasts degrade bone matrix and release $Ca^{2+}$ into blood.

↑PTH

matrix. The bones become very fragile, and spontaneous fractures begin to occur. _____+

## Did You Get It?

11. How are the thyroid and parathyroid glands linked anatomically?
12. What hormone increases the blood calcium ion level, and which endocrine gland produces this hormone?
13. What hormone reduces the blood calcium ion level, and which endocrine gland produces this hormone?

**For answers, see Appendix A.**

 **A:** Blood calcium levels would drop because the bones would no longer be exposed to the bone-digesting stimulus of PTH.

## Thymus

The **thymus** is located in the upper thorax, posterior to the sternum. Large in infants and children, it decreases in size throughout adulthood. By old age, it is composed mostly of fibrous connective tissue and fat. The thymus produces a hormone called **thymosin** (thi′mo-sin) and others that appear to be essential for normal development of a special group of white blood cells (*T lymphocytes*) and the immune response. (We describe the role of the thymus and its hormones in immunity in Chapter 12.)

## Adrenal Glands

The two **adrenal glands** curve over the top of the kidneys like triangular hats. Although each

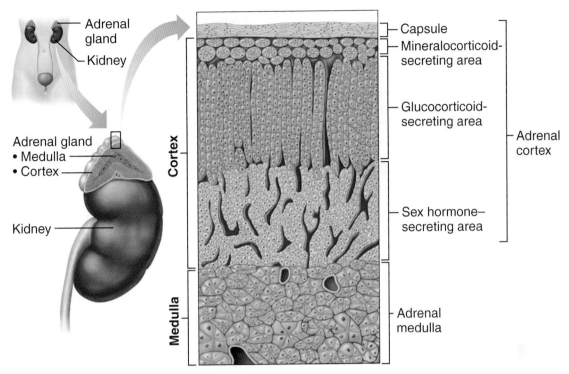

**Figure 9.8 Microscopic structure of the adrenal gland.** Diagram of the three regions of the adrenal cortex and part of the adrenal medulla.

adrenal gland looks like a single organ, it is structurally and functionally two endocrine organs in one. Much like the pituitary gland, it has parts made of glandular (cortex) and neural tissue (medulla). The central medulla region is enclosed by the adrenal cortex, which contains three separate layers of cells **(Figure 9.8)**.

### Hormones of the Adrenal Cortex

The **adrenal cortex** produces three major groups of steroid hormones, which are collectively called **corticosteroids**   (kor″ti-ko-ster′oidz)—mineralocorticoids, glucocorticoids, and sex hormones.

The **mineralocorticoids**, mainly **aldosterone** (al″dos-ter′ōn), are produced by the outermost adrenal cortex cell layer. As their name suggests, the mineralocorticoids are important in regulating the mineral (or salt) content of the blood, particularly the concentrations of sodium and potassium ions. These hormones target the kidney tubules that selectively reabsorb the minerals or allow them to be flushed out of the body in urine. When the blood level of aldosterone rises, the kidney tubule cells reabsorb increasing amounts of sodium ions and secrete more potassium ions into the urine. When sodium is reabsorbed, water follows. Thus, the mineralocorticoids help regulate both

water and electrolyte balance in body fluids. The release of aldosterone is stimulated by humoral factors, such as fewer sodium ions or more potassium ions in the blood (and, to a lesser degree, by ACTH, a hormonal stimulus) **(Figure 9.9**, p. 348). **Renin**, an enzyme produced by the kidneys when blood pressure drops, also causes the release of aldosterone by triggering a series of reactions that form **angiotensin II**, a potent stimulator of aldosterone release. A hormone released by the heart, **atrial natriuretic** (na″tre-u-ret′ik) **peptide (ANP)**, prevents aldosterone release, its goal being to *reduce* blood volume and blood pressure.

The middle cortical layer mainly produces **glucocorticoids**, which include **cortisone** and **cortisol**. Glucocorticoids promote normal cell metabolism and help the body to resist *long-term stressors*, primarily by increasing the blood glucose level. When blood levels of glucocorticoids are high, fats and even proteins are broken down by body cells and converted to glucose, which is released to the blood. For this reason, glucocorticoids are said to be *hyperglycemic hormones*. Glucocorticoids also seem to control the more unpleasant effects of inflammation by decreasing edema, and they reduce pain by inhibiting the pain-causing prostaglandins (see Table 9.2, p. 356).

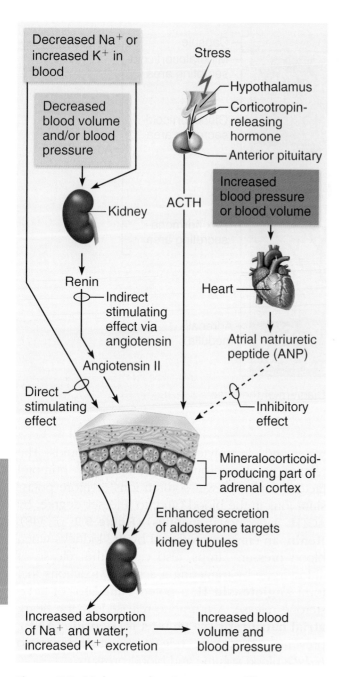

**Figure 9.9 Major mechanisms controlling aldosterone release from the adrenal cortex.**

Because of their anti-inflammatory properties, glucocorticoids are often prescribed as drugs to suppress inflammation for patients with rheumatoid arthritis. Glucocorticoids are released from the adrenal cortex in response to a rising blood level of ACTH.

In both men and women, the adrenal cortex produces both male and female **sex hormones** throughout life in relatively small amounts. The bulk of the sex hormones produced by the innermost cortex layer are **androgens** (male sex

hormones), but some **estrogens** (female sex hormones) are also formed.

### ⚖ Homeostatic Imbalance 9.6

A generalized hyposecretion of all the adrenal cortex hormones leads to **Addison's disease**, characterized by a peculiar bronze tone of the skin, which can resemble a suntan. Because the aldosterone level is low, sodium and water are lost from the body, which leads to problems with electrolyte and water balance. This, in turn, causes the muscles to become weak, and shock is a possibility. Other signs and symptoms of Addison's disease result from deficient levels of glucocorticoids, such as hypoglycemia, a lessened ability to cope with stress (burnout), and suppression of the immune system (and thereby increased susceptibility to infection). A complete lack of glucocorticoids is incompatible with life.

Hypersecretion problems may result from an ACTH-releasing tumor of the pituitary or from adrenal cortical tumors. The resulting condition depends on the cortical area involved. Hyperactivity of the outermost cortical area results in **hyperaldosteronism** (hi″per-al″dos′-terōn-izm). Excessive water and sodium ions are retained, leading to high blood pressure and edema. Potassium ions are lost to such an extent that the activity of the heart and nervous system may be disrupted. When the tumor is in the middle cortical area or the patient has been receiving pharmacological doses (amounts higher than those released in the body) of glucocorticoids to counteract inflammatory disease, **Cushing's syndrome** occurs. Excessive glucocorticoids result in

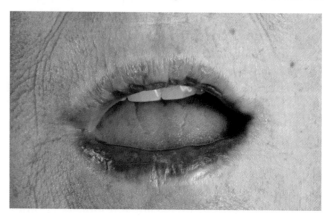

Lips of this individual show the hyperpigmentation of Addison's disease.

a swollen "moon face" and the appearance of a "buffalo hump" of fat on the upper back. Other common and undesirable effects include high blood pressure, hyperglycemia (steroid diabetes), weakening of the bones (as protein is withdrawn to be converted to glucose), and severe depression of the immune system.

Hypersecretion of the sex hormones leads to **masculinization**, regardless of sex. In adult males these effects may be masked, but in females the results are often dramatic. A beard develops, and a masculine pattern of body hair distribution occurs, among other effects. _____ ✚

## Did You Get It?

14. How do thymosin and other thymus hormones help to protect the body?
15. What hormone stimulates the kidneys to reabsorb more sodium ions?
16. Which group of hormones produced by the adrenal cortex have anti-inflammatory effects and participate in the long-term stress response?

For answers, see Appendix A.

### Hormones of the Adrenal Medulla

The **adrenal medulla**, like the posterior pituitary, is a knot of nervous tissue. When the medulla is stimulated by sympathetic nervous system neurons, its cells release two similar hormones, **epinephrine** (ep″ĭ-nef′rin), also called **adrenaline**, and **norepinephrine (noradrenaline)**, into the bloodstream. Collectively, these hormones are called **catecholamines** (kat″ĕ-kol′ah-mēnz). Because some sympathetic neurons also release norepinephrine as a neurotransmitter, the adrenal medulla is often thought of as a "misplaced sympathetic nervous system ganglion."

When you are (or feel) threatened physically or emotionally, your sympathetic nervous system brings about the "fight-or-flight" response to help you cope with the stressful situation. One of the organs it stimulates is the adrenal medulla, which literally pumps its hormones into the bloodstream to enhance and prolong the effects of the neurotransmitters of the sympathetic nervous system. The catecholamines increase heart rate, blood pressure, and the blood glucose level and dilate the small passageways of the lungs. These events result in more oxygen and glucose in the blood and a faster circulation of blood to the body organs (most importantly, to the brain, muscles, and heart). Thus, the body is better able to deal with a short-term stressor, whether the job at hand is to fight, begin the inflammatory process, or alert you so you think more clearly (**Figure 9.10**, p. 350).

The catecholamines of the adrenal medulla prepare the body to cope with short-term stressful situations and cause the so-called *alarm stage* of the stress response. Glucocorticoids, by contrast, are produced by the adrenal cortex and are important when coping with prolonged or continuing stressors, such as dealing with the death of a family member or having a major operation. Glucocorticoids operate primarily during the *resistance stage* of the stress response. If they are successful in protecting the body, the problem will eventually be resolved without lasting damage. When the stress continues on and on, the adrenal cortex may simply "burn out," which is usually fatal. (The roles of glucocorticoids in the stress response are also shown in Figure 9.10.)

### ⚖ Homeostatic Imbalance 9.7

Damage to or destruction of the adrenal medulla has no major effects as long as the sympathetic nervous system neurons continue to function normally. However, hypersecretion of catecholamines leads to symptoms such as rapid heartbeat, high blood pressure, and a tendency to perspire and be very irritable. Surgical removal of the catecholamine-secreting cells corrects this condition. _____ ✚

## Pancreatic Islets

The **pancreas**, located close to the stomach in the abdominal cavity (see Figure 9.3), is a mixed gland. The **pancreatic islets**, also called the **islets of Langerhans** (lahng′er-hanz), are little masses of endocrine (hormone-producing) tissue scattered among the exocrine (enzyme-producing) tissue of the pancreas. The exocrine, or *acinar*, part of the pancreas acts as part of the digestive system (see Chapter 14). Here we will consider only the pancreatic islets.

Although there are more than a million islets separated by exocrine cells, each of these tiny clumps of cells works like an organ within an organ as it busily manufactures its hormones. Two

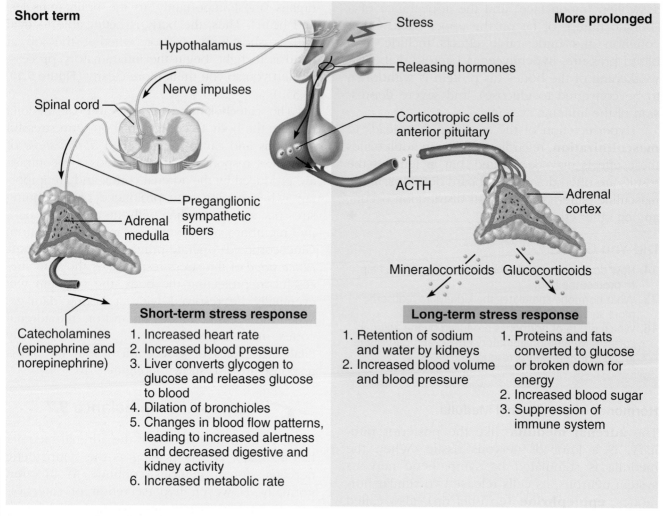

**Figure 9.10 Roles of the hypothalamus, adrenal medulla, and adrenal cortex in the stress response.** (Note that ACTH is only a weak stimulator of mineralocorticoid release under normal conditions.)

important hormones produced by the islet cells are **insulin** and **glucagon** (gloo'kah-gon).

Islet cells act as fuel sensors, secreting insulin and glucagon appropriately during fed and fasting states. A high level of glucose in the blood stimulates the release of insulin from the **beta** (ba'tah) **cells (Figure 9.11)** of the islets. Insulin acts on just about all body cells, increasing their ability to import glucose across their plasma membranes. Once inside the cells, glucose is oxidized for energy or converted to glycogen or fat for storage. Insulin also speeds up these "use it" or "store it" activities. Because insulin sweeps the glucose out of the blood, its effect is said to be *hypoglycemic*. As the blood glucose level falls, the stimulus for insulin release ends—another classic case of nega-

tive feedback control. Many hormones have hyperglycemic effects (glucagon, glucocorticoids, and epinephrine, to name a few), but insulin is the only hormone that decreases the blood glucose level. Insulin is absolutely necessary for the use of glucose by body cells. Without it, essentially no glucose can get into the cells to be used.

### Homeostatic Imbalance 9.8

Without insulin, the blood level of glucose (which normally ranges from 80 to 120 mg/100 ml of blood) rise to a dramatically high level (for example, 600 mg/100 ml of blood). In such instances, glucose begins to spill into the urine because the kidney tubule cells cannot reabsorb it fast enough.

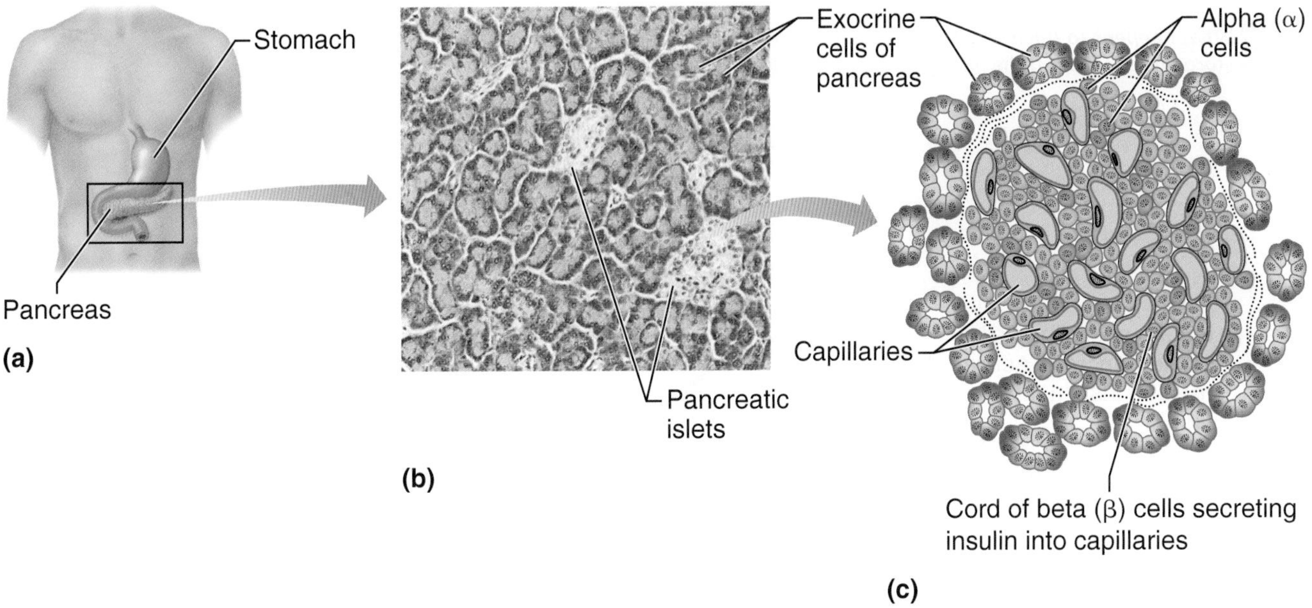

Stomach

Pancreas

**(a)**

Exocrine cells of pancreas

Pancreatic islets

**(b)**

Alpha (α) cells

Capillaries

Cord of beta (β) cells secreting insulin into capillaries

**(c)**

**Figure 9.11 Pancreatic tissue. (a)** Location of pancreas relative to the stomach and small intestine. **(b)** Photomicrograph of pancreas with exocrine and endocrine (islets) areas clearly visible (140×). **(c)** Diagrammatic view of a pancreatic islet. Beta cells produce insulin; alpha cells produce glucagon.

View PAL Histology
Mastering A&P®

As glucose flushes from the body, water follows, leading to dehydration. The clinical name for this condition is **diabetes mellitus** (mě-li′tus), which literally means that something sweet (*mel* = honey) is passing through, or siphoning (*diabetes* = Greek "siphon"), from the body. Because cells do not have access to glucose, fats and even proteins are broken down and used to meet the energy requirements of the body. As a result, body weight begins to decline. Loss of body proteins leads to a decreased ability to fight infections, so diabetics must maintain good hygiene and care for even small cuts and bruises. When large amounts of fats (instead of sugars) are used for energy, the blood becomes very acidic (**acidosis** [as″ĭ-do′sis]) as *ketones* (intermediate products of fat breakdown) appear in the blood. This type of acidosis is referred to as **ketosis**. Unless corrected, coma and death result. The three cardinal signs of diabetes mellitus are (1) **polyuria** (pol″e-u′re-ah)—excessive urination to flush out the glucose and ketones; (2) **polydipsia** (pol″e-dip′se-ah)—excessive thirst resulting from water loss; and (3) **polyphagia** (pol″e-fa′je-ah)—

hunger due to inability to use sugars and the loss of fat and proteins from the body.

People with mild cases of diabetes mellitus (most cases of *type 2*, or *adult-onset, diabetes*) produce insulin, but for some reason their insulin receptors are unable to respond to it, a situation called **insulin resistance**. Type 2 diabetics are treated with special diets or oral hypoglycemic medications that prod the sluggish islets into action and increase the sensitivity of the target tissues to insulin and of beta cells to the stimulating effects of glucose. To regulate the blood glucose level in the more severe *type 1 (juvenile) diabetic*, insulin is infused continuously by an insulin pump worn externally or by a regimen of carefully planned insulin injections is administered throughout the day. _____ ✚

Glucagon acts as an antagonist of insulin; that is, it helps to regulate the blood glucose level but in a way opposite that of insulin (**Figure 9.12**, p. 352). Its release by the **alpha** (al′fah) **cells** (see Figure 9.11c) of the islets is stimulated by a low blood level of glucose. Its action is basically

**Q:** *What happens to the liver's ability to synthesize and store glycogen when the glucagon blood level rises?*

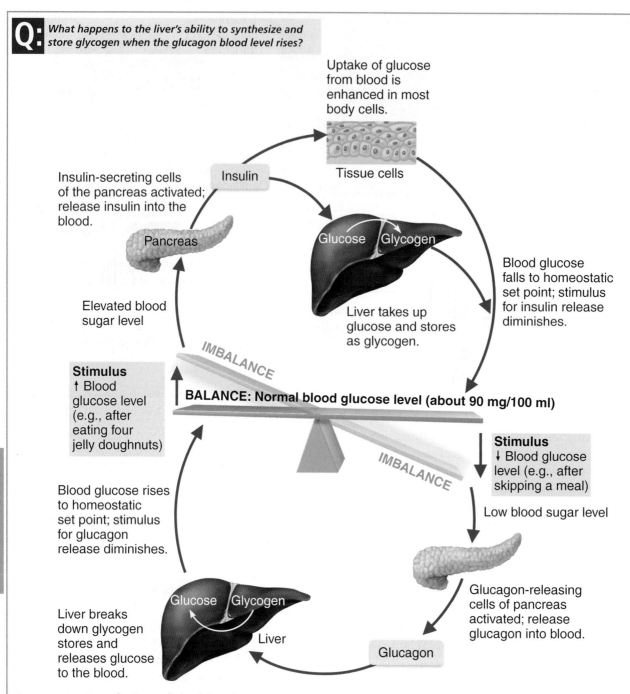

Uptake of glucose from blood is enhanced in most body cells.

Tissue cells

Insulin

Insulin-secreting cells of the pancreas activated; release insulin into the blood.

Pancreas

Glucose    Glycogen

Blood glucose falls to homeostatic set point; stimulus for insulin release diminishes.

Elevated blood sugar level

Liver takes up glucose and stores as glycogen.

IMBALANCE

**Stimulus**
↑ Blood glucose level (e.g., after eating four jelly doughnuts)

**BALANCE: Normal blood glucose level (about 90 mg/100 ml)**

IMBALANCE

**Stimulus**
↓ Blood glucose level (e.g., after skipping a meal)

Low blood sugar level

Blood glucose rises to homeostatic set point; stimulus for glucagon release diminishes.

Liver breaks down glycogen stores and releases glucose to the blood.

Glucose    Glycogen

Liver

Glucagon-releasing cells of pancreas activated; release glucagon into blood.

Glucagon

**Figure 9.12 Regulation of the blood glucose level by a negative feedback mechanism involving pancreatic hormones.**

hyperglycemic. Its primary target organ is the liver, which it stimulates to break down stored glycogen to glucose and to release the glucose into the blood. No important disorders resulting from hypo- or hypersecretion of glucagon are known.

### Did You Get It?

17. Mrs. Bellamy's husband has suffered a heart attack and is hospitalized. Would you expect her blood glucose level to be elevated, normal, or lower than normal? Why?
18. Insulin and glucagon are both pancreatic hormones. Which stimulates cellular uptake of glucose?

**For answers, see Appendix A.**

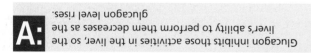

**A:** Glucagon inhibits those activities in the liver, so the liver's ability to perform them decreases as the glucagon level rises.

## Gonads

The female and male *gonads* (see Figure 9.3) produce sex cells (an exocrine function). They also produce sex hormones that are identical to those produced by adrenal cortex cells. The major differences from the adrenal sex hormone production are the source and relative amounts of hormones produced.

### Hormones of the Ovaries

The female gonads, or **ovaries**, are paired, slightly larger than almond-sized organs located in the pelvic cavity. Besides producing female sex cells (*ova*, or eggs), ovaries produce two groups of steroid hormones, **estrogens** and **progesterone**.

Alone, the estrogens are responsible for the development of sex characteristics in women (primarily growth and maturation of the reproductive organs) and the appearance of secondary sex characteristics (for example, hair in the pubic and axillary regions) at puberty. Acting with progesterone, estrogens promote breast development and cyclic changes in the uterine lining (the **menstrual cycle**, or **menstruation**).

Progesterone (pro-jes′tĕ-rōn), as already noted, acts with estrogen to bring about the menstrual cycle. During pregnancy, it quiets the muscles of the uterus so that an implanted embryo will not be aborted and helps prepare breast tissue for lactation.

Ovaries are stimulated to release their estrogens and progesterone in a cyclic way by the anterior pituitary gonadotropic hormones. We give more detail on this feedback cycle and on the structure, function, and controls of the ovaries in the reproductive system chapter (Chapter 16), but keep in mind that hyposecretion of the ovarian hormones severely hampers a woman's ability to conceive and bear children.

### Hormones of the Testes

The paired oval **testes** of the male are suspended in a sac, the *scrotum*, outside the pelvic cavity. In addition to male sex cells, or *sperm*, the testes also produce male sex hormones, or **androgens**, of which **testosterone** (tes-tos′tĕ-rōn) is the most important. At puberty, testosterone promotes the growth and maturation of the reproductive system organs to prepare the young man for reproduction. It also causes the male's secondary sex characteristics (growth of facial hair, development of heavy bones and muscles, and lowering of the voice) to appear and stimulates the male sex drive.

In adults, testosterone is necessary for continuous production of sperm. In cases of hyposecretion, the man becomes sterile; such cases are usually treated by testosterone injections. Testosterone production is specifically stimulated by LH (luteinizing hormone). (Chapter 16, which deals with the reproductive system, contains more information on the structure and exocrine function of the testes.)

Table 9.1 (pp. 354–355) summarizes the major endocrine glands and some of their hormones.

### Did You Get It?

19. Which gonadal hormone causes a young girl's reproductive organs to mature at puberty—estrogen or progesterone?

For the answer, see Appendix A.

# Other Hormone-Producing Tissues and Organs

→ **Learning** Objective

☐ **Indicate the endocrine role of the kidneys, the stomach and intestine, the heart, and the placenta.**

Besides the major endocrine organs, pockets of hormone-producing cells are found in fatty tissue and in the walls of the small intestine, stomach, kidneys, and heart—organs whose chief functions have little to do with hormone production. Because we describe most of these hormones in later chapters, we summarize only their chief characteristics in Table 9.2 (p. 356).

The **placenta** (plah-sen′tah) is a remarkable organ formed temporarily in the uterus of pregnant women. In addition to its roles as the respiratory, excretory, and nutrition-delivery systems for the fetus, it also produces several protein and steroid hormones that help to maintain the pregnancy and pave the way for delivery of the baby.

During very early pregnancy, a hormone called **human chorionic** (ko″re-on′ik) **gonadotropin (hCG)** is produced by the developing embryo and then by the fetal part of the placenta. Similar to LH (luteinizing hormone), hCG stimulates the ovaries

*(Text continues on p. 357.)*

Table **9.1** **Major Endocrine Glands and Some of Their Hormones**

| Gland | | Hormone | Chemical class* | Major actions | Regulated by |
|---|---|---|---|---|---|
| Hypothalamus | | Hormones released by the posterior pituitary; releasing and inhibiting hormones that regulate the anterior pituitary (see below) | | | |
| Pituitary gland | | | | | |
| • Posterior lobe (releases hormones made by the hypothalamus) | | Oxytocin | Peptide | Stimulates contraction of uterus and the milk "let-down" reflex | Nervous system (hypothalamus), in response to uterine stretching and/or suckling of a baby |
| | | Antidiuretic hormone (ADH) | Peptide | Promotes retention of water by kidneys | Hypothalamus, in response to water/salt imbalance |
| • Anterior lobe | | Growth hormone (GH) | Protein | Stimulates growth (especially of bones and muscles) and metabolism | Hypothalamic releasing and inhibiting hormones |
| | | Prolactin (PRL) | Protein | Stimulates milk production | Hypothalamic hormones |
| | | Follicle-stimulating hormone (FSH) | Protein | Stimulates production of ova and sperm | Hypothalamic hormones |
| | | Luteinizing hormone (LH) | Protein | Stimulates ovaries and testes | Hypothalamic hormones |
| | | Thyroid-stimulating hormone (TSH) | Protein | Stimulates thyroid gland | Thyroxine in blood; hypothalamic hormones |
| | | Adrenocorticotropic hormone (ACTH) | Protein | Stimulates adrenal cortex to secrete glucocorticoids | Glucocorticoids; hypothalamic hormones |
| Pineal gland | | Melatonin | Amine | Involved in biological rhythms (daily and seasonal) | Light/dark cycles |
| Thyroid gland | | Thyroxine ($T_4$) and triiodothyronine ($T_3$) | Amine | Stimulates metabolism | TSH |
| | | Calcitonin | Peptide | Reduces blood calcium ion level | Calcium ion level in blood |
| Parathyroid glands | | Parathyroid hormone (PTH) | Peptide | Raises blood calcium ion level | Calcium ion level in blood |

Table **9.1**    *(continued)*

| Gland | | Hormone | Chemical class* | Major actions | Regulated by |
|---|---|---|---|---|---|
| Thymus | | Thymosin | Peptide | "Programs" T lymphocytes | Not known |
| Adrenal glands | | | | | |
| • Adrenal medulla | | Epinephrine and norepinephrine | Amines | Raise blood glucose level; increase rate of metabolism; constrict certain blood vessels | Nervous system (sympathetic division) |
| • Adrenal cortex | | Glucocorticoids | Steroids | Increase blood glucose level | ACTH |
| | | Mineralocorticoids | Steroids | Promote reabsorption of Na$^+$ and excretion of K$^+$ (potassium) in kidneys | Changes in blood volume or blood pressure; K$^+$ or Na$^+$ level in blood |
| | | Androgens and estrogens (see entry under gonads) | | | |
| Pancreas | | Insulin | Protein | Reduces blood glucose level | Glucose level in blood |
| | | Glucagon | Protein | Raises blood glucose level | Glucose level in blood |
| Gonads | | | | | |
| • Testes | | Androgens | Steroids | Support sperm formation; development and maintenance of male secondary sex characteristics | FSH and LH |
| • Ovaries | | Estrogens | Steroids | Stimulate uterine lining growth; development and maintenance of female secondary sex characteristics | FSH and LH |
| | | Progesterone | Steroids | Promotes growth of uterine lining | FSH and LH |

* Any class not indicated as steroid is amino acid based.

**Table 9.2  Hormones Produced by Organs Other Than the Major Endocrine Organs**

| Hormone | Chemical composition | Source | Stimulus for secretion | Target organ/Effects |
|---|---|---|---|---|
| Prostaglandins (PGs); several groups indicated by letters A–I (PGA–PGI) | Derived from fatty acid molecules | Plasma membranes of virtually all body cells | Various (local irritation, hormones, etc.) | Have many targets but act locally at site of release. Examples of effects include the following: increase blood pressure by acting as vasoconstrictors; cause constriction of respiratory passageways; stimulate muscle of the uterus, promoting menstrual pain and labor; enhance blood clotting; promote inflammation and pain; increase output of digestive secretions by stomach; cause fever. |
| Gastrin | Peptide | Stomach | Food | *Stomach:* stimulates glands to release hydrochloric acid (HCl). |
| Intestinal gastrin | Peptide | Duodenum (first segment) of small intestine | Food, especially fats | *Stomach:* stimulates gastric glands and motility. |
| Secretin | Peptide | Duodenum | Food | *Pancreas:* stimulates release of bicarbonate-rich juice. *Liver:* increases release of bile. *Stomach:* reduces secretions and motility. |
| Cholecystokinin (CCK) | Peptide | Duodenum | Food | *Pancreas:* stimulates release of enzyme-rich juice. *Gallbladder:* stimulates expulsion of stored bile. *Duodenal papilla* (entry point at the small intestine): causes sphincter to relax, allowing bile and pancreatic juice to enter duodenum. |
| Erythropoietin | Glycoprotein | Kidney | Hypoxia (low oxygen level) | *Bone marrow:* stimulates production of red blood cells. |
| Active vitamin $D_3$ | Steroid | Kidney (activates provitamin D made by epidermal cells) | PTH | *Intestine:* stimulates active transport of dietary calcium ions across intestinal cell membranes. |
| Atrial natriuretic peptide (ANP) | Peptide | Heart | Stretching of atria (receiving chambers) of heart | *Kidney:* inhibits sodium ion reabsorption and renin release. *Adrenal cortex:* inhibits secretion of aldosterone, thereby decreasing blood volume and blood pressure. |
| Leptin | Peptide | Adipose tissue | Fatty foods | *Brain:* suppresses appetite and increases energy expenditure. |
| Resistin | Peptide | Adipose tissue | Unknown | *Fat, muscle, liver:* antagonizes insulin's action on liver cells. |

to *continue* producing estrogen and progesterone so that the lining of the uterus is not sloughed off in menses. (The home pregnancy tests sold over the counter test for the presence of hCG in the woman's urine.) In the third month, the placenta assumes the job of producing estrogen and progesterone, and the ovaries become inactive for the rest of the pregnancy. The high estrogen and progesterone blood levels maintain the lining of the uterus (thus, the pregnancy) and prepare the breasts for producing milk. **Human placental lactogen (hPL)** works cooperatively with estrogen and progesterone in preparing the breasts for lactation. **Relaxin**, another placental hormone, causes the mother's pelvic ligaments and the pubic symphysis to relax and become more flexible, which eases birth passage.

# Developmental Aspects of the Endocrine System

### → Learning Objective

☐ Describe the effect of aging on the endocrine system and body homeostasis.

The embryonic development of the endocrine glands varies. The pituitary gland is derived from epithelium of the oral cavity and a neural tissue projection of the hypothalamus. The pineal gland is entirely neural tissue. Most strictly epithelial glands develop as little saclike outpocketings of the mucosa of the digestive tract. These are the thyroid, thymus, and pancreas. Formation of the gonads and the adrenal and parathyroid glands is much more complex and is not considered here.

Barring outright malfunctions of the endocrine glands, most endocrine organs seem to operate smoothly until old age. In late middle age, the efficiency of the ovaries begins to decline, causing **menopause** ("the change of life"). During this period, a woman's reproductive organs begin to atrophy, and the ability to bear children ends.

Problems associated with estrogen deficiency begin to occur, such as arteriosclerosis, osteoporosis, decreased skin elasticity, and changes in the operation of the sympathetic nervous system that result in "hot flashes." In addition, fatigue, nervousness, and mood changes such as depression are common. No such dramatic changes seem to happen in men. In fact, many men remain fertile throughout their life span, indicating that testosterone is still being produced in adequate amounts.

The efficiency of the endocrine system as a whole gradually declines in old age. Striking changes in aging women are due to decreasing levels of female hormones, and there is no question that growth hormone output by the anterior pituitary declines, which partially explains muscle atrophy in old age. Elderly persons are less able to resist stress and infection. This decreased resistance may result from overproduction or underproduction of the defensive hormones, because both "derail" the stress defense equilibrium and alter general body metabolism. Additionally, exposure to pesticides, industrial chemicals, dioxin, and other soil and water pollutants diminishes endocrine function, which may explain the higher cancer rates among older adults in certain areas of the country. Older people are often mildly hypothyroid. All older people have some decline in insulin production, and type 2 diabetes is most common in this age group.

## Did You Get It?

20. Which two digestive system organs are important sources of hormones associated with digestion?
21. What temporary organ produces the same hormones as the ovaries?
22. Failure of which endocrine organ(s) leads to menopause in women?
23. In the elderly, the decline in the level of which hormone is associated with muscle atrophy? With osteoporosis in women?

**For answers, see Appendix A.**

# Homeostatic Relationships between the **Endocrine System** and Other Body Systems

## Endocrine System

### Nervous System
- Many hormones (growth hormone, thyroxine, sex hormones) influence normal maturation and function of the nervous system
- Hypothalamus controls anterior pituitary function and secretes two hormones

### Lymphatic System/Immunity
- Lymphocytes "programmed" by thymic hormones seed the lymph nodes; glucocorticoids depress the immune response and inflammation
- Lymph provides a route for transport of hormones

### Respiratory System
- Epinephrine influences ventilation (dilates bronchioles)
- Respiratory system provides oxygen; disposes of carbon dioxide; converting enzyme in lungs converts angiotensin I to angiotensin II

### Cardiovascular System
- Several hormones influence blood volume, blood pressure, and heart contractility; erythropoietin stimulates red blood cell production
- Blood is the main transport medium of hormones; heart produces atrial natriuretic peptide

### Digestive System
- Local gastrointestinal (GI) hormones influence GI function; activated vitamin D necessary to absorb calcium from diet; catecholamines influence digestive system activity
- Digestive system provides nutrients to endocrine organs

### Reproductive System
- Hypothalamic, anterior pituitary, and gonadal hormones direct reproductive system development and function; oxytocin and prolactin are involved in birth and breastfeeding
- Gonadal hormones feed back to influence endocrine system function

### Urinary System
- Aldosterone and ADH influence renal function; erythropoietin released by kidneys promotes red blood cell formation
- Kidneys activate vitamin D (considered a hormone)

### Integumentary System
- Androgens activate sebaceous glands; estrogen increases skin hydration
- Skin produces a precursor of vitamin D (cholecalciferol or provitamin D)

### Muscular System
- Growth hormone is essential for normal muscular development; other hormones (thyroxine and catecholamines) influence muscle metabolism
- Muscular system mechanically protects some endocrine glands; muscular activity promotes catecholamine release

### Skeletal System
- PTH is important in regulating calcium blood levels; growth hormone, $T_3$, $T_4$, and sex hormones are necessary for normal skeletal development
- The skeleton protects some endocrine organs, especially those in brain, chest, and pelvis

# Summary

## The Endocrine System and Hormone Function— An Overview (pp. 335–338)

1. The endocrine system is a major controlling system of the body. Through hormones, it stimulates such long-term processes as growth and development, metabolism, reproduction, and body defense.

2. Endocrine organs are small and widely separated in the body. Some are part of mixed glands (both endocrine and exocrine in function). Others produce only hormones.

3. Nearly all hormones are amino acid–based hormones or steroids.

4. Endocrine organs are activated to release their hormones into the blood by hormonal, humoral, or neural stimuli. Negative feedback is important in regulating blood hormone levels.

5. Bloodborne hormones alter the metabolic activities of their target organs. The ability of a target organ to respond to a hormone depends on the presence of receptors in or on its cells to which the hormone binds or attaches.

6. Amino acid–based hormones act through second messengers. Steroid hormones can directly influence the target cell's DNA or act via second messenger.

## The Major Endocrine Organs (pp. 338–353)

1. Pituitary gland
   a. The pituitary gland hangs from the hypothalamus of the brain by a stalk and is enclosed by bone. It consists of a glandular portion (anterior lobe) and a neural portion (posterior lobe).
   b. Except for growth hormone and prolactin, hormones of the anterior pituitary are all tropic hormones.
      (1) Growth hormone (GH): An anabolic and protein-conserving hormone that promotes total body growth. Its most important effect is on skeletal muscles and bones. Untreated hyposecretion during childhood results in pituitary dwarfism; hypersecretion produces gigantism (in childhood) and acromegaly (in adulthood).
      (2) Prolactin (PRL): Stimulates production of breast milk.
      (3) Adrenocorticotropic hormone (ACTH): Stimulates the adrenal cortex to release its hormones, mainly glucocorticoids.
      (4) Thyroid-stimulating hormone (TSH): Stimulates the thyroid gland to release thyroid hormone.
      (5) Gonadotropic hormones
         (a) Follicle-stimulating hormone (FSH): Beginning at puberty, stimulates follicle development and estrogen production by the female ovaries; promotes sperm production in the testes.
         (b) Luteinizing hormone (LH): Beginning at puberty, stimulates ovulation, causes the ruptured ovarian follicle to produce progesterone; stimulates testes to produce testosterone.
   c. The posterior pituitary stores and releases hypothalamic hormones on command.
      (1) Oxytocin: Stimulates powerful uterine contractions and causes milk ejection in nursing women.
      (2) Antidiuretic hormone (ADH): Causes kidney tubule cells to reabsorb and conserve body water and increases blood pressure by constricting blood vessels. Hyposecretion leads to diabetes insipidus.
   d. Releasing and inhibiting hormones made by the hypothalamus regulate release of hormones made by the anterior pituitary. The hypothalamus also makes two hormones that are transported to the posterior pituitary for storage and later release.

2. The pineal gland, located posterior to the third ventricle of the brain, releases melatonin, which affects sleep/wake cycles.

3. Thyroid gland
   a. The thyroid gland is located in the anterior throat.
   b. Thyroid hormone, which has two forms (thyroxine [$T_4$] and triiodothyronine [$T_3$]), is released from the thyroid follicles when the blood level of TSH rises. Thyroid hormone is the body's major metabolic hormone. It increases the rate at which cells oxidize glucose and is necessary for normal growth and development. Lack of iodine leads to goiter. Hyposecretion of thyroxine results in cretinism in children and myxedema in adults. Hypersecretion results from Graves' disease or other forms of hyperthyroidism.

9

c. Calcitonin is released by parafollicular cells surrounding the thyroid follicles in response to a high blood level of calcium ions. It causes calcium to be deposited in bones.

4. Parathyroid glands

   a. The parathyroid glands are four small glands located on the posterior aspect of the thyroid gland.

   b. A low blood level of calcium stimulates the parathyroid glands to release parathyroid hormone (PTH). It causes liberation of calcium from bone to blood. Hyposecretion of PTH results in tetany; hypersecretion leads to extreme bone wasting and fractures.

5. The thymus, located in the upper thorax, functions during youth but atrophies in old age. Its hormone, thymosin, promotes maturation of T lymphocytes, important in body defense.

6. Adrenal glands

   a. The adrenal glands are paired glands perched on the kidneys. Each gland has two functional endocrine portions, the cortex and the medulla.

   b. Adrenal cortex hormones include the following:

      (1) Mineralocorticoids, primarily aldosterone, regulate sodium ion ($Na^+$) reabsorption and potassium ion ($K^+$) secretion by the kidneys. Their release is stimulated primarily by low $Na^+$ and/or high $K^+$ levels in blood.

      (2) Glucocorticoids enable the body to resist long-term stress by increasing the blood glucose level and depressing the inflammatory response.

      (3) Sex hormones (mainly androgens) are produced in small amounts throughout life.

   c. Generalized hypoactivity of the adrenal cortex results in Addison's disease. Hypersecretion can result in hyperaldosteronism, Cushing's syndrome, and/or masculinization.

   d. The adrenal medulla produces catecholamines (epinephrine and norepinephrine) in response to sympathetic nervous system stimulation. Its catecholamines enhance and prolong the effects of the "fight-or-flight" (sympathetic nervous system) response to short-term stress. Hypersecretion leads to symptoms typical of sympathetic nervous system overactivity.

7. Pancreatic islets

   a. Located in the abdomen close to the stomach, the pancreas is both an exocrine and endocrine gland. The endocrine islets release insulin and glucagon to blood.

   b. Insulin is released when the blood level of glucose is high. It increases the rate of glucose uptake and metabolism by body cells. Hyposecretion of insulin results in diabetes mellitus, which severely disturbs body metabolism. Cardinal signs are polyuria, polydipsia, and polyphagia.

   c. Glucagon, released when the blood level of glucose is low, stimulates the liver to release glucose to blood, thus increasing the blood glucose level.

8. Gonads

   a. The ovaries of the female release:

      (1) Estrogens: Release of estrogens by ovarian follicles begins at puberty under the influence of FSH. Estrogens stimulate maturation of the female reproductive organs and female secondary sex characteristics. With progesterone, they cause the menstrual cycle.

      (2) Progesterone: Progesterone is released from the ovary in response to a high blood level of LH. It works with estrogens in establishing the menstrual cycle.

   b. The testes of the male begin to produce testosterone at puberty in response to LH stimulation. Testosterone promotes maturation of the male reproductive organs, male secondary sex characteristics, and production of sperm by the testes.

   c. Hyposecretion of gonadal hormones results in sterility in both females and males.

## Other Hormone-Producing Tissues and Organs (pp. 353–357)

1. Several organs that are generally nonendocrine in overall function, such as the stomach, small intestine, kidneys, and heart, have cells that secrete hormones.

2. The placenta is a temporary organ formed in the uterus of pregnant women. Its primary endocrine role is to produce estrogen and progesterone, which maintain pregnancy and ready the breasts for lactation.

## Developmental Aspects of the Endocrine System (p. 357)

1. Decreasing function of female ovaries at menopause leads to such symptoms as osteoporosis,

increased chance of arteriosclerosis, and possible mood changes.

2. Efficiency of all endocrine glands gradually decreases with aging, which leads to a generalized increase in incidence of diabetes mellitus, immune system depression, lower metabolic rate, and, in some areas, cancer rates.

## Review Questions

Access additional practice questions using your smartphone, tablet, or computer:
Mastering A&P®> Study Area > Practice Tests & Quizzes

### Multiple Choice

*More than one choice may apply.*

1. The major endocrine organs of the body
   a. tend to be very large organs.
   b. are closely connected with each other.
   c. all contribute to the same function (digestion).
   d. tend to lie near the midline of the body.

2. Which is (are) true of the hormone melatonin?
   a. It is a releasing hormone.
   b. Its level is the highest at night.
   c. It is a sleep trigger.
   d. Iodine is needed to produce it.

3. Which of the following hormones is (are) secreted by neurons?
   a. Oxytocin          c. ADH
   b. Insulin           d. Cortisol

4. ANP, the hormone secreted by the heart, has exactly the opposite function to this hormone secreted by the outermost zone of the adrenal cortex.
   a. Epinephrine       c. Aldosterone
   b. Cortisol          d. Testosterone

5. Hormones that act directly or indirectly to elevate the blood glucose level include which of the following?
   a. GH
   b. Cortisol
   c. Insulin
   d. ACTH

6. Hypertension may result from hypersecretion of
   a. thyroxine.
   b. cortisol.
   c. aldosterone.
   d. ADH.

7. Which of the following is (are) associated with the short-term stress response?
   a. Increased blood sugar
   b. Mineralocorticoids
   c. Epinephrine
   d. Increased metabolic rate

8. Which of the following is given as a drug to reduce inflammation?
   a. Epinephrine       c. Aldosterone
   b. Cortisol          d. ADH

9. The element needed for thyroid gland function is
   a. potassium.        c. calcium.
   b. iodine.           d. manganese.

### Short Answer Essay

10. Explain how the nervous and endocrine systems differ in (a) the rate of their control, (b) the way in which they communicate with body cells, and (c) the types of body processes they control.

11. Which endocrine organs are actually mixed (endocrine and exocrine) glands? Which are purely endocrine?

12. Briefly describe how a steroid hormone affects the activity of the target cell.

13. Provide one example for each way endocrine glands are stimulated to release their hormones.

14. Define *negative feedback.*

15. Explain why not all organs are target organs for all hormones.

16. Describe the body location of each of the following endocrine organs: anterior pituitary, pineal gland, thymus, pancreas, ovaries, testes. Then, for each organ, name its hormones and their effect(s) on body processes. Finally, for each hormone, list the important results of its hypersecretion or hyposecretion.

17. What anatomical feature makes it possible for the releasing hormones from the hypothalamus to influence the activity of the anterior pituitary?

18. What is the primary function of antidiuretic hormone? Why is it also called "vasopressin"?

19. What is the most common cause of hypersecretion by endocrine organs?

20. What is the role of alpha and beta cells in the regulation of blood glucose?

21. The adrenal cortex produces corticosteroids. What is the difference between mineralocorticoids and glucocorticoids with regard to their production sites? What major functions do they monitor?

22. What is insulin resistance?

23. In general, the endocrine system becomes less efficient as we age. List some examples of problems that elderly individuals have as a result of decreasing hormone production.

 ## Critical Thinking and Clinical Application Questions

24. Jenny's body mass index (BMI) is too high, and she wants to lose some weight. Is there a hormone that can suppress appetite? Explain.

25. The parents of 14-year-old Megan are concerned about her height because she is only 4 feet tall and they are both close to 6 feet tall. After tests by their doctor, certain hormones are prescribed for the girl. What is the probable diagnosis? What hormones are prescribed, and why might the girl expect to reach normal height?

26. Shannon, a 28-year-old, has been in the first stage of labor for 15 hours. Her uterine contractions are weak, and her labor is not progressing normally. Shannon and her doctor desire a vaginal delivery, so the physician orders that Pitocin (a synthetic oxytocin) be infused. What is the effect of this hormone?

27. Mr. Flores brings his wife to the clinic, concerned about her nervousness, heart palpitations, and excessive sweating. Tests show hyperglycemia and hypertension. What hormones are probably being hypersecreted? What is the cause? What physical factors allow you to rule out thyroid problems?

28. What are the possible harmful effects of using anabolic steroids to increase muscle mass and strength?

29. Melissa, age 40, comes to the clinic, troubled by swelling in her face and unusual fat deposition on her back and abdomen. She reports that she bruises easily. Blood tests show an elevated glucose level. What is your diagnosis, and which glands might be causing the problem?

30. Multiple births are observed commonly in women treated for infertility with drugs that stimulate the release of gonadotropic hormones. What could be the reason?

31. Ryan had symptoms of excessive secretion of PTH (a high blood calcium ion level), and his physicians were certain he had a parathyroid gland tumor. Yet when surgery was performed on his neck, the surgeon could not find the parathyroid glands at all. Where should the surgeon look next to find the tumorous parathyroid gland?

# 10 Blood

**WHAT**

Blood transports substances such as oxygen and nutrients throughout the body and participates in processes such as clotting and fighting infections.

**HOW**

Blood is moved through blood vessels by the pumping action of the heart. This fluid contains red blood cells to carry oxygen, clotting proteins to stop bleeding, and white blood cells to fight infections.

**WHY**

Transportation via blood is the only way substances can be moved to distant body locations. In addition, clotting proteins are found only in blood—without them, a minor cut could be life-threatening!

**INSTRUCTORS**

New **Building Vocabulary Coaching** Activities for this chapter are assignable in Mastering A&P°

Blood is the "river of life." **Blood** transports everything that must be carried from one place to another within the body—nutrients, hormones, wastes (headed for elimination from the body), and body heat—through blood vessels. Long before modern medicine, blood was viewed as magical because when it drained from the body, life departed as well.

In this chapter, we consider the composition and function of this life-sustaining fluid. In the cardiovascular system chapter (Chapter 11), we discuss the means by which blood is propelled throughout the body.

## Composition and Functions of Blood

→ **Learning Objectives**

☐ Describe the composition and volume of whole blood.

☐ Describe the composition of plasma, and discuss its importance in the body.

Blood is unique: It is the only *fluid* tissue in the body. Although blood appears to be a thick, homogeneous liquid, the microscope reveals that it has both solid and liquid components.

## Components

Blood is a complex connective tissue in which living blood cells, the **formed elements**, are suspended in **plasma** (plaz'mah), a nonliving fluid matrix. The collagen and elastin fibers typical of other connective tissues are absent from blood; instead, dissolved proteins become visible as fibrin strands during blood clotting.

If a sample of blood is separated, the plasma rises to the top, and the formed elements, being heavier, fall to the bottom **(Figure 10.1)**. Most of the reddish "pellet" at the bottom of the tube is *erythrocytes* (ĕ-rith'ro-sīts; *erythro* = red), or *red blood cells*, the formed elements that function in oxygen transport. There is a thin, whitish layer called the **buffy coat** at the junction between the erythrocytes and the plasma (just barely visible in Figure 10.1). This layer contains the remaining formed elements, *leukocytes* (lu'ko-sīts; *leuko* = white), *white blood cells* that act in various ways to protect the body; and *platelets*, cell fragments that help stop bleeding. Erythrocytes normally account for about 45 percent of the total volume of a blood sample, a percentage known as the **hematocrit** ("blood fraction"). White blood cells and platelets contribute less than 1 percent, and plasma makes up most of the remaining 55 percent of whole blood.

## Physical Characteristics and Volume

Blood is a sticky, opaque fluid that is heavier than water and about five times thicker, or more viscous, largely because of its formed elements. Depending on the amount of oxygen it is carrying, the color of blood varies from scarlet (oxygen-rich) to a dull red or purple (oxygen-poor). Blood has a characteristic metallic, salty taste (something we often discover as children). Blood is slightly alkaline, with a pH between 7.35 and 7.45. Its temperature (38°C, or 100.4°F) is always slightly higher than body temperature because of the friction produced as blood flows through the vessels. Blood accounts for approximately 8 percent of body weight, and its volume in healthy adults is 5 to 6 liters, or about 6 quarts.

## Plasma

Plasma, which is approximately 90 percent water, is the liquid part of the blood. Over 100 different substances are dissolved in this straw-colored fluid. Examples of dissolved substances include nutrients, salts (electrolytes), respiratory gases, hormones, plasma proteins, and various wastes and products of cell metabolism. (Other substances found in plasma are listed in Figure 10.1).

*Plasma proteins* are the most abundant solutes in plasma. Except for antibodies and protein-based hormones, the liver makes most plasma proteins. The plasma proteins serve a variety of functions. For instance, **albumin** (al-bu'min) acts as a carrier to shuttle certain molecules through the circulation, is an important blood buffer, and contributes to the osmotic pressure of blood, which acts to keep water in the bloodstream. Clotting proteins help stem blood loss when a blood vessel is injured, and antibodies help protect the body from pathogens. Plasma proteins are *not* taken up by cells to be used as food fuels or metabolic nutrients, as are other solutes such as glucose, fatty acids, and oxygen.

The composition of plasma varies continuously as cells exchange substances with the blood. Assuming a healthy diet, however, the composition of plasma is kept relatively constant by various homeostatic mechanisms of the body. For example, when blood proteins drop to undesirable levels, the liver is stimulated to make more proteins, and when the blood starts to become too acid (*acidosis*) or too basic (*alkalosis*), both the respiratory and urinary systems are called into action to restore it to its normal, slightly alkaline pH range of 7.35 to 7.45. Various body organs make dozens of adjustments day in and day out to maintain the many plasma solutes at life-sustaining levels. Besides transporting various substances around the body, plasma helps to distribute body heat, a by-product of cellular metabolism, evenly throughout the body.

### Did You Get It?

1. Which body organ plays the main role in producing plasma proteins?
2. What are the three major categories of formed elements?
3. What determines whether blood is bright red (scarlet) or dull red?

**For answers, see Appendix A.**

**Q:** *How would a decrease in the amount of plasma proteins affect plasma volume?*

**Figure 10.1 The composition of blood.**

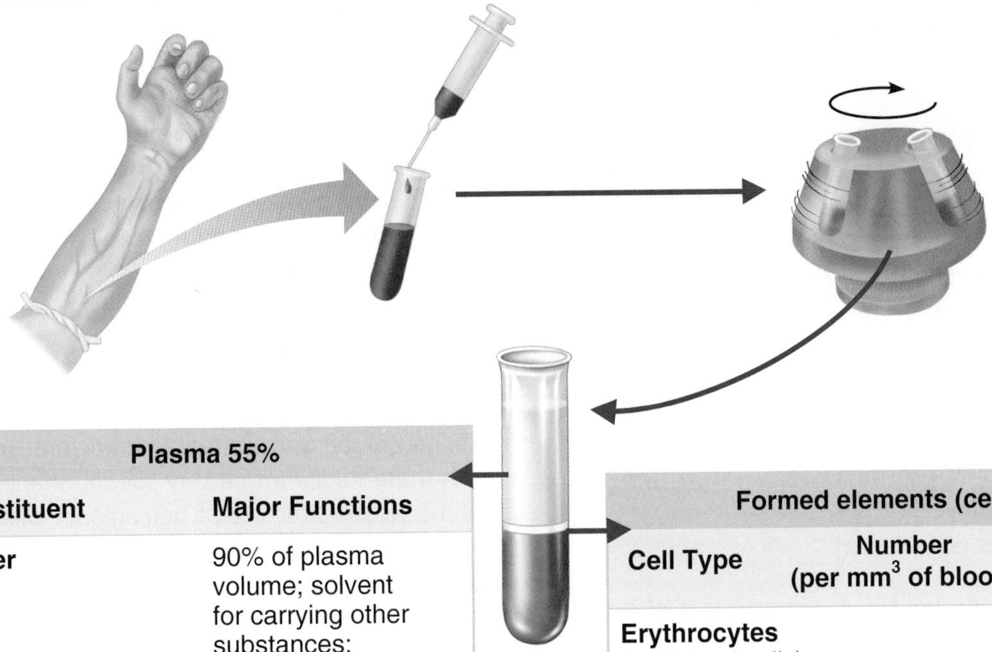

| Plasma 55% | |
|---|---|
| **Constituent** | **Major Functions** |
| **Water** | 90% of plasma volume; solvent for carrying other substances; absorbs heat |
| **Salts (electrolytes)** Sodium Potassium Calcium Magnesium Chloride Bicarbonate | Osmotic balance, pH buffering, regulation of membrane permeability |
| **Plasma proteins** Albumin | Osmotic balance, pH buffering |
| Fibrinogen | Clotting of blood |
| Globulins | Defense (antibodies) and lipid transport |
| **Substances transported by blood** Nutrients (glucose, fatty acids, amino acids, vitamins) Waste products of metabolism (urea, uric acid) Respiratory gases ($O_2$ and $CO_2$) Hormones (steroids and thyroid hormone are carried by plasma proteins) | |

| Formed elements (cells) 45% | | |
|---|---|---|
| **Cell Type** | **Number (per mm³ of blood)** | **Functions** |
| **Erythrocytes** (red blood cells) 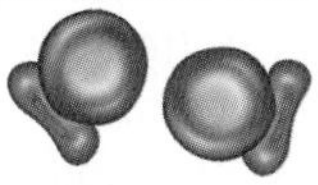 | 4–6 million | Transport oxygen and help transport carbon dioxide |
| **Leukocytes** (white blood cells) | 4,800–10,800 | Defense and immunity |
| Basophil | Eosinophil | Lymphocyte |
| Neutrophil | | Monocyte |
| **Platelets** 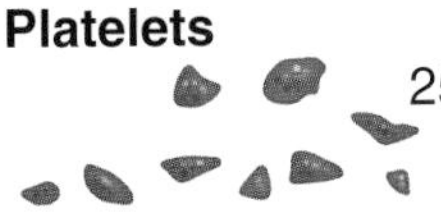 | 250,000–400,000 | Blood clotting |

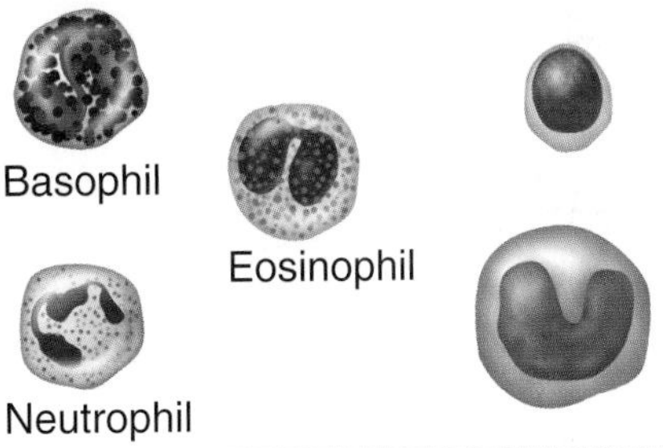

**10**

**A:** Plasma proteins create the osmotic pressure that helps to maintain plasma volume and draws leaked fluid back into circulation. Hence, a decrease in the amount of plasma proteins would result in a reduced plasma volume.

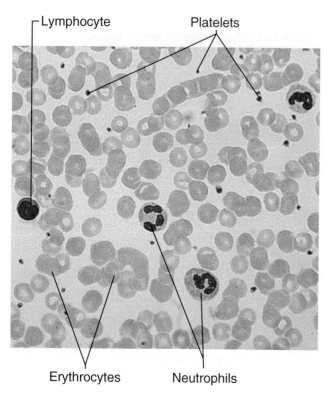

Lymphocyte          Platelets

Erythrocytes        Neutrophils

**Figure 10.2 Photomicrograph of a blood smear.**
Most of the cells in this view are erythrocytes (red blood cells). Two kinds of leukocytes (white blood cells) are also present: lymphocytes and neutrophils. Also note the platelets.

View PAL Histology
Mastering A&P®

## Formed Elements

→ **Learning** Objectives

☐ **List the cell types making up the formed elements, and describe the major functions of each type.**

☐ **Define** *anemia, polycythemia, leukopenia,* **and** *leukocytosis,* **and list possible causes for each condition.**

If you observe a stained smear of human blood under a light microscope, you will see disc-shaped red blood cells, a variety of gaudily stained spherical white blood cells, and some scattered platelets that look like debris **(Figure 10.2)**. However, erythrocytes vastly outnumber the other types of formed elements. (Look ahead to Table 10.2 on p. 370, which provides a summary of the important characteristics of the various formed elements.)

### Erythrocytes

**Erythrocytes**, or **red blood cells (RBCs)**, function primarily to ferry oxygen to all cells of the body. They are superb examples of the link between structure and function. RBCs differ from other blood cells because they are *anucleate* (a-nu'kle-at); that is, they lack a nucleus. They also contain very few organelles. In fact, mature RBCs circulating in the blood are literally "bags" of hemoglobin molecules. **Hemoglobin** (he"mo-glo'bin) **(Hb)**, an iron-bearing protein, transports most of the oxygen that is carried in the blood. (It also binds with a small amount of carbon dioxide.)

### CONCEPTLINK

Recall that hemoglobin is an example of a globular protein (look back at Figure 2.19b, p. 76). Globular, or functional, proteins have tertiary structure, meaning that they are folded into a very specific shape. In this case, the folded structure of hemoglobin allows it to perform the specific function of binding and carrying oxygen. The structure of globular proteins is also very vulnerable to pH changes and can be denatured (unfolded) by a pH that is too low (acidic); denatured hemoglobin is unable to bind oxygen. ←

Moreover, because erythrocytes lack mitochondria and make ATP by anaerobic mechanisms, they do not use up any of the oxygen they are transporting, making them very efficient oxygen transporters indeed.

Erythrocytes are small, flexible cells shaped like biconcave discs—flattened discs with depressed centers on both sides (see the photo on p. 367). Because of their thinner centers, erythrocytes look like miniature doughnuts when viewed with a microscope. Their small size and peculiar shape provide a large surface area relative to their volume, making them ideally suited for gas exchange.

RBCs outnumber white blood cells by about 1,000 to 1 and are the major factor contributing to blood viscosity. Although the numbers of RBCs in the circulation do vary, there are normally about 5 million cells per cubic millimeter of blood. (A cubic millimeter [$mm^3$] is a very tiny drop of blood, almost too small to be seen.) When the number of RBC/$mm^3$ increases, blood viscosity, or thickness, increases. Similarly, as the number of RBCs decreases, blood thins and flows more rapidly.

Although the numbers of RBCs are important, it is the amount of hemoglobin in the bloodstream at any time that really determines how well the erythrocytes are performing their role of oxygen transport. The more hemoglobin molecules the RBCs contain, the more oxygen they will be able to carry. A single red blood cell contains about 250 million hemoglobin molecules, each capable of binding 4 molecules of oxygen, so each of these tiny cells can carry about 1 billion molecules of oxygen! However, much more important clinically is the fact that normal blood contains 12–18 grams (g) of hemoglobin per 100 milliliters (ml) of blood. The hemoglobin content is slightly higher in men (13–18 g/ml) than in women (12–16 g/ml).

## Homeostatic Imbalance 10.1

A decrease in the oxygen-carrying ability of the blood, whatever the reason, is called **anemia** (ah-ne′me-ah; "lacking blood"). Anemia may be the result of (1) a lower-than-normal *number* of RBCs or (2) abnormal or deficient *hemoglobin content* in the RBCs. There are several types of anemia (classified and described briefly in Table 10.1, p. 368), but one of these, *sickle cell anemia*, deserves a little more attention because people with this genetic disorder are frequently seen in hospital emergency rooms.

In **sickle cell anemia (SCA)**, the body does not form normal hemoglobin (as in the RBC shown in part (a) of the figure). Instead, abnormal hemoglobin is formed that becomes spiky and sharp (see part (b) of the figure) when either oxygen is unloaded or the oxygen content in the blood decreases below normal. This change in hemoglobin causes the RBCs to become sickled (crescent-shaped), to rupture easily, and to dam up small blood vessels. These events interfere with oxygen delivery (leaving victims gasping for air) and cause extreme pain. It is amazing that this havoc results from a change in just *one* of the amino acids in two of the four polypeptide chains of the hemoglobin molecule!

Sickle cell anemia occurs chiefly in dark-skinned people who live in the malaria belt of Africa and among their descendants. Apparently, the same gene that causes sickling makes red blood cells infected by the malaria-causing parasite stick to the capillary walls and then lose potassium, an essential nutrient for survival of the

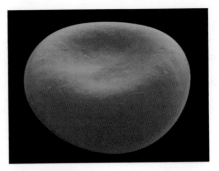

**(a) Normal RBC and part of the amino acid sequence of its hemoglobin**

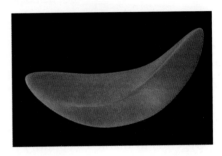

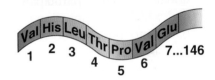

**(b) Sickled RBC and part of its hemoglobin sequence**

Comparison of **(a)** a normal erythrocyte to **(b)** a sickled erythrocyte (6,550×).

parasite. Hence, the malaria-causing parasite is prevented from multiplying within the red blood cells, and individuals with the sickle cell gene have a better chance of surviving where malaria is prevalent. Only individuals carrying two copies of the defective gene have sickle cell anemia. Those carrying just one sickling gene have **sickle cell trait (SCT)**; they generally do not display the symptoms but can pass on the sickling gene to their offspring.

An excessive or abnormal increase in the number of erythrocytes is **polycythemia** (pol″e-si-the′me-ah). Polycythemia may result from bone marrow cancer (*polycythemia vera*). It may also be a normal physiologic (homeostatic) response to

**10**

Table **10.1**  **Types of Anemia**

| Direct cause | Resulting from | Leading to |
|---|---|---|
| Decrease in RBC number | Sudden hemorrhage | Hemorrhagic anemia |
| | Lysis of RBCs as a result of bacterial infections | Hemolytic (he"mo-lit'ik) anemia |
| | Lack of vitamin $B_{12}$ (usually due to lack of intrinsic factor required for absorption of the vitamin; intrinsic factor is formed by stomach mucosa cells) | Pernicious (per-nish'us) anemia |
| | Depression/destruction of bone marrow by cancer, radiation, or certain medications | Aplastic anemia |
| Inadequate hemoglobin content in RBCs | Lack of iron in diet or slow/prolonged bleeding (such as heavy menstrual flow or bleeding ulcer), which depletes iron reserves needed to make hemoglobin; RBCs are small and pale because they lack hemoglobin | Iron-deficiency anemia |
| Abnormal hemoglobin in RBCs | Genetic defect leads to abnormal hemoglobin, which becomes sharp and sickle-shaped under conditions of increased oxygen use by body; occurs mainly in people of African descent | Sickle cell anemia |

living at high altitudes, where the air is thinner and less oxygen is available (*secondary polycythemia*). Occasionally, athletes participate in illegal "blood doping," which is the infusion of a person's own RBCs back into their bloodstream to artificially raise oxygen-carrying capacity. The major problem that results from excessive numbers of RBCs is increased blood viscosity, which causes blood to flow sluggishly in the body and impairs circulation. _____ ✚

## Leukocytes

Although **leukocytes**, or **white blood cells (WBCs)**, are far less numerous than red blood cells, they are crucial to body defense. On average, there are 4,800 to 10,800 WBCs/mm³ of blood, and they account for less than 1 percent of total blood volume. White blood cells contain nuclei and the usual organelles, which makes them the only complete cells in blood.

Leukocytes form a protective, movable army that helps defend the body against damage by bacteria, viruses, parasites, and tumor cells. As such, they have some very special characteristics. Red blood cells are confined to the bloodstream. White blood cells, by contrast, are able to slip into and out of the blood vessels—a process called

**diapedesis** (di"ah-pĕ-de'sis; "leaping across"). The circulatory system is simply their means of transportation to areas of the body where their services are needed for inflammatory or immune responses (as described in Chapter 12).

In addition, WBCs can locate areas of tissue damage and infection in the body by responding to certain chemicals that diffuse from the damaged cells. This capability is called **positive chemotaxis** (ke"mo-tax'is). Once they have "caught the scent," the WBCs move through the tissue spaces by **amoeboid** (ah-me'boid) **motion** (they form flowing cytoplasmic extensions that help move them along). By following the diffusion gradient, they pinpoint areas of tissue damage and rally round in large numbers to destroy microorganisms and dispose of dead cells.

Whenever WBCs mobilize for action, the body speeds up their production, and as many as twice the normal number of WBCs may appear in the blood within a few hours. A total WBC count above 11,000 cells/mm³ is referred to as **leukocytosis** (lu"ko-si-to'sis) (*cytosis* = an increase in cells). Leukocytosis generally indicates that a bacterial or viral infection is stewing in the body. The opposite condition, **leukopenia** (lu"ko-pe'ne-ah), is an abnormally low WBC count (*penia* = deficiency). It is commonly caused by certain drugs, such as corticosteroids and anti-cancer agents.

 ## Homeostatic Imbalance 10.2

Leukocytosis is a normal and desirable response to infectious threats to the body. By contrast, the excessive production of abnormal WBCs that occurs in infectious mononucleosis and leukemia is distinctly pathological. In **leukemia** (lu-ke′me-ah), literally "white blood," the bone marrow becomes cancerous, and huge numbers of WBCs are turned out rapidly. Although this might not appear to present a problem, the "newborn" WBCs are immature and incapable of carrying out their normal protective functions. Consequently, the body becomes easy prey for disease-causing bacteria and viruses. Additionally, because other blood cell lines are crowded out, severe anemia and bleeding problems result. _____ +

WBCs are classified into two major groups—*granulocytes* and *agranulocytes*—depending on whether or not they contain visible granules in their cytoplasm. Specific characteristics and microscopic views of the leukocytes are presented in Table 10.2, p. 370.

**Granulocytes** (gran′u-lo-sītz″) are granule-containing WBCs. They have lobed nuclei, which typically consist of several rounded nuclear areas connected by thin strands of nuclear material. The granules in their cytoplasm stain specifically with Wright's stain. The granulocytes include neutrophils (nu′tro-filz), eosinophils (e″o-sin′o-filz), and basophils (ba′so-filz).

- **Neutrophils** are the most numerous WBCs. They have a multilobed nucleus and very fine granules that respond to both acidic and basic stains. Consequently, the cytoplasm as a whole stains pink. Neutrophils are avid phagocytes at sites of acute infection. They are particularly partial to bacteria and fungi, which they kill during a *respiratory burst* that deluges the phagocytized invaders with a potent brew of oxidizing substances (bleach, hydrogen peroxide, and others).

- **Eosinophils** have a blue-red nucleus that resembles earmuffs and brick-red cytoplasmic granules. Their number increases rapidly during infections by parasitic worms (tapeworms,

etc.) ingested in food such as raw fish or entering through the skin. When eosinophils encounter a parasitic worm, they gather around and release enzymes from their cytoplasmic granules onto the parasite's surface, digesting it away.

- **Basophils**, the rarest of the WBCs, have large histamine-containing granules that stain dark blue. **Histamine** is an inflammatory chemical that makes blood vessels leaky and attracts other WBCs to the inflamed site.

The second group of WBCs, **agranulocytes**, lack *visible* cytoplasmic granules. Their nuclei are closer to the norm—that is, they are spherical, oval, or kidney-shaped. The agranulocytes include lymphocytes (lim′fo-sītz) and monocytes (mon′o-sītz).

- **Lymphocytes** have a large, dark purple nucleus that occupies most of the cell volume. Only slightly larger than RBCs, lymphocytes tend to take up residence in lymphatic tissues, such as the tonsils, where they play an important role in the immune response. They are the second most numerous leukocytes in the blood.

- **Monocytes** are the largest of the WBCs. Except for their more abundant cytoplasm and distinctive U- or kidney-shaped nucleus, they resemble large lymphocytes. When they migrate into the tissues, they change into macrophages (*macro* = large; *phage* = one that eats) with huge appetites. Macrophages are important in fighting chronic infections, such as tuberculosis, and in activating lymphocytes.

Students are often asked to list the WBCs in order of relative abundance in the blood—from most to least. The following phrase may help you with this task: **N**ever **l**et **m**onkeys **e**at **b**ananas (neutrophils, lymphocytes, monocytes, eosinophils, basophils).

### Platelets

**Platelets** are not technically cells. They are fragments of bizarre multinucleate cells called **megakaryocytes** (meg″ah-kar′e-o-sītz), which pinch off thousands of anucleate platelet "pieces" that quickly seal themselves off from the surrounding fluids. The platelets appear as darkly staining, irregularly shaped bodies scattered among the other blood cells. The normal platelet count in

**10**

Table **10.2**  **Characteristics of Formed Elements of the Blood**

| Cell type | Occurrence in blood (cells per mm³) | Cell anatomy* | Function |
|---|---|---|---|
| *Erythrocytes* (red blood cells) | 4–6 million | Salmon-colored biconcave disks; anucleate; literally, sacs of hemoglobin; most organelles have been ejected | Transport oxygen bound to hemoglobin molecules; also transport small amount of carbon dioxide |
| *Leukocytes* (white blood cells) | 4,800–10,800 | | |
| Granulocytes | | | |
| • Neutrophils | 3,000–7,000 (40–70% of WBCs) | Cytoplasm stains pale pink and contains fine granules, which are difficult to see; deep purple nucleus consists of three to seven lobes connected by thin strands of nucleoplasm | Active phagocytes; number increases rapidly during short-term or acute infections |
| • Eosinophils | 100–400 (1–4% of WBCs) | Red coarse cytoplasmic granules; figure-8 or bilobed nucleus stains blue-red | Kill parasitic worms by deluging them with digestive enzymes; play a complex role in allergy attacks |
| • Basophils | 20–50 (0–1% of WBCs) | Cytoplasm has a few large blue-purple granules; U- or S-shaped nucleus with constrictions, stains dark blue | Release histamine (vasodilator chemical) at sites of inflammation; contain heparin, an anticoagulant |
| Agranulocytes | | | |
| • Lymphocytes | 1,500–3,000 (20–45% of WBCs) | Cytoplasm pale blue and appears as thin rim around nucleus; spherical (or slightly indented) dark purple-blue nucleus | Part of immune system; B lymphocytes produce antibodies; T lymphocytes are involved in graft rejection and in fighting tumors and viruses via direct cell attack |
| • Monocytes | 100–700 (4–8% of WBCs) | Abundant gray-blue cytoplasm; dark blue-purple nucleus often U- or kidney-shaped | Active phagocytes that become macrophages in the tissues; long-term "cleanup team"; increase in number during chronic infections; activate lymphocytes during immune response |
| *Platelets* | 150,000–400,000 | Essentially irregularly shaped cell fragments; stain deep purple | Needed for normal blood clotting; initiate clotting cascade by clinging to torn area |

*Appearance when stained with Wright's stain.

blood is about 300,000 cells per mm³. Platelets are needed for the clotting process that stops blood loss from broken blood vessels (see Table 10.2). (We explain this process on pp. 373–374.)

## Did You Get It?

4. What is the role of hemoglobin in the red blood cell?
5. Which white blood cells are most important in body immunity?
6. If you had a severe infection, would you expect your WBC count to be closest to 5,000, 10,000, or 15,000 per mm³?
7. Little Lisa is pale and fatigued. What disorder of erythrocytes might she be suffering from?

**For answers, see Appendix A.**

## Hematopoiesis (Blood Cell Formation)

### → Learning Objective

☐ **Explain the role of the hemocytoblast.**

Blood cell formation, or **hematopoiesis** (hem″ah-to-po-e′sis), occurs in red bone marrow, or *myeloid* tissue. In adults, this tissue is found chiefly in the axial skeleton, pectoral and pelvic girdles, and proximal epiphyses of the humerus and femur. Each type of blood cell is produced in different numbers in response to changing body needs and different stimuli. After they mature, they are discharged into the blood vessels surrounding the area. On average, the red marrow turns out an ounce of new blood containing 100 billion new cells every day.

All the formed elements arise from a common *stem cell*, the **hemocytoblast** (he″mo-si′to-blast; "blood cell former"), which resides in red bone marrow. Their development differs, however, and once a cell is committed to a specific blood pathway, it cannot change. The hemocytoblast forms two types of descendants—the *lymphoid stem cell*, which produces lymphocytes, and the *myeloid stem cell*, which can produce all other classes of formed elements **(Figure 10.3)**.

### Formation of Red Blood Cells

Because they are anucleate, RBCs are unable to synthesize proteins, grow, or divide. As they age, RBCs become rigid and begin to fall apart in 100 to 120 days. Their remains are eliminated by phagocytes in the spleen, liver, and other body tissues. RBC components are salvaged. Iron is bound to protein as ferritin, and the balance of the heme

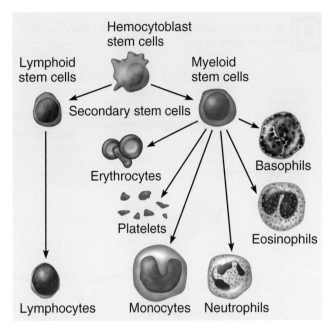

**Figure 10.3 The development of blood cells.** All blood cells differentiate from hemocytoblast stem cells in red bone marrow. The population of stem cells renews itself by mitosis. Some daughter cells become lymphoid stem cells, which develop into two classes of lymphocytes that function in the immune response. All other blood cells differentiate from myeloid stem cells.

group is degraded to bilirubin, which is then secreted into the intestine by liver cells. There it becomes a brown pigment called stercobilin that leaves the body in feces. Globin is broken down to amino acids, which are released into the circulation.

Lost blood cells are replaced more or less continuously by the division of hemocytoblasts in the red bone marrow. The developing RBCs divide many times and then begin synthesizing huge amounts of hemoglobin. When enough hemoglobin has been accumulated, the nucleus and most organelles are ejected, and the cell collapses inward. The result is the young RBC, called a *reticulocyte* (rĕ-tik′u-lo-sīt) because it still contains some rough endoplasmic reticulum (ER). The reticulocytes enter the bloodstream to begin their task of transporting oxygen. Within 2 days of release, they have ejected the remaining ER and have become fully functioning erythrocytes. The entire developmental process from hemocytoblast to mature RBC takes 3 to 5 days.

The rate of erythrocyte production is controlled by a hormone called **erythropoietin** (ĕ-rith″ro-po-e′tin). Normally a small amount of erythropoietin circulates in the blood at all times,

**10**

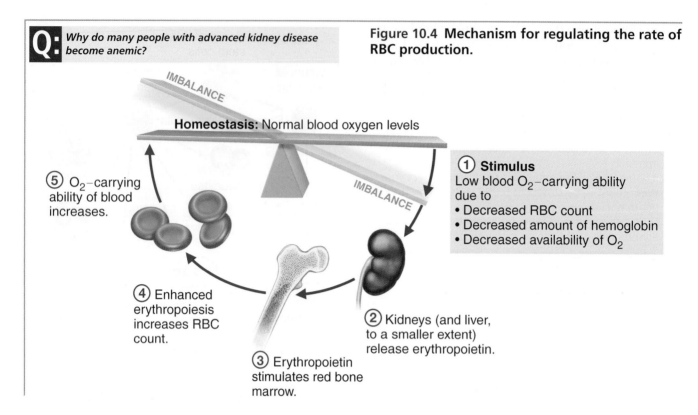

**Q:** *Why do many people with advanced kidney disease become anemic?*

**Figure 10.4 Mechanism for regulating the rate of RBC production.**

IMBALANCE

**Homeostasis:** Normal blood oxygen levels

IMBALANCE

⑤ $O_2$–carrying ability of blood increases.

① **Stimulus**
Low blood $O_2$–carrying ability due to
• Decreased RBC count
• Decreased amount of hemoglobin
• Decreased availability of $O_2$

④ Enhanced erythropoiesis increases RBC count.

② Kidneys (and liver, to a smaller extent) release erythropoietin.

③ Erythropoietin stimulates red bone marrow.

and red blood cells are formed at a fairly constant rate. Although the liver produces some, the kidneys play the major role in producing this hormone. When the blood level of oxygen begins to decline for any reason, the kidneys step up their release of erythropoietin. Erythropoietin targets the bone marrow, prodding it into "high gear" to turn out more RBCs. (Follow this sequence of events in **Figure 10.4**.)

### CONCEPT**LINK**

Recall the concept of negative feedback control (see Chapter 1, p. 45). In this case, erythropoietin is released in response to a low blood oxygen level, which stimulates the bone marrow to produce more red blood cells. With their numbers increased, the red cells carry more oxygen, increasing the blood oxygen level and reducing the initial stimulus. ←

As you might expect, an overabundance of erythrocytes, or an excessive amount of oxygen in

the bloodstream, depresses erythropoietin release and red blood cell production. However, RBC production is controlled *not* by the relative number of RBCs in the blood, but by the ability of the available RBCs to transport enough oxygen to meet the body's demands.

### Formation of White Blood Cells and Platelets

Like erythrocyte production, the formation of leukocytes and platelets is stimulated by hormones. These *colony stimulating factors (CSFs)* and *interleukins* not only prompt red bone marrow to turn out leukocytes, but also enhance the ability of mature leukocytes to protect the body. Apparently, they are released in response to specific chemical signals in the environment, such as inflammatory chemicals and certain bacteria or their toxins. The hormone *thrombopoietin* accelerates the production of platelets from megakaryocytes, but little is known about how that process is regulated.

When bone marrow problems or a disease condition such as leukemia is suspected, a special needle is used to withdraw a small sample of red marrow from one of the flat bones (ilium or sternum) close to the body surface. This procedure provides cells for a microscopic examination called a *bone marrow biopsy*.

**A:** The kidneys produce most of the body's erythropoietin, which stimulates red blood cell production by the bone marrow.

# Hemostasis

→ **Learning Objectives**

☐ **Describe the blood-clotting process.**

☐ **Name some factors that may inhibit or enhance the blood-clotting process.**

Normally, blood flows smoothly past the intact lining (endothelium) of blood vessel walls. But if a blood vessel wall breaks, a series of reactions starts the process of **hemostasis** (*hem* = blood; *stasis* = standing still), or stopping the bleeding. This response, which is fast and localized, involves many substances normally present in plasma, as well as some that are released by platelets and injured tissue cells.

## Phases of Hemostasis

Hemostasis involves three major phases, which occur in rapid sequence: **vascular spasms**, **platelet plug formation**, and **coagulation**, or **blood clotting**. Blood loss at the site is prevented when fibrous tissue grows into the clot and seals the hole in the blood vessel.

Hemostasis occurs as follows **(Figure 10.5)**:

① **Vascular spasms occur.** The immediate response to blood vessel injury is vasoconstriction, which causes blood vessel spasms. The spasms narrow the blood vessel, decreasing blood loss until clotting can occur. (Other factors causing vessel spasms include direct injury to the smooth muscle cells, stimulation of local pain receptors, and release of serotonin by anchored platelets.)

② **Platelet plug forms.** Platelets are repelled by an intact endothelium, but when the underlying collagen fibers of a broken vessel are exposed, the platelets become "sticky" and cling to the damaged site. Anchored platelets release chemicals that enhance the vascular spasms and attract more platelets to the site. As more and more platelets pile up, a *platelet plug* forms.

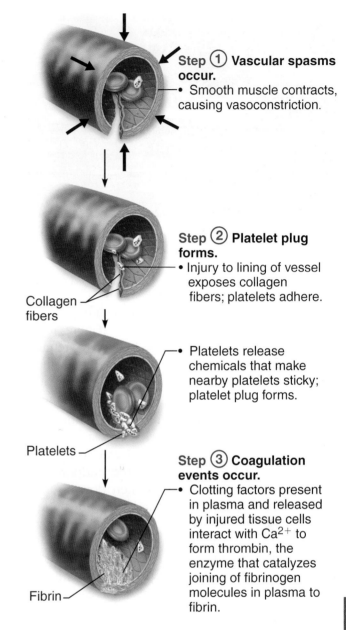

Step ① **Vascular spasms occur.**
• Smooth muscle contracts, causing vasoconstriction.

Step ② **Platelet plug forms.**
• Injury to lining of vessel exposes collagen fibers; platelets adhere.

Collagen fibers

• Platelets release chemicals that make nearby platelets sticky; platelet plug forms.

Platelets

Step ③ **Coagulation events occur.**
• Clotting factors present in plasma and released by injured tissue cells interact with $Ca^{2+}$ to form thrombin, the enzyme that catalyzes joining of fibrinogen molecules in plasma to fibrin.

Fibrin

• Fibrin forms a mesh that traps red blood cells and platelets, forming the clot.

**Figure 10.5 Events of hemostasis.**

③ **Coagulation events occur.** At the same time, the injured tissues are releasing **tissue factor (TF)**, which interacts with **$PF_3$** (platelet factor 3), a phospholipid that coats the surfaces of the platelets. This combination interacts with other clotting factors and calcium ions ($Ca^{2+}$), which are essential for many steps in the clotting process, to form an activator that leads to the formation of **thrombin**, an enzyme. Thrombin then joins soluble **fibrinogen** (fi-brin′o-jen) proteins into long, hairlike molecules of insoluble

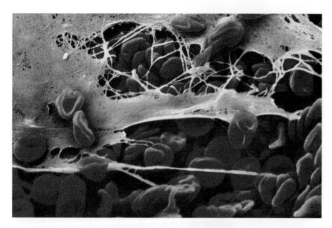

**Figure 10.6 Fibrin clot.** Scanning electron micrograph (artificially colored) of red blood cells trapped in a mesh of fibrin threads.

**fibrin**. Fibrin forms a meshwork that traps RBCs and forms the basis of the *clot* **(Figure 10.6)**. Within the hour, the clot begins to retract, squeezing **serum** (plasma minus the clotting proteins) from the mass and pulling the ruptured edges of the blood vessel closer together.

Normally, blood clots within 3 to 6 minutes. As a rule, once the clotting cascade has started, the triggering factors are rapidly inactivated to prevent widespread clotting. Eventually, the endothelium regenerates, and the clot is broken down. Once these events of the clotting cascade were understood, it became clear that placing sterile gauze over a cut or applying pressure to a wound would speed up the clotting process. The gauze provides a rough surface to which the platelets can adhere, and the pressure fractures cells, increasing the release of tissue factor locally.

## Disorders of Hemostasis

### Homeostatic Imbalance 10.3

The major disorders of hemostasis include undesirable clot formation and bleeding disorders.

### Undesirable Clotting

Despite the body's safeguards against abnormal clotting, undesirable clots sometimes form in unbroken blood vessels, particularly in the legs. A clot that develops and persists in an unbroken blood vessel is called a **thrombus** (throm′bus). If the thrombus is large enough, it may prevent blood from flowing to the cells beyond the blockage. For example, if the blockage forms in the blood vessels serving the lungs (*pulmonary thrombosis*, as shown in the figure), the consequences may be death of lung tissue and fatal *hypoxia* (inadequate oxygen delivery to body tissues). If a thrombus breaks away from the vessel wall and floats freely in the bloodstream, it becomes an **embolus** (em′bo-lus; plural *emboli*). An embolus is usually no problem unless or until it lodges in a blood vessel too narrow for it to pass through. For example, a *cerebral embolus* may cause a *stroke*, in which brain tissue dies.

Undesirable clotting may be caused by anything that roughens the endothelium of a blood vessel and encourages clinging of platelets, such as severe burns, physical blows, or an accumulation of fatty material. Slowly flowing blood, or blood pooling, is another risk factor, especially in immobilized patients. In this case, clotting factors are not washed away as usual and accumulate so that clot formation becomes possible. A number of anticoagulants, the most important of which are aspirin, heparin, and warfarin, are used clinically for thrombus-prone patients.

### Bleeding Disorders

The most common causes of abnormal bleeding are platelet deficiency (*thrombocytopenia*) and deficits of some of the clotting factors, such as might result from impaired liver function or certain genetic conditions.

**Thrombocytopenia** results from an insufficient number of circulating platelets. It can arise from any condition that suppresses the bone marrow, such as bone marrow cancer, radiation, or certain drugs. In this disorder, even normal movements cause spontaneous bleeding from small blood vessels. This is evidenced by many small purplish blotches, called **petechiae** (pĕ-te′ke-e), that resemble a rash on the skin.

A thrombus, or clot, occluding a small pulmonary blood vessel in a human lung.

When the liver is unable to synthesize its usual supply of clotting factors, abnormal and often severe bleeding episodes occur. If vitamin K (needed by the liver cells to produce the clotting factors) is deficient, the problem is easily corrected with vitamin K supplements. However, when liver function is severely impaired (as in hepatitis and cirrhosis), only whole blood transfusions are helpful. Transfusions of concentrated platelets provide temporary relief from bleeding.

The term **hemophilia** (he″mo-fil′e-ah) applies to several different hereditary bleeding disorders that result from a lack of any of the factors needed for clotting. The hemophilias have similar signs and symptoms that begin early in life. Even minor tissue trauma results in prolonged bleeding and can be life-threatening. Repeated bleeding into joints causes them to become disabled and painful.

When a bleeding episode occurs, hemophiliacs are given a transfusion of fresh plasma or injections of the purified clotting factor they lack. Because hemophiliacs are absolutely dependent on one or the other of these therapies, some have become the victims of blood-transmitted viral diseases such as hepatitis and AIDS. (AIDS, acquired immune deficiency syndrome, is a condition of depressed immunity and is described in Chapter 12.) These problems have been largely resolved because of the availability of genetically engineered clotting factors and hepatitis vaccines. _____ ✚

## Did You Get It?

11. What factors enhance the risk of thrombus formation in intact blood vessels?

**For the answer, see Appendix A.**

# Blood Groups and Transfusions

## → Learning Objectives

☐ Describe the ABO and Rh blood groups.

☐ Explain the basis for a transfusion reaction.

As we have seen, blood is vital for transporting substances through the body. When blood is lost, the blood vessels constrict, and the bone marrow steps up blood cell formation in an attempt to maintain circulation. However, the body can compensate for a loss of blood volume only up to a certain limit. Losses of 15 to 30 percent lead to pallor and weakness. Loss of over 30 percent causes severe shock, which can be fatal.

Whole blood transfusions are routinely given to replace substantial blood loss and to treat severe anemia or thrombocytopenia. The usual blood bank procedure involves collecting blood from a donor and mixing it with an anticoagulant to prevent clotting. The treated blood can be stored (refrigerated at 4°C, or 39.2°F) until needed for about 35 days.

## Human Blood Groups

Although whole blood transfusions can save lives, people have different blood groups, and transfusing incompatible or mismatched blood can be fatal. How so? The plasma membranes of RBCs, like those of all other body cells, bear genetically determined proteins (antigens), which identify each person as unique. An **antigen** (an′tĭ-jen) is a substance that the body recognizes as foreign; it stimulates the immune system to mount a defense against it. Most antigens are foreign proteins, such as those that are part of viruses or bacteria that have managed to invade the body. Although each of us tolerates our own cellular (self) antigens, one person's RBC proteins will be recognized as foreign if transfused into another person with different RBC antigens. The "recognizers" are **antibodies** present in plasma that attach to RBCs bearing surface antigens different from those on the patient's (recipient's) RBCs. Binding of the antibodies causes the foreign RBCs to clump, a phenomenon called **agglutination*** (ah-gloo″tĭ-na′shun), which leads to the clogging of small blood vessels throughout the body. During the next few hours, the foreign RBCs are lysed (ruptured), and their hemoglobin is released into the bloodstream.

Although the transfused blood is unable to deliver the increased oxygen-carrying capacity hoped for and some tissue areas may be deprived of blood, the most devastating consequence of severe transfusion reactions is that the freed hemoglobin molecules may block the kidney tubules, causing kidney failure and death. Transfusion reactions can also cause fever, chills, nausea, and vomiting, but in the absence of kidney shutdown these reactions are rarely fatal. Treatment is aimed at preventing kidney damage by infusing fluids to dilute and dissolve the hemoglobin and diuretics to flush it out of the body in urine.

10

---

*The RBC antigens that promote this clumping are sometimes called **agglutinogens** (ag″loo-tin′o-jenz), and the antibodies that bind them together are called **agglutinins** (ah-gloo′tĭ-ninz).

# Phlebotomy Technician

Phlebotomists must know where all the arteries and veins are located in the body.

"Phlebotomy is the most important procedure done for a medical laboratory," says Michael Coté, who supervises the phlebotomy staff at Palo Alto Veterans Administration Hospital in California. "To make accurate diagnosis and effective treatment possible, it's vital to draw a good blood sample, place it in a sterile container, and process it accurately in the lab. Without a high-quality specimen, none of this can happen."

*Phlebotomy* is not exactly a household word. It derives from the Greek terms for "vein" and "to cut." A phlebotomy technician, or phlebotomist, is trained to collect and process blood samples that will be subjected to laboratory analysis.

Coté appreciates how important it is for phlebotomists to understand anatomy and physiology. "Anatomy is a key requirement in phlebotomy training," he says, "because you have to learn where all the arteries and veins are located in the body. Some patients' veins are easy to find, but others have veins that are practically invisible. You need to know the right place to insert that needle. Although 90 percent of the blood we draw comes from the antecubital region inside the elbow, we may also draw blood from the cephalic vein in the forearm, or from veins in the hands."

Coté notes that knowledge of physiology is also important. "I have to be able to assess patients' overall health and physical condition because this affects their ability to give an adequate blood sample and may demand that a different needle size be used to draw the blood sample. People who are dehydrated can be difficult

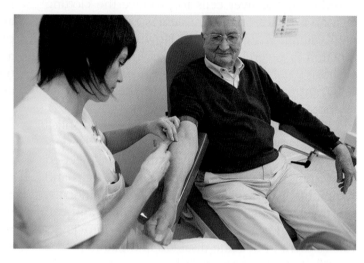

because their blood pressure is lower, and venous return is impaired. Patients with poor circulation are also harder to work with. The blood tends to stay in the body trunk rather than flowing freely into the extremities because they're cold, and it's difficult to get a good return from veins in the arms. Cancer patients often show increased sensitivity to pain, so we have to be very gentle and use the smallest needle possible." Patients with a history of drug abuse pose other challenges. "Frequent 'sticking' with needles causes scar tissue to form. You can tell people who have used intravenous drugs—their veins feel rock-hard and are much more difficult to penetrate with a needle."

> ## Phlebotomy is the most important procedure done for a medical laboratory.

Coté says that a good phlebotomist must also possess effective interpersonal skills: "People are apprehensive about being stuck with needles, so you have to be patient and be able to put them at ease. Above all, you have to be confident. If the phlebotomist is nervous, the patient will sense it and get nervous too."

To become certified, a phlebotomy technician must be a high school graduate, complete a phlebotomy training program or acquire equivalent experience, and pass a certification exam offered by the American Society for Clinical Pathology (ASCP).

Accreditation procedures for phlebotomists vary from state to state. For more information, contact the American Society for Clinical Pathology at

33 West Monroe Street
Suite 1600
Chicago, IL 60603
(312) 541-4999
https://www.ascp.org

For additional information on this career and others, click the Focus on Careers link at Mastering A&P®.

## Table **10.3**  **ABO Blood Groups**

| Blood group | RBC antigens (agglutinogens) | Illustration | Plasma antibodies (agglutinins) | Blood that can be received | Frequency (% of U.S. population) | | | |
|---|---|---|---|---|---|---|---|---|
| | | | | | White | Black | Asian | Native American |
| AB | A, B | | None | A, B, AB, O "Universal recipient" | 4 | 4 | 5 | <1 |
| B | B | | Anti-A (a) | B, O | 11 | 20 | 27 | 4 |
| A | A | | Anti-B (b) | A, O | 40 | 27 | 28 | 16 |
| O | None | | Anti-A (a) Anti-B (b) | O "Universal donor" | 45 | 49 | 40 | 79 |

There are over 30 common RBC antigens in humans, so each person's blood cells can be classified into several different blood groups. However, it is the antigens of the *ABO* and *Rh blood groups* that cause the most vigorous transfusion reactions. We describe these two blood groups here.

The **ABO blood groups** are based on which of two antigens, type A or type B, a person inherits (Table 10.3). Absence of both antigens results in type O blood, presence of both antigens leads to type AB, and the presence of either A or B antigen yields type A or B blood, respectively. In the ABO blood group, antibodies form during infancy against the ABO antigens *not* present on your own RBCs. As shown in the table, a baby with neither the A nor the B antigen (group O) forms both anti-A and anti-B antibodies; those with type A antigens (group A) form anti-B antibodies, and so on. To keep this idea straight, remember that antibodies against a person's own blood type will *not* be produced.

The **Rh blood groups** are so named because one of the eight Rh antigens (agglutinogen D) was originally identified in **Rh**esus monkeys. Later the same antigen was discovered in humans. Most Americans are $Rh^+$ ("Rh positive"), meaning that their RBCs carry Rh antigen. Unlike the antibodies of the ABO system, anti-Rh antibodies are *not* automatically formed by $Rh^-$ ("Rh negative") individuals. However, if an $Rh^-$ person receives $Rh^+$ blood, shortly after the transfusion his or her immune system becomes sensitized and begins producing anti-$Rh^+$ antibodies against the foreign blood type.

**Hemolysis** (rupture of RBCs) does not occur in an $Rh^-$ person with the first transfusion of $Rh^+$ blood because it takes time for the body to react and start making antibodies. However, the second time and every time thereafter, a typical transfusion reaction occurs in which the patient's antibodies attack and rupture the donor's $Rh^+$ RBCs.

An important Rh-related problem occurs in pregnant $Rh^-$ women who are carrying $Rh^+$ babies. The *first* such pregnancy usually results in the delivery of a healthy baby. But because the mother is sensitized by $Rh^+$ antigens that have passed through the placenta into her bloodstream, she will form anti-$Rh^+$ antibodies unless treated with

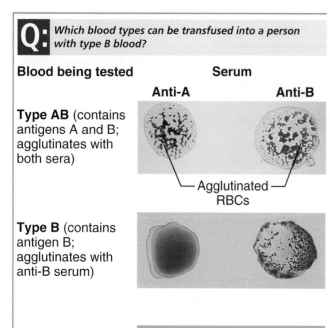

**Q:** *Which blood types can be transfused into a person with type B blood?*

| Blood being tested | Serum | |
| --- | --- | --- |
| | **Anti-A** | **Anti-B** |

**Type AB** (contains antigens A and B; agglutinates with both sera)

Agglutinated RBCs

**Type B** (contains antigen B; agglutinates with anti-B serum)

**Type A** (contains antigen A; agglutinates with anti-A serum)

**Type O** (contains no antigens; does not agglutinate with either serum)

**Figure 10.7 Blood typing of ABO blood groups.** When serum containing anti-A or anti-B antibodies is added to a blood sample diluted with saline, agglutination will occur between the antibody and the corresponding antigen (if present).

RhoGAM in the 28th week of pregnancy and again shortly after giving birth. RhoGAM is an immune serum that prevents this sensitization and subsequent immune response. If she is not treated and becomes pregnant again with an Rh⁺ baby, her antibodies will cross through the placenta and destroy the baby's RBCs, producing a condition known as *hemolytic disease of the newborn*. The baby is anemic and becomes hypoxic and *cyanotic* (the skin takes on a blue cast). Brain damage and even death may result unless fetal transfusions are done before birth to provide more RBCs for oxygen transport.

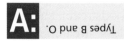

**A:** Types B and O.

## Blood Typing

The importance of determining the blood group of both the donor and the recipient *before* blood is transfused is glaringly obvious. The general procedure for determining ABO blood type essentially involves testing the blood by mixing it with two different types of immune serum—anti-A and anti-B **(Figure 10.7)**. Agglutination occurs when RBCs of a group A person are mixed with the anti-A serum but not when they are mixed with the anti-B serum. Likewise, RBCs of type B blood are clumped by anti-B serum but not by anti-A serum. In order to double check compatibility, cross matching is also done. *Cross matching* involves testing for agglutination of donor RBCs by the recipient's serum and of the recipient's RBCs by the donor serum. Typing for the Rh factors is done in the same manner as ABO blood typing.

### Did You Get It?

12. What are the classes of human blood groups based on?
13. What is the probable result of infusing mismatched blood?
14. Cary is bleeding profusely after being hit by a truck as he was pedaling his bike home. At the hospital, the nurse asked him whether he knew his blood type. He told her that he "had the same blood as most other people." What is his ABO blood type?
15. What is the difference between an antigen and an antibody?

For answers, see Appendix A.

# Developmental Aspects Of Blood

→ **Learning Objectives**

☐ **Explain the basis of physiologic jaundice seen in some newborn babies.**

☐ **Indicate blood disorders that increase in frequency in the aged.**

In the embryo, the entire circulatory system develops early. Before birth, there are many sites of blood cell formation—the fetal liver and spleen, among others—but by the seventh month of development, the fetus's red marrow has become the chief site of hematopoiesis, and it remains so throughout life. Generally, embryonic blood cells are circulating in the newly formed blood vessels by day 28 of development. Fetal hemoglobin (HbF) differs from the hemoglobin formed after birth. It has a greater ability to pick up oxygen, a characteristic that is highly desirable because fetal blood is less oxygen rich than that of the mother. After birth, fetal blood cells are gradually replaced by RBCs that contain the more typical hemoglobin A (HbA). In situations in which fetal RBCs are destroyed so fast that the immature liver cannot rid the body of hemoglobin breakdown products fast enough, the infant becomes *jaundiced* (jawn'dist). This type of jaundice generally causes no major problems and is referred to as **physiologic jaundice**, to distinguish it from more serious disease conditions.

 ## Homeostatic Imbalance 10.4

Various congenital diseases result from genetic factors (such as hemophilia and sickle cell anemia) and from interactions with maternal blood factors (such as hemolytic disease of the newborn). Dietary factors can lead to abnormalities in blood cell formation and hemoglobin production. Iron-deficiency anemia is especially common in women because of their monthly blood loss during menses. The young and the old are particularly at risk for leukemia.

With increasing age, chronic types of leukemias, anemias, and diseases involving undesirable clot formation are more prevalent. However, these are usually secondary to disorders of the heart, blood vessels, or immune system. The elderly are particularly at risk for pernicious anemia, caused by a lack of vitamin $B_{12}$, because the stomach mucosa (which produces the intrinsic factor required to absorb vitamin $B_{12}$) atrophies with age. _____ **+**

### Did You Get It?

16. How does fetal hemoglobin differ from that of the adult?
17. What blood-related disorders are particularly common in the elderly?

**For answers, see Appendix A.**

## Summary

### Composition and Functions of Blood (pp. 363–373)

1. Blood is composed of a nonliving fluid matrix (plasma) and formed elements. It is scarlet to dull red, depending on the amount of oxygen carried. Normal adult blood volume is 5 to 6 liters.

2. Dissolved in plasma (primarily water) are nutrients, gases, hormones, wastes, proteins, salts, and so on. Plasma composition changes as body cells remove or add substances to it, but homeostatic mechanisms act to keep it relatively constant. Plasma makes up 55 percent of whole blood.

3. Formed elements, the living blood cells that make up about 45 percent of whole blood, include the following:

    a. Erythrocytes, or RBCs—disc-shaped, anucleate cells that transport oxygen bound to their hemoglobin molecules. Their life span is 100 to 120 days.

    b. Leukocytes, or WBCs—amoeboid cells involved in protecting the body.

    c. Platelets—cell fragments that act in blood clotting.

4. Anemia is a decrease in oxygen-carrying ability of blood. Possible causes are bleeding, a decrease in the number of functional RBCs, or a decrease in the amount of hemoglobin they contain. Polycythemia is an excessive number of RBCs that may result from bone marrow cancer or a move to a location where less oxygen is available in the air (at high altitude, for example).

5. Leukocytes are nucleated cells, classed into two groups:

    a. Granulocytes include neutrophils, eosinophils, and basophils.

    b. Agranulocytes include monocytes and lymphocytes.

6. When bacteria, viruses, or other foreign substances invade the body, WBCs increase in number (leukocytosis) and fight them in various ways.

7. An abnormal decrease in number of WBCs is leukopenia. An abnormal increase in WBCs is seen in infectious mononucleosis and leukemia (cancer of blood-forming bone marrow).

8. All formed elements arise in red bone marrow from a common stem cell, the hemocytoblast. However, their developmental pathways differ. The stimulus for hematopoiesis is hormonal (erythropoietin, in the case of RBCs).

### Hemostasis (pp. 373–375)

1. Stoppage of blood loss from an injured blood vessel, or hemostasis, involves three steps: vascular spasms, platelet plug formation, and blood clot formation.

10

2. Hemostasis is started by a tear or interruption in the blood vessel lining. Vascular spasms and accumulation of platelets at the site temporarily stop or slow blood loss. Platelet PF$_3$ and tissue factor initiate the clotting cascade, leading to formation of fibrin threads. Fibrin traps RBCs as they flow past, forming a clot.

3. Clots are digested once a vessel has been repaired. An attached clot that forms or persists in an unbroken blood vessel is a thrombus; a clot traveling in the bloodstream is an embolus.

4. Abnormal bleeding may reflect a deficit of platelets (thrombocytopenia), genetic factors (hemophilia), or inability of the liver to make clotting factors.

## Blood Groups and Transfusions (pp. 375–378)

1. Blood groups are classified on the basis of proteins (antigens) on RBC membranes. Complementary antibodies may or may not be present in blood. Antibodies act to agglutinate (clump) and mark foreign RBCs for lysis.

2. The blood group most commonly typed for is ABO. Type O is most common; least common is AB. ABO antigens are accompanied by preformed antibodies, which act against RBCs that have "foreign" antigens.

3. Rh factor is found in most Americans. Rh$^-$ people do not have preformed antibodies to Rh$^+$ RBCs but form them once exposed (sensitized) to Rh$^+$ blood.

## Developmental Aspects of Blood (pp. 378–379)

1. Congenital blood defects include various types of hemolytic anemias and hemophilia. Incompatibility between maternal and fetal blood can lead to fetal cyanosis, resulting from destruction of fetal blood cells.

2. Fetal hemoglobin (HbF) binds more readily with oxygen than does HbA.

3. Physiologic jaundice in a newborn reflects immaturity of the infant's liver.

4. Leukemias are most common in the very young and very old. Older adults are also at risk for anemia and clotting disorders.

# Review Questions

Access additional practice questions using your smartphone, tablet, or computer:
Mastering A&P®> Study Area > Practice Tests & Quizzes

## Multiple Choice

*More than one choice may apply.*

1. Which would lead to increased erythropoiesis?
   a. Chronic bleeding ulcer
   b. Reduction in respiratory ventilation
   c. Decreased level of physical activity
   d. Reduced blood flow to the kidneys

2. In a person with sickle cell anemia, sickling of RBCs can be induced by
   a. blood loss.           c. stress.
   b. vigorous exercise.    d. fever.

3. A child is diagnosed with sickle cell anemia. This means that
   a. one parent had sickle cell anemia.
   b. one parent carried the sickle cell gene.
   c. both parents had sickle cell anemia.
   d. both parents carried the sickle cell gene.

4. Polycythemia vera will result in
   a. overproduction of WBCs.
   b. exceptionally high blood volume.
   c. abnormally high blood viscosity.
   d. abnormally low hematocrit.

5. Which of the following are normally associated with leukocytes?
   a. Positive chemotaxis    c. Diapedesis
   b. Phototaxis             d. Hemostasis

6. Which of the following cell types are granulocytes?
   a. Lymphocytes    c. Eosinophils
   b. Platelets      d. Basophils

7. A person with blood group A can receive blood from a person with blood group
   a. B.    c. AB.
   b. A.    d. O.

8. A condition resulting from thrombocytopenia is
   a. thrombus formation.    c. petechiae.
   b. embolus formation.     d. hemophilia.

9. Which of the following can cause problems in a transfusion reaction?
   a. Donor antibodies attacking recipient RBCs
   b. Clogging of small vessels by agglutinated clumps of RBCs
   c. Lysis of donated RBCs
   d. Blockage of kidney tubules

10. If an Rh⁻ mother becomes pregnant, when can hemolytic disease of the newborn *not possibly* occur in the child?
    a. If the child is Rh⁻
    b. If the child is Rh⁺
    c. If the father is Rh⁺
    d. If the father is Rh⁻

11. Hematocrit is determined by the percentage of
    a. plasma.
    b. leukocytes.
    c. platelets.
    d. erythrocytes.

12. Clotting proteins
    a. stem blood loss after injury.
    b. transport certain molecules.
    c. help keep water in the bloodstream.
    d. protect the body from pathogens.

## Short Answer Essay

13. What is the blood volume of an average-sized adult?

14. Name as many different categories of substances carried in plasma as you can.

15. Define *formed elements*. Which category is most numerous? Which makes up the buffy coat?

16. Define *anemia*, and give three possible causes.

17. What is positive chemotaxis? What kind of motion is involved in this?

18. Name the formed elements that arise from myeloid stem cells. Name those arising from lymphoid stem cells.

19. What is the function of platelets?

20. Describe the process of hemostasis. Indicate what starts the process.

21. What is the difference between a thrombus and an embolus?

22. What are agglutinins?

23. What factors induce the red bone marrow to produce leukocytes?

24. What is a transfusion reaction? Why does it happen?

25. Explain why an Rh⁻ person does not have a transfusion reaction on the first exposure to Rh⁺ blood. Why is there a transfusion reaction the second time he or she receives the Rh⁺ blood?

26. Why is phlebotomy often considered the most important procedure for a medical laboratory?

# Critical Thinking and Clinical Application Questions

27. Following a surgery for hip replacement, Albert has been experiencing severe chest pain. He also feels dizzy, sweats heavily, and coughs up blood often. What is the likely diagnosis?

28. A bone marrow biopsy of Mr. Lee, a man on long-term drug therapy, shows an abnormally high percentage of nonhematopoietic connective tissue. What condition does this indicate? If the symptoms are critical, what short-term and long-term treatments are indicated? Which treatment is he more likely to be given: infusion of whole blood or of packed red cells?

29. A woman comes to the clinic complaining of fatigue, shortness of breath, and chills. Blood tests show anemia, and a bleeding ulcer is diagnosed. What type of anemia is this?

30. A patient is diagnosed with bone marrow cancer and has a hematocrit of 70 percent. What is this condition called?

31. A middle-aged college professor from Boston is in the Swiss Alps studying astronomy. He arrived two days ago and plans to stay the entire year. However, he notices that he is short of breath when he walks up steps and that he tires easily with any physical activity. His symptoms gradually disappear; after two months, he feels fine. Upon returning to the United States, he has a complete physical exam and is told that his erythrocyte count is higher than normal. (a) Attempt to explain this finding. (b) Will his RBC count remain at this higher-than-normal level? Why or why not?

32. Why is someone more likely to bleed to death when an artery is cleanly severed than when it is crushed and torn?

33. What could be wrong with Linda's newborn baby whose skin has taken on a purplish-blue color?

34. Shao-Mei and Elisha, who are good friends, decide to donate blood. Shao-Mei discovers that her blood group is AB Rh⁻ and Elisha's blood group is O Rh⁻. Explain how Elisha could help Shao-Mei if she ever needed a blood transfusion. Why cannot Shao-Mei help Elisha in the same way?

35. Mr. Malone is going into shock because of blood loss, so paramedics infuse a saline solution. Why would this help?

10

# 11 The Cardiovascular System

When most people hear the term *cardiovascular system*, they immediately think of the heart. We have all felt our own heart "pound" from time to time when we are nervous. The crucial importance of the heart has been recognized for ages. However, the **cardiovascular system** is much more than just the heart, and from a scientific and medical standpoint, it is important to understand *why* this system is so vital to life.

Night and day, minute after minute, our trillions of cells take up nutrients and excrete wastes. Although the pace of these exchanges slows during sleep, they must go on continuously: when they stop, we die. Cells can make such exchanges only with the interstitial fluid in their immediate vicinity. Thus, some means of changing and "refreshing" these fluids is necessary to renew the nutrients and prevent pollution caused by the buildup of wastes. Like a bustling factory, the body must have a transportation system to carry its various "cargoes" back and forth. Instead of roads, railway tracks, and subways, the body's delivery routes are its hollow blood vessels.

Most simply stated, the major function of the cardiovascular system is transportation. Using blood as the transport vehicle, the system carries oxygen, nutrients, cell wastes, hormones, and many other substances vital for body homeostasis to and from the cells. The force to move the blood

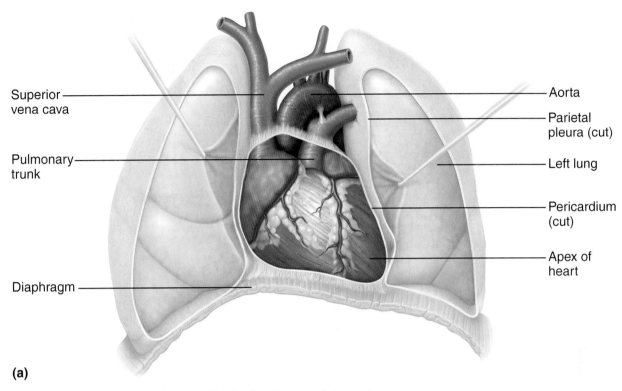

Superior vena cava

Pulmonary trunk

Diaphragm

Aorta

Parietal pleura (cut)

Left lung

Pericardium (cut)

Apex of heart

**(a)**

**Figure 11.1 Location of the heart within the thorax. (a)** Relationship of the heart and great vessels to the lungs.

*(Figure continues on page 384.)*

around the body is provided by the beating heart and by blood pressure.

The cardiovascular system includes a muscular pump equipped with one-way valves and a system of large and small "plumbing" tubes within which the blood travels. (We discussed blood, the substance transported, in Chapter 10.) Here we will consider the heart (the pump) and the blood vessels (the "plumbing").

# The Heart

## Anatomy of the Heart

### → Learning Objective

☐ **Describe the location of the heart in the body, and identify its major anatomical areas on an appropriate model or diagram.**

### Size, Location, and Orientation

The modest size and weight of the heart give few hints of its incredible strength. Approximately the size of a person's fist, the hollow, cone-shaped heart weighs less than a pound. Snugly enclosed within the inferior **mediastinum** (me″de-as-ti′num), the medial section of the thoracic cavity, the heart is flanked on each side by the lungs **(Figure 11.1)**. Its pointed **apex** is directed toward the left hip and rests on the diaphragm, approximately at the level of the fifth intercostal space. (This is exactly where one would place a stethoscope to count the heart rate for an apical pulse.) Its broad posterosuperior aspect, or **base**, from which the great vessels of the body emerge, points toward the right shoulder and lies beneath the second rib.

### Coverings and Walls of the Heart

The heart is enclosed by a sac called the **pericardium** (per″i-kar′de-um) that is made up of three layers: an outer fibrous layer and an inner serous membrane pair. The loosely fitting superficial part of this sac is referred to as the **fibrous pericardium**. This fibrous layer helps protect the heart and anchors it to surrounding structures, such as the diaphragm and sternum. Deep to the fibrous pericardium is the slippery, two-layered **serous pericardium**. The parietal layer of the serous pericardium, or **parietal pericardium**, lines the interior of the fibrous pericardium. At the superior aspect of the heart, this parietal layer attaches to the large arteries leaving the heart and then makes

**11**

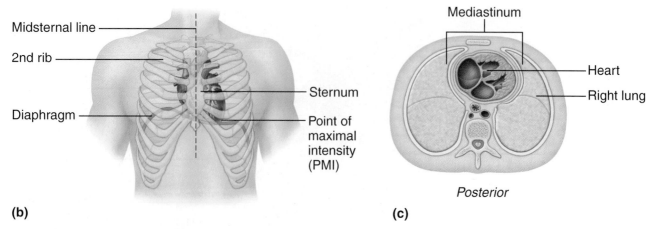

**Figure 11.1** *(continued)* **Location of the heart within the thorax.** **(b)** Relationship of the heart to the sternum and ribs. **(c)** Cross-sectional view showing relative position of the heart in the thorax.

a U-turn and continues inferiorly over the heart surface. The visceral layer of the serous pericardium, or **visceral pericardium**, also called the **epicardium**, is part of the heart wall **(Figure 11.2)**. In other words, the epicardium is the innermost layer of the pericardium and the outermost layer of the heart wall. Lubricating serous fluid is produced by the serous pericardial membranes and collects in the *pericardial cavity* between these serous layers. This fluid allows the heart to beat easily in a relatively frictionless environment as the serous pericardial layers slide smoothly across each other.

### Homeostatic Imbalance 11.1

Inflammation of the pericardium, **pericarditis** (per″ĭ-kar-di′tis), often results in a decrease in the already small amount of serous fluid. This causes the pericardial layers to rub, bind, and stick to each other, forming painful *adhesions* that interfere with heart movements. _____ +

The heart walls are composed of three layers: the outer *epicardium* (the visceral pericardium just described), the *myocardium*, and the innermost *endocardium* (see Figure 11.2). The **myocardium** (mi″o-kar′de-um) consists of thick bundles of cardiac muscle twisted and whorled into ringlike arrangements (see Figure 6.2b, p. 210). It is the layer that actually contracts. Myocardial cells are linked together by intercalated discs, which contain both desmosomes and gap junctions. The gap junctions at the intercalated discs allow ions to flow from cell to cell, carrying a wave of excitement across the heart. The

myocardium is reinforced internally by a network of dense fibrous connective tissue called the "skeleton of the heart." The **endocardium** (en″do-kar′de-um) is a thin, glistening sheet of endothelium that lines the heart chambers. It is continuous with the linings of the blood vessels leaving and entering the heart. (**Figure 11.3** shows two views of the heart—an external anterior view and a frontal section. As the anatomical areas of the heart are described in the next section, keep referring to Figure 11.3 to locate each of the heart structures or regions.)

## Chambers and Associated Great Vessels

### → Learning Objectives
- ☐ **Trace the pathway of blood through the heart.**
- ☐ **Compare the pulmonary and systemic circuits.**

The heart has four hollow cavities, or chambers—two **atria** (a′tre-ah; singular *atrium*) and two **ventricles** (ven′trĭ-kulz). Each of these chambers is lined with endocardium, which helps blood flow smoothly through the heart. The superior atria are primarily *receiving chambers*. As a rule, they are not important in the pumping activity of the heart. Instead, they assist with filling the ventricles. Blood flows into the atria under low pressure from the veins of the body and then continues on to fill the ventricles. The inferior, thick-walled ventricles are the *discharging chambers*, or actual pumps of the heart. When they contract, blood is propelled out of the heart and into circulation. The right ventricle forms most of the heart's anterior surface; the left ventricle forms its apex (Figure 11.3a). The

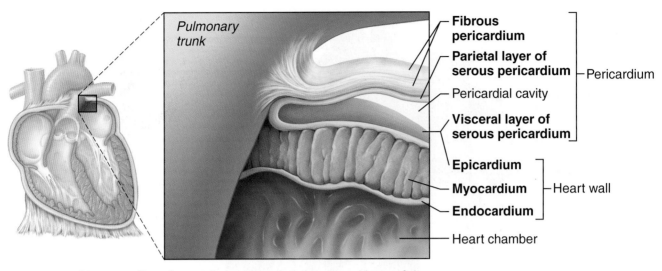

**Figure 11.2 Heart wall and coverings.** Note that the visceral layer of the pericardium and the epicardium of the heart wall are the same structure.

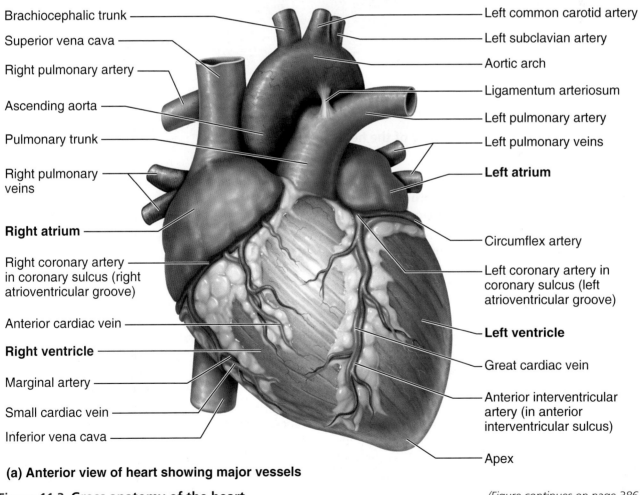

**(a) Anterior view of heart showing major vessels**

**Figure 11.3 Gross anatomy of the heart.**

*(Figure continues on page 386.)*

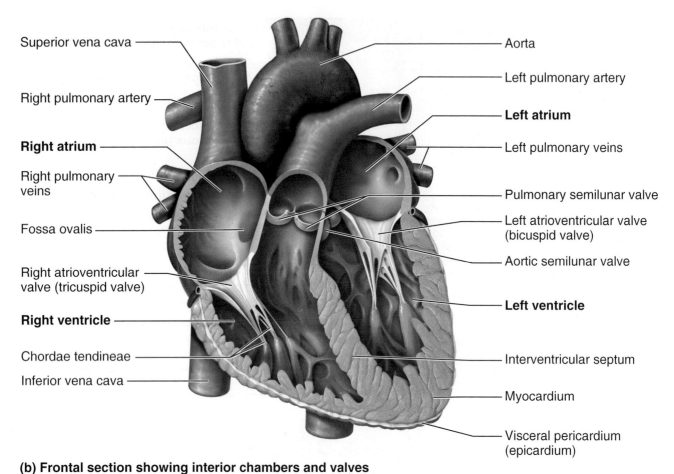

Superior vena cava

Right pulmonary artery

**Right atrium**

Right pulmonary veins

Fossa ovalis

Right atrioventricular valve (tricuspid valve)

**Right ventricle**

Chordae tendineae

Inferior vena cava

Aorta

Left pulmonary artery

**Left atrium**

Left pulmonary veins

Pulmonary semilunar valve

Left atrioventricular valve (bicuspid valve)

Aortic semilunar valve

**Left ventricle**

Interventricular septum

Myocardium

Visceral pericardium (epicardium)

**(b) Frontal section showing interior chambers and valves**

**Figure 11.3** *(continued)* **Gross anatomy of the heart.**

septum that divides the heart longitudinally is referred to as the **interatrial septum** where it divides the atria and the **interventricular septum** where it divides the ventricles.

Although it is a single organ, the heart functions as a double pump, with arteries carrying blood away from and veins carrying blood toward the heart. The right side works as the pulmonary circuit pump. It receives oxygen-poor blood from the veins of the body through the large **superior vena cava** and **inferior vena cava** (plural *venae cavae*; ka′ve) and pumps it out through the **pulmonary trunk**. The pulmonary trunk splits into the right and left **pulmonary arteries**, which carry blood to the lungs, where oxygen is picked up and carbon dioxide is unloaded. Oxygen-rich blood drains from the lungs and is returned to the left side of the heart through the four **pulmonary veins**. This circuit, from the right ventricle (the pump) to the lungs and back to the left atrium (receiving chamber), is called the **pulmonary circulation (Figure 11.4)**. Its only function is to carry blood to the lungs for gas exchange

(oxygen enters the blood and carbon dioxide enters the lungs) and then return it to the heart.

Oxygen-rich blood returned to the left atrium flows into the left ventricle and is pumped out into the **aorta** (a-or′tah), from which the systemic arteries branch to supply essentially all body tissues. After oxygen is delivered to tissues, oxygen-poor blood circulates from the tissues back to the right atrium via the systemic veins, which finally empty their cargo into either the superior or inferior vena cava. This second circuit, from the left ventricle through the body tissues and back to the right atrium, is called the **systemic circulation** (see Figure 11.4). It supplies oxygen- and nutrient-rich blood to all body organs. Because the left ventricle pumps blood over the much longer systemic pathway through the body, its walls are substantially thicker than those of the right ventricle **(Figure 11.5)**, and it is a much more powerful pump.

### Did You Get It?

1. What is the location of the heart in the thorax?

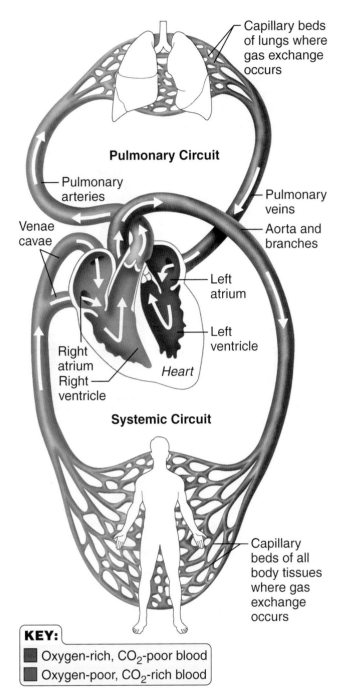

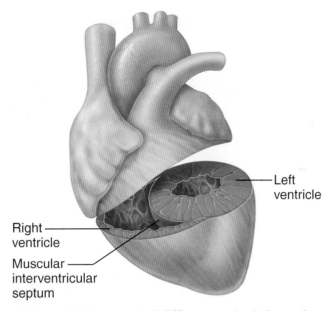

**Figure 11.5 Anatomical differences in right and left ventricles.** The left ventricle has a thicker wall, and its cavity is basically circular. The right ventricle cavity is crescent-shaped and wraps around the left ventricle.

## Heart Valves

### → Learning Objective

☐ **Explain the operation of the heart valves.**

The heart is equipped with four valves, which allow blood to flow in only one direction through the heart chambers—from the atria through the ventricles and out the great arteries leaving the heart (see Figure 11.3b). The **atrioventricular (AV) valves** (a″tre-o-ven-trik′u-lar) are located between the atria and ventricles on each side. These valves prevent backflow into the atria when the ventricles contract. The left AV valve—the **bicuspid valve**, also called the **mitral** (mi′tral) **valve**—consists of two flaps, or cusps, of endocardium. The right AV valve, the **tricuspid valve**, has three cusps. Tiny white cords, the **chordae tendineae** (kor′de ten-din′e)—literally, "tendinous cords" (think of them as "heart strings")—anchor the cusps to the walls of the ventricles. When the heart is relaxed and blood is passively filling its chambers, the AV valve cusps hang limply into the ventricles (**Figure 11. 6a**, p. 388).

As the ventricles contract, they press on the blood in their chambers, and the pressure inside the ventricles (intraventricular pressure) begins to rise. This forces the AV valve cusps upward, closing the valves. At this point the chordae tendineae tighten and anchor the cusps in a closed position.

**Figure 11.4 The systemic and pulmonary circulations.** The left side of the heart is the systemic pump; the right side is the pulmonary circuit pump. (Although there are two pulmonary arteries, one each to the right and left lung, for simplicity only one is shown.)

2. Which heart chamber has the thickest walls? What is the functional significance of this structural difference?
3. How does the function of the systemic circulation differ from that of the pulmonary circulation?

**For answers, see Appendix A.**

## (a) Operation of the AV valves

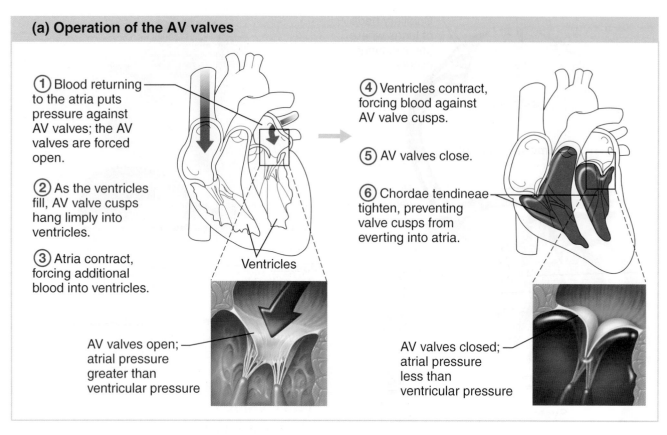

① Blood returning to the atria puts pressure against AV valves; the AV valves are forced open.

② As the ventricles fill, AV valve cusps hang limply into ventricles.

③ Atria contract, forcing additional blood into ventricles.

Ventricles

AV valves open; atrial pressure greater than ventricular pressure

④ Ventricles contract, forcing blood against AV valve cusps.

⑤ AV valves close.

⑥ Chordae tendineae tighten, preventing valve cusps from everting into atria.

AV valves closed; atrial pressure less than ventricular pressure

## (b) Operation of the semilunar valves

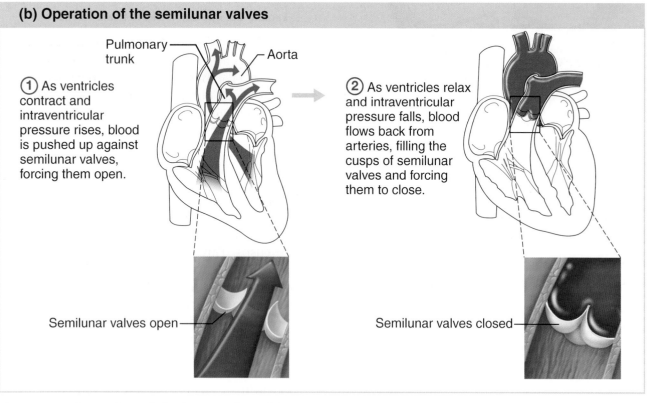

Pulmonary trunk

Aorta

① As ventricles contract and intraventricular pressure rises, blood is pushed up against semilunar valves, forcing them open.

② As ventricles relax and intraventricular pressure falls, blood flows back from arteries, filling the cusps of semilunar valves and forcing them to close.

Semilunar valves open

Semilunar valves closed

**Figure 11.6 Operation of the heart valves. (a)** Atrioventricular (AV) valves. **(b)** Semilunar valves.

If the cusps were unanchored, they would blow upward into the atria like an umbrella being turned inside out by a gusty wind. In this manner, the AV valves prevent backflow into the atria when the ventricles are contracting.

The second set of valves, the **semilunar** (sem″ĭ-lu′nar) **valves**, guards the bases of the two large arteries leaving the ventricular chambers. Thus, they are known as the **pulmonary semilunar valve** and **aortic semilunar valve** (see Figure 11.3b). Each semilunar valve has three cusps that fit tightly together when the valves are closed. When the ventricles are contracting and forcing blood out of the heart, the cusps are forced open and flattened against the walls of the arteries by the tremendous force of rushing blood (Figure 11.6b). Then, when the ventricles relax, the blood begins to flow backward toward the heart, and the cusps fill with blood like a parachute filling with air, closing the valves. This prevents arterial blood from reentering the heart.

Each set of valves operates at a different time. The AV valves are open during heart relaxation and closed when the ventricles are contracting. The semilunar valves are closed during heart relaxation and are forced open when the ventricles contract. The valves force blood to continually move forward through the heart by opening and closing in response to pressure changes in the heart.

## Homeostatic Imbalance 11.2

Heart valves are simple devices, and the heart—like any mechanical pump—can function with "leaky" valves as long as the damage is not too great. However, severely deformed valves can seriously hamper cardiac function. For example, an *incompetent valve* forces the heart to pump and repump the same blood because the valve does not close properly, so blood backflows. In *valvular stenosis*, the valve cusps become stiff, often because of repeated bacterial infection of the endocardium (**endocarditis**). This forces the heart to contract more vigorously than normal to create enough pressure to drive blood through the narrowed valve. In each case, the heart's workload increases, and ultimately the heart weakens and may fail. Under such conditions, the faulty valve is replaced with a synthetic valve (see photo), a cryopreserved human valve, or a chemically treated valve taken from a pig heart.

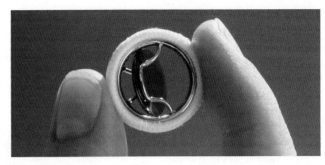

Prosthetic aortic heart valve.

+

### Cardiac Circulation

→ **Learning Objective**

☐ **Name the functional blood supply of the heart.**

Although the heart chambers are bathed with blood almost continuously, the blood contained in the heart does *not* nourish the myocardium. The *functional blood supply* that oxygenates and nourishes the myocardium is provided by the right and left coronary arteries. The **coronary arteries** branch from the base of the aorta and encircle the heart in the **coronary sulcus (atrioventricular groove)** at the junction of the atria and ventricles (see Figure 11.3a). The coronary arteries and their major branches (the **anterior interventricular artery** and **circumflex artery** on the left, and the **posterior interventricular artery** and **marginal artery** on the right) are compressed (flow is inhibited, not stopped completely) when the ventricles are contracting and fill when the heart is relaxed. The myocardium is drained by several **cardiac veins**, which empty into an enlarged vessel on the posterior of the heart called the **coronary sinus**. The coronary sinus, in turn, empties into the right atrium.

## Homeostatic Imbalance 11.3

When the heart beats at a very rapid rate, the myocardium may receive an inadequate blood supply because the relaxation periods (when the blood is able to flow to the heart tissue) are shortened. Situations in which the myocardium is deprived of oxygen often result in crushing chest pain called **angina pectoris** (an-ji′nah pek′tor-is). This pain is a warning that should *never* be ignored, because if angina is prolonged, the oxygen-deprived heart cells may die, forming an area called an **infarct**. The resulting **myocardial infarction** (in-fark′shun), or MI, is commonly called a "heart attack" or a "coronary." _____+

11

**Figure 11.7 The intrinsic conduction system of the heart.** The depolarization wave initiated by the sinoatrial (SA) node passes successively through the atrial myocardium to the atrioventricular (AV) node, the AV bundle, the right and left bundle branches, and the Purkinje fibers in the ventricular walls.

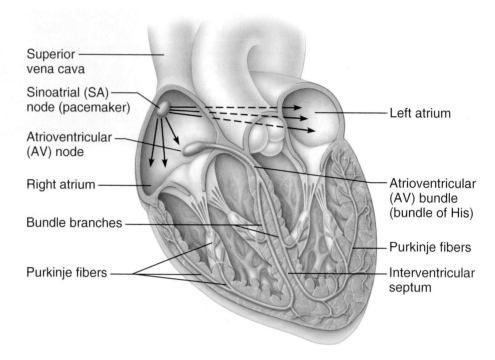

Superior vena cava

Sinoatrial (SA) node (pacemaker)

Atrioventricular (AV) node

Right atrium

Bundle branches

Purkinje fibers

Left atrium

Atrioventricular (AV) bundle (bundle of His)

Purkinje fibers

Interventricular septum

## Did You Get It?

4. Why are the heart valves important?
5. Why might a thrombus in a coronary artery cause sudden death?

**For answers, see Appendix A.**

## Physiology of the Heart

As the heart beats, or contracts, the blood makes continuous round-trips—into and out of the heart, through the rest of the body, and then back to the heart—only to be sent out again. The amount of work that a heart does is almost too incredible to believe. In one day it pushes the body's supply of 6 quarts or so of blood (6 liters [L]) through the blood vessels over 1,000 times, meaning that it actually pumps about 6,000 quarts of blood (1500 gallons) in a single day!

### Intrinsic Conduction System of the Heart: Setting the Basic Rhythm

→ **Learning Objectives**

☐ **Name the elements of the intrinsic conduction system of the heart, and describe the pathway of impulses through this system.**

☐ **Explain what information can be gained from an electrocardiogram.**

What makes the heart beat? Unlike skeletal muscle cells, which must be stimulated by nerve impulses before they will contract, cardiac muscle cells can and do contract spontaneously and independently, even if all nervous connections are severed. Moreover, these spontaneous contractions occur in a regular and continuous way. Although cardiac muscle *can* beat independently, the muscle cells in different areas of the heart have different rhythms. Atrial cells beat about 60 times per minute, but ventricular cells contract more slowly (20–40 times per minute). Therefore, without some type of unifying control system, the heart would be an uncoordinated and inefficient pump.

Two systems act to regulate heart activity. One of these involves the nerves of the autonomic nervous system, which act like brakes and gas pedals to decrease or increase the heart rate, depending on which division is activated. We consider this topic later (see p. 394). The second system is the **intrinsic conduction system**, or **nodal system**, that is built into the heart tissue **(Figure 11.7)** and sets its basic rhythm like a drummer sets the beat for a rock band playing a song. The intrinsic conduction system is composed of a special tissue found nowhere else in the body; it is much like a cross between muscle and nervous tissue. This system causes heart muscle depolarization in only one direction—from the atria to the ventricles.

### CONCEPTLINK

This is very similar to the one-way generation of an action potential as it travels down the axon of a neuron like a wave (Chapter 7, pp. 260–262). The signals that stimulate cardiac muscle contraction also travel one way throughout the intrinsic conduction system. ←

In addition, the intrinsic conduction system enforces a contraction rate of approximately 75 beats per minute on the heart; thus, the heart beats as a coordinated unit.

One of the most important parts of the intrinsic conduction system is a crescent-shaped node of tissue called the **sinoatrial** (si″no-a′tre-al) **(SA) node**, located in the right atrium. Other components include the **atrioventricular (AV) node** at the junction of the atria and ventricles, the **atrioventricular (AV) bundle (bundle of His)** and the right and left **bundle branches** located in the interventricular septum, and finally the **Purkinje** (pur-kin′je) **fibers**, which spread within the myocardium of the ventricle walls.

The SA node is a tiny cell mass with a mammoth job. Because it has the highest rate of depolarization in the whole system, it starts each heartbeat and sets the pace for the whole heart. Consequently, the SA node is often called the **pacemaker**. From the SA node, the impulse spreads through the atria to the AV node, and then the atria contract. At the AV node, the impulse is delayed briefly to give the atria time to finish contracting. It then passes rapidly through the AV bundle, the bundle branches, and the Purkinje fibers, resulting in a "wringing" contraction of the ventricles that begins at the heart apex and moves toward the atria. This contraction effectively ejects blood superiorly into the large arteries leaving the heart. "A Closer Look" (on p. 393) describes *electrocardiography*, the clinical procedure for mapping the electrical activity of the heart.

## Homeostatic Imbalance 11.4

Because the atria and ventricles are separated from one another by "insulating" connective tissue, which is part of the fibrous skeleton of the heart, depolarization waves can reach the ventricles only by traveling through the AV node. Thus, any damage to the AV node can partially or totally block the ventricles from the control of the SA node. When this occurs, the ventricles begin to beat at their own rate, which is much slower, some or all of the time. This condition is called **heart block**.

There are other conditions that can interfere with the regular conduction of impulses across the heart—for example, damage to the SA node results in a slower heart rate. When this is a problem, artificial pacemakers are usually installed surgically.

**Ischemia** (is-ke′me-ah), or lack of an adequate blood supply to the heart muscle, may lead to **fibrillation**—a rapid, uncoordinated quivering of the ventricles (it looks like a bag of wiggling worms). Fibrillation makes the heart unable to pump any blood and so is a major cause of death from heart attacks in adults. Many businesses train their employees in the use of AEDs (*automatic external defibrillators*), which has proven to be lifesaving in many cases. _____ ✚

**Tachycardia** (tak″e-kar′de-ah) is a rapid heart rate (over 100 beats per minute). **Bradycardia** (brad″e-kar′de-ah) is a heart rate that is substantially slower than normal (less than 60 beats per minute). Neither condition is pathological, but prolonged tachycardia may progress to fibrillation.

### Cardiac Cycle and Heart Sounds

→ **Learning Objective**

☐ Define *systole, diastole, stroke volume, cardiac cycle, heart sounds,* and *murmur.*

In a healthy heart, the atria contract simultaneously. Then, as they start to relax, the ventricles begin to contract. **Systole** (sis′to-le) and **diastole** (di-as′to-le) mean heart *contraction* and *relaxation*, respectively. Because most of the pumping work is done by the ventricles, these terms refer to the contraction and relaxation of the *ventricles* unless otherwise stated.

The term **cardiac cycle** refers to the events of one complete heartbeat, during which both atria and ventricles contract and then relax. The average heart beats approximately 75 times per minute, so the length of the cardiac cycle is normally about 0.8 second. We will consider the cardiac cycle in terms of events occurring during five periods (**Figure 11.8**, p. 392).

① **Atrial diastole (ventricular filling).** Our discussion begins with the heart completely relaxed. Pressure in the heart is low, the AV valves are open, and blood is flowing passively through the atria into the ventricles. The semilunar valves are closed.

② **Atrial systole.** The ventricles remain in diastole as the atria contract, forcing blood into the ventricles to complete ventricular filling.

**11**

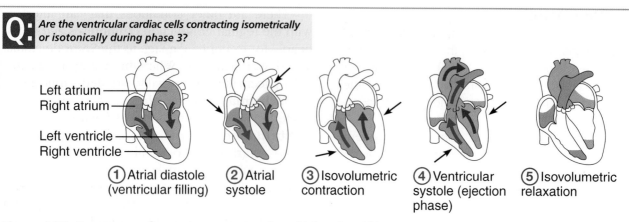

**Q:** *Are the ventricular cardiac cells contracting isometrically or isotonically during phase 3?*

Left atrium
Right atrium

Left ventricle
Right ventricle

① Atrial diastole (ventricular filling)　② Atrial systole　③ Isovolumetric contraction　④ Ventricular systole (ejection phase)　⑤ Isovolumetric relaxation

**Figure 11.8 Summary of events occurring during the cardiac cycle.** Small black arrows indicate the regions of the heart that are contracting; thick red and blue arrows indicate direction of blood flow. During the *isovolumetric* (literally "same volume measurement") phases in periods 3 and 5, the ventricles are closed chambers, and the volume of blood they contain is unchanging.

③ **Isovolumetric contraction.** Atrial systole ends, and ventricular systole begins. The initial rise in intraventricular pressure closes the AV valves, preventing backflow of blood into the atria. For a moment, the ventricles are completely closed chambers.

④ **Ventricular systole (ejection phase).** The ventricles continue to contract, causing the intraventricular pressure to surpass the pressure in the major arteries leaving the heart. This causes the semilunar valves to open and blood to be ejected from the ventricles. During this phase, the atria are again relaxed and filling with blood.

⑤ **Isovolumetric relaxation.** As ventricular diastole begins, the pressure in the ventricles falls below that in the major arteries, and the semilunar valves close to prevent backflow into the ventricles. For another moment, the ventricles are completely closed chambers and intraventricular pressure continues to decrease. Meanwhile, the atria have been in diastole, filling with blood. When atrial pressure increases above intraventricular pressure, the AV valves open, and the cycle repeats.

When using a stethoscope, you can hear two distinct sounds during each cardiac cycle. These **heart sounds** are often described by the two syllables "lub" and "dup," and the sequence is lub-dup, pause, lub-dup, pause, and so on. The first heart sound (lub) is caused by the closing of the AV valves. The second heart sound (dup) occurs when the semilunar valves close at the end of ventricular systole. The first heart sound is longer and louder than the second heart sound, which tends to be short and sharp.

 **Homeostatic Imbalance 11.5**

Abnormal or unusual heart sounds are called **heart murmurs**. Blood flows silently as long as the flow is smooth and uninterrupted. If it strikes obstructions, its flow becomes turbulent and generates sounds that can be heard with a stethoscope. Heart murmurs are fairly common in young children (and some elderly people) with perfectly healthy hearts, probably because their heart walls are relatively thin and vibrate with rushing blood. However, murmurs in patients who do not fall into either of these groups most often indicate valve problems. For example, if a valve does not close tightly (is *incompetent*), a swishing sound will be heard *after* that valve has (supposedly) closed, as the blood flows back through the partially open valve. Distinct sounds also can be heard when blood flows turbulently through *stenosed* (narrowed) valves. _____ ✚

**Did You Get It?**

6. What is the function of the intrinsic conduction system of the heart?
7. To which heart chambers do the terms *systole* and *diastole* usually apply?

 **A:** The cells contract isometrically until they have enough force to overcome the back pressure of the blood against the semilunar valves, at which point their contraction becomes isotonic.

# Electrocardiography: (Don't) Be Still My Heart

When impulses pass through the heart, electrical currents are generated that spread throughout the body. These impulses can be detected on the body surface and recorded with an *electrocardiograph*. The recording that is made, the **electrocardiogram (ECG)**, traces the flow of current through the heart. The illustration shows a normal ECG tracing.

The typical ECG has three recognizable waves. The first wave, which follows the firing of the SA node, is the **P wave**. The P wave is small and signals the depolarization of the atria immediately before they contract. The large **QRS complex**, which results from the depolarization of the ventricles, has a complicated shape. It precedes the contraction of the ventricles. The **T wave** results from currents flowing during the repolarization of the ventricles. (Atrial repolarization is generally hidden by the large QRS complex, which is being recorded at the same time.)

Abnormalities in the shape of the waves and changes in their timing

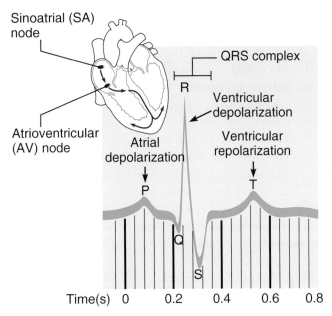

An electrocardiogram tracing showing the three normally recognizable deflection waves—P, QRS, and T.

signal that something may be wrong with the nodal system, or they may indicate a *myocardial infarct* (present or past). A myocardial infarct (heart attack) is an area of heart tissue in which the cardiac cells have died; it is generally a result of *ischemia* (inadequate blood flow). During *fibrillation* (quivering of the heart), the normal pattern of the ECG is totally lost, and the heart ceases to pump blood.

---

8. During isovolumetric contraction of the cardiac cycle, which chambers are relaxing, and which are contracting?
9. What causes the lub-dup sounds heard with a stethoscope?

**For answers, see Appendix A.**

## Cardiac Output

→ **Learning Objective**

☐ **Describe the effect of each of the following on heart rate: stimulation by the vagus nerve, exercise, epinephrine, and various ions.**

**Cardiac output (CO)** is the amount of blood pumped out by *each* side of the heart (actually each ventricle) in 1 minute. It is the product of the **heart rate (HR)** and the **stroke volume (SV)**. Stroke volume is the volume of blood pumped out by a ventricle with each heartbeat. In general, stroke volume increases as the force of ventricular contraction increases. If we use the normal resting values for heart rate (75 beats per minute) and stroke volume (70 ml per beat), the average adult cardiac output can be easily calculated:

$$CO = HR \text{ (75 beats/min)} \times SV \text{ (70 ml/beat)}$$
$$CO = 5250 \text{ ml/min} = 5.25 \text{ L/min}$$

The normal adult blood volume is about 6,000 ml, so nearly the entire blood supply passes through the body once each minute. Cardiac output varies with the demands of the body. It rises when the stroke volume is increased or the heart beats faster or both; it drops when either or both of these factors decrease. Let's take a look at how stroke volume and heart rate are regulated.

***Regulation of Stroke Volume*** A healthy heart pumps out about 60 percent of the blood present in its ventricles. As noted previously, this is approximately 70 ml (about 2 ounces) with each heartbeat. According to *Starling's law of the heart*, the critical factor controlling stroke volume is how much the cardiac muscle cells are stretched by the filling of the chambers just before they contract. The more they are stretched, the stronger the contraction will be. The important factor stretching the heart muscle is *venous return*, the amount of blood entering the heart and distending its ventricles. If one side of the heart suddenly begins to pump more blood than the other, the increased venous return to the opposite ventricle will force it to pump out an equal amount, thus preventing backup of blood in the circulation.

Anything that increases the volume or speed of venous return also increases stroke volume and force of contraction **(Figure 11.9)**. For example, a slow heartbeat allows more time for the ventricles to fill. Exercise speeds venous return because it results in increased heart rate and force; the enhanced squeezing action of active skeletal muscles on the veins helps return blood to the heart. This so-called *muscular pump* also plays a major role in increasing the venous return. In contrast, low venous return, such as might result from severe blood loss or an abnormally rapid heart rate, decreases stroke volume, causing the heart to beat less forcefully.

***Factors Modifying Basic Heart Rate*** In healthy people, stroke volume tends to be relatively constant. However, when blood volume drops suddenly or when the heart has been seriously weakened, stroke volume declines, and cardiac output is maintained by a faster heartbeat. Although heart contraction does not depend on the nervous system, its rate *can* be changed temporarily by the autonomic nerves. Indeed, the most important external influence on heart rate is the activity of the autonomic nervous system. Several chemicals, hormones, and ions also modify heart rate. Some of these factors are discussed next (see also Figure 11.9).

1. **Neural (ANS) controls.** During times of physical or emotional stress, the nerves of the *sympathetic division* of the autonomic nervous system more strongly stimulate the SA and AV nodes and the cardiac muscle itself. As a result, the heart beats more rapidly. This is a familiar phenomenon to anyone who has ever been frightened or has had to run to catch a bus. As fast as the heart pumps under ordinary conditions, it really speeds up when special demands are placed on it. Because a faster blood flow increases the rate at which fresh blood reaches body cells, more oxygen and glucose are made available to them during periods of stress. When demand declines, the heart adjusts accordingly. *Parasympathetic nerves*, primarily vagus nerve fibers, slow and steady the heart, giving it more time to rest during noncrisis times.

2. **Hormones and ions.** Various hormones and ions can have a dramatic effect on heart activity. Both *epinephrine*, which mimics sympathetic nerves and is released in response to sympathetic nerve stimulation, and *thyroxine*, a thyroid hormone, increase heart rate. Electrolyte imbalances pose a real threat to the heart. For example, recall that calcium ions are required for muscle contraction. A reduced level of ionic calcium in the blood depresses the heartbeat, whereas an excessive level of blood calcium ions causes such prolonged contractions that the heart may stop entirely. Either excess or lack of needed ions such as sodium and potassium also modifies heart activity. A deficit of potassium ions in the blood, for example, causes the heart to beat feebly, and abnormal heart rhythms appear.

3. **Physical factors.** A number of physical factors, including age, gender, exercise, and body temperature, influence heart rate. Resting heart rate is fastest in the fetus (140–160 beats per minute) and then gradually decreases throughout life. The average adult heart rate is faster in females (72–80 beats per minute) than in males (64–72 beats per minute). Heat increases heart rate by boosting the metabolic rate of heart cells. This explains the rapid, pounding heartbeat you feel when you have a

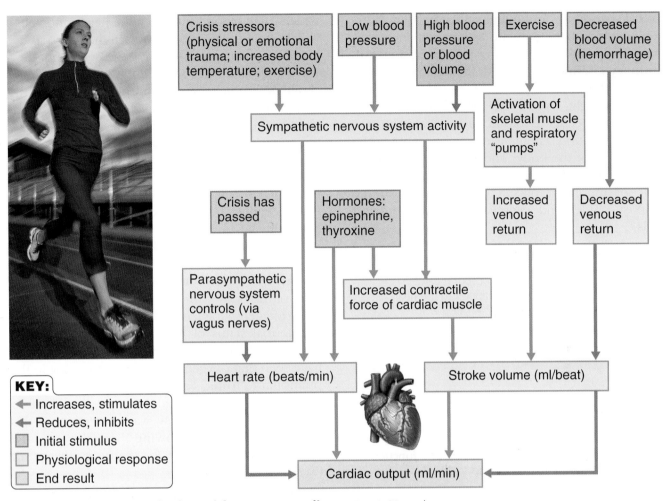

**Figure 11.9 Influence of selected factors on cardiac output.** Note that any decrease in heart rate or stroke volume will result in a corresponding decrease in cardiac output.

high fever and accounts in part for the effect of exercise on heart rate (remember, working muscles generate heat). Cold has the opposite effect; it directly decreases heart rate. As noted previously, exercise acts through nervous system controls (sympathetic division) to increase heart rate (and also, through the action of the muscular pump, to increase stroke volume).

### Homeostatic Imbalance 11.6

The pumping action of the healthy heart maintains a balance between cardiac output and venous return. But when the pumping efficiency of the heart is reduced so that circulation is inadequate to meet tissue needs, **congestive heart failure (CHF)** occurs. Congestive heart failure is usually a progressive condition that reflects weakening of the heart by *coronary atherosclerosis* (clogging of the coronary vessels with fatty buildup), hypertensive heart disease, or multiple myocardial infarctions (repaired with noncontracting scar tissue). In these patients, the heart pumps weakly and is nearly "worn out." The weak contractions of a heart in CHF result in a lower stroke volume. For those patients, the drug digitalis is routinely prescribed. It enhances contractile force and stroke volume of the heart, resulting in greater cardiac output.

Because the heart is a double pump, each side can fail independently of the other. If the left heart fails, *pulmonary congestion* occurs. The right side of the heart continues to propel blood to the lungs, but the left side is unable to eject the returning blood into the systemic circulation. As blood "backs up" in the lungs, they become swollen with blood, the pressure within them increases, and fluid leaks into the lung tissue, causing **pulmonary edema**. If untreated, the person "drowns" in these fluids.

If the right side of the heart fails, *peripheral congestion* occurs as blood backs up in the systemic circulation. Edema is most noticeable in the distal parts of the body: The feet, ankles, and fingers become swollen and puffy. Failure of one side of the heart puts a greater strain on the opposite side, and eventually the whole heart fails. _____ ✚

## Did You Get It?

10. What does the term *cardiac output* mean?
11. What would you expect to happen to the heart rate of an individual with a fever? Why?
12. What is the most important factor affecting stroke volume?

For answers, see Appendix A.

# Blood Vessels

## → Learning Objective

☐ **Compare and contrast the structure and function of arteries, veins, and capillaries.**

Blood circulates inside the blood vessels, which form a closed transport system called the **vascular system**. The idea that blood circulates, or "makes rounds," through the body is only about 300 years old. The ancient Greeks believed that blood moved through the body like an ocean tide, first moving out from the heart and then ebbing back to it in the same vessels to get rid of its impurities in the lungs. It was not until the seventeenth century that William Harvey, an English physician, proved that blood did, in fact, move in circles.

Like a system of roads, the vascular system has its freeways, secondary roads, and alleys. As the heart beats, it propels blood into the large **arteries** leaving the heart. As the large arteries branch, blood moves into successively smaller and smaller arteries and then into the **arterioles** (ar-ter′e-ōlz), which feed the **capillary** (kap′ĭ-lar″e) **beds** in the tissues. Capillary beds are drained by **venules** (ven′ulz), which in turn empty into **veins** that merge and finally empty into the great veins (venae cavae) entering the heart. Thus arteries,

which carry blood away from the heart, and veins, which drain the tissues and return the blood to the heart, are simply conducting vessels—the freeways and secondary roads. Only the tiny hairlike capillaries, which extend and branch through the tissues and connect the smallest arteries (arterioles) to the smallest veins (venules), directly serve the needs of the body cells. The capillaries are the side streets or alleys that intimately intertwine among the body cells and provide access to individual "homes." It is only through their walls that exchanges between tissue cells and the blood can occur.

Notice that we routinely depict arteries in red and veins in blue. By convention, red indicates oxygen-rich blood, the normal status of blood in most of the body's arteries, and blue indicates relatively oxygen-depleted, carbon dioxide–rich blood, the normal status of blood in most of the veins. However, there are exceptions. For instance, we have seen that oxygen-poor blood is carried in the pulmonary trunk, an artery, while oxygen-rich blood is transported back to the heart in pulmonary veins. An easy way to remember this difference is the following: *Arteries are red and veins are blue, but for the lungs there's an exception of two.*

## Microscopic Anatomy of Blood Vessels

### Tunics

Except for the microscopic capillaries (which have only one layer), the walls of blood vessels have three layers, or *tunics* **(Figure 11.10)**. The **tunica intima** (tu′nĭ-kah in-tim′ah), which lines the *lumen*, or interior, of the vessels, is a thin layer of endothelium (squamous epithelial cells) resting on a basement membrane. Its cells fit closely together and form a slick surface that decreases friction as blood flows through the vessel lumen.

The **tunica media** (me′de-ah) is the bulky middle layer, made up mostly of smooth muscle and elastic fibers. Some of the larger arteries have *elastic laminae*, sheets of elastic tissue, in addition to the scattered elastic fibers. The smooth muscle, which is controlled by the sympathetic nervous system, is active in changing the diameter of the vessels. As the vessels constrict or dilate, blood pressure increases or decreases, respectively.

### Figure 11.10 Structure of blood vessels.

**(a)** Light photomicrograph of a muscular artery and the corresponding vein in cross section (85×). **(b)** The walls of arteries and veins are composed of three tunics: the tunica intima, tunica media, and tunica externa. Capillaries—between arteries and veins in the circulatory pathway—are composed only of the tunica intima. Notice that the tunica media is thick in arteries and relatively thin in veins.

View **PAL** Histology
Mastering A&P®

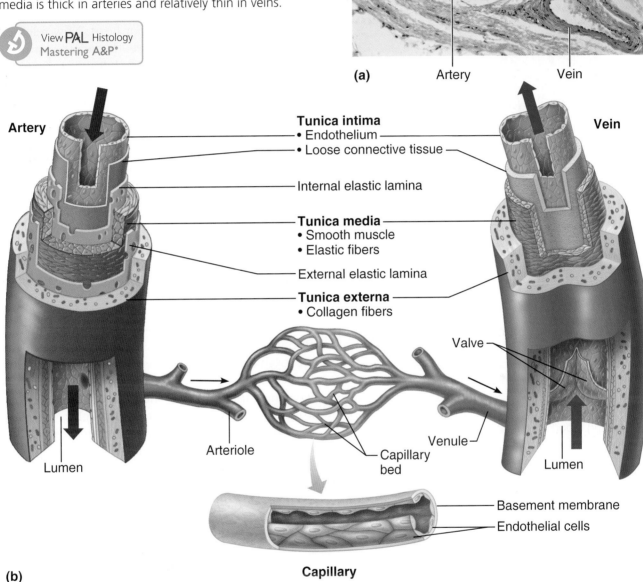

(a)    Artery    Vein

(b)    Capillary

The **tunica externa** (eks'tern-ah) is the outermost tunic. This layer is composed largely of fibrous connective tissue, and its function is to support and protect the vessels.

### Structural Differences in Arteries, Veins, and Capillaries

The walls of arteries are usually much thicker than those of veins. The arterial tunica media, in particular, tends to be much heavier. This structural difference is related to a difference in function of these two types of vessels. Arteries, which carry blood away from the heart, must be able to expand as blood is forced into them and then recoil passively as the blood flows off into the circulation during diastole. Their walls must be strong and stretchy enough to take these continuous

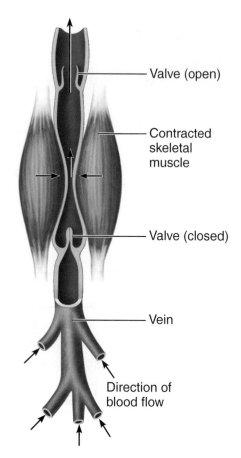

Valve (open)

Contracted skeletal muscle

Valve (closed)

Vein

Direction of blood flow

**Figure 11.11 Operation of the muscular pump.**
When skeletal muscles contract and press against the flexible veins, the valves proximal to the area of contraction are forced open, and blood is squeezed toward the heart like toothpaste from a tube. The valves distal to the point of contraction are closed by the backflowing blood.

changes in pressure without stretching out permanently (see Figure 11.19).

Veins, in contrast, carry blood back toward the heart, and the pressure in them tends to be low all the time. Thus veins have thinner walls. However, because the blood pressure in veins is usually too low to force the blood back to the heart, and because blood returning to the heart often flows against gravity (up the legs, for example), veins are modified to ensure that the amount of blood returning to the heart (venous return) equals the amount being pumped out of the heart (cardiac output) at any time. The lumens of veins tend to be much larger than those of corresponding arteries, and they tend to have a thinner tunica media but a thicker tunica externa. The larger veins have **valves** that prevent backflow of blood like those in the heart (see Figure 11.10).

- To see the effect of venous valves, perform the following simple experiment on yourself: Allow one hand to hang by your side for a minute or two, until the blood vessels on its dorsal aspect become distended (swollen) with blood. Place two fingertips side by side on top of and perpendicular to one of the distended veins. Then, pressing firmly, move your proximal finger along the vein toward your heart. Now release that finger. As you can see, the vein remains collapsed in spite of gravity because your proximal finger pushed the blood past a valve. Now remove your distal finger, and watch the vein fill rapidly with blood.

Skeletal muscle activity, known as the *muscular pump*, also enhances venous return. As the muscles surrounding the veins contract and relax, the blood is squeezed, or "milked," through the veins toward the heart **(Figure 11.11)**. Finally, the drop in pressure that occurs in the thorax just before we inhale causes the large veins near the heart to fill. Thus, the *respiratory pump* also helps return blood to the heart (see Figure 11.9).

The transparent walls of the capillaries are only one cell layer thick—just the tunica intima. Because of this exceptional thinness, substances are exchanged easily between the blood and the tissue cells. The tiny capillaries tend to form interweaving capillary beds. The flow of blood from an arteriole to a venule—that is, through a capillary bed—is called **microcirculation**. In most body regions, a capillary bed consists of two types of vessels: (1) a **vascular shunt**, a vessel that directly connects the arteriole and venule at opposite ends of the bed, and (2) true capillaries, the actual *exchange vessels* **(Figure 11.12)**.

The *true capillaries* number 10 to 100 per capillary bed, depending on the organ or tissues served. They usually branch off the proximal end of the shunt and return to the distal end, but occasionally they spring from the **terminal arteriole** and empty directly into the **postcapillary venule**. A cuff of smooth muscle fibers, called a **precapillary sphincter**, surrounds the root of each true capillary and acts as a valve to regulate the flow of blood into the capillary. Blood flowing through a terminal arteriole may take one of two routes: through the true capillaries or through the shunt. When the precapillary sphincters are relaxed (open), blood flows through the true capillaries and takes part in exchanges with tissue cells. When the sphincters are contracted (closed), blood flows through the shunts and bypasses the tissue cells in that region.

 ## Homeostatic Imbalance 11.7

**Varicose veins** are common in people who stand for long periods of time (for example, cashiers and hairdressers) and in obese (or pregnant) individuals. The common factors are the pooling of blood in the feet and legs and inefficient venous return resulting from inactivity or pressure on the veins. In any case, the overworked valves give way, and the veins become twisted and dilated. A serious complication of varicose veins is **thrombophlebitis** (throm″bo-flĕ-bi′tis), inflammation of a vein that results when a clot forms in a vessel with poor circulation. Because all venous blood must pass through the pulmonary circulation before traveling through the body tissues again, a common consequence of thrombophlebitis is clot detachment and **pulmonary embolism**, which is a life-threatening condition in which the embolism lodges in a vessel in the lung. _____ ✚

### Did You Get It?

13. Assume you are viewing a blood vessel under the microscope. It has a large, lopsided lumen, relatively thick tunica externa, and a relatively thin tunica media. Which kind of blood vessel is this?
14. Arteries lack valves, but veins have them. How is this structural difference related to blood pressure?
15. How is the structure of capillaries related to their function in the body?

For answers, see Appendix A.

## Gross Anatomy of Blood Vessels

→ **Learning** Objective

☐ **Identify the body's major arteries and veins, and name the body region supplied by each.**

### Major Arteries of the Systemic Circulation

The **aorta** is the largest artery of the body, and it is a truly splendid vessel. In adults, the aorta is about the size of a garden hose (with an internal diameter about equal to the diameter of your thumb) where it leaves the left ventricle of the heart. It decreases only slightly in diameter as it runs to its terminus. Different parts of the aorta are named for either their location or their shape. The aorta springs upward from the left ventricle of the

 *Assume the capillary bed depicted here is in the biceps brachii muscle of your arm. What condition would the capillary bed be in, (a) or (b), if you were doing push-ups at the gym?*

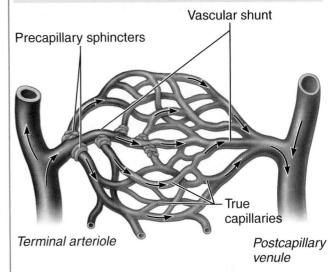

**(a) Sphincters open; blood flows through true capillaries.**

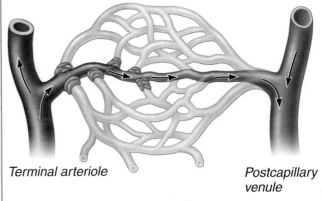

**(b) Sphincters closed; blood flows through vascular shunt.**

**Figure 11.12 Anatomy of a capillary bed.** The vascular shunt bypasses the true capillaries when precapillary sphincters controlling blood entry into the true capillaries are constricted.

heart as the **ascending aorta**, arches to the left as the **aortic arch**, and then plunges downward through the thorax, following the spine (**thoracic aorta**) finally to pass through the diaphragm into the abdominopelvic cavity, where it becomes the **abdominal aorta** (**Figure 11.13**, p. 401).

 **A:** Condition (a). The true capillaries would be flushed with blood to serve the working muscle cells.

The major branches of the aorta and the organs they serve are listed next in sequence from the heart. (Figure 11.13 shows the course of the aorta and its major branches.) As you locate the arteries on the figure, use what you already know to make your learning easier. In many cases the name of the artery tells you the body region or organs served (for example, renal artery, brachial artery, and coronary artery) or the bone followed (femoral artery and ulnar artery).

### Arterial Branches of the Ascending Aorta

- The only branches of the ascending aorta are the **right (R.) coronary artery** and **left (L.) coronary artery**, which serve the heart.

### Arterial Branches of the Aortic Arch

- The **brachiocephalic** (bra″ke-o-sĕ-fal′ik) **trunk** (the first branch off the aortic arch) splits into the **R. common carotid** (kah-rŏ′tid) **artery**, which further branches into the R. internal and R. external carotid arteries, and the **R. subclavian** (sub-kla′ve-an) **artery**. (See same-named vessels on the left side of the body for organs served.)

- The **L. common carotid artery** is the second branch off the aortic arch. It divides, forming the **L. internal carotid**, which serves the brain, and the **L. external carotid**, which serves the skin and muscles of the head and neck.

- The third branch of the aortic arch, the **L. subclavian artery**, gives off an important branch—the **vertebral artery**, which serves part of the brain. In the axilla, the subclavian artery becomes the **axillary artery** and then continues into the arm as the **brachial artery**, which supplies the arm. At the elbow, the brachial artery splits to form the **radial artery** and **ulnar artery**, which serve the forearm.

### Arterial Branches of the Thoracic Aorta

- The *intercostal arteries* (10 pairs) supply the muscles of the thorax wall. Other branches of the thoracic aorta supply the lungs (*bronchial arteries*), the esophagus (*esophageal arteries*), and the diaphragm (*phrenic arteries*). (These arteries are not illustrated in Figure 11.13.)

### Arterial Branches of the Abdominal Aorta

- The **celiac trunk** is the first branch of the abdominal aorta. It is a single vessel that has three branches: the L. gastric artery, which supplies the stomach; the splenic artery, which supplies the spleen; and the common hepatic artery, which supplies the liver.

- The unpaired **superior mesenteric** (mes″enter′ik) **artery** supplies most of the small intestine and the first half of the large intestine, or colon.

- The **renal** (R. and L.) **arteries** serve the kidneys.

- The **gonadal** (R. and L.) **arteries** supply the gonads. They are called the *ovarian arteries* in females (serving the ovaries) and the *testicular arteries* in males (serving the testes).

- The *lumbar arteries* (not illustrated in Figure 11.13) are several pairs of arteries serving the heavy muscles of the abdomen and trunk walls.

- The **inferior mesenteric artery** is a small, unpaired artery supplying the second half of the large intestine.

- The **common iliac** (R. and L.) **arteries** are the final branches of the abdominal aorta. Each divides into an **internal iliac artery**, which supplies the pelvic organs (bladder, rectum, and so on), and an **external iliac artery**, which enters the thigh, where it becomes the **femoral artery**. The femoral artery and its branch, the **deep artery of the thigh**, serve the thigh. At the knee, the femoral artery becomes the **popliteal artery**, which then splits into the **anterior tibial artery** and **posterior tibial artery**, which supply the leg and foot. The anterior tibial artery terminates in the **dorsalis pedis artery**, which via the **arcuate artery** supplies the dorsum of the foot. (The dorsalis pedis is often palpated in patients with circulatory problems of the legs to determine whether the distal part of the leg has adequate circulation.)

## Major Veins of the Systemic Circulation

Although arteries are generally located in deep, well-protected body areas, many veins are more superficial, and some are easily seen and palpated on the body surface. Most deep veins follow the course of the major arteries, and with a few exceptions, the naming of these veins is identical to that of their companion arteries. Major systemic arteries branch off the aorta, whereas the veins converge on the venae cavae, which enter the right atrium of the heart. Veins draining the head and arms empty

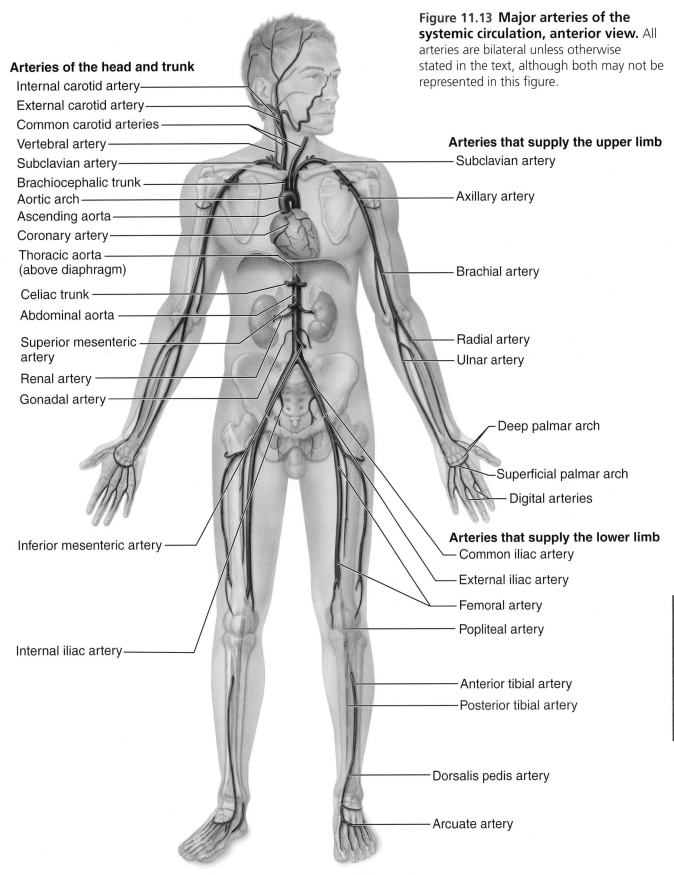

**Figure 11.13 Major arteries of the systemic circulation, anterior view.** All arteries are bilateral unless otherwise stated in the text, although both may not be represented in this figure.

**Arteries of the head and trunk**

Internal carotid artery

External carotid artery

Common carotid arteries

Vertebral artery

Subclavian artery

Brachiocephalic trunk

Aortic arch

Ascending aorta

Coronary artery

Thoracic aorta
(above diaphragm)

Celiac trunk

Abdominal aorta

Superior mesenteric
artery

Renal artery

Gonadal artery

Inferior mesenteric artery

Internal iliac artery

**Arteries that supply the upper limb**

Subclavian artery

Axillary artery

Brachial artery

Radial artery

Ulnar artery

Deep palmar arch

Superficial palmar arch

Digital arteries

**Arteries that supply the lower limb**

Common iliac artery

External iliac artery

Femoral artery

Popliteal artery

Anterior tibial artery

Posterior tibial artery

Dorsalis pedis artery

Arcuate artery

11

into the **superior vena cava**, and those draining the lower body empty into the **inferior vena cava**. (These veins are described next and shown in **Figure 11.14**. As before, locate the veins on the figure as you read their descriptions.)

### Veins Draining into the Superior Vena Cava

Veins draining into the superior vena cava are listed in a distal-to-proximal direction; that is, in the same direction the blood flows into the superior vena cava.

- The **radial vein** and **ulnar vein** are deep veins draining the forearm. They unite to form the deep **brachial vein**, which drains the arm and empties into the **axillary vein** in the axillary region.

- The **cephalic** (se-fal′ik) **vein** provides for the superficial drainage of the lateral aspect of the arm and empties into the axillary vein.

- The **basilic** (bah-sil′ik) **vein** is a superficial vein that drains the medial aspect of the arm and empties into the brachial vein proximally. The basilic and cephalic veins are joined at the anterior aspect of the elbow by the **median cubital vein**. (The median cubital vein is often chosen as the site for withdrawing blood for the purpose of blood testing.)

- The **subclavian vein** receives venous blood from the arm through the axillary vein and from the skin and muscles of the head through the **external jugular vein**.

- The **vertebral vein** drains the posterior part of the head.

- The **internal jugular vein** drains the dural sinuses of the brain.

- The **brachiocephalic** (R. and L.) **veins** are large veins that receive venous drainage from the subclavian, vertebral, and internal jugular veins on their respective sides. The brachiocephalic veins join to form the superior vena cava, which enters the heart.

- The *azygos* (āz′ĭ-gos) *vein* is a single vein that drains the thorax and enters the superior vena cava just before it joins the heart. (This vein is not illustrated in Figure 11.14.)

### Veins Draining into the Inferior Vena Cava

The inferior vena cava, which is much longer than the superior vena cava, returns blood to the heart from all body regions inferior to the diaphragm.

As before, we will trace the venous drainage in a distal-to-proximal direction.

- The **anterior tibial vein** and **posterior tibial vein** and the **fibular vein** drain the leg (calf and foot). (The fibular vein is not shown in Figure 11.14.) The posterior tibial vein becomes the **popliteal vein** at the knee and then the **femoral vein** in the thigh. The femoral vein becomes the **external iliac vein** as it enters the pelvis.

- The **great saphenous** (sah-fe′nus) **veins** are the longest veins in the body. They receive the superficial drainage of the leg. They begin at the **dorsal venous arch** in the foot and travel up the medial aspect of the leg to empty into the femoral vein in the thigh.

- Each **common iliac** (R. and L.) **vein** is formed by the union of the **external iliac vein** and the **internal iliac vein** (which drains the pelvis) on its own side. The common iliac veins join to form the inferior vena cava, which then ascends superiorly in the abdominal cavity.

- The **R. gonadal vein** drains the right ovary in females and the right testicle in males. (The **L. gonadal vein** empties into the left renal vein superiorly.) (The gonadal veins are not illustrated in Figure 11.14.)

- The **renal** (R. and L.) **veins** drain the kidneys.

- The **hepatic portal vein** is a single vein that drains the digestive tract organs and carries this blood through the liver before it enters the systemic circulation. (We discuss the hepatic portal circulation in the next section.)

- The **hepatic** (R. and L.) **veins** drain the liver.

## Did You Get It?

16. In what part of the body are the femoral, popliteal, and arcuate arteries found?
17. In what part of the body are the axillary, cephalic, and basilic veins located?

**For answers, see Appendix A.**

### Special Circulations

→ **Learning Objective**

☐ Discuss the unique features of the arterial circulation of the brain, and hepatic portal circulation.

### Arterial Supply of the Brain and the Circle of Willis

Because a lack of blood for even a few minutes

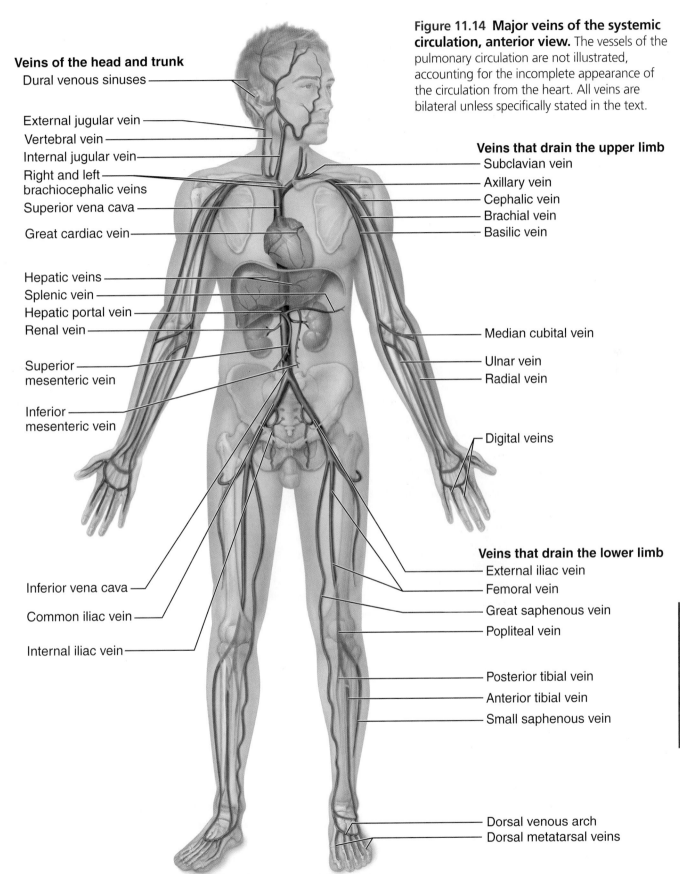

**Figure 11.14 Major veins of the systemic circulation, anterior view.** The vessels of the pulmonary circulation are not illustrated, accounting for the incomplete appearance of the circulation from the heart. All veins are bilateral unless specifically stated in the text.

**Veins of the head and trunk**
Dural venous sinuses
External jugular vein
Vertebral vein
Internal jugular vein
Right and left brachiocephalic veins
Superior vena cava
Great cardiac vein
Hepatic veins
Splenic vein
Hepatic portal vein
Renal vein
Superior mesenteric vein
Inferior mesenteric vein
Inferior vena cava
Common iliac vein
Internal iliac vein

**Veins that drain the upper limb**
Subclavian vein
Axillary vein
Cephalic vein
Brachial vein
Basilic vein
Median cubital vein
Ulnar vein
Radial vein
Digital veins

**Veins that drain the lower limb**
External iliac vein
Femoral vein
Great saphenous vein
Popliteal vein
Posterior tibial vein
Anterior tibial vein
Small saphenous vein
Dorsal venous arch
Dorsal metatarsal veins

11

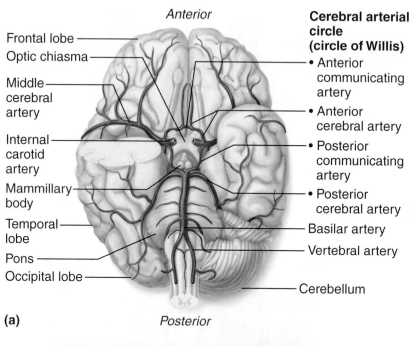

*Anterior*

Frontal lobe

Optic chiasma

Middle cerebral artery

Internal carotid artery

Mammillary body

Temporal lobe

Pons

Occipital lobe

**Cerebral arterial circle (circle of Willis)**
- Anterior communicating artery
- Anterior cerebral artery
- Posterior communicating artery
- Posterior cerebral artery

Basilar artery

Vertebral artery

Cerebellum

**(a)**

*Posterior*

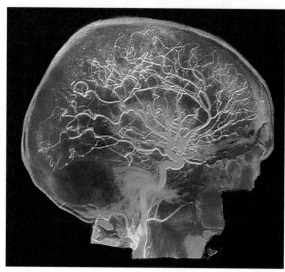

**(b)**

**Figure 11.15 Arterial supply of the brain. (a)** Major arteries of the brain. The four arteries composing the circle of Willis are indicated by bullet points. (The cerebellum is shown only on the left side of the brain.) **(b)** Colorized arteriograph of the brain's arteries.

causes the delicate brain cells to die, a continuous blood supply to the brain is crucial. The brain is supplied by two pairs of arteries, the internal carotid arteries and the vertebral arteries **(Figure 11.15)**.

The **internal carotid arteries**, branches of the common carotid arteries, run through the neck and enter the skull through the temporal bone. Once inside the cranium, each divides into the **anterior cerebral artery** and **middle cerebral artery**, which supply most of the cerebrum.

The paired **vertebral arteries** pass upward from the subclavian arteries at the base of the neck. Within the skull, the vertebral arteries join to form the single **basilar artery**. This artery serves the brain stem and cerebellum as it travels upward. At the base of the cerebrum, the basilar artery divides to form the **posterior cerebral arteries**, which supply the posterior part of the cerebrum.

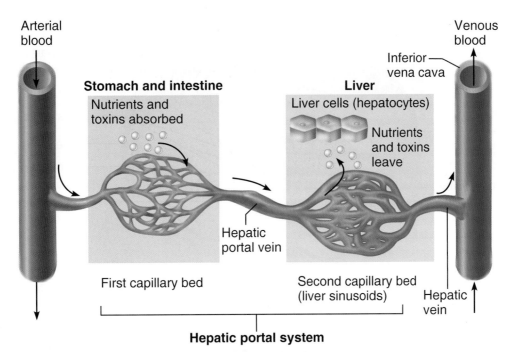

**Figure 11.16 The basic scheme of the hepatic portal system.** Note the presence of two capillary beds within the portal system.

Nutrients and toxins picked up from capillaries in the stomach and intestine are transported to the liver for processing. From the liver

sinusoids, the blood continues into the hepatic veins and inferior vena cava.

The anterior and posterior blood supplies of the brain are united by small *communicating arterial branches*. The result is a complete circle of connecting blood vessels called the **cerebral arterial circle** or the **circle of Willis**, which surrounds the base of the brain. The cerebral arterial circle protects the brain by providing more than one route for blood to reach brain tissue in case of a clot or impaired blood flow anywhere in the system.

*Hepatic Portal Circulation* The veins of the **hepatic portal circulation** drain the digestive organs, spleen, and pancreas and deliver this blood to the liver through the **hepatic portal vein (Figure 11.16)**. When you have just eaten, the hepatic portal blood contains large amounts of nutrients. Because the liver is a key body organ involved in maintaining the proper glucose, fat, and protein concentrations in the blood, this system allows blood to "take a detour" to ensure that the liver processes these substances before they enter the systemic circulation. The liver also helps detoxify blood by removing and processing toxins absorbed by the stomach and intestines. As blood flows slowly through the liver, some of the nutrients are removed to be

stored or processed in various ways for later release to the blood. The liver is drained by the hepatic veins that enter the inferior vena cava.

### CONCEPT**LINK**

Like the portal circulation that links the hypothalamus of the brain and the anterior pituitary gland (Chapter 9, p. 339), the hepatic portal circulation is a unique and unusual circulation. Normally, arteries feed capillary beds, which in turn drain into veins. In the hepatic portal circulation, *veins* feed the liver circulation (Figure 11.16). ←

The major vessels composing the hepatic portal circulation (**Figure 11.17**, p. 406) include the inferior and superior mesenteric veins, the splenic vein, and the left gastric vein. The **inferior mesenteric vein**, draining the terminal part of the large intestine, drains into the **splenic vein**, which itself drains the spleen, pancreas, and the left side of the stomach. The splenic vein and **superior mesenteric vein** (which drains the small intestine and the first part of the colon) join to form the hepatic portal vein. The **L. gastric vein**, which drains the right side of the stomach, drains directly into the hepatic portal vein.

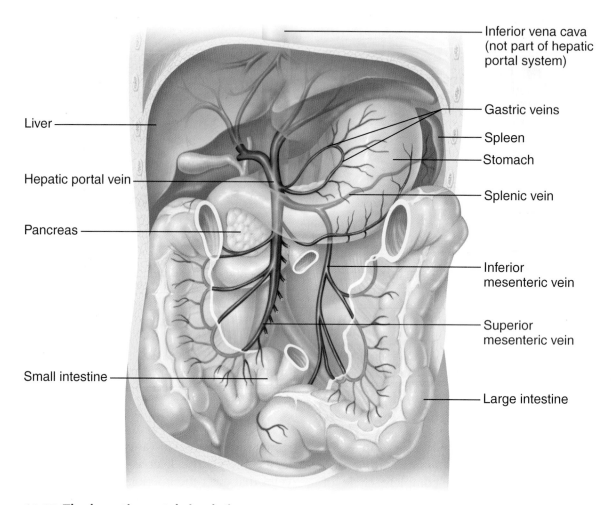

**Figure 11.17 The hepatic portal circulation.**

## Did You Get It?

18. Which vessel—the hepatic portal vein, hepatic vein, or hepatic artery—has the highest content of nutrients after a meal?
19. In what two important ways is the pulmonary circulation different from the systemic circulation?

> **For answers, see Appendix A.**

## Physiology of Circulation

→ **Learning Objective**

☐ **Define *pulse*, and name several pulse points.**

A fairly good indication of the efficiency of a person's circulatory system can be obtained by taking arterial pulse and blood pressure measurements. These measurements, along with those of respiratory rate and body temperature, are referred to collectively as **vital signs** in clinical settings.

### Arterial Pulse

The alternating expansion and recoil of an artery that occurs with each beat of the left ventricle creates a pressure wave—a **pulse**—that travels through the entire arterial system. Normally the pulse rate (pressure surges per minute) equals the heart rate (beats per minute). The pulse averages 70 to 76 beats per minute in a healthy resting person. It is influenced by activity, postural changes, and emotions.

You can feel a pulse in any artery lying close to the body surface by compressing the artery against firm tissue; this provides an easy way of counting heart rate. Because it is so accessible, the point where the radial artery surfaces at the wrist (the radial pulse) is routinely used to take a pulse measurement, but there are several other clinically important arterial pulse points **(Figure 11.18)**.

Because these same points are compressed to stop blood flow into distal tissues during significant blood loss or hemorrhage, they are also called **pressure points**. For example, if you seriously cut your hand, you can stop the bleeding somewhat by compressing the brachial artery.

- Palpate each of the pulse points shown in Figure 11.18 by placing the tips of your first two or three fingers of one hand over the artery at the site indicated. Do not use your thumb, because it has its own pulse. Compress the artery firmly as you begin and then immediately ease up on your pressure slightly. In each case, notice the regularity and relative strength of the pulse.

## Did You Get It?

**20.** Which artery is palpated at the wrist? At the groin? At the side of the neck?

For the answer, see Appendix A.

## Blood Pressure

### → Learning Objectives

☐ Define *blood pressure*, and list factors affecting and/or determining blood pressure.

☐ Define *hypertension* and *atherosclerosis*, and describe possible health consequences of these conditions.

Any system equipped with a pump that forces fluid through a one-way network of closed tubes operates under pressure. **Blood pressure** is the pressure the blood exerts against the inner walls of the blood vessels, and it is the force that keeps blood circulating continuously even between heartbeats. Unless stated otherwise, the term *blood pressure* in this discussion is understood to mean the pressure within the large systemic arteries near the heart.

***Blood Pressure Gradient*** When the ventricles contract, they force blood into large, thick-walled elastic arteries close to the heart that expand as the blood is pushed into them.

### CONCEPT**LINK**

As you remember, in the passive process of filtration, substances move from areas of high pressure to areas of low pressure through a filter (Chapter 3, p. 102). Blood flow is driven by these same differences in pressure, but without a filter. ←

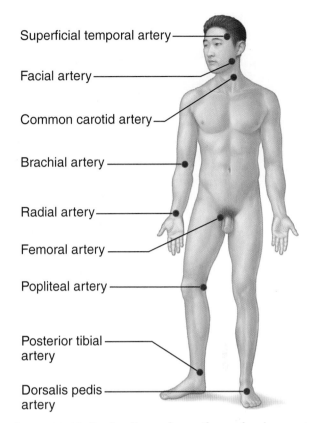

**Figure 11.18 Body sites where the pulse is most easily palpated.** (Most of the specific arteries indicated are discussed on p. 400.)

The high pressure in these elastic arteries forces the blood to continuously move into areas where the pressure is lower. The pressure is highest in the large arteries closest to the heart and continues to drop throughout the systemic pathway, reaching zero at the right atrium **(Figure 11.19)**. Recall that the blood flows into the smaller arteries, then arterioles, capillaries, venules, veins, and finally back to the large venae cavae entering the right atrium of the heart. It flows continuously along a pressure gradient (from high to low pressure) as it makes its circuit, day in and day out. The valves in the large veins, the milking activity of the skeletal muscles, and pressure changes in the thorax are so important because together they ensure blood flows back to the heart to be pumped out to the body again. The pressure differences between arteries and veins become very clear when these vessels are cut. If a vein is cut, the blood flows evenly from the wound; a lacerated artery produces rapid spurts of blood. A similar pressure drop occurs as blood flows through the pulmonary pathway.

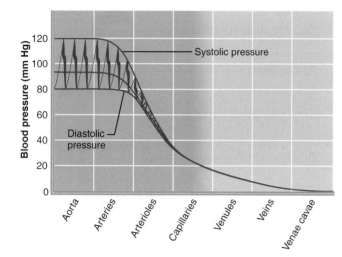

**Figure 11.19 Blood pressure in the systemic circuit of the cardiovascular system.**

Continuous blood flow absolutely depends on the stretchiness of the larger arteries and their ability to recoil and keep exerting pressure on the blood as it flows into the rest of the vascular system. Think of a garden hose with relatively hard walls. When the water is turned on, the water spurts out under high pressure because the hose walls don't expand. However, when the water faucet is suddenly turned off, the flow of water stops just as abruptly. The reason is that the walls of the hose cannot recoil to keep pressure on the water; therefore, the pressure drops, and the flow of water stops. The importance of the elasticity of the arteries is best appreciated when it is lost, as happens in *arteriosclerosis*. Arteriosclerosis is also called "hardening of the arteries" (see "A Closer Look" on pp. 412–413).

## Did You Get It?

21. How does blood pressure change throughout the systemic circulatory pathway?

**For the answer, see Appendix A.**

***Measuring Blood Pressure*** The off-and-on flow of blood into the arteries as the heart alternately contracts and relaxes causes the blood pressure to rise and fall during each beat. Thus, two arterial blood pressures are usually measured: **systolic** (sis-tŏ′lik) **pressure**, the pressure in the arteries at the peak of ventricular contraction, and **diastolic** (di″us-tŏ′lik) **pressure**, the pressure when the ventricles are relaxing. Blood pressures are reported in millimeters of mercury (mm Hg), with

the higher systolic pressure written first—120/80 (read "120 over 80") translates to a systolic pressure of 120 mm Hg and a diastolic pressure of 80 mm Hg. Most often, systemic arterial blood pressure is measured indirectly by the **auscultatory** (os-kul′tuh-tor-e) **method**. This procedure is used to measure blood pressure in the brachial artery of the arm (**Figure 11.20**).

### Effects of Various Factors on Blood Pressure

Arterial blood pressure (BP) is directly related to cardiac output (CO; the amount of blood pumped out of the left ventricle per minute) and peripheral resistance (PR). This relationship is expressed by the equation BP = CO × PR. We have already considered regulation of cardiac output, so we will concentrate on peripheral resistance here.

**Peripheral resistance** is the amount of friction the blood encounters as it flows through the blood vessels. Many factors increase peripheral resistance, but probably the most important is the constriction, or narrowing, of blood vessels, especially arterioles, as a result of either sympathetic nervous system activity or atherosclerosis. Increased blood volume or increased blood viscosity (thickness) also raises peripheral resistance. Any factor that increases either cardiac output or peripheral resistance causes an almost immediate rise in blood pressure. Many factors can alter blood pressure—age, weight, time of day, exercise, body position, emotional state, and various drugs, to name a few. The influence of a few of these factors is discussed next.

• **Neural factors: the autonomic nervous system.** The parasympathetic division of the autonomic nervous system has little or no effect on blood pressure, but the sympathetic division is important. The major action of the sympathetic nerves on the vascular system is to cause **vasoconstriction** (vās″o-kon-strik′shun), or narrowing of the blood vessels, which increases the blood pressure. The sympathetic center in the medulla of the brain is activated to cause vasoconstriction in many different circumstances (**Figure 11.21**, p. 410). For example, when we stand up suddenly after lying down, gravity causes blood to pool very briefly in the vessels of the legs and feet, and blood pressure drops. This activates *pressoreceptors*, also called *baroreceptors*

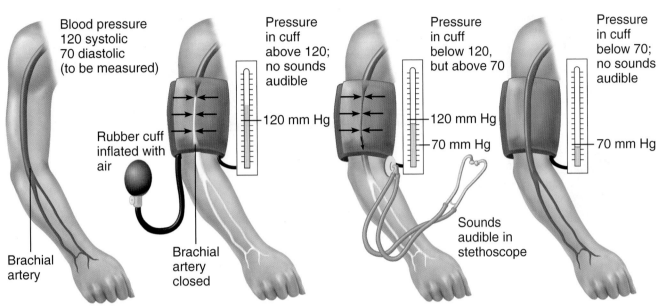

Blood pressure
120 systolic
70 diastolic
(to be measured)

Brachial
artery

Rubber cuff
inflated with
air

Brachial
artery
closed

Pressure
in cuff
above 120;
no sounds
audible

120 mm Hg

Pressure
in cuff
below 120,
but above 70

120 mm Hg

70 mm Hg

Sounds
audible in
stethoscope

Pressure
in cuff
below 70;
no sounds
audible

70 mm Hg

**(a)** The course of the brachial artery of the arm. Assume a blood pressure of 120/70 in a young, healthy person.

**(b)** The blood pressure cuff is wrapped snugly around the arm just above the elbow and inflated until the cuff pressure exceeds the systolic blood pressure. At this point, blood flow into the arm is stopped, and a brachial pulse cannot be felt or heard.

**(c)** The pressure in the cuff is gradually reduced while the examiner listens (auscultates) for sounds in the brachial artery with a stethoscope. The pressure read as the first soft tapping sounds are heard (the first point at which a small amount of blood is spurting through the constricted artery) is recorded as the systolic pressure.

**(d)** As the pressure is reduced still further, the sounds become louder and more distinct; when the artery is no longer constricted and blood flows freely, the sounds can no longer be heard. The pressure at which the sounds disappear is recorded as the diastolic pressure.

**Figure 11.20 Measuring blood pressure.**

(*baro* = pressure), in the large arteries of the neck and chest. They send off warning signals that result in reflexive vasoconstriction, quickly increasing blood pressure back to the homeostatic level.

When blood volume suddenly decreases, as in hemorrhage, blood pressure drops, and the heart begins to beat more rapidly as it tries to compensate. However, because blood loss reduces venous return, the heart also beats weakly and inefficiently. In such cases, the sympathetic nervous system causes vasoconstriction to increase the blood pressure so that (it is hoped) venous return increases and circulation can continue. This phenomenon also happens during severe dehydration.

The final example concerns sympathetic nervous system activity when we exercise vigorously or are frightened and have to make a hasty escape. Under these conditions, generalized vasoconstriction occurs *except* in the skeletal muscles. The vessels of the skeletal muscles dilate to increase the blood flow to the working muscles. (However, note that the sympathetic nerves *never* cause vasoconstriction of blood vessels of the heart or brain.)

• **Renal factors: the kidneys.** The kidneys play a major role in regulating arterial blood pressure by altering blood volume. As blood pressure (and/or blood volume) increases above normal, the kidneys allow more water

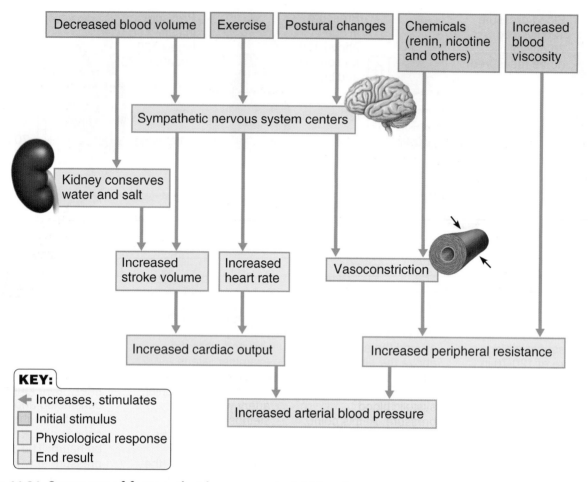

**Figure 11.21 Summary of factors that increase arterial blood pressure.**

to leave the body in the urine. Because the source of this water is the bloodstream, blood volume decreases, which in turn decreases blood pressure. However, when arterial blood pressure falls, the kidneys retain body water, maintaining blood volume and blood pressure (see Figure 11.21). In order to increase blood volume and blood pressure, fluids must be ingested or administered intravenously.

In addition, when arterial blood pressure is low, certain kidney cells release the enzyme *renin* into the blood. Renin triggers a series of chemical reactions that result in the formation of *angiotensin II*, a potent vasoconstrictor chemical. Angiotensin also stimulates the adrenal cortex to release aldosterone, a hormone that enhances sodium ion reabsorption by the kidneys. As sodium ions move into

the blood, water follows. Thus, blood volume and blood pressure both rise in response to aldosterone.

- **Temperature.** In general, cold has a *vasoconstricting* effect. This is why your exposed skin feels cold to the touch on a winter day and why cold compresses are recommended to prevent swelling of a bruised area. Heat has a *vasodilating* effect. This explains why skin reddens during exercise as body temperature increases and why warm compresses are used to speed the circulation into an inflamed area.

- **Chemicals.** The effects of chemical substances, many of which are drugs, on blood pressure are widespread and well known in many cases. We will give just a few examples here. **Epinephrine** increases both heart rate and blood pressure.

**CONCEPTLINK**
Recall that epinephrine is the "fight-or-flight" hormone, which is produced by the adrenal medulla and
helps us deal with short-term stress (Chapter 9,
p. 349). ←

*Nicotine* increases blood pressure by causing vasoconstriction. Both *alcohol* and *histamine* cause vasodilation and decrease blood
pressure. The reason a person who has "one
too many" becomes flushed is that alcohol dilates the skin vessels.

- **Diet.** Although medical opinions tend to
  change and are at odds from time to time, it is
  generally thought that a diet low in salt, saturated fats, and cholesterol helps to prevent *hypertension*, or high blood pressure.

## Did You Get It?

22. What is the effect of hemorrhage on blood pressure?
    Why?

For the answer, see Appendix A.

***Variations in Blood Pressure***  In normal adults at
rest, systolic blood pressure varies between 110
and 140 mm Hg, and diastolic pressure between
70 and 80 mm Hg—but blood pressure varies
considerably from one person to another and
cycles over a 24-hour period, peaking in the
morning. What is normal for you may not be normal for your grandfather or your neighbor. Blood
pressure varies with age, weight, race, mood,
physical activity, and posture. Nearly all these
variations can be explained in terms of the factors
affecting blood pressure that we have already
discussed.

**Hypotension**, or low blood pressure, is generally considered to be a systolic blood pressure
below 100 mm Hg. In many cases, it simply
reflects individual differences and is no cause for
concern. In fact, low blood pressure is an
expected result of physical conditioning and is
often associated with long life and an old age free
of illness.

## Homeostatic Imbalance 11.8

Elderly people may experience temporary low
blood pressure and dizziness when they rise suddenly from a reclining or sitting position—a condition called **orthostatic hypotension**. Because an
aging sympathetic nervous system reacts more
slowly to postural changes, blood pools briefly in
the lower limbs, reducing blood pressure and,
consequently, blood delivery to the brain. Making
postural changes more slowly to give the nervous
system time to make the necessary adjustments
usually prevents this problem. _____ ✚

Chronic hypotension (not explained by physical conditioning) may hint at poor nutrition and
inadequate levels of blood proteins. Because
blood viscosity is low, blood pressure is also lower
than normal. Acute hypotension is one of the most
important warnings of **circulatory shock**, a condition in which the blood vessels are inadequately
filled and blood cannot circulate normally. The
most common cause is blood loss.

A brief elevation in blood pressure is a normal
response to fever, physical exertion, and emotional upset, such as anger or fear. Persistent
**hypertension (high blood pressure)**, is pathological and is defined as a condition of sustained
elevated arterial pressure of 140/90 or higher.

## Homeostatic Imbalance 11.9

Chronic hypertension is a common and dangerous disease that warns of increased peripheral
resistance. Although it progresses without symptoms for the first 10 to 20 years, it slowly and
surely strains the heart and damages the arteries.
For this reason, hypertension is often called the
"silent killer." Because the heart is forced to pump
against increased resistance, it must work harder,
and in time, the myocardium enlarges. When
finally strained beyond its capacity to respond,
the heart weakens and its walls become flabby.
Hypertension also ravages blood vessels, causing
small tears in the endothelium that accelerate the
progress of atherosclerosis (the early stage of arteriosclerosis).

Although hypertension and atherosclerosis are
often linked, it is difficult to blame hypertension on
any distinct anatomical pathology. In fact, about 90
percent of hypertensive people have **primary
(essential) hypertension**, which cannot be attributed to any specific organic cause. However, factors
such as diet, obesity, heredity, race, and stress appear
to be involved. For instance, more women than men
and more blacks than whites are hypertensive.

**11**

# Atherosclerosis?
# Get Out the Cardiovascular Drāno!

When arteries are narrowed by **atherosclerosis**, the clogging process begins on the inside: the walls of the vessels thicken and then protrude into the vessel lumen. Once this happens, a roaming blood clot or arterial spasms can close the vessel completely. All blood vessels are susceptible to atherosclerosis, but for some unknown reason the aorta and the coronary arteries are most often affected.

## Onset and Stages of Atherosclerosis

What triggers this scourge of blood vessels that indirectly causes half of the deaths in the Western world? The initial event is damage to the tunica intima caused by bloodborne chemicals such as carbon monoxide (present in cigarette smoke or auto exhaust); by bacteria or viruses; or by physical factors such as a blow or persistent hypertension. Once a break has occurred, blood platelets cling to the injured site and initiate clotting to prevent blood loss. The injured endothelium summons the immune system to repair the damage. Most plaques grow slowly, through a series of injuries that heal, only to be ruptured again and again. As the plaque grows, the injured endothelial cells release chemicals that make the endothelium more permeable, allowing fats and cholesterol to take up residence just deep to the tunica intima. Monocytes attracted to the area migrate beneath the endothelium, where they become macrophages that gorge themselves on the fat in particular. These cells can become so filled with oxidized fats that they are transformed into "foam cells" that lose their ability to function. Soon they are joined by smooth muscle cells migrating from the tunica media of the blood vessel wall. These cells deposit collagen and elastin fibers in the area and also take in fat, becoming foam cells. The result is the erroneously named **fatty streak stage**, characterized by thickening of the tunica intima by lesions called **fibrous plaques** or **atherosclerotic plaques**. When these small, fatty

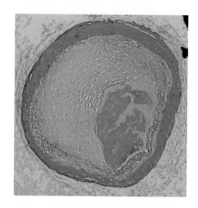

Atherosclerotic plaques nearly close a human artery.

mounds of muscle begin to protrude into the vessel wall (and ultimately the vessel lumen), the condition is called *atherosclerosis* (see photo).

**Arteriosclerosis** is the end stage of the disease. As enlarging plaques hinder diffusion of nutrients from the blood to the deeper tissues of the artery wall, smooth muscle cells in the tunica media die, and the elastic fibers deteriorate and are gradually replaced by nonelastic scar tissue. Then, calcium salts are deposited in the

Hypertension runs in families. The child of a hypertensive parent is twice as likely to develop high blood pressure as is a child of parents with normal blood pressure. High blood pressure is common in obese people because the total length of their blood vessels is greater than that in thinner individuals. For each pound of fat, miles of additional blood vessels are required, making the heart work harder to pump blood over longer distances. _____ ✚

## Capillary Exchange of Gases and Nutrients

→ **Learning Objective**

☐ **Describe the exchanges that occur across capillary walls.**

Capillaries form an intricate network among the body's cells, and no substance has to diffuse very far to enter or leave a cell. The substances to be exchanged diffuse through the **interstitial fluid (tissue fluid)** between cells.

Substances tend to move to and from body cells according to their concentration gradients. Thus, oxygen and nutrients leave the blood and move into the tissue cells, and carbon dioxide and other wastes exit the tissue cells and enter the blood. Basically, substances entering or leaving the blood may take one of four routes across the plasma membranes of the single layer of endothelial cells forming the capillary wall **(Figure 11.22)**.

lesions, forming **complicated plaques**. Collectively, these events cause the arterial wall to fray and ulcerate, conditions that encourage thrombus formation. The increased rigidity of the vessels leads to hypertension. Together, these events increase the risk of myocardial infarctions, strokes, and aneurysms.

The popular view that most heart attacks are the consequence of severe vessel narrowing and hardening is now being challenged. It appears that the body's defense system betrays it. The inflammatory process that occurs in the still soft, unstable, cholesterol-rich plaques changes the biology of the vessel wall and makes the plaques susceptible to rupture, exploding off fragments that trigger massive clots that can cause lethal heart attacks.

### Treatment and Prevention

Some medical centers test heart patients for elevated levels of cholesterol and C-reactive protein, a marker of inflammation. Electron beam CT scans may be able to identify people at risk by detecting calcium deposits in their coronary arteries. Antibiotics and anti-inflammatory drugs are being tested as preventive measures, and

*statins*, cholesterol-reducing drugs, show promise. Even the humble aspirin is gaining new respect, and many cardiologists recommend that people at high risk take one "baby aspirin" (81 mg) daily.

So what can help when the damage is done and the heart is at risk because of atherosclerotic coronary vessels? In the past, the only choice has been coronary artery bypass surgery, in which vessels removed from the legs or thoracic cavity are implanted in the heart to restore circulation. A more recently developed technique, *balloon angioplasty*, uses a catheter with a balloon packed into its tip. When the catheter reaches the blockage, the balloon is inflated, and the fatty mass is compressed against the vessel wall. However, this procedure is useful to clear only very localized obstructions. A newer catheter device uses a laser beam to vaporize the arterial clogs. Although these intravascular devices are faster, cheaper, and less risky than bypass surgery, they carry with them the same major shortcoming: they do nothing to stop the underlying disease, and in time new blockages occur in 30 to 50 percent of cases.

Sometimes after angioplasty, a metal-mesh tube called a *stent* is placed in the artery to keep it open.

When a blood clot is trapped by the diseased vessel walls, the answer may be a *clot-dissolving agent*, for example, *tissue plasminogen activator (tPA)*, a naturally occurring substance now being produced by genetic engineering techniques. Intravenously injecting tPA can restore blood flow quickly and put an early end to a heart attack in progress.

There is little doubt that lifestyle factors—emotional stress, smoking, obesity, high-fat and high-cholesterol diets, and lack of exercise—contribute to both atherosclerosis and hypertension. With these changeable risk factors, why not just have patients change their lifestyle? This is easier said than done. Although taking antioxidants (vitamins C and E and beta carotene) and exercising more may "undo" some of the damage, old habits die hard, and North Americans like their burgers and butter. If atherosclerosis could be reversed to give the heart a longer and healthier life, many more people with diseased arteries may be more willing to trade lifelong habits for a healthy old age!

1. **Direct diffusion through membrane.** As with all cells, substances can diffuse directly through (cross) their plasma membranes if the substances are lipid-soluble (such as the respiratory gases oxygen and carbon dioxide).

2. **Diffusion through intercellular clefts.** Limited passage of fluid and small solutes is allowed by **intercellular clefts** (gaps between cells in the capillary wall). Most of our capillaries have intercellular clefts, except for brain capillaries, which are entirely secured together by tight junctions (the basis of the blood-brain barrier, described in Chapter 7).

3. **Diffusion through pores.** Very free passage of small solutes and fluids is allowed by

**fenestrated capillaries**. These unique capillaries are found where absorption is a priority (intestinal capillaries or capillaries serving endocrine glands) or where filtration occurs (the kidney). A fenestra is an oval pore, or opening (*fenestra* = window), and is usually covered by a delicate membrane (see Figure 11.22). Even so, a fenestra is much more permeable than other regions of the plasma membrane.

4. **Transport via vesicles.** Certain lipid-insoluble substances may enter or leave the blood and/or pass through the plasma membranes of endothelial cells within vesicles, that is, by endocytosis or exocytosis.

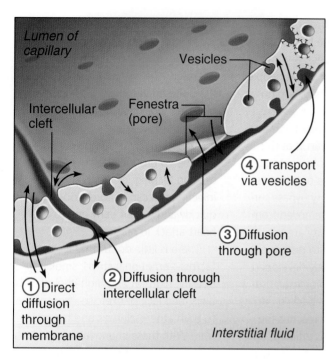

**Figure 11.22 Capillary transport mechanisms.** The four possible pathways or routes of transport across the wall of an endothelial cell of a capillary. (The endothelial cell is illustrated as if cut in cross section.)

Only substances unable to pass by one of these routes are prevented from leaving (or entering) the capillaries. These include protein molecules (in plasma or interstitial fluid) and blood cells.

## Fluid Movements at Capillary Beds

In addition to the exchanges made via passive diffusion through capillary endothelial cell plasma membranes, clefts, or fenestrations, and via vesicles, there are active forces operating at capillary beds. Because of their intercellular clefts and fenestrations, some capillaries are leaky, and bulk fluid flow (fluid moving all at once) occurs across their plasma membranes. Hence, blood pressure tends to force fluid (and solutes) out of the capillaries, and osmotic pressure tends to draw fluid into them because blood has a higher solute concentration (due to its plasma proteins) than does interstitial fluid. Whether fluid moves out of or into a capillary depends on the difference between the two pressures. As a rule, blood pressure is higher than osmotic pressure at the arterial end of the capillary bed, and lower than osmotic pressure at the venous end. Consequently, fluid moves out of the capillaries at the beginning of the bed and is reclaimed at the opposite (venule) end **(Figure 11.23)**. However,

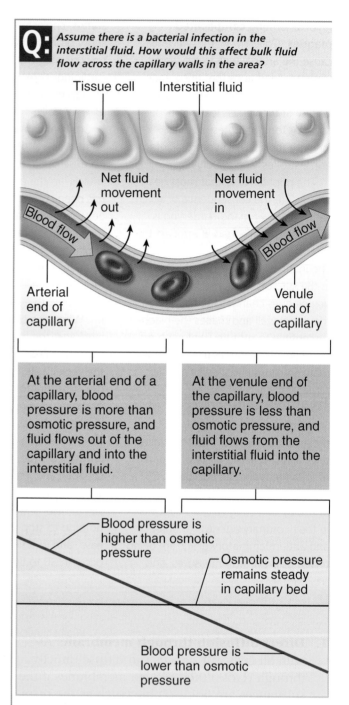

**Figure 11.23 Bulk fluid flow across capillary walls.** Fluid flow depends largely on the difference between the blood pressure and the osmotic pressure at different regions of the capillary bed.

Bacterial infection would increase fluid flow because the osmotic pressure of the interstitial fluid would rise as inflammatory molecules and debris accumulated in the area.

not quite all of the fluid forced out of the blood is reclaimed at the venule end. Returning fluid left in tissues to the blood is the chore of the lymphatic system (discussed in Chapter 12).

## Did You Get It?

23. Would you expect fluid to be entering or leaving the capillaries at the venous end of a capillary bed?

**For the answer, see Appendix A.**

# Developmental Aspects of the Cardiovascular System

→ **Learning Objectives**

☐ **Briefly describe the development of the cardiovascular system.**

☐ **Name the fetal vascular modifications, or "fetal shunts," and describe their function before birth.**

☐ **Describe changes in the cardiovascular system that occur with aging, and list several factors that help maintain cardiovascular health.**

The heart begins as a simple tube in the embryo. It is beating and busily pumping blood by the fourth week of pregnancy. During the next 3 weeks, the heart continues to change and mature, finally becoming a four-chambered structure capable of acting as a double pump—all without missing a beat!

Because the lungs and digestive system are immature and not functioning in a fetus, all nutrient, excretory, and gas exchanges occur through the placenta. Nutrients and oxygen move from the mother's blood into the fetal blood, and fetal wastes move in the opposite direction. The *umbilical cord* contains three blood vessels: one large **umbilical vein** and two smaller **umbilical arteries** (**Figure 11.24**, p. 416). The umbilical vein carries blood rich in nutrients and oxygen to the fetus. The umbilical arteries carry blood laden with carbon dioxide and metabolic waste products from the fetus to the placenta. As blood flows superiorly toward the heart of the fetus, most of it bypasses the immature liver through the **ductus venosus** (duk′tus ve-no′sus) and enters the inferior vena cava, which carries the blood to the right atrium of the heart.

Because the nonfunctional fetal lungs are collapsed, two shunts (diverting vessels) make sure that they are almost entirely bypassed. Some of the blood entering the right atrium is shunted directly into the left atrium through the **foramen ovale** (fo-ra′men o-val′e), a flaplike opening in the interatrial septum. Blood that does manage to enter the right ventricle is pumped out to the pulmonary trunk, where it meets a second shunt, the **ductus arteriosus** (ar-ter″e-o′sus), a short vessel that connects the aorta and the pulmonary trunk. Because the collapsed lungs are a high-pressure area, blood tends to enter the systemic circulation through the ductus arteriosus. The aorta carries blood to the tissues of the fetal body and ultimately back to the placenta through the umbilical arteries.

At birth, or shortly after, the foramen ovale closes. Its remnant, the **fossa ovalis**, is visible in the right atrium (see Figure 11.3b). The ductus arteriosus collapses and is converted to the fibrous **ligamentum arteriosum** (lig″ah-men′tum ar-ter″e-o′sum) (see Figure 11.3a). As blood stops flowing through the umbilical vessels, they collapse, and the circulatory pattern converts to that of an adult.

## ⚖ Homeostatic Imbalance 11.10

Congenital heart defects account for about half of infant deaths resulting from all congenital defects. Environmental interferences, such as maternal infection and drugs ingested during the first 3 months of pregnancy (when the embryonic heart is forming), seem to be the major causes of such problems. Congenital heart defects may include a ductus arteriosus that does not close, septal openings, and other structural abnormalities of the heart. Such problems can usually be corrected surgically. ____ ✚

In the absence of congenital heart problems, the heart usually functions smoothly throughout a long lifetime for most people. Homeostatic mechanisms are so effective that we rarely are aware of when the heart is working harder. The heart will hypertrophy and its cardiac output will increase substantially if we exercise regularly and aerobically (that is, vigorously enough to force it to beat at a higher-than-normal rate for extended periods of time). The heart becomes not only a more powerful pump but also a more efficient one: pulse rate and blood pressure decrease. An added benefit of aerobic exercise is that it clears fatty deposits from the blood vessel walls, helping to slow the progress of atherosclerosis. However, let's raise a

*(Text continues on page 418.)*

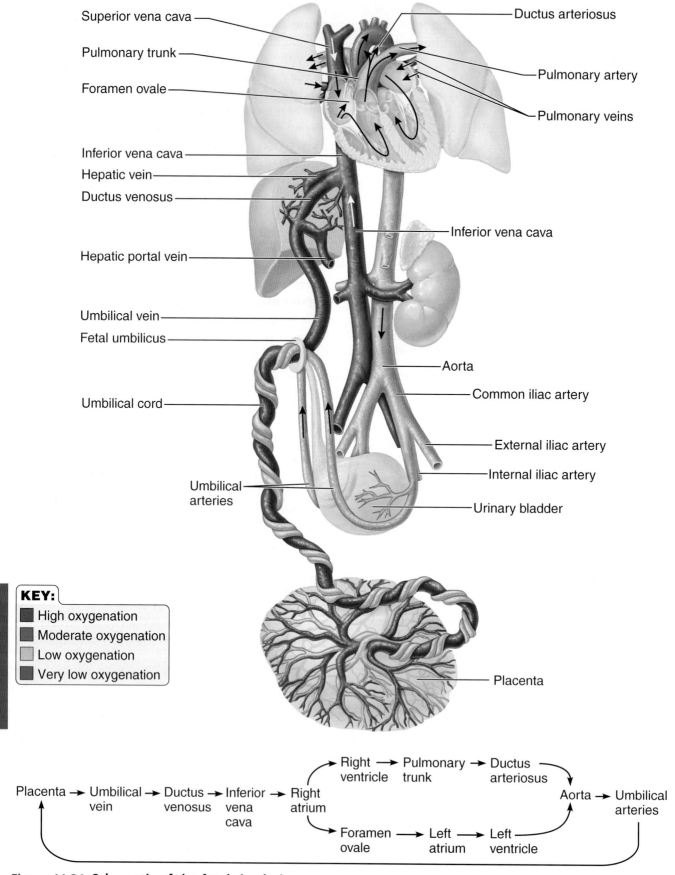

**KEY:**
- ■ High oxygenation
- ■ Moderate oxygenation
- ■ Low oxygenation
- ■ Very low oxygenation

Labels:
Superior vena cava
Pulmonary trunk
Foramen ovale
Inferior vena cava
Hepatic vein
Ductus venosus
Hepatic portal vein
Umbilical vein
Fetal umbilicus
Umbilical cord
Umbilical arteries
Ductus arteriosus
Pulmonary artery
Pulmonary veins
Inferior vena cava
Aorta
Common iliac artery
External iliac artery
Internal iliac artery
Urinary bladder
Placenta

Placenta → Umbilical vein → Ductus venosus → Inferior vena cava → Right atrium → Right ventricle → Pulmonary trunk → Ductus arteriosus → Aorta → Umbilical arteries

Right atrium → Foramen ovale → Left atrium → Left ventricle → Aorta

**Figure 11.24 Schematic of the fetal circulation.**

# Homeostatic Relationships between the **Cardiovascular System** and Other Body Systems

### Endocrine System

- The cardiovascular system delivers oxygen and nutrients; carries away wastes; blood serves as a transport vehicle for hormones
- Several hormones influence blood pressure (epinephrine, ANP, thyroxine, ADH); estrogen maintains vascular health in women

### Lymphatic System/Immunity

- The cardiovascular system delivers oxygen and nutrients to lymphoid organs, which house immune cells; transports lymphocytes and antibodies; carries away wastes
- The lymphatic system picks up leaked fluid and plasma proteins and returns them to the cardiovascular system; its immune cells protect cardiovascular organs from specific pathogens

### Digestive System

- The cardiovascular system delivers oxygen and nutrients; carries away wastes
- The digestive system provides nutrients to the blood including iron and B vitamins essential for RBC (and hemoglobin) formation

### Urinary System

- The cardiovascular system delivers oxygen and nutrients; carries away wastes; blood pressure maintains kidney function
- The urinary system helps regulate blood volume and pressure by altering urine volume and releasing renin

### Muscular System

- The cardiovascular system delivers oxygen and nutrients; carries away wastes
- Aerobic exercise enhances cardiovascular efficiency and helps prevent arteriosclerosis; the muscle "pump" aids venous return

### Nervous System

- The cardiovascular system delivers oxygen and nutrients; removes wastes
- ANS regulates cardiac rate and force; sympathetic division maintains blood pressure and controls blood distribution according to need

### Respiratory System

- The cardiovascular system delivers oxygen and nutrients; carries away wastes
- The respiratory system carries out gas exchange: loads oxygen and unloads carbon dioxide from the blood; respiratory "pump" aids venous return

## Cardiovascular System

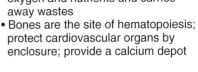

### Reproductive System

- The cardiovascular system delivers oxygen and nutrients; carries away wastes
- Estrogen maintains vascular health in women

### Integumentary System

- The cardiovascular system delivers oxygen and nutrients; carries away wastes
- The skin's blood vessels provide an important blood reservoir and a site for heat loss from body

### Skeletal System

- The cardiovascular system delivers oxygen and nutrients and carries away wastes
- Bones are the site of hematopoiesis; protect cardiovascular organs by enclosure; provide a calcium depot

caution flag here: The once-a-month or once-a-year tennis player or snow shoveler has not built up this type of heart endurance and strength. When such an individual pushes his or her heart too much, it may not be able to cope with the sudden demand. This is why many "weekend athletes" are myocardial infarction (heart attack) victims.

As we get older, more and more signs of cardiovascular system disturbances start to appear. In some people, the venous valves weaken, and purple, snakelike varicose veins appear. Not everyone has varicose veins, but we all have progressive atherosclerosis. Some say the process begins at birth, and even children's arteries show atherosclerotic plaques. There's an old saying, "You are only as old as your arteries," referring to this degenerative process. The gradual loss in elasticity of the blood vessels leads to hypertension and hypertensive heart disease. The insidious filling of the blood vessels with fatty, calcified deposits leads most commonly to **coronary artery disease**. Also, the roughening of the vessel walls encourages thrombus formation (see Chapter 10). At least 30 percent of the population in the United States has hypertension by the age of 50, and cardiovascular disease causes more than one-half of the deaths in people over age 65. Although the aging process itself contributes to changes in the walls of the blood vessels that can lead to strokes or myocardial infarctions, most researchers feel that diet, not aging, is the single most important contributing factor to cardiovascular diseases. There is some agreement that the risk is lowered if people eat less animal fat, cholesterol, and salt. Other recommendations include avoiding stress, eliminating cigarette smoking, and taking part in a regular, moderate exercise program.

# Summary

### The Heart (pp. 383–396)

1. The heart, located in the thorax, is flanked laterally by the lungs and enclosed in a multi-layered pericardium.

2. The bulk of the heart wall (myocardium) is composed of cardiac muscle. The heart has four hollow chambers—two atria (receiving chambers) and two ventricles (discharging chambers), each lined with endocardium. (The heart is divided longitudinally by a septum.)

3. The heart functions as a double pump. The right ventricle is the pulmonary pump (right ventricle to lungs to left atrium). The left ventricle is the systemic pump (left ventricle to body tissues to right atrium).

4. Four valves prevent backflow of blood in the heart. The AV valves (mitral, or bicuspid, and tricuspid) prevent backflow into the atria when the ventricles are contracting. The semilunar valves (pulmonary and aortic) prevent backflow into the ventricles when the heart is relaxing. The valves open and close in response to pressure changes in the heart.

5. The myocardium is nourished by the coronary circulation, which branches off the aorta and consists of the right and left coronary arteries and their branches, and is drained by the cardiac veins and the coronary sinus.

6. Cardiac muscle is able to initiate its own contraction in a regular way, but its rate is influenced by both intrinsic and extrinsic factors. The intrinsic conduction system increases the rate of heart contraction and ensures that the heart beats as a unit. The SA node is the heart's pacemaker. Extrinsic factors include neural and hormonal stimuli.

 Complete an interactive tutorial: Mastering A&P®> Study Area > Interactive Physiology > Cardiovascular System > Electrical Activity of the Heart

7. The time and events occurring from one heartbeat to the next are the cardiac cycle.

8. As the heart beats, sounds resulting from the closing of the valves ("lub-dup") can be heard. Faulty valves reduce the efficiency of the heart as a pump and result in abnormal heart sounds (murmurs).

 Complete an interactive tutorial: Mastering A&P®> Study Area > Interactive Physiology > Cardiovascular System > Cardiac Cycle

9. Cardiac output, the amount of blood pumped out by each ventricle in one minute, is the product of heart rate (HR) × stroke volume (SV). SV is the amount of blood ejected by a ventricle with each beat.

 Complete an interactive tutorial: Mastering A&P®> Study Area > Interactive Physiology > Cardiovascular System > Cardiac Output

10. SV rises or falls with the volume of venous return. HR is influenced by the nerves of the autonomic nervous system, drugs (and other chemicals), and ion levels in the blood.

## Blood Vessels (pp. 396–415)

1. Arteries, which transport blood away from the heart, and veins, which carry blood back to the heart, are conducting vessels. Only capillaries play a role in actual exchanges with tissue cells.

2. Except for capillaries, blood vessels are composed of three tunics: The tunica intima forms a friction-reducing lining for the vessel. The tunica media is the bulky middle layer of muscle and elastic tissue. The tunica externa is the protective, outermost connective tissue layer. Capillary walls are formed of the tunica intima only.

3. Artery walls are thick and strong to withstand blood pressure fluctuations. They expand and recoil as the heart beats. Vein walls are thinner, their lumens are larger, and they are equipped with valves. These modifications reflect the low pressure of the blood flowing in veins.

4. Capillary beds have two types of vessels—a vascular shunt and true capillaries. Blood flow into true capillaries is controlled by precapillary sphincters. Exchanges with tissue cells occur across the walls of the true capillaries. When the precapillary sphincters are closed, blood flows directly through the vascular shunt.

5. Varicose veins, a structural defect due to incompetent valves, is a common vascular problem, especially in people who are obese or who stand for long hours. It is a predisposing factor for thrombophlebitis.

6. All the major arteries of the systemic circulation are branches of the aorta, which leaves the left ventricle. They branch into smaller arteries and then into the arterioles, which feed the capillary beds of the body tissues. (For the names and locations of the systemic arteries, see pp. 399–401.)

7. The major veins of the systemic circulation ultimately converge on one of the venae cavae. All veins superior to the diaphragm drain into the superior vena cava, and those inferior to the diaphragm drain into the inferior vena cava. Both venae cavae enter the right atrium of the heart. (See pp. 400–403 for the names and locations of the systemic veins.)

8. The arterial circulation of the brain is formed by branches of paired vertebral and internal carotid arteries. The cerebral arterial circle provides alternate routes for blood flow in case of a blockage in the brain's arterial supply.

9. The hepatic portal circulation is formed by veins draining the digestive organs, which empty into the hepatic portal vein. The hepatic portal vein carries the nutrient-rich blood to the liver, where it is processed before the blood is allowed to enter the systemic circulation.

10. The pulse is the alternating expansion and recoil of a blood vessel wall (the pressure wave) that occurs as the heart beats. It may be felt easily over any superficial artery; such sites are called pressure points.

11. Blood pressure is the pressure that blood exerts on the walls of the blood vessels. It is the force that causes blood to flow down its pressure gradient in the blood vessels. It is high in the arteries, lower in the capillaries, and lowest in the right atrium. Both systolic and diastolic pressures are recorded when measuring blood pressure.

12. Arterial blood pressure is directly influenced by heart activity (increased heart rate leads to increased blood pressure) and by resistance to blood flow. The most important factors increasing the peripheral resistance are a decrease in the diameter or stretchiness of the arteries and arterioles and an increase in blood viscosity.

13. Many factors influence blood pressure, including the activity of the sympathetic nerves and kidneys, drugs, and diet.

 Complete an interactive tutorial: Mastering A&P®> Study Area > Interactive Physiology > Cardiovascular System > Factors that Affect Blood Pressure

14. Hypertension, which reflects an increase in peripheral resistance, strains the heart and damages blood vessels. In most cases, the precise cause is unknown.

15. Substances move to and from the blood and tissue cells through capillary walls. Some substances are transported in vesicles, but most move by diffusion—directly through the endothelial cell

11

plasma membranes, through intercellular clefts, or through fenestrations. Fluid is forced from the bloodstream by blood pressure and drawn back into the blood by osmotic pressure.

## Developmental Aspects of the Cardiovascular System (pp. 415–418)

1. The heart begins as a tubelike structure that is beating and pumping blood by the fourth week of embryonic development.

2. The fetal circulation is a temporary circulation seen only in the fetus. It consists primarily of three special vessels: the single umbilical vein that carries nutrient- and oxygen-laden blood to the fetus from the placenta, and the two umbilical arteries that carry carbon dioxide and waste-laden blood from the fetus to the placenta. Shunts bypassing the lungs and liver are also present.

3. Congenital heart defects account for half of all infant deaths resulting from congenital problems.

4. Arteriosclerosis is an expected consequence of aging. Gradual loss of elasticity in the arteries leads to hypertension and hypertensive heart disease, and clogging of the vessels with fatty substances leads to coronary artery disease and stroke. Cardiovascular disease is an important cause of death in individuals over age 65.

5. A healthy diet low in fat, cholesterol, and salt; stopping smoking; and regular aerobic exercise may help to reverse the atherosclerotic process and prolong life.

# Review Questions

Access additional practice questions using your smartphone, tablet, or computer: Mastering A&P®> Study Area > Practice Tests & Quizzes

## Multiple Choice

*More than one choice may apply.*

1. Both venae cavae deliver oxygen-poor blood from the body to the
   a. right ventricle.
   b. left ventricle.
   c. right atrium.
   d. left atrium.

2. Given a volume of 150 ml at the end of diastole, a volume of 50 ml at the end of systole, and a heart rate of 60 bpm, the cardiac output is
   a. 600 ml/min.
   b. 6 liters/min.
   c. 1200 ml/min.
   d. 3 liters/min.

3. Which of the following depolarizes next after the AV node?
   a. Atrial myocardium
   b. Ventricular myocardium
   c. Bundle branches
   d. AV bundle

4. During atrial diastole,
   a. the atrial pressure exceeds the ventricular pressure.
   b. blood from the pulmonary and systemic circuits fills atria.
   c. the semilunar valves are forced open.
   d. the SA node depolarizes.

5. Atrial repolarization coincides in time with the
   a. P wave.
   b. T wave.
   c. QRS wave.
   d. P-Q interval.

6. Soon after the onset of ventricular systole, the
   a. AV valves close.
   b. semilunar valves open.
   c. first heart sound is heard.
   d. aortic pressure increases.

7. To count the heart rate for an apical pulse, you would place your stethoscope approximately at the level of the
   a. third intercostal space.
   b. fourth intercostal space.
   c. fifth intercostal space.
   d. sixth intercostal space.

8. In comparing a parallel artery and vein, you would find that
    a. the artery wall is thicker.
    b. the artery diameter is greater.
    c. the artery lumen is smaller.
    d. the artery endothelium is thicker.

9. Which of these vessels is bilaterally symmetrical (i.e., one vessel of the pair occurs on each side of the body)?
    a. Internal carotid artery
    b. Brachiocephalic trunk
    c. Azygos vein
    d. Renal vein

10. A stroke that occludes a posterior cerebral artery will most likely affect
    a. hearing.
    b. vision.
    c. smell.
    d. higher thought processes.

11. Most of the small intestine is supplied by the
    a. superior mesenteric artery.
    b. inferior mesenteric artery.
    c. lumbar arteries.
    d. celiac trunk.

12. Which layer of the artery wall thickens most in atherosclerosis?
    a. Tunica media
    b. Tunica intima
    c. Tunica adventitia
    d. Tunica externa

13. Which of the following are associated with aging?
    a. Increasing blood pressure
    b. Weakening of venous valves
    c. Arteriosclerosis
    d. Collapse of the ductus arteriosus

14. What effects do the kidneys have on blood pressure?
    a. They decrease pressure by causing vasodilation.
    b. They increase pressure by increasing blood volume.
    c. They decrease pressure by decreasing blood volume.
    d. They increase pressure by decreasing blood volume.

15. The most external part of the pericardium is the
    a. parietal layer of serous pericardium.
    b. fibrous pericardium.
    c. visceral layer of serous pericardium.
    d. epicardium.

16. Which of the following are features of capillaries?
    a. They allow direct diffusion of nutrients.
    b. They allow transport by vesicles.
    c. They have large pores to allow transport.
    d. They rely on valves to allow transport.

17. How many cusps does the semilunar valve at the base of the aorta have?
    a. Two
    b. Three
    c. Four
    d. Six

18. Which layer of the heart wall is an endothelium?
    a. Endocardium
    b. Myocardium
    c. Epicardium
    d. Pericardium

## Short Answer Essay

19. Draw a diagram of the heart showing the three layers composing its wall and its four chambers. Label each. Show where the AV and semilunar valves are, and name them. Show and label all blood vessels entering and leaving the heart chambers.

20. Trace one drop of blood from the time it enters the right atrium of the heart until it enters the left atrium. What is this circuit called?

21. What is the function of the fluid that fills the pericardial sac?

22. Define *systole* and *diastole*.

23. Define *stroke volume* and *cardiac cycle*.

24. How does the heart's ability to contract differ from that of other muscles of the body?

25. Name the elements of the intrinsic conduction system, *in order*, beginning with the pacemaker.

26. Which system is responsible for decreasing heart rate after a crisis has passed?

**11**

27. What is congestive heart failure? What is the difference between pulmonary congestion and peripheral congestion?

28. Describe the structure of capillary walls.

29. How is blood supply to the myocardium organized?

30. Name three factors that are important in promoting venous return.

31. Arteries are often described as vessels that carry oxygen-rich blood, and veins are said to carry oxygen-poor (carbon dioxide–rich) blood. Name two sets of exceptions to this rule that were discussed in this chapter.

32. Trace a drop of blood from the left ventricle of the heart to the wrist of the right hand and back to the heart. Now trace it to the dorsum of the right foot and back to the right heart.

33. What is the function of the hepatic portal circulation? In what way is a portal circulation a "strange" circulation?

34. In a fetus, the liver and lungs are almost entirely bypassed by blood. Why is this? Name the vessel that bypasses the liver. Name two lung bypasses. Three vessels travel in the umbilical cord; which of these carries oxygen- and nutrient-rich blood?

35. Define *pulse*. Palpate your pulse. Which pulse point did you use?

36. Which artery is palpated at the inner elbow? At the dorsum of the foot?

37. Define *systolic pressure* and *diastolic pressure*.

38. Which part of the nervous system plays an important role in regulating blood pressure? By which action does it increase blood pressure?

39. In which position—sitting, lying down, or standing—is the blood pressure normally highest? Lowest? Explain why.

40. Microcirculation is the flow of blood through a capillary bed consisting of a vascular shunt and true capillaries. What is the difference between a vascular shunt and a true capillary?

41. What are varicose veins? What factors seem to promote their formation?

42. Explain why blood flow in arteries is pulsatile and blood flow in veins is not.

43. What is the relationship between cross-sectional area of a blood vessel and velocity (speed) of blood flow in that vessel?

44. Which type of blood vessel is most important in regulating vascular resistance, and how does it achieve this?

45. What is the foramen ovale, and what is its function?

## Critical Thinking and Clinical Application Questions

46. John is a 30-year-old man who is overweight and smokes. He has been diagnosed with *hypertension* and *arteriosclerosis*. Define each of these conditions. How are they often related? Why is hypertension called the "silent killer"? Name three changes in your lifestyle that might help prevent cardiovascular disease in your old age.

47. Following a viral infection, George has been feeling a sharp stabbing pain in the retrosternal region. The pain gets worse when he breathes deeply and also when he lies down. A physical examination reveals that he has pericarditis. Can you explain why pericarditis causes this severe pain?

48. Hannah, a 14-year-old girl undergoing a physical examination before being admitted to summer camp, was found to have a loud heart murmur at the second intercostal space on the left side of the sternum. The murmur takes the form of a swishing sound with no high-pitched whistle. What, exactly, is producing the murmur?

49. Colin is in the hospital to have a coronary angiography so that a possible coronary stenosis can be detected. A catheter is inserted through the right radial artery. What route must the catheter travel to reach the coronary arteries?

50. Stuart, a 49-year-old man, has been diagnosed with coronary artery disease. He needs a bypass graft in

his circumflex artery and right coronary artery. Which parts of his heart muscle are most likely to see an improved blood flow after the procedure?

51. Mr. Grimaldi was previously diagnosed as having a posterior pituitary tumor that causes hypersecretion of ADH. He comes to the clinic regularly to have his blood pressure checked. Would you expect his blood pressure to be chronically elevated or depressed? Why?

52. Grandma tells Hailey not to swim for 30 minutes after eating. Explain why taking a vigorous swim immediately after lunch is more likely to cause indigestion than cramping of your muscles.

53. The guards at the royal palace in London stand at attention while on duty. On a very hot day, it is not unusual for one (or more) to become lightheaded and faint. Explain this phenomenon.

11

# 12

# The Lymphatic System and Body Defenses

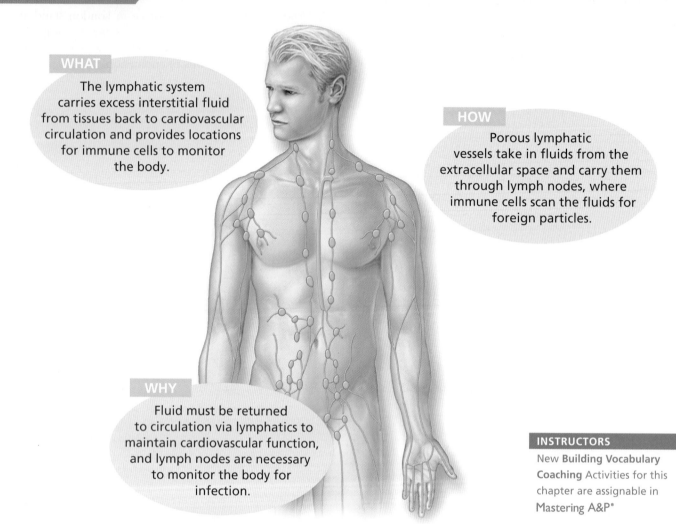

**WHAT**

The lymphatic system carries excess interstitial fluid from tissues back to cardiovascular circulation and provides locations for immune cells to monitor the body.

**HOW**

Porous lymphatic vessels take in fluids from the extracellular space and carry them through lymph nodes, where immune cells scan the fluids for foreign particles.

**WHY**

Fluid must be returned to circulation via lymphatics to maintain cardiovascular function, and lymph nodes are necessary to monitor the body for infection.

**INSTRUCTORS**

New **Building Vocabulary Coaching** Activities for this chapter are assignable in Mastering A&P®

## PART I: THE LYMPHATIC SYSTEM

→ **Learning Objectives**

☐ **Explain how the lymphatic system is functionally related to the cardiovascular system and the immune defenses.**

☐ **Name the two major types of structures composing the lymphatic system.**

☐ **Describe the source of lymph, and explain its formation and transport.**

When we mentally tick off the names of the body's organ systems, the lymphatic (lim-fat′ik) system is probably not the first that comes to mind. Yet without this system, our cardiovascular system would stop working, and our immune system would be hopelessly impaired. The **lymphatic system** consists of two semi-independent parts: (1) a meandering network of *lymphatic vessels* and (2) various *lymphoid tissues* and *organs* scattered throughout the body. The lymphatic vessels transport fluids that have escaped from the blood back to the cardiovascular system. The lymphoid tissues and organs house *phagocytic cells* and *lymphocytes*, which play essential roles in body defense and resistance to disease.

# Lymphatic Vessels

As blood circulates through the body, nutrients, wastes, and gases are exchanged between the blood and the interstitial fluid.

### CONCEPTLINK

Recall that the hydrostatic and osmotic pressures operating at capillary beds force fluid out of the blood at the arterial ends of the beds ("upstream") and cause most of the expelled fluid to be reabsorbed at the venous ends ("downstream") (Chapter 11, p. 414). ⬅

The fluid that remains behind in the tissue spaces, as much as 3 L daily, becomes part of the interstitial fluid. This excess tissue fluid, as well as any plasma proteins that escape from the blood, must be carried back to the blood if the vascular system is to have sufficient blood volume to continue to operate properly. If it does not, fluid accumulates in the tissues, producing **edema**, or swelling. Excessive edema impairs the ability of cells to make exchanges with the interstitial fluid and ultimately the blood. The function of the **lymphatic vessels** is to form an elaborate drainage system that picks up this excess interstitial fluid, now called **lymph** (*lymph* = clear water), and returns it to the blood **(Figure 12.1)**.

The lymphatic vessels, also called **lymphatics**, form a one-way system, and lymph flows only toward the heart. The microscopic **lymph capillaries** weave between the tissue cells and blood capillaries in the loose connective tissues of the body **(Figure 12.2a**, p. 426) and absorb the leaked fluid. Although similar to blood capillaries, lymphatic capillaries are so remarkably permeable that they were once thought to be open at one end like a straw. Not so—instead, the edges of the endothelial cells forming their walls loosely overlap one another, forming flaplike *minivalves* (Figure 12.2b) that act as one-way swinging doors. The flaps, anchored by fine collagen fibers to surrounding structures, gape open when the fluid pressure is higher in the interstitial space, allowing fluid to enter the lymphatic capillary. However, when the pressure is higher inside the lymphatic vessels, the endothelial cell flaps are forced together, preventing the lymph

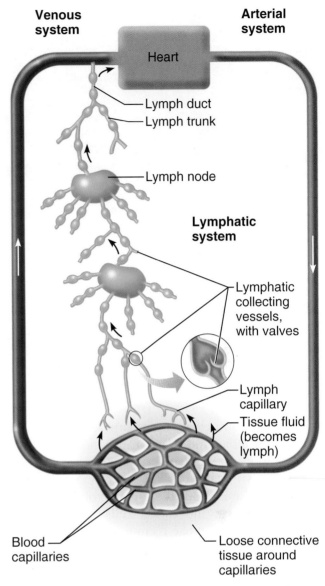

**Figure 12.1 Relationship of lymphatic vessels to blood vessels.** Beginning at the bottom of this figure, we see that lymph, which begins as tissue fluid derived from blood capillaries, enters the lymph capillaries, travels through the lymphatic vessels and lymph nodes, and enters the bloodstream via the great veins at the root of the neck.

from leaking back out and forcing it along the vessel.

### CONCEPTLINK

This is very similar to the way that valves in veins work to ensure blood returns to the heart, despite being under low pressure (Chapter 11, p. 398). ⬅

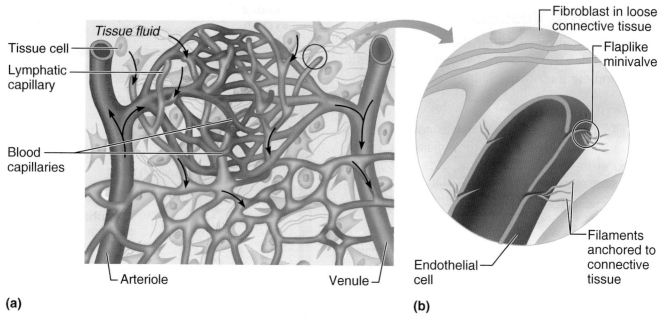

**Figure 12.2 Special structural features of lymphatic capillaries. (a)** Structural relationship between blood capillaries and lymph capillaries. Black arrows indicate direction of fluid movement. **(b)** Lymph capillaries begin as blind-ended tubes. The endothelial cells forming their walls overlap one another, forming flaplike minivalves.

Proteins, and even larger particles such as cell debris, bacteria, and viruses, are normally prevented from entering blood capillaries, but they enter the lymphatic capillaries easily, particularly in inflamed areas. But there is a problem here—bacteria, viruses, and cancer cells that enter the lymphatics can then use them to travel throughout the body. This dilemma is partly resolved by the fact that white blood cells (WBCs) can also travel in lymph and lymph takes "detours" through the lymph nodes, where it is cleansed of debris and "examined" by cells of the immune system, as we explain in more detail shortly.

Lymph is transported from the lymph capillaries through successively larger lymphatic vessels, referred to as **lymphatic collecting vessels**, until it is finally returned to the venous system through two large ducts in the thoracic region. The **right lymphatic duct** drains lymph from the right arm and the right side of the head and thorax. The large **thoracic duct** receives lymph from the rest of the body **(Figure 12.3)**. Both ducts empty the lymph into the subclavian vein on their own side of the body.

Like veins of the cardiovascular system, lymphatic vessels are thin walled, and the larger ones have valves. The lymphatic system is a low-pressure, pumpless system. Lymph is transported by the same mechanisms that aid return of venous blood—the milking action of the skeletal muscles and pressure changes in the thorax during breathing (that is, the muscular and respiratory "pumps"). In addition, smooth muscle in the walls of the larger lymphatics contracts rhythmically, helping to "pump" the lymph along.

### Did You Get It?

1. What is the function of lymphatic vessels?
2. How do lymphatic capillaries and blood capillaries differ structurally from each other?

For answers, see Appendix A.

## Lymph Nodes

→ **Learning Objective**

☐ Describe the function(s) of lymph nodes, tonsils, the thymus, Peyer's patches, and the spleen.

In addition to returning tissue fluid to circulation, the lymphatic system plays a major role in immunity. Cells in **lymph nodes** in particular help protect the body by removing foreign material such as bacteria and tumor cells from the lymphatic stream and by producing lymphocytes that function in the immune response.

**Q:** *What would be the consequence of blockage of the thoracic duct?*

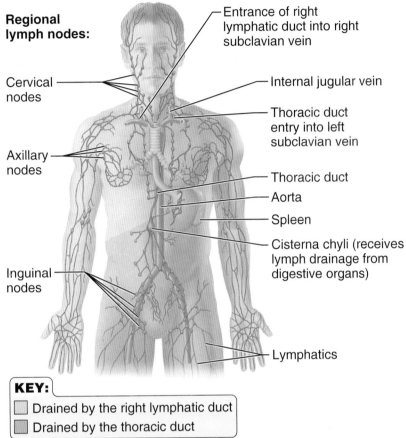

**Regional lymph nodes:**

Cervical nodes

Axillary nodes

Inguinal nodes

Entrance of right lymphatic duct into right subclavian vein

Internal jugular vein

Thoracic duct entry into left subclavian vein

Thoracic duct

Aorta

Spleen

Cisterna chyli (receives lymph drainage from digestive organs)

Lymphatics

**KEY:**
- ▢ Drained by the right lymphatic duct
- ▢ Drained by the thoracic duct

**Figure 12.3 Distribution of lymphatic vessels and lymph nodes.** Note that until recently, scientists thought that the CNS had no lymphatics. However, we now know the CNS is drained by lymphatics near the venous dural sinuses.

As lymph is transported toward the heart, it is filtered through the thousands of lymph nodes that cluster along the lymphatic vessels (see Figure 12.1). Particularly large clusters are found in the inguinal, axillary, and cervical regions of the body (see Figure 12.3). Within the lymph nodes are **macrophages** (mak′ro-fāj-ez), which engulf and destroy bacteria, viruses, and other foreign substances in the lymph before it is returned to the blood. Collections of **lymphocytes** (a type of white blood cell, covered in Chapter 10) are also strategically located in the lymph nodes and

respond to foreign substances in the lymphatic stream. Although we are not usually aware of the protective nature of the lymph nodes, most of us have had "swollen glands" during an active infection. What are actually swollen in these cases are not glands at all but rather lymph nodes (the term *swollen glands* is a misnomer), which swell as a result of the trapping function of the nodes.

Lymph nodes vary in shape and size, but most are kidney-shaped, about 1 centimeter long, and "buried" in the connective tissue that surrounds them. Each node is surrounded by a fibrous *capsule* from which connective tissue strands called *trabeculae* (trah-bek′yu-le) extend inward to divide the node into a number of compartments (**Figure 12.4**, p. 428). The internal framework is a

**A:** Edema in the areas that drain into the thoracic duct.

12

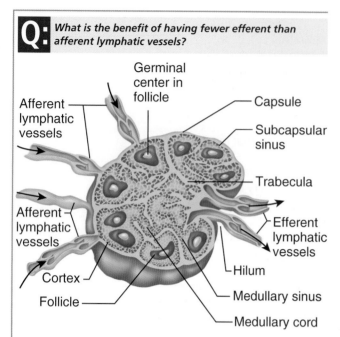

Germinal center in follicle

Afferent lymphatic vessels

Capsule

Subcapsular sinus

Trabecula

Afferent lymphatic vessels

Efferent lymphatic vessels

Cortex

Hilum

Follicle

Medullary sinus

Medullary cord

**Figure 12.4 Structure of a lymph node.** Longitudinal section of a lymph node and associated lymphatics. Notice that several afferent lymphatics enter the node, whereas fewer efferent lymphatics exit at its hilum.

network of soft reticular connective tissue, like "cellular bleachers," that provides a place for the continually changing population of lymphocytes to "sit" as they monitor the lymphatic stream. Lymphocytes arise from the red bone marrow but then migrate to the lymph nodes and other lymphoid organs, where they reproduce further (see Chapter 10).

The outer part of the node, its **cortex**, contains collections of lymphocytes called **follicles**, many of which have dark-staining centers called **germinal centers**. These centers enlarge when specific B lymphocytes (the *B cells*) are generating daughter cells called *plasma cells*, which release *antibodies*. The rest of the cortical cells are lymphocytes "in transit," the so-called *T cells* that circulate continuously between the blood, lymph nodes, and lymphatic stream, performing their

surveillance role. The *medullary cords* are inward extensions of cortical tissue that contain both B and T cells. Phagocytic macrophages are located in the central **medulla** of the lymph node.

Lymph enters the convex side of a lymph node through **afferent lymphatic vessels**. It then flows through a number of **sinuses** (spaces) that meander through the lymph node and finally exits from the node at its indented region, the **hilum** (hi'lum), via **efferent lymphatic vessels**. Because there are fewer efferent vessels draining the node than afferent vessels feeding it, the flow of lymph through the node is very slow, kind of like sand flowing through an hourglass. This allows time for the lymphocytes and macrophages to perform their protective functions.

### Homeostatic Imbalance 12.1

Lymph nodes help rid the body of infectious agents and cancer cells, but sometimes they are overwhelmed by the very agents they are trying to destroy. For example, when large numbers of bacteria or viruses are trapped in the nodes, the nodes become inflamed, swollen, and tender to the touch. The lymph nodes can also become secondary cancer sites, particularly in cancers that use lymphatic vessels to spread throughout the body. Swelling is temporary in lymph nodes fighting infection. In contrast, swelling persists or nodes enlarge over time if they contain cancer cells. _____**+**

## Other Lymphoid Organs

Lymph nodes are just one of the many types of **lymphoid organs** in the body. Others include the spleen, thymus, tonsils, Peyer's patches, and the appendix **(Figure 12.5)**, as well as bits of lymphoid tissue scattered in epithelial and connective tissues. All these organs have a predominance of reticular connective tissue and lymphocytes. Although all lymphoid organs have roles in protecting the body, only the lymph nodes filter lymph.

The **spleen** is a soft organ located in the left side of the abdominal cavity, just beneath the diaphragm, that curls around the anterolateral aspect of the stomach. Instead of filtering lymph, the spleen filters and cleanses blood of bacteria, viruses, and other debris. As with the other lymphoid organs, the spleen provides a site for lymphocyte proliferation and immune surveillance, but its most important function is to destroy worn-out red blood cells and return some of their breakdown products to the liver. For example, iron is used again for making hemoglobin, and the rest of the hemoglobin molecule is secreted in bile.

Other functions of the spleen include storing platelets and acting as a blood reservoir (like the liver). During hemorrhage, both the spleen and liver contract and empty their blood into circulation to help bring the blood volume back to its normal level. In the fetus, the spleen is an important hematopoietic (blood cell–forming) site, but as a rule the adult spleen produces only lymphocytes.

The **thymus**, which functions at peak level only during youth, is a lymphoid mass found in the anterior mediastinum overlying the heart.

### CONCEPT**LINK**

Remember that the thymus produces hormones, *thymosin* and others, that function in the programming of T lymphocytes so they can carry out their protective roles in the body (Chapter 9, p. 346). ←

The **tonsils** are small masses of lymphoid tissue deep to the mucosa surrounding the pharynx (throat). Their job is to trap and remove bacteria or other foreign pathogens entering the throat. They carry out this function so efficiently that sometimes they become congested with bacteria and become red, swollen, and sore, a condition called **tonsillitis**.

**Peyer's patches**, which resemble tonsils, are found in the wall of the distal small intestine. Lymphoid follicles are also located in the wall of the **appendix**, a tubelike offshoot of the proximal large intestine. The macrophages of Peyer's patches and the appendix are in an ideal position to capture and destroy harmful bacteria (always present in tremendous numbers in the intestine), thereby preventing them from penetrating the intestinal wall. Peyer's patches, the appendix, and the tonsils are part of the collection of small lymphoid tissues referred to as **mucosa-associated lymphoid tissue (MALT)**. Collectively, MALT acts as a sentinel to protect the upper respiratory and digestive tracts from the constant attacks of foreign matter entering those cavities.

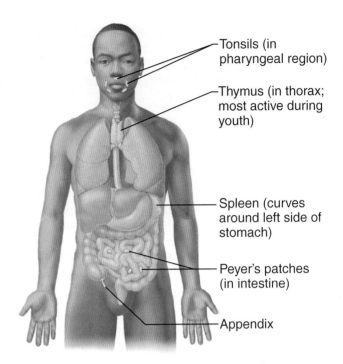

**Figure 12.5 Lymphoid organs.** Locations of the tonsils, spleen, thymus, Peyer's patches, and appendix.

### Did You Get It?

3. In which three regions of the body are the lymph nodes most densely located?
4. What anatomical characteristic ensures that lymph flows through the lymph nodes slowly?
5. Which lymphoid organ gets rid of aged red blood cells?
6. What is MALT?

**For answers, see Appendix A.**

## PART II: BODY DEFENSES

Every second of every day, an army of hostile microbes swarms on our skin and invades our inner passageways—yet we remain healthy most of the time. The body seems to have developed a single-minded approach toward such foes—if you're not with us, you're against us!

Two types of mechanisms defend the body against these tiny but mighty enemies: *innate* and *adaptive defense mechanisms* **(Figure 12.6)**. Together they make up the **immune system**. Although certain body organs (lymphoid organs and blood vessels) are intimately involved with the immune response, the immune system is a *functional system* rather than an organ system in an anatomical sense. Its "structures" are a variety

| The Immune System | | |
| --- | --- | --- |
| **Innate (nonspecific) defense mechanisms** | | **Adaptive (specific) defense mechanisms** |
| First line of defense | Second line of defense | Third line of defense |
| • Skin<br>• Mucous membranes<br>• Secretions of skin and mucous membranes | • Phagocytic cells<br>• Natural killer cells<br>• Antimicrobial proteins<br>• The inflammatory response<br>• Fever | • Lymphocytes<br>• Antibodies<br>• Macrophages and other antigen-presenting cells |

**Figure 12.6 An overview of the body's defenses.**

of molecules and trillions of immune cells that inhabit lymphoid tissues and organs and circulate in body fluids. The most important of the immune cells are *lymphocytes*, *dendritic cells*, and *macrophages*. Macrophages actually play an important role in both innate and adaptive mechanisms. The **innate defense system**, also called the **nonspecific defense system**, responds immediately to protect the body from all foreign substances, whatever they are. We are born with our innate defenses, which include intact skin and mucous membranes, the inflammatory response, and a number of proteins produced by body cells. These innate mechanisms reduce the workload of the adaptive defense mechanisms by generally preventing the entry and spread of microorganisms throughout the body.

The **adaptive defense system**, or **specific defense system**, fights invaders that get past the innate defenses by mounting an attack against one or more *particular* foreign substances.

When our immune system is operating effectively, it protects us from most bacteria, viruses, transplanted organs or grafts, and even our own cells that have "turned against" us (cancer cells). The immune system does this both directly, by cell attack, and indirectly, by releasing mobilizing chemicals and protective antibody molecules. The resulting highly specific resistance to disease is called **immunity** (*immun* = free).

Unlike the innate defenses, which are always prepared to defend the body, the adaptive system must first "meet," or be sensitized by, an initial

exposure to a foreign substance (antigen) before it can protect the body against the invader. Another important feature of the adaptive response is that it "remembers" which invaders it has fought. Nonetheless, what the adaptive system lacks in speed it makes up for in the precision of its counterattacks. Although we consider them separately, keep in mind that innate and adaptive defenses always work hand-in-hand to protect the body.

**Did You Get It?**

7. How do the innate and adaptive defenses differ?

For the answer, see Appendix A.

# Innate Body Defenses

Some innate resistance to disease is inherited. For instance, there are certain diseases that humans never get, such as some forms of tuberculosis that affect birds. Most often, however, the term *innate*, or *nonspecific*, *body defense* refers to the mechanical barriers that cover body surfaces and to the cells and chemicals that act on the initial battlefronts to protect the body from invading **pathogens** (harmful microorganisms). Table 12.1 provides a detailed summary of the innate defenses.

## Surface Membrane Barriers

→ **Learning Objective**

☐ Describe the protective functions of skin and mucous membranes.

The body's *first line of defense* against the invasion of disease-causing microorganisms includes the

Table **12.1**   **Summary of Innate (Nonspecific) Body Defenses**

| Category and associated elements | Protective mechanism |
|---|---|
| **Surface membrane barriers—first line of defense** | |
| Intact skin (epidermis) | Forms mechanical barrier that prevents entry of pathogens and other harmful substances into body. |
| • Acid mantle | Skin secretions make epidermal surface acidic, which inhibits bacterial growth; sebum also contains bacteria-killing chemicals. |
| • Keratin | Provides resistance against acids, alkalis, and bacterial enzymes. |
| Intact mucous membranes | Form mechanical barrier that prevents entry of pathogens. |
| • Mucus | Traps microorganisms in respiratory and digestive tracts. |
| • Nasal hairs | Filter and trap microorganisms and other airborne particles in nasal passages. |
| • Cilia | Propel debris-laden mucus away from lower respiratory passages. |
| • Gastric juice | Contains concentrated hydrochloric acid and protein-digesting enzymes that destroy pathogens in stomach. |
| • Acid mantle of vagina | Inhibits growth of bacteria and fungi in female reproductive tract. |
| • Lacrimal secretion (tears); saliva | Continuously lubricate and cleanse eyes (tears) and oral cavity (saliva); contain lysozyme, an enzyme that destroys microorganisms. |
| **Cellular and chemical defenses—second line of defense** | |
| Phagocytes | Engulf and destroy pathogens that breach surface membrane barriers; macrophages also contribute to immune response. |
| Natural killer cells | Promote cell lysis by direct cell attack against virus-infected or cancerous body cells; do not depend on specific antigen recognition. |
| Inflammatory response | Prevents spread of injurious agents to adjacent tissues, disposes of pathogens and dead tissue cells, and promotes tissue repair; releases chemical mediators that attract phagocytes (and immune cells) to the area. |
| Antimicrobial chemicals | |
| • Complement | Group of plasma proteins that lyses microorganisms, enhances phagocytosis by opsonization, and intensifies inflammatory response. |
| • Interferons | Proteins released by virus-infected cells that protect uninfected tissue cells from viral takeover; mobilize immune system. |
| • Fluids with acid pH | Normally acid pH inhibits bacterial growth; urine cleanses the lower urinary tract as it flushes from the body. |
| Fever | Systemic response triggered by pyrogens; high body temperature inhibits multiplication of bacteria and enhances body repair processes. |

12

*skin* and *mucous membranes*. As long as the skin is unbroken, its keratinized epidermis is a strong physical barrier to most microorganisms that swarm on the skin. Intact mucous membranes provide similar mechanical barriers within the body. Recall that mucous membranes line all body cavities open to the exterior. Besides serving as physical barriers, these membranes produce a variety of protective secretions:

- The acidic pH of skin secretions (called the *acid mantle*) and usually of urine (pH of 4.5 to 6) inhibits bacterial growth, and sebum contains chemicals that are toxic to bacteria. The vaginal secretions of adult females are also very acidic.

- Sticky mucus traps many microorganisms that enter digestive and respiratory passageways.

- The stomach mucosa secretes gastric juice containing hydrochloric acid and protein-digesting enzymes that both kill pathogens.

- Saliva and lacrimal secretions (tears) contain **lysozyme**, an enzyme that destroys bacteria.

Some mucosae also have structural modifications that fend off potential invaders. Mucus-coated hairs inside the nasal cavity trap inhaled particles, and the respiratory tract mucosa is ciliated. The cilia sweep dust- and bacteria-laden mucus (think "sticky trap") superiorly toward the mouth, preventing it from entering the lungs, where the warm, moist environment provides ideal conditions for bacterial growth.

Although the surface barriers are quite effective, some microbes may get through and they *are* broken from time to time by small nicks and cuts resulting, for example, from shaving or a paper cut. When this happens and microorganisms do invade deeper tissues, the internal innate mechanisms come into play.

## Internal Defenses: Cells and Chemicals

For its *second line of defense*, the body uses an enormous number of cells and chemicals. These defenses rely on the destructive powers of cells called *phagocytes* and *natural killer cells*, on the inflammatory response, and on a variety of chemical substances that kill pathogens and help repair tissue. Fever is another nonspecific protective response.

### Natural Killer Cells

→ **Learning Objective**

☐ Explain the role of natural killer cells.

**Natural killer (NK) cells** roam the body in blood and lymph. They are a unique group of aggressive lymphocytes that can lyse (burst) and kill cancer cells, virus-infected body cells, and some other nonspecific targets well before the adaptive arm of the immune system is enlisted in the fight. Natural killer cells can act spontaneously against *any* such target by recognizing certain sugars on the "intruder's" surface as well as its lack of certain "self" cell surface molecules. NK cells are not phagocytic. They attack the target cell's membrane and release lytic chemicals called *perforin* and *granzymes* (enzymes), which degrade target cell contents. Shortly thereafter, the target cell dies as its membrane and nucleus disintegrate. NK cells also release powerful inflammatory chemicals.

### Inflammatory Response

→ **Learning Objective**

☐ Describe the inflammatory process.

The **inflammatory response** is a nonspecific response that is triggered whenever body tissues are injured **(Figure 12.7)**. For example, it occurs in response to physical trauma, intense heat, irritating chemicals, and infection by viruses and bacteria. The four most common indicators, or *cardinal signs*, of acute inflammation are *redness*, *heat* (*inflamm* = set on fire), *pain*, and *swelling* (edema). It is easy to understand why these signs and symptoms occur, once you understand the events of the inflammatory response.

The inflammatory process begins with a chemical "alarm." When cells are damaged, they release inflammatory chemicals, including **histamine** and **kinins** (ki′ninz), that (1) cause blood vessels in the area to dilate, (2) make capillaries leaky, and (3) attract phagocytes and white blood cells to the area. (This third phenomenon is called **positive chemotaxis** because the cells are moving toward a high concentration of signaling molecules.) Dilation of the blood vessels increases the blood flow to the area, accounting for the redness and heat observed. Increased permeability of the capillaries allows fluid to leak from the blood into the

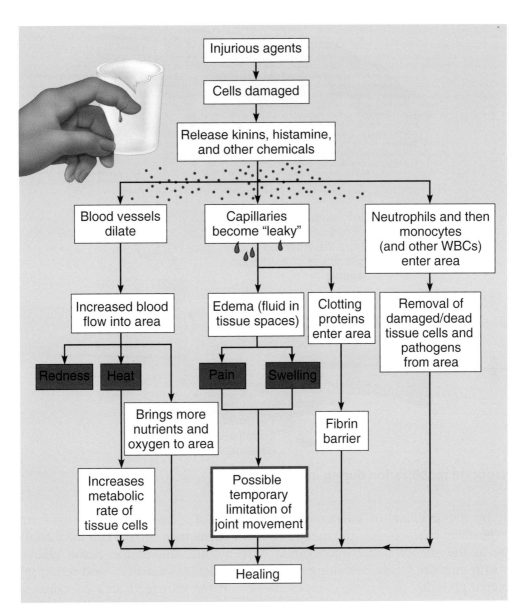

**Figure 12.7 Flowchart of inflammatory events.** The four cardinal signs (most common indicators) of acute inflammation are shown in red boxes. Limitation of joint movement (box outlined in red) occurs in some cases and is considered to be the fifth cardinal sign of acute inflammation.

tissue spaces, causing local edema (swelling) that also activates pain receptors in the area as pressure in the tissues builds. If the swollen, painful area is a joint, its function (movement) may be impaired temporarily. This forces the injured part to rest, which aids healing. Some authorities consider limitation of joint movement to be the fifth cardinal sign of inflammation.

The inflammatory response prevents the spread of damaging agents to nearby tissues, disposes of cell debris and pathogens, and sets the stage for repair. Let's look at how it accomplishes these tasks by mobilizing phagocytes (**Figure 12.8**, p. 434). Once the inflammatory process has begun:

① Neutrophils, responding to a gradient of diffusing inflammatory chemicals, enter the blood from the bone marrow and roll along the blood vessel walls, following the "scent."

② At the point where the chemical signal is the strongest, they flatten out and squeeze through the capillary walls, a process called **diapedesis**.

12

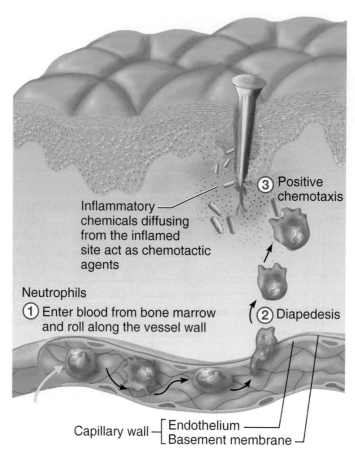

**Figure 12.8 Phagocyte mobilization during inflammation.**

③ Still drawn by the gradient of inflammatory chemicals (positive chemotaxis), the neutrophils gather at the site of tissue injury, and within an hour, they are busily devouring any foreign material present.

As the counterattack continues, monocytes follow neutrophils into the inflamed area. Monocytes are fairly poor phagocytes, but within 12 hours after entering the tissues they become macrophages with insatiable appetites. The macrophages continue to wage the battle, replacing the short-lived neutrophils at the site of damage. Macrophages are the central actors in the final disposal of cell debris as the inflammation subsides.

Besides phagocytosis, other protective events are also occurring at the inflamed site. Clotting proteins, leaked into the area from the blood, are activated and begin to wall off the damaged area with fibrin to prevent the spread of harmful agents to neighboring tissues. The fibrin mesh (think of a small fish net) also forms scaffolding for permanent repair. The local heat increases the metabolic rate of the tissue cells, speeding up their defensive actions and repair processes.

If the damaged area contains pathogens that have previously invaded the body, the third line of defense also comes into play—the adaptive response mediated by lymphocytes.

## Homeostatic Imbalance 12.2

In severely infected areas, the battle takes a considerable toll on both sides, and creamy, yellow pus may form in the wound. **Pus** is a mixture of dead or dying neutrophils, broken-down tissue cells, and living and dead pathogens. If the inflammatory mechanism fails to fully clear the area of debris, the sac of pus may become walled off, forming an *abscess*. Surgical drainage of abscesses is often necessary before healing can occur.

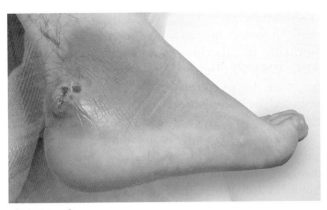

Abscess on foot.

## Did You Get It?

8. What are the four common indicators of inflammation?

**For the answer, see Appendix A.**

## Phagocytes

### → Learning Objective

☐ **Explain the importance of phagocytes.**

Pathogens that make it through the mechanical barriers are confronted by **phagocytes** (fag′o-sītz″; *phago* = eat) in nearly every body organ. A phagocyte, such as a *macrophage* or *neutrophil*, engulfs a foreign particle by the process of *phagocytosis* (**Figure 12.9**). (Recall what you learned in Chapter 3.) Flowing cytoplasmic extensions bind to the particle and then pull it inside, forming a phagocytic vesicle. The vesicle then fuses with a *lysosome*, where enzymes digest its contents.

### Antimicrobial Proteins

### → Learning Objective

☐ **Name several antimicrobial substances produced by the body that act in innate body defense.**

A variety of **antimicrobial proteins** enhance the innate defenses either by attacking microorganisms directly or by hindering their ability to reproduce. The most important of these are *complement* and *interferon*.

***Complement*** The term **complement** refers to a group of at least 20 plasma proteins that circulate in the blood in an inactive state, much like inactive

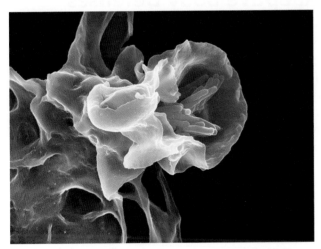

**(a) A macrophage (purple) uses its cytoplasmic extensions to ingest bacillus-shaped bacteria (pink) by phagocytosis.** Scanning electron micrograph.

**Figure 12.9 Phagocytosis by a macrophage.**

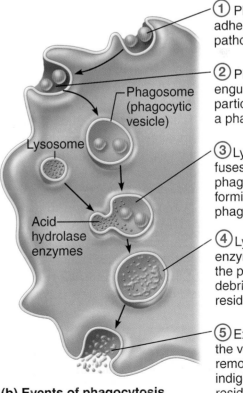

① Phagocyte adheres to pathogens.

② Phagocyte engulfs the particles, forming a phagosome.

Phagosome (phagocytic vesicle)

Lysosome

③ Lysosome fuses with the phagocytic vesicle, forming a phagolysosome.

Acid hydrolase enzymes

④ Lysosomal enzymes digest the pathogens or debris, leaving a residual body.

⑤ Exocytosis of the vesicle removes indigestible and residual material.

**(b) Events of phagocytosis**

clotting proteins. However, when complement becomes attached, or *fixed*, to foreign cells such as bacteria, fungi, or mismatched red blood cells, it is activated and becomes a major factor in the fight against foreign cells **(Figure 12.10)**. This **complement fixation** occurs when complement proteins bind to certain sugars or proteins (such as antibodies) on the foreign cell's surface. One result of complement fixation is ① the formation of **membrane attack complexes (MAC)** that produce holes, or pores, in the foreign cell's surface. ② These pores allow water to rush into the cell, ③ causing it to burst, or lyse. Activated complement also amplifies the inflammatory response. Some of the molecules released during the activation process are *vasodilators*, and some are *chemotaxis chemicals* that attract neutrophils and macrophages into the region. Still others cause the cell membranes of the foreign cells to become sticky so they are easier to phagocytize; this effect is called *opsonization*. Complement is a nonspecific defensive mechanism that "complements," or enhances, the effectiveness of *both* innate and adaptive defenses. Complement proteins must be activated in a particular sequence called a *cascade*, which insures that complement is not accidentally activated.

**Interferon**  Viruses are unique, acellular particles—essentially nucleic acids surrounded by a protein coat—that lack the cellular machinery required to generate ATP or make proteins. They do their "dirty work" in the body by taking over target cells (different cells for different viruses), then using cellular machinery and nutrients to reproduce themselves. Although the virus-infected cells can do little to save themselves, they help defend cells that have not yet been infected by secreting small proteins called **interferons** (in-ter-feer'onz). The interferon molecules diffuse to nearby cells and bind to their membrane receptors. This binding stimulates the synthesis of proteins that "interfere" with the ability of viruses to multiply within these still-healthy cells, reducing the spread of the virus. Interferons do not assist with fighting bacterial or fungal infections.

## Fever

### → Learning Objective

☐ **Describe how fever helps protect the body.**

**Fever**, or abnormally high body temperature, is a systemic response to invading microorganisms.

Body temperature is regulated by the hypothalamus, referred to as the body's "thermostat." Normally the thermostat is set at approximately 37°C (98.6°F), but it can be reset upward in response to **pyrogens** (*pyro* = fire), chemicals secreted by white blood cells and macrophages exposed to foreign cells or substances in the body.

Although high fevers are dangerous because excess heat "scrambles" (destroys protein structure) enzymes and other body proteins, rendering them nonfunctional, mild or moderate fever seems to benefit the body. Bacteria require large amounts of iron and zinc to multiply, but during a fever the liver and spleen gather up these nutrients, making them less available. Fever also increases the metabolic rate of tissue cells in general, speeding up repair processes.

### Did You Get It?

9. How does complement cause lysis of a pathogenic microorganism?
10. Which type of infectious microorganism causes the body's level of interferons to rise?

For answers, see Appendix A.

# Adaptive Body Defenses

Most of us would find it wonderfully convenient if we could walk into a single clothing store and buy custom outfits for every occasion—that fit us "to a T." We know that such a service is nearly impossible to find—yet we take for granted our built-in *specific defense system*, which stalks and eliminates with nearly equal precision almost any type of pathogen that intrudes into the body.

The **immune response** to a threat involves tremendously increased internal nonspecific defenses (inflammatory responses and others) and also provides protection that is carefully targeted against *specific* antigens. Furthermore, the initial exposure to an antigen "primes" (sensitizes) the body to react more vigorously to later meetings with the same antigen.

Referred to as the body's *third line of defense* (see Figure 12.6), the adaptive, or specific, defense mechanism is a functional system that recognizes foreign molecules called *antigens* and acts to inactivate or destroy them. Normally it protects us from a wide variety of pathogens, as well as from abnormal body cells. When it fails, malfunctions, or is disabled, some of the most devastating

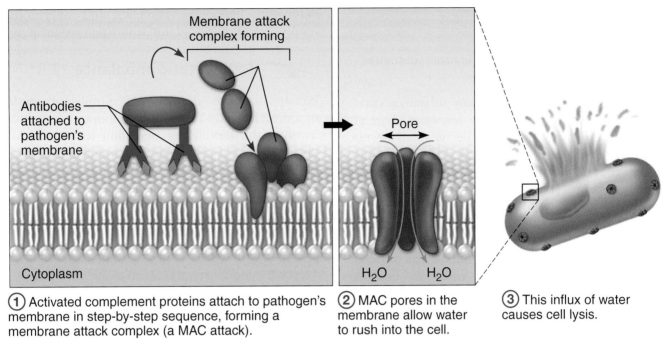

① Activated complement proteins attach to pathogen's membrane in step-by-step sequence, forming a membrane attack complex (a MAC attack).

② MAC pores in the membrane allow water to rush into the cell.

③ This influx of water causes cell lysis.

**Figure 12.10 Activation of complement, resulting in lysis of a target cell.**

diseases—such as cancer, rheumatoid arthritis, and AIDS—may result.

Although *immunology*, the study of immunity, had not been established yet, the ancient Greeks knew that once someone had suffered a certain infectious disease, that person was unlikely to have the same disease again. The basis of this immunity was revealed in the late 1800s, when it was shown that animals surviving a serious bacterial infection have "factors" in their blood that protect them from future attacks by the same pathogen. (These factors are now known to be unique proteins called *antibodies*.) Furthermore, it was demonstrated that if antibody-containing serum from the surviving animals (immune serum) were injected into animals that had not been exposed to the pathogen, those animals would also be protected. This was exciting news. But then, in the mid-1900s, it was discovered that injection of serum containing antibodies did not always protect a recipient from diseases the donor had survived. In such cases, however, injection of the donor's lymphocytes *did* provide immunity. These landmark experiments revealed three important aspects of adaptive defense:

- **It is antigen specific**—it recognizes and acts against *particular* pathogens or foreign substances.

- **It is systemic**—immunity is not restricted to the initial infection site.

- **It has "memory"**—it recognizes and mounts even stronger attacks on previously encountered pathogens.

As the pieces began to fall into place, two separate but overlapping arms of the adaptive defense system were recognized. **Humoral (hu′mor-al) immunity**, also called **antibody-mediated immunity**, is provided by antibodies (immune proteins) present in the body's "humors," or fluids. When lymphocytes themselves defend the body, the immunity is called **cellular immunity** or **cell-mediated immunity** because the protective factor is living cells. The cellular arm also has cellular targets—virus-infected cells, cancer cells, and cells of foreign grafts. The lymphocytes act against such targets either *directly*, by lysing the foreign cells, or *indirectly*, by releasing chemicals that enhance the inflammatory response or activate other immune cells. However, before we describe the humoral and cellular responses individually, we will consider both the remarkable cells involved in these immune responses and the antigens that trigger their activity.

12

## Antigens

→ **Learning Objective**

☐ Define *antigen* and *hapten*, and name substances that act as complete antigens.

An **antigen** (an'tĭ-jen; **Ag**) is any substance capable of provoking an immune response. Foreign antigens are large, complex molecules that are not normally present in our bodies. Consequently, they are foreign intruders, or **nonself**. An almost limitless variety of substances can act as antigens, including virtually all foreign proteins, nucleic acids, many large carbohydrates, and some lipids. Of these, proteins provoke the strongest responses. Pollen grains and microorganisms such as bacteria, fungi, and virus particles are *antigenic* because their surfaces bear such foreign molecules.

Our own cells are richly studded with a variety of protein and carbohydrate molecules. As our immune system develops, it takes inventory of all these molecules so that, thereafter, they are recognized as self. Although these **self-antigens** do not trigger an immune response in us, they *are* strongly antigenic to other people. This helps explain why our bodies reject cells of transplanted organs or foreign grafts unless special measures (drugs and others) are taken to cripple or stifle the immune response. In the unusual situations where self-antigens do elicit an immune response, autoimmune disease results.

### CONCEPTLINK

Likewise, this explains why only people of compatible blood types can donate blood to one another. Recall our discussion of blood antigens (Chapter 10, p. 378). Depending on your individual blood type, one or more blood antigen types may be self to your body, and one or more may be nonself. For example, if you are blood type A, type A antigen is self, but type B antigen is nonself. However, in a person with type AB blood, both A and B antigens are self-antigens. ←

As a rule, small molecules are not antigenic, but when they link up with our own proteins, the immune system may recognize the combination as foreign and mount an attack that is harmful rather than protective. In such cases, the troublesome small molecule is called a **hapten** (*haptein* = to grasp) or **incomplete antigen**. Besides certain drugs, chemicals that act as haptens are found in poison ivy, animal dander, and even in some detergents, hair dyes, cosmetics, and other commonly used household and industrial products.

 **Homeostatic Imbalance 12.3**

Perhaps the most dramatic and familiar example of a drug hapten's provoking an immune response involves the binding of penicillin to blood proteins, which causes a **penicillin reaction** in some people. The combination of blood protein and penicillin is recognized as foreign. In such cases, the immune system mounts such a vicious attack that the person's life is endangered. _____ ✚

### Did You Get It?

11. What is the difference between an antigen and a self-antigen?

For the answer, see Appendix A.

## Cells of the Adaptive Defense System: An Overview

→ **Learning Objectives**

☐ Name the two arms of the adaptive defense system, and relate each to a specific lymphocyte type (B or T cell).

☐ Compare and contrast the development of B and T cells.

☐ State the roles of B cells and T cells.

☐ Explain the importance of antigen-presenting cells in immunity.

The crucial cells of the adaptive system are lymphocytes and antigen-presenting cells (APCs). We already discussed the unique lymphocytes called NK cells as part of the innate defenses. Lymphocytes exist in two other major "flavors." The **B lymphocytes**, or **B cells**, produce antibodies and oversee humoral immunity, whereas the **T lymphocytes**, or **T cells**, constitute the cell-mediated arm of the adaptive defenses and do not make antibodies. T lymphocytes can recognize and eliminate *specific* virus-infected or tumor cells, and B lymphocytes can target *specific* extracellular antigens. In contrast, APCs do not respond to specific antigens but instead play an essential role in activating the lymphocytes that do.

### Lymphocytes

Like all blood cells, lymphocytes originate from hemocytoblasts in red bone marrow. The immature (called *naive*) lymphocytes released from the marrow are essentially identical. Whether a

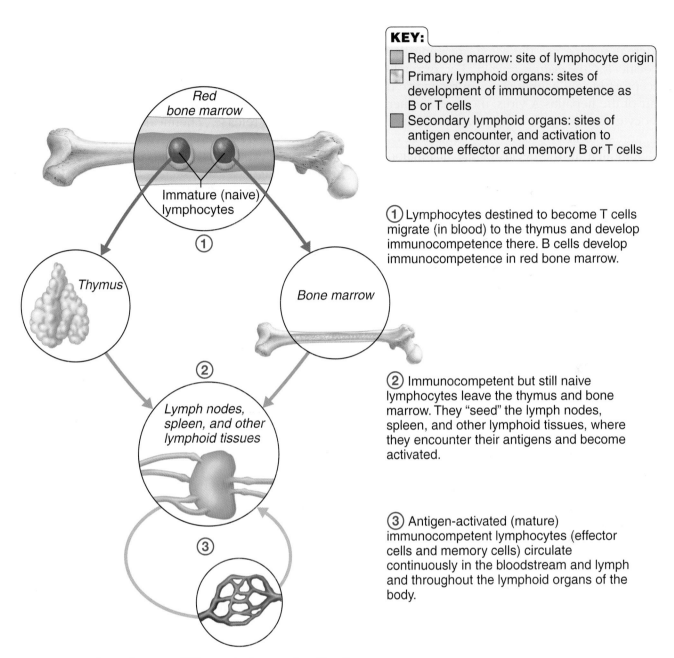

**KEY:**
- ■ Red bone marrow: site of lymphocyte origin
- □ Primary lymphoid organs: sites of development of immunocompetence as B or T cells
- ■ Secondary lymphoid organs: sites of antigen encounter, and activation to become effector and memory B or T cells

① Lymphocytes destined to become T cells migrate (in blood) to the thymus and develop immunocompetence there. B cells develop immunocompetence in red bone marrow.

② Immunocompetent but still naive lymphocytes leave the thymus and bone marrow. They "seed" the lymph nodes, spleen, and other lymphoid tissues, where they encounter their antigens and become activated.

③ Antigen-activated (mature) immunocompetent lymphocytes (effector cells and memory cells) circulate continuously in the bloodstream and lymph and throughout the lymphoid organs of the body.

**Figure 12.11  Lymphocyte differentiation and activation.**

given lymphocyte matures into a B cell or a T cell depends on where in the body it becomes **immunocompetent**, that is, capable of responding to a specific antigen by binding to it with antigen-specific receptors that appear on the lymphocyte's surface. **T** cells arise from lymphocytes that migrate to the **t**hymus (**Figure 12.11** ①). There they undergo a maturation process lasting 2 to 3 days, directed by thymic hormones (*thymosin* and others). Within the thymus, the immature lymphocytes divide rapidly and their numbers increase enormously, but only those maturing T cells with

the sharpest ability to identify *foreign* antigens survive. Lymphocytes capable of binding strongly with *self-antigens* (and of acting against body cells) are vigorously weeded out and destroyed. Thus, the development of **self-tolerance** for the body's own cells is an essential part of a lymphocyte's "education." This is true not only for T cells but also for B cells. **B** cells develop immunocompetence in **b**one marrow, but less is known about the factors that regulate B cell maturation.

Once a lymphocyte is immunocompetent, it will be able to react to one and only one distinct antigen, because *all* the antigen receptors on its external surface are the same, meaning they are specific for the same antigen. For example, the receptors of one lymphocyte can recognize only a part of the hepatitis A virus, those of another lymphocyte can recognize only pneumococcus bacteria, and so forth.

Although all the details of the maturation process are still being worked out, we know that lymphocytes become immunocompetent *before* meeting the antigens they may later attack. Thus, *it is our genes, not antigens, that determine what foreign substances our immune system will be able to recognize and resist.* Only some of the possible antigens our lymphocytes are programmed to resist will ever invade our bodies. Consequently, only some members of our army of immunocompetent cells will be mobilized during our lifetime.

After they become immunocompetent, both T cells and B cells migrate to the lymph nodes and spleen (and loose connective tissues), where their encounters with antigens will occur (Figure 12.11 ②). Then, when the lymphocytes bind with recognized antigens, they complete their differentiation from naive cells into fully mature T cells and B cells. Mature lymphocytes, especially T cells, circulate continuously through the body (Figure 12.11 ③). This makes sense because circulating greatly increases a lymphocyte's chance of coming into contact with antigens, as well as with huge numbers of macrophages and other lymphocytes along the route.

### Antigen-Presenting Cells

The major role of **antigen-presenting cells (APCs)** in immunity is to engulf antigens and then present fragments of them, like signal flags, on their own surfaces, where they can be recognized by T cells. In other words, they *present antigens* to the cells that will actually deal with the antigens. The major types of cells acting as APCs are dendritic cells, macrophages, and B lymphocytes.

All of these cell types are in sites that make it easy to encounter and process antigens. Dendritic cells are present in connective tissues and in the epidermis (where they are also called *Langerhans cells*). These cells are therefore at the body's frontiers, best situated to act as mobile sentinels. Macrophages are widely distributed throughout the lymphoid organs and connective tissues, where they act as phagocytes in the innate defense system. Macrophages tend to remain fixed in the lymphoid organs (as if waiting for antigens to come to them).

When they present antigens, dendritic cells and macrophages activate T cells. Activated T cells, in turn, release chemicals (see Table 12.3, pp. 450–451) that prod macrophages to become *activated macrophages*, true "killers" that are insatiable phagocytes and secrete bactericidal (bacteria-killing) chemicals. As you will see, interactions between lymphocytes, and between lymphocytes and APCs, play roles in virtually all phases of the immune response.

In addition to T cell circulation and passive delivery of antigens to lymphoid organs by lymphatics, there is a third antigen capture and delivery system—migration of dendritic cells to secondary lymphoid organs. With their long, wispy extensions, dendritic cells are very efficient antigen catchers. Once they have engulfed antigens by phagocytosis, they enter nearby lymphatics to get to the lymphoid organ where they will present the antigens to T cells. Indeed, dendritic cells are the most effective antigen presenters known—it's their *only* job. Dendritic cells are a key link between innate and adaptive immunity. They initiate adaptive immune responses particularly tailored to the type of pathogen that they have encountered.

To summarize, the adaptive immune system is a two-fisted defensive system, with a humoral arm and a cellular arm. It uses lymphocytes, APCs, and specific molecules to identify and destroy all substances—both living and nonliving—that are in the body but are not recognized as self. The immune system's ability to respond to such threats depends on the ability of its cells (1) to recognize foreign substances (antigens) in the body by binding to them and (2) to communicate with one another so that the system as a whole mounts a response specific to those antigens.

### Did You Get It?

12. What are the two types of lymphocytes involved in adaptive immune responses, and how do their functions in body protection differ?
13. Where does the "programming" phase leading to immunocompetence occur for T cells?
14. What is the essential role of macrophages and dendritic cells in adaptive immunity?

For answers, see Appendix A.

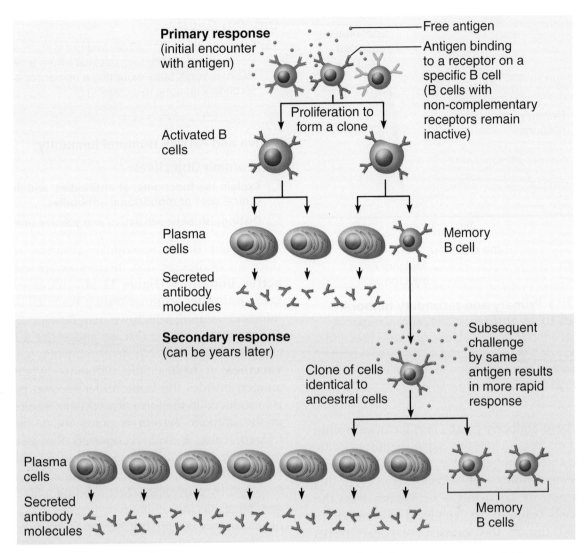

**Figure 12.12 Clonal selection of a B cell.** The initial meeting with the antigen stimulates the primary response, in which the B cell divides rapidly, forming many clones (clonal selection), most of which become antibody-producing plasma cells. Cells that do not differentiate into plasma cells become memory cells, which are primed to respond to subsequent exposures to the same antigen. Should such a meeting occur, the memory cells quickly produce more memory cells and larger numbers of effector plasma cells with the same antigen specificity. Responses generated by memory cells are called *secondary responses.*

## Humoral (Antibody-Mediated) Immune Response

→ **Learning Objectives**

☐ **Define *humoral immunity*.**

☐ **State the role of plasma cells.**

An immunocompetent but as yet immature B lymphocyte is stimulated to complete its development (from naive to fully mature B cell) when antigens bind to its surface receptors. This binding event *sensitizes*, or *activates*, the lymphocyte to "switch on" and undergo **clonal selection**. In this process, the lymphocyte begins to grow and then multiplies rapidly to form an army of cells exactly like itself and bearing the same antigen-specific receptors **(Figure 12.12)**. The resulting family of identical cells descended from the *same* ancestor cell are called **clones**, and clone formation is the **primary humoral response** to that antigen. (As described later, a special group of T cells called helper T cells also influence B cell activation.)

Most of the B cell clone members, or descendants, become **plasma cells**. After an initial lag

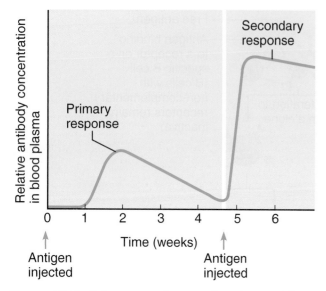

**Figure 12.13 Primary and secondary humoral responses to an antigen.** In the primary response, the level of antibodies in the blood gradually rises and then rapidly declines. The secondary response is both more rapid and more intense, and the antibody level remains high for a much longer time.

period, these antibody-producing "factories" swing into action, producing the same highly specific antibodies at an unbelievable rate of about 2,000 antibody molecules per second. However, this flurry of activity lasts only 4 or 5 days; then the plasma cells begin to die. The level of antibody in the blood during this primary response peaks about 10 days after the response begins and then slowly declines **(Figure 12.13)**.

B cell clone members that do not become plasma cells become long-lived **memory cells** capable of responding to the same antigen if they "see" it again. Memory cells are responsible for the **immunological memory** mentioned earlier. These later immune responses, called **secondary humoral responses**, are produced much faster, are more prolonged, and are more effective than the events of the primary response because all the preparations for this attack have already been made. Within hours after recognition of the "old-enemy" antigen, a new army of plasma cells is being generated, and antibodies flood into the bloodstream. Within 2 to 3 days, the blood antibody level peaks (at a much higher level than seen in the primary response), and the level remains high for weeks to months. We will describe how antibodies work shortly.

## Did You Get It?

**15.** Suzy had the flu last winter and got it again this year. However, her doctor says this year's flu is a new strain. Is Suzy's body mounting a primary or a secondary immune response?

For the answer, see Appendix A.

### Active and Passive Humoral Immunity

→ **Learning Objectives**

☐ **Explain the function(s) of antibodies, and describe clinical uses of monoclonal antibodies.**

☐ **Distinguish between active and passive immunity.**

When your B cells encounter antigens and produce antibodies against them, you are exhibiting **active immunity (Figure 12.14)**. Active immunity is (1) *naturally acquired* during bacterial and viral infections, during which we may develop the signs and symptoms of the disease and suffer a little (or a lot), and (2) *artificially acquired* when we receive **vaccines**. It makes little difference whether the antigen invades the body under its own power or is introduced in the form of a vaccine. The response of the immune system is pretty much the same. Indeed, once it was recognized that secondary responses are so much more vigorous, the race was on to develop vaccines to "prime" the immune response by providing a first meeting with the antigen. Most vaccines contain pathogens that are dead or *attenuated* (living, but extremely weakened).

We receive two benefits from vaccines: (1) they spare us most of the signs and symptoms (and discomfort) of the disease that would otherwise occur during the primary response and (2) the weakened antigens are still able to stimulate antibody production and promote immunological memory. So-called *booster shots*, which may intensify the immune response at later meetings with the same antigen, are also available. Vaccines have virtually wiped out smallpox and are currently available against microorganisms that cause pneumonia, polio, tetanus, diphtheria, whooping cough, measles, and many other diseases. In the United States, active immunization programs have dramatically reduced the incidence of many potentially serious childhood diseases. The more children who are vaccinated, the better the *herd immunity*, a phenomenon in which a population of people are generally protected because most of a given population is immune to a disease or infection. Herd immunity prevents an outbreak of the disease or infection and thus helps to protect individuals who have not been immunized.

**Passive immunity** is quite different from active immunity, both in the antibody source and in the degree of protection it provides (see Figure 12.14). Instead of being made by your plasma cells, the antibodies are obtained from the serum of an immune human or animal donor. As a result, your B cells are *not* challenged by the antigen, immunological memory does *not* occur, and the temporary protection provided by the "donated antibodies" ends when they naturally degrade in the body.

Passive immunity is conferred *naturally* on a fetus when the mother's antibodies cross the placenta and enter the fetal circulation, and after birth during breastfeeding. For several months after birth, the baby is protected from all the antigens to which the mother has been exposed.

Passive immunity is *artificially* conferred when a person receives immune serum or gamma globulin (donated antibodies). Gamma globulin is commonly administered after exposure to hepatitis. Other immune sera are used to treat poisonous snakebites (an *antivenom*), botulism, rabies, and tetanus (an *antitoxin*) because these diseases will kill a person before active immunity can be established. The donated antibodies provide immediate protection, but their effect is short-lived (2 to 3 weeks). Meanwhile, however, the body's own defenses take over.

In addition to their use to provide passive immunity, antibodies are prepared commercially for use in research, in clinical testing for diagnostic purposes, and in treating certain cancers. **Monoclonal antibodies** used for such purposes are descendants of a single cell and are pure antibody preparations that exhibit specificity for one, and only one, antigen. Besides their use in delivering cancer-fighting drugs to cancerous tissue, monoclonal antibodies are being used for early cancer diagnosis and to track cancers hidden deep within the body. They are also used for diagnosing pregnancy, hepatitis, and rabies.

## Antibodies

### → Learning Objectives

☐ Describe the structure of an antibody monomer.

☐ List the five antibody classes, and describe their specific roles in immunity.

☐ Describe several ways in which antibodies act against antigens.

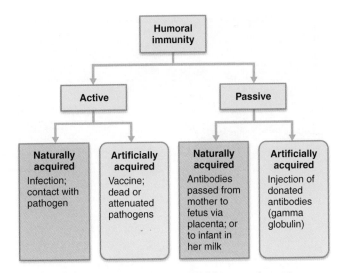

**Figure 12.14 Types of humoral immunity.** There are two overall types of humoral immunity, active and passive. Darker green boxes signify naturally acquired immunity, whereas lighter green boxes show artificially acquired immunity.

**Antibodies**, also referred to as **immunoglobulins** (im″mu-no-glob′u-linz; **Igs**), constitute the *gamma globulin* part of blood proteins. Antibodies are soluble proteins secreted by activated B cells or by their plasma-cell offspring in response to an antigen, and they are capable of binding specifically with that antigen.

Antibodies are formed in response to a huge number of different antigens. Despite their variety, they all have a similar basic anatomy that allows them to be grouped into five Ig classes, each slightly different in structure and function.

***Basic Antibody Structure*** Regardless of its class, every antibody has a basic structure consisting of four polypeptide chains linked together by *disulfide* (sulfur-to-sulfur) *bonds* (**Figure 12.15**, p. 444). Two of the four chains are identical; these are the *heavy chains*. The other two chains, the *light chains*, are also identical to each other but are only about half as long as the heavy chains. When the four chains are combined, the antibody molecule formed has two identical halves, each consisting of a heavy and a light chain, and the molecule as a whole is commonly described as being Y-shaped.

When scientists began investigating antibody structure, they discovered something very peculiar. Each of the four chains forming an antibody had a **variable (V) region** at one end and a **constant (C) region** at the other end. Antibodies responding

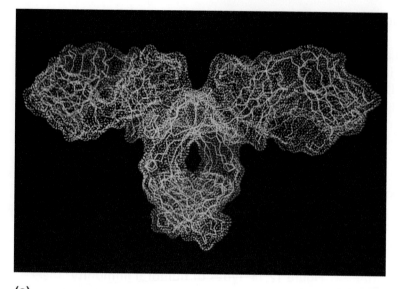

**(a)**

**(b)**

**Figure 12.15 Basic antibody structure. (a)** Computer-generated image. **(b)** Simplified diagram. The monomer of each type of antibody is formed by four polypeptide chains (two light chains and two heavy chains) that are joined by disulfide bonds. Each chain has a variable (V) region (antigen-binding site that differs in different antibodies) and a constant (C) region (essentially identical in different antibodies of the same class).

to different antigens had different variable regions, but their constant regions were the same or nearly so. This made sense when it was discovered that the variable regions of the heavy and light chains combine their efforts to form an **antigen-binding site** (see Figure 12.15) uniquely shaped to "fit" a specific antigen. Hence, each antibody has two such antigen-binding regions.

The constant regions that form the "stem" of an antibody can be compared to the handle of a key. A key handle has a common function for all keys: it allows you to hold the key and place its tumbler-moving portion into the lock. Similarly, the constant regions of the antibody chains serve common functions in all antibodies: they determine the type of antibody formed (antibody class), as well as how the antibody class will carry out its immune roles in the body, and the cell types or chemicals with which the antibody can bind.

## Did You Get It?

16. What is the importance of the variable region of antibodies?

For the answer, see Appendix A.

***Antibody Classes***    There are five major immunoglobulin classes—IgM, IgA, IgD, IgG, and IgE.

(Remember the name MADGE to recall the five Ig types.) Antibodies IgD, IgG, and IgE have the same basic Y-shaped structure described previously and are referred to as *monomers* (Table 12.2). IgA antibodies occur in both monomer and *dimer* (two linked monomers) forms. (Only the dimer form is shown in the table.) Because they are constructed of five linked monomers, IgM antibodies are called *pentamers* (*penta* = five).

The antibodies of each class have slightly different biological roles and locations in the body. For example, IgG is the most abundant antibody in blood plasma and is the only type that can cross the placental barrier. Hence, the passive immunity that a mother transfers to her fetus is "with the compliments" of her IgG antibodies. Only IgM and IgG can fix complement. The IgA dimer, *secretory IgA*, is found mainly in secretions that bathe body surfaces, such as mucus and tears. It plays a major role in preventing pathogens from gaining entry into the body. IgE antibodies are the "troublemaker" antibodies involved in allergies.

## Did You Get It?

17. Which class of antibody is found in saliva and tears?

For the answer, see Appendix A.

| Table **12.2** | Immunoglobulin Classes | | |
|---|---|---|---|
| **Class** | **Generalized structure** | **Where found** | **Biological function** |
| IgM | | Attached to B cell; free in plasma | When bound to B cell membrane, serves as antigen receptor; first Ig class *released* by plasma cells during primary response; potent agglutinating agent; fixes complement. |
| IgA | | Some (monomer) in plasma; dimer in secretions such as saliva, tears, intestinal juice, and milk | Bathes and protects mucosal surfaces from attachment of pathogens. |
| IgD | | Almost always attached to B cell | Believed to be cell surface receptor of immunocompetent B cell; important in activation of B cell. |
| IgG | | Most abundant antibody in plasma; represents 75–85% of circulating antibodies | Main antibody of both primary and secondary responses; crosses placenta and provides passive immunity to fetus; fixes complement. |
| IgE | | Secreted by plasma cells in skin, mucosae of gastrointestinal and respiratory tracts, and tonsils | Binds to mast cells and basophils and triggers release of histamine and other chemical mediators of inflammation and some allergic responses. |

***Antibody Function*** Antibodies inactivate antigens in a number of ways—by complement fixation, neutralization, agglutination, opsonization, and precipitation (**Figure 12.16**, p. 446). Of these, complement fixation and neutralization are most important to body protection.

Complement is the chief antibody ammunition used against cellular antigens, such as bacteria or mismatched red blood cells. As noted earlier, complement is fixed (activated) during innate defenses and when it binds to antibodies attached to cellular targets (Look back at Figure 12.10, p. 437).

**Neutralization** occurs when antibodies bind to specific sites (usually at or close to the site where a cell would bind) on bacterial *exotoxins* (toxic chemicals secreted by bacteria) or on viruses that can cause cell injury. In this way, they block the harmful effects of the exotoxin or virus by preventing them from binding to body cells.

Because antibodies have more than one antigen-binding site, they can bind to more than one antigen at a time; consequently, *antigen-antibody complexes* can be cross-linked into large lattices. When the cross-linking involves cell-bound antigens, the process causes clumping of the foreign cells, a process called **agglutination**. This type of antigen-antibody reaction occurs when mismatched blood is transfused (the foreign red blood cells are clumped) and is the basis of tests used for blood typing. (Look back at Figure 10.7, p. 378.) When the cross-linking process involves soluble antigenic molecules, the resulting antigen-antibody complexes are so large that they become insoluble and settle out of solution. This cross-linking reaction is more precisely called **precipitation**. Such agglutinated bacteria and immobilized (precipitated) antigen molecules are much more easily captured and engulfed by the body's phagocytes than are freely moving antigens.

12

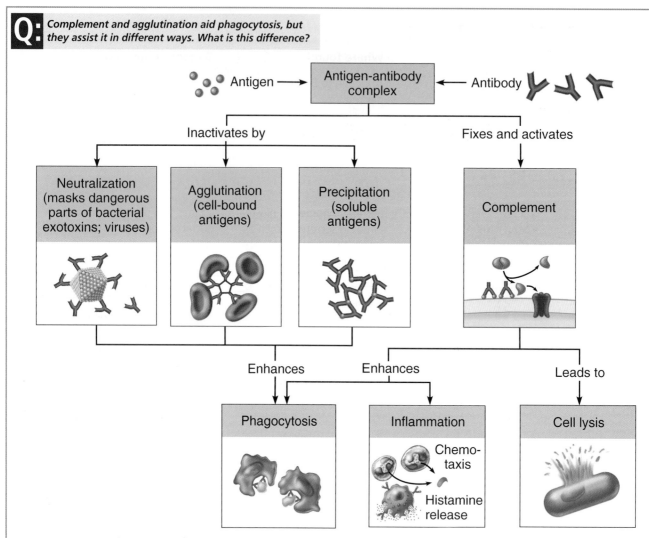

**Q:** *Complement and agglutination aid phagocytosis, but they assist it in different ways. What is this difference?*

**Figure 12.16 Mechanisms of antibody action.** Note that opsonization, the process by which antibodies "mark" antigens for phagocytes, is not pictured here.

## Did You Get It?

18. Regarding the action of antibodies, what is neutralization?
19. Distinguish among precipitation, agglutination, and opsonization.

**For the answers, see Appendix A.**

**A:** Complement fixation opsonizes, or provides "knobs" for the phagocyte to attach to, whereas agglutination binds antigens (typically microorganisms) together into large complexes that become nonmobile and easier to destroy.

## Cellular (Cell-Mediated) Immune Response

→ **Learning Objective**

☐ **Distinguish the roles of helper, regulatory, and cytotoxic T cells.**

The main difference between the two arms of the adaptive response is that B cells secrete their antibody "weapons," whereas T cells fight their antigens directly. Like B cells, immunocompetent T cells are activated to form a clone by binding with a "recognized" antigen (look ahead to Figure 12.19, p. 449). However, unlike B cells, T cells are not able to bind with *free* antigens. Instead, the antigens must be "presented" by a macrophage (or other antigen-presenting cell), and a *double recognition*

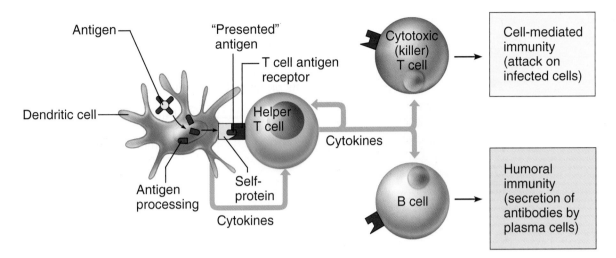

**Figure 12.17 T cell activation and interactions with other cells of the immune response.** Dendritic cells are important antigen-presenting cells (APCs). After they ingest an antigen, they display parts of it on their surface, where it can be recognized by a helper T cell specific for the same antigen. During the binding process, the helper T cell binds simultaneously to the antigen and to the APC (self) receptor, which leads to helper T cell activation and cloning (not illustrated). If the APC is a macrophage, it releases cytokines, which enhance helper T cell activation. Activated helper T cells release cytokines, which stimulate proliferation and activity of other helper T cells and help activate B cells and cytotoxic T cells.

must occur. The APC engulfs an antigen and processes it internally. Parts of the processed antigen are then displayed on the external surface of the presenting cell in combination with one of the APC's own proteins. Note that there are different classes of *effector* T cells, those that carry out direct activities. Two types that are involved in the activation process are called *helper T cells* and *cytotoxic T cells* **(Figure 12.17)**. We will discuss more about the functions of these cells as we work through this process.

A T cell must recognize "nonself," the antigen fragment presented by the APC, and "self," by coupling with a specific glycoprotein on the APC's surface at the same time. Antigen binding alone is not enough to sensitize T cells. They must be "spoon-fed" the antigens by APCs, and something like a "double handshake" must occur. This process of **antigen presentation** is now known to be essential for the activation and clonal selection of T cells. Without the "presenters," the immune response is severely impaired. **Cytokine** chemicals released by macrophages and dendritic cells also play important roles in the immune response (**Table 12.3**, pp. 450–451).

**Cytotoxic T cells** specialize in killing virus-infected, cancer, or foreign graft cells directly (**Figure 12.18**, p. 448). One way a cytotoxic T cell accomplishes this is by ① binding tightly to a foreign cell and ② releasing toxic chemicals called *perforin* and *granzymes* from its granules. The perforin ③ enters the foreign cell's plasma membrane (delivering the so-called *lethal hit*). Shortly thereafter, pores appear in the target cell's membrane, allowing ④ the granzymes (protein-digesting enzymes) to enter and kill the foreign cell. The cytotoxic T cell then ⑤ detaches and seeks other foreign prey to attack.

**Helper T cells** are the T cells that act as the "directors" or "managers" of the adaptive immune response. Once activated, they circulate through the body, recruiting other cells to fight the invaders. For example, helper T cells interact directly with B cells (that have already attached to antigens), prodding the B cells into more rapid division (clone production) and then, like the "boss" of an assembly line, signaling for antibody formation to begin. The helper T cells also release a variety of cytokines (see Table 12.3) that act indirectly to rid the body of antigens by (1) stimulating

12

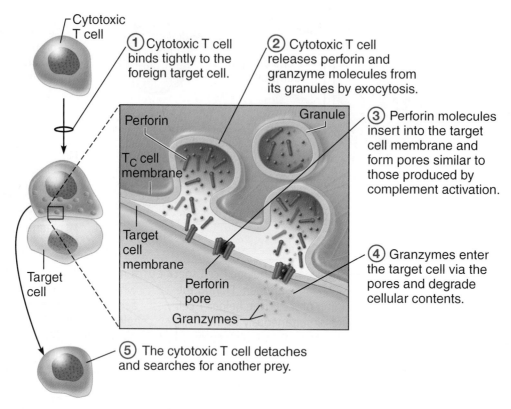

**Figure 12.18 A proposed mechanism by which cytotoxic T cells kill target cells.**

cytotoxic T cells and B cells to grow and divide; (2) attracting other types of protective white blood cells, such as neutrophils, into the area; and (3) enhancing the ability of macrophages to engulf and destroy microorganisms. (Actually, macrophages are good phagocytes even in the absence of cytokines, but in the presence of cytokines they develop an even bigger appetite.) As the released cytokines summon more and more cells into the battle, the immune response gains momentum, and the antigens are often overwhelmed by the sheer numbers of immune elements acting against them.

Another T cell population, **regulatory T cells** (formerly called *suppressor T cells*), release chemicals that suppress the activity of both T and B cells. Regulatory T cells are vital for winding down and stopping the immune response after an antigen has been successfully inactivated or destroyed. This helps prevent uncontrolled or unnecessary immune system activity.

Most of the T cells enlisted to fight in a particular immune response are dead within a few days. However, a few members of each clone are long-lived **memory cells** that remain behind to provide the immunological memory for each antigen encountered and enable the body to respond quickly to subsequent invasions.

The major elements of the immune response are summarized in **Figure 12.19**.

### Did You Get It?

20. T cells must take part in what is sometimes called the "double handshake" in order to be activated. What does this mean?
21. How is the lethal hit accomplished?
22. What is the role of regulatory T cells in the adaptive immune response?
23. In which branch of the adaptive response do helper T cells participate?

For answers, see Appendix A.

## Organ Transplants and Rejection

For people suffering with end-stage heart or kidney disease, organ transplants are a desirable treatment option. However, organ transplants have had mixed success because the immune system is ever vigilant, and rejection is a real problem.

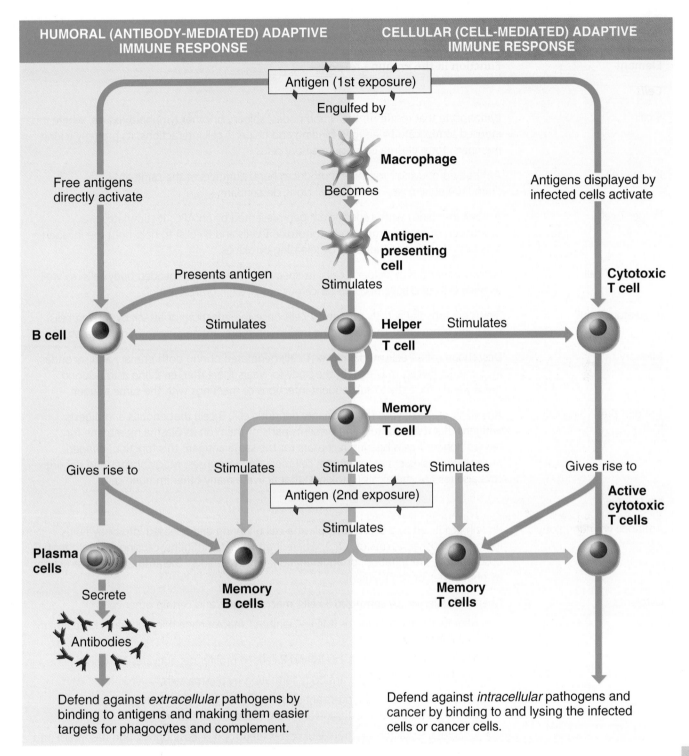

HUMORAL (ANTIBODY-MEDIATED) ADAPTIVE IMMUNE RESPONSE

CELLULAR (CELL-MEDIATED) ADAPTIVE IMMUNE RESPONSE

Antigen (1st exposure)

Engulfed by

Macrophage

Free antigens directly activate

Becomes

Antigens displayed by infected cells activate

Antigen-presenting cell

Presents antigen

Stimulates

Cytotoxic T cell

B cell    Stimulates    Helper T cell    Stimulates

Gives rise to    Stimulates    Stimulates    Stimulates    Gives rise to

Memory T cell

Antigen (2nd exposure)

Stimulates

Active cytotoxic T cells

Plasma cells

Secrete

Memory B cells

Memory T cells

Antibodies

Defend against *extracellular* pathogens by binding to antigens and making them easier targets for phagocytes and complement.

Defend against *intracellular* pathogens and cancer by binding to and lysing the infected cells or cancer cells.

**Figure 12.19  A summary of the adaptive immune responses.** In this flowchart, green arrows track the primary response, and blue arrows track the secondary response.

12

Table **12.3**  **Functions of Cells and Molecules Involved in Immunity**

| Element | Function in the immune response |
| --- | --- |
| *Cells* | |
| B cell | Lymphocyte that resides in the lymph nodes, spleen, or other lymphoid tissues, where it is induced to replicate by antigen-binding and helper T cell interactions; its progeny (clone members) form plasma cells and memory cells. |
| Plasma cell | Antibody-producing "machine"; produces huge numbers of the same antibody (immunoglobulin); specialized B cell clone descendant. |
| Helper T cell | A T cell that binds with a specific antigen presented by an APC; it stimulates the production of other immune cells (cytotoxic T cells and B cells) to help fight the invader; acts both directly and indirectly by releasing cytokines. |
| Cytotoxic T cell | Activity enhanced by helper T cells; its specialty is killing virus-invaded body cells, as well as body cells that have become cancerous; involved in graft rejection. |
| Regulatory T cell | Slows or stops the activity of B and T cells once the infection (or attack by foreign cells) has been conquered. Thought to be important in preventing autoimmune diseases. |
| Memory cell | Descendant of an activated B cell or T cell; generated during both primary and secondary immune responses; may exist in the body for years thereafter, enabling it to respond quickly and efficiently to subsequent infections or meetings with the same antigen. |
| Antigen-presenting cell (APC) | Any of several cell types (macrophage, dendritic cell, B cell) that engulfs and digests antigens that it encounters and presents parts of them on its plasma membrane for recognition by T cells bearing receptors for the same antigen; this function, *antigen presentation*, is essential for normal cell-mediated responses. Macrophages and dendritic cells also release chemicals (cytokines) that activate many other immune cells. |
| *Molecules* | |
| Antibody (immunoglobulin) | Protein produced by a B cell or its plasma-cell offspring and released into body fluids (blood, lymph, saliva, mucus, etc.), where it attaches to antigens, causing neutralization, opsonization, precipitation, or agglutination, which "marks" the antigens for destruction by phagocytes or complement. |
| Cytokines | Chemicals released by sensitized T cells, macrophages, and certain other cells:<br>• Migration inhibiting factor (MIF)—"inhibits" macrophage migration and keeps them in the local area.<br>• Interleukin 2—stimulates T cells and B cells to proliferate; activates NK cells.<br>• Helper factors—enhance antibody formation by plasma cells.<br>• Suppressor factors—suppress antibody formation or T cell–mediated immune responses (interleukin-10 transforming growth factor and others).<br>• Chemotactic factors—attract leukocytes (neutrophils, eosinophils, and basophils) into inflamed area.<br>• Gamma interferon—secreted by lymphocytes; helps make tissue cells resistant to viral infection; activates macrophages and NK cells; enhances maturation of cytotoxic T cells. |
| Tumor necrosis factor (TNF) | Like perforin, causes cell killing; attracts granulocytes; activates T cells and macrophages. |
| Complement | Group of bloodborne proteins activated after binding to antibody-covered antigens; when activated, complement causes lysis of the microorganism and enhances inflammatory response. |

Table **12.3** *(continued)*

| Element | Function in the immune response |
| --- | --- |
| Antigen | Substance capable of provoking an immune response; typically a large, complex molecule not normally present in the body. |
| Cytotoxins | Perforin, granzymes—cell toxins released by cytotoxic T cells and NK cells. |

There are four major types of transplants, or *grafts*:

- **Autografts** are tissue grafts transplanted from one site to another in the *same person*.

- **Isografts** are tissue grafts donated by a *genetically identical person*, the only example being an identical twin.

- **Allografts** are tissue grafts taken from *a person other than an identical twin*.

- **Xenografts** are tissue grafts harvested from a *different animal species*, such as a porcine (pig) heart valve transplanted into a human.

Autografts and isografts are ideal donor organs or tissues and are just about always successful given an adequate blood supply and no infection. Although pig heart valves have been transplanted with success, xenografts of whole organs are virtually never successful. The graft type most used is an allograft taken from a recently deceased person.

Before an allograft is even attempted, the ABO and other blood group antigens of both the donor and recipient must be determined and must match. Then, the cell membrane antigens of their tissue cells are typed to determine how closely they match. At least a 75 percent match is needed to attempt a graft; as you might guess, good tissue matches between unrelated people are hard to find.

After surgery, to prevent rejection, the patient receives **immunosuppressive therapy**, including one or more of the following: corticosteroids to suppress inflammation, antiproliferative drugs, radiation (X-ray) therapy, and immunosuppressor drugs. Many of these drugs kill rapidly dividing cells (such as activated lymphocytes), and all of them have severe side effects. However, the major problem with immunosuppressive therapy is that while the immune system is suppressed, it cannot protect the body against other foreign agents. Explosive infection (bacterial, fungal, viral) is the most frequent cause of death in these patients. Even under the best conditions, roughly 50 percent of patients have rejected their transplant within 10 years of receiving it. In rare cases, transplant patients have been able to "ditch the drugs," and finding a way to induce such tolerance is the goal of many current research projects.

### Did You Get It?

24. Sheila is receiving a kidney transplant. The donor is her fraternal twin. What name is given to this type of graft?

For the answer, see Appendix A.

## Disorders of Immunity

→ **Learning** Objective

☐ **Describe allergies, autoimmune diseases, and immunodeficiencies.**

 **Homeostatic Imbalance 12.4**

The most important disorders of the immune system are allergies, autoimmune diseases, and immunodeficiencies.

### Allergies

At first, the immune response was thought to be purely protective. However, it was not long before its dangerous potentials were discovered. **Allergies**, or **hypersensitivities**, are abnormally vigorous immune responses in which the immune system causes tissue damage as it fights off a perceived "threat" that would otherwise be harmless to the body. The term *allergen* (*allo* = altered; *erg* = reaction) is used to distinguish this type of antigen from those producing normal immune responses. Most of the time, people do not die of allergies; they are just miserable with them.

12

There are several different types of allergies, including reactions where cells are lysed, such as during a transfusion reaction, and inflammation due to antigen-antibody complexes, such as those of glomerulonephritis. The most common type is **immediate hypersensitivity**, or **acute hypersensitivity**. After sensitization to a particular allergen, this type of response is triggered when that allergen is encountered again. This secondary response starts with the release of a flood of histamine when IgE antibodies bind to *mast cells*. Histamine causes small blood vessels in the area to become dilated and leaky and is largely to blame for the best-recognized symptoms of allergy: a runny nose, watery eyes, and itchy, reddened skin (hives). When the allergen is inhaled, symptoms of asthma appear because smooth muscle in the walls of the bronchioles contracts, constricting the passages and restricting air flow. Over-the-counter (OTC) anti-allergy drugs contain *antihistamines* that counteract these effects. Most of these reactions begin within seconds after contact with the allergen and last about half an hour.

Fortunately, the bodywide, or systemic, acute allergic response known as **anaphylactic** (an″ah-fǐ-lak′tik) **shock** is fairly rare. Anaphylactic shock occurs when the allergen directly enters the blood and circulates rapidly through the body, as might happen with certain bee stings, spider bites, or an injection of a foreign substance (such as horse serum, penicillin, or other drugs that act as haptens) into susceptible individuals. Food allergies (peanut or wheat allergies) may also trigger anaphylaxis, and can be fatal. The mechanism of anaphylactic shock is essentially the same as that of local responses, but when the entire body is involved, the outcome is life-threatening. For example, it is difficult to breathe when the smooth muscles of lung passages contract, and the sudden vasodilation (and fluid loss) that occurs may cause circulatory collapse and death within minutes. Epinephrine, found in emergency EpiPen® shots, is the drug of choice to reverse these histamine-mediated effects.

**Delayed hypersensitivities**, mediated mainly by a special subgroup of helper T cells, cytotoxic T cells, and macrophages, take much longer to appear (1 to 3 days) than any of the acute reactions produced by antibodies. Instead of histamine, the chemicals mediating these reactions are

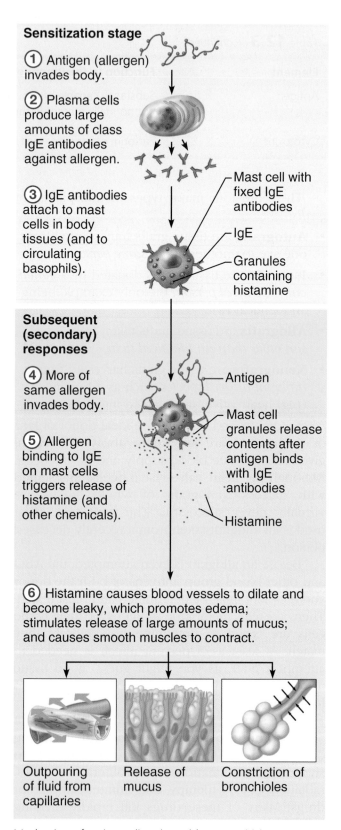

**Sensitization stage**

① Antigen (allergen) invades body.

② Plasma cells produce large amounts of class IgE antibodies against allergen.

③ IgE antibodies attach to mast cells in body tissues (and to circulating basophils).

Mast cell with fixed IgE antibodies

IgE

Granules containing histamine

**Subsequent (secondary) responses**

④ More of same allergen invades body.

⑤ Allergen binding to IgE on mast cells triggers release of histamine (and other chemicals).

Antigen

Mast cell granules release contents after antigen binds with IgE antibodies

Histamine

⑥ Histamine causes blood vessels to dilate and become leaky, which promotes edema; stimulates release of large amounts of mucus; and causes smooth muscles to contract.

Outpouring of fluid from capillaries

Release of mucus

Constriction of bronchioles

Mechanism of an immediate (acute) hypersensitivity response.

cytokines released by the activated T cells. Hence, antihistamine drugs are *not* helpful against the delayed types of allergies. Corticosteroid drugs are used to provide relief.

The most familiar examples of delayed hypersensitivity reactions are those classed as **allergic contact dermatitis**, which follow skin contact with poison ivy, some heavy metals (lead, mercury, and others), and certain cosmetic and deodorant chemicals. These agents act as haptens; after diffusing through the skin and attaching to body proteins, they are perceived as foreign by the immune system. The *Mantoux* and *tine tests*, skin tests for detecting tuberculosis, depend on delayed hypersensitivity reactions. When the tubercle antigens are injected just under the skin, a small, hard lesion forms if the person has been sensitized to the antigen.

## Autoimmune Diseases

Occasionally, the immune system loses its ability to distinguish friend from foe, that is, to tolerate self-antigens while still recognizing and attacking foreign antigens. When this happens, the body produces antibodies (*auto-antibodies*) and sensitized T cells that attack and damage its own tissues. This puzzling phenomenon is called **autoimmune disease** because it is a person's own immune system that produces the disorder.

Some 5 percent of adults in North America—two-thirds of them women—are afflicted with autoimmune disease. Most common are the following:

- **Rheumatoid arthritis (RA)**, which systematically destroys joints (see pp. 198–199)

- **Myasthenia gravis**, which impairs communication between nerves and skeletal muscles (see p. 244)

- **Multiple sclerosis (MS)**, which destroys the white matter (myelin sheaths) of the brain and spinal cord (see p. 257)

- **Graves' disease**, in which the thyroid gland produces excessive amounts of thyroxine (see p. 345) in response to autoantibodies that mimic TSH

- **Type 1 diabetes mellitus**, which destroys pancreatic beta cells, resulting in deficient production of insulin (see p. 351)

- **Systemic lupus erythematosus (SLE)**, a systemic disease that occurs mainly in young women and particularly affects the kidneys, heart, lungs, and skin

- **Glomerulonephritis**, a severe impairment of kidney function due to acute inflammation (see p. 559)

Current therapies include treatments that depress certain aspects of the immune response.

How does the normal state of self-tolerance break down? It appears that one of the following may trigger it:

- **New self-antigens appear.** Such "hidden" antigens are found in sperm cells, the eye lens, and certain proteins in the thyroid gland. In addition, "new" self-antigens may appear as a result of gene mutations that change the structure of self-proteins or as a result of alterations in self-proteins by hapten attachment or by bacterial or viral damage.

- **Foreign antigens resemble self-antigens.** For instance, antibodies produced during an infection caused by streptococcus bacteria (such as strep throat or scarlet fever) are known to cross-react with heart antigens, causing damage to both the heart muscle and its valves, as well as to the joints and kidneys. This age-old disease is called **rheumatic fever**.

## Immunodeficiencies

The **immunodeficiencies** include both congenital and acquired conditions in which the production or function of immune cells or complement is abnormal. The most devastating congenital condition is **severe combined immunodeficiency disease (SCID)**, in which there is a marked deficit of both B and T cells. Because T cells are absolutely required for normal operation of *both* arms of the adaptive response, afflicted children have essentially no protection against pathogens of any type. Minor infections easily shrugged off by most children are lethal to those with SCID. Bone marrow transplants and umbilical cord blood, which provide normal lymphocyte stem cells, have helped some SCID victims. Without such treatment, the only hope for survival is living behind protective barriers (in a plastic "bubble") that keep out all infectious agents.

Currently, the most important and most devastating of the acquired immunodeficiencies is **acquired immune deficiency syndrome (AIDS)**. AIDS cripples the immune system by interfering with the activity of helper T cells (see "A Closer Look" on pp. 454–455). _____✚

# AIDS: An Ongoing Pandemic

A *pandemic* is a widespread epidemic, or ongoing outbreak of disease. The AIDS pandemic has been going on for over 30 years.

Although AIDS was first identified in the United States in 1981 among homosexual men and intravenous drug users of both sexes, it had begun afflicting the heterosexual population of Africa several years earlier. AIDS is characterized by severe weight loss, night sweats, swollen lymph nodes, and increasingly frequent infections, including a fungal brain infection called *cryptococcal meningitis*, a rare type of pneumonia called *pneumocystis pneumonia*, an eye infection called *cytomegalovirus retinitis*, and the bizarre malignancy *Kaposi's sarcoma*, a cancerlike condition of blood vessels evidenced by purple lesions of the skin. Some AIDS victims develop slurred speech and severe dementia. The course of untreated AIDS is grim and thus far nearly inescapable, finally ending in complete debilitation and death from cancer or overwhelming infection.

AIDS is caused by a virus transmitted in blood, breast milk, semen, and vaginal secretions. Most commonly, the virus enters the body via blood transfusions or blood-contaminated needles and during intimate sexual contact in which the mucosa is torn or where open lesions caused by sexually transmitted infections allow the virus access to the blood. It is *not* transmitted by casual contact, because the virus dries and dies when exposed to air.

The virus, named *HIV (human immunodeficiency virus)*, specifically targets and destroys helper T cells, severely depressing the adaptive immune response. Without helper T cells, B cells and cytotoxic T cells cannot be activated or maintained, and the whole immune system is turned topsy-turvy. It is now clear that the virus multiplies steadily in the lymph nodes throughout most of the chronic asymptomatic period. Symptomatic AIDS appears when the lymph nodes can no longer contain the virus and the immune system collapses. This process can take 10 years or more. The virus also invades the brain, which accounts for the dementia of some AIDS patients.

As of 2014, nearly 37 million people worldwide were living with HIV. Since the pandemic began, 39 million people have died. Sub-Saharan Africa

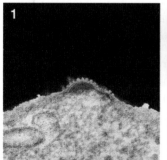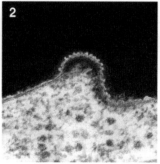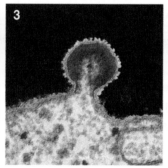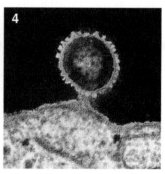

A new HIV virus emerges from an infected human cell.

## Did You Get It?

25. What is an allergy?
26. What causes the difficulty in breathing that occurs with anaphylactic shock?
27. What is the principal problem common to all immunodeficiency diseases?
28. What are two possible causes of autoimmune disease?

For answers, see Appendix A.

has the highest population of infected individuals, accounting for nearly 70 percent of the world HIV cases. Because there is a 6-month "window" during which antibodies may develop after exposure to HIV, there are probably 100 asymptomatic carriers of the virus for every newly diagnosed case. Moreover, the disease has a long incubation period (from a few months to 10 years) between exposure and the appearance of clinical symptoms. The Centers for Disease Control and Prevention (CDC) estimates that over 1.2 million U.S. residents are HIV infected and that one of every eight of those people does not realize he or she is infected.

The "face of AIDS" in the United States is changing. Victims have begun to include people who do not belong to the original high-risk groups. Before reliable testing of donated blood was available, some people contracted the virus from blood transfusions. Hemophiliacs have been especially vulnerable because the blood-clotting factors they need are isolated from pooled blood donations. Although manufacturers began taking measures to kill the virus in 1984, by then an estimated 60 percent of the hemophiliacs in the United States were already infected. The virus can also be transmitted from an infected mother to her fetus. Homosexual men still account for the bulk of cases transmitted by sexual contact, but more and more heterosexuals are contracting this disease. Particularly disturbing is the near-epidemic increase in diagnosed cases among teenagers and young adults; among adolescents aged 10–19, the death rate from AIDS has tripled since 2000. As of 2013, AIDS is the eighth leading killer of all Americans ages 25 to 34 and ranks ninth for Americans 35 to 44.

Large-city hospitals are seeing more HIV/AIDS patients daily, and statistics reporting the number of cases in the inner city, where intravenous drug use is the chief means of HIV transmission, are alarming. Currently, about 13 percent of the 12.2 million IV drug users are thought to have HIV. Furthermore, 75 percent of HIV/AIDS cases in newborns occur where drug abuse abounds.

Diagnostic tests to identify carriers of the AIDS virus are increasingly sophisticated. For example, a urine test provides a painless alternative to the standard HIV tests. However, no sure cure has yet been found. Over 100 drugs are now in the U.S. Food and Drug Administration pipeline, and several vaccines are undergoing clinical trials.

Clinically available are several **a**nti**r**etroviral **t**herapies (ART)—with drugs named like alphabet soup—that act by inhibiting the enzymes that the HIV virus needs to multiply in body cells. The *reverse transcriptase inhibitors*, such as AZT, were early on the scene. AZT was followed by others, including ddC and 3TC. These were followed by *protease inhibitors* (saquinavir, ritonavir, and others) and *fusion inhibitors*, which block the virus's entry into the cell in the first place. New drugs are constantly appearing. In the early days, treatment required taking many pills at varying times, and the schedule for taking them was difficult to follow. Now most patients take one pill daily that contains three drugs. Combination therapy using drugs from each class delivers a one-two punch to the HIV virus, postpones drug resistance (a problem with single-drug therapy using AZT alone), and substantially reduces the amount of HIV in the blood. Earlier treatment and more effective combination drug therapy have turned AIDS into a controllable illness with a life expectancy well beyond 8 years. However, preventing infection is the preferred way to go. Perhaps, as urged in the media, the best defense is to practice "safer sex" by using condoms and knowing one's sexual partner's history, but the only foolproof method is sexual abstinence.

# PART III: DEVELOPMENTAL ASPECTS OF THE LYMPHATIC SYSTEM AND BODY DEFENSES

→ **Learning Objectives**

☐ **Describe the origin of the lymphatic vessels.**

☐ **Describe the effects of aging on immunity.**

The lymphatic vessels, which bud from the veins of the blood vascular system, and the main clusters of lymph nodes are obvious by the fifth week of development. Except for the thymus and the spleen, the lymphoid organs are poorly developed before birth. However, soon after birth, they become heavily populated with lymphocytes as the immune system begins to function.

# Homeostatic Relationships between the
# Lymphatic System and Other Body Systems

## Nervous System

- The lymphatic vessels pick up leaked plasma fluid, proteins, and immune cells from both central and peripheral nervous system structures; immune cells protect nervous system structures from specific pathogens
- The nervous system innervates larger lymphatic vessels; lymphatics drain excess fluid from the brain

## Endocrine System

- Lymphatic vessels pick up leaked fluids and proteins; lymph distributes hormones; immune cells protect endocrine organs from pathogens
- The thymus produces hormones that promote development of lymphoid organs and "program" T lymphocytes

## Lymphatic System/Immunity

## Respiratory System

- Lymphatic vessels pick up leaked fluid and proteins from respiratory organs; immune cells protect respiratory organs from specific pathogens; plasma cells in the respiratory mucosa secrete IgA to prevent pathogen invasion of deeper tissues
- The lungs provide oxygen needed by lymphoid/immune cells and eliminate carbon dioxide; the pharynx houses tonsils; the respiratory "pump" aids lymph flow

## Digestive System

- Lymphatic vessels pick up leaked fluids and proteins from digestive organs; lymph transports some products of fat digestion to the blood; lymphoid nodules in the wall of the intestine prevent invasion of pathogens
- The digestive system digests and absorbs nutrients needed by cells of lymphoid organs; gastric acidity inhibits pathogens' entry into blood

## Muscular System

- Lymphatic vessels pick up leaked fluid and proteins from skeletal muscle; immune cells protect muscles from specific pathogens
- The skeletal muscle "pump" aids the flow of lymph; muscles protect superficial lymph nodes

## Cardiovascular System

- Lymphatic vessels pick up leaked plasma and proteins; spleen destroys aged RBCs, stores iron, and removes debris from blood; immune cells protect cardiovascular organs from specific pathogens
- Blood is the source of lymph; lymphatics develop from veins; blood provides the route for circulation of immune elements

## Urinary System

- Lymphatic vessels pick up leaked fluid and proteins from urinary organs; immune cells protect urinary organs from specific pathogens
- Urinary system eliminates wastes and maintains homeostatic water/acid-base/electrolyte balance of the blood for immune cell functioning; urine flushes some pathogens out of the body

## Reproductive System

- Lymphatic vessels pick up leaked fluid and proteins from reproductive organs; immune cells protect the organs from specific pathogens
- Acidity of vaginal secretions prevents bacterial multiplication

## Integumentary System

- Lymphatic vessels pick up leaked plasma fluid and proteins from the dermis; lymphocytes in lymph enhance the skin's protective role by defending against specific pathogens
- The skin's keratinized epithelium provides a mechanical barrier to pathogens; acid pH of skin secretions inhibits growth of bacteria on skin

## Skeletal System

- Lymphatic vessels pick up leaked plasma fluid and proteins from the periostea; immune cells protect bones from specific pathogens
- The bones house hematopoietic tissue (red marrow), which produces the cells that populate the lymphoid organs and provide body immunity

Lymphatic system problems are relatively uncommon, but when they do occur they are painfully obvious. For example, when the lymphatic vessels are blocked (as in *elephantiasis*, a tropical disease in which the lymphatics become clogged with parasitic worms) or when lymphatics are removed (as in radical breast surgery), severe edema may result. However, lymphatic vessels removed surgically do grow back in time.

Stem cells of the immune system originate in the spleen and liver during the first month of embryonic development. Later, the bone marrow becomes the predominant source of stem cells (hemocytoblasts), and it persists in this role into adult life. In late fetal life and shortly after birth, the young lymphocytes develop self-tolerance and immunocompetence in their "programming organs" (thymus and bone marrow) and then populate the other lymphoid tissues. Upon meeting "their" specific antigens, the T and B cells complete their development to fully mature immune cells.

Although the ability of our adaptive immune system to recognize foreign substances is determined by our genes, the nervous system may help to control the activity of the immune response. The immune response is definitely impaired in individuals who are under severe stress. Our immune system normally serves us well throughout our lifetime, until old age. But during the later years, its efficiency begins to wane. As a result, the body becomes less able to fight infections and destroy cells that have become cancerous. Additionally, we become more susceptible to both autoimmune and immunodeficiency diseases.

## Summary

### PART I: THE LYMPHATIC SYSTEM (pp. 424–429)

1. The lymphatic system consists of the lymphatic vessels, lymph nodes, and certain other lymphoid organs in the body.

2. Extremely porous, blind-ended lymphatic capillaries pick up excess fluid leaked from the blood capillaries. The fluid (lymph) flows into the larger lymphatics and finally into the blood vascular system through the right lymphatic duct and the left thoracic duct.

3. Lymph transport is aided by the muscular and respiratory pumps and by contraction of smooth muscle in the walls of the lymphatic vessels.

4. Lymph nodes are clustered along lymphatic vessels, and the lymphatic stream flows through them. Lymph nodes serve as multiplication sites for lymphocytes; phagocytic cells within them remove bacteria, viruses, and the like from the lymph stream before it is returned to the blood.

5. Other lymphoid organs include the tonsils, which remove bacteria trying to enter the digestive or respiratory tracts; the thymus, a programming region for T lymphocytes; Peyer's patches, which prevent bacteria in the intestine from penetrating deeper into the body; and the spleen, a red blood cell graveyard and blood reservoir.

### PART II: BODY DEFENSES (pp. 429–455)

#### Innate Body Defenses (pp. 430–436)

1. Surface membranes (skin and mucous membranes) provide mechanical barriers to pathogens. Some produce secretions and/or have structural modifications that enhance their defensive effects; the skin's acidity, lysozyme, mucus, keratin, and ciliated cells are examples.

2. The inflammatory response prevents the spread of harmful agents, disposes of pathogens and dead tissue cells, and promotes healing. Protective leukocytes enter the area, fibrin walls off the area, and tissue repair occurs.

3. Natural killer cells are lymphocytes that act nonspecifically to lyse virus-infected and malignant cells.

4. Phagocytes (macrophages and neutrophils) engulf and destroy pathogens that penetrate epithelial barriers. This process is enhanced when the pathogen's surface is altered by attachment of antibodies and/or complement.

5. When complement (a group of plasma proteins) becomes fixed on the membrane of a foreign cell, lysis of the target cell occurs. Complement also enhances phagocytosis and the inflammatory and immune responses.

6. Interferons are a group of proteins synthesized by virus-infected cells. They prevent viruses from multiplying in other body cells.

**12**

7. Fever enhances the fight against infectious micro-organisms by increasing metabolism (which speeds up repair processes) and by causing the liver and spleen to store iron and zinc (which bacteria need for multiplication).

## Adaptive Body Defenses (pp. 436–455)

1. The immune system recognizes something as foreign and acts to inactivate or remove it. The adaptive immune response is antigen specific, is systemic, and has memory. The two arms are humoral immunity (mediated by antibodies) and cellular immunity (mediated by living cells, T lymphocytes).

2. Antigens

   a. Antigens are large, complex molecules (or parts of them) recognized as foreign by the body. Foreign proteins are the strongest antigens.

   b. Complete antigens provoke an immune response and bind with products of that response (antibodies or sensitized lymphocytes).

   c. Incomplete antigens, or haptens, are small molecules that are unable to cause an immune response by themselves but do so when they bind to body proteins and the complex is recognized as foreign.

3. Cells of the adaptive defense system: an overview

   a. Two main cell populations, lymphocytes and antigen-presenting cells, are involved in adaptive defense.

   b. Lymphocytes arise from hemocytoblasts of bone marrow. T cells develop immunocompetence in the thymus and oversee cell-mediated immunity. B cells develop immunocompetence in bone marrow and provide humoral immunity. Immunocompetent lymphocytes seed lymphoid organs, where antigen challenge occurs, and circulate through blood, lymph, and lymphoid organs.

   c. Immunocompetence is signaled by the appearance of antigen-specific receptors on the surfaces of lymphocytes.

   d. Macrophages arise from monocytes produced in bone marrow. They and other APCs phagocytize pathogens and present parts of the antigens on their surfaces, for recognition by T cells.

4. Humoral (antibody-mediated) immune response

   a. Clonal selection of B cells occurs when antigens bind to their receptors, causing them to proliferate. Most clone members become plasma cells, which secrete antibodies. This is called the *primary humoral response.*

   b. Other clone members become memory B cells, capable of mounting a rapid attack against the same antigen in subsequent meetings (*secondary humoral responses*). These cells provide immunological "memory."

   c. Active humoral immunity is acquired during an infection or via vaccination and provides immunological memory. Passive immunity is conferred when a donor's antibodies are injected into the bloodstream or when the mother's antibodies cross the placenta. It does not provide immunological memory.

   d. Basic antibody structure

      (1) Antibodies are proteins produced by sensitized B cells or plasma cells in response to an antigen, and they are capable of binding specifically with that antigen.

      (2) An antibody is composed of four polypeptide chains (two heavy and two light) that form a Y-shaped molecule.

      (3) Each polypeptide chain has a variable and a constant region. Variable regions form antigen-binding sites, one on each arm of the Y. Constant regions determine antibody function and class.

      (4) Five classes of antibodies exist: IgM, IgA, IgD, IgG, and IgE. They differ structurally and functionally.

      (5) Antibody functions include agglutination, complement fixation, neutralization, opsonization, and precipitation.

      (6) Monoclonal antibodies are pure preparations of a single antibody type useful in diagnosing various infectious disorders and cancer and in treating certain cancers.

5. Cellular (cell-mediated) immune response

   a. T cells are sensitized by binding simultaneously to an antigen and a self-protein displayed on the surface of a macrophage or another type of antigen-presenting cell. Clonal selection occurs, and clone members differentiate into effector T cells or memory T cells.

   b. There are several different classes of T cells. Cytotoxic (killer) T cells directly attack and lyse infected and cancerous cells. Helper T cells interact directly with B cells bound to antigens. They also release cytokines, chemicals that enhance the killing activity of macrophages, attract other leukocytes, or act as helper factors that stimulate activity of B cells and cytotoxic T cells. Regulatory T cells "turn off" the normal immune response by releasing suppressor chemicals.

6. Organ transplants include autografts, isografts, allografts, and xenografts. The most common graft is an allograft. Blood group and tissue matching are done to ensure the best match possible, and organ transplant is followed by immunosuppressive therapy.

7. Disorders of immunity

   a. Autoimmune disease occurs when the body's self-tolerance breaks down and antibodies and/or T cells attack the body's own tissues. Most forms of autoimmune disease result from the appearance of formerly hidden self-antigens or changes in the structure of self-antigens, and antibodies formed against foreign antigens that resemble self-antigens.

   b. In hypersensitivity (allergy), the immune system overreacts to an otherwise harmless antigen, and tissue destruction occurs. Immediate (acute) hypersensitivity, as seen in hay fever, hives, and anaphylaxis, is due to IgE antibodies. Delayed hypersensitivity (for example, contact dermatitis) reflects activity of T cells, macrophages, and cytokines.

   c. Immunodeficiencies result from abnormalities in any immune element. Most serious is severe combined immunodeficiency disease (a congenital disease) and AIDS, an acquired immunodeficiency disease caused by a virus that attacks and cripples helper T cells.

## PART III: DEVELOPMENTAL ASPECTS OF THE LYMPHATIC SYSTEM AND BODY DEFENSES (pp. 455–457)

1. Lymphatic vessels form by budding off veins. The thymus and the spleen are the first lymphoid organs to appear in the embryo. Other lymphoid organs remain relatively undeveloped until after birth.

2. The immune response develops around the time of birth.

3. The ability of immunocompetent cells to recognize foreign antigens is genetically determined. Stress appears to interfere with normal immune response.

4. Efficiency of the immune response wanes in old age, and infections, cancer, immunodeficiencies, and autoimmune diseases become more prevalent.

## Review Questions

 Access additional practice questions using your smartphone, tablet, or computer: Mastering A&P®> Study Area > Practice Tests & Quizzes

### Multiple Choice

*More than one choice may apply.*

1. The body's largest lymphatic vessel is the
   a. lymph capillary.
   b. lymphatic collecting vessel.
   c. right lymphatic duct.
   d. thoracic duct.

2. Which parts of the lymph node show increased activity when antibody production is high?
   a. Germinal centers
   b. Outer follicle
   c. Medullary cords
   d. Medullary sinuses

3. Which of the following drain into the right lymphatic duct?
   a. Inguinal nodes
   b. Cisterna chyli
   c. Cervical nodes
   d. Axillary nodes

4. Which of the following are part of MALT?
   a. Tonsils
   b. Thymus
   c. Peyer's patches
   d. Any lymphoid tissue along the digestive tract

5. Developmentally, embryonic lymphatic vessels are most closely associated with the
   a. veins.
   b. arteries.
   c. nerves.
   d. thymus gland.

6. Which of the following are among the most common indicators of inflammation?
   a. Phagocytosis
   b. Edema
   c. Leukocytosis
   d. Pain

7. Chemical mediators of inflammation include
   a. interferon.          c. histamine.
   b. complement.          d. antibodies.

8. The second line of defense of the innate defense mechanisms includes
   a. the mucous membranes.
   b. the macrophages.
   c. fever.
   d. the natural killer cells.

9. Which of these antibody classes is usually arranged as a pentamer?
   a. IgG        c. IgA
   b. IgM        d. IgD

10. A damaged cell releases
    a. perforin.
    b. kinins.
    c. histamine.
    d. granzymes.

11. Which of the following antibody capabilities causes a transfusion reaction with A or B erythrocyte antigens?
    a. Neutralization
    b. Precipitation
    c. Complement fixation
    d. Agglutination

12. Which of the following is/are examples of auto-immune disease?
    a. Type 1 diabetes
    b. Multiple sclerosis
    c. Graves' disease
    d. Rheumatoid arthritis

13. Helper T cells directly stimulate
    a. B cells.
    b. cytotoxic T cells.
    c. memory B cells.
    d. plasma cells.

## Short Answer Essay

14. What is the most important function of the lymph nodes?

15. Compare and contrast blood, interstitial fluid, and lymph.

16. Where would you find Peyer's patches and the appendix? What are their functions?

17. Besides acting as mechanical barriers, the skin and mucosae of the body contribute to body protection in other ways. Cite the common body locations and the importance of mucus, lysozyme, keratin, acid pH, and cilia.

18. What is complement? Besides bacterial lysis, what are some of the roles of complement?

19. Why is the term "swollen glands" a misnomer in the case of an infection?

20. Define *immune response.*

21. What happens to pathogens during phagocytosis?

22. Differentiate clearly between humoral and cellular immunity.

23. Although the immune system has two arms, it has been said, "No T cells, no immunity." How is this so?

24. Define *immunocompetence.* What indicates that a B cell or T cell has developed immunocompetence? Where does the "programming phase" occur in the case of B cells?

25. Binding of antigens to receptors of immunocompetent lymphocytes leads to clonal selection. Describe the process of clonal selection. What nonlymphocyte cell is a central actor in this process, and what is its function?

26. What is the connection between T cells and memory cells?

27. Describe the specific roles of helper, cytotoxic, and regulatory T cells in cell-mediated immunity. Which is thought to be disabled in AIDS?

28. Compare and contrast a primary and a secondary immune response. Which is more rapid, and why?

29. Describe the structure of an antibody, and explain the importance of its variable and constant regions.

30. Name the five classes of immunoglobulins. Which is most likely to be found attached to a B cell membrane? Which is most abundant in plasma? Which is important in allergic responses? Which is the first Ig to be released during the primary response? Which can cross the placental barrier?

31. How can mild fever be beneficial to the body?

32. Distinguish between immediate types of allergy and delayed allergic reactions in terms of cause and consequences.

33. What events can result in the loss of self-tolerance and autoimmune disease?

34. Would lack of memory B cells for a particular antigen impact the primary or the secondary humoral response?

35. Contrast lysozyme, perforin and granzymes, and a membrane attack complex. Include which cell(s) or proteins secrete or produce each.

 ## Critical Thinking and Clinical Application Questions

36. As an infant receives her first dose of oral polio vaccine, the nurse explains to her parents that the vaccine is a preparation of weakened virus. What type of immunity will the infant develop?

37. Some people with a deficit of IgA exhibit recurrent paranasal sinus and respiratory tract infections. Explain these symptoms.

38. Mr. James, an 80-year-old man, is grumbling about having to receive a flu shot every year. Flu viruses have a high mutation rate (undergo rapid genetic changes), which results in the appearance of new proteins on the flu virus's "coat." How does this help explain the need to get a flu shot each year?

39. Mrs. Morrow, a 59-year-old woman, has undergone a left radical mastectomy (removal of the left breast and left axillary lymph nodes and vessels). Her left arm is severely swollen and painful, and she is unable to raise it more than shoulder height. (a) Explain her signs and symptoms. (b) Can she expect relief from these symptoms in time? How so?

40. Lymphocytes continuously circulate through the body using blood and lymph as their transport vehicles. What is the importance of this recirculation behavior?

41. Capillary permeability increase and plasma proteins leak into the interstitial fluid as part of the inflammatory process. Why is this desirable?

# The Respiratory System

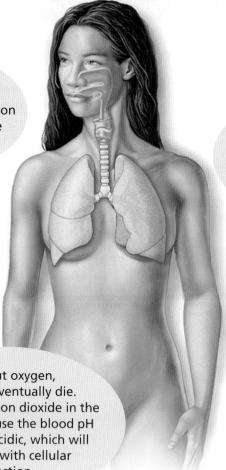

**WHAT**

The respiratory system provides oxygen to the body, disposes of carbon dioxide, and helps reguate blood pH.

**HOW**

Gas exchange occurs in the air sacs of the lungs, called *alveoli*, and at capillary beds around the body.

**WHY**

Without oxygen, cells will eventually die. Too much carbon dioxide in the blood will cause the blood pH to become acidic, which will interfere with cellular function.

**INSTRUCTORS**

New **Building Vocabulary Coaching** Activities for this chapter are assignable in Mastering A&P®

The trillions of cells in the body require an abundant and endless supply of oxygen to carry out their vital functions. We cannot "do without" oxygen for even a little while, as we can without food or water. Furthermore, as cells use oxygen, they give off carbon dioxide, a waste product the body must get rid of.

The *cardiovascular* and *respiratory systems* share responsibility for supplying the body with oxygen and disposing of carbon dioxide. The respiratory system organs oversee the gas exchanges that occur between the blood and the external environment. Using blood as the transporting fluid, the cardiovascular system organs transport respiratory gases between the lungs and the cells in the rest of the body. If either system fails, cells begin to die from oxygen starvation and accumulation of carbon dioxide.

## Functional Anatomy of the Respiratory System

→ **Learning Objectives**

☐ **Name the organs forming the respiratory passageway from the nasal cavity to the alveoli of the lungs (or identify them on a diagram or model), and describe the function of each.**

☐ **Describe several protective mechanisms of the respiratory system.**

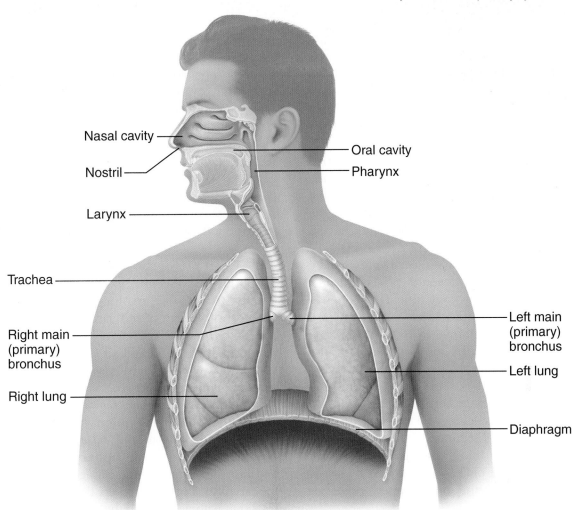

**Figure 13.1 The major respiratory organs shown in relation to surrounding structures.**

The organs of the **respiratory system** include the nose, pharynx, larynx, trachea, bronchi and their smaller branches, and the lungs, which contain the *alveoli* (al-ve′o-li), or terminal air sacs **(Figure 13.1)**. Because gas exchange with the blood happens *only* in the alveoli, the other respiratory system structures are really just conducting passageways that carry air through the lungs. The passageways from the nose to the larynx are called the *upper respiratory tract*, and those from the trachea to the alveoli are the *lower respiratory tract*. These conducting passageways also purify, humidify, and warm incoming air. Thus, the air finally reaching the lungs has fewer irritants (such as dust or bacteria) than the air outside, and it is warm and damp.

## The Nose

The **nose**, whether "button" or "hooked" in shape, is the only externally visible part of the respiratory system. During breathing, air enters the nose by passing through the **nostrils**, or **nares**. The interior of the nose consists of the **nasal cavity**, divided by a midline **nasal septum**. The *olfactory receptors* for the sense of smell are located in the mucosa in the slitlike superior part of the nasal cavity, just beneath the ethmoid bone.

### CONCEPT**LINK**

Recall that any area open to the outside of the body, including respiratory passages, is lined with mucous membrane (mucosa), which is a "wet," or moist, membrane (Chapter 4, p. 136).

The rest of the mucosa lining the nasal cavity, called the *respiratory mucosa*, rests on a rich network of thin-walled veins that warms the air as it flows past. In addition, the sticky mucus produced by this mucosa's glands moistens the air and traps

13

incoming bacteria and other foreign debris, and lysozyme enzymes in the mucus destroy bacteria chemically. The ciliated cells of the nasal mucosa create a gentle current that moves the sheet of contaminated mucus posteriorly toward the throat (pharynx), where it is swallowed and digested by stomach juices. We are usually unaware of this important ciliary action, but when the external temperature is extremely cold, these cilia become sluggish, allowing mucus to accumulate in the nasal cavity and to dribble outward through the nostrils. This helps explain why you might have a "runny" nose on a crisp, wintry day.

The lateral walls of the nasal cavity are uneven, owing to three mucosa-covered projections, or lobes, called **conchae** (kong′ke). The conchae greatly increase the surface area of the mucosa exposed to the air (as shown in **Figure 13.2**). The conchae also increase the air turbulence in the nasal cavity. As the air swirls through the twists and turns, inhaled particles are deflected onto the mucus-coated surfaces, where they are trapped and prevented from reaching the lungs.

The nasal cavity is separated from the oral cavity below by a partition, the **palate** (pal′et). Anteriorly, where the palate is supported by bone, is the **hard palate**; the unsupported posterior part is the **soft palate**.

### Homeostatic Imbalance 13.1

The genetic defect **cleft palate** (failure of the bones forming the palate to fuse medially) results in breathing difficulty as well as problems with oral cavity functions, such as nursing and speaking. _____ +

The nasal cavity is surrounded by a ring of **paranasal sinuses** located in the frontal, sphenoid, ethmoid, and maxillary bones. (See Figure 5.13, p. 177.) The sinuses lighten the skull and act as resonance chambers for speech. They also produce mucus, which drains into the nasal cavities. The suctioning effect created by nose blowing helps to drain the sinuses. The *nasolacrimal ducts*, which drain tears from the eyes, also empty into the nasal cavities.

### Homeostatic Imbalance 13.2

Cold viruses and various allergens can cause **rhinitis** (ri-ni′tis), inflammation of the nasal mucosa. The excessive mucus produced results in nasal congestion and postnasal drip. Because the nasal mucosa is continuous throughout the respiratory tract and extends tentacle-like into the nasolacrimal (tear) ducts and paranasal sinuses, nasal cavity infections often spread to those regions as well. **Sinusitis**, or sinus inflammation, is difficult to treat and can cause marked changes in voice quality. When the passageways connecting the sinuses to the nasal cavity are blocked with mucus or infectious matter, the air in the sinus cavities is absorbed. The result is a partial vacuum and a **sinus headache** localized over the inflamed area. _____ +

## The Pharynx

The **pharynx** (far′inks) is a muscular passageway about 13 cm (5 inches) long that vaguely resembles a short length of red garden hose. Commonly called the *throat*, the pharynx serves as a common passageway for food and air (see Figure 13.2). It is continuous with the nasal cavity anteriorly via the **posterior nasal aperture**.

The pharynx has three regions (Figure 13.2a). Air enters the superior portion, the **nasopharynx** (na″zo-far′inks), from the nasal cavity and then descends through the **oropharynx** (o″ro-far′inks) and **laryngopharynx** (lah-ring″go-far′inks) to enter the larynx below. Food enters the mouth, then travels along with air through the oropharynx and laryngopharynx. But instead of entering the larynx, food is directed into the *esophagus* (ĕ-sof′ah-gus) posteriorly by a flap called the *epiglottis* (ep″ĭ-glot′is).

The *pharyngotympanic tubes*, which drain the middle ears, open into the nasopharynx. The mucosae of these two regions are continuous, so ear infections such as *otitis media* (o-ti′tis me′de-ah) may follow a sore throat or other types of pharyngeal infections.

Clusters of lymphatic tissue called *tonsils* are also found in the pharynx. The single **pharyngeal** (far-rin′je-al) **tonsil**, often called the *adenoid*, is located high in the nasopharynx. The two **palatine tonsils** are in the oropharynx at the end of the soft palate, as are the two **lingual tonsils**, which lie at the base of the tongue. The tonsils also play a role in protecting the body from infection (see Chapter 12).

### Homeostatic Imbalance 13.3

If the pharyngeal tonsil becomes inflamed and swollen (as during a bacterial infection), it obstructs

**Figure 13.2 Basic anatomy of the upper respiratory tract, sagittal section.**

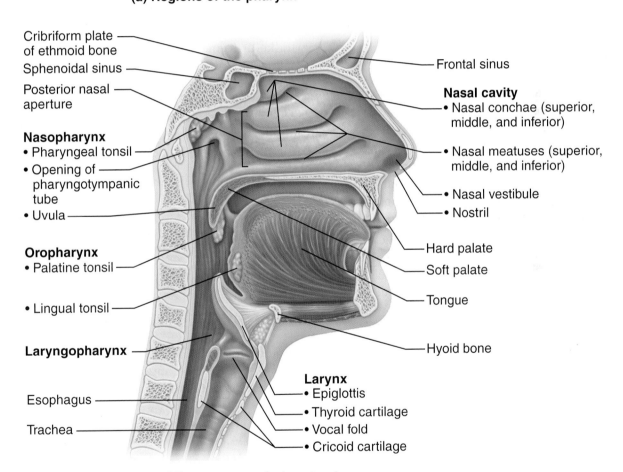

**Pharynx**
• Nasopharynx
• Oropharynx
• Laryngopharynx

**(a) Regions of the pharynx**

Cribriform plate of ethmoid bone

Sphenoidal sinus

Posterior nasal aperture

**Nasopharynx**
• Pharyngeal tonsil
• Opening of pharyngotympanic tube
• Uvula

**Oropharynx**
• Palatine tonsil

• Lingual tonsil

**Laryngopharynx**

Esophagus

Trachea

Frontal sinus

**Nasal cavity**
• Nasal conchae (superior, middle, and inferior)

• Nasal meatuses (superior, middle, and inferior)

• Nasal vestibule
• Nostril

Hard palate

Soft palate

Tongue

Hyoid bone

**Larynx**
• Epiglottis
• Thyroid cartilage
• Vocal fold
• Cricoid cartilage

**(b) Detailed anatomy of the upper respiratory tract**

the nasopharynx and forces the person to breathe through the mouth. In mouth breathing, air is not properly moistened, warmed, or filtered before reaching the lungs. Many children routinely get this condition, called **tonsillitis**. Years ago the belief was that the tonsils, though protective, were more trouble than they were worth, and they were routinely removed. Now, with the widespread use of antibiotics, this is no longer necessary. _____✚

## The Larynx

The **larynx** (lar′inks), or *voice box*, routes air and food into the proper channels and plays a role in speech. Located inferior to the pharynx (Figure 13.2b), it is formed by eight rigid hyaline cartilages and a spoon-shaped flap of elastic cartilage, the **epiglottis**. The largest of the hyaline cartilages is the shield-shaped **thyroid cartilage**, which protrudes anteriorly and is commonly called the *Adam's apple*.

Sometimes referred to as the "guardian of the airway," the epiglottis protects the superior opening of the larynx. During regular breathing, the epiglottis allows the passage of air into the lower respiratory passages. When we swallow food or fluids, the situation changes dramatically; the larynx is pulled upward, and the epiglottis tips, forming a lid over the larynx's opening. This routes food into the esophagus, which leads to the stomach, posteriorly. If anything other than air enters the larynx, a *cough reflex* is triggered to prevent the substance from continuing into the lungs. Because this protective reflex does *not* work when we are unconscious, never try to give fluids to an unconscious person when attempting to revive him or her.

- Palpate your larynx by placing your hand midway on the anterior surface of your neck. Swallow. Can you feel the larynx rising as you swallow?

Part of the mucous membrane of the larynx forms a pair of folds, called the **vocal folds**, or **true vocal cords**, which vibrate with expelled air. This ability of the vocal folds to vibrate allows us to speak. The vocal folds and the slitlike passageway between them are called the **glottis**.

## The Trachea

Air entering the **trachea** (tra′ke-ah), or *windpipe*, from the larynx travels down its length (10–12 cm, or about 4 inches) to the level of the fifth thoracic vertebra, which is approximately midchest (see Figure 13.1).

The trachea is fairly rigid because its walls are reinforced with C-shaped rings of **hyaline cartilage (Figure 13.3a)**. These rings serve a double purpose. The open parts of the rings abut the *esophagus* and allow it to expand anteriorly when we swallow a large piece of food. The solid portions support the trachea walls and keep it *patent*, or open, in spite of the pressure changes that occur during breathing. The *trachealis* muscle lies next to the esophagus and completes the wall of the trachea posteriorly.

 **Homeostatic Imbalance 13.4**

Because the trachea is the only way air can enter the lungs, tracheal obstruction is life-threatening. Many people have suffocated after choking on a piece of food that suddenly closed off the trachea (or the glottis of the larynx). The **Heimlich maneuver**, a procedure in which the air in a person's own lungs is used to "pop out," or expel, an obstructing piece of food, has saved many people from becoming victims of choking. The Heimlich maneuver is simple to learn and easy to do. However, it is best learned by demonstration because cracked ribs are a distinct possibility when it is done incorrectly. In some cases of obstructed breathing, an emergency *tracheostomy* (tra′ke-ost′o-me; surgical opening of the trachea) is done to provide an alternative route for air to reach the lungs. Individuals with tracheostomy tubes in place form huge amounts of mucus the first few days because of irritation to the trachea. Thus, they must be suctioned frequently during this time to prevent the mucus from pooling in their lungs. _____ ✚

The trachea is lined with a ciliated mucosa (Figure 13.3b). The cilia beat continuously in a superior direction. They are surrounded by *goblet cells* that produce mucus. The cilia propel this mucus, loaded with dust particles and other debris, away from the lungs to the throat, where it can be swallowed or spat out.

 **Homeostatic Imbalance 13.5**

Smoking inhibits and ultimately destroys the cilia. Without these cilia, coughing is the only means of preventing mucus from accumulating in the lungs. Smokers with respiratory congestion should avoid medications that inhibit the cough reflex. _____ ✚

### Did You Get It?

1. Why is nose breathing preferable to mouth breathing?
2. What is the specific protective function of cilia in the trachea?

For answers, see Appendix A.

## The Main Bronchi

The right and left **main (primary) bronchi** (brong′ki) are formed by the division of the trachea. Each main bronchus runs obliquely before it

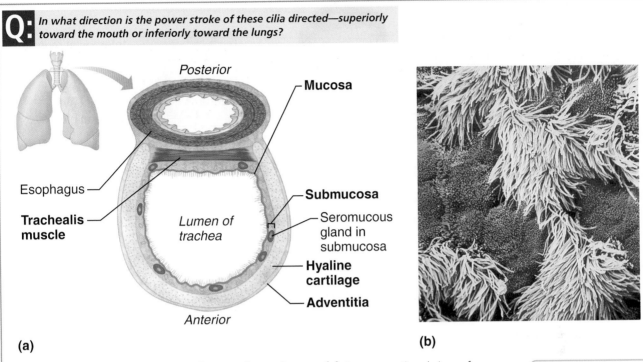

**Q:** *In what direction is the power stroke of these cilia directed—superiorly toward the mouth or inferiorly toward the lungs?*

**Figure 13.3 Anatomy of the trachea and esophagus. (a)** A cross-sectional view of the structural relationship between the trachea and esophagus. **(b)** Cilia in the trachea. The cilia are the yellow, grasslike projections surrounded by the mucus-secreting goblet cells, which exhibit short microvilli (orange). (Scanning electron micrograph, 1,800×.)

View **PAL** Histology
Mastering A&P°

plunges into the medial depression (*hilum*) of the lung on its own side (see Figure 13.1). The right main bronchus is wider, shorter, and straighter than the left. Consequently, it is the more common site for an inhaled foreign object to become lodged. By the time incoming air reaches the bronchi, it is warm, cleansed of most impurities, and humid. The smaller subdivisions of the main bronchi within the lungs are direct routes to the air sacs.

## The Lungs

→ **Learning** Objectives

☐ **Describe the structure and function of the lungs and the pleural coverings.**

☐ **Describe the structure of the respiratory membrane.**

The **lungs** are fairly large organs. They occupy the entire thoracic cavity except for the most central area, the **mediastinum** (me″de-as-ti′num), which houses the heart, the great blood vessels, bronchi, the esophagus, and other organs (**Figure 13.4**, p. 468). The narrow superior portion of each lung, the **apex**, is just deep to the clavicle. The broad lung area resting on the diaphragm is the **base**. Each lung is divided into lobes by fissures; the left lung has two lobes, and the right lung has three.

The surface of each lung is covered with its own visceral serosa, called the **pulmonary pleura** or **visceral pleura** (ploor′ah), and the walls of the thoracic cavity are lined by the **parietal pleura**. The pleural membranes produce *pleural fluid*, slippery serous fluid, which allows the lungs to glide easily over the thorax wall during breathing and causes the two pleural layers to cling together. The pleurae can slide easily from side to side across one another, but they strongly resist being pulled apart. Consequently, the lungs are held tightly to the thorax wall, and the **pleural space** is more of a potential space than an actual one. As we describe shortly, this tight adherence of the pleural membranes is absolutely essential for normal breathing. (Figure 13.4a shows the position of the pleurae on the lungs and the thorax wall.)

**A:** Superiorly toward the mouth, to prevent unwanted substances from entering the lungs.

13

**Figure 13.4 Anatomical relationships of organs in the thoracic cavity.**
Part **(a)** includes a close-up illustration of the pleurae. In **(b)**, the size of the pleural (and pericardial) cavity is exaggerated for clarity.

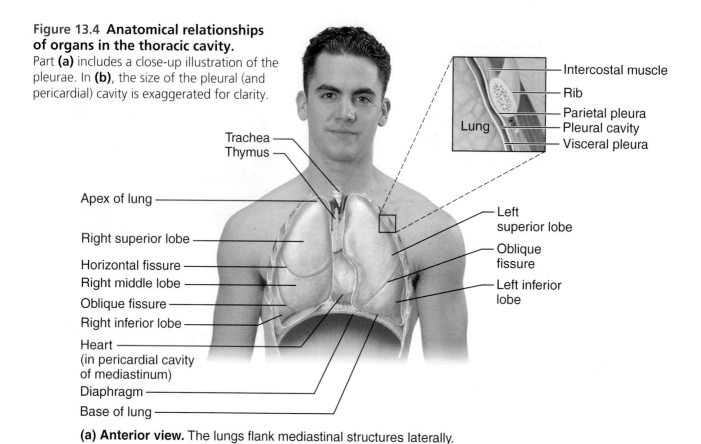

Intercostal muscle
Rib
Parietal pleura
Pleural cavity
Visceral pleura
Lung

Trachea
Thymus

Apex of lung

Right superior lobe

Horizontal fissure

Right middle lobe

Oblique fissure

Right inferior lobe

Heart
(in pericardial cavity
of mediastinum)

Diaphragm

Base of lung

Left superior lobe

Oblique fissure

Left inferior lobe

**(a) Anterior view.** The lungs flank mediastinal structures laterally.

*Posterior*

Vertebra

Esophagus
(in posterior mediastinum)

**Root of lung at hilum**
• Left main bronchus
• Left pulmonary artery
• Left pulmonary vein

Right lung

Parietal pleura

Visceral pleura

Pleural cavity

Left lung

Thoracic wall

Pulmonary trunk

Pericardial membranes

Heart (in mediastinum)

Sternum

Anterior mediastinum

*Anterior*

**(b) Transverse section through the thorax, viewed from above**

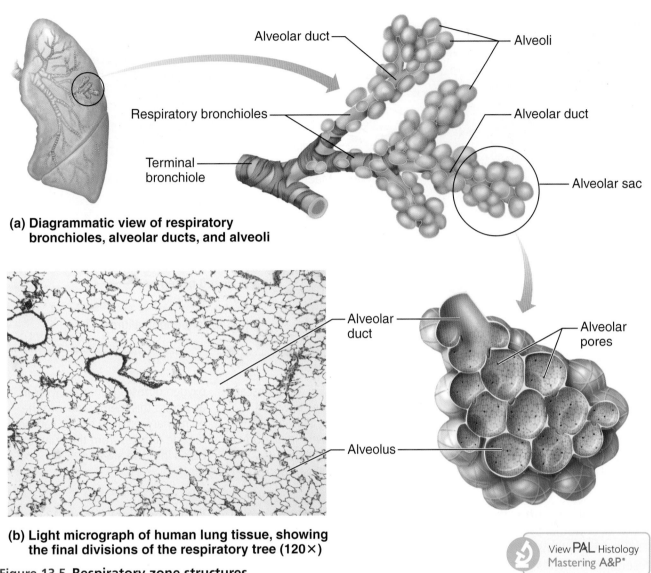

(a) **Diagrammatic view of respiratory bronchioles, alveolar ducts, and alveoli**

(b) **Light micrograph of human lung tissue, showing the final divisions of the respiratory tree (120×)**

View PAL Histology
Mastering A&P®

**Figure 13.5 Respiratory zone structures.**

### Homeostatic Imbalance 13.6

**Pleurisy** (ploo′rĭ-se), inflammation of the pleurae, can be caused by insufficient secretion of pleural fluid. The pleural surfaces become dry and rough, resulting in friction and stabbing pain with each breath. Conversely, the pleurae may produce excessive amounts of fluid, which exerts pressure on the lungs. This type of pleurisy hinders breathing movements, but it is much less painful than the dry rubbing type. _____ +

### The Bronchial Tree

After entering the lungs, the main bronchi subdivide into smaller and smaller branches (secondary and tertiary bronchi, and so on), finally ending in the small-est of the conducting passageways, the **bronchioles** (brong′ke-ōlz). Because of this branching and rebranching of the respiratory passageways within the lungs, the network formed is often referred to as the *bronchial*, or *respiratory*, *tree*. All but the smallest branches have reinforcing cartilage in their walls.

### Respiratory Zone Structures and the Respiratory Membrane

The *terminal bronchioles* lead into *respiratory zone structures*, even smaller conduits that eventually terminate in **alveoli** (al-ve′o-li; *alveol* = small cavity), or air sacs. The **respiratory zone**, which includes the *respiratory bronchioles, alveolar ducts, alveolar sacs*, and *alveoli*, is the only site of gas exchange **(Figure 13.5)**. All other respiratory

13

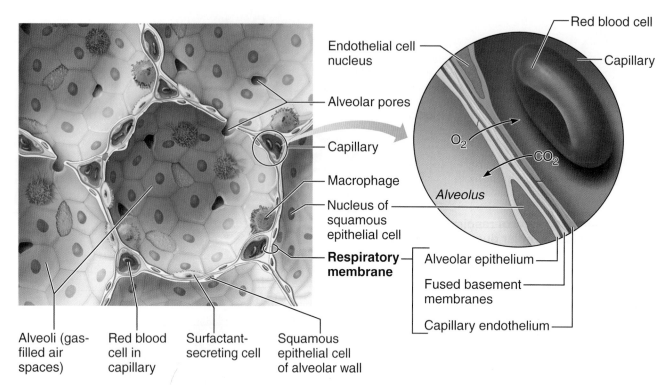

**Figure 13.6 Functional anatomy of the respiratory membrane (air-blood barrier).** As shown in the left illustration, the respiratory membrane is composed of squamous epithelial cells of the alveoli, the capillary endothelium, and the scant basement membranes between. Surfactant-secreting cells in the alveoli are also shown. Small pores connect neighboring alveoli. As shown in the right illustration, oxygen diffuses from the alveolar air across this membrane into the pulmonary capillary blood; carbon dioxide diffuses from the pulmonary blood into the alveolus.

passages are **conducting zone structures** that serve as conduits to and from the respiratory zone. There are millions of the clustered alveoli, which resemble bunches of grapes, and they make up the bulk of the lungs. Consequently, the lungs are mostly air spaces. The balance of the lung tissue, its *stroma*, is mainly elastic connective tissue that allows the lungs to stretch and recoil (spring back) as we breathe. Thus, in spite of their relatively large size, the lungs weigh only about 2½ pounds, and they are soft and spongy.

The walls of the alveoli are composed largely of a single, thin layer of simple squamous epithelial cells. The thinness of their walls is hard to imagine, but a sheet of tissue paper is much thicker. *Alveolar pores* connect neighboring air sacs and provide alternative routes for air to reach alveoli whose feeder bronchioles have been clogged by mucus or otherwise blocked. The external surfaces of the alveoli are covered with a "cobweb" of pulmonary capillaries. Together, the alveolar and capillary walls, their fused basement membranes, and occasional elastic fibers construct the **respiratory**

**membrane (air-blood barrier) (Figure 13.6)**. The respiratory membrane has gas (air) flowing past on one side and blood flowing past on the other. Gas exchange occurs by simple diffusion through the respiratory membrane—oxygen passes from the alveolar air into the capillary blood, and carbon dioxide leaves the blood to enter the alveoli (Figure 13.6). The total gas exchange surface provided by the alveolar walls of a healthy man is estimated to be 50 to 70 square meters, or approximately 40 times greater than the surface area of his skin!

The final line of defense for the respiratory system is in the alveoli. Remarkably efficient **alveolar macrophages**, sometimes called "dust cells," wander in and out of the alveoli picking up bacteria, carbon particles, and other debris. Also scattered amid the epithelial cells that form most of the alveolar walls are cuboidal surfactant-secreting cells, which look very different from the squamous epithelial cells. These cells produce a lipid (fat) molecule called *surfactant*, which coats the gas-exposed alveolar surfaces and is very important in lung function (as described on p. 483).

## Did You Get It?

3. Name the order of the following parts of the human respiratory system from the site where air enters the nostrils to the site where air reaches the end passages of the lungs—bronchi, larynx, nasal cavity, alveoli, trachea, pharynx, bronchioles.

4. Which main bronchus is the most likely site for an inhaled object to become lodged? Why?

5. The lungs are mostly passageways and elastic tissue. What is the role of the passageways? Of the elastic tissue?

6. Name the four structures that make up the respiratory zone.

**For answers, see Appendix A.**

# Respiratory Physiology

→ **Learning Objectives**

☐ Define *cellular respiration, external respiration, internal respiration, pulmonary ventilation, expiration,* and *inspiration.*

☐ **Explain how the respiratory muscles cause volume changes that lead to air flow into and out of the lungs (breathing).**

The major function of the respiratory system is to supply the body with oxygen and to dispose of carbon dioxide. To do this, at least four distinct events, collectively called **respiration**, must occur:

1. **Pulmonary ventilation.** Air must move into and out of the lungs so that the gases in the alveoli of the lungs are continuously refreshed. This process of pulmonary ventilation is commonly called **breathing**.

2. **External respiration.** Gas exchange (oxygen loading and carbon dioxide unloading) between the pulmonary blood and alveoli must take place. Remember that in **ex**ternal respiration, gas exchanges are being made between the blood and the body *exterior*.

3. **Respiratory gas transport.** Oxygen and carbon dioxide must be transported to and from the lungs and tissue cells of the body via the bloodstream.

4. **Internal respiration.** At systemic capillaries, gas exchange occurs between the blood and cells *inside* the body.

Although only the first two processes are the special responsibility of the respiratory system, all four processes are necessary for gas exchange to occur. Note that **cellular respiration**, the *use* of oxygen to produce ATP and carbon dioxide, is the cornerstone of all energy-producing chemical reactions and occurs in all cells (see Chapter 14).

## Mechanics of Breathing

Breathing, or pulmonary ventilation, is a mechanical process that depends on volume changes occurring in the thoracic cavity. Keep this rule in mind about the mechanics of breathing: *Volume changes lead to pressure changes, which lead to the flow of gases to equalize the pressure.*

A gas, like a liquid, always conforms to the shape of its container. However, unlike a liquid, a gas *fills* its container. Therefore, in a large volume, the gas molecules will be far apart, and the pressure (created by the gas molecules hitting each other and the walls of the container) will be low. Assuming the amount of gas remains constant, if the volume is reduced, the gas molecules will be closer together, and the pressure will rise.

> **CONCEPTLINK**
> Recall that pressure changes also drive other processes in the body, such as filtration (passive transport; Chapter 3, pp. 102–103) and blood flow (Chapter 11). In these processes, substances move from high to low pressure and achieve a specific function, such as membrane transport or circulation.

Let's see how volume changes relate to the two phases of breathing—**inspiration**, when air is flowing into the lungs, and **expiration**, when air is leaving the lungs.

### Inspiration

When the inspiratory muscles, the **diaphragm** and **external intercostals**, contract, the size of the thoracic cavity increases. As the dome-shaped diaphragm contracts inferiorly, the superior-inferior dimension (height) of the thoracic cavity increases (**Figure 13.7a**, p. 472). Contraction of the external intercostals lifts the rib cage and thrusts the sternum forward, which increases the antero-posterior and lateral dimensions of the thorax. The lungs adhere tightly to the thorax walls (because of the surface tension of the fluid between the pleural membranes), so they are stretched to the new, larger size of the thorax. As **intrapulmonary volume** (the volume within the lungs) increases, the gases within the lungs spread out to fill the larger space. The resulting decrease in gas

**13**

**Changes in anterior-posterior and superior-inferior dimensions**

**Changes in lateral dimensions**

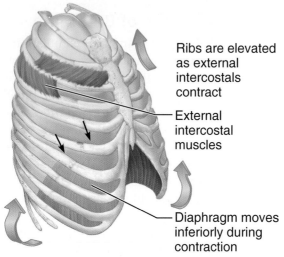

Ribs are elevated as external intercostals contract

External intercostal muscles

Diaphragm moves inferiorly during contraction

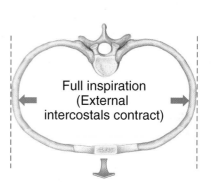

Full inspiration (External intercostals contract)

**(a) Inspiration: Air (gases) flows into the lungs**

Ribs are depressed as external intercostals relax

External intercostal muscles

Diaphragm moves superiorly as it relaxes

Expiration (External intercostals relax)

**(b) Expiration: Air (gases) flows out of the lungs**

**Figure 13.7 Rib cage and diaphragm positions during breathing.** Black arrows represent the direction of diaphragm movement. **(a)** At the end of a normal inspiration: Chest is expanded laterally, rib cage is elevated, and diaphragm is depressed. Lungs are stretched to the larger thoracic volume, causing the intrapulmonary pressure to fall and air to flow into the lungs. **(b)** At the end of a normal expiration: The chest is depressed, and the lateral dimension is reduced; the rib cage is descended; and the diaphragm is elevated and dome-shaped. Lungs recoil to a smaller volume, intrapulmonary pressure rises, and air flows out of the lungs.

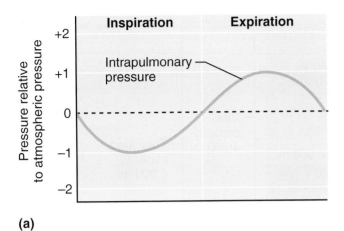

**(a)**

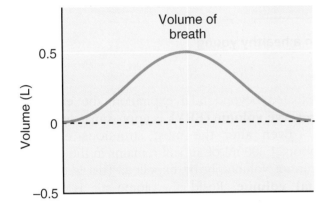

**(b)**

**Figure 13.8 Changes in (a) intrapulmonary pressure and (b) air flow during inspiration and expiration.**

pressure in the lungs produces a partial vacuum (pressure less than atmospheric pressure outside the body), which causes air to flow into the lungs. Air continues to move into the lungs until the intrapulmonary pressure equals atmospheric pressure **(Figure 13.8a)**. This series of events is called *inspiration* (inhalation).

### Expiration

*Expiration* (exhalation) in healthy people is largely a passive process that depends more on the natural elasticity of the lungs than on muscle contraction. As the inspiratory muscles relax and resume their initial resting length, the rib cage descends, the diaphragm relaxes superiorly, and the lungs recoil (Figure 13.7b). Thus, both the thoracic and intrapulmonary volumes decrease. As the intrapulmonary volume decreases, the gases inside the lungs are forced more closely together, and the intrapulmonary pressure rises to a point higher

than atmospheric pressure (see Figure 13.8a). This causes the gases to passively flow out to equalize the pressure with the outside.

However, if the respiratory passageways are narrowed by spasms of the bronchioles (as in *asthma*) or clogged with mucus or fluid (as in *chronic bronchitis* or *pneumonia*), expiration becomes an active process. In such cases of *forced expiration*, the internal intercostal muscles are activated to help depress the rib cage, and the abdominal muscles contract and help to force air from the lungs by squeezing the abdominal organs upward against the diaphragm.

Normally the pressure within the pleural space, the **intrapleural pressure**, is *always* negative. This is the major factor preventing lung collapse. If for any reason the intrapleural pressure becomes equal to the atmospheric pressure, the lungs immediately recoil and collapse.

### Homeostatic Imbalance 13.7

During **atelectasis** (a″teh-lek′tuh-sis), or lung collapse, the lung is useless for ventilation. This phenomenon occurs when air enters the pleural space through a chest wound, but it may also result from a rupture of the visceral pleura, which allows air to enter the pleural space from the respiratory tract. The presence of air in the intrapleural space, which disrupts the fluid bond between the pleurae, is referred to as **pneumothorax** (nu″mo-tho′raks).

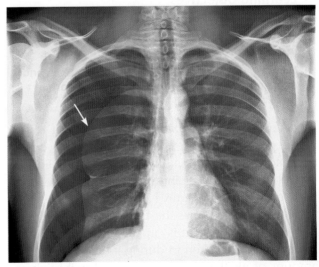

A colored chest X-ray film showing a pneumothorax, or collapsed lung. The patient's right lung (on left side of X-ray film, bright blue) has collapsed because of a buildup of air (purple) between the lung and chest wall.

13

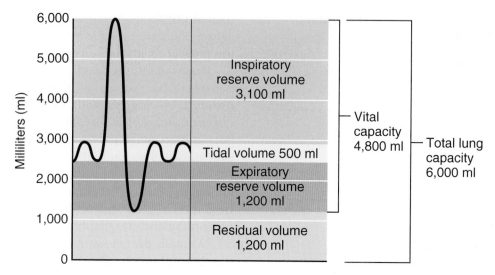

**Figure 13.9 Graph of the various respiratory volumes in a healthy young adult male.**

Pneumothorax is reversed by drawing air out of the intrapleural space with a chest tube, which allows the lung to reinflate and resume its normal function. _____ ✛

### Did You Get It?

7. What is the most basic function of respiration?
8. What causes air to flow out of the lungs during expiration?

For answers, see Appendix A.

## Respiratory Volumes and Capacities

→ **Learning** Objectives

☐ **Define the following respiratory volumes:** *tidal volume, vital capacity, expiratory reserve volume, inspiratory reserve volume,* and *residual air.*

☐ **Name several nonrespiratory air movements, and explain how they modify or differ from normal respiratory air movements.**

Many factors affect *respiratory capacity*—for example, a person's size, sex, age, and physical condition. Normal quiet breathing moves approximately 500 ml of air (about a pint) into and out of the lungs with each breath (Figure 13.8b). This respiratory volume is referred to as the **tidal volume (TV)**.

As a rule, a person is capable of inhaling much more air than is taken in during a tidal breath. The amount of air that can be taken in forcibly above the tidal volume is the **inspiratory reserve volume (IRV)**, which is around 3,100 ml.

Similarly, after a normal expiration, more air can be exhaled. The amount of air that can be forcibly exhaled beyond tidal expiration, the **expiratory reserve volume (ERV)**, is approximately 1,200 ml.

Even after the most strenuous expiration, about 1,200 ml of air still remains in the lungs and cannot voluntarily be expelled. This is the **residual volume**. Residual volume air is important because it allows gas exchange to go on continuously even between breaths and helps to keep the alveoli open (inflated).

The total amount of exchangeable air (around 4,800 ml in healthy young men and 3,100 ml in healthy young women) is the **vital capacity (VC)**. The vital capacity is the sum of the tidal volume plus the inspiratory and expiratory reserve volumes. Respiratory volumes of a male are summarized in **Figure 13.9**.

Note that some of the air that enters the respiratory tract remains in the conducting zone passageways and never reaches the alveoli. This is called the **dead space volume** and during a normal tidal breath is about 150 ml. The functional volume—air that actually reaches the respiratory zone and contributes to gas exchange—is about 350 ml.

Respiratory capacities are measured with a *spirometer* (spi-rom′ĕ-ter). As a person breathes, the volumes of air exhaled can be read on an indicator, which shows the changes in air volume inside the apparatus. Spirometer testing is useful for evaluating losses in respiratory functioning and in following the course of some respiratory diseases. In pneumonia, for example, inspiration is obstructed, and the IRV and VC decrease. In emphysema, where expiration is hampered, the

| Table **13.1** | Nonrespiratory Air (Gas) Movements |
|---|---|
| **Movement** | **Mechanism and result** |
| Cough | Taking a deep breath, closing glottis, and forcing air superiorly from lungs against glottis. Then, glottis opens suddenly, and a blast of air rushes upward. Coughs act to clear the lower respiratory passageways. |
| Sneeze | Similar to a cough, except that expelled air is directed through nasal cavities instead of through oral cavity. The uvula (u'vu-lah), a dangling tag of tissue hanging from the soft palate, becomes depressed and closes oral cavity off from pharynx, routing air through nasal cavities. Sneezes clear upper respiratory passages. |
| Crying | Inspiration followed by release of air in a number of short expirations. Primarily an emotionally induced mechanism. |
| Laughing | Essentially same as crying in terms of the air movements produced. Also an emotionally induced response. |
| Hiccups | Sudden inspirations resulting from spasms of diaphragm; initiated by irritation of diaphragm or phrenic nerves, which serve diaphragm. The sound occurs when inspired air hits vocal folds of closed glottis. |
| Yawn | Very deep inspiration, taken with jaws wide open; ventilates all alveoli (some alveoli may remain collapsed during normal quiet breathing). |

ERV is much lower than normal, and the residual volume is higher.

## Nonrespiratory Air Movements

Many situations other than breathing move air into or out of the lungs and may modify the normal respiratory rhythm. Coughs and sneezes clear the air passages of debris or collected mucus. Laughing and crying reflect our emotions. For the most part, these **nonrespiratory air movements** are a result of reflex activity, but some may be produced voluntarily. (Examples of nonrespiratory movements are given in **Table 13.1**.)

## Respiratory Sounds

As air flows into and out of the respiratory tree, it produces two recognizable sounds that can be picked up with a stethoscope. **Bronchial sounds** are produced by air rushing through the large respiratory passageways (trachea and bronchi). **Vesicular** (vě-sik'u-lar) **breathing sounds** occur as air fills the alveoli. The vesicular sounds are soft murmurs that resemble a muffled breeze.

### Homeostatic Imbalance 13.8

Diseased respiratory tissue, mucus, or pus can produce abnormal sounds such as **crackle** (a bubbling sound), **wheezing** (a whistling sound), and *rales*. **Rales** are abnormal bronchial sounds produced by the presence of mucus or exudate in the lung passages or by thickening of the bronchial walls. ____✚

### Did You Get It?

9. Which is the largest respiratory volume—ERV, IRV, TV, or VC? Which is the smallest?
10. Dead space volume accounts for about 150 ml of tidal volume. How much of a tidal breath actually reaches the alveoli?
11. Jimmy broke a left rib when he fell from his bike. The rib punctured the chest wall. What happened to his left lung? Why?

For answers, see Appendix A.

## External Respiration, Gas Transport, and Internal Respiration

→ **Learning Objectives**

☐ **Describe the process of gas exchanges in the lungs and tissues.**

☐ **Describe how oxygen and carbon dioxide are transported in the blood.**

As explained earlier, *external respiration* is the actual exchange of gases between the alveoli and the blood (pulmonary gas exchange), and *internal respiration* is the gas exchange process that occurs between the blood and the tissue cells (systemic capillary gas exchange). Keep in mind that all gas exchanges obey the laws of diffusion; that is, movement occurs *toward* the area of lower concentration of the diffusing substance. The relative amounts of $O_2$ and $CO_2$ in the alveolar tissues, and in the arterial and venous blood, are illustrated in **Figure 13.10** (p. 476).

**13**

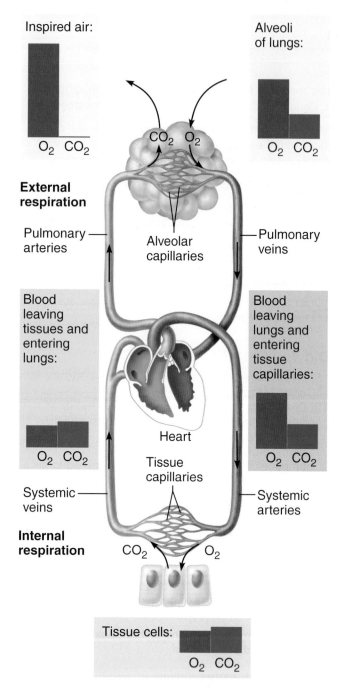

**Figure 13.10 Gas exchanges in external and internal respiration.** Note that these exchanges occur in the body according to the laws of diffusion.

## External Respiration

During external respiration, dark red blood flowing through the pulmonary circuit is transformed into the scarlet river that is returned to the heart for distribution to the systemic circuit. Although this color change is due to oxygen pickup by hemoglobin in the lungs, carbon dioxide is being unloaded from the blood equally fast. Because

body cells continually remove oxygen from blood, there is always more oxygen in the alveoli than in the blood. Thus, oxygen tends to diffuse from the air of the alveoli through the respiratory membrane into the more oxygen-poor blood of the pulmonary capillaries.

In contrast, as tissue cells remove oxygen from the blood in the systemic circulation, they release carbon dioxide into the blood. Because the concentration of carbon dioxide is much higher in the pulmonary capillaries than it is in the alveolar air, it will diffuse from the blood into the alveoli and be flushed out of the lungs during expiration. Relatively speaking, blood draining from the lungs into the pulmonary veins is rich in oxygen and poor in carbon dioxide.

## Gas Transport in the Blood

Oxygen is transported in the blood in two ways. Most attaches to hemoglobin molecules inside the red blood cells (RBCs) to form **oxyhemoglobin** (ok″se-he″mo-glo′bin)—$HbO_2$ **(Figure 13.11a)**. A very small amount of oxygen is carried dissolved in the plasma.

Carbon dioxide is twenty times more soluble in plasma compared to oxygen. As a result, most carbon dioxide is transported in plasma as **bicarbonate ion** ($HCO_3^-$), which plays a very important role in buffering blood pH. Carbon dioxide is enzymatically converted to bicarbonate ion within red blood cells; then the newly formed bicarbonate ions diffuse into the plasma.

### CONCEPT**LINK**

Remember that blood pH should remain between 7.35 and 7.45 (Chapter 10, p. 364). Buffers, such as bicarbonate ion, minimize changes in pH in order to maintain homeostasis.

A smaller amount of the transported $CO_2$ (between 20 and 30 percent) is carried inside the RBCs bound to hemoglobin. Carbon dioxide binds to hemoglobin at a different site from oxygen, so it does not interfere with oxygen transport.

Before carbon dioxide can diffuse out of the blood into the alveoli, it must first be released from its bicarbonate ion form. For this to occur, bicarbonate ions ($HCO_3^-$) must enter the red blood cells, where they combine with hydrogen ions ($H^+$) to form carbonic acid ($H_2CO_3$). Carbonic acid quickly splits to form water and carbon dioxide, and carbon dioxide then diffuses from the blood into the alveoli.

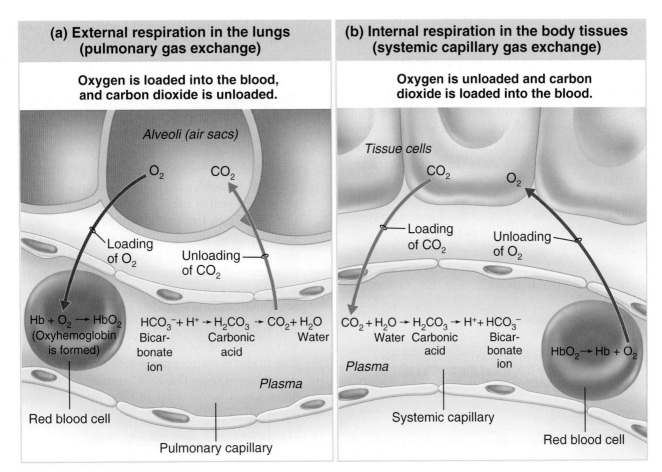

### (a) External respiration in the lungs (pulmonary gas exchange)

**Oxygen is loaded into the blood, and carbon dioxide is unloaded.**

*Alveoli (air sacs)*

$O_2$  $CO_2$

Loading of $O_2$  Unloading of $CO_2$

$Hb + O_2 \rightarrow HbO_2$ (Oxyhemoglobin is formed)

$HCO_3^- + H^+ \rightarrow H_2CO_3 \rightarrow CO_2 + H_2O$
Bicarbonate ion  Carbonic acid  Water

*Plasma*

Red blood cell

Pulmonary capillary

### (b) Internal respiration in the body tissues (systemic capillary gas exchange)

**Oxygen is unloaded and carbon dioxide is loaded into the blood.**

*Tissue cells*

$CO_2$  $O_2$

Loading of $CO_2$  Unloading of $O_2$

$CO_2 + H_2O \rightarrow H_2CO_3 \rightarrow H^+ + HCO_3^-$
Water  Carbonic acid  Bicarbonate ion

*Plasma*

$HbO_2 \rightarrow Hb + O_2$

Systemic capillary

Red blood cell

**Figure 13.11 The loading and unloading of oxygen ($O_2$) and carbon dioxide ($CO_2$) in the body.** Note that although the conversion of $CO_2$ to bicarbonate ion and the reverse reaction are shown occurring in the plasma in (b), most such conversions occur within the red blood cells. Additionally, though not illustrated, some $CO_2$ is carried within red blood cells bound to hemoglobin, and very small amounts of $O_2$ and $CO_2$ are carried dissolved in blood plasma.

---

### Homeostatic Imbalance 13.9

*Impaired oxygen transport:* Whatever the cause, inadequate oxygen delivery to body tissues is called **hypoxia** (hi-pok′se-ah). This condition is easy to recognize in light-skinned people because their skin and mucosae take on a bluish cast (become **cyanotic**; si″ah-not′ik). In dark-skinned individuals, this color change can be observed only in the mucosae and nailbeds. Hypoxia may be the result of anemia, pulmonary disease, or impaired or blocked blood circulation.

**Carbon monoxide poisoning** represents a unique type of hypoxia. Carbon monoxide (CO) is an odorless, colorless gas that competes vigorously with oxygen for the same binding sites on hemoglobin. Moreover, because hemoglobin binds to CO more readily than to oxygen, carbon monoxide is a very successful competitor—so much so that it crowds out or displaces oxygen.

Carbon monoxide poisoning is the leading cause of death from fire. It is particularly dangerous because it kills its victims softly and quietly. It does not produce the characteristic signs of hypoxia—cyanosis and respiratory distress. Instead, the victim becomes confused and has a throbbing headache. In rare cases, the skin becomes cherry red (the color of the hemoglobin-CO complex), which is often interpreted as a healthy "blush." People with CO poisoning are given 100 percent oxygen until the carbon monoxide has been cleared from the body. _____+

### Internal Respiration

Internal respiration, the exchange of gases between the blood and the tissue cells, is the opposite of what occurs in the lungs. In this process, oxygen leaves and carbon dioxide enters the blood (see Figure 13.11b). In the blood, carbon dioxide combines with water to form carbonic acid ($H_2CO_3$), which quickly releases bicarbonate ions. As previously mentioned, most of the conversion of carbon dioxide to bicarbonate ions occurs *inside* the RBCs, where a special enzyme (carbonic anhydrase) speeds up this reaction. Then the bicarbonate ions diffuse out into plasma, where they are transported. At the same time, oxygen is released from hemoglobin, and the oxygen diffuses quickly out of the blood to enter the cells. As a result of these exchanges, venous blood in the systemic circulation is much poorer in oxygen and richer in carbon dioxide than blood leaving the lungs.

## Did You Get It?

12. Which type of cellular transport moves respiratory gases between the blood and the body's cells?
13. What is the major form in which $CO_2$ is transported in the blood?
14. What is cyanosis?

For answers, see Appendix A.

## Control of Respiration

### ➜ Learning Objectives

☐ **Name the brain areas involved in control of respiration.**

☐ **Name several physical factors that influence respiratory rate.**

☐ **Explain the relative importance of oxygen and carbon dioxide in modifying breathing rate and depth.**

☐ **Explain why it is not possible to stop breathing voluntarily.**

☐ **Define *apnea*, *hyperventilation*, and *hypoventilation*.**

Although our tidelike breathing seems so beautifully simple, its control is fairly complex. We will cover only the most basic aspects of the respiratory controls.

### Neural Regulation: Setting the Basic Rhythm

The activity of the respiratory muscles, the diaphragm and external intercostals, is regulated by nerve impulses transmitted from the brain by the **phrenic nerves** and **intercostal nerves**. Neural centers that control respiratory rhythm and depth are located mainly in the *medulla* and *pons*

(Figure 13.12). Much remains to be discovered about neural regulation of breathing. A brief summary of what we know follows.

- The medulla contains two respiratory centers. The first, the *ventral respiratory group (VRG)*, contains both inspiratory and expiratory neurons that alternately send impulses to control the rhythm of breathing. The inspiratory neurons stimulate the diaphragm and external intercostal muscles via the phrenic and intercostal nerves, respectively, during quiet breathing. Impulses from the expiratory neurons *stop* the stimulation of the diaphragm and external intercostal muscles, allowing passive exhalation to occur. Impulses from the VRG maintain a normal quiet breathing rate of 12 to 15 respirations/minute, a rate called eupnea (ūp-ne′ah).

- The other medullary center, the *dorsal respiratory group (DRG)*, integrates sensory information from chemoreceptors and peripheral stretch receptors. The DRG communicates this information to the VRG to help modify breathing rhythms.

- The pons respiratory centers, which also communicate with the VRG, help to smooth the transitions (modify timing) between inhalation and exhalation during activities such as singing, sleeping or exercising.

The bronchioles and alveoli have stretch receptors that respond to extreme overinflation (which might damage the lungs) by initiating protective reflexes. In the case of overinflation, the vagus nerves send impulses from the stretch receptors to the medulla; soon thereafter, inspiration ends and expiration occurs. This is one example of DRG integration during respiratory control.

During exercise, we breathe more vigorously and deeply because the brain centers send more impulses to the respiratory muscles. This respiratory pattern is called **hyperpnea** (hy-perp′ne-ah). After strenuous exercise, expiration becomes active, and the abdominal muscles and any other muscles capable of depressing the ribs are used to aid expiration.

### ⚖ Homeostatic Imbalance 13.10

If the medullary centers are completely suppressed (as with an overdose of sleeping pills, morphine, or alcohol), respiration stops completely, and death occurs. _____✚

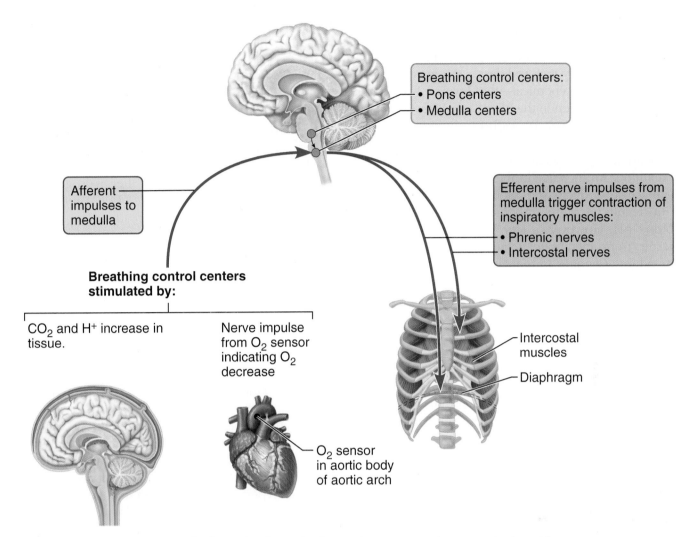

**Figure 13.12 Neural control of respiration.** This figure shows sensory inputs to the breathing control centers, which send signals through effector nerves to the respiratory muscles.

### Nonneural Factors Influencing Respiratory Rate and Depth

***Physical Factors***   Although the medulla's respiratory centers set the basic rhythm of breathing, physical factors such as talking, coughing, and exercising can modify both the rate and depth of breathing. We have already examined some of these factors in our discussion of nonrespiratory air movements. Increased body temperature also causes an increase in the rate of breathing.

***Volition (Conscious Control)***   We all have consciously controlled our breathing pattern at one time or another. During singing and swallowing, breath control is extremely important, and many of us have held our breath for short periods to swim underwater. However, voluntary control of breathing is limited, and the respiratory centers will simply ignore messages from the cortex (our wishes) when the oxygen supply in the blood is getting low or blood pH is falling. All you need do to prove this is to try to talk normally or to hold your breath after running at breakneck speed for a few minutes. It simply cannot be done. Many toddlers try to manipulate their parents by holding their breath "to death." Even though this threat causes many parents to become anxious, they need not worry because the involuntary controls take over and normal respiration begins again.

***Emotional Factors***   Emotional factors also modify the rate and depth of breathing. Have you ever watched a horror movie with bated (held) breath or been so scared by what you saw that you were nearly panting? Have you ever touched something cold and clammy and gasped? All of these result

**13**

from reflexes initiated by emotional stimuli acting through centers in the hypothalamus.

***Chemical Factors*** Although many factors can modify respiratory rate and depth, the most important factors are chemical—the levels of carbon dioxide and oxygen in the blood (see Figure 13.12). An increased level of carbon dioxide and a decreased blood pH are the most important stimuli leading to an increase in the rate and depth of breathing. An increase in the carbon dioxide level can cause a decreased blood pH because high $CO_2$ results in more carbonic acid, which lowers blood pH. However, a low blood pH could also result from metabolic activities independent of breathing. Changes in the carbon dioxide concentration or $H^+$ ion concentration (which affects pH) in brain tissue seem to act directly on the medulla centers by influencing the pH of local tissues in the brain stem (see Figure 13.12).

Conversely, changes in oxygen concentration in the blood are detected by peripheral chemoreceptor regions in the aorta (aortic body in the aortic arch) and in the fork of the common carotid artery (the carotid body). These, in turn, send impulses to the medulla when the blood oxygen level is dropping. When oxygen levels are low, these same chemoreceptors are also able to detect high carbon dioxide levels. Although every cell in the body must have oxygen to live, it is the body's need to rid itself of carbon dioxide that is the *most* important stimulus for breathing. A decrease in the oxygen level becomes an important stimulus only when the level is dangerously low.

## Homeostatic Imbalance 13.11

In people who retain carbon dioxide, such as people with emphysema, chronic bronchitis, or other chronic lung diseases, the brain no longer recognizes an increased level of carbon dioxide as important. In such cases, a dropping oxygen level becomes the respiratory stimulus. This interesting fact explains why such patients are always given a low level of oxygen; it helps to maintain the prevailing stimulus for breathing (a low oxygen level). _____✚

The respiratory system in healthy individuals has homeostatic mechanisms. As carbon dioxide or other sources of acids begin to accumulate in the blood and tissues, and pH starts to drop, you begin to breathe more deeply and more rapidly. **Hyperventilation** is an increase in the rate and depth of breathing that

exceeds the body's need to remove carbon dioxide. In other words, during hyperventilation, we exhale more $CO_2$ than we should, resulting in elevated blood pH (there is less carbonic acid).

By contrast, when blood starts to become slightly alkaline, or basic, breathing slows and becomes shallow. Slower breathing allows carbon dioxide to accumulate in the blood and brings the blood pH back into the normal range.

Indeed, control of breathing during rest is aimed primarily at regulating the hydrogen ion concentration in the brain. *Hypoventilation* (extremely slow or shallow breathing) or *hyperventilation* can dramatically change the amount of carbonic acid in the blood. Carbonic acid increases dramatically during hypoventilation and decreases substantially during hyperventilation. In both situations, the buffering ability of the blood is likely to be overwhelmed; the result is *acidosis* or *alkalosis*.

## Homeostatic Imbalance 13.12

Hyperventilation, often brought on by anxiety attacks, frequently leads to brief periods of **apnea** (ap'ne-ah), cessation of breathing, until the carbon dioxide builds up in the blood again. If breathing stops for an extended time, cyanosis may occur as a result of insufficient oxygen in the blood. In addition, the hyperventilating person may get dizzy and faint because the resulting alkalosis causes cerebral blood vessels to constrict. Such attacks can be prevented by having the hyperventilating person breathe into a paper bag. Because exhaled air contains more carbon dioxide than atmospheric air, it upsets the normal diffusion gradient that causes carbon dioxide to be unloaded from the blood and leave the body. As a result, the carbon dioxide (and thus carbonic acid) level begins to rise in the blood, ending alkalosis. _____✚

### Did You Get It?

15. Which brain area is most important for setting the basic respiratory rate and rhythm?
16. What chemical factor in blood normally provides the most powerful stimulus to breathe?

For answers, see Appendix A.

# Respiratory Disorders

## Learning Objective

☐ **Describe symptoms and probable causes of COPD and lung cancer.**

## Homeostatic Imbalance 13.13

The respiratory system is particularly vulnerable to infections because it is open to airborne pathogens. We have already considered some of these inflammatory conditions, such as rhinitis and tonsillitis. Now we will turn our attention to the most disabling respiratory disorders, the group of diseases collectively referred to as *chronic obstructive pulmonary disease (COPD)* and *lung cancer*. These disorders are "living proof" of cigarette smoking's devastating effects on the body. Long known to promote cardiovascular disease, cigarettes are perhaps even more effective at destroying the lungs.

### Chronic Obstructive Pulmonary Disease (COPD)

The **chronic obstructive pulmonary diseases**, exemplified by *chronic bronchitis* and *emphysema* (em"fĭ-se'mah), are a major cause of death and disability in the United States. These diseases have certain features in common: (1) Patients almost always have a history of smoking; (2) **dyspnea** (disp'ne-ah), difficult or labored breathing, often referred to as "air hunger," occurs and becomes progressively worse; (3) coughing and frequent pulmonary infections are common; and (4) most COPD victims are hypoxic, retain carbon dioxide and have respiratory acidosis, and ultimately develop respiratory failure (see the figure).

In **chronic bronchitis**, the mucosa of the lower respiratory passages becomes severely inflamed and produces excessive mucus. The pooled mucus impairs ventilation and gas exchange and dramatically increases the risk of lung infections, including pneumonias. Chronic bronchitis patients are sometimes called "blue bloaters" because hypoxia and carbon dioxide retention occur early in the disease and cyanosis is common.

In **emphysema**, the walls of some alveoli are destroyed, causing the remaining alveoli to be enlarged. In addition, chronic inflammation promotes fibrosis of the lungs. As the lungs become less elastic, the airways collapse during expiration and obstruct outflow of air. As a result, these patients use an incredible amount of energy to exhale, and they are always exhausted. Because air is retained in the lungs, oxygen exchange is surprisingly efficient, and cyanosis does not usually appear until late in the disease. Consequently, emphysema sufferers are sometimes referred to as "pink puffers." However, overinflation of the lungs leads to a permanently expanded barrel chest.

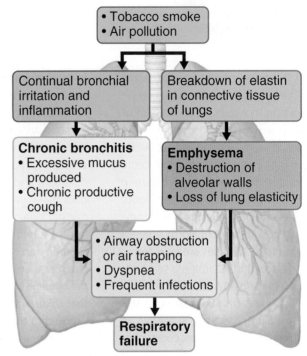

The pathogenesis of COPD.

### Lung Cancer

Lung cancer is the leading cause of cancer death for both men and women in North America, causing more deaths than breast, prostate, and colorectal cancer combined. This is tragic, because lung cancer is largely preventable—nearly 90 percent of lung cancers result from smoking. Lung cancer is aggressive and metastasizes rapidly and widely, so most cases are not diagnosed until they are well advanced. The cure rate for lung cancer is notoriously low; most victims die within 1 year of diagnosis. The overall 5-year survival of people with lung cancer is about 17 percent.

Ordinarily, nasal hairs, sticky mucus, and the action of cilia do a fine job of protecting the lungs from irritants, but smoking overwhelms these cleansing devices, and they eventually stop functioning. Continuous irritation prompts the production of more mucus, but smoking slows the movements of cilia that clear this mucus and depresses lung macrophages. One result is a pooling of mucus in the lower respiratory tract and an increased frequency of pulmonary infections, including pneumonia and COPD. However, it is the irritating effects of the "cocktail" of toxic chemicals in tobacco smoke that ultimately lead to lung cancer.

The three most common types of lung cancer are (1) **adenocarcinoma** (40 percent of cases), which originates as solitary nodules in peripheral

**13**

# Too Clean for Our Own Good?

Ah, spring is here! You can tell by the sounds of wheezing and sneezing, not to mention the glut of advertisements for allergy medications. It seems that just about everyone is allergic to something. A century ago, allergies were rare—a disease of rich city folk who had the best food, water, and medical care that money could buy. In cities of North America and Europe, allergies are now commonplace, and asthma is a serious problem for many children.

Asthma is chronic inflammation of small lung passages. In allergic asthma, allergens trigger constriction of these passages, making it difficult to breathe. Over the past 20 years, the cases of asthma have increased by about 75 percent in the United States and Canada, yet asthma and allergy are still rare in developing countries.

What is behind this? The cause can't be genetic, because the increase in allergy and asthma has been much too rapid. Clearly, there is something about living in an industrialized society that increases the risk of allergy. The observation that the increase in allergy is correlated with the decrease in childhood diseases (smallpox, measles, whooping cough, and others) has led to the *hygiene hypothesis*. The twentieth century saw remarkable advances in public health. Improvements in public sanitation and clean water supplies together with vaccination and the use of antibiotics have kept us living longer, healthier lives and have drastically reduced infant mortality in the industrialized world. The cost? Allergy.

Back in the "good old days" before improvements in sanitation, newborns fell prey to a wide variety of childhood diseases. Those that survived childhood diseases ended up with an immune system that had received a well-rounded education. Today, children are immunized against childhood infections and are often given antibiotics that wipe out beneficial symbiotic bacteria of the gut. Kept inside houses sterilized by antiseptic soaps, children don't play in (and eat) dirt as much as they used to. According to the hygiene hypothesis, the result is that children's immune systems aren't exposed to the large numbers of pathogens and toxins they once were. Voilà! Allergy.

This simple story doesn't tell all, though. First, it isn't quite true that people in developing countries don't have allergies—although they make IgE, they don't respond when exposed to allergens. It turns out that many people in developing countries are infected by parasites, particularly intestinal worms. When cured of the worm infections, they get allergies. The worms seem to actively suppress the immune system.

A second problem with the hygiene hypothesis is that allergy isn't the only immune disorder on the rise. Autoimmune diseases in children and young adults, such as type 1 diabetes and multiple sclerosis, have also increased in industrial societies.

Normally, as regulatory T cells (a group of lymphocytes that help control the immune response) become established during infancy, the host becomes *tolerant* to the antigens these cells have been exposed to (meaning they will not be attacked), preventing allergy and autoimmune

disorders. Regulatory T cell development may also depend on the presence of the very bacteria depleted by antibiotics and excessive cleanliness. Thus, allergy and autoimmune disorders may be the result of faulty regulatory T cell development.

Perhaps, in our haste to rid ourselves of household dust, dirt, bacteria, and parasites, we have eliminated the signals our immune systems need to fully mature. Scientists are now seeking ways to imitate these lost signals.

Instead of relying on injections of synthetic "dirt," perhaps we need to rethink our obsession with cleanliness. Clearly, no one wants to return to the "good old days" when 25 percent of infants died, but it may be that allowing babies to get a little dirtier early in life should be looked at as a golden opportunity to prevent allergy and asthma later on!

lung areas and develops from bronchial glands and alveolar cells; (2) **squamous cell carcinoma** (25–30 percent), which arises in the epithelium of the larger bronchi and tends to form masses that hollow out and bleed; and (3) **small cell carcinoma** (about 20 percent), which contains lymphocyte-like cells that originate in the main bronchi and grow aggressively in small grapelike clusters within the mediastinum, a site from which metastasis is especially rapid.

The most effective treatment for lung cancer is complete removal of the diseased lung lobes in an attempt to halt metastasis. However, removal is an option only if metastasis has not already occurred. In most cases, radiation therapy and chemotherapy are the only options, but most lung cancers are resistant to these treatments.

Fortunately, there are several new therapies on the horizon. These include (1) antibodies that target specific molecules required by the tumor or that deliver toxic agents directly to the tumor, (2) cancer vaccines to stimulate the immune system, and (3) various forms of gene therapy to replace the defective genes that make tumor cells divide continuously. As clinical trials progress, we will learn about the effectiveness of these approaches. However, prevention is worth a pound of cure. Quitting smoking is a valuable goal. _____ +

## Did You Get It?

17. Alvin, a smoker, sees his doctor because he has a persistent cough and is short of breath after very little exertion. He has a barrel chest and explains that it is difficult for him to exhale. What diagnosis will the doctor make?

For the answer, see Appendix A.

# Developmental Aspects of the Respiratory System

## Learning Objective

☐ **Describe normal changes that occur in respiratory system functioning from infancy to old age.**

In the fetus, the lungs are filled with fluid, and all respiratory exchanges are made by the placenta. At birth, the fluid-filled pathway is drained, and the respiratory passageways fill with air. The alveoli inflate and begin to function in gas exchange, but the lungs are not fully inflated for 2 weeks. The success of this change—that is, from nonfunc-

tional to functional respiration—depends on the presence of **surfactant** (sur-fak'tant), a fatty molecule made by the cuboidal alveolar cells (see Figure 13.6). Surfactant lowers the surface tension of the film of water lining each alveolar sac so that the alveoli do not collapse between each breath. Surfactant is not usually present in large enough amounts to accomplish this function until late in pregnancy (between 28 and 30 weeks).

 ## Homeostatic Imbalance 13.14

Infants who are born prematurely (before week 28) or in whom surfactant production is inadequate (as in many infants born to diabetic mothers) have **infant respiratory distress syndrome (IRDS)**. These infants have dyspnea within a few hours after birth and use tremendous amounts of energy just to reinflate their alveoli, which collapse after each breath. Although IRDS still accounts for over 20,000 newborn deaths a year, many babies with IRDS survive now because of the use of equipment that supplies a positive pressure continuously and keeps the alveoli open until the maturing lungs produce enough surfactant.

Important birth defects of the respiratory system include cleft palate and cystic fibrosis. **Cystic fibrosis (CF)**, the most common lethal genetic disease in the United States, strikes in 1 out of every 2,400 births, and every day two children die of it. CF causes oversecretion of a thick mucus that clogs the respiratory passages and puts the child at risk for fatal respiratory infections. It affects other secretory processes as well. Most important, it impairs food digestion by clogging ducts that deliver pancreatic enzymes and bile to the small intestine. Also, sweat glands produce an extremely salty perspiration. At the heart of CF is a faulty gene that codes for the CFTR protein. CFTR works as a chloride ion ($Cl^-$) channel to control the flow of $Cl^-$ into and out of cells. In people with the mutated gene, CFTR gets "stuck" in the endoplasmic reticulum and is unable to reach the plasma membrane to perform its normal role. Consequently, less $Cl^-$ is secreted from the cells and less water follows, resulting in the thick mucus typical of CF. The goal of CF research is to restore normal salt and water movement. Conventional therapy for CF is mucus-dissolving drugs, "clapping" the chest to loosen the thick mucus, and antibiotics to prevent infection. A surprisingly simple approach is to inhale hypertonic saline droplets. _____ +

**13**

# Homeostatic Relationships between the **Respiratory System** and Other Body Systems

## Endocrine System
- Respiratory system provides oxygen; disposes of carbon dioxide
- Epinephrine dilates the bronchioles; testosterone promotes laryngeal enlargement in males at puberty

## Lymphatic System/Immunity
- Respiratory system provides oxygen; disposes of carbon dioxide; tonsils in pharynx house immune cells
- Lymphatic system helps to maintain blood volume required for respiratory gas transport; immune system protects respiratory organs from pathogens and cancer

## Digestive System
- Respiratory system provides oxygen; disposes of carbon dioxide
- Digestive system provides nutrients needed by respiratory system

## Urinary System
- Respiratory system provides oxygen; disposes of carbon dioxide
- Kidneys dispose of metabolic wastes (other than carbon dioxide) of respiratory system organs

## Muscular System
- Respiratory system provides oxygen needed for muscle activity; disposes of carbon dioxide
- The diaphragm and intercostal muscles produce volume changes necessary for breathing; regular exercise increases respiratory efficiency

## Nervous System
- Respiratory system provides oxygen needed for normal neural activity; disposes of carbon dioxide
- Medullary and pons centers regulate respiratory rate/depth; stretch receptors in lungs and chemoreceptors in large arteries provide feedback

## Respiratory System

## Cardiovascular System
- Respiratory system provides oxygen; disposes of carbon dioxide; carbon dioxide present in blood as $HCO_3^-$ and $H_2CO_3$ contributes to blood buffering
- Blood transports respiratory gases

## Reproductive System
- Respiratory system provides oxygen; disposes of carbon dioxide

## Integumentary System
- Respiratory system provides oxygen; disposes of carbon dioxide
- Skin protects respiratory system organs by forming surface barriers

## Skeletal System
- Respiratory system provides oxygen; disposes of carbon dioxide
- Bones enclose and protect lungs and bronchi

The respiratory rate is highest in newborn infants, about 40 to 80 respirations per minute. It continues to drop through life: In the infant it is around 30 per minute, at 5 years it is around 25 per minute, and in adults it is 12 to 18 per minute. However, the rate often increases again in old age. The lungs mature throughout childhood, and more alveoli are formed until young adulthood. But in people who begin smoking during the early teens, the lungs never completely mature, and those additional alveoli are lost forever.

 ## Homeostatic Imbalance 13.15

In **sudden infant death syndrome (SIDS)**, also called *crib death*, apparently healthy infants stop breathing and die in their sleep. Believed to be a problem with neural control of respiration, most cases occur in infants placed on their tummy to sleep. Since 1992, the recommendation to place infants on their back to sleep has resulted in a 40 percent decline in SIDS cases in the United States. _____ ✛

Except for sneezes or coughs (responses to irritants) and the occasional common cold that blocks the upper respiratory passageways with mucus, the respiratory system works so efficiently and smoothly that we are mostly not even aware of it. Most problems that occur are a result of external factors—for example, obstruction of the trachea by a piece of food or aspiration of food particles or vomitus (which leads to aspiration pneumonia). Some unfortunate individuals are plagued by **asthma**, caused by chronically inflamed, hypersensitive bronchial passages that respond to many irritants (such as dust mite and cockroach droppings, dog dander, and fungi) with dyspnea, coughing, and wheezing. "A Closer Look" (see p. 482) discusses how hygiene may be related to recent increases in cases of asthma.

For many years, tuberculosis and pneumonia were the worst killers in the United States. Antibiotics have largely decreased their lethal threat, but they are still dangerous diseases. Newly diagnosed, and frequently drug-resistant, tuberculosis cases in AIDS patients are increasing by leaps and bounds the world over, but *at present* the most damaging and disabling respiratory diseases are still those described previously, COPD and lung cancer.

As we age, the chest wall becomes more rigid and the lungs begin to lose their elasticity, resulting in a slowly decreasing ability to ventilate the lungs. Vital capacity decreases by about one-third by the age of 70. In addition, the blood oxygen level decreases, and sensitivity to the stimulating effects of carbon dioxide decreases, particularly in a reclining or supine position. As a result, many old people tend to become hypoxic during sleep and exhibit **sleep apnea**.

Additionally, many of the respiratory system's protective mechanisms also become less efficient with age. Ciliary activity of the mucosa decreases, and the phagocytes in the lungs become sluggish. The net result is that the elderly population is more at risk for respiratory tract infections, particularly pneumonia and influenza.

### Did You Get It?

18. What happens to the alveoli if surfactant is not produced in a newborn baby? What name is given to this condition?

For the answer, see Appendix A.

## Summary

### Functional Anatomy of the Respiratory System (pp. 462–471)

1. The nasal cavity (the chamber within the nose) is divided medially by a nasal septum and separated from the oral cavity by the palate. The nasal cavity is lined with mucosa, which warms, filters, and moistens incoming air. The mucosa also contains receptors for sense of smell. Paranasal sinuses and nasolacrimal ducts drain into the nasal cavity.

2. The pharynx (throat) is a mucosa-lined, muscular tube with three regions—nasopharynx, oropharynx, and laryngopharynx. The nasopharynx functions in respiration only; the others serve both respiratory and digestive functions. The pharynx contains tonsils, which act as part of the body's defense system.

3. The larynx (voice box) is a cartilaginous structure; most prominent is the thyroid cartilage (Adam's

**13**

apple). The larynx connects the pharynx with the trachea inferiorly. The laryngeal opening (glottis) is hooded by the epiglottis, which prevents entry of food or drink into respiratory passages when swallowing. The larynx contains the vocal folds (true vocal cords), which produce sounds used in speech.

4. The trachea (windpipe) extends from the larynx to the main bronchi. The trachea is a smooth-muscle tube lined with a ciliated mucosa and reinforced with C-shaped cartilaginous rings, which keep the trachea patent (open).

5. Right and left main (primary) bronchi result from division of the trachea. Each plunges into the hilum of the lung on its side.

6. The lungs are paired organs flanking the mediastinum in the thoracic cavity. Each lung is covered with pulmonary (visceral) pleura; the thorax wall is lined with parietal pleura. Pleural fluid decreases friction during breathing. The lungs are primarily elastic tissue and passageways of the bronchial tree. The smallest passageways end in clusters of air sacs called alveoli.

7. The conducting zone includes all respiratory passages from the nasal cavity to the terminal bronchioles; they conduct air to and from the alveoli of the lungs. Respiratory bronchioles, alveolar ducts and sacs, and alveoli—which have thin walls through which all gas exchange occurs with pulmonary capillary blood—are respiratory zone structures.

## Respiratory Physiology (pp. 471–480)

1. Mechanics of breathing: Gas travels from high-pressure to low-pressure areas. Pressure outside the body is atmospheric pressure; pressure inside the lungs is intrapulmonary pressure; pressure between the pleurae is intrapleural pressure (which is always negative). Movement of air into and out of the lungs is called pulmonary ventilation, or breathing. When inspiratory muscles contract, intrapulmonary volume increases, its pressure decreases, and air rushes in (inspiration). When inspiratory muscles relax, the lungs recoil and air rushes out (expiration). Expansion of the lungs is helped by cohesion between pleurae and by the presence of surfactant in alveoli.

Complete an interactive tutorial: Mastering A&P®> Study Area > Interactive Physiology > Respiratory System > Pulmonary Ventilation

2. Respiratory volumes and capacities: Air volumes exchanged during breathing are TV, IRV, ERV, and VC (see p. 474 for values). Residual volume, which is nonexchangeable respiratory volume, allows gas exchange to go on continually.

3. Nonrespiratory air movements: These are voluntary or reflex activities that move air into or out of the lungs. They include coughing, sneezing, laughing, crying, hiccuping, and yawning.

4. Respiratory sounds: Bronchial sounds are sounds of air passing through large respiratory passageways. Vesicular breathing sounds occur as air fills alveoli.

5. External respiration, gas transport, and internal respiration: Gases move according to the laws of diffusion. Oxygen moves from alveolar air into pulmonary blood. Most oxygen is transported bound to hemoglobin inside RBCs. Carbon dioxide moves from pulmonary blood into alveolar air. Most carbon dioxide is transported as bicarbonate ion in plasma. At body tissues, oxygen moves from blood to the tissues, whereas carbon dioxide moves from the tissues to blood.

6. Control of respiration

   a. Nervous control: Neural centers for control of respiratory rhythm are in the medulla and pons. The medulla is the respiratory rate "pacemaker." Reflex arcs initiated by stretch receptors in the lungs also play a role in respiration by notifying neural centers of excessive overinflation.

   b. Physical factors: Increased body temperature, exercise, speech, singing, and nonrespiratory air movements modify both rate and depth of breathing.

   c. Volition: To a degree, breathing may be consciously controlled if it does not interfere with homeostasis.

   d. Emotional factors: Some emotional stimuli can modify breathing. Examples are fear, anger, and excitement.

   e. Chemical factors: Changes in carbon dioxide levels are the most important stimuli affecting respiratory rhythm and depth. Carbon dioxide acts directly on the medulla via its effect on reducing the pH of blood and brainstem tissue. A rising level of carbon dioxide or a drop in pH independent of $CO_2$ in the medulla results in faster, deeper breathing; falling levels lead to shallow, slow breathing. Hyperventilation may result in apnea and dizziness due to alkalosis.

Oxygen, monitored by peripheral chemoreceptors, is less important as a respiratory stimulus in healthy people. It *is* the stimulus for people whose systems have become accustomed to a high level of carbon dioxide as a result of disease.

## Respiratory Disorders (pp. 480–483)

1. The major respiratory disorders are COPD (emphysema and chronic bronchitis) and lung cancer. Cigarette smoking is a significant cause of both.

2. Emphysema is characterized by permanent destruction and enlargement of alveoli. The lungs lose their elasticity, and expiration becomes an active process.

3. Chronic bronchitis is characterized by excessive production and pooling of mucus in lower respiratory passageways, which severely impairs ventilation and gas exchange. Patients may become cyanotic as a result of chronic hypoxia.

4. Lung cancer is extremely aggressive and metastasizes rapidly. The three most common lung cancers are adenocarcinoma, squamous cell carcinoma, and small cell carcinoma.

## Developmental Aspects of the Respiratory System (pp. 483–485)

1. Premature infants have problems keeping their lungs inflated because of lack of surfactant in their alveoli. (Surfactant is formed late in pregnancy.)

2. The most important birth defects of the respiratory system are cleft palate and cystic fibrosis.

3. The lungs continue to mature until young adulthood. Smoking in the teen years can lead to irreparable damage.

4. During youth and middle age, most respiratory system problems are a result of external factors, such as infections and substances that physically block respiratory passageways.

5. In old age, the thorax becomes more rigid and lungs become less elastic, leading to decreased vital capacity. Protective mechanisms of the respiratory system become less effective in elderly persons, predisposing them to more respiratory tract infections.

---

## Review Questions

Access additional practice questions using your smartphone, tablet, or computer: Mastering A&P® > Study Area > Practice Tests & Quizzes

### Multiple Choice

*More than one choice may apply.*

1. When you exhale, air flows through respiratory structures in which sequence?
   a. Alveolus, bronchiole, bronchus, larynx, trachea, pharynx, nasal cavity
   b. Alveolus, trachea, bronchus, bronchiole, larynx, pharynx, nasal cavity
   c. Alveolus, bronchus, bronchiole, trachea, larynx, pharynx, nasal cavity
   d. Alveolus, bronchiole, bronchus, trachea, larynx, pharynx, nasal cavity

2. The structures that open into the nasal cavity include the
   a. sphenoidal sinus.
   b. esophagus.
   c. pharyngotympanic tubes.
   d. nasolacrimal ducts.

3. Which of the following is (are) located in the larynx?
   a. Glottis
   b. Epiglottis
   c. Hyoid bone
   d. Thyroid cartilage

4. Lung collapse is prevented by
   a. high surface tension of alveolar fluid.
   b. adhesion of the pleural membranes.
   c. high pressure in the pleural cavities.
   d. high elasticity of lung tissue.

5. Disorders classified as COPDs include
   a. pneumonia.          c. bronchitis.
   b. emphysema.          d. sleep apnea.

6. Which of the following changes will accompany the loss of lung elasticity associated with aging?
   a. Increase in tidal volume
   b. Increase in inspiratory reserve volume
   c. Increase in residual volume
   d. Increase in vital capacity

7. Which of the following is (are) *not* present in the alveoli?
   a. Surfactant-secreting cells
   b. Dust cells
   c. Red blood cells
   d. Squamous epithelial cells
   e. Capillary endothelial cells

**13**

## Short Answer Essay

8. Clearly explain the difference between external and internal respiration.

9. Trace the route of air from the nares to an alveolus.

10. Why is it important that the trachea be reinforced with cartilaginous rings? What is the advantage of the fact that the rings are incomplete posteriorly?

11. Name the lymphatic tissue present in the oropharynx.

12. What distinguishes the mucosa in the slitlike superior part of the nasal cavity from that in the rest of the nasal cavity?

13. In terms of general health, what is the importance of the fact that the pharyngotympanic tubes and the sinuses drain into the nasal cavities and nasopharynx?

14. What is it about the structure of the alveoli that makes them an ideal site for gas exchange?

15. What do TV, ERV, and VC mean?

16. Name several nonrespiratory air movements, and explain how each differs from normal breathing.

17. The contraction of the diaphragm and the external intercostal muscles begins inspiration. What happens, in terms of volume and pressure changes in the lungs, when these muscles contract?

18. How can you recognize whether an isolated lung is a right or a left lung?

19. What determines in which direction carbon dioxide and oxygen will diffuse in the lungs? In the tissues?

20. Name the two major brain areas involved in the nervous control of breathing.

21. Why are elderly people more at risk for respiratory tract infections?

22. Name two chemical factors that modify respiratory rate and depth. Which is usually more important?

23. Define *hyperventilation*. If you hyperventilate, do you retain or expel more carbon dioxide? What effect does hyperventilation have on blood pH? On breathing rate?

24. Compare and contrast the signs and symptoms of emphysema and chronic bronchitis.

25. Contrast the stimulus for and the result of hyperventilation with the stimulus for and result of hyperpnea.

## Critical Thinking and Clinical Application Questions

26. Anne has a bad cold, and she is also suffering from a continuing headache. An X-ray image of her head shows that her nasal sinuses have a cloudy white aspect, indicating the presence of mucus or infectious matter. Does this explain her headache? Justify your answer.

27. Why doesn't Laney have to worry when her 3-year-old son Ethan threatens to "hold his breath till he dies"?

28. Mr. Alvarez bumped a bee's nest while making repairs on his roof. Not surprisingly, he was promptly stung several times. Because he knew he was allergic to bee stings, he rushed to the hospital. While waiting, he went into a state of shock and had extreme difficulty breathing. Examination showed his larynx to be edematous, and a tracheostomy was performed. Why is edema of the larynx likely to obstruct the airway? What is a tracheostomy, and what purpose does it serve?

29. As a result of a stroke, Mrs. Minnick's swallowing is uncoordinated. What detrimental effect might this have on her ability to breathe?

30. An elderly lady living alone is found dead on her bathroom floor. There are no signs of violence, but the rescue team notices a gas-powered heating system and the lack of proper ventilation in the room. What is a likely cause of the death of this lady?

31. Nineteen-year-old Tyler stumbled into the drugstore gasping for breath. Blood was oozing from a small hole in his chest wall. When the paramedics arrived they said that Tyler had been shot and suffered a pneumothorax and atelectasis. What do both of these terms mean, and how do you explain his respiratory distress? How will it be treated?

# 14 The Digestive System and Body Metabolism

**WHAT**

The digestive system breaks down the food you eat into nutrients needed for metabolic processes, such as making ATP, and rids the body of materials that cannot be used, such as fiber.

**HOW**

Chewing breaks down food into small pieces easy for enzymes to access; enzymes then chemically digest food into nutrients that are actively transported into blood and delivered to cells around the body.

**WHY**

The digestive system is essential for providing the body with the energy and building blocks it requires to maintain life.

**INSTRUCTORS**

New **Building Vocabulary Coaching** Activities for this chapter are assignable in Mastering A&P®

Children are fascinated by the workings of the digestive system: They relish crunching a potato chip, delight in making "mustaches" with milk, and giggle when their stomach "growls." As adults, we know that a healthy digestive system is essential for good health because it converts food into the raw materials that build and fuel our body's cells. Specifically, the digestive system *ingests* food (takes it in), *digests* it (breaks it down) into nutrient molecules, *absorbs* the nutrients into the bloodstream, and then *defecates* (excretes) to rid the body of the indigestible wastes.

## PART I: ANATOMY AND PHYSIOLOGY OF THE DIGESTIVE SYSTEM

## Anatomy of the Digestive System

→ **Learning Objectives**

☐ **Name the organs of the alimentary canal and accessory digestive organs, and identify each on an appropriate diagram or model.**

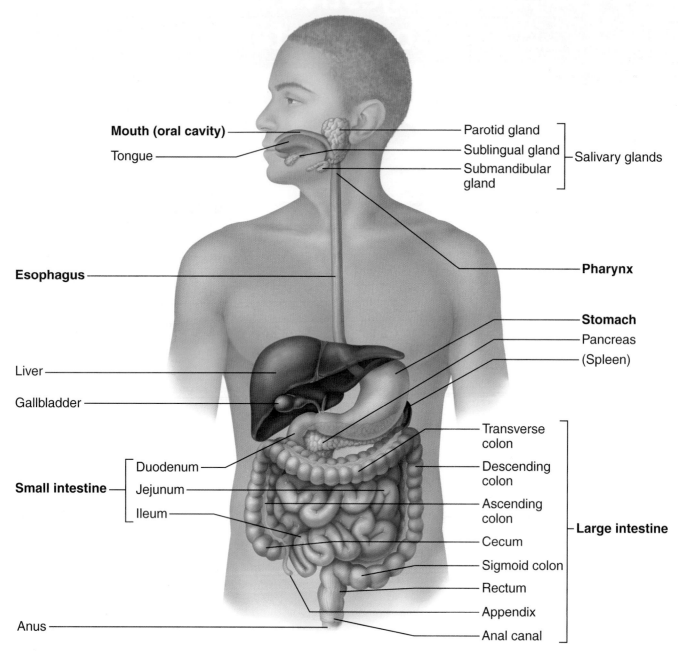

**Figure 14.1 The human digestive system: Alimentary canal and accessory organs.** The organs of the alimentary canal have boldfacedlabels, whereas accessory organs do not. The liver and gallbladder are reflected superiorly and to the right side of the body.

☐ **Identify the overall function of the digestive system as digestion and absorption of foodstuffs, and describe the general activities of each digestive system organ.**

We can separate the organs of the digestive system into two main groups: those forming the *alimentary* (al″ĕ-men′tar-e; *aliment* = nourish) *canal* and the *accessory digestive organs* **(Figure 14.1)**. The alimentary canal performs the whole menu of

digestive functions (ingests, digests, absorbs, and defecates) as it propels the foodstuffs along its length. The accessory organs (teeth, tongue, and several large digestive glands) assist digestion in various ways, as described shortly.

## Organs of the Alimentary Canal

The **alimentary canal**, also called the **gastrointestinal (GI) tract** or *gut*, is a continuous, coiled,

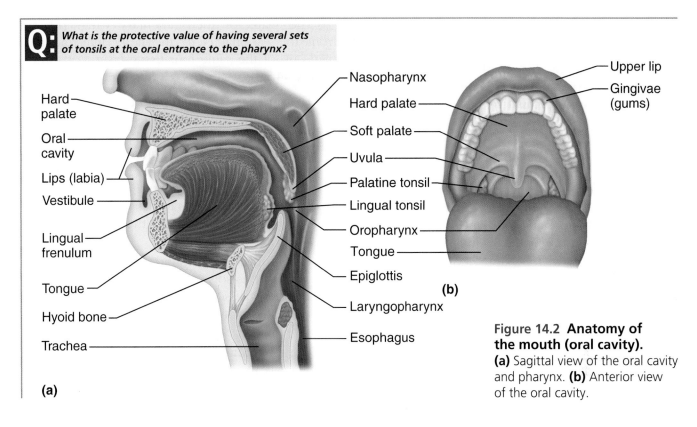

Hard palate
Oral cavity
Lips (labia)
Vestibule
Lingual frenulum
Tongue
Hyoid bone
Trachea

Nasopharynx
Hard palate
Soft palate
Uvula
Palatine tonsil
Lingual tonsil
Oropharynx
Tongue
Epiglottis
Laryngopharynx
Esophagus

**(a)**

Upper lip
Gingivae (gums)

**(b)**

**Figure 14.2 Anatomy of the mouth (oral cavity).** **(a)** Sagittal view of the oral cavity and pharynx. **(b)** Anterior view of the oral cavity.

hollow muscular tube that winds through the ventral body cavity from mouth to anus. Its organs are the *mouth, pharynx, esophagus, stomach, small intestine,* and *large intestine.* The large intestine leads to the terminal opening, or *anus.* In a cadaver, the alimentary canal is approximately 9 m (about 30 feet) long, but in a living person, it is considerably shorter because of its muscle tone. Food material within this tube is technically outside the body, because it has contact only with cells lining the tract and the tube is open to the external environment at both ends. This relationship becomes clearer if you think of your alimentary canal as an elongated doughnut. Then it's easy to see that if you stick your finger through a doughnut hole, your finger is not really *in* the doughnut. Each organ of the alimentary canal is described next (find the organs in Figure 14.1).

## Mouth

Food enters the digestive tract through the **mouth**, or **oral cavity**, a mucous membrane–lined cavity **(Figure 14.2)**. The **lips**, or **labia**, protect its anterior opening; the **cheeks** form its lateral walls; the **hard palate** forms its anterior roof; and the **soft palate** forms its posterior roof. The **uvula** (u′vu-lah) is a fleshy fingerlike projection of the soft palate, which dangles from the posterior edge of the soft palate. The space between the lips and cheeks externally and the teeth and gums internally is the **vestibule**. The area contained by the teeth is the **oral cavity proper**. The muscular **tongue** occupies the floor of the mouth. The tongue has several bony attachments—two of these are to the hyoid bone and the styloid processes of the skull. The **lingual frenulum** (ling′gwal fren′u-lum), a fold of mucous membrane, secures the tongue to the floor of the mouth and limits its posterior movements (see Figure 14.2a).

### Homeostatic Imbalance 14.1

Children born with an extremely short lingual frenulum are often referred to as "tongue-tied" because movement of the tongue is restricted, leading to distorted speech. This congenital

14

condition can be corrected surgically by cutting the frenulum. _____ ✚

At the posterior end of the oral cavity are paired masses of lymphatic tissue, the **palatine tonsils**. The **lingual tonsil** covers the base of the tongue just beyond. The tonsils, along with other lymphatic tissues, are part of the body's defense system (look back at Figure 13.2, p. 465). When the tonsils become inflamed and enlarge, they partially block the entrance into the throat (pharynx), making swallowing difficult and painful, as described in Homeostatic Imbalance 13.3, p. 464.

As food enters the mouth, it is mixed with saliva and *masticated* (chewed). The cheeks and closed lips hold the food between the teeth during chewing. The nimble tongue continuously mixes food with saliva and initiates swallowing. Thus, the breakdown of food begins before it has even left the mouth.

### CONCEPT**LINK**

Recall that *papillae* containing taste buds, or taste receptors, are found on the tongue surface (see Chapter 8, p. 326). So, besides its food-manipulating function, the tongue allows us to enjoy and appreciate the food we eat.

### Pharynx

From the mouth, food passes posteriorly into the *oropharynx* and *laryngopharynx*, both of which are common passageways for food, fluids, and air. The pharynx is subdivided into the *nasopharynx*, part of the respiratory passageway; the **oropharynx**, posterior to the oral cavity; and the **laryngopharynx**, which is continuous with the esophagus inferiorly (see Chapter 13).

The walls of the pharynx contain two skeletal muscle layers. The cells of the outer layer run longitudinally; those of the inner layer (the constrictor muscles) run around the wall in a circular fashion. Alternating contractions of these two muscle layers propel food through the pharynx inferiorly into the esophagus. Later we describe this propelling mechanism, called *peristalsis* (per″ĭ-stal′sis).

### Esophagus

The **esophagus** (ĕ-sof′ah-gus), or *gullet*, runs from the pharynx through the diaphragm to the stomach. About 25 cm (10 inches) long, it is essentially a passageway that conducts food (by peristalsis) to the stomach.

The walls of the alimentary canal organs from the esophagus to the large intestine are made up of the same four tissue layers, or tunics **(Figure 14.3)**:

1. The **mucosa** is the innermost layer, a moist mucous membrane that lines the hollow cavity, or **lumen**, of the organ. It consists primarily of *surface epithelium* plus a small amount of connective tissue (*lamina propria*) and a scanty *smooth muscle layer*. Beyond the esophagus, which has a friction-resisting stratified squamous epithelium, the epithelium is mostly simple columnar.

2. The **submucosa** is found just beneath the mucosa. It is soft connective tissue containing blood vessels, nerve endings, mucosa-associated lymphoid tissue (MALT), and lymphatic vessels.

3. The **muscularis externa** is a muscle layer typically made up of an inner *circular layer* and an outer *longitudinal layer* of smooth muscle cells.

4. The **serosa** is the outermost layer of the wall. As half of a serous membrane pair, the **visceral peritoneum** (per″ĭ-to-ne′um) consists of a single layer of flat, serous fluid–producing cells. The visceral peritoneum is continuous with the slippery **parietal peritoneum**, which lines the abdominopelvic cavity by way of a membrane extension, the **mesentery** (mes′en-ter″e). (These relationships are illustrated in Figure 14.5 on p. 495.)

### ⚖ Homeostatic Imbalance 14.2

When the peritoneum is infected, a condition called **peritonitis** (per″ĭ-to-ni′tis), the peritoneal membranes tend to stick together around the infection site. This helps to seal off and localize many intraperitoneal infections (at least initially), providing time for macrophages in the lymphatic tissue to mount an attack. _____ ✚

The alimentary canal wall contains two important *intrinsic nerve plexuses*—the **submucosal nerve plexus** and the **myenteric** (mi-en′ter-ik; "intestinal muscle") **nerve plexus**. These networks of nerve fibers are actually part of the autonomic nervous system. They help regulate the mobility and secretory activity of GI tract organs.

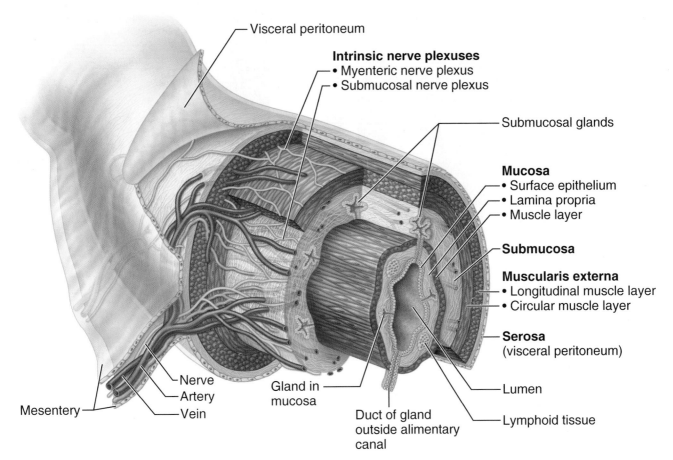

Visceral peritoneum

**Intrinsic nerve plexuses**
• Myenteric nerve plexus
• Submucosal nerve plexus

Submucosal glands

**Mucosa**
• Surface epithelium
• Lamina propria
• Muscle layer

**Submucosa**

**Muscularis externa**
• Longitudinal muscle layer
• Circular muscle layer

**Serosa**
(visceral peritoneum)

Lumen

Lymphoid tissue

Duct of gland
outside alimentary
canal

Gland in
mucosa

Nerve
Artery
Vein

Mesentery

**Figure 14.3  Basic structure of the alimentary canal wall.**

## Stomach

The C-shaped **stomach** (**Figure 14.4a** and **b**, p. 494) is on the left side of the abdominal cavity, nearly hidden by the liver and diaphragm. Different regions of the stomach have been named. The *cardial region*, or *cardia* (named for its position near the heart), surrounds the **cardioesophageal** (kar″de-o-ĕ-sof″ah-je′al) **sphincter**, through which food enters the stomach from the esophagus. The *fundus* is the expanded part of the stomach lateral to the cardial region. The *body* is the midportion of the stomach; in the body, the convex lateral surface is the **greater curvature**, and its concave medial surface is the **lesser curvature**. As it narrows inferiorly, the body becomes the *pyloric antrum* and then the funnel-shaped *pylorus* (pi-lor′us), the terminal part of the stomach. The pylorus is continuous with the small intestine through the **pyloric sphincter**, or **pyloric valve**.

### CONCEPTLINK

Recall that valves control the flow of fluids, including blood flow in veins and through the heart (see Chapter 11, pp. 387–389). The valves of the digestive system (formed by sphincter muscles) control the flow of food and digestive juices through the GI tract.

The stomach varies from 15 to 25 cm (6 to 10 inches) in length, but its diameter and volume depend on how much food it contains. When it is full, it can hold about 4 liters (1 gallon) of food. When it is empty, it collapses inward on itself, and its mucosa is thrown into large folds called **rugae** (roo′ge; *ruga* = wrinkle, fold).

The **lesser omentum** (o-men′tum), a double layer of peritoneum, extends from the liver to the lesser curvature of the stomach. The **greater omentum**, another extension of the peritoneum, drapes downward and covers the abdominal organs like a lacy apron before attaching to the posterior body wall (**Figure 14.5a**, p. 495). The

14

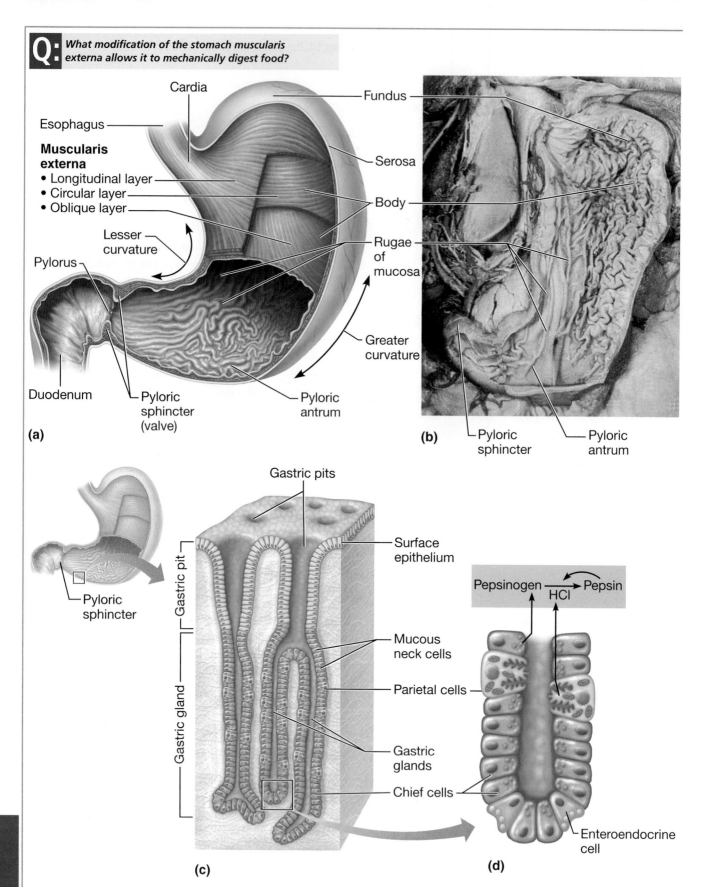

**Q:** *What modification of the stomach muscularis externa allows it to mechanically digest food?*

**(a)**

**(b)**

**(c)**

**(d)**

**Figure 14.4 Anatomy of the stomach.** Gross internal anatomy (frontal section). **(a)** Diagram. **(b)** Photo. **(c)** Enlarged view of gastric pits and glands (longitudinal section). **(d)** Pepsinogen produced by the chief cells is activated (converted to pepsin) by HCl secreted by the parietal cells.

**A:** Its third layer of smooth muscle, the oblique layer, which allows it to knead or pummel the food.

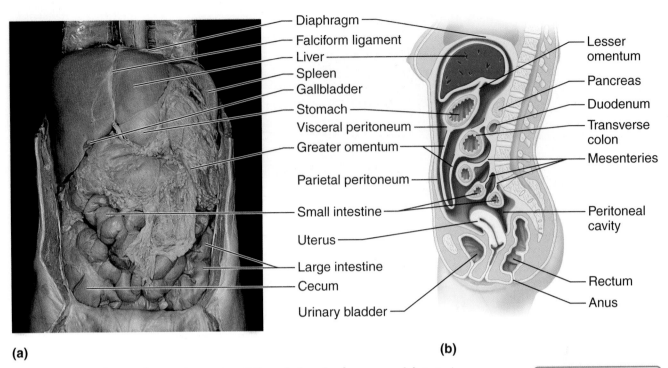

**(a)**                                                                                              **(b)**

**Figure 14.5 Peritoneal attachments of the abdominal organs. (a)** Anterior
view; the greater omentum is shown in its normal position, covering the abdominal
viscera. **(b)** Sagittal view of the abdominopelvic cavity of a female.

greater omentum is riddled with fat, which helps to insulate, cushion, and protect the abdominal organs. It also has large collections of lymphoid follicles containing macrophages and defensive cells of the immune system.

The stomach acts as a temporary "storage tank" for food as well as a site for food breakdown. Besides the usual longitudinal and circular muscle layers, its wall contains a third, obliquely arranged layer in the *muscularis externa* (see Figure 14.4a). This arrangement allows the stomach not only to move food along the tract, but also to churn, mix, and pummel the food, physically breaking it down into smaller fragments. In addition, chemical breakdown of proteins begins in the stomach.

The mucosa of the stomach is a simple columnar epithelium composed entirely of mucous cells. They produce a protective layer of bicarbonate-rich alkaline mucus that clings to the stomach mucosa and protects the stomach wall from being damaged by acid and digested by enzymes. This otherwise smooth lining is dotted with millions of deep **gastric pits**, which lead into **gastric glands** (Figure 14.4c) that secrete the components of **gastric juice**. For example, some stomach cells produce **intrinsic factor**, a substance needed for absorption of vitamin $B_{12}$ from the small intestine.

The **chief cells** produce inactive protein-digesting enzymes, mostly **pepsinogens**. The **parietal cells** produce corrosive hydrochloric acid (HCl), which makes the stomach contents acidic and activates the enzymes, as in the conversion of pepsinogen to pepsin (shown in Figure 14.4d). The *mucous neck cells* produce a thin acidic mucus with an unknown function that is quite different from that secreted by the mucous cells of the mucosa. Still other cells, the **enteroendocrine cells** (*entero =* gut), produce local hormones, such as *gastrin*, that are important in regulating the digestive activities of the stomach (see Table 14.1 on p. 497).

Most digestive activity occurs in the pyloric region of the stomach. After food has been processed in the stomach, it is thick like heavy cream and is called **chyme** (kīm). The chyme enters the small intestine through the pyloric sphincter.

## Did You Get It?

1. What is the sequential order (mouth to anus) of the digestive organs making up the alimentary canal?
2. In which organ of the alimentary canal does protein digestion begin?
3. The stomach epithelium secretes several substances, including alkaline mucus and intrinsic factor. What is the function of each of these two secretions?

**For answers, see Appendix A.**

14

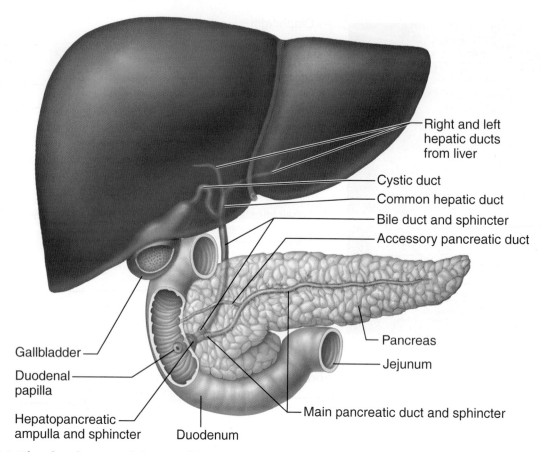

**Right and left hepatic ducts from liver**

**Cystic duct**

**Common hepatic duct**

**Bile duct and sphincter**

**Accessory pancreatic duct**

**Pancreas**

**Jejunum**

**Main pancreatic duct and sphincter**

**Gallblader**

**Duodenal papilla**

**Hepatopancreatic ampulla and sphincter**

**Duodenum**

**Figure 14.6 The duodenum of the small intestine and related organs.**

## Small Intestine

### → Learning Objective

☐ **Explain how villi aid digestive processes in the small intestine.**

The **small intestine** is the body's major digestive organ. Within its twisted passageways, usable nutrients are finally prepared for their journey into the cells of the body. The small intestine is a muscular tube extending from the pyloric sphincter to the large intestine (see Figure 14.1, p. 490). It is the longest section of the alimentary tube, with an average length of 2 to 4 m (7 to 13 feet) in a living person. Except for the initial part of the small intestine, which mostly lies in a retroperitoneal position (posterior to the parietal peritoneum), the small intestine hangs in sausagelike coils in the abdominal cavity, suspended from the posterior abdominal wall by the fan-shaped mesentery (Figure 14.5b, p. 495). The large intestine encircles and frames it in the abdominal cavity.

The small intestine has three subdivisions: the **duodenum** (doo″ uh-de′num; "twelve finger widths long"), the **jejunum** (jĕ-joo′num; "empty"), and the **ileum** (il′e-um; "twisted intestine"), which contribute 5 percent, nearly 40 percent, and almost 60 percent of the length of the small intestine, respectively (see Figure 14.1). The ileum joins the large intestine at the **ileocecal** (il″e-o-se′kal) **valve** (look forward to Figure 14.8, p. 499).

Chemical digestion of foods begins in earnest in the small intestine. The small intestine is able to process only a small amount of food at one time. The *pyloric sphincter* (literally, "gatekeeper") controls the movement of chyme into the small intestine from the stomach and prevents the small intestine from being overwhelmed. In the C-shaped duodenum, some enzymes are produced by the intestinal cells. More important are enzymes that are produced by the pancreas and then delivered to the duodenum through the **pancreatic ducts**, where they complete the chemical breakdown of foods in the small intestine. *Bile* (formed by the liver) also enters the duodenum through the **bile duct** in the same area **(Figure 14.6)**. The main pancreatic and bile ducts join at the duodenum to

Table **14.1** Hormones and Hormonelike Products That Act in Digestion

| Hormone | Source | Stimulus for secretion | Action |
| --- | --- | --- | --- |
| Gastrin | Stomach | Food in stomach, particularly partially digested proteins; ACh released by nerve fibers | • Stimulates release of gastric juice<br>• Stimulates stomach emptying |
| Intestinal gastrin | Duodenum | Food in stomach | • Stimulates gastric secretion and emptying |
| Histamine | Stomach | Food in stomach | • Activates parietal cells to secrete hydrochloric acid |
| Somatostatin | Stomach and duodenum | Food in stomach stimulated by sympathetic nerve fibers | • Inhibits secretion of gastric juice and pancreatic juice<br>• Inhibits emptying of stomach and gallbladder |
| Secretin | Duodenum | Acidic chyme and partially digested foods in duodenum | • Increases output of pancreatic juice rich in bicarbonate ions<br>• Increases bile output by liver<br>• Inhibits gastric mobility and gastric gland secretion |
| Cholecystokinin (CCK) | Duodenum | Fatty chyme and partially digested proteins in duodenum | • Increases output of enzyme-rich pancreatic juice<br>• Stimulates gallbladder to expel stored bile<br>• Relaxes sphincter of duodenal papilla to allow bile and pancreatic juice to enter the duodenum |
| Gastric inhibitory peptide (GIP) | Duodenum | Food in duodenum | • Inhibits secretion of gastric juice<br>• Stimulates insulin release |

form the flasklike *hepatopancreatic ampulla* (hĕ-pă"to-pan-kre-ă'tik am-pu'lah), literally, the "liver-pancreatic enlargement." From there, the bile and pancreatic juice travel through the *duodenal papilla* and enter the duodenum together.

Nearly all nutrient absorption occurs in the small intestine. The small intestine is well suited for its function. Its wall has three structures that increase the absorptive surface tremendously—villi, microvilli, and circular folds (**Figure 14.7**, p. 498). **Villi** are fingerlike projections of the mucosa that give it a velvety appearance and feel, much like the soft nap of a towel. Within each villus is a rich capillary bed and a modified lymphatic capillary called a **lacteal**. The nutrients are absorbed through the mucosal cells into both the capillaries and the lacteal (indicated in Figure 14.13 on p. 505). **Microvilli** (mi"kro-vil'i) are tiny projections of the plasma membrane of the mucosa cells that give the cell surface a fuzzy

appearance, sometimes referred to as the **brush border**. The plasma membranes bear enzymes (*brush border enzymes*) that complete the digestion of proteins and carbohydrates in the small intestine. **Circular folds**, also called **plicae circulares** (pli'se ser-ku-la'res), are deep folds of both mucosa and submucosa layers. Unlike the rugae of the stomach, the circular folds do not disappear when food fills the small intestine. Instead, they form an internal "corkscrew slide" to increase surface area and force chyme to travel slowly through the small intestine so nutrients can be absorbed efficiently. All these structural modifications, which increase the surface area, decrease in number toward the end of the small intestine. In contrast, local collections of lymphatic tissue (called **Peyer's patches**) found in the submucosa increase in number toward the end of the small intestine. This reflects the fact that the remaining (undigested) food residue in

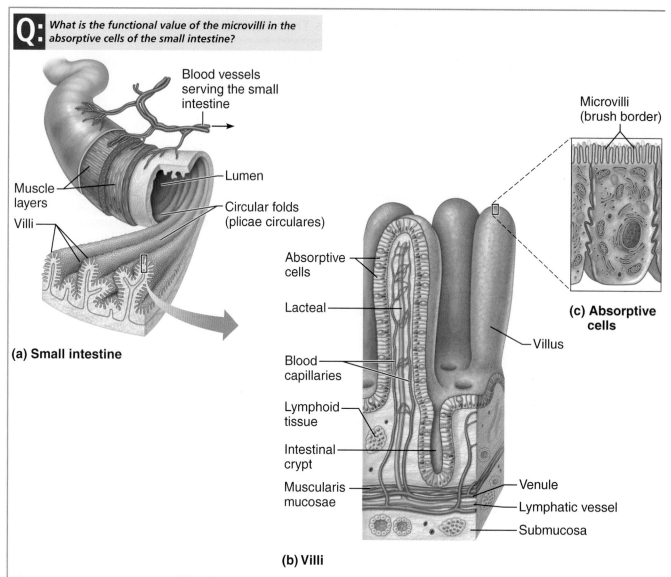

**Q:** *What is the functional value of the microvilli in the absorptive cells of the small intestine?*

Blood vessels serving the small intestine

Muscle layers

Villi

Lumen

Circular folds (plicae circulares)

**(a) Small intestine**

Absorptive cells

Lacteal

Blood capillaries

Lymphoid tissue

Intestinal crypt

Muscularis mucosae

Villus

Venule

Lymphatic vessel

Submucosa

**(b) Villi**

Microvilli (brush border)

**(c) Absorptive cells**

**Figure 14.7 Structural modifications of the small intestine. (a)** Several circular folds (plicae circulares), seen on the inner surface of the small intestine. **(b)** Enlargement of a villus extension of the circular fold. **(c)** Enlargement of an absorptive cell to show microvilli (brush border).

the intestine contains huge numbers of bacteria, which must be prevented from entering the bloodstream if at all possible.

### Did You Get It?

4. Which muscular sphincter regulates the flow of chyme into the small intestine?
5. What are villi, and why are they important?

**For answers, see Appendix A.**

**A:** They tremendously increase the surface area available for absorption of digested foodstuffs.

### Large Intestine

The **large intestine** is much larger in diameter than the small intestine (thus its name) but shorter in length. About 1.5 m (5 feet) long, it extends from the ileocecal valve to the anus (**Figure 14.8**, p. 499). Its major functions are to dry out the indigestible food residue by absorbing water and to eliminate these residues from the body as **feces** (fe′sēz). It frames the small intestine on three sides and has these subdivisions: *cecum* (se′kum), *appendix, colon, rectum*, and *anal canal*.

The saclike **cecum** is the first part of the large intestine. Hanging from the cecum is the wormlike

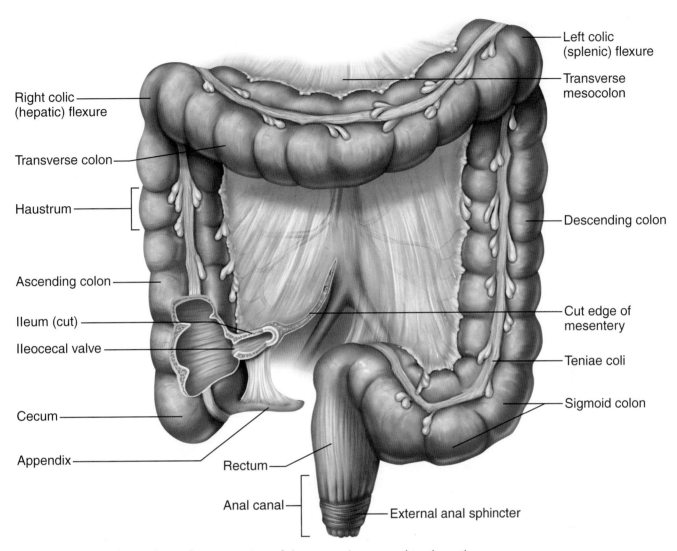

Right colic (hepatic) flexure

Transverse colon

Haustrum

Ascending colon

Ileum (cut)

Ileocecal valve

Cecum

Appendix

Rectum

Anal canal

Left colic (splenic) flexure

Transverse mesocolon

Descending colon

Cut edge of mesentery

Teniae coli

Sigmoid colon

External anal sphincter

**Figure 14.8 The large intestine.** A section of the cecum is removed to show the ileocecal valve.

**appendix**, a potential trouble spot. Because it is usually twisted, it is an ideal location for bacteria to accumulate and multiply. Inflammation of the appendix, **appendicitis**, is the usual result.

The **colon** is divided into several distinct regions. The **ascending colon** travels up the right side of the abdominal cavity and makes a turn, the *right colic* (or *hepatic*) *flexure*, to travel across the abdominal cavity as the **transverse colon**. It then turns again at the *left colic* (or *splenic*) *flexure* and continues down the left side as the **descending colon** to enter the pelvis, where it becomes the S-shaped **sigmoid** (sig′moyd) **colon**. The sigmoid colon, **rectum**, and **anal canal** lie in the pelvis.

The anal canal ends at the **anus** (a′nus), which opens to the exterior. The anal canal has two valves: the **external anal sphincter**, composed of skeletal muscle, is voluntary, and the **internal anal sphincter**, formed by smooth muscle, is involuntary. These sphincters, which act rather like purse strings to open and close the anus, are ordinarily closed except during defecation, when feces are eliminated from the body.

Because most nutrients have been absorbed before the large intestine is reached, no villi are present in the large intestine, but there are tremendous numbers of *goblet cells* in its mucosa that produce alkaline (bicarbonate-rich) mucus. The mucus lubricates the passage of feces to the end of the digestive tract.

In the large intestine, the longitudinal layer of the muscularis externa is reduced to three bands

14

of muscle called *teniae coli* (ten'ne-e ko'li; "ribbons of the colon"). Because these muscle bands usually display some degree of tone (are partially contracted), they cause the wall to pucker into small pocketlike sacs called **haustra** (haws'trah).

## Did You Get It?

6. What are the two main functions of the large intestine?

For the answer, see Appendix A.

## Accessory Digestive Organs

### ➔ Learning Objectives

☐ List the accessory digestive organs, and describe the general function of each.

☐ Name the deciduous and permanent teeth, and describe the basic anatomy of a tooth.

☐ Describe the composition and functions of saliva.

☐ Name the main digestive products of the pancreas and of the liver.

## Teeth

The role teeth play in processing food needs little introduction. We **masticate**, or *chew*, by opening and closing our jaws and moving them from side to side while continuously using our tongue and cheek muscles to keep the food between our teeth. In the process, the teeth tear and grind the food, breaking it down into smaller fragments.

Ordinarily, by the age of 21, two sets of teeth have been formed (**Figure 14.9**, p. 500). The first set are **deciduous** (de-sid'u-us) **teeth**, also called **baby teeth** or **milk teeth**. The deciduous teeth begin to erupt around 6 months; the first teeth to appear are the lower central incisors. A baby has a full set (20 teeth) by the age of 2 years.

As the second set of teeth, the deeper **permanent teeth**, enlarge and develop, the roots of the milk teeth are reabsorbed, and between the ages of 6 and 12 years they loosen and fall out. All of the permanent teeth but the third molars have erupted by the end of adolescence. The third molars, also called *wisdom teeth*, emerge between the ages of 17 and 25. Although there are 32 permanent teeth in a full set, the wisdom teeth often fail to erupt; sometimes they are completely absent.

### 🔨 Homeostatic Imbalance 14.3

When teeth remain embedded in the jawbone, they are said to be *impacted*. Impacted teeth can

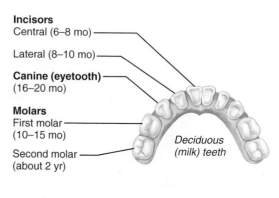

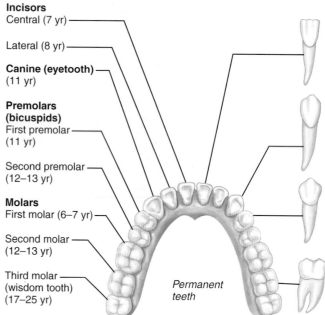

**Figure 14.9 Deciduous and permanent teeth.** Approximate time of tooth eruption is shown in parentheses. The same number and arrangement of teeth exist in both upper and lower jaws, so only the lower jaw is shown in each case. The shapes of individual teeth are shown on the right.

exert pressure and cause a good deal of pain, so they usually must be removed surgically. Wisdom teeth are the most commonly impacted. _____ ✚

We classify the teeth according to shape and function as incisors, canines, premolars, and molars (see Figure 14.9). The chisel-shaped **incisors** are adapted for cutting; the fanglike **canines** (eyeteeth) are for tearing or piercing. The **premolars** (bicuspids) and **molars** have broad crowns with rounded cusps (tips) and are best suited for crushing and grinding.

A tooth consists of two major regions, the *crown* and the *root* (**Figure 14.10**). The enamel-covered **crown** is the exposed part of the tooth

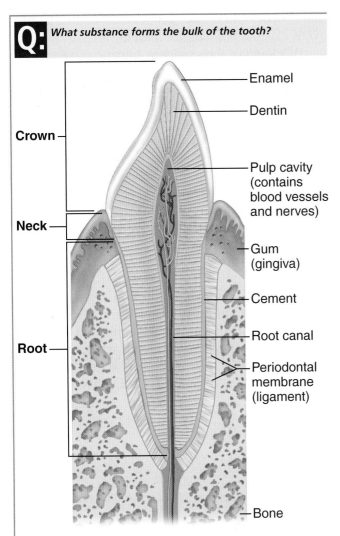

**Q:** *What substance forms the bulk of the tooth?*

Enamel

Dentin

Crown

Pulp cavity (contains blood vessels and nerves)

Neck

Gum (gingiva)

Cement

Root canal

Root

Periodontal membrane (ligament)

Bone

**Figure 14.10 Longitudinal section of a canine tooth.**

**A:** Dentin.

above the **gingiva** (jin′-jĭvah), or **gum**. The crown is covered with **enamel**, a ceramic-like substance as thick as a dime, that directly bears the force of chewing. It is the hardest substance in the body and is fairly brittle because it is heavily mineralized with calcium salts.

The portion of the tooth embedded in the jawbone is the **root**; the root and crown are connected by a region called the **neck**. The outer surface of the root is covered by a substance called **cement**, which attaches the tooth to the **periodontal** (per″e-o-don′tal) **membrane (ligament)**. This ligament holds the tooth in place in

the bony jaw. **Dentin**, a bonelike material, underlies the enamel and forms the bulk of the tooth. It surrounds a central **pulp cavity**, which contains a number of structures (connective tissue, blood vessels, and nerve fibers) collectively called **pulp**. Pulp supplies nutrients to the tooth tissues and provides for tooth sensations. Where the pulp cavity extends into the root, it becomes the **root canal**, which provides a route for blood vessels, nerves, and other pulp structures to enter the pulp cavity of the tooth.

## Did You Get It?

7. What is the general function of each of the four types of teeth?

For the answer, see Appendix A.

### Salivary Glands

Three pairs of **salivary glands** empty their secretions into the mouth. The large **parotid** (pah-rot′id) **glands** lie anterior to the ears. **Mumps**, a common childhood disease, is an inflammation of the parotid glands. If you look at the location of the parotid glands (see Figure 14.1), you can readily understand why people with mumps complain that it hurts to open their mouth or chew.

The **submandibular glands** and the small **sublingual** (sub-ling′gwal) **glands** empty their secretions into the floor of the mouth through tiny ducts. The product of the salivary glands, **saliva**, is a mixture of mucus and serous fluids. The mucus moistens and helps to bind food together into a mass called a *bolus* (bo′lus), which makes chewing and swallowing easier. The clear serous portion contains an enzyme, **salivary amylase** (am′ĭ-lās), in a bicarbonate-rich (alkaline) juice that begins the process of starch digestion in the mouth. Saliva also contains substances such as lysozyme and antibodies (IgA) that inhibit bacteria; therefore, it has a protective function as well. Last but not least, saliva dissolves food chemicals so they can be tasted.

### Pancreas

The **pancreas** is a soft, pink, triangular gland that extends across the abdomen from the spleen to the duodenum (look back at Figure 14.6). Most of the pancreas lies posterior to the parietal peritoneum; hence its location is referred to as *retroperitoneal*.

Only the pancreas produces enzymes (described later) that break down all categories of digestible

14

foods. The pancreatic enzymes are secreted into the duodenum in an alkaline fluid that neutralizes the acidic chyme coming in from the stomach. The pancreas also has an endocrine function; it produces the hormones insulin and glucagon (as explained in Chapter 9).

### Liver and Gallbladder

The **liver** is the largest gland in the body. Located under the diaphragm, more to the right side of the body, it overlies and almost completely covers the stomach (look back at Figure 14.5a). The liver has four lobes and is suspended from the diaphragm and abdominal wall by a delicate mesentery cord, the **falciform** (fal'sĭ-form) **ligament**.

The liver has many critical metabolic and regulatory roles; however, its digestive function is to produce **bile**. Bile leaves the liver through the **common hepatic duct** and enters the duodenum through the *bile duct* (look back at Figure 14.6).

Bile is a yellow-to-green, watery solution containing bile salts, bile pigments (chiefly bilirubin, a breakdown product of hemoglobin), cholesterol, phospholipids, and a variety of electrolytes. Of these components, only the bile salts (derived from cholesterol) and phospholipids help the digestive process. Bile does not contain enzymes, but its bile salts *emulsify* fats by physically breaking large fat globules into smaller ones, thus providing more surface area for the fat-digesting enzymes to work on.

The **gallbladder** is a small, thin-walled green sac that snuggles in a shallow fossa in the inferior surface of the liver (look back at Figure 14.6). When food digestion is not occurring, bile backs up the **cystic duct** and enters the gallbladder to be stored. While in the gallbladder, bile is concentrated by the removal of water. Later, when fatty food enters the duodenum, a hormonal stimulus prompts the gallbladder to contract and spurt out stored bile, making it available to the duodenum.

### Homeostatic Imbalance 14.4

If bile is stored in the gallbladder for too long or too much water is removed, the cholesterol it contains may crystallize, forming **gallstones**. Because gallstones tend to be quite sharp, agonizing pain

may occur when the gallbladder contracts (an event called a *gallbladder attack*).

Blockage of the common hepatic or bile ducts (for example, by wedged gallstones) prevents bile from entering the small intestine, and it begins to accumulate and eventually backs up into the liver. This exerts pressure on the liver cells, and bile salts and bile pigments begin to enter the bloodstream. As the bile pigments circulate through the body, the tissues become yellow, or **jaundiced**. Blockage of the ducts is just one cause of jaundice. More often it results from actual liver problems such as **hepatitis**, an inflammation of the liver, or **cirrhosis** (sir-ro'sis), a chronic inflammatory condition in which the liver is severely damaged and becomes hard and fibrous. Hepatitis is most often due to viral infection resulting from drinking contaminated water or transmitted in blood via transfusion or contaminated needles. Cirrhosis is almost guaranteed when someone drinks alcoholic beverages in excess for many years, and it is a common consequence of severe hepatitis. _____ ✚

### Did You Get It?

8. Mary has a dry mouth—very little saliva is being secreted. Digestion of which type of food will be affected (decreased) by this situation?
9. What is the digestive role of bile? What organ produces bile?
10. Which organ is the only one that produces enzymes capable of digesting all groups of food?

For answers, see Appendix A.)

# Functions of the Digestive System

→ **Learning Objectives**

☐ **List and describe the six main activities of the digestive system.**

☐ **Describe how foodstuffs in the digestive tract are mixed and moved along the tract.**

☐ **Describe the function of local hormones in digestion.**

☐ **List the major enzymes or enzyme groups involved in digestion, and name the foodstuffs on which they act.**

☐ **Describe the mechanisms of swallowing, vomiting, and defecation.**

☐ **Name the end products of protein, fat, and carbohydrate digestion.**

# Overview of Gastrointestinal Processes and Controls

The major functions of the digestive tract are usually summarized in two words—*digestion* and *absorption*. However, many of its specific activities (such as smooth muscle activity) and certain regulatory events are not really covered by either term. To describe digestive system processes a little more accurately, we need to consider a few more functional terms. The essential activities of the GI tract include the following six processes (summarized in **Figure 14.11**).

1.  **Ingestion.** Food must be placed into the mouth before it can be acted on. This is an active, voluntary process called *ingestion*.

2.  **Propulsion.** To be processed by more than one digestive organ, foods must be propelled from one organ to the next. Swallowing is one example of food movement that depends largely on the propulsive process called **peristalsis**. Peristalsis is involuntary and involves alternating waves of contraction and relaxation of the longitudinal muscles in the organ wall **(Figure 14.12a)**. The net effect is to squeeze the food along the tract.

3.  **Food breakdown: Mechanical breakdown.** Mechanical breakdown physically fragments food into smaller particles, increasing surface area and preparing food for further degradation by enzymes. Chewing and mixing of food in the mouth by the teeth and tongue, and churning of food in the stomach are examples of processes contributing to mechanical food breakdown. In addition, **segmentation** (Figure 14.12b) in the small intestine moves food back and forth across the internal wall of the organ, mixing it with the digestive juices. Although segmentation may also help to propel foodstuffs through the small intestine, it is more an example of mechanical digestion than of propulsion.

4.  **Food breakdown: Digestion.** The sequence of steps in which large food molecules are chemically broken down to their building blocks by enzymes (protein molecules that act as catalysts) is called *digestion*.

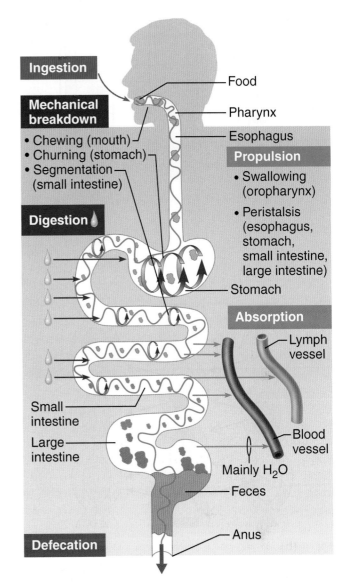

**Figure 14.11 Schematic summary of gastrointestinal tract activities.** Gastrointestinal tract activities include ingestion, mechanical breakdown, chemical (enzymatic) breakdown or digestion, propulsion, absorption, and defecation. Sites of chemical digestion are also sites that produce enzymes or that receive enzymes or other secretions made by accessory organs. The entire GI tract mucosa secretes mucus, which protects and lubricates.

### CONCEPT**LINK**

Recall that these enzymatic reactions are called *hydrolysis* reactions because a water molecule is added to each bond to be broken (see Chapter 2, p. 69). Water is also necessary as a dissolving medium and a softening agent for food digestion.

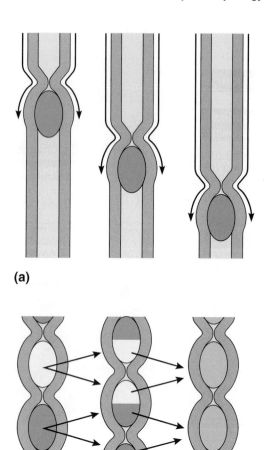

**(a)**

**(b)**

**Figure 14.12 Peristaltic and segmental movements of the digestive tract. (a)** In peristalsis, adjacent or neighboring segments of the intestine (or other alimentary canal organs) alternately contract and relax, thereby moving food distally along the tract. **(b)** In segmentation, single segments of the intestine alternately contract and relax. Because active segments are separated by inactive ones, the food is moved forward and then backward. Thus the food is mixed rather than simply propelled along the tract.

Because each of the major food groups has very different building blocks, let's take a little time to review these organic molecules (which we first introduced in Chapter 2). The building blocks, or units, of *carbohydrate* foods are *monosaccharides* (mon″o-sak′ah-rīdz), or simple sugars. We need to remember only three of these that are common in our diet—*glucose, fructose,* and *galactose.* Glucose is by far the most important, and

when we talk about blood sugar level, we are referring to glucose. Fructose is the most abundant sugar in fruits, and galactose is found in milk.

The only carbohydrates that our digestive system digests, or breaks down to simple sugars, are *sucrose* (table sugar), *lactose* (milk sugar), *maltose* (malt sugar), and *starch*. Sucrose, maltose, and lactose are referred to as *disaccharides*, or double sugars, because each consists of two simple sugars linked together. Starch is a *polysaccharide* (literally, "many sugars") formed by linking hundreds of glucose units. Although we eat foods containing other polysaccharides, such as *cellulose*, we do not have enzymes capable of breaking them down. The indigestible polysaccharides do not provide us with any nutrients, but they help move the foodstuffs along the gastrointestinal tract by providing bulk, or *fiber*, in our diet.

*Proteins* are digested to their building blocks, which are *amino* (ah-me′no) *acids*. The intermediate product of protein digestion is polypeptides.

When *lipids* (fats) are digested, they yield two different types of building blocks—*fatty acids* and an alcohol called *glycerol* (glis′er-ol). The digestion, or chemical breakdown, of carbohydrates, proteins, and fats is summarized in **Figure 14.13**, and we describe it in more detail shortly.

5. **Absorption.** Absorption is the transport of digestive end products from the lumen of the GI tract to the blood or lymph. For absorption to occur, the digested foods must first enter the mucosal cells by active or passive transport processes. The small intestine is the major absorptive site.

6. **Defecation.** Defecation is the elimination of indigestible residues from the GI tract via the anus in the form of feces.

Some of these processes are the job of a single organ. For example, only the mouth ingests, and only the large intestine defecates. But most digestive system activities occur bit by bit as food is moved along the tract. Thus, in one sense, the digestive tract can be viewed as a "disassembly line" in which food is carried from one stage of its processing to the next and its nutrients are made available to the cells in the body along the way.

Throughout this book we have stressed the body's drive to maintain a constant internal environment, particularly in terms of homeostasis of the blood, which comes into contact with all body

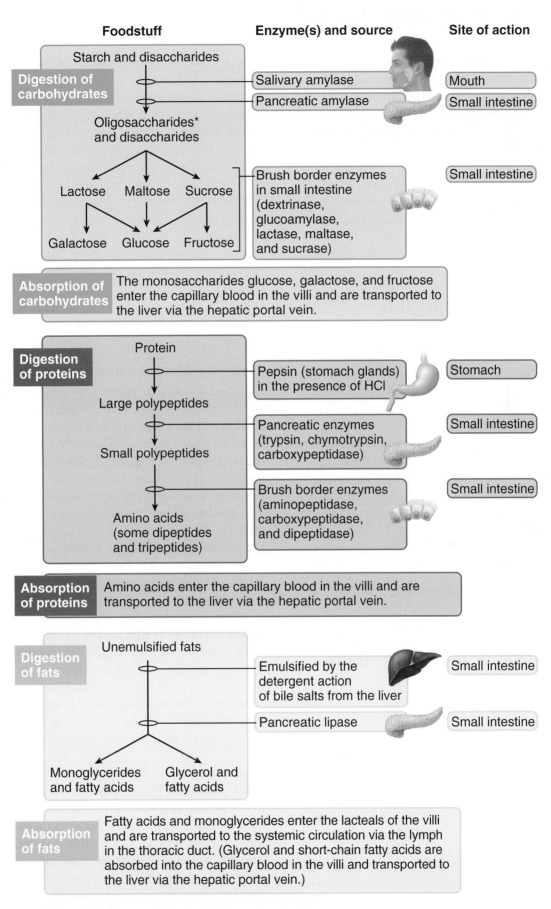

| Foodstuff | Enzyme(s) and source | Site of action |
|---|---|---|

**Digestion of carbohydrates**

Starch and disaccharides
- Salivary amylase — Mouth
- Pancreatic amylase — Small intestine

Oligosaccharides* and disaccharides

Lactose   Maltose   Sucrose

Galactose   Glucose   Fructose
- Brush border enzymes in small intestine (dextrinase, glucoamylase, lactase, maltase, and sucrase) — Small intestine

**Absorption of carbohydrates**
The monosaccharides glucose, galactose, and fructose enter the capillary blood in the villi and are transported to the liver via the hepatic portal vein.

**Digestion of proteins**

Protein
- Pepsin (stomach glands) in the presence of HCl — Stomach

Large polypeptides
- Pancreatic enzymes (trypsin, chymotrypsin, carboxypeptidase) — Small intestine

Small polypeptides
- Brush border enzymes (aminopeptidase, carboxypeptidase, and dipeptidase) — Small intestine

Amino acids (some dipeptides and tripeptides)

**Absorption of proteins**
Amino acids enter the capillary blood in the villi and are transported to the liver via the hepatic portal vein.

**Digestion of fats**

Unemulsified fats
- Emulsified by the detergent action of bile salts from the liver — Small intestine
- Pancreatic lipase — Small intestine

Monoglycerides and fatty acids    Glycerol and fatty acids

**Absorption of fats**
Fatty acids and monoglycerides enter the lacteals of the villi and are transported to the systemic circulation via the lymph in the thoracic duct. (Glycerol and short-chain fatty acids are absorbed into the capillary blood in the villi and transported to the liver via the hepatic portal vein.)

*Oligosaccharides consist of a few linked monosaccharides.

**Figure 14.13 Flowchart of digestion and absorption of foodstuffs.**

14

cells. The digestive system, however, creates an optimal environment for itself to function in the lumen (cavity) of the alimentary canal, an area that is actually *outside* the body. Conditions in that lumen are controlled so that digestive processes occur efficiently.

Digestive activity is mostly controlled by reflexes via the parasympathetic division of the autonomic nervous system. (This division is the "rest-and-digest" arm. See Chapter 7.) The sensors (mechanoreceptors, chemoreceptors) involved in these reflexes are located in the walls of the alimentary canal organs and respond to a number of stimuli, the most important being stretch of the organ by food in its lumen, pH of the contents, and presence of certain breakdown products of digestion. When these receptors are activated, they trigger reflexes that activate or inhibit (1) the glands that secrete digestive juices into the lumen or hormones into the blood, and (2) the smooth muscles of the muscularis that mix and propel the food along the tract.

Now that we have summarized some points that apply to the function of the digestive organs as a group, we are ready to look at their special capabilities.

### Did You Get It?

11. What is the proper order of the following stages of food processing—defecation, absorption, digestion, ingestion?

12. How do mechanical breakdown and digestion differ from each other?

For answers, see Appendix A.

## Activities Occurring in the Mouth, Pharynx, and Esophagus

### Food Ingestion and Breakdown

Once food has been placed in the mouth, both mechanical and digestive (chemical) processing begins. First the food is *physically* broken down into smaller particles by chewing. Then, as the food is mixed with saliva, salivary amylase begins the digestion of starch, chemically breaking it down into maltose (see Figure 14.13). The next time you eat a piece of bread, chew it for a few minutes before swallowing it. You will notice that it begins to taste sweet as the sugars are released.

Saliva is normally secreted continuously to keep the mouth moist, but when food enters the mouth, much larger amounts of saliva pour out. However, the simple pressure of anything put in the mouth and chewed, such as rubber bands or sugarless gum, will also stimulate the release of saliva. Some emotional stimuli can also cause salivation. For example, the mere thought of a hot fudge sundae will make many a mouth water. All these reflexes, though initiated by different stimuli, are brought about by parasympathetic fibers in cranial nerves VII and IX.

Essentially no food absorption occurs in the mouth. (However, some drugs, such as nitroglycerine, are absorbed easily through the oral mucosa.) The pharynx and esophagus have no digestive function; they simply provide passageways to carry food to the next processing site, the stomach.

### Food Propulsion—Swallowing and Peristalsis

For food to be sent on its way from the mouth, it must first be swallowed. **Deglutition** (de"gloo-tish'un), or **swallowing**, is a complex process that involves the coordinated activity of several structures (tongue, soft palate, pharynx, and esophagus). It has two major phases, the buccal and the pharyngeal-esophageal. The first phase, the voluntary **buccal phase**, occurs in the mouth. Once the food has been chewed and well mixed with saliva, the **bolus** (food mass) is forced into the pharynx by the tongue. As food enters the pharynx, it passes out of our control and into the realm of reflex activity.

The second phase, the involuntary **pharyngeal-esophageal phase**, transports food through the pharynx and esophagus. The parasympathetic division of the autonomic nervous system (primarily the vagus nerve) controls this phase and promotes the mobility of the digestive organs from this point on. All routes that the food might take, except the desired route further into the digestive tract, are blocked off. The tongue blocks off the mouth, and the soft palate closes off the nasal passages. The larynx rises so that its opening (into the respiratory passageways) is covered by the flaplike epiglottis. Food is moved through the pharynx and then into the esophagus inferiorly by wavelike peristaltic contractions of their muscular walls—first the longitudinal muscles contract, and then the circular muscles contract. (The events of the swallowing process are illustrated in **Figure 14.14**.)

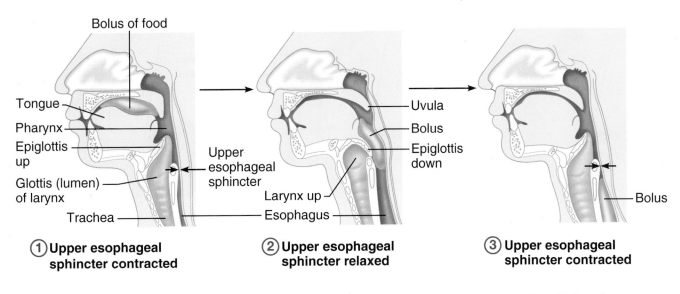

① **Upper esophageal sphincter contracted**

② **Upper esophageal sphincter relaxed**

③ **Upper esophageal sphincter contracted**

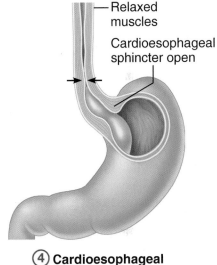

④ **Cardioesophageal sphincter relaxed**

**Figure 14.14 Swallowing.** ① With the upper esophageal sphincter contracted, the tongue pushes the food bolus posteriorly and against the soft palate. ② The soft palate rises to close off the nasal passages as the bolus enters the pharynx. The larynx rises so that the epiglottis covers its opening as peristalsis carries the food through the pharynx and into the esophagus. The upper esophageal sphincter relaxes to allow food entry. ③ The upper esophageal sphincter contracts again as the larynx and epiglottis return to their former positions and peristalsis moves the food bolus inferiorly to the stomach. ④ The cardioesophageal sphincter opens, and food enters the stomach.

If we try to talk or laugh while swallowing, we confuse our bodies with mixed messages, and as a result food may enter the respiratory passages. This triggers still another protective reflex—coughing—during which air rushes upward from the lungs in an attempt to expel the food.

Once food reaches the distal end of the esophagus, it presses against the cardioesophageal sphincter, causing it to open, and the food enters the stomach. The movement of food through the pharynx and esophagus is so automatic that a person can swallow and food will reach the stomach even if he is standing on his head. Gravity plays no part in the transport of food once it has left the mouth, which explains why astronauts in the zero gravity of outer space can still swallow and get nourishment.

## Activities of the Stomach

### Food Breakdown

Secretion of **gastric juice** is regulated by both neural and hormonal factors. The sight, smell, and taste of food stimulate parasympathetic nervous system reflexes, which increase the secretion of gastric juice by the gastric (stomach) glands. In addition, the presence of food and a rising pH in the stomach stimulate the stomach cells to release the hormone **gastrin**. Gastrin prods the gastric glands to produce still more of the protein-digesting enzymes (such as pepsinogen), mucus, and hydrochloric acid. Under normal conditions, 2 to 3 liters of gastric juice are produced every day.

Hydrochloric acid makes the stomach contents very acidic. This can be dangerous because both

14

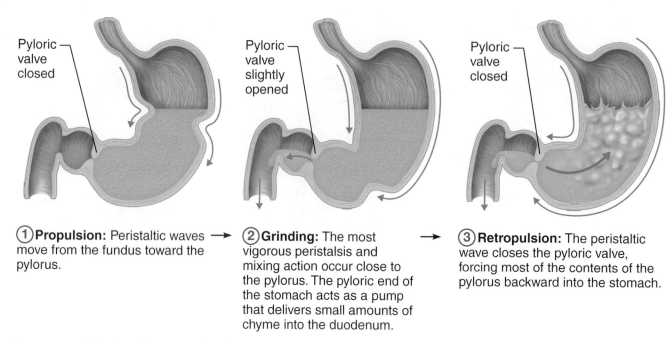

① **Propulsion:** Peristaltic waves move from the fundus toward the pylorus. →

② **Grinding:** The most vigorous peristalsis and mixing action occur close to the pylorus. The pyloric end of the stomach acts as a pump that delivers small amounts of chyme into the duodenum. →

③ **Retropulsion:** The peristaltic wave closes the pyloric valve, forcing most of the contents of the pylorus backward into the stomach.

**Figure 14.15 Peristaltic waves in the stomach.**

hydrochloric acid and the protein-digesting enzymes have the ability to digest the stomach itself, causing *ulcers* (see "A Closer Look" on p. 512). However, as long as enough mucus is made, the stomach is protected.

### ⚖ Homeostatic Imbalance 14.5

Occasionally, the cardioesophageal sphincter fails to close tightly and gastric juice backs up into the esophagus, which has little mucus protection. This results in a characteristic pain known as **heartburn**, which, if uncorrected, leads to inflammation of the esophagus (**esophagitis** [ĕ-sof′ah-ji′tis]) and perhaps even to ulceration of the esophagus. A common cause is a **hiatal hernia**, a structural abnormality in which the superior part of the stomach protrudes slightly above the diaphragm. Because the diaphragm no longer reinforces the relatively weak cardioesophageal sphincter, gastric juice flows into the unprotected esophagus. Conservative treatment involves restricting food intake after the evening meal, taking antacids, and sleeping with the head elevated. _____ ✚

The extremely acidic environment that hydrochloric acid provides *is* necessary, because it activates *pepsinogen* to **pepsin**, the active protein-digesting enzyme. *Rennin*, the second protein-digesting enzyme produced by the stomach, works primarily on milk protein and converts it to a substance that looks like sour milk. Many parents mistakenly think that the curdy substance infants spit up after feeding is milk that has soured in their stomach. Rennin, produced in large amounts in infants, is the same enzyme used to make milk curdle into cheese. It is not believed to be produced in adults.

Other than the beginning of protein digestion, little chemical digestion occurs in the stomach. Except for aspirin and alcohol (which seem somehow to have a "special pass"), virtually no absorption occurs through the stomach walls.

As food enters and fills the stomach, its wall begins to stretch (at the same time the gastric juices are being secreted, as just described). Then the three muscle layers of the stomach wall become active. They compress and pummel the food, breaking it apart physically, all the while continuously mixing the food with the enzyme-containing gastric juice so that the thick fluid chyme is formed. The process looks something like the preparation of a cake mix, in which the floury mixture is repeatedly folded on itself and mixed with the liquid until it reaches a uniform texture.

#### Food Propulsion

Once the food has been well mixed, a rippling peristalsis begins in the upper half of the stomach (**Figure 14.15**, ①). The contractions increase in force as the food approaches the pyloric valve, grinding the food into chyme. The pylorus of the stomach, which holds about 30 ml of chyme, acts like a meter that allows only liquids and very small particles to pass through the pyloric sphincter. Because the pyloric sphincter barely opens, each contraction of the stomach muscle squirts 3 ml or less of chyme into the small intestine (Figure 14.15, ②). The contraction also *closes* the valve, so the rest of the chyme (about 27 ml) is propelled backward into the stomach for more mixing, a process called *retropulsion* (Figure 14.15, ③).

When the duodenum is filled with chyme and its wall is stretched, a nervous reflex, the *enterogastric* (en"ter-o-gas'trik) *reflex*, occurs. This reflex "puts the brakes on" gastric activity. It slows the emptying of the stomach by inhibiting the vagus nerve and tightening the pyloric sphincter, thus allowing time for intestinal processing to catch up. Generally, it takes about 4 hours for the stomach to empty completely after the person eats a well-balanced meal and 6 hours or more if the meal has a high fat content.

 **Homeostatic Imbalance 14.6**

Local irritation of the stomach, such as occurs with bacterial food poisoning, may activate the *emetic* (ē-met'ik) center in the brain (medulla). The emetic center, in turn, causes **vomiting**, or **emesis**. Vomiting is essentially a reverse peristalsis occurring in the stomach (and perhaps the small intestine), accompanied by contraction of the abdominal muscles and the diaphragm, which increases the pressure on the abdominal organs. The emetic center may also be activated through other pathways; disturbance of the equilibrium apparatus of the inner ear during a boat ride on rough water is one example. _____✚

## Did You Get It?

13. What happens during the buccal stage of swallowing?
14. How does segmentation differ from peristalsis in terms of moving substances along the GI tract?
15. What does it mean when we say that food "went down the wrong tube"?
16. Why is it necessary for the stomach contents to be so acidic?

**For answers, see Appendix A.**

## Activities of the Small Intestine

### Chyme Breakdown and Absorption

Chyme reaching the small intestine is only partially digested. Carbohydrate and protein digestion has begun, but virtually no fats or nucleic acids have been digested up to this point. Here the process of chemical digestion is accelerated as the food now takes a rather wild 3- to 6-hour journey through the looping coils and twists of the small intestine. By the time the food reaches the end of the small intestine, digestion will be complete, and nearly all food absorption will have occurred.

As mentioned earlier, the microvilli of small intestine cells bear several important enzymes, the so-called **brush border enzymes**, that break down double sugars into simple sugars and complete protein digestion (see Figure 14.13). *Intestinal juice* itself is relatively enzyme poor, and protective mucus is probably the most important intestinal gland secretion. However, foods entering the small intestine are literally deluged with enzyme-rich **pancreatic juice** delivered via a duct from the pancreas, as well as bile from the liver.

Pancreatic juice contains enzymes that (1) along with brush border enzymes, complete the digestion of starch (*pancreatic amylase*); (2) carry out about half of protein digestion (via the action of *trypsin, chymotrypsin, carboxypeptidase,* and others); (3) are totally responsible for fat digestion, because the pancreas is essentially the only source of *lipases*; and (4) digest nucleic acids (*nucleases*).

In addition to enzymes, pancreatic juice contains a rich supply of bicarbonate ions, which makes it very basic (about pH 8). When pancreatic juice reaches the small intestine, it neutralizes the acidic chyme coming in from the stomach and provides the proper environment for activation and activity of intestinal and pancreatic digestive enzymes.

 **Homeostatic Imbalance 14.7**

**Pancreatitis** (pan"kre-ah-ti'tis) is a rare but extremely serious inflammation of the pancreas that results from activation of pancreatic enzymes in the pancreatic duct. Because pancreatic enzymes break down all categories of biological molecules, they cause digestion of the pancreatic tissue and

14

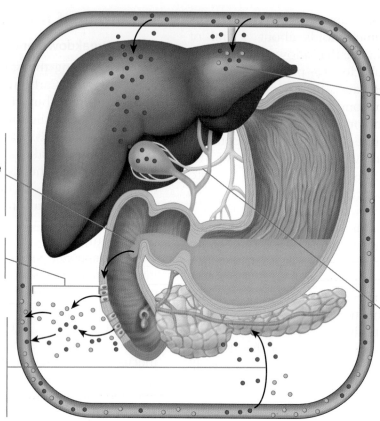

① Chyme entering duodenum causes duodenal enteroendocrine cells to release cholecystokinin (CCK) and secretin.

② CCK (red dots) and secretin (blue dots) enter the bloodstream.

③ Upon reaching the pancreas, CCK induces secretion of enzyme-rich pancreatic juice; secretin causes secretion of bicarbonate-rich pancreatic juice.

④ Secretin causes the liver to secrete more bile; CCK stimulates the gallbladder to release stored bile and the hepatopancreatic sphincter to relax (allows bile from both sources to enter the duodenum).

⑤ Stimulation by vagal nerve fibers causes release of pancreatic juice and weak contractions of the gallbladder.

**Figure 14.16 Regulation of pancreatic juice and bile secretion and release.** Hormonal controls exerted by secretin and cholecystokinin ①–④, are the main regulatory factors. Neural control ⑤ is mediated by the vagus nerves.

duct. This painful condition can lead to nutritional deficiencies because pancreatic enzymes are essential for digestion in the small intestine. _____ ✚

The release of pancreatic juice into the duodenum is stimulated by both the vagus nerve and local hormones. When chyme enters the small intestine, it stimulates the mucosa cells to produce several hormones (Table 14.1, p. 511). Two of these hormones, **secretin** (se-kre′tin) and **cholecystokinin** (ko″le-sis″to-kin′in) **(CCK)**, influence the release of pancreatic juice and bile.

The hormones enter the blood and circulate to their target organs, the pancreas, liver, and gallbladder. Both hormones work together to stimulate the pancreas to release its enzyme- and bicarbonate-rich product (**Figure 14.16**, p. 510). In addition, secretin causes the liver to increase its output of bile, and cholecystokinin causes the gallbladder to contract and release stored bile into the bile duct so that bile and pancreatic juice enter the

small intestine together. As mentioned before, bile is not an enzyme. Instead, it acts like a detergent to emulsify, or mechanically separate, large fat globules into thousands of tiny ones, providing a much greater surface area for the pancreatic lipases to work on. Bile is also necessary for absorption of fats—and the fat-soluble vitamins (K, D, E, and A) that are absorbed along with them—from the intestinal tract.

### Homeostatic Imbalance 14.8

If *either* bile or pancreatic juice is absent, essentially no fat digestion or absorption goes on, and fatty, bulky stools are the result. In such cases, blood-clotting problems also occur because the liver needs fat-soluble vitamin K to make prothrombin, an important clotting factor. _____ ✚

Absorption of water and of the end products of digestion occurs all along the length of the small intestine. Most substances are absorbed through the intestinal cell plasma membranes by the process of *active transport*. Then they enter the capillary beds in the villi to be transported in the blood to the liver via the hepatic portal vein. The exception seems to be lipids, or fats, which are absorbed passively by the process of *diffusion*. Lipid breakdown products enter both the capillary beds and the lacteals in the villi and are carried to the liver by both blood and lymphatic fluids.

At the end of the ileum, all that remains is some water, indigestible food materials (plant fibers such as cellulose), and large amounts of bacteria. This debris enters the large intestine through the ileocecal valve. (The complete process of food digestion and absorption is summarized in Figure 14.13 on p. 505.)

### Chyme Propulsion

As mentioned previously, *peristalsis* is the major means of propelling chyme through the digestive tract. It involves waves of contraction that move along the length of the intestine, followed by waves of relaxation. The net effect is that the food is moved through the small intestine in much the same way that toothpaste is squeezed from a tube. Rhythmic segmental movements produce local constrictions of the intestine (see Figure 14.12b) that mix the chyme with the digestive juices and help to propel food through the intestine.

## Activities of the Large Intestine

### Nutrient Breakdown and Absorption

What is finally delivered to the large intestine contains few nutrients, but that residue still has 12 to 24 hours more to spend there. The colon itself produces no digestive enzymes. However, the "resident" bacteria that live in its lumen metabolize some of the remaining nutrients, releasing gases (methane and hydrogen sulfide) that contribute to flatulence and the odor of feces. About 500 ml of gas (flatus) is produced each day, much more when certain carbohydrate-rich foods (such as beans) are eaten.

Bacteria residing in the large intestine also make some vitamins (vitamin K and some B vitamins). Absorption by the large intestine is limited to the absorption of these vitamins, some ions, and most of the remaining water. Feces, the more or less solid product delivered to the rectum, contains undigested food residues, mucus, millions of bacteria, and just enough water to allow its smooth passage.

### Propulsion of Food Residue and Defecation

When presented with food residue, the colon begins contractions, but they are sluggish or short-lived. The movements most seen in the colon are **haustral contractions**, slow segmenting movements lasting about 1 minute that occur every 30 minutes or so. As a haustrum fills with food residue, the distension stimulates its muscle to contract, which propels the luminal contents into the next haustrum. These movements also mix the residue, which aids in water absorption.

**Mass movements** are long, slow-moving but powerful contractile waves that move over large areas of the colon three or four times daily and force the contents toward the rectum. Typically, they occur during or just after eating, when food begins to fill the stomach and small intestine. Bulk, or fiber, in the diet increases the strength of colon contractions and softens the stool, allowing the colon to perform its function more effectively.

### Homeostatic Imbalance 14.9

When the diet lacks bulk, the colon narrows and its circular muscles contract more powerfully, increasing the pressure on its walls. This encourages formation of **diverticula** (di″ver-tik′u-lah), in which the mucosa protrudes through the colon walls, a condition called **diverticulosis**. **Diverticulitis**, a condition in which the diverticula become inflamed, can be life-threatening if ruptures occur. _____ ✚

The rectum is generally empty, but when feces are forced into it by mass movements and its wall is stretched, the **defecation reflex** is initiated. The defecation reflex is a spinal (sacral region) reflex that causes the walls of the sigmoid colon and the rectum to contract and the anal sphincters to relax. As the feces are forced through the anal canal, messages reach the brain giving us time to decide whether the external voluntary sphincter should

**14**

# Peptic Ulcers: "Something Is Eating at Me"

Archie, a 53-year-old factory worker, began to experience a burning pain in his upper abdomen an hour or two after each meal. At first, he blamed his home cooking, but his abdominal pain became markedly worse during a hectic week when he worked 15 hours of overtime. Archie consulted a physician and was told that he had a peptic ulcer.

Peptic ulcers affect one of every eight Americans. A peptic ulcer is typically a round, sharply defined crater 1 to 4 cm in diameter in the mucosa of any part of the GI tract exposed to secretions of the stomach (see photo a). A few peptic ulcers occur in the lower esophagus, following the reflux of stomach contents, but most (98 percent) occur in the pyloric part of the stomach (gastric ulcers) or the first part of the duodenum (duodenal ulcers). Peptic ulcers may appear at any age, but they develop most frequently between ages 50 and 70. They tend to recur—healing, then flaring up periodically—for the rest of a person's life if not treated.

Duodenal ulcers are about three times more common than gastric ulcers. Gastric and duodenal ulcers may produce a gnawing or burning epigastric pain that seems to bore through to your back. Other symptoms include loss of appetite, bloating, nausea, and vomiting. Not all people with ulcers experience these symptoms, however, and some exhibit no symptoms at all.

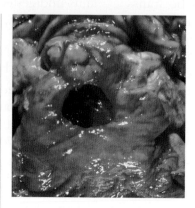

(a) A peptic ulcer lesion

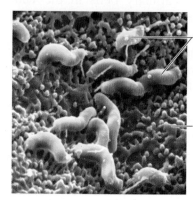

Bacteria

Mucosa layer of stomach

(b) *H. pylori* bacteria

For years the stereotypical ulcer patient was the overworked business executive. Although a stressful lifestyle does seem to *aggravate* existing ulcers, recent studies indicate that most such ulcers are caused by a strain of acid-resistant bacteria (*Helicobacter pylori*) that burrow through the mucus coat of the stomach, leaving bare areas. This corkscrew-shaped bacterium is found in the stomachs of 50 percent of healthy people and 70 to 90 percent of people with ulcers (see photo b).

Peptic ulcers can produce serious complications. In about 20 percent of cases, eroded blood vessels bleed into the GI tract. In such cases, anemia may result from severe blood loss. In 5 to 10 percent of patients, scarring within the stomach obstructs the pyloric opening, blocking digestion. About 5 percent of peptic ulcers *perforate*, leaking stomach or duodenal contents into the peritoneal cavity. A perforated ulcer is a life-threatening condition.

In spite of these potential complications, most peptic ulcers heal readily and respond well to treatment. The first steps are to avoid smoking, alcohol (especially wine), ibuprofen, and aspirin, all of which aggravate ulcers. Antacid drugs are suggested to neutralize the stomach acids. In ulcers colonized by *H. pylori*, a simple 2-week regimen of antibiotics permanently cures nearly all such patients. For active ulcers, a histamine blocker (a medication that suppresses acid production in the stomach) can be combined with the antibiotic therapy to speed recovery.

The relatively few peptic ulcers not caused by *H. pylori* generally result from long-term use of NSAIDs, such as ibuprofen. Treatment with histamine-blocker drugs (such as Zantac) is the therapy of choice.

Recent animal trials using a newly developed vaccine against *H. pylori* have been successful. Preventive vaccination, coupled with antibiotic cures, may eradicate peptic ulcers within the next 25 years.

remain open or be constricted to stop passage of feces. If it is not convenient, defecation (bowel movement) can be delayed temporarily. Within a few seconds, the reflex contractions end, and the rectal walls relax. With the next mass movement, the defecation reflex is initiated again.

 ## Homeostatic Imbalance 14.10

Watery stools, or **diarrhea** (di″ah-re′ah), result from any condition that rushes food residue through the large intestine before that organ has had sufficient time to absorb the water (as in irritation of the colon by bacteria). Because fluids and ions are lost from the body, prolonged diarrhea may result in dehydration and electrolyte imbalance, which can be fatal if severe.

If food residue remains in the large intestine for extended periods, too much water is absorbed, and the stool becomes hard and difficult to pass. This condition, called **constipation**, may result from lack of fiber in the diet, poor bowel habits ("failing to heed the call"), narcotic pain medications such as hydrocodone, and laxative abuse. Stool softeners can help increase water content of material in the colon to ease passage during defecation. _____ ✚

### Did You Get It?

17. What are the building blocks (and digestion products) of proteins?
18. What is the role of CCK in digestion?
19. What are brush border enzymes?

For answers, see Appendix A.

# PART II: NUTRITION AND METABOLISM

### → Learning Objectives

☐ Define *nutrient* and *kilocalorie*.

☐ List the six nutrient categories. Note important dietary sources and their main cellular uses.

Although it seems at times that people can be divided into two camps—those who live to eat and those who eat to live—we all recognize the vital importance of food for life. The common saying "You are what you eat" is true in that part of the food we eat is converted to our living flesh. In other words, a certain fraction of nutrients is used to build cellular molecules and structures and to replace worn-out parts. However, most foods are used as metabolic fuels. That is, they are oxidized and transformed into *adenosine triphosphate (ATP)*, the chemical energy form body cells need to drive their many activities. The energy value of foods is measured in units called **kilocalories (kcal)**, or Calories (with a capital C), the units conscientiously counted by dieters.

We have just considered how foods are digested and absorbed. But what happens to these foods once they enter the blood? Why do we need bread, meat, and fresh vegetables? Why does everything we eat seem to turn to fat? We will try to answer these questions in this section.

# Nutrition

A **nutrient** is a substance in food that the body uses to promote normal growth, maintenance, and repair. The nutrients divide neatly into six categories. The **major nutrients**—carbohydrates, lipids, and proteins—make up the bulk of what we eat. **Minor nutrients**—vitamins and minerals—although equally crucial for health, are required in very small amounts. Water, which accounts for about 60 percent of the volume of the food we eat, is also considered to be a major nutrient. However, because we described its importance as a solvent and in many other aspects of body functioning in the chapter on basic chemistry (see Chapter 2), we will consider only the other five groups of nutrients here.

Most foods offer a combination of nutrients. For example, a bowl of cream of mushroom soup contains all of the major nutrients plus some vitamins and minerals. A diet consisting of foods selected from each of the five food groups (Table 14.2, p. 514), that is, grains, fruits, vegetables, meats and meat alternatives, and milk products, normally guarantees adequate amounts of all of the needed nutrients.

## Dietary Recommendations

Several types of detailed food guide pyramids have been developed in the United States to provide dietary recommendations. The version issued in 1992 by Walter Willett, called the **Healthy Eating Pyramid (Figure 14.17a**, p. 515), looks at six major food groups, subdividing some of them further. It uses the traditional (horizontal) orientation of food groups; emphasizes eating whole-grain foods and

**14**

Table **14.2**   **Five Basic Food Groups and Some of Their Major Nutrients**

| Group | Example foods | Major nutrients supplied in significant amounts | |
|---|---|---|---|
| | | **By all in group** | **By only some in group** |
| **Fruits** | Apples, bananas, dates, oranges, tomatoes | Carbohydrate<br>Water | Vitamins: A, C, folic acid<br>Minerals: iron, potassium<br>Fiber |
| **Vegetables** | Broccoli, cabbage, green beans, lettuce, potatoes | Carbohydrate<br>Water | Vitamins: A, C, E, K, and B vitamins except $B_{12}$<br>Minerals: calcium, magnesium, iodine, manganese, phosphorus<br>Fiber |
| **Grain products** (preferably whole grain; otherwise, enriched or fortified) | Breads, rolls, bagels; cereals, dry and cooked; pasta; rice, other grains; tortillas, pancakes, waffles; crackers; popcorn | Carbohydrate Protein<br>Vitamins: thiamine ($B_1$), niacin | Water<br>Fiber<br>Minerals: iron, magnesium, selenium |
| **Milk products** | Milk, yogurt; cheese; ice cream, ice milk, frozen yogurt | Protein<br>Fat<br>Vitamins: riboflavin, $B_{12}$<br>Minerals: calcium, phosphorus<br>Water | Carbohydrate<br>Vitamins: A, D |
| **Meats and meat alternatives** | Meat, fish, poultry; eggs; seeds; nuts, nut butters; soybeans, tofu; other legumes (peas and beans) | Protein<br>Vitamins: niacin, $B_6$<br>Minerals: iron, zinc | Carbohydrate<br>Fat<br>Vitamins: $B_{12}$, thiamine ($B_1$)<br>Water<br>Fiber |

lots of fruits and vegetables; and recommends substituting plant oils for animal fats and restricting red meat, sweets, and starchy foods.

**MyPlate** is a more recent food guide from 2011 that uses a different symbol—a round dinner plate. Issued by the United States Department of Agriculture (USDA), it shows food categories in healthy proportions in sections of a place setting rather than as segments of a pyramid (Figure 14.17b). Occupying half the plate are fruits and vegetables (more vegetable than fruit). The other half shows grains and proteins (more grain than protein). A glass represents dairy, the fifth food group. Links provide details on healthy choices in each food group. You can personalize your MyPlate diet by age, sex, and activity level by going to the MyPlate website (www.choosemyplate.gov).

Nutrition advice is constantly in flux and often mired in the self-interest of food companies. Nonetheless, basic dietary principles have not changed in years and are not in dispute: Eat less overall; eat plenty of fruits, vegetables, and whole grains; avoid junk food; and exercise regularly.

## Dietary Sources of the Major Nutrients

### Carbohydrates

Except for milk sugar (lactose) and small amounts of glycogen in meats, all the **carbohydrates**—sugars

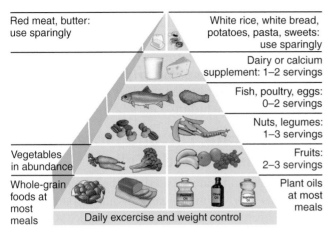

**(a) Healthy Eating Pyramid**

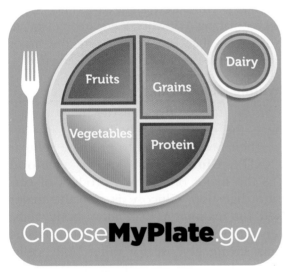

**(b) USDA's MyPlate**

**Figure 14.17 Two visual food guides.**

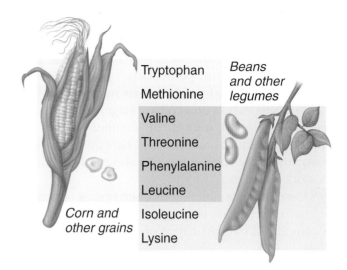

**Figure 14.18 The eight essential amino acids.** Vegetarian diets must be carefully constructed to provide all essential amino acids. A meal of corn and beans fills the bill: Corn provides the essential amino acids not in beans, and vice versa.

and starches—we ingest are derived from plants. Sugars come mainly from fruits, sugar cane, and milk. The polysaccharide starch is found in grains, legumes, and root vegetables. The polysaccharide cellulose, which is plentiful in most vegetables, is not digested by humans, but it provides fiber that increases the bulk of the stool and aids defecation.

## Lipids

Although we also ingest cholesterol and phospholipids, most dietary **lipids** are triglycerides (**neutral fats**). We eat saturated fats in animal products such as meat and dairy foods and in a few plant products, such as coconut. Unsaturated fats are present in seeds, nuts, and most vegetable oils.

Major sources of cholesterol are egg yolk, meats, and milk products.

## Proteins

Animal products contain the highest-quality **proteins**, molecules that are amino acid polymers. Eggs, milk, fish, and most meat proteins are *complete proteins* that meet all of the body's amino acid requirements for tissue maintenance and growth.

Legumes (beans and peas), nuts, and cereals are also protein rich, but their proteins are nutritionally incomplete because they are low in one or more of the essential amino acids. The **essential amino acids** are the eight amino acids (listed in **Figure 14.18**, p. 516) that our bodies cannot make. Hence we must obtain these amino acids from our diet. As you can see, strict vegetarians must carefully plan their diets to obtain all the essential amino acids and prevent protein malnutrition. Cereal grains and legumes when eaten together provide all the needed amino acids, and some variety of this combination is found in the diets of all cultures (for example, rice and beans in the Mexican tradition).

## Vitamins

**Vitamins** are organic nutrients of various forms that the body requires in small amounts. Although vitamins are found in all major food groups, no one food contains all the required vitamins. Thus, a balanced diet is the best way to ensure a full vitamin complement, particularly because certain vitamins (A, C, and E) appear to have anticancer effects. Diets rich in broccoli, cabbage, and brussels sprouts (all good sources of vitamins A and C) appear to reduce cancer risk. However, controversy abounds concerning the ability of vitamins to work wonders.

Most vitamins function as **coenzymes** (or parts of coenzymes); that is, they act with an enzyme to accomplish a particular type of catalysis.

### Minerals

The body also requires adequate supplies of seven **minerals** (which are inorganic substances)—calcium, phosphorus, potassium, sulfur, sodium, chloride, and magnesium—as well as trace amounts of about a dozen others.

Fats and sugars have practically no minerals, and cereals and grains are poor sources. The most mineral-rich foods are vegetables, legumes, milk, and some meats.

The section on metabolism discusses the main uses of the major nutrients in the body. (Appendix D details some important roles of vitamins and minerals in the body.)

### Did You Get It?

20. What is the major source of carbohydrates in our diet?
21. Why is it important to include cellulose in a healthy diet even though we cannot digest it?
22. Are oils saturated or unsaturated lipids?
23. What is the most significant role that vitamins play in the body?

**For answers, see Appendix A.**

# Metabolism

### → Learning Objectives

☐ Define *metabolism*, *anabolism*, and *catabolism*.

☐ Recognize the uses of carbohydrates, fats, and proteins in cell metabolism.

**Metabolism** (mĕ-tab′o-lizm; *metabol* = change) is a broad term referring to all chemical reactions that are necessary to maintain life. It involves **catabolism** (kah-tab′o-lizm), the breakdown of substances to simpler substances, and **anabolism** (ah-nab′o-lizm), the building of larger molecules or structures from smaller ones. During catabolism, bond energy of foods is released and captured to make **adenosine triphosphate (ATP)**, the energy-rich molecule used to energize all cellular activities, including catabolic reactions.

## Carbohydrate, Fat, and Protein Metabolism in Body Cells

Not all foodstuffs are treated in the same way by body cells. For example, carbohydrates, particularly glucose, are usually broken down to make ATP. Fats are used to build cell membranes, make myelin sheaths around neurons, and insulate the body with a fatty cushion. They are also used as the body's main fuel for making ATP when there are inadequate carbohydrates in the diet. Proteins are the major structural materials used for building cell structures.

### Carbohydrate Metabolism

Just as an oil furnace uses oil (its fuel) to produce heat, the cells of the body use carbohydrates as their preferred fuel to produce cellular energy (ATP). **Glucose**, also known as **blood sugar**, is the major breakdown product of carbohydrate digestion. Glucose is also the major fuel used for making ATP in most body cells. The liver is an exception; it routinely uses fats as well, thus saving glucose for other body cells. Essentially, glucose is broken apart piece by piece, and some of the chemical energy released when its bonds are broken is captured and used to bind phosphate to ADP molecules to make ATP.

The carbon atoms released from glucose leave the cells as carbon dioxide, and the hydrogen atoms removed (which contain energy-rich electrons) are eventually combined with oxygen to form water. (The overall reaction is summed up simply in **Figure 14.19**.) These oxygen-using events are referred to collectively as **cellular respiration**. The three main metabolic pathways involved in cellular respiration are *glycolysis*, the *citric acid cycle*, and the *electron transport chain* (shown schematically in **Figure 14.20**).

Oxidation via the removal of hydrogen atoms (which are temporarily passed to vitamin-containing coenzymes) is a major role of glycolysis and the citric acid cycle. **Glycolysis**, which takes place in the cytosol, also energizes each glucose molecule

**Figure 14.19 Summary equation for cellular respiration.**

so that it can be split into two pyruvic acid molecules and yield a small amount of ATP in the process (Figure 14.20). The **citric acid cycle** occurs in the mitochondria and produces virtually all the carbon dioxide that results during cell respiration. Like glycolysis, it yields a small amount of ATP by transferring high-energy phosphate groups directly

from phosphorylated substances to ADP, a process called *substrate-level phosphorylation*. Free oxygen is not involved.

The **electron transport chain** is where the action is for ATP production. The hydrogen atoms removed during the first two metabolic phases are loaded with energy. These hydrogens are delivered

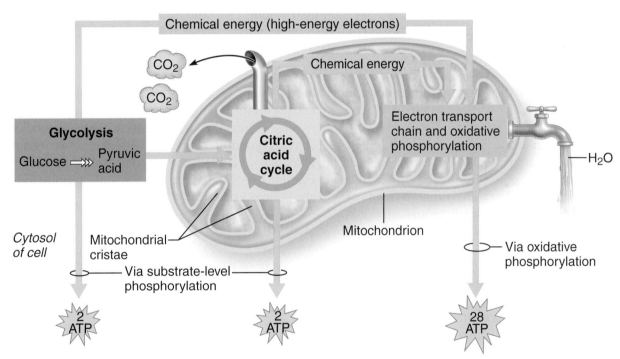

① During glycolysis, each glucose molecule is broken down into two molecules of pyruvic acid as hydrogen atoms containing high-energy electrons are removed.

② The pyruvic acid enters the mitochondrion, where citric acid cycle enzymes remove more hydrogen atoms and decompose it to $CO_2$. During glycolysis and the citric acid cycle, small amounts of ATP are formed.

③ Energy-rich electrons picked up by coenzymes are transferred to the electron transport chain, built into the cristae membrane. The electron transport chain carries out oxidative phosphorylation, which accounts for most of the ATP generated by cellular respiration, and finally unites the removed hydrogen with oxygen to form water.

**Figure 14.20 The formation of ATP in the cytosol and the mitochondria during cellular respiration.** The maximum energy yield per glucose molecule is 32 ATP.

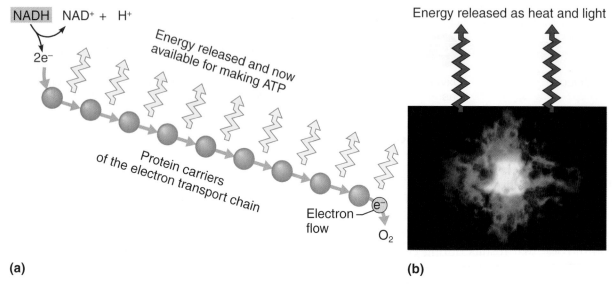

**(a)**

**(b)**

**Figure 14.21 Energy release in the electron transport chain versus one-step reduction of oxygen. (a)** In cellular respiration, cascading electrons release energy in small steps and finally reduce $O_2$ and form water. The energy released is in quantities easily used to form ATP. (NADH is a niacin-containing coenzyme that delivers electrons to the electron transport chain.) **(b)** When $O_2$ is reduced (combined with hydrogen) in one step, the result is an explosion.

by the coenzymes to the protein carriers of the electron transport chain, which form part of the mitochondrial cristae membranes **(Figure 14.21a)**. There the hydrogen atoms are split into hydrogen ions ($H^+$) and electrons ($e^-$). The electrons "fall down an energy hill," going from each carrier to a carrier of lower energy. They give off their "load" of energy in a series of steps in small enough amounts to enable the cell to attach phosphate to ADP and make ATP. Ultimately, free oxygen is reduced (the electrons and hydrogen ions are united with molecular oxygen), forming water and a large amount of ATP. This more complicated process of ATP formation is called *oxidative phosphorylation*. The beauty of this system is that, unlike the explosive reaction (Figure 14.21b) that usually happens when fuels are burned ($O_2$ is combined with hydrogen), relatively small amounts of energy are lost as heat (and light).

Because glucose is the major fuel for making ATP, homeostasis of the blood glucose level is critically important. If there is an excessively high level of glucose in the blood (**hyperglycemia** [hi″per-gli-se′me-ah]), some of the excess is stored in body cells (particularly liver and muscle cells) as glycogen. If the blood glucose level is still too high, excesses are converted to fat. There is no question that eating large amounts of empty-calorie foods such as candy and other sugary sweets causes a rapid deposit of fat in the body's adipose tissues. When the blood glucose level is too low (**hypoglycemia**), the liver breaks down stored glycogen and releases glucose to the blood for cellular use. (These various fates of carbohydrates are shown in **Figure 14.22a**.)

## Did You Get It?

24. What name is given to the process in which glucose is combined with oxygen to yield $CO_2$, $H_2O$, and ATP?
25. What are the major end products of the citric acid cycle?
26. What are the major products of the electron transport chain?

For answers, see Appendix A.

### Fat Metabolism

As we will describe shortly, the liver handles most lipid, or fat, metabolism that goes on in the body. The liver cells use some fats to make ATP for their own use; use some to synthesize lipoproteins, thromboplastin (a clotting protein), and cholesterol; and then release the rest to the blood in the form of relatively small fat-breakdown products. Body cells remove the fat products and cholesterol from the blood and build them into their membranes or steroid hormones as needed. Fats are also used to form myelin sheaths (see Chapter 7)

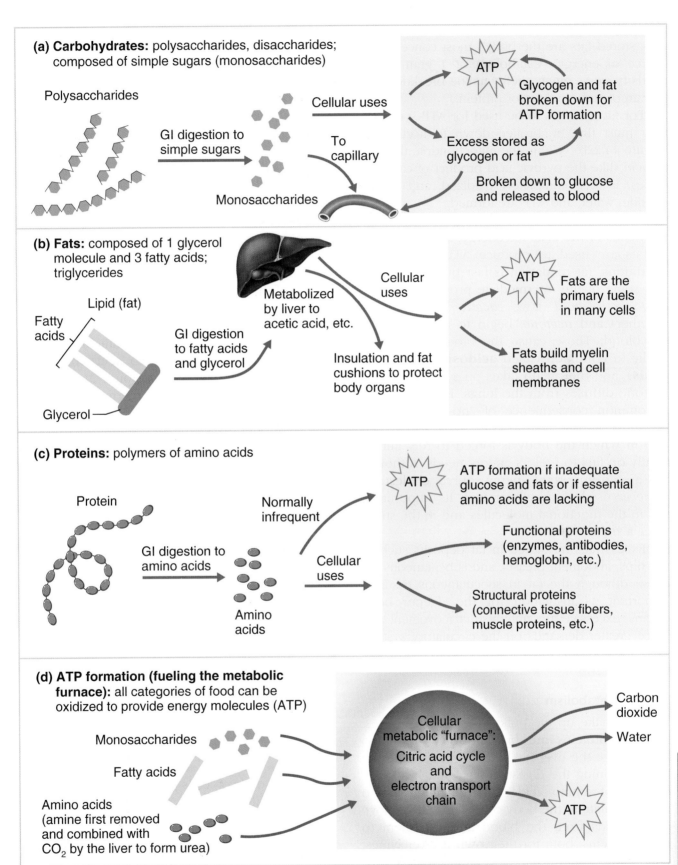

**(a) Carbohydrates:** polysaccharides, disaccharides; composed of simple sugars (monosaccharides)

Polysaccharides

GI digestion to simple sugars

Monosaccharides

Cellular uses

To capillary

ATP

Glycogen and fat broken down for ATP formation

Excess stored as glycogen or fat

Broken down to glucose and released to blood

**(b) Fats:** composed of 1 glycerol molecule and 3 fatty acids; triglycerides

Lipid (fat)

Fatty acids

Glycerol

GI digestion to fatty acids and glycerol

Metabolized by liver to acetic acid, etc.

Cellular uses

Insulation and fat cushions to protect body organs

ATP

Fats are the primary fuels in many cells

Fats build myelin sheaths and cell membranes

**(c) Proteins:** polymers of amino acids

Protein

GI digestion to amino acids

Amino acids

Normally infrequent

Cellular uses

ATP

ATP formation if inadequate glucose and fats or if essential amino acids are lacking

Functional proteins (enzymes, antibodies, hemoglobin, etc.)

Structural proteins (connective tissue fibers, muscle proteins, etc.)

**(d) ATP formation (fueling the metabolic furnace):** all categories of food can be oxidized to provide energy molecules (ATP)

Monosaccharides

Fatty acids

Amino acids (amine first removed and combined with $CO_2$ by the liver to form urea)

Cellular metabolic "furnace":

Citric acid cycle and electron transport chain

Carbon dioxide

Water

ATP

**Figure 14.22 Metabolism by body cells. (a)** Carbohydrate metabolism. **(b)** Fat metabolism. **(c)** Protein metabolism. **(d)** ATP formation.

14

and fatty cushions around body organs. In addition, stored fats are the body's most concentrated source of energy. (Catabolism of 1 gram of fat yields twice as much energy as the breakdown of 1 gram of carbohydrate or protein.)

For fat products to be used for ATP synthesis, they must first be broken down to acetic acid (Figure 14.22b). Within the mitochondria, the acetic acid (like the pyruvic acid product of carbohydrates) is then completely oxidized, and carbon dioxide, water, and ATP are formed.

When there is not enough glucose to fuel the needs of the cells for energy, larger amounts of fats are used to produce ATP. Under such conditions, fat oxidation is fast but incomplete, and some of the intermediate products, such as *acetoacetic acid* (two *acetic acids* linked together) and *acetone*, begin to accumulate in the blood. These cause the blood to become acidic (a condition called **acidosis** or **ketoacidosis**), and the breath takes on a fruity odor as acetone diffuses from the lungs. Ketoacidosis is a common consequence of "no-carbohydrate" diets, uncontrolled diabetes mellitus, and starvation in which the body is forced to rely almost totally on fats to fuel its energy needs. Although fats are an important energy source, cholesterol is *never* used as a cellular fuel. Its importance lies in the functional molecules and in the structures it helps to form.

Excess fats are stored in fat deposits such as the hips, abdomen, breasts, and subcutaneous tissues. Although the fat in subcutaneous tissue is important as insulation for the deeper body organs, excessive amounts restrict movement and place greater demands on the circulatory system. (The metabolism and uses of fats are shown in Figure 14.22b.)

### Protein Metabolism

Proteins make up the bulk of cellular structures. Ingested proteins are broken down to amino acids. Once the liver has finished processing the blood draining the digestive tract and has taken its "fill" of amino acids, the remaining amino acids circulate to the body cells. The cells remove amino acids from the blood and use them to build proteins, both for their own use (enzymes, membranes, mitotic spindle proteins, muscle proteins) and for export (mucus, hormones, and others). Cells take few chances with their amino

acid supply. They use ATP to actively transport amino acids into their interior even though in many cases they may contain more of those amino acids than are in the blood flowing past them. Even though this may appear to be "cellular greed," there is an important reason for this active uptake of amino acids. Cells cannot build their proteins unless *all* the needed amino acids, which number around 20, are present. As mentioned earlier, because cells cannot make the essential amino acids, they are available to the cells only through the diet. This helps explain the avid accumulation of amino acids, which ensures that all amino acids needed will be available for present and (at least some) future protein-building needs of the cells (Figure 14.22c).

Amino acids are used to make ATP only when proteins are overabundant and/or when carbohydrates and fats are not available. When it is necessary to oxidize amino acids for energy (Figure 14.22d), their amine groups are removed as *ammonia*, and the rest of the molecule enters the citric acid cycle pathway in the mitochondria. The ammonia that is released during this process is toxic to body cells, especially nerve cells. We will discuss how the liver comes to the rescue in this situation shortly.

### Did You Get It?

27. Other than for ATP production, list two uses of fats in the body.

For the answer, see Appendix A.

## The Central Role of the Liver in Metabolism

### → Learning Objective

☐ Describe the metabolic roles of the liver.

The liver is one of the most versatile and complex organs in the body. Without it, we would die within 24 hours. Its role in digestion (manufacturing bile) is important to the digestive process to be sure, but it is only one of the many functions of liver cells. The liver cells detoxify drugs and alcohol, degrade hormones, and make many substances vital to the body as a whole (cholesterol, blood proteins such as albumin and clotting proteins, and lipoproteins). They play a central role in metabolism as they process nearly every class of nutrient. Because of the liver's key roles, nature has provided us with a surplus of liver tissue. We have much more than we need, and even if part of it is damaged or removed, it is one of the few body organs that can regenerate rapidly and easily.

A unique circulation, the *hepatic portal circulation*, brings nutrient-rich blood draining from the digestive viscera directly to the liver (see Chapter 11). The liver is the body's major metabolic organ, and this detour that nutrients take through the liver ensures that the liver's needs will be met first. As blood circulates slowly through the liver, liver cells remove amino acids, fatty acids, and glucose from the blood. These nutrients are stored for later use or processed in various ways. At the same time, the liver's phagocytic cells remove and destroy bacteria that have managed to get through the walls of the digestive tract and into the blood.

## General Metabolic Functions

The liver is vitally important in helping to maintain the blood glucose level within the normal range (around 100 mg glucose/100 ml of blood). After a carbohydrate-rich meal, thousands of glucose molecules are removed from the blood and combined to form the large polysaccharide molecules called **glycogen** (gli'ko-jen), which are then stored in the liver. This process is **glycogenesis** (gli"ko-jen'ĕ-sis), literally, "glycogen formation" (*genesis* = beginning).

Later, as body cells continue to remove glucose from the blood to meet their needs, the blood glucose level begins to drop. At this time, liver cells break down the stored glycogen by a process called **glycogenolysis** (gli"ko-jen-ol'ĭ-sis), which means "glycogen splitting." The liver cells then release glucose bit by bit to the blood to maintain blood glucose homeostasis.

If necessary, the liver can also make glucose from noncarbohydrate substances, such as fats and proteins. This process is **gluconeogenesis** (glu"ko-ne"o-jen'ĕ-sis), which means "formation of new sugar" (**Figure 14.23**, p. 522). Hormones, such as thyroxine, insulin, and glucagon are vitally important in controlling the blood sugar level and in the handling of glucose in all body cells (see Chapter 9).

Some of the fats and fatty acids picked up by the liver cells are oxidized for energy (to make ATP) for use by the liver cells themselves. The rest are broken down to simpler substances such as acetic acid and acetoacetic acid and then are released into the blood or stored as fat reserves in the liver. The liver also makes cholesterol and secretes cholesterol's breakdown products in bile.

All blood proteins made by the liver are built from the amino acids its cells pick up from the blood. The completed proteins are then released back into the blood to travel throughout the circulation. *Albumin*, the most abundant protein in blood, holds fluid in the bloodstream. When insufficient albumin is present in blood, fluid leaves the bloodstream and accumulates in the tissue spaces, causing edema. (In Chapter 10, we discussed the role of the protective *clotting proteins* made by the liver.) Liver cells also synthesize nonessential amino acids and, as mentioned earlier, detoxify ammonia. In this process, ammonia is combined with carbon dioxide to form **urea** (u-re'ah). Urea, which is not harmful to the body cells, is then flushed from the body in urine.

Nutrients not needed by the liver cells, as well as the products of liver metabolism, are released into the blood and drain from the liver in the hepatic vein. These materials then enter the systemic circulation, where they become available to other body cells.

## Cholesterol Metabolism and Transport

Although it's a very important lipid in the diet, as we mentioned, **cholesterol** is not used as an energy fuel. Instead, it serves as the structural basis of steroid hormones and vitamin D and is a major building block of plasma membranes. Because we hear so much about "cutting down our cholesterol intake" in the media, it may be surprising to learn that only about 15 percent of blood cholesterol comes from the diet. The other 85 percent or so is made by the liver. Cholesterol is lost from the body when it is broken down and secreted in bile salts, which eventually leave the body in feces.

Fatty acids, fats, and cholesterol are insoluble in water, so they cannot circulate freely in the bloodstream. Instead they are transported bound to the small lipid-protein complexes called *lipoproteins*. Although the entire story is complex, the important thing to know is that the **low-density lipoproteins**, or **LDLs**, transport cholesterol and other lipids *to* body cells, where they are used in various ways. If large amounts of LDLs are circulating, the chance that fatty substances will be deposited on the arterial walls, initiating atherosclerosis, is high. Because of this

14

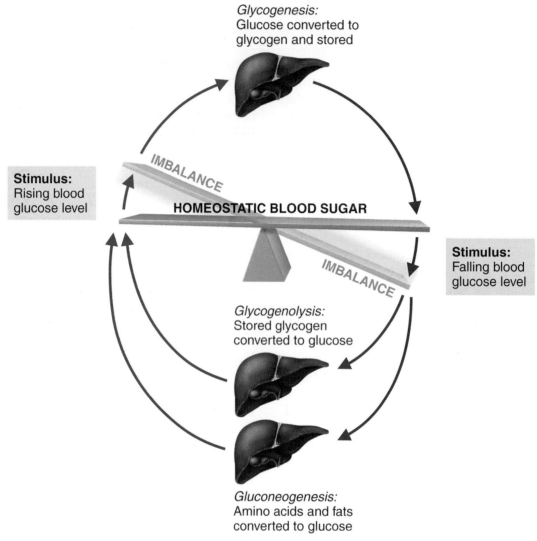

*Glycogenesis:*
Glucose converted to
glycogen and stored

**IMBALANCE**

**Stimulus:**
Rising blood
glucose level

**HOMEOSTATIC BLOOD SUGAR**

**Stimulus:**
Falling blood
glucose level

**IMBALANCE**

*Glycogenolysis:*
Stored glycogen
converted to glucose

*Gluconeogenesis:*
Amino acids and fats
converted to glucose

**Figure 14.23 Metabolic events occurring in the liver as the blood glucose level rises and falls.** When the blood glucose level is rising, the liver removes glucose from the blood and stores it as glycogen (glycogenesis). When the blood glucose level falls, the liver breaks down stored glycogen (glycogenolysis) and makes new glucose from amino acids and fats (gluconeogenesis). The glucose is then released to the blood to restore homeostasis of blood sugar.

possibility, the LDLs are unkindly tagged as "bad lipoproteins." By contrast, the lipoproteins that transport cholesterol *from* the tissue cells (or arteries) to the liver for disposal in bile are **high-density lipoproteins**, or **HDLs**. A high HDL level is considered "good" because the cholesterol is destined to be broken down and eliminated from the body. Obviously both LDLs and HDLs are "good and necessary"; it is just their relative ratio in the blood that determines whether or not potentially lethal cholesterol deposits are likely to be laid down on the artery walls. In general, aerobic exercise, a diet low in saturated fats and cholesterol, and abstaining from smoking and drinking coffee all appear to favor a desirable HDL/LDL ratio.

### Did You Get It?

28. What is gluconeogenesis?
29. What happens to the ammonia released when amino acids are "burned" for energy?
30. If you had your choice, would you prefer to have high blood levels of HDLs or LDLs? Explain your answer.

**For answers, see Appendix A.**

# Body Energy Balance

## → Learning Objective

☐ **Explain the importance of energy balance in the body, and indicate consequences of energy imbalance.**

When any fuel is burned, it consumes oxygen and liberates heat. The "burning" of food fuels by body cells is no exception. Energy cannot be created or destroyed—it can only be converted from one form to another (see Chapter 2). If we apply this principle to cellular metabolism, it means that a dynamic balance exists between the body's energy intake and its energy output:

Energy intake = total energy output
(heat + work + energy storage)

**Energy intake** is the energy liberated during food oxidation—that is, during the reactions of glycolysis, the citric acid cycle, and the electron transport chain. **Energy output** includes the energy we immediately lose as heat (about 60 percent of the total), plus energy used to do work (driven by ATP), plus energy that is stored in the form of fat or glycogen. Energy storage is important only during periods of growth and during net fat deposit.

## Regulation of Food Intake

When energy intake and energy output are balanced, body weight remains stable. When they are not, we either gain or lose weight. Because body weight in most people is surprisingly stable, mechanisms that control food intake or heat production or both must exist. Unhappily for many people, the body's weight-controlling systems appear to be designed more to protect us against weight loss than weight gain. (Some weight-control methods used by people who are obese are discussed in "A Closer Look" on pp. 529–530.)

But how is food intake controlled? That is a difficult question, and one that is still not fully answered. For example, what type of receptor could sense the body's total calorie content and alert one to start eating or to put down that fork? Despite heroic research efforts, no such single receptor type has been found.

We have known for some time that the hypothalamus releases several peptides that influence feeding behavior. Current theories of how feeding behavior and hunger are regulated focus most importantly on neural signals from the digestive tract, bloodborne signals related to body energy stores, and hormones. To a smaller degree, body temperature and psychological factors also seem to play a role. All these factors appear to operate through feedback signals to the feeding centers of the brain. Brain receptors include thermoreceptors, chemoreceptors (for glucose, insulin, and others), and receptors that respond to leptin and other polypeptides. Sensors in peripheral locations have also been suggested, with the liver and gut itself (alimentary canal) the prime candidates.

## Metabolic Rate and Body Heat Production

## → Learning Objective

☐ **List several factors that influence metabolic rate, and indicate each one's effect.**

When nutrients are broken down to produce cellular energy (ATP), they yield different amounts of energy. As mentioned earlier, the energy value of food is measured in a unit called the *kilocalorie (kcal)*. In general, carbohydrates and proteins each yield 4 kcal/gram, and fats yield 9 kcal/gram when they are broken down for energy production. Most meals, and even many individual foods, are mixtures of carbohydrates, fats, and proteins. To determine the caloric value of a meal, we must know how many grams of each type of foodstuff it contains. For most of us, this is a difficult chore indeed, but approximations can easily be made with the help of a simple app such as MyFitnessPal.

*Basal Metabolic Rate*  The amount of energy used by the body is also measured in kilocalories. The **basal metabolic rate (BMR)** is the amount of heat produced by the body per unit of time when it is under basal conditions—that is, at rest. It reflects the energy supply a person's body needs just to perform essential life activities such as breathing, maintaining the heartbeat, and kidney function. An average 70-kg (154-pound) adult has a BMR of about 60 to 72 kcal/hour.

Many factors influence BMR, including surface area and gender. For example, small, thin males tend to have a higher BMR than large, obese females (Table 14.3). Age is also important; children and adolescents require large amounts of energy for growth and have relatively high BMRs.

14

Table **14.3**    **Factors Determining the Basal Metabolic Rate (BMR)**

| Factor | Variation | Effect on BMR |
|---|---|---|
| Surface area | Large surface area in relation to body volume, as in thin, small individuals | Increased |
| | Small surface area in relation to body volume, as in large, heavy individuals | Decreased |
| Sex | Male | Increased |
| | Female | Decreased |
| Thyroxine production | Increased | Increased |
| | Decreased | Decreased |
| Age | Young, rapid growth | Increased |
| | Aging, elderly | Decreased |
| Strong emotions (anger or fear) and infections | | Increased |

In old age, the BMR decreases dramatically as the muscles begin to atrophy.

The amount of **thyroxine** produced by the thyroid gland is probably the most important factor in determining a person's BMR; hence, thyroxine has been dubbed the "metabolic hormone." The more thyroxine produced, the higher the oxygen consumption and ATP use, and the higher the metabolic rate. In the past, most BMR tests were done to determine whether sufficient thyroxine was being made. Today, thyroid activity is more easily assessed by blood tests.

### Homeostatic Imbalance 14.11

**Hyperthyroidism** causes a host of effects due to the excessive metabolic rate it produces. The body catabolizes stored fats and tissue proteins, and despite increased hunger and food intake, the person often loses weight. Bones weaken, and body muscles, including the heart, atrophy. In contrast, **hypothyroidism** results in slowed metabolism, obesity, and diminished thought processes. ___+

*Total Metabolic Rate*    When we are active, the body must oxidize more glucose to provide energy for the additional activities. Digesting food and even modest physical activity increase the body's caloric requirements dramatically. These additional fuel requirements are above and beyond the energy required to maintain the body in the basal state. **Total metabolic rate (TMR)** refers to the total amount of kilocalories the body must consume to fuel all ongoing activities. Muscular work is the major body activity that increases the TMR. Even slight increases in skeletal muscle activity cause remarkable leaps in metabolic rate. When a well-trained athlete exercises vigorously for several minutes, the TMR may increase to 15 to 20 times normal, and it remains elevated for several hours afterward.

When the total number of kilocalories consumed is equal to the TMR, homeostasis is maintained, and our weight remains constant. However, if we eat more than we need to sustain our activities, excess kilocalories appear in the form of fat deposits. Conversely, if we are extremely active and do not properly feed the "metabolic furnace," we begin to break down fat reserves and even tissue proteins to satisfy our TMR. This principle is used in every good weight-loss diet. (The total kilocalories needed are calculated on the basis of body size and age. Then, 20 percent or more of the requirements are cut from the daily diet.) If the dieting person exercises regularly, weight drops off even more quickly because the TMR increases above the person's former rate.

### Did You Get It?

31. Which of the following would you expect to yield a relatively high BMR: old age, large surface area relative to body volume, female sex, deficient thyroxine production?

For the answer, see Appendix A.

## Body Temperature Regulation

→ **Learning Objective**

☐ **Describe how body temperature is regulated.**

Although we have been emphasizing that foods are "burned" to produce ATP, remember that ATP is not the only product of cell catabolism. Most of the energy released during food oxidation escapes as heat. Less than 40 percent of available food energy is actually captured to form ATP. The heat released warms the tissues and, more important, the blood, which circulates to all body tissues, keeping them at homeostatic temperatures. This allows metabolism to occur efficiently.

Body temperature reflects the balance between heat production and heat loss. The body's thermostat is in the *hypothalamus* of the brain. Through autonomic nervous system pathways, the hypothalamus continuously regulates body temperature around a set point of 37°C, fluctuating within the range of 35.6° to 37.8°C (96° to 100°F). The hypothalamus does this by initiating heat-promoting mechanisms or heat loss mechanisms (**Figure 14.24**, p. 526).

***Heat-Promoting Mechanisms*** When the environmental temperature is low, the body must produce more heat to maintain normal body temperature. And if, for whatever reason, the temperature of circulating blood falls, body heat must be conserved and more heat generated to restore normal body (blood) temperature. Short-term means of accomplishing this are **vasoconstriction** of blood vessels of the skin and **shivering**.

When the skin vasculature constricts, the blood temporarily bypasses the skin and is rerouted to the deeper, more vital body organs. When this happens, the temperature of the exposed skin drops to that of the external environment.

 ## Homeostatic Imbalance 14.12

Restriction of blood delivery to the skin is no problem for brief periods of time. But if it is extended, the skin cells, chilled by internal ice crystals and deprived of oxygen and nutrients, begin to die. This condition, called **frostbite**, can cause the loss of extremities such as fingers and toes. _____ ✚

When the *core* body temperature (the temperature of the deep organs) drops to the point beyond which simple constriction of skin capillaries can handle the situation, shivering begins. Shivering, involuntary shuddering contractions of the voluntary muscles, is very effective in increasing the body temperature because skeletal muscle activity produces large amounts of heat.

## Homeostatic Imbalance 14.13

Extremely low body temperature resulting from prolonged exposure to cold is **hypothermia**. In hypothermia, the individual's vital signs (respiratory rate, blood pressure, heart rate) decrease. The person becomes drowsy and oddly comfortable, even though previously he or she felt extremely cold. Uncorrected, the situation progresses to coma and finally death as metabolic processes grind to a stop. _____✚

***Heat Loss Mechanisms*** Just as the body must be protected from becoming too cold, it must also be protected from excessively high temperatures. Most heat loss occurs through the skin via **radiation** or **evaporation**. When body temperature increases above the normal range, the blood vessels serving the skin dilate, and capillary beds in the skin become flushed with warm blood. As a result, heat radiates from the skin surface. However, if the external environment is as hot as or hotter than the body, heat cannot be lost by radiation. In such cases, the only means of getting rid of excess heat is by the evaporation of perspiration off the skin surface. This is an efficient means of liberating body heat as long as the air is dry. If it is humid, evaporation occurs at a much slower rate. Our heat-liberating mechanisms don't work well, and we feel miserable and irritable.

## Homeostatic Imbalance 14.14

When normal heat loss processes become ineffective, the resulting **hyperthermia**, or elevated body temperature, depresses the activity of the hypothalamus. As a result, a vicious positive feedback cycle occurs: The soaring body temperature increases the metabolic rate, which in turn increases heat production. The skin becomes hot and dry; and, as the temperature continues to spiral upward, permanent

**14**

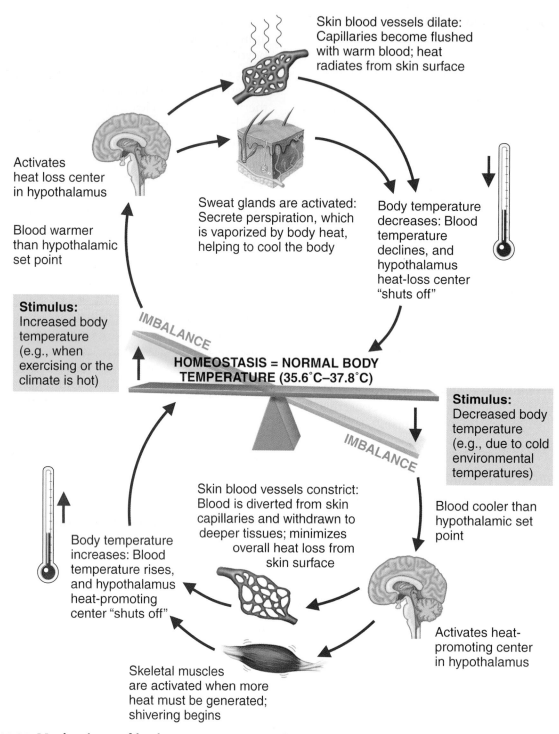

**Figure 14.24 Mechanisms of body temperature regulation.**

Skin blood vessels dilate: Capillaries become flushed with warm blood; heat radiates from skin surface

Activates heat loss center in hypothalamus

Blood warmer than hypothalamic set point

Sweat glands are activated: Secrete perspiration, which is vaporized by body heat, helping to cool the body

Body temperature decreases: Blood temperature declines, and hypothalamus heat-loss center "shuts off"

**Stimulus:** Increased body temperature (e.g., when exercising or the climate is hot)

*IMBALANCE*

**HOMEOSTASIS = NORMAL BODY TEMPERATURE (35.6°C–37.8°C)**

*IMBALANCE*

**Stimulus:** Decreased body temperature (e.g., due to cold environmental temperatures)

Blood cooler than hypothalamic set point

Skin blood vessels constrict: Blood is diverted from skin capillaries and withdrawn to deeper tissues; minimizes overall heat loss from skin surface

Body temperature increases: Blood temperature rises, and hypothalamus heat-promoting center "shuts off"

Activates heat-promoting center in hypothalamus

Skeletal muscles are activated when more heat must be generated; shivering begins brain damage becomes a distinct possibility. This condition, called **heat stroke**, can be fatal unless rapid corrective measures are taken immediately (immersion in cool water and administration of fluids).

**Heat exhaustion** is the term used to describe the heat-associated collapse of an individual during or following vigorous physical activity. Heat exhaustion results from excessive loss of body fluids (dehydration) and is evidenced by low blood pressure, a rapid heartbeat, and cool, clammy skin. In contrast to heat stroke, heat loss mechanisms still operate in heat exhaustion. _____

*Fever* is controlled hyperthermia. Most often, it results from infection somewhere in the body, but it may be caused by other conditions (cancer, allergic reactions, and CNS injuries). Macrophages, white blood cells, and injured tissue cells release chemical substances called *pyrogens* (pi'ro-jenz) that act directly on the hypothalamus, causing its thermostat to be set to a higher temperature (*pyro* = fire). After thermostat resetting, heat-promoting mechanisms are initiated. Vasoconstriction causes the skin to become cool, and shivering begins to generate heat. This situation, called "the chills," is a sure sign that body temperature is rising. Body temperature is allowed to rise until it reaches the new setting. Then, it is maintained at the "fever setting" until natural body defense processes or antibiotics reverse the disease process. At that point, the thermostat is reset again to the lower normal level, causing heat loss mechanisms to swing into action—the individual begins to sweat, and the skin becomes flushed and warm. Physicians have long recognized that these signs signal a turn for the better, saying that the patient's fever has "broken" once body temperature begins falling.

Fever, by increasing the metabolic rate, helps speed the various healing processes and also appears to inhibit bacterial growth (see Chapter 12). The danger of fever is that if the body thermostat is set too high, body proteins may be denatured (unfolded), and permanent brain damage may occur.

### Did You Get It?

32. What are two means of either maintaining or increasing body temperature?
33. How does vasodilation of skin blood vessels affect body temperature on a hot day?

**For answers, see Appendix A.**

# PART III: DEVELOPMENTAL ASPECTS OF THE DIGESTIVE SYSTEM AND METABOLISM

→ **Learning Objectives**

☐ **Name important congenital disorders of the digestive system and significant inborn errors of metabolism.**

☐ **Describe the effect of aging on the digestive system.**

The very young embryo is flat and pancake-shaped. However, it soon folds to form a cylindrical body, and its internal cavity becomes the cavity of the alimentary canal. By the fifth week of development, the alimentary canal is a continuous tube-like structure extending from the mouth to the anus. Shortly after, the digestive glands (salivary glands, liver, and pancreas) bud out from the mucosa of the alimentary tube. These glands retain their connections (ducts) and can easily empty their secretions into the alimentary canal to promote its digestive functions.

### Homeostatic Imbalance 14.15

The digestive system is susceptible to many congenital defects that interfere with feeding. The most common is the **cleft palate/cleft lip defect**.

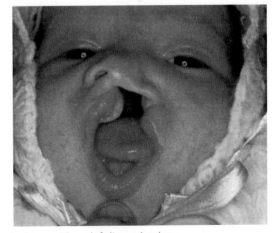

A baby born with a cleft lip and palate.

Of the two, cleft palate is more serious because the child is unable to suck properly. Another relatively common congenital defect is **tracheoesophageal fistula** (tra"ke-o-ĕ-sof'ah-je-al fis'tu-lah), an abnormal connection between the esophagus and the trachea. In addition, the esophagus often (but not always) ends in a blind sac and does not

**14**

connect to the stomach. The baby chokes, drools, and becomes cyanotic during feedings because food is entering the respiratory passageways. All three defects can be corrected surgically.

There are many types of genetic problems that interfere with metabolism, but perhaps the two most common are *cystic fibrosis* and *phenylketonuria* (fen″il-ke′to-nu″re-ah). **Cystic fibrosis (CF)** primarily affects the lungs, but it also significantly impairs the activity of the pancreas. In CF, huge amounts of mucus are produced, which block the passages of involved organs. Blockage of the pancreatic duct prevents pancreatic fluid from reaching the small intestine. As a result, fats and fat-soluble vitamins are not digested or absorbed, and bulky, fat-laden stools result. This condition is usually handled by administering pancreatic enzymes with meals.

**Phenylketonuria (PKU)** involves an inability of tissue cells to use phenylalanine (fen″il-al′ah-nin), an amino acid present in all protein foods. In such cases, brain damage and intellectual dysfunction occur unless a special diet low in phenylalanine is prescribed. _____ **+**

The developing infant receives all its nutrients through the placenta, and at least at this period of life, obtaining and processing nutrients is no problem (assuming the mother is adequately nourished). Obtaining nutrition is the most important activity of the newborn baby, and several reflexes present at this time help in this activity. For example, the *rooting reflex* helps the infant find the nipple (mother's or bottle), and the *sucking reflex* helps him or her to hold on to the nipple and swallow. The stomach of a newborn infant is very small, so feeding must be frequent (every 3 to 4 hours). Peristalsis is rather inefficient at this time, and vomiting is not at all unusual.

Teething begins around age 6 months and continues until about the age of 2 years. During this interval, the infant progresses to more and more solid foods and usually is eating an adult diet by toddlerhood. Appetite decreases in the elementary school–aged child and then increases again during the rapid growth of adolescence. (Parents of adolescents can expect higher grocery bills!)

All through childhood and into adulthood, the digestive system usually operates with relatively few problems unless there are abnormal interferences, such as contaminated food or extremely spicy or irritating foods (which may cause inflam-

mation of the gastrointestinal tract, or **gastroenteritis** [gas″tro-en-ter-i′tis]). Inflammation of the appendix, **appendicitis**, is particularly common in teenagers for some unknown reason.

Between middle age and early old age, the metabolic rate decreases by 5 to 8 percent in every 10-year period. This is the time of life when the weight seems to creep up, and obesity often becomes a fact of life. To maintain desired weight, we must be aware of this gradual change and be prepared to reduce caloric intake. Two distinctly middle-age digestive problems are *ulcers* (see "A Closer Look" on p. 512) and *gallbladder problems* (inflammation of the gallbladder or gallstones). For some unknown reason, *irritable bowel syndrome (IBS)*, characterized by alternating bouts of diarrhea and constipation, is becoming much more prevalent in adults, but its cause has been difficult to pin down.

During old age, activity of the GI tract declines. Fewer digestive juices are produced, and peristalsis slows. Taste and smell become less acute, and periodontal disease often develops. Many elderly individuals live alone or on a reduced income. These factors, along with increasing physical disability, tend to make eating less appealing, and nutrition is inadequate in many of our elderly citizens. Diverticulosis, the bulging of the colon wall into the lumen, and cancer of the gastrointestinal tract are fairly common problems. Cancer of the stomach and colon rarely show early signs and often progress to an inoperable stage because they spread to distant parts of the body before the individual seeks medical attention. However, when detected early, both diseases are treatable. A diet high in plant fiber and low in fat may help to decrease the incidence of colon cancer. Additionally, because most colorectal cancers derive from initially benign mucosal tumors called **polyps**, and because the incidence of polyp formation increases with age, colonoscopy screenings should begin at the age of 50 and be repeated every 10 years. However, very sensitive fecal occult blood tests, which look for blood in the stool, should be completed annually.

### Did You Get It?

34. How does cystic fibrosis affect digestion?
35. What dietary changes must be made to prevent brain damage in children with PKU?
36. What occurs when your total caloric intake exceeds your TMR?

**For answers, see Appendix A.**

# Obesity: Magical Solution Wanted

How fat is too fat? What distinguishes an obese person from one who is merely overweight? The bathroom scale is a poor guide, because body weight alone cannot reveal body composition. Dense bones and well-developed muscles may make a person technically overweight.

The most common view of obesity is that it is a condition of excessive fat storage. A body fat content that is 18 to 22 percent of body weight (males and females, respectively) is considered normal for adults. Anything over that is defined as obesity.

Body fat content can be estimated using the *body mass index (BMI)* scale, an index of a person's weight relative to height. To estimate BMI, multiply weight in pounds (lb) by 703 and then divide by your height in inches squared:

$$BMI = \frac{wt(lb)}{(height\ in\ inches)^2} \times 703$$

A BMI of 25 or more is defined as *overweight*, and a BMI of 30 or above is defined as *obese*. Although BMI calculations are simple and inexpensive, it is important to note that they are not diagnostic. In other words, BMI does not measure body composition directly, so a high BMI does not necessarily represent a health crisis. For example, a healthy athlete with a high percentage of lean muscle mass may have a BMI in the overweight or even obese range. In order to verify high body fat content, other testing would need to be done. Let's look at three such tests.

- DEXA, or dual-energy X-ray absorption, is a special X-ray

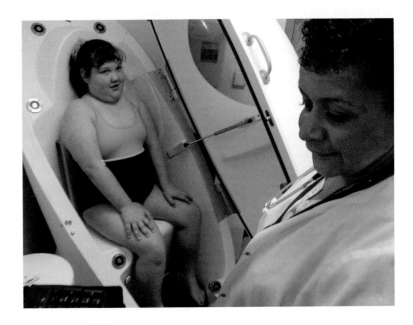

technique that measures bone mineral content and uses that information to estimate the amounts of lean tissue and fat.

- Air displacement plethysmography technology, known widely as the Bod Pod, is a way of measuring body composition using air displacement to determine body volume and thus body density (*density* is the mass, or weight, divided by volume).

- Underwater weighing is similar to the Bod Pod but is performed with a water-filled rather than air-filled chamber.

Of course, the most accurate method for determining body composition is dissection, which can only be performed after death. Although the three techniques described here are more accurate than BMI, they are also more expensive and less readily available to the general population.

However it's defined, obesity is a perplexing and poorly understood disease. Despite its adverse effects on health (a higher incidence of arteriosclerosis, hypertension, coronary artery disease, and diabetes mellitus), it is the most common health problem in the United States. According to the Centers for Disease Control and Prevention (CDC), nearly 35% of adults and about 17% of all youth ages 2–19 are obese in the United States. Overeating, consuming fried foods and processed foods, and sedentary activities, such as video gaming or watching television, have replaced healthy eating and exercise.

With all its health and social problems, it's a fair bet that few people choose to be obese. So what causes obesity? Let's look at three theories.

- **Overeating during childhood.** Some researchers believe that the "clean your plate" syndrome early in life sets the stage for adult obesity by increasing the number of fat cells formed during childhood. Then, during adulthood, the more cells there are, the more fat can be stored.

- **Greater fuel efficiency in obese people.** Obese people may be more fuel efficient and more effective "fat storers." Although it is often assumed that obese people eat more than other people, this is not necessarily true—many actually eat less than people of normal weight.

  Fat packs more wallop per calorie (is more fattening) than proteins or carbohydrates because very little energy is used to process fat. For example, when someone ingests 100 excess *carbohydrate* calories, 23 calories are used in metabolic processing, and 77 are stored. However, if the excess calories come from fat, only 3 calories are "burned," and 97 are stored. These facts apply to everyone, but for obese individuals, the picture is even bleaker. Fat cells of overweight people "sprout" more receptors that favor fat accumulation, and their lipoprotein lipase enzyme, which unloads fat from the blood, is exceptionally efficient.

- **Genetic predisposition.** Morbid obesity is the fate of people inheriting two obesity genes. However, a true genetic predisposition for "fatness" accounts for only about 5 percent of obesity in the United States. These people, given excess

calories, will always deposit them as fat.

## False and Risky Cures

Many people have "tried everything" to lose weight. Poor choices for fighting obesity include diuretics; amphetamines; over-the-counter medications containing ephedra; and drugs that interfere with fat, and thus fat-soluble vitamin, absorption. All of these have serious health risks such as electrolyte imbalance, heart valve malfunction, cardiovascular problems, liver damage, stroke, seizures, and, in some cases, death. Because weight-loss supplements are not monitored by the U.S. Food and Drug Administration (FDA), the true extent of the damage they cause is not known.

Diet products and books sell well, but which ones really work? At the present time, many dieters find themselves choosing between two "dueling" schools of thought: those promoting low-carbohydrate (high-protein and high-fat) diets such as the Atkins and South Beach diets, and those espousing the traditional low-fat diet (high in complex carbohydrates). Clinical studies show that people on low-carbohydrate diets lose weight more quickly at first but tend to plateau at 6 months. Among long-term dieters, people on low-fat diets lost just as much weight as did those on the low-carbohydrate diets.

## Surgery

In desperation, some people turn to surgery. Options for bariatric surgery

(to reduce stomach size) mainly focus on reducing food intake, reducing the amount of food absorbed, or both. Procedures such as *gastric bypass* or *sleeve gastrectomy* can make the stomach smaller through placement of a gastric band or surgical removal of part of the stomach. These solutions help patients feel full after a small meal. To reduce food absorption, one approach is to route food away from a portion of the small intestine, as in *biliopancreatic diversion (BPD)*. During BPD surgery, the lower half of the small intestine is connected directly to the stomach in such a way that bile and pancreatic juices do not pass through this "new" segment of the small intestine. Because pancreatic enzymes are absent, fewer nutrients are available for reabsorption.

*Liposuction* directly removes fat. In this procedure, suction is used to remove adipose tissue in order to reshape the body. Like BPD, it carries all the risks of surgery, and unless the patient changes his or her eating habits, the remaining fat deposits elsewhere in the body will overfill.

## Eat Less, Exercise More

Unfortunately, there is no magical solution for obesity. The best way for most of us to lose weight is to take in fewer fat calories and increase physical activity. Resistance exercise also helps because it increases muscle mass, and muscle consumes more energy at rest compared to fat. The only way to keep the weight off is to make healthy eating and exercise lifelong habits.

# Homeostatic Relationships between the Digestive System and Other Body Systems

**Endocrine System**

- Liver removes hormones from blood, ending their activity; digestive system provides nutrients needed for energy fuel, growth, and repair; pancreas has hormone-producing cells
- Local hormones help regulate digestive function

**Lymphatic System/Immunity**

- Digestive system provides nutrients for normal functioning; HCl of stomach provides nonspecific protection against bacteria
- Lacteals drain fatty lymph from digestive tract organs and convey it to blood; Peyer's patches and lymphoid tissue in mesentery house macrophages and immune cells that protect digestive tract organs against infection

**Digestive System**

**Urinary System**

- Digestive system provides nutrients for energy fuel, growth, and repair
- Kidneys transform vitamin D to its active form, which is needed for calcium absorption; excrete some bilirubin produced by the liver

**Muscular System**

- Digestive system provides nutrients for energy fuel, growth, and repair; liver removes lactic acid, resulting from muscle activity, from the blood
- Skeletal muscle activity increases contractions of GI tract

**Nervous System**

- Digestive system provides nutrients for normal neural functioning
- Neural controls of digestive function; in general, parasympathetic fibers accelerate and sympathetic fibers inhibit digestive activity; reflex and voluntary controls of defecation

**Respiratory System**

- Digestive system provides nutrients for energy metabolism, growth, and repair
- Respiratory system provides oxygen and carries away carbon dioxide produced by digestive system organs

**Cardiovascular System**

- Digestive system provides nutrients to heart and blood vessels; absorbs iron needed for hemoglobin synthesis; absorbs water necessary for normal blood volume
- Cardiovascular system transports nutrients absorbed by alimentary canal to all tissues of the body; distributes hormones of the digestive tract

**Reproductive System**

- Digestive system provides nutrients for energy fuel, growth, and repair and extra nutrition needed to support fetal growth

**Integumentary System**

- Digestive system provides nutrients for energy fuel, growth, and repair; supplies fats that provide insulation in subcutaneous tissues
- The skin synthesizes vitamin D needed for calcium absorption from the intestine; protects by enclosure

**Skeletal System**

- Digestive system provides nutrients for energy fuel, growth, and repair; absorbs calcium needed for bone salts
- Skeletal system protects some digestive organs by bone; cavities store some nutrients (e.g., calcium, fats)

# Summary

## PART I: ANATOMY AND PHYSIOLOGY OF THE DIGESTIVE SYSTEM (pp. 489–513)

### Anatomy of the Digestive System (pp. 489–502)

1. The digestive system consists of the alimentary canal (a hollow tube extending from mouth to anus) and several accessory digestive organs. The wall of the alimentary canal has four main tissue layers—mucosa, submucosa, muscularis externa, serosa. The serosa (visceral peritoneum) is continuous with the parietal peritoneum, which lines the abdominal cavity wall.

2. Organs of the alimentary canal:

   a. The mouth, or oral cavity, contains teeth and tongue and is bound by lips, cheeks, and palate. Tonsils guard its posterior margin.

   b. The pharynx is a muscular tube that provides a passageway for food and air.

   c. The esophagus is a muscular tube that completes the passageway from the pharynx to the stomach.

   d. The stomach is a C-shaped organ located on the left side of the abdomen beneath the diaphragm. Food enters it through the cardioesophageal sphincter and leaves it to enter the small intestine through the pyloric sphincter. The stomach has a third oblique layer of muscle in its wall that allows it to perform mixing or churning movements. Gastric glands produce hydrochloric acid, pepsin, rennin, mucus, gastrin, and intrinsic factor. Mucus protects the stomach itself from being digested.

   e. The tubelike small intestine is suspended from the posterior body wall by the mesentery. Its subdivisions are the duodenum, jejunum, and ileum. Food digestion and absorption are completed here. Pancreatic juice and bile enter the duodenum through a sphincter at the distal end of the bile duct. Microvilli, villi, and circular folds increase the surface area of the small intestine for enhanced absorption.

   f. The large intestine frames the small intestine. Subdivisions are the cecum; appendix; ascending, transverse, and descending colon; sigmoid colon; rectum; and anal canal. The large intestine delivers undigested food residue (feces) to the body exterior.

3. Two sets of teeth are formed. The first set consists of 20 deciduous teeth that begin to appear at 6 months and are lost by 12 years of age. Permanent teeth (32 of them) begin to replace deciduous teeth around age 7. A typical tooth consists of crown covered with enamel and root covered with cement. Most of the tooth is bonelike dentin. The pulp cavity contains blood vessels and nerves.

4. Salivary glands (three pairs—parotid, submandibular, and sublingual) secrete saliva into the oral cavity. Saliva contains mucus and serous fluids. The serous component contains salivary amylase.

5. Several accessory organs deposit substances into the alimentary tube via ducts.

   a. The pancreas is a soft gland lying in the mesentery between the stomach and small intestine. Pancreatic juice contains enzymes (which digest all categories of foods) in an alkaline fluid.

   b. The liver is a four-lobed organ overlying the stomach. Its digestive function is to produce bile, which it ducts into the small intestine.

   c. The gallbladder is a muscular sac that stores and concentrates bile. When fat digestion is not occurring, the continuously made bile backs up the cystic duct and enters the gallbladder.

### Functions of the Digestive System (pp. 502–513)

1. Foods must be broken down to their building blocks to be absorbed. Building blocks of carbohydrates are simple sugars, or monosaccharides. Building blocks of proteins are amino acids. Building blocks of fats, or lipids, are fatty acids and glycerol, and building blocks of nucleic acids are nucleotides.

2. Both mechanical (chewing) and chemical food breakdown begin in the mouth. Saliva contains mucus, which helps bind food together into a bolus, and salivary amylase, which begins the chemical breakdown of starch. Saliva is secreted in response to food in the mouth, mechanical pressure, and psychic stimuli. Essentially no food absorption occurs in the mouth.

3. Swallowing has two phases: The buccal phase is voluntary; the tongue pushes the bolus into the pharynx. The involuntary pharyngeal-esophageal phase involves the closing off of nasal and respiratory passages and the conduction of food to the stomach by peristalsis.

4. When food enters the stomach, gastric secretion is stimulated by the vagus nerve and by gastrin (a local hormone). Hydrochloric acid activates the protein-digesting enzyme pepsin, and digestion of proteins

begins. Food is also mechanically broken down by the churning activity of stomach muscles. Movement of chyme into the small intestine is controlled by the enterogastric reflex.

5. Digestion of fats, proteins, and carbohydrates is completed in the small intestine by intestinal enzymes and, more important, pancreatic enzymes. Alkaline pancreatic juice neutralizes acidic chyme and provides the proper environment for the operation of its enzymes. Both pancreatic juice (the only source of lipases) and bile (formed by the liver) are necessary for normal fat breakdown and absorption. Bile acts as a fat emulsifier. Secretin and cholecystokinin, hormones produced by the small intestine, stimulate release of bile and pancreatic juice. Segmental movements mix foods; peristaltic movements move foodstuffs along the small intestine.

6. Most nutrient absorption occurs by active transport into the capillary blood of the villi. Fats are absorbed by diffusion into both capillary blood and lacteals in the villi.

7. The large intestine receives bacteria-laden indigestible food residue. Activities of the large intestine are absorption of water, of salts, and of vitamins made by resident bacteria. When feces are delivered to the rectum by mass peristalsis, the defecation reflex is initiated.

# PART II: NUTRITION AND METABOLISM

(pp. 513–527)

## Nutrition (pp. 513–516)

1. Most foods are used as fuels to form ATP; proteins and cholesterol are exceptions.

2. A nutrient is a substance in food used to promote growth, maintenance, and repair of the body.

3. The major nutrients are carbohydrates, lipids, proteins, and water. Vitamins and minerals are required in very small amounts.

4. Dietary carbohydrates (sugars and starch) are included in fruits and vegetables (plant products).

5. Dietary lipids are found in meats, dairy products, and vegetable oils.

6. Eggs, milk, meats, poultry, and fish are rich sources of protein.

7. Vitamins, found mainly in fruits, vegetables, and milk, function mainly as coenzymes in the body.

8. Minerals, most plentiful in vegetables and legumes, are mainly important for enzyme activity. Iron is important to make hemoglobin, and calcium is important for building bone, blood clotting, nerve impulses, muscle contraction, and secretory activities.

## Metabolism (pp. 516–527)

1. Metabolism includes all chemical breakdown (catabolic) and building (anabolic) reactions needed to maintain life.

2. Carbohydrates, the most important of which is glucose, are the body's major energy fuel. As glucose is oxidized, carbon dioxide, water, and ATP are formed. The sequential pathways of glucose catabolism are glycolysis (which occurs in the cytosol) and the citric acid cycle and electron transport chain (which function in the mitochondria). During hyperglycemia, glucose is stored as glycogen or converted to fat. In hypoglycemia, glycogenolysis, gluconeogenesis, and fat breakdown occur to restore the normal blood glucose level.

3. Fats insulate the body, protect organs, build some cell structures (membranes and myelin sheaths), and provide reserve energy. When carbohydrates are in limited supply, more fats are oxidized to produce ATP. Excessive fat breakdown causes blood to become acidic. Excess dietary fat is stored in subcutaneous tissue and other fat deposits.

4. Proteins form the bulk of cell structure and most functional molecules. They are carefully conserved by body cells. Amino acids are actively taken up from blood by tissue cells; those that cannot be made by body cells are called *essential amino acids.* Amino acids are oxidized to form ATP when other fuel sources are not available. Ammonia is released as amino acids are catabolized.

5. The liver is the body's key metabolic organ. Its cells remove nutrients from hepatic portal blood. It performs glycogenesis, glycogenolysis, and gluconeogenesis to maintain homeostasis of the blood glucose level. Its cells make blood proteins and other substances and release them to blood. Liver cells also detoxify ammonia by combining it with carbon dioxide to form urea. Liver cells burn fats to provide some of their energy (ATP); excesses are stored or released to blood in simpler forms that can be used by other tissue cells. Phagocytic cells remove bacteria from hepatic portal blood. Most cholesterol is made by the liver; cholesterol breakdown products are secreted in bile. Fats and cholesterol are transported in the blood by

14

lipoproteins. LDLs transport cholesterol to body cells; HDLs carry cholesterol to the liver for degradation. Cholesterol is used to make functional molecules and for some structural purposes; it is not used for ATP synthesis.

6. A dynamic balance exists between energy intake and total energy output (heat + work + energy storage). Interference with this balance results in obesity or in malnutrition leading to body wasting.

7. When the three major types of foods are oxidized for energy, they yield different amounts of energy. Carbohydrates and proteins yield 4 kcal/gram; fats yield 9 kcal/gram. Basal metabolic rate (BMR) is the total amount of energy used by the body in a basal (resting) state. Age, sex, body surface area, and amount of thyroxine produced influence BMR.

8. Total metabolic rate (TMR) is the number of calories used by the body to accomplish all ongoing daily activities. It increases dramatically as muscle activity increases. When TMR equals total caloric intake, weight remains constant.

9. As foods are catabolized to form ATP, more than 60 percent of energy released escapes as heat, warming the body. The hypothalamus initiates heat loss processes (radiation of heat from skin and evaporation of sweat) or heat-promoting processes (vasoconstriction of skin blood vessels and shiv-ering) as necessary to maintain body temperature within the normal range. Fever (hyperthermia) represents body temperature regulated at a higher-than-normal level.

## PART III: DEVELOPMENTAL ASPECTS OF THE DIGESTIVE SYSTEM AND METABOLISM (pp. 527–531)

1. The alimentary canal forms as a hollow tube. Accessory glands form as outpocketings from this tube.

2. Common congenital defects include cleft palate, cleft lip, and tracheoesophageal fistula, all of which interfere with normal nutrition. Common inborn errors of metabolism are cystic fibrosis (CF) and phenylketonuria (PKU).

3. Various inflammatory conditions plague the digestive system throughout life. Appendicitis is common in adolescents; gastroenteritis and food poisoning can occur at any time (if irritating factors are present); and ulcers and gallbladder problems increase in middle age. Obesity and diabetes mellitus are bothersome during later middle age.

4. Efficiency of all digestive system processes decreases in the elderly. Gastrointestinal cancers, such as stomach and colon cancer, appear with increasing frequency in an aging population.

## Review Questions

 Access additional practice questions using your smartphone, tablet, or computer:
Mastering A&P®> Study Area > Practice Tests & Quizzes

### Multiple Choice

*More than one choice may apply.*

1. Which of the following digestive organs are accessory organs?

   a. Esophagus          c. Pancreas

   b. Stomach            d. Liver

2. Ducts that are part of the digestive system include the

   a. bile duct.

   b. pancreatic duct.

   c. thoracic duct.

   d. ejaculatory duct.

   e. cystic duct.

3. The GI tube tissue layer responsible for the actions of segmentation and peristalsis is the

   a. serosa.

   b. mucosa.

   c. muscularis externa.

   d. submucosa.

4. Match the digestive organ listed in column B with the function listed in column A.

| Column A | Column B |
| --- | --- |
| _____ 1. produces bile | a. salivary glands |
| _____ 2. absorbs water | b. esophagus |
| _____ 3. churning occurs here | c. stomach |
| _____ 4. muscular tube connecting the laryngopharynx with the stomach | d. small intestine |
| | e. liver |
| _____ 5. produces both endocrine and exocrine secretions | f. gallbladder |
| | g. pancreas |
| _____ 6. secretes a substance that initiates carbohydrate digestion | h. large intestine |
| _____ 7. stores bile | |
| _____ 8. segmentation occurs here | |

5. Which of the following is (are) true concerning the stomach?

   a. It is located within the peritoneum.

   b. The pylorus is continuous with the large intestine.

   c. It covers the liver.

   d. The breakdown of fats starts in it.

6. Which of these organs lie in the left hypochondriac region of the abdomen?

   a. Spleen

   b. Stomach

   c. Cecum

   d. Duodenum

7. Release of secretin leads to

   a. contraction of smooth muscle in the duodenal papilla.

   b. increased secretory activity of liver cells.

   c. contraction of the gallbladder wall.

   d. release of bicarbonate-rich fluid by the pancreas.

8. The pH of chyme entering the duodenum is adjusted by

   a. bile.

   b. intestinal juice.

   c. enzyme secretions from the pancreas.

   d. bicarbonate-rich secretions from the pancreas.

9. A 3-year-old girl is rewarded with a hug because she is now completely toilet trained. Which muscle is one that she has learned to control?

   a. Levator ani

   b. Internal anal sphincter

   c. Internal and external obliques

   d. External anal sphincter

10. Hormones that act to decrease the blood glucose level include

   a. insulin.

   c. epinephrine

   b. glucagon.

   d. growth hormone.

11. Which events occur shortly after eating?

   a. Use of amino acids as a major source of energy

   b. Lipogenesis (and fat deposit)

   c. Breakdown of fat reserves

   d. Increased uptake of glucose by skeletal muscles and other body tissues

12. What is the approximate time of eruption of the third molars, also called wisdom teeth?

   a. 6–7 years

   c. 17–25 years

   b. 12–13 years

   d. They do not erupt sometimes.

13. Complete this statement. In glycolysis, _____ is oxidized, and _____ is reduced.

   a. vitamin-containing coenzyme; glucose

   b. ATP; ADP

   c. glucose; oxygen

   d. glucose; vitamin-containing coenzyme

## Short Answer Essay

14. Make a simple line drawing of the organs of the alimentary tube, and label each organ.

15. Add three labels to your drawing—salivary glands, liver, and pancreas—and use arrows to show where each of these organs empties its secretion into the alimentary tube.

16. Name the layers of the alimentary tube wall from the lumen outward.

17. What is the mesentery? The peritoneum?

18. Name the subdivisions of the small intestine beginning near the stomach. Do the same for the subdivisions of the large intestine, beginning closest to the small intestine.

19. Name the two major regions of a tooth. Briefly describe them. What connects these two regions?

20. Describe how salivary secretions assist the digestive process.

21. Assume you have been chewing a piece of bread for five or six minutes. How would you expect its taste to change during this time? Why?

22. Name the three subdivisions of the pharynx, and briefly describe their locations.

23. What hormones and hormonelike products are secreted by the stomach?

24. Bile emulsifies fat. Define *emulsify*.

25. Where are CCK and secretin produced? What is the stimulus for their release?

26. Describe the two phases of swallowing.

27. How do segmental and peristaltic movements differ?

28. A cream cheese and jelly sandwich contains proteins, carbohydrates, and fats. Describe what happens to the sandwich when you eat it relative to events occurring in ingestion, digestion, absorption, and defecation.

29. Which structures greatly increase the absorptive surface of the small intestine?

30. How is the passage of feces facilitated?

31. Define *defecation reflex, constipation,* and *diarrhea.*

32. Define *metabolism*, *anabolism*, and *catabolism*.

33. Define *gluconeogenesis*, *glycogenolysis*, and *glycogenesis*.

34. What are the major food groups?

35. What is the harmful result when excessive amounts of fats are burned to produce ATP? Name two conditions that might lead to this result.

36. How many calories are produced when 1 gram of carbohydrate is oxidized? 1 gram of protein? 1 gram of fat? If you just ate 100 grams of food that was 20 percent protein, 30 percent carbohydrate, and 10 percent fat, how many kilocalories did you consume?

37. In terms of surface area, which kinds of individuals will have a high basal metabolic rate (BMR)?

38. Name two ways in which heat is lost from the body.

39. What is fever? What does it indicate?

40. Why must the feeding of a newborn baby be frequent? Which reflexes help a newborn to obtain nutrition?

# Critical Thinking and Clinical Application Questions

41. After chopping wood for about two hours on a hot but breezy afternoon, John stumbled into the house and then fainted. His T-shirt was wringing wet with perspiration, and his pulse was faint and rapid. Was he suffering from heat stroke or heat exhaustion? Explain your reasoning, and note what you should do to help John recover.

42. When Serge went to his doctor with the reports of his annual health checkup, the doctor was not happy with the HDL/LDL ratio in Serge's blood. The LDL level being far too high, he was concerned that Serge might have atherosclerosis. How can an undesirable HDL/LDL ratio lead to atherosclerosis?

43. A young woman is put through an extensive battery of tests to determine the cause of her stomach pains. She is diagnosed with gastric ulcers. An antihistamine drug is prescribed, and she is sent home. What is the mechanism of her medication? What life-threatening problems can result from a poorly managed ulcer? Why did the clinic doctor warn the woman not to take aspirin?

44. Continuing from the previous question, the woman's ulcer got worse. She started complaining of back pain. The physician discovered that the back pain occurred because the pancreas was now damaged. Use logic to deduce how a perforating gastric ulcer could come to damage the pancreas.

45. Ryan, a 6-year-old child who is allergic to milk, has extremely bowed legs. What condition do you suspect, and what is the connection to not drinking milk?

46. Kathy's baby is not thriving because she has problems sucking. Outwardly, no abnormalities can be observed, but a close examination of the baby's mouth reveals a possible cause of her condition. Explain what the cause is, and how this condition can be corrected.

47. An anorexic girl shows a high level of acetone in her blood. What is this condition called, and what has caused it?

48. Every year dozens of elderly people are found dead in their unheated apartments—victims of hypothermia. What is hypothermia, and how does it kill?

49. Charlie has been vomiting ever since he ate last night's fast food for breakfast. What could be a possible cause of his emesis?

50. Kayla has been on antibiotics for an extended period. A visit to her physician reveals that she has a vitamin K deficiency. What is the cause-and-effect relationship here?

# 15 The Urinary System

The kidneys, which maintain the purity and constancy of our internal fluids, are perfect examples of homeostatic organs. Like sanitation workers who keep a city's water supply drinkable and dispose of its waste, the kidneys are usually unappreciated until there is a malfunction and "internal garbage" piles up. Every day, the kidneys filter gallons of fluid from the bloodstream. They then process this *filtrate*, allowing wastes and excess ions to leave the body in *urine* while returning needed substances to the blood in just the right proportions. Although the lungs and the skin also play roles in excretion, the kidneys bear the major responsibility for eliminating nitrogenous (nitrogen-containing) wastes, toxins, and drugs from the body.

Disposing of wastes and excess ions is only part of the kidneys' work. As they perform these excretory functions, the kidneys also regulate blood volume and maintain the proper balance between water and salts, and between acids and bases.

The kidneys have other regulatory functions, too:

- By producing the enzyme *renin*, they help regulate blood pressure.

- The hormone *erythropoietin*, released by the kidneys, stimulates red blood cell production in bone marrow (see Chapter 10).

- Kidney cells convert vitamin D to its active form.

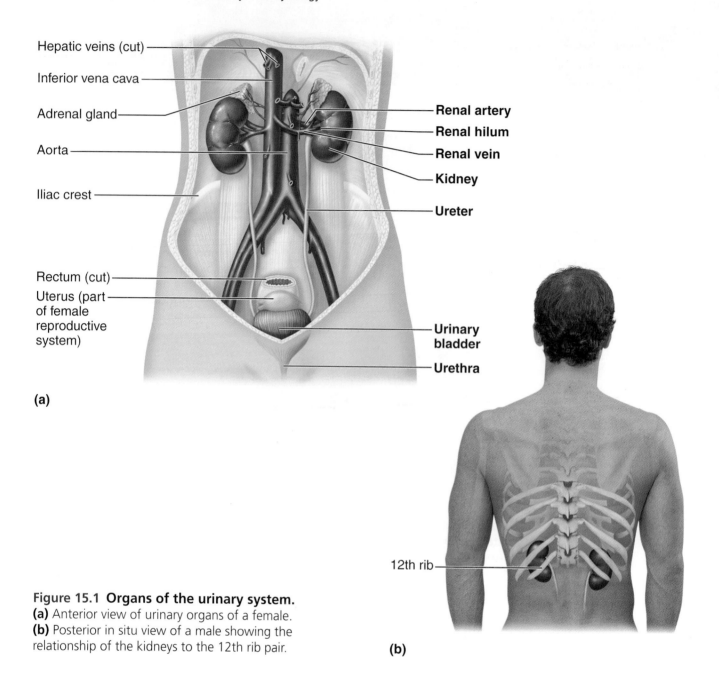

Hepatic veins (cut)

Inferior vena cava

Adrenal gland

Aorta

Iliac crest

Rectum (cut)

Uterus (part
of female
reproductive
system)

**Renal artery**

**Renal hilum**

**Renal vein**

**Kidney**

**Ureter**

**Urinary
bladder**

**Urethra**

**(a)**

12th rib

**(b)**

**Figure 15.1 Organs of the urinary system.**
**(a)** Anterior view of urinary organs of a female.
**(b)** Posterior in situ view of a male showing the
relationship of the kidneys to the 12th rib pair.

The kidneys alone perform the functions just described and manufacture urine in the process. The other organs of the **urinary system**—the paired *ureters* and the single *urinary bladder* and *urethra* **(Figure 15.1a)**—provide temporary storage for urine or serve as transportation channels to carry it from the kidneys to the outside of the body.

# Kidneys

## Location and Structure

→ **Learning Objectives**

☐ Describe the location of the kidneys in the body.

☐ Identify the following regions of a kidney (longitudinal section): hilum, cortex, medulla, medullary pyramids, calyces, pelvis, and renal columns.

Although many people believe the **kidneys** are located in the lower back, this is *not* the case. These small, dark red organs with a kidney-bean shape lie against the dorsal body wall in a *retroperitoneal* position (behind the parietal peritoneum) in the *superior* lumbar region (Figure 15.1b). The kidneys extend from the $T_{12}$ to the $L_3$ vertebra; thus they receive some protection from the lower part of the rib cage. Because it is crowded by the liver, the right kidney is slightly lower than the left.

## Kidney Structure

An adult kidney is about 12 cm (5 inches) long, 6 cm (2.5 inches) wide, and 3 cm (1 inch) thick, or about the size of a large bar of soap. It is convex laterally and has a medial indentation called the *renal hilum*. Several structures, including the ureters, the renal blood vessels, and nerves, enter or exit the kidney at the hilum (**Figure 15.2**, p. 540). Sitting atop each kidney is an *adrenal gland*, which is part of the endocrine system and is a separate organ.

The kidney has three protective layers. Deep to superficial, they are as follows:

- A transparent *fibrous capsule* encloses each kidney and gives it a glistening appearance.

- A fatty mass, the *perirenal fat capsule*, surrounds each kidney and cushions it against blows.

- The *renal fascia*, the most superficial layer made of dense fibrous connective tissue, anchors the kidney and adrenal gland to surrounding structures.

### ◢ Homeostatic Imbalance 15.1

The fat surrounding the kidneys is important in holding them in their normal body position. If the amount of fatty tissue dwindles (as with rapid weight loss), the kidneys may drop to a lower position, a condition called *ptosis* (to´sis; "a fall"). Ptosis creates problems if the ureters, which drain urine from the kidneys, become kinked. When this happens, urine that can no longer pass through the ureters backs up and exerts pressure on the kidney tissue. This condition, called **hydronephrosis** (hi˝dro-ne˘-fro´sis), can severely damage the kidney. \_\_\_\_✚

When a kidney is cut lengthwise, three distinct regions become apparent (see Figure 15.2). The outer region, which is light in color, is the **renal cortex** (*cortex* = bark). Deep to the cortex is a darker reddish brown area, the **renal medulla**. The medulla has triangular regions with a striped appearance, the **renal pyramids**, or **medullary** (med´u-lar˝e), **pyramids**. The broad *base* of each pyramid faces toward the cortex; its tip, the *apex*, points toward the inner region of the kidney. The pyramids are separated by extensions of cortexlike tissue, called **renal columns**.

Lateral to the hilum is a flat, funnel-shaped tube, the **renal pelvis**. The pelvis is continuous with the ureter leaving the hilum (see Figure 15.2b). Extensions of the pelvis, **calyces** (kal´˘i-sēz; singular *calyx*), form cup-shaped "drains" that enclose the tips of the pyramids. The calyces collect urine, which continuously drains from the tips of the pyramids into the renal pelvis. Urine then flows from the pelvis into the ureter, which transports it to the bladder for temporary storage until it leaves the body.

### Blood Supply

The kidneys continuously cleanse the blood and adjust its composition, so it is not surprising that they have a rich blood supply (see Figure 15.2b and c). Approximately one-quarter of the total blood supply of the body passes through the kidneys each minute. The artery supplying each kidney is the **renal artery**. As the renal artery approaches the hilum, it divides into **segmental arteries**, each of which gives off several branches called **interlobar arteries**, which travel through the renal columns to reach the cortex. At the cortex-medulla junction, interlobar arteries give off the **arcuate** (ar´ku-at) **arteries**, which arch over the medullary pyramids. Small **cortical radiate arteries** then branch off the arcuate arteries to supply the renal cortex. Venous blood draining from the kidney flows through veins that trace the pathway of the arterial supply but in a reverse direction—**cortical radiate veins** to **arcuate veins** to **interlobar veins** to the **renal vein**, which emerges from the kidney hilum and empties into the inferior vena cava. (There are no segmental veins.)

### Did You Get It?

1. The kidneys are retroperitoneal. What does that mean?
2. Mary lost lots of weight, and suddenly she was having problems with her urine flow. What would you guess happened, and what caused the problem?
3. From the most superficial aspect of a kidney to its ureter, name its three major regions.

**For answers, see Appendix A.**

## Nephrons

### → Learning Objective

☐ **Recognize that the nephron is the structural and functional unit of the kidney, and describe its anatomy.**

**Nephrons** (nef´ronz) are the structural and functional units of the kidneys and are responsible for forming urine. Each kidney contains over a million of these tiny filters. In addition, there are thousands of *collecting ducts*, each of which collects fluid from several nephrons and conveys it to the renal pelvis. **Figure 15.3a** and **b**, p. 541, show the
*(Text continues on page 542.)*

15

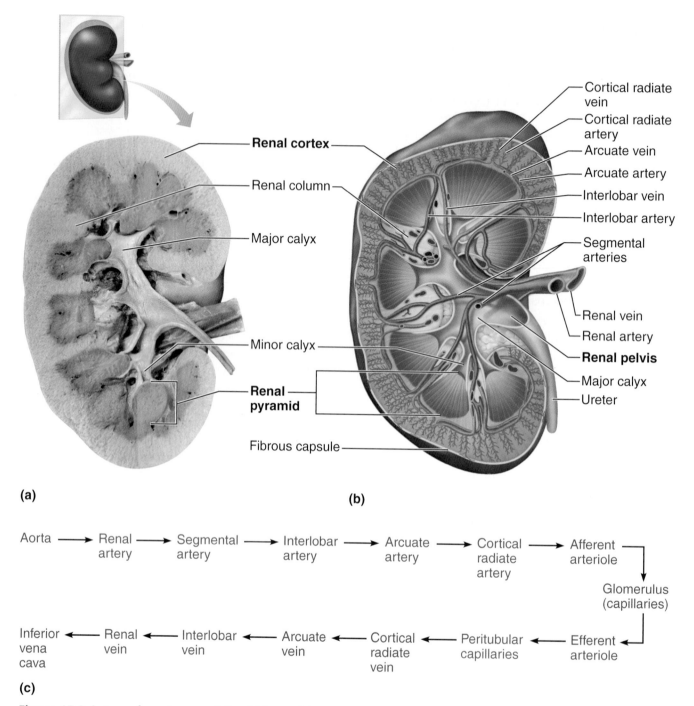

(a)

(b)

(c)

**Figure 15.2 Internal anatomy of the kidney. (a)** Photograph of a coronally sectioned kidney. **(b)** Diagram of coronal kidney section illustrating major blood vessels. **(c)** Summary of the pathway of renal blood vessels.

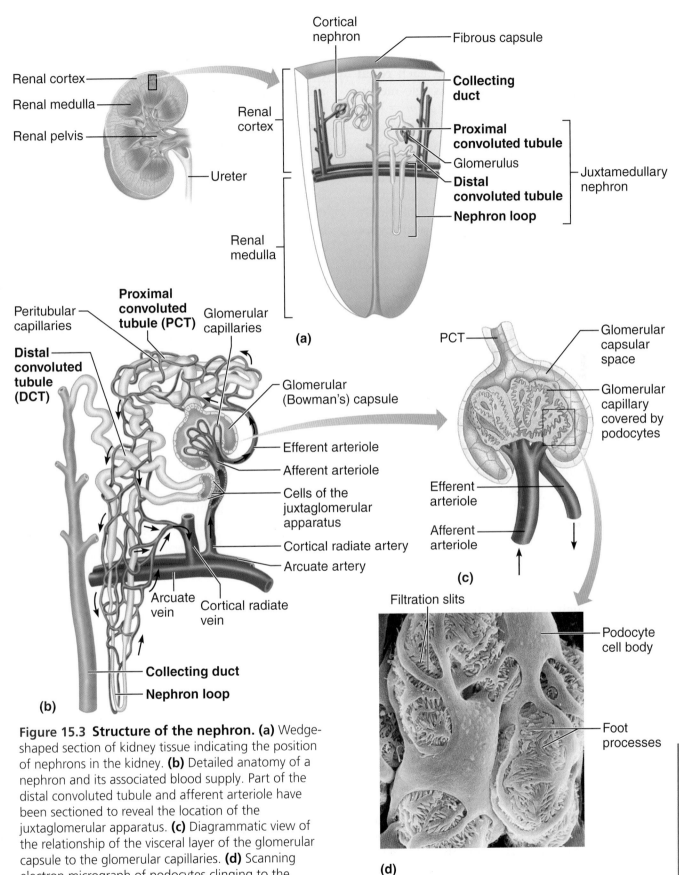

Renal cortex
Renal medulla
Renal pelvis
Ureter

Cortical nephron
Fibrous capsule
Renal cortex
Renal medulla
**Collecting duct**
**Proximal convoluted tubule**
Glomerulus
**Distal convoluted tubule**
**Nephron loop**
Juxtamedullary nephron

**(a)**

Peritubular capillaries
**Proximal convoluted tubule (PCT)**
Glomerular capillaries
**Distal convoluted tubule (DCT)**
Glomerular (Bowman's) capsule
Efferent arteriole
Afferent arteriole
Cells of the juxtaglomerular apparatus
Cortical radiate artery
Arcuate artery
Arcuate vein
Cortical radiate vein
**Collecting duct**
**Nephron loop**

**(b)**

PCT
Glomerular capsular space
Glomerular capillary covered by podocytes
Efferent arteriole
Afferent arteriole

**(c)**

Filtration slits
Podocyte cell body
Foot processes

**(d)**

**Figure 15.3 Structure of the nephron. (a)** Wedge-shaped section of kidney tissue indicating the position of nephrons in the kidney. **(b)** Detailed anatomy of a nephron and its associated blood supply. Part of the distal convoluted tubule and afferent arteriole have been sectioned to reveal the location of the juxtaglomerular apparatus. **(c)** Diagrammatic view of the relationship of the visceral layer of the glomerular capsule to the glomerular capillaries. **(d)** Scanning electron micrograph of podocytes clinging to the glomerular capillaries.

15

anatomy and relative positioning of nephrons and collecting ducts in each kidney.

Each nephron consists of two main structures: a *renal corpuscle* and a *renal tubule*. Each **renal corpuscle** consists of a **glomerulus** (glo-mer'u-lus), which is a knot of capillaries (*glom* = little ball), and a cup-shaped hollow structure that completely surrounds the glomerulus like a well-worn baseball glove encloses a ball. This portion of the renal corpuscle is called the **glomerular capsule** or **Bowman's capsule**. The inner (visceral) layer of the capsule is made up of highly modified octopus-like cells called **podocytes** (pod'o-sītz; Figure 15.3c and d). Podocytes have long branching extensions called *foot processes* that intertwine with one another and cling to the glomerulus. Openings called *filtration slits* between these processes allow the podocytes to form a porous, or "holey," membrane around the glomerulus ideal for filtration.

The **renal tubule**, which makes up the rest of the nephron, is about 3 cm (approximately 1.25 inches) long. As it extends from the glomerular capsule, it coils and twists before forming a hairpin loop and then again becomes coiled and twisted before entering a collecting duct. These different regions of the tubule have specific names (see Figure 15.3a and b); in order from the glomerular capsule, they are the **proximal convoluted tubule (PCT)**; the **nephron loop**, or *loop of Henle* (hen'le); and the **distal convoluted tubule (DCT)**. The surfaces of the tubule cells exposed to the filtrate in the proximal convoluted tubules are covered with dense microvilli, which increases their surface area tremendously. Microvilli also occur on the tubule cells in other parts of the tubule, but in much reduced numbers.

Most nephrons are called **cortical nephrons** because they are located almost entirely within the cortex. In a few cases, the nephrons are called **juxtamedullary nephrons** because they are situated close to the cortex-medulla junction, and their nephron loops dip deep into the medulla (see Figure 15.3a). The **collecting ducts**, each of which receives urine from many nephrons, run downward through the medullary pyramids, giving the pyramids a striped appearance. They deliver the final urine product into the calyces and renal pelvis.

Each and every nephron is associated with two capillary beds—the glomerulus (mentioned earlier) and the *peritubular* (per"ĭ-tu'bu-lar) *capillary bed* (see Figure 15.3b). The glomerulus is fed by the **afferent arteriole**, which arises from a cortical radiate artery. It is the "feeder vessel," and the **efferent arteriole** receives the blood that has passed through the glomerulus. The glomerulus, specialized for filtration, differs from any other capillary bed in the entire body because it is both fed *and* drained by *arterioles*. Also, the afferent arteriole has a larger diameter than the efferent, resulting in a much higher blood pressure in the glomerular capillaries compared to other capillary beds. This high pressure forces fluid and small solutes out of the blood into the glomerular capsule. Most of this filtrate (99 percent) is eventually reclaimed by the renal tubule cells and returned to the blood in the peritubular capillary beds.

The second capillary bed, the **peritubular capillaries**, arises from the efferent arteriole that drains the glomerulus. Unlike the high-pressure glomerulus, these capillaries are low-pressure, porous vessels adapted for absorption instead of filtration. They cling closely to the whole length of the renal tubule, where they are in an ideal position to receive solutes and water from the tubule cells as these substances are reabsorbed from the filtrate percolating through the tubule. The peritubular capillaries ultimately drain into interlobar veins leaving the cortex.

## Urine Formation and Characteristics

### → Learning Objectives

☐ **Describe the process of urine formation, identifying the areas of the nephron that are responsible for filtration, reabsorption, and secretion.**

☐ **Describe the function of the kidneys in excretion of nitrogen-containing wastes.**

☐ **Define *anuria* and *oliguria*.**

Urine formation is a result of three processes—*glomerular filtration*, *tubular reabsorption*, and *tubular secretion* **(Figure 15.4)**.

### Glomerular Filtration

As just described, the glomerulus acts as a filter. **Glomerular filtration** is a nonselective, passive process in which fluid passes from the blood into the glomerular capsule part of the renal tubule.

## CONCEPT**LINK**

Recall that filtration, as a passive process, requires a pressure gradient (Chapter 3, p. 104). The capillaries of the glomerulus are under higher pressure compared to the glomerular capsule; as a result, fluids move down the pressure gradient, from the blood into the glomerular capsule.

Once in the capsule, the fluid is called **filtrate**; it is essentially blood plasma without blood proteins. Both proteins and blood cells are normally too large to pass through the filtration membrane, and when either of these appears in the urine, there is usually a problem with the glomerular filters. As long as the systemic blood pressure is normal, filtrate will be formed. If arterial blood pressure drops too low, glomerular pressure becomes inadequate to force substances out of the blood into the tubules, and filtrate formation stops.

### Homeostatic Imbalance 15.2

An abnormally low urinary output is called **oliguria** (ol″i-gu′re-ah) if it is between 100 and 400 ml/day, and **anuria** (ah-nu′re-ah) if it is less than 100 ml/day. Low urinary output usually indicates that glomerular blood pressure is too low to cause filtration, but anuria may also result from transfusion reactions and acute inflammation or from crushing injuries to the kidneys. _____+

### Tubular Reabsorption

Besides wastes and excess ions that must be removed from the blood, the filtrate contains many useful substances (including water, glucose, amino acids, and ions), which must be reclaimed from the filtrate and returned to the blood. **Tubular reabsorption** begins as soon as the filtrate enters the proximal convoluted tubule (**Figure 15.5**, p. 544). The tubule cells are "transporters," taking up needed substances from the filtrate and then passing them out their posterior aspect into the extracellular space, from which they are absorbed into peritubular capillary blood. Some reabsorption is done passively (for example, water passes by osmosis), but reabsorption of most substances depends on active transport processes, which use membrane carriers, require ATP, and are very selective. There is an abundance of carriers for substances that need to be retained, and few or no carriers for substances of no use to the body.

**Q:** *How would liver disease, in which the liver is unable to make many of the blood proteins, affect process 1? (Reviewing Chapter 10 might help.)*

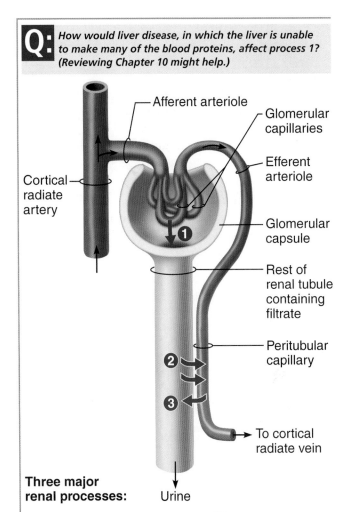

**Three major renal processes:**

**1** ➞ **Glomerular filtration:** Water and solutes smaller than proteins are forced through the capillary walls and pores of the glomerular capsule into the renal tubule.

**2** ➞ **Tubular reabsorption:** Water, glucose, amino acids, and needed ions are transported out of the filtrate into the tubule cells and then enter the capillary blood.

**3** ➞ **Tubular secretion:** H⁺, K⁺, creatinine, and drugs are removed from the peritubular blood and secreted by the tubule cells into the filtrate.

**Figure 15.4 The kidney depicted schematically as a single large, uncoiled nephron.** A kidney actually has millions of nephrons acting in parallel. Three processes in the kidneys adjust the composition of plasma.

**A:** The amount of renal filtrate formed is a function of filtration (blood) pressure and blood osmotic pressure (exerted largely by blood proteins). In this situation, more filtrate than normal will be formed because the blood pressure is opposed by lower-than-normal osmotic pressure of the blood, caused by low blood protein levels.

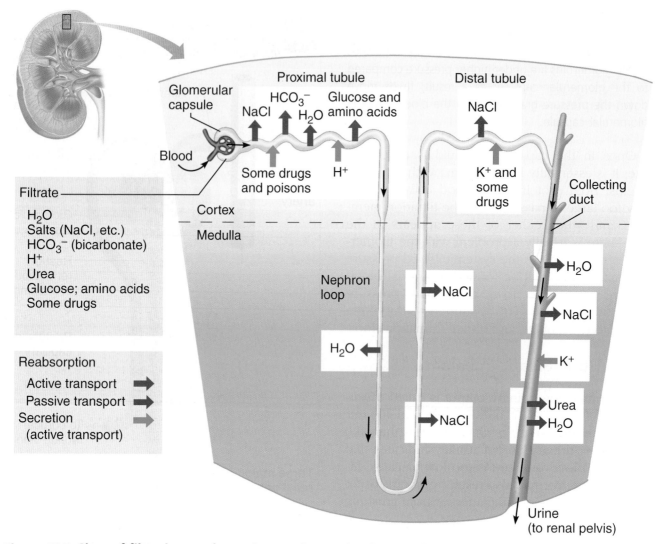

**Figure 15.5 Sites of filtration, reabsorption, and secretion in a nephron.**

Needed substances (for example, glucose and amino acids) are usually entirely removed from the filtrate. Most reabsorption occurs in the proximal convoluted tubules, but the distal convoluted tubule and the collecting duct are also active.

## Tubular Secretion

**Tubular secretion** is essentially tubular reabsorption in reverse. Some substances, such as hydrogen and potassium ions ($H^+$ and $K^+$) and creatinine, also move from the blood of the peritubular capillaries through the tubule cells or from the tubule cells themselves into the filtrate to be eliminated in urine. This process seems to be important for getting rid of substances not already in the filtrate, such as certain drugs or excess potassium ions, or as an additional means for controlling blood pH (see Figure 15.5).

### CONCEPTLINK

Recall that pH is a measure of hydrogen ion ($H^+$) concentration (see Chapter 2, p. 67). When the body experiences a high level of hydrogen ions, which can lower pH, the kidneys help by eliminating excess hydrogen ions from the body via the urine.

### Nitrogenous Wastes

**Nitrogenous waste products** are poorly reabsorbed, if at all. Tubule cells have few membrane carriers to reabsorb these substances because we do not need them. They tend to remain in the filtrate and are found in high concentrations in urine excreted from the body. Various ions are reabsorbed or allowed to go out in the urine, according to what is needed at a particular time to maintain the proper pH and *electrolyte* (solute) composition of the blood. Common nitrogenous wastes include the following:

- **Urea** (u-re′ah), formed by the liver as an end product of protein breakdown when amino acids are used to produce energy

- **Uric acid**, released when nucleic acids are metabolized

- **Creatinine** (kre-at′ĭ-nin), associated with creatine (kre′ah-tēn) metabolism in muscle tissue

## Did You Get It?

4. What is the structural and functional unit of the kidney?
5. What are two functions of the renal tubule and one function of the peritubular capillaries?
6. How does a decrease in blood pressure affect glomerular pressure?

**For answers, see Appendix A.**

## Characteristics of Urine

### → Learning Objectives

☐ **Describe the composition of normal urine.**

☐ **List abnormal urinary components.**

In 24 hours, the marvelously complex kidneys filter some 150 to 180 liters of blood plasma through their glomeruli into the tubules, which process the filtrate by taking substances out of it (reabsorption) and adding substances to it (secretion). In the same 24 hours, only about 1.0 to 1.8 liters of urine are produced. Therefore, urine and filtrate are quite different. Filtrate contains everything that blood plasma does (except proteins), but by the time it reaches the collecting ducts, the filtrate has lost most of its water and just about all of its nutrients and necessary ions. What remains, **urine**, contains nitrogenous wastes and unneeded or excess substances. Assuming we are healthy, our kidneys can keep our blood composition fairly constant despite wide variations in diet and cell activity.

Freshly voided urine is generally clear and pale to deep yellow. The normal yellow color is due to *urochrome* (u′ro-krōm), a pigment that results from the body's destruction of hemoglobin. The more solutes are in the urine, the deeper yellow its color. Dilute urine is a pale, straw color. At times, urine may be a color other than yellow. This might be a result of eating certain foods (beets, for example) or the presence of bile or blood in the urine.

When formed, urine is sterile, and its odor is slightly aromatic. If it is allowed to stand, it takes on an ammonia odor caused by the action of bacteria on the urine solutes. Some drugs, vegetables (such as asparagus), and various diseases (such as diabetes mellitus) alter the usual odor of urine.

Urine pH is usually slightly acid (around 6), but changes in body metabolism and certain foods may cause it to be much more acidic or basic. For example, a diet with large amounts of protein (eggs and cheese) and whole-wheat products causes urine to become quite acidic. Conversely, a vegetarian diet makes urine quite alkaline as the kidneys excrete excess bases. Bacterial infection of the urinary tract also may cause the urine to be alkaline.

Because urine is water plus solutes, urine is denser (weighs more) than distilled water. **Specific gravity** compares how *much* heavier urine is than distilled water. Whereas the specific gravity of pure water is 1.0, the specific gravity of urine usually ranges from 1.001 to 1.035 (dilute to concentrated urine, respectively). Urine is generally *dilute* (has a low specific gravity) when a person drinks excessive fluids, uses diuretics (drugs that increase urine output), or has chronic renal failure, a condition in which the kidney loses its ability to concentrate urine (see "A Closer Look" on p. 549). Conditions that produce *concentrated* urine with a high specific gravity include inadequate fluid intake, fever, and kidney inflammation called *pyelonephritis* (pi″ĕ-lo-nĕ-fri′tis).

Solutes normally found in urine include sodium ions ($Na^+$) and potassium ions ($K^+$), urea, uric acid, creatinine, ammonia, bicarbonate ions, and various other ions, depending on blood composition. With certain diseases, urine composition can change dramatically, and the presence of abnormal substances in urine is often helpful in diagnosing the problem. This is why a routine urinalysis should always be part of a physical examination.

Substances *not* normally found in urine are glucose, blood proteins, red blood cells, hemoglobin, white blood cells (pus), and bile. Table 15.1, p. 546 lists conditions that may involve abnormal urinary constituents and volumes.

15

Table **15.1**    **Abnormal Urinary Constituents**

| Substance | Name of condition | Possible causes |
| --- | --- | --- |
| Glucose | Glycosuria (gli"ko-su're-ah) | Nonpathological: Excessive intake of sugary foods<br>Pathological: Diabetes mellitus |
| Proteins | Proteinuria (pro"te-ĭ-nu're-ah) (also called *albuminuria*) | Nonpathological: Physical exertion, pregnancy<br>Pathological: Glomerulonephritis, hypertension |
| Pus (WBCs and bacteria) | Pyuria (pi-u're-ah) | Urinary tract infection |
| RBCs | Hematuria (he"mah-tu're-ah) | Bleeding in the urinary tract (due to trauma, kidney stones, infection) |
| Hemoglobin | Hemoglobinuria (he"mo-glo-bĭ-nu're-ah) | Various: Transfusion reaction, hemolytic anemia |
| Bile pigment | Bilirubinuria (bil"ĭ-roo-bĭ-nu're-ah) | Liver disease (hepatitis) |

## Did You Get It?

7. Is the specific gravity of urine higher or lower than that of water? Why?

For the answer, see Appendix A.

# Ureters, Urinary Bladder, and Urethra

→ **Learning Objectives**

☐ **Describe the general structure and function of the ureters, urinary bladder, and urethra.**

☐ **Compare the course and length of the male urethra to that of the female.**

☐ **Define *micturition*.**

☐ **Describe the difference in control of the external and internal urethral sphincters.**

## Ureters

The **ureters** (yer'ĕ-terz) are two slender tubes each 25 to 30 cm (10 to 12 inches) long and 6 mm (¼ inch) in diameter. Each ureter runs behind the peritoneum from the renal hilum to the posterior aspect of the bladder, which it enters at a slight angle **(Figure 15.6)**. The superior end of each ureter is continuous with the renal pelvis, and its mucosal lining is continuous with the mucosa lining the renal pelvis and the bladder inferiorly.

The ureters carry urine from the kidneys to the bladder. Although it might appear that gravity could help drain urine to the bladder, it would not transport urine while a person is lying down. Urine reaches the bladder even when standing on your head because the ureters *do* play an active role in urine transport. Smooth muscle layers in their walls contract to propel urine by peristalsis. Once urine has entered the bladder, it is prevented from flowing back into the ureters by small valve-like folds of bladder mucosa that cover the ureter openings.

 **Homeostatic Imbalance 15.3**

When urine becomes extremely concentrated, solutes such as uric acid salts form crystals that precipitate in the renal pelvis. These crystals are called **renal calculi** (kal'ku-li; *calculus* = little stone) or *kidney stones* (see photo).

Excruciating pain that radiates to the flank (lateral, posterior lower back) occurs when the ureter walls close in on the sharp calculi as they are being eased through the ureter by peristalsis or when the calculi become wedged in a ureter. Frequent bacterial infections of the urinary tract, urinary retention, and alkaline urine all favor calculi formation. Surgery is one treatment option, but a noninvasive procedure (*lithotripsy*) that uses ultrasound waves to shatter the calculi is also available. The pulverized remnants of the calculi are then painlessly eliminated in urine.

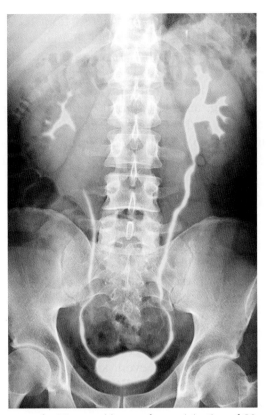

A urogram of a 35-year-old man after an injection of 80 ml of iodized contrast medium. The contrast medium (in yellow) travels from the kidneys toward the bladder. The presence of a kidney stone is highlighted in red.

## Urinary Bladder

The **urinary bladder** is a smooth, collapsible, muscular sac that stores urine temporarily. It is located retroperitoneally in the pelvis just posterior to the pubic symphysis. A scan of the interior of the bladder reveals three openings—the two ureter openings (*ureteral orifices*), and the single opening of the *urethra* (u-re′thrah) (the *internal urethral orifice*), which drains the bladder (see Figure 15.6). The smooth triangular region of the bladder base outlined by these three openings is called the **trigone** (tri′gon). The trigone is important clinically because infections tend to persist in this region. In males, the *prostate* (part of the male reproductive system) surrounds the neck of the bladder where it empties into the urethra.

The bladder wall contains three layers of smooth muscle, collectively called the *detrusor muscle*, and its mucosa is a special type of epithelium, *transitional epithelium* (see Chapter 3, p. 119). Both of these structural features make the bladder uniquely suited for its function of urine storage. When the bladder is empty, it is collapsed, 5 to 7.5 cm (2 to 3 inches) long at most, and its walls are thick and thrown into folds. As urine accumulates, the bladder expands and rises superiorly in the abdominal cavity (**Figure 15.7**, p. 549). Its muscular wall stretches, and the transitional epithelial

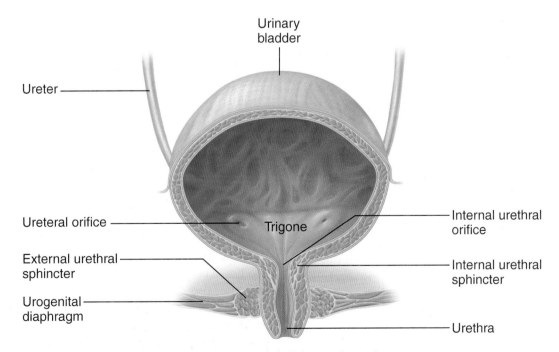

Figure 15.6 **Basic structure of the female urinary bladder and urethra.**

Urinary bladder

Ureter

Ureteral orifice

Trigone

Internal urethral orifice

External urethral sphincter

Internal urethral sphincter

Urogenital diaphragm

Urethra

15

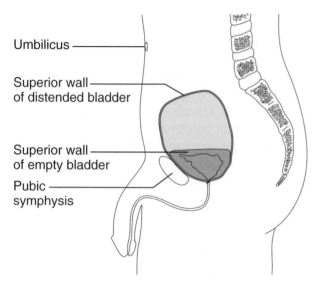

**Figure 15.7 Position and shape of a distended and an empty urinary bladder in an adult male.**

layer thins, increasing its volume and allowing the bladder to store more urine without substantially increasing its internal pressure. A moderately full bladder is about 12.5 cm (5 inches) long and holds about 500 ml (1 pint) of urine, but it is capable of holding more than twice that amount. When the bladder is really distended, or stretched by urine, it becomes firm and pear-shaped and may be felt just above the pubic symphysis. Urine is formed continuously by the kidneys, but it is stored in the bladder until its release is convenient.

## Urethra

The **urethra** is a thin-walled tube that carries urine by peristalsis from the bladder to the outside of the body. At the bladder-urethra junction, a thickening of the smooth muscle forms the **internal urethral sphincter** (see Figure 15.6), an involuntary sphincter (think "**in**ternal is **in**voluntary") that keeps the urethra closed when urine is not being passed. A second sphincter, the **external urethral sphincter**, is formed by skeletal muscle as the urethra passes through the pelvic floor. This sphincter is voluntarily controlled.

The length and relative function of the urethra differ in the two sexes. In men, the urethra is approximately 20 cm (8 inches) long and has three named regions (look ahead to Figure 16.2a, p. 566): the *prostatic*, *membranous*, and *spongy* (or *penile*) *urethrae*. The urethra opens at the tip of the penis after traveling down its length. The urethra of the male has a double function. It carries both urine and sperm (in semen) from the body, but never at the same time. Thus, in males, the urethra is part of both the urinary and reproductive systems.

In women, the urethra is about 3 to 4 cm (1½ inches) long, and its external orifice, or opening, lies anterior to the vaginal opening (look ahead to Figure 16.8a, p. 575). Its only function is to conduct urine from the bladder to the body exterior.

### Homeostatic Imbalance 15.4

The female urinary orifice is close to the anal opening, so improper toileting habits (wiping from back to front rather than from front to back) can carry fecal bacteria into the urethra. And because the mucosa of the urethra is continuous with that of the rest of the urinary tract organs, inflammation of the urethra, or **urethritis** (u″re-thri′tis), can easily ascend the tract to cause bladder inflammation (**cystitis**) or even kidney inflammation (**pyelonephritis**, or **pyelitis**). Symptoms of **urinary tract infection (UTI)** include *dysuria* (painful urination), urinary *urgency* and *frequency*, fever, and sometimes cloudy or blood-tinged urine. When the kidneys are involved, back pain and a severe headache are common. _____ **+**

## Micturition

**Micturition** (mik″tu-rish′un), or **voiding**, is the act of emptying the bladder. As noted, two sphincters, or valves—the internal urethral sphincter (more superiorly located) and the external urethral sphincter (more inferiorly located)—control the flow of urine from the bladder (see Figure 15.6). Ordinarily, the bladder continues to collect urine until about 200 ml have accumulated. At this point, stretching of the bladder wall activates stretch receptors. Impulses transmitted to the sacral region of the spinal cord and then back to the bladder via the *pelvic splanchnic nerves* cause the bladder to go into reflex contractions. As the contractions become stronger, stored urine is forced past the internal urethral sphincter (the smooth muscle, involuntary sphincter) into the upper part of the urethra. The person will then feel the urge to void. Because the lower external sphincter is skeletal muscle and is controlled voluntarily, we can choose to keep it closed and postpone bladder emptying temporarily. However, if it is convenient, the external sphincter can be relaxed so that urine is flushed

# Renal Failure and the Artificial Kidney

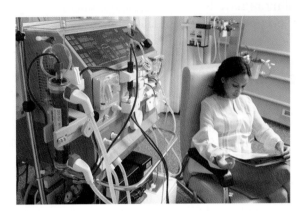

**A renal patient undergoing hemodialysis.**

One out of every 10 Americans has some degree of kidney dysfunction, and nearly 4 percent require dialysis or a kidney transplant just to stay alive. The incidence of kidney disease is growing fastest among adults age 60 and above.

Warning signs of kidney trouble may include high blood pressure, frequent urination, painful urination, puffy eyes, or swollen hands or feet. But all too often, there are few noticeable symptoms until a significant percentage of kidney function has already been lost. Urinalysis can detect *proteinuria*, a sensitive and early marker of kidney damage, and a simple blood test can determine the creatinine level, which can be used to estimate the rate of blood filtration.

In adults, chronic renal disease often develops together with other chronic health conditions. The leading causes are diabetes mellitus, which accounts for approximately 44 percent of new cases each year, and hypertension, accounting for about 28 percent of cases. Hypertension is both a cause and a symptom: High blood pressure impairs kidney function by damaging renal blood vessels and reducing circulation to the organs, even as hypertension-battered kidneys push blood pressure upward. Atherosclerosis compounds the problem by further impairing circulation.

Other possible causes of renal failure include the following:

- Repeated kidney infections
- Physical trauma to the kidneys (crush injury and others)
- Chemical poisoning of the tubule cells by heavy metals (mercury or lead) or organic solvents (dry-cleaning fluids, paint thinner)
- Prolonged pressure on skeletal muscles (causes release of myoglobin, a muscle pigment that can clog renal tubules)

In renal failure, filtrate formation decreases significantly or stops completely. Because toxic wastes accumulate quickly in the blood, *dialysis* (*dialys* = separate) is necessary to cleanse the blood while the kidneys are shut down. In *hemodialysis*, which uses an "artificial kidney" apparatus (see photo), the patient's blood is passed through tubing that is permeable only to selected substances, and the tubing is immersed in a bathing solution that differs slightly from normal "cleansed" plasma.

As blood circulates through the tubing, nitrogenous wastes and potassium ions ($K^+$) present in the blood (but not in the bath) diffuse out of the blood through the tubing into the surrounding solution. Substances to be added to the blood (mainly buffers for hydrogen ions and glucose for malnourished patients) move from the bathing solution into the blood. In this way, needed substances are retained in the blood or added to it, while wastes and ion excesses are removed. Hemodialysis is routinely done three times weekly, and each session takes 4 to 8 hours, depending on patient size and how well the kidneys are functioning. Hemodialysis patients sometimes encounter serious problems such as thrombosis, infection, and ischemia at the shunt site. Hemorrhage is an added risk because the blood must be treated with an anticoagulant, such as *heparin*, to prevent clotting during hemodialysis.

When renal damage is nonreversible, as in chronic renal failure, a kidney transplant is the only answer. Unless the new kidney comes from an identical twin, recipients must take immunosuppressive drugs for the rest of their lives to prevent rejection. If this sounds bleak, consider that 40 years ago people who reached end-stage renal failure survived for only a few days. Today, they can live for years or even decades.

from the body. When a person chooses not to void, the reflex contractions of the bladder stop within a minute or so, and urine collection continues. After 200 to 300 ml more have been collected, the micturition reflex occurs again. Eventually, micturition occurs whether the person wills it or not.

### ⚓ Homeostatic Imbalance 15.5

**Incontinence** (in-kon'tĭ-nens) occurs when a person is unable to voluntarily control the external sphincter. Incontinence is normal in children 2 years old or younger because they have not yet gained control over their voluntary sphincter. It may also occur in older children who sleep so soundly that they are not awakened by the stimulus. However, after the toddler years, incontinence is usually a result of emotional problems, pressure (as in pregnancy), or nervous system problems (stroke or spinal cord injury).

**Urinary retention** is essentially the opposite of incontinence. In this condition, the bladder is unable to expel its contained urine. There are various causes for urinary retention. It often occurs after surgery in which general anesthesia has been given, because it takes a little time for the smooth muscles to regain their activity. Another cause of urinary retention, occurring primarily in older men, is enlargement, or **hyperplasia**, of the prostate, which surrounds the neck of the bladder. As the prostate gland enlarges, it narrows the urethra, making it very difficult to void. When urinary retention is prolonged, a slender flexible drainage tube called a *catheter* (kath'ĭ-ter) must be inserted through the urethra to drain the urine and prevent bladder trauma from excessive stretching. _____ ✚

### Did You Get It?

8. A kidney stone blocking a ureter would interfere with the flow of urine to which organ?
9. Why is the presence of transitional epithelium in the urinary bladder important?
10. Structurally and functionally, how does the male urethra differ from the female urethra?
11. What is a synonym for *micturition*?

For answers, see Appendix A.

# Fluid, Electrolyte, and Acid-Base Balance

The composition of the blood depends on three major factors: diet, cellular metabolism, and urine output. In general, the kidneys have four major roles to play, which help keep the blood composition relatively constant:

- Excreting nitrogen-containing wastes
- Maintaining water balance of the blood
- Maintaining electrolyte balance of the blood
- Ensuring proper blood pH

Excretion of nitrogenous wastes has already been considered; the other roles of the kidneys are discussed next.

## Maintaining Water Balance of Blood

### → Learning Objectives

☐ Name and localize the three main fluid compartments of the body.

☐ Explain the role of antidiuretic hormone (ADH) in the regulation of water balance by the kidney.

☐ Explain the role of aldosterone in sodium ion and potassium ion balance of the blood.

☐ Define *diuresis* and *polyuria*.

### Body Fluids and Fluid Compartments

If you are a healthy young adult, water probably accounts for half or more of your body weight—50 percent in women and about 60 percent in men. These differences reflect the fact that women have relatively less muscle and a larger amount of body fat (of all body tissues, fat contains the least water). Babies, with little fat and low bone mass, are about 75 percent water. This accounts for their "dewy" skin, like that of a freshly picked peach. After infancy, total body water content declines through life, finally accounting for only about 45 percent of body mass in old age. Water is the universal body solvent within which all solutes (including the very important electrolytes) are dissolved. (We described the importance of water to the functioning of the body and its cells in Chapter 2.)

Water occupies three main locations within the body, referred to as *fluid compartments* (**Figure 15.8**). About two-thirds of body fluid, the so-called **intracellular fluid (ICF)**, is contained within the living cells. The remainder, called **extracellular fluid (ECF)**, includes all body fluids located outside the cells. ECF includes blood plasma, interstitial fluid (IF), lymph, and *transcellular fluid*, which includes cerebrospinal and serous fluids, the humors of the eye, and others.

As plasma circulates throughout the body delivering substances, it serves as the "highway" that links the external and internal environments

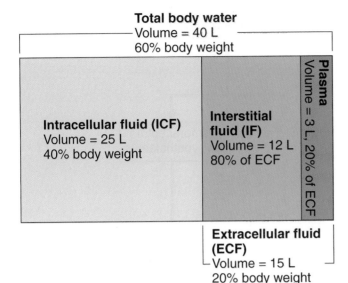

Figure 15.8 **The major fluid compartments of the body.** Approximate values are noted for a 70-kg (154-pound) man. For simplicity, this figure shows only interstitial fluid and plasma in the ECF.

**(Figure 15.9)**. Exchanges occur almost continuously in the lungs, gastrointestinal tract, and kidneys. Although these exchanges alter plasma composition and volume, adjustments in the other two fluid compartments follow quickly so that balance is restored.

### The Link between Water and Electrolytes

Water certainly accounts for nearly the entire volume of body fluids, which are all similar, regardless of type. But there is more to *fluid balance* than just water. Recall from Chapter 2 that electrolytes are charged particles (ions) that conduct an electrical current in an aqueous solution. The types and amounts of solutes in body fluids, especially electrolytes such as sodium, potassium, and calcium ions, are also vitally important to overall body homeostasis. Water and electrolyte balance, though we consider them separately here, are tightly linked as the kidneys continuously process the blood.

### Regulation of Water Intake and Output

If the body is to remain properly hydrated, we cannot lose more water than we take in. Most

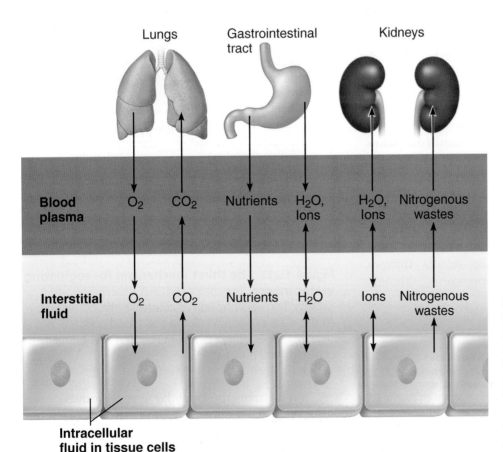

Figure 15.9 **The continuous mixing of body fluids.** Nutrients and wastes are exchanged between the intracellular fluid and the plasma through the interstitial fluid. Blood plasma transports nutrients and wastes between the cells and the external environment of the body.

15

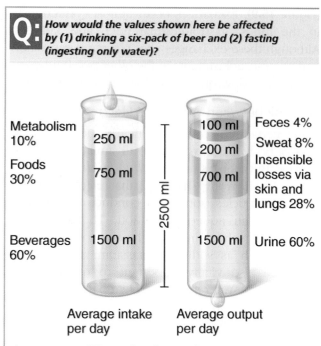

| Metabolism 10% | 250 ml | | 100 ml | Feces 4% |
| Foods 30% | 750 ml | 2500 ml | 700 ml | Sweat 8% / Insensible losses via skin and lungs 28% |
| Beverages 60% | 1500 ml | | 1500 ml | Urine 60% |

Average intake per day | Average output per day

**Figure 15.10 Water intake and output.** Major sources of body water and routes of water loss from the body are shown. When intake and output are in balance, the body is adequately hydrated.

water intake results from fluids and foods we ingest in our diet **(Figure 15.10)**. However, a small amount (about 10 percent) is produced during cellular metabolism (explained in Chapter 14).

The **thirst mechanism** is the driving force for water intake. An increase in plasma solute content of only 2 to 3 percent excites highly sensitive cells in the hypothalamus called **osmoreceptors** (oz″mo-re-sep′torz), which in turn activate the hypothalamic *thirst center* **(Figure 15.11)**. The mouth also becomes dry because the salivary glands obtain the water they require from the blood. When less fluid leaves the bloodstream, less saliva is produced, reinforcing the thirst response. The same response results from a decline in blood volume (or pressure), although this is a less potent stimulus.

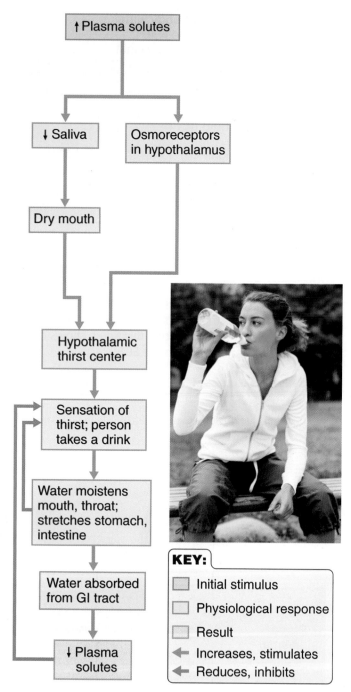

↑Plasma solutes

↓Saliva

Osmoreceptors in hypothalamus

Dry mouth

Hypothalamic thirst center

Sensation of thirst; person takes a drink

Water moistens mouth, throat; stretches stomach, intestine

Water absorbed from GI tract

↓Plasma solutes

**KEY:**
- Initial stimulus
- Physiological response
- Result
- ← Increases, stimulates
- ← Reduces, inhibits

**Figure 15.11 The thirst mechanism for regulating water intake.** The major stimulus is increased osmolarity (concentration of solutes) of blood plasma. (A decrease in plasma volume and blood pressure, not shown, is a less important stimulus that acts primarily to control blood pressure.)

Water leaves the body by several routes. Some water vaporizes out of the lungs (insensible water loss), some is lost in perspiration, and some leaves the body in the stool (see Figure 15.10). The job of the kidneys is like that of a juggler. If large amounts of water are lost in other ways, the kidneys compensate by producing less urine to conserve body water. However, when water intake is excessive, the kidneys excrete generous amounts of water in urine to reduce fluid volume.

In addition to regulating fluid volume, the kidneys ensure that proper concentrations of the various electrolytes are present in both intracellular and extracellular fluids. Most electrolytes enter the body in foods and "hard" (mineral-rich) water. Although very small amounts are lost in perspiration and in feces, the major factor regulating the electrolyte composition of body fluids is the kidneys.

Reabsorption of water and electrolytes by the kidneys is regulated primarily by hormones. When blood volume drops for any reason (for example, as a result of hemorrhage or excessive water loss through sweating or diarrhea), arterial blood pressure drops, which in turn decreases the amount of filtrate formed by the kidneys. In addition, the hypothalamic osmoreceptors react to the change in blood composition (that is, less water and more solutes) by becoming more active. The result is that nerve impulses are sent to the posterior pituitary (**Figure 15.12**, p. 554), which then releases **antidiuretic hormone (ADH)**. (The term *antidiuretic* is derived from *diuresis* [di″u-re′sis], which means "flow of urine from the kidney," and *anti*, which means "against.") As you might guess, this hormone prevents excessive water loss in the urine.

### CONCEPT**LINK**

Remember the concept of interrelationships among organ systems (see Figure 1.3, p. 34), and notice that the interdependent events regulating sodium ion and water balance (see Figure 15.12, p. 554) involve four body systems: urinary, nervous, endocrine, and cardiovascular.

ADH travels in the blood to its main target, the kidney's collecting ducts, where it causes the duct cells to reabsorb more water. As more water is returned to the bloodstream, blood volume and blood pressure increase to normal levels, and only a small amount of concentrated urine is formed. ADH is released more or less continuously unless the solute concentration of the blood drops too low. When this happens, the osmoreceptors become "quiet," and excess water is allowed to leave the body in the urine.

 ## Homeostatic Imbalance 15.6

When ADH is *not* released (perhaps because of injury or destruction of the hypothalamus or posterior pituitary gland), huge amounts of very dilute urine (up to 25 liters/day) flush from the body day after day. This condition, **diabetes insipidus** (in-sip′ĭ-dus), can lead to severe dehydration and electrolyte imbalances. Affected individuals are always thirsty and have to drink fluids almost continuously to maintain normal fluid balance. ____✚

## Maintaining Electrolyte Balance

Very small changes in the electrolyte concentrations in the various fluid compartments cause water to move from one compartment to another. This movement not only alters blood volume and blood pressure, but also can severely impair the activity of irritable cells, such as nerve and muscle cells. For example, when the level of potassium ions in the ECF is low, muscle cells may not be able to repolarize properly, which can lead to muscle cramps.

Besides ADH, **aldosterone** (al″dos′ter-on) is a second hormone that helps to regulate blood composition and blood volume by acting on the kidney. Aldosterone is the major factor regulating sodium ion content of the ECF and in the process helps regulate the concentration of other ions (potassium, chloride [Cl−], and magnesium [Mg$^{2+}$]) as well.

Sodium ions are the electrolytes most responsible for osmotic water flow. When too few sodium ions are in the blood, water will leave the blood and enter tissues, causing edema. If serious, this situation can cause the circulatory system to shut down. Regardless of whether aldosterone is present or not, about 80 percent of the sodium ions in the filtrate are reabsorbed in the proximal convoluted tubules of the kidneys. When the aldosterone concentration is high, most of the remaining sodium ions are actively reabsorbed in the distal convoluted tubules and the collecting ducts. Generally speaking, for each sodium ion reabsorbed, a chloride ion follows, and a potassium ion is secreted into the filtrate. Thus, as the sodium ion content of the blood increases, the potassium ion concentration decreases, bringing these two ions back to their normal balance in the blood. Still another effect of aldosterone is to increase water reabsorption by the tubule cells; as sodium ions are reclaimed, water follows it passively back into the blood. Here is a little rule to keep in mind: *Water follows salt.*

**15**

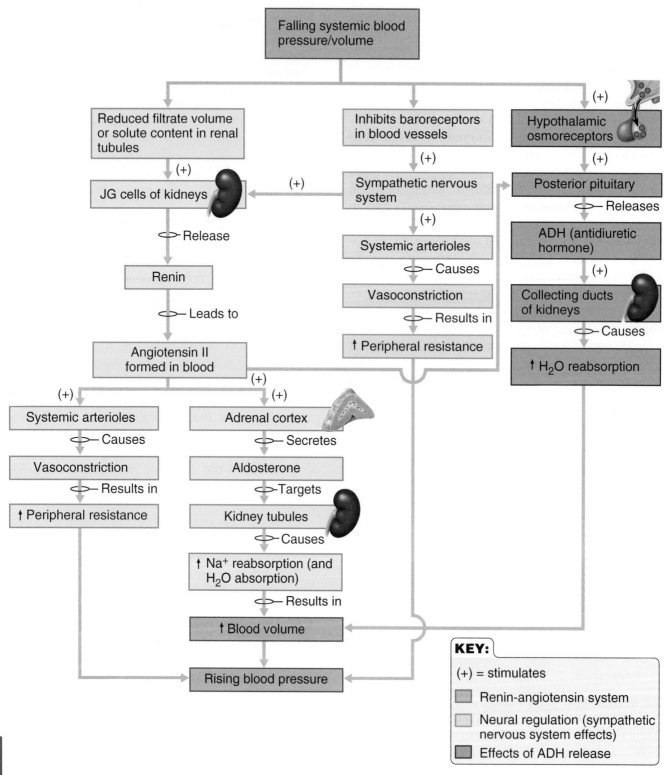

**Figure 15.12 Flowchart of mechanisms regulating sodium ion and water balance to help maintain blood pressure homeostasis.**

Recall that aldosterone is produced by the adrenal cortex. Although a rising potassium ion level or a falling sodium ion level in the ECF directly stimulates the adrenal cells to release aldosterone, the most important trigger for aldosterone release is the **renin-angiotensin mechanism** (see Figure 15.12) mediated by the *juxtaglomerular (JG) apparatus* of the renal tubules. The juxtaglomerular apparatus (look back at Figure 15.3b) consists of a complex of modified smooth muscle cells (JG cells) in the afferent arteriole plus some modified epithelial cells forming part of the distal convoluted tubule. The naming of this cell cluster reflects its location close to (*juxta*) the glomerulus. When the cells of the JG apparatus are stimulated by low blood pressure in the afferent arteriole or changes in solute content of the filtrate, they respond by releasing the enzyme **renin** (re'nin) into the blood. (Notice the different spelling and pronunciation of this enzyme from *rennin* (rĕn'in), an enzyme secreted by stomach glands.) Renin initiates the series of reactions that produce angiotensin II. Angiotensin II in turn acts directly on the blood vessels to cause vasoconstriction (leading to an increase in peripheral resistance) and on the adrenal cortical cells to promote aldosterone release. As a result, blood volume and blood pressure increase (see Figure 15.12). The renin-angiotensin mechanism is extremely important for regulating blood pressure.

When pressure drops, baroreceptors in large blood vessels are also excited. These baroreceptors alert sympathetic nervous system centers of the brain to cause vasoconstriction (via release of epinephrine and norepinephrine) (see Figure 15.12). However, this neural mechanism's major focus is blood pressure regulation, not water or electrolyte balance.

### ⚖ Homeostatic Imbalance 15.7

People with **Addison's disease** (hypoaldosteronism) have **polyuria** (excrete large volumes of urine) and so lose tremendous amounts of salt and water to urine. As long as adequate amounts of salt and fluids are ingested, people with this condition can avoid problems, but they are constantly teetering on the brink of dehydration. _____ ✚

### Did You Get It?

12. What are the four main functions of the kidneys?
13. What is the chief mechanism prompting water intake?
14. What is the role of aldosterone in water balance?

15. Where are osmoreceptors located, and to what stimulus do they respond?

**For answers, see Appendix A.**

## Maintaining Acid-Base Balance of Blood

### Learning Objective

☐ Compare and contrast the relative speed of buffers, the respiratory system, and the kidneys in maintaining the acid-base balance of the blood.

For the cells of the body to function properly, blood pH must be maintained between 7.35 and 7.45, a very narrow range. Whenever the pH of arterial blood rises above 7.45, a person is said to have **alkalosis** (al"kah-lo'sis). A drop in arterial pH to below 7.35 results in **acidosis** (as"ĭ-do'sis). Because a pH of 7.0 is neutral, any pH between 7.0 and 7.35 is not acidic, chemically speaking; however, it represents a lower-than-optimal pH for the functioning of most body cells. Therefore, any arterial pH in this range is called **physiological acidosis**.

Although small amounts of acidic substances enter the body in ingested foods, most hydrogen ions originate as by-products of cellular metabolism, which continuously adds substances to the blood that tend to disturb its **acid-base balance**. Many different acids are produced (for example, phosphoric acid, lactic acid, and many types of fatty acids). In addition, carbon dioxide, which is released during energy production, forms carbonic acid. Ammonia and other basic substances are also released to the blood as cells go about their usual "business." Although the chemical buffers in the blood can temporarily "tie up" excess acids and bases, and the lungs have the chief responsibility for eliminating carbon dioxide from the body, the kidneys assume most of the load for maintaining acid-base balance of the blood. Before describing how the kidneys function in acid-base balance, let's look at how blood buffers and the respiratory system help regulate pH.

### Blood Buffers

Chemical buffers are systems of one or two molecules that act to prevent dramatic changes in the hydrogen ion concentration when acids or bases are added. They do this by binding to hydrogen ions whenever the pH drops and by releasing hydrogen ions when the pH rises. The chemical buffers act within a fraction of a second, so they are the first line of defense in resisting pH changes.

15

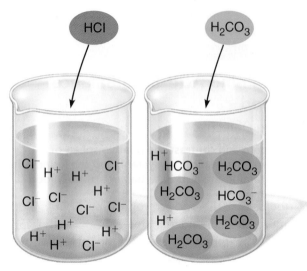

**(a) A strong acid such as HCl dissociates completely into its ions.**

**(b) A weak acid such as H₂CO₃ does *not* dissociate completely.**

**Figure 15.13 Dissociation of strong and weak acids in water.** Undissociated molecules are shown in colored ovals.

To better understand how a chemical buffer system works, let's review strong and weak acids and bases (covered in Chapter 2). Acids are proton (H⁺) donors, and the acidity of a solution reflects only the *free* hydrogen ions, not those still bound to anions. *Strong acids* such as hydrochloric acid (HCl) dissociate completely and liberate all their hydrogen ions (H⁺) in water **(Figure 15.13a)**. Consequently they can cause large changes in pH. By contrast, *weak acids* such as carbonic acid (H₂CO₃) release only some of their H⁺ (partial dissociation) and so have a lesser effect on pH (Figure 15.13b). However, weak acids are very effective at preventing pH changes because they are forced to dissociate and release more H⁺ when the pH rises over the desirable pH range. This feature allows them to play a very important role in the chemical buffer systems.

Also recall that bases are proton, or hydrogen ion, acceptors. *Strong bases* such as hydroxides dissociate easily in water and quickly tie up H⁺, but *weak bases* such as bicarbonate ion (HCO₃⁻) and ammonia (NH₃) are slower to accept H⁺. However, as pH drops, the weak bases become "stronger" and begin to tie up more H⁺. Thus, like weak acids, they are valuable members of the chemical buffer systems.

The three major chemical buffer systems of the body are the *bicarbonate, phosphate,* and *protein buffer systems,* each of which helps to maintain the pH in one or more of the fluid compartments. They all work together, and anything that causes a shift in hydrogen ion concentration in one compartment also causes changes in the others. Thus, any abnormal pH level is resisted by the entire buffering system. Because all three systems operate in a similar way, we will examine just one, the bicarbonate buffer system, which is important in preventing changes in blood pH.

The **bicarbonate buffer system** is a mixture of *carbonic acid* (H₂CO₃) and its salt, *sodium bicarbonate* (NaHCO₃). Carbonic acid is a weak acid, so it does not dissociate much in neutral or acidic solutions. Thus, when a strong acid such as hydrochloric acid is added, most of the carbonic acid remains intact. However, the *bicarbonate ions* (HCO₃⁻) of the salt act as bases to tie up the hydrogen ions released by the stronger acid, forming more carbonic acid:

$$HCl + NaHCO_3 \longrightarrow H_2CO_3 + NaCl$$

strong acid   weak base   weak acid   salt

Because the strong acid is (effectively) changed to a weak one, it lowers the pH of the solution only very slightly.

Similarly, if a strong base such as sodium hydroxide (NaOH) is added to a solution containing the bicarbonate buffer system, NaHCO₃ will not dissociate further under such alkaline conditions. However, carbonic acid (H₂CO₃) will be forced to dissociate further by the presence of the strong base—releasing more H⁺ to bind with the OH⁻ released by NaOH.

$$NaOH + H_2CO_3 \longrightarrow NaHCO_3 + H_2O$$

strong acid   weak acid   weak base   water

The net result is replacement of a strong base by a weak one, so that the pH of the solution rises very little.

### Respiratory Mechanisms

The respiratory system eliminates carbon dioxide from the blood while it "loads" oxygen into the blood. When carbon dioxide (CO₂) enters the blood from the tissue cells, most of it enters the red blood cells, where it is converted to bicarbonate ion (HCO₃⁻) for transport in the plasma, as shown by the equation

$$\underset{\substack{\text{carbonic} \\ \text{anhydrase}}}{} \quad CO_2 + H_2O \rightleftharpoons H_2CO_3 \rightleftharpoons H^+ + HCO_3^-$$

$$\underset{\substack{\text{carbon} \\ \text{dioxide}}}{} \quad \underset{\text{water}}{} \quad \underset{\substack{\text{carbonic} \\ \text{acid}}}{} \quad \underset{\substack{\text{hydrogen} \\ \text{ion}}}{} \quad \underset{\substack{\text{bicarbonate} \\ \text{ion}}}{}$$

The double arrows reveal that this reaction goes both ways. However, an increase in carbon dioxide pushes the reaction to the right, producing more carbonic acid (refer back to Figure 13.11, p. 477). Likewise, an increase in hydrogen ions pushes the reaction to the left, producing more carbonic acid.

In healthy people, carbon dioxide is expelled from the lungs at the same rate as it is formed in the tissues. Thus the $H^+$ released when $CO_2$ is loaded into the blood is not allowed to accumulate, because it is tied up in $H_2O$ when $CO_2$ is unloaded in the lungs. So, under normal conditions, the $H^+$ produced by $CO_2$ transport has essentially no effect on blood pH.

However, when $CO_2$ accumulates in the blood (for example, during restricted breathing) or more $H^+$ is released to the blood by metabolic processes, the chemoreceptors in the respiratory control centers of the brain (or in peripheral blood vessels) are activated. As a result, breathing rate and depth increase, and the excess $H^+$ is "blown off" as more $CO_2$ is removed from the blood.

In contrast, when blood pH begins to rise (alkalosis), the respiratory center is depressed. Consequently, the respiratory rate and depth fall, allowing $CO_2$ (and hence, $H^+$) to accumulate in the blood. Again blood pH is restored to the normal range. Generally, these respiratory system corrections of blood pH (via regulation of $CO_2$ content of the blood) are accomplished within a minute or so.

### Renal Mechanisms

Chemical buffers can tie up excess acids or bases temporarily, but they cannot eliminate them from the body. And although the lungs can dispose of carbonic acid by eliminating carbon dioxide, only the kidneys can rid the body of other acids generated during metabolism. Additionally, only the kidneys have the power to regulate blood levels of alkaline substances. Thus, although the kidneys act slowly and require hours or days to change blood pH, they are the most potent of the pH-regulating mechanisms.

The most important means by which the kidneys maintain acid-base balance of the blood are by excreting bicarbonate ions and reabsorbing or generating new bicarbonate ions. Look back again at the equation showing how the carbonic acid–bicarbonate buffer system operates. Notice that losing a bicarbonate ion ($HCO_3^-$) from the body has the same effect as gaining a hydrogen ion ($H^+$) because it pushes the equation to the right (that is, it leaves a free $H^+$). By the same token, reabsorbing or generating new $HCO_3^-$ is the same as losing $H^+$ because it tends to combine with an $H^+$ and pushes the equation to the left.

Renal mechanisms undertake these adjustments: As blood pH rises, bicarbonate ions are excreted and hydrogen ions are retained by the tubule cells. Conversely, when blood pH falls, bicarbonate is reabsorbed and generated, and hydrogen ions are secreted. Urine pH varies from 4.5 to 8.0, which reflects the ability of the renal tubules to excrete basic or acid ions to maintain blood pH homeostasis.

### Did You Get It?

16. Why is blood pH between 7.0 and 7.35 considered acidic even though basic chemistry defines any pH above 7.0 as basic?
17. To minimize the pH shift that occurs when a strong acid is added to water, would it be better to add a strong base or a weak base? Why?
18. Brian's ultralight plane has crashed in the desert. It is hot, and he neglected to bring a canteen of water. Assume he is there all day. How would you expect this situation to affect the solute concentration of his blood?
19. What are the two main ways the kidneys maintain the acid-base balance of the blood?

For answers, see Appendix A.

# Developmental Aspects of the Urinary System

→ Learning Objectives

☐ Describe three common congenital problems of the urinary system.

☐ Describe the effect of aging on urinary system functioning.

When you trace the embryonic development of the kidneys, it almost seems as though they can't "make up their mind" about whether to come or go. The first tubule system forms and then begins to degenerate as a second, lower, set appears. The second set, in turn, degenerates as a third set

15

# Licensed Practical Nurse (LPN)

**Knowledge of anatomy and physiology helps licensed practical nurses care for their patients.**

Originally, the term *practical nurse* was used to describe health workers who went into private homes to provide bedside nursing. Some licensed practical nurses (LPNs) still provide care in patients' homes, but others play important roles in hospitals, doctors' offices, and health care clinics.

Linda Davila, an LPN employed by Stein Hospice in Sandusky, Ohio, specializes in caring for the terminally ill. "Most of our patients, perhaps 60 percent, have cancer," she says. "Others have life-threatening kidney, lung, or cardiac conditions. Most hospice patients have been diagnosed as having 6 months or less to live, and they have chosen to forgo aggressive medical treatments. My job is to help them stay as comfortable and pain free as possible for the remaining days of their lives and to help maintain quality of life for them and their families."

Hospice care provides a multidisciplinary approach. Davila's team includes social workers, spiritual advisers, registered nurses (RNs), and other LPNs. Explains Davila, "I provide personal care for our patients by helping them with bathing and grooming and monitoring their medications. I also work closely with their families and teach them how to care for the patient: how to empty catheters, bathe someone in bed, and change the sheets for someone who's bedridden."

Davila's clinical skills and anatomy/physiology training enable her to monitor a patient's condition closely and alert team members to any changes. "Hospice patients take many medications, including narcotics," she notes. The narcotics also slow down the digestive process and can make patients constipated. "I monitor patients for side effects and report any symptoms immediately to the RN on my team so we can adjust their medications."

In cancer patients, Davila can often gauge the extent of the disease and recognize when it spreads. "In class we learned that many types of cancer spread to the bone. Sometimes patients will complain of sudden pain in their ribs, which could indicate their cancer has progressed. With lung cancer patients, Davila monitors for shortness of breath, a bluish tinge in skin and fingernails, and mental confusion—signs of reduced oxygen, indicating that the cancer is impairing lung function. She promptly alerts team members so appropriate treatment can begin.

LPN training applicants must hold a high school diploma or the equivalent and pass the school's entrance examination. A one-year program that includes anatomy and physiology, as well as mathematics and other general educational requirements, is typical.

> ## Davila's anatomy and physiology training enable her to monitor patients closely.

After graduating from training, an LPN must also pass a state licensing examination and earn continuing education credits to maintain licensure. Accreditation procedures vary from state to state. For more information about licensed practical nursing and testing requirements, contact:

National League for Nursing
The Watergate
2600 Virginia Ave, NW
Eighth Floor
Washington, DC 20037
800-669-1656
www.nln.org

For additional information on this career and others, click the Focus on Careers link at Mastering A&P.

makes its appearance. This third set develops into the functional kidneys, which are excreting urine by the third month of fetal life. However, the fetal kidneys do not work nearly as hard as they will after birth, because exchanges with the mother's blood through the placenta allow her system to clear many of the undesirable substances from the fetal blood.

## Homeostatic Imbalance 15.8

There are many congenital abnormalities of the urinary system. Two of the most common are polycystic kidney and hypospadias.

**Adult polycystic** (pol″e-sis′tik) **kidney disease** is a degenerative condition that appears to run in families. One or both kidneys enlarge, sometimes to the size of a football, and have many blisterlike sacs (cysts) containing urine. These cysts interfere with renal function by obstructing but initially not stopping urine drainage. Currently, little can be done for this condition except to prevent further kidney damage by avoiding infection. Renal failure is the eventual outcome, but kidney transplants improve chances for survival.

In the rarer, infantile form of the disease, the kidney has blind pouches into which the filtrate flows, totally blocking drainage. The disease progresses rapidly, resulting in death by 2 years of age.

**Hypospadias** (hi″po-spa′de-as) is a condition found in male babies only. It occurs when the urethral orifice is located on the ventral surface of the penis. Corrective surgery is generally done when the child is around 12 months old. _____+

Because the bladder is very small and the kidneys are unable to concentrate urine for the first 2 months, a newborn baby voids from 5 to 40 times per day, depending on fluid intake. By 2 months of age, the infant is voiding approximately 400 ml/day, and the amount steadily increases until adolescence, when adult urine output (about 1500 ml/day) is achieved.

Control of the voluntary urethral sphincter goes hand in hand with nervous system development. By 15 months, most toddlers are aware when they have voided. By 18 months they can hold urine in their bladder for about 2 hours, which is the first sign that potty training can begin. Daytime control usually occurs well before nighttime control is achieved. Complete nighttime control usually does not occur before the child is 4 years old.

During childhood and through late middle age, most urinary system problems are infectious or inflammatory conditions. Many types of bacteria may invade the urinary tract to cause urethritis, cystitis, or pyelonephritis. *Escherichia coli* (esh″er-ĭke-ah ko′li) bacteria are normal residents of the digestive tract and generally cause no problems there, but they act as pathogens (disease-causing agents) in the sterile environment of the urinary tract and account for 80 percent of urinary tract infections. Bacteria and viruses responsible for *sexually transmitted infections (STIs)*, which are primarily reproductive tract infections, may also invade and cause inflammation in the urinary tract, which leads to the clogging of some of its ducts.

## Homeostatic Imbalance 15.9

Childhood streptococcal (strep″to-kok′al) infections, such as strep throat and scarlet fever, may cause inflammatory damage to the kidneys if the original infections are not treated promptly and properly. A common sequel to untreated childhood strep infections is **glomerulonephritis** (glomer″u-lo-ně-fri′tis), in which the glomerular filters become clogged with antigen-antibody complexes resulting from the strep infection. _____+

As we age, kidney size and function progressively decline. By age 70, the rate of filtrate formation is only about half that of the middle-aged adult. This is believed to result from impaired renal circulation due to atherosclerosis, which affects the entire circulatory system of the aging person. In addition to a decrease in the number of functional nephrons, the tubule cells become less efficient in their ability to concentrate urine.

Another consequence of aging is bladder shrinkage and loss of bladder tone, causing many elderly individuals to experience **urgency** (a feeling that it is necessary to void) and **frequency** (frequent voiding of small amounts of urine). **Nocturia** (nok-tu′re-ah), the need to get up during the night to urinate, plagues almost two-thirds of this population. In many, incontinence is the final outcome of the aging process. Urinary retention is another common problem; most often it is a

# Homeostatic Relationships between the **Urinary System** and Other Body Systems

## Nervous System
- Kidneys dispose of nitrogenous wastes; maintain fluid, electrolyte, and acid-base balance of blood; renal regulation of $Na^+$, $K^+$, and $Ca^{2+}$ content in ECF is essential for normal neural function
- Neural controls are involved in micturition; sympathetic nervous system activity triggers the renin-angiotensin mechanism

## Endocrine System
- Kidneys dispose of nitrogenous wastes; maintain fluid, electrolyte, and acid-base balance of blood; produce the hormone erythropoietin; renal regulation of $Na^+$ and water balance is essential for blood pressure homeostasis
- ADH, aldosterone, ANP, and other hormones help regulate renal reabsorption of water and electrolytes

## Respiratory System
- Kidneys dispose of nitrogenous wastes; maintain fluid, electrolyte, and acid-base balance of blood
- Respiratory system provides oxygen required by kidney; disposes of carbon dioxide; cells in the lungs convert angiotensin I to angiotensin II

## Lymphatic System/Immunity
- Kidneys dispose of nitrogenous wastes; maintain fluid, electrolyte, and acid-base balance of blood
- By returning leaked plasma fluid to cardiovascular system, lymphatic vessels help maintain normal systemic blood pressure needed for kidney function; immune cells protect urinary organs from infection, cancer, and other foreign substances

## Cardiovascular System
- Kidneys dispose of nitrogenous wastes; maintain fluid, electrolyte, and acid-base balance of blood; renal regulation of $Na^+$ and water balance is essential for blood pressure homeostasis. $Na^+$, $K^+$, and $Ca^{2+}$ regulation help maintain normal heart function
- Systemic arterial blood pressure is the driving force for glomerular filtration; heart secretes atrial natriuretic peptide (ANP); blood vessels transport nutrients, oxygen, etc. to urinary organs

## Digestive System
- Kidneys dispose of nitrogenous wastes; maintain fluid, electrolyte, and acid-base balance of blood; also, metabolize vitamin D to the active form needed for calcium absorption
- Digestive organs provide nutrients needed for kidney health; liver synthesizes most urea, a nitrogenous waste that must be excreted by the kidneys

## Reproductive System
- Kidneys dispose of nitrogenous wastes; maintain fluid, electrolyte, and acid-base balance of blood

# Urinary System

## Integumentary System
- Kidneys dispose of nitrogenous wastes; maintain fluid, electrolyte, and acid-base balance of blood
- Skin provides external protective barrier; serves as site for vitamin D synthesis and water loss (via perspiration)

## Muscular System
- Kidneys dispose of nitrogenous wastes; maintain fluid, electrolyte, and acid-base balance of blood; renal regulation of $Na^+$, $K^+$, and $Ca^{2+}$ content in ECF is crucial for muscle activity
- Muscles of pelvic diaphragm and external urethral sphincter function in voluntary control of micturition; muscle metabolism generates creatinine, a nitrogenous waste product that must be excreted by the kidneys

## Skeletal System
- Kidneys dispose of nitrogenous wastes; maintain fluid, electrolyte, and acid-base balance of blood
- Bones of rib cage provide some protection to kidneys

result of hypertrophy of the prostate gland in males. Some of the problems of incontinence and retention can be avoided by a regular regimen of activity that keeps the body as a whole in optimum condition and promotes alertness to elimination signals.

## Did You Get It?

20. What is hypospadias?
21. What is urgency as related to the urinary system? What is frequency?
22. What is the difference between polycystic kidney disease and glomerulonephritis?

For answers, see Appendix A.

## Summary

### Kidneys (pp. 538–546)

1. The paired kidneys are retroperitoneal in the superior lumbar region. Each kidney has a medial indentation (hilum), where the renal artery, renal vein, and ureter attach. Each kidney is enclosed in a tough fibrous capsule. A fatty cushion holds the kidneys against the body trunk wall.

2. A longitudinal section of a kidney reveals an outer cortex, deeper medulla, and medial pelvis. Extensions of the pelvis (calyces) surround the tips of medullary pyramids and collect urine draining from them.

3. The renal artery, which enters the kidney, breaks up into segmental and then interlobar arteries that travel outward through the medulla. Interlobar arteries split into arcuate arteries, which branch to produce cortical radiate arteries, which serve the cortex.

4. Nephrons are structural and functional units of the kidneys. Each consists of a glomerulus and a renal tubule. Subdivisions of the renal tubule (from the glomerulus onward) are glomerular capsule, proximal convoluted tubule, nephron loop, and distal convoluted tubule. A second (peritubular) capillary bed is also associated with each nephron.

5. Nephron functions include filtration, reabsorption, and secretion. Filtrate formation is the role of the high-pressure glomerulus. Filtrate is essentially plasma without blood proteins. In reabsorption, done by tubule cells, needed substances are removed from filtrate (amino acids, glucose, water, some ions) and returned to blood. The tubule cells also secrete additional substances into filtrate. Secretion is important to rid the body of drugs and excess ions and to maintain acid-base balance of blood.

6. Urine is clear, yellow, and usually slightly acidic, but its pH value varies widely. Substances normally found in urine are nitrogenous wastes, water, various ions (always including sodium and potassium). Substances normally absent from urine include glucose, blood proteins, blood, pus (WBCs), and bile.

### Ureters, Urinary Bladder, and Urethra (pp. 546–550)

1. The ureters are slender tubes that conduct urine by peristalsis from kidney to bladder.

2. The bladder is a muscular sac posterior to the pubic symphysis. It has two inlets (ureters) and one outlet (urethra). In males, the prostate gland surrounds its outlet. The function of the bladder is to temporarily store urine.

3. The urethra carries urine from the bladder to the body exterior. In females, it is 3 to 4 cm long and conducts only urine. In males, it is 20 cm long and conducts both urine and sperm. The internal sphincter of smooth muscle is at the bladder-urethra junction. The external sphincter of skeletal muscle is located more inferiorly.

4. Micturition is emptying the bladder. The micturition reflex causes the involuntary internal sphincter to open when stretch receptors in the bladder wall are stimulated. The external sphincter is voluntarily controlled, so micturition can ordinarily be temporarily delayed. Incontinence is the inability to control micturition.

### Fluid, Electrolyte, and Acid-Base Balance (pp. 550–557)

1. Blood composition depends on diet, cellular metabolism, and urinary output. To maintain blood composition, the kidneys must do the following:

   a. Allow nitrogen-containing wastes (urea, ammonia, creatinine, uric acid) to leave the body in the urine.

   b. Maintain water and electrolyte balance by absorbing more or less water and reclaiming ions in

**15**

response to hormonal signals. ADH increases water reabsorption and conserves body water. Aldosterone increases reabsorption of sodium and water and decreases potassium reabsorption.

c. Maintain acid-base balance by actively secreting bicarbonate ions (and retaining hydrogen ions [$H^+$]) or by absorbing bicarbonate ions (and secreting $H^+$). Chemical buffers tie up excess $H^+$ or bases temporarily; respiratory centers modify blood pH by retaining $CO_2$ (decreases pH) or by eliminating more $CO_2$ from the blood (increases blood pH). Only the kidney can remove metabolic acids and excess bases from the body.

## Developmental Aspects of the Urinary System (pp. 557–561)

1. The kidneys begin to develop in the first few weeks of embryonic life and are excreting urine by the third month.

2. Common congenital abnormalities include polycystic kidney and hypospadias.

3. Common urinary system problems in children and young to middle-aged adults are infections caused by fecal microorganisms, microorganisms causing sexually transmitted infections, and *Streptococcus*.

4. Renal failure is an uncommon, but extremely serious, problem in which kidneys are unable to concentrate urine. Dialysis must be done to maintain chemical homeostasis of blood.

5. With age, filtration rate decreases, and tubule cells become less efficient at concentrating urine, leading to urgency, frequency, and incontinence. In men, urinary retention is another common problem.

---

# Review Questions

Access additional practice questions using your smartphone, tablet, or computer:
Mastering A&P®> Study Area > Practice Tests & Quizzes

## Multiple Choice

*More than one choice may apply.*

1. Microscopic examination of a section of the kidney shows a thick-walled vessel with glomeruli scattered in the tissue on one side of the vessel but not on the other side. What vessel is this?

   a. Segmental vein

   b. Cortical radiate artery

   c. Interlobar vein

   d. Arcuate artery

2. What is the glomerulus?

   a. The same as the renal tubule

   b. The same as Bowman's capsule

   c. The same as the nephron

   d. Capillaries

3. Which of the following substances should not be present in normal urine?

   a. Red blood cells     c. Creatinine

   b. Uric acid     d. Glucose

4. Effects of aldosterone include

   a. increase in sodium ion excretion.

   b. increase in water retention.

   c. increase in potassium ion concentration in the urine.

   d. higher blood pressure.

5. Which of the following is dependent on tubular secretion?

   a. Clearing penicillin from the blood

   b. Removal of nitrogenous wastes that have been reabsorbed

   c. Removal of excess potassium ions from the blood

   d. Control of blood pH

6. What is the length of the urethra in women?

   a. 25–30 cm     c. 3–4 cm

   b. 18–20 cm     d. 10–15 cm

7. The smallest fluid compartment is the

   a. intracellular fluid.     c. plasma.

   b. extracellular fluid.     d. interstitial fluid.

8. In the carbonic acid–bicarbonate buffer system, strong acids are buffered by

   a. carbonic acid.

   b. water.

   c. bicarbonate ion.

   d. the salt of the strong acid.

## Short Answer Essay

9. Name the organs of the urinary system, and describe the general function of each organ.

10. Describe the location of the kidneys in the body.

11. Make a diagram of a nephron. Identify and label the glomerulus, proximal convoluted tubule, distal convoluted tubule, and nephron loop.

12. About 150 L of plasma filters into the renal tubules each day, but only 1–2 L of urine are formed. What happens to the rest of the plasma fluid?

13. What is the function of the collecting ducts? What do they contain?

14. What is the function of the glomerulus?

15. Where do the peritubular capillaries arise from, and how do they differ from the glomerular capillaries?

16. Which part of the glomerulus is under the highest pressure? What is the significance of this?

17. How does aldosterone modify the chemical composition of urine?

18. What is the juxtaglomerular apparatus, and what is its role in the regulation of blood pressure?

19. Through which parts of the nephron is NaCl transported out? Is this active transport or passive transport?

20. In which condition presence of proteins can be observed in urine? Is this condition pathological or nonpathological? Explain.

21. Describe the role of the respiratory system in controlling acid-base balance.

22. Why is urinalysis a routine part of any good physical examination?

23. How do the internal and external urethral sphincters differ structurally and functionally?

24. Define *micturition*, and describe the micturition reflex.

25. What sometimes happens when urine becomes too concentrated or remains too long in the bladder?

26. Define *urinary retention*.

27. Contrast the following homeostatic imbalances: oliguria, anuria, polyuria, and nocturia.

28. Why is the trigone of the urinary bladder important clinically?

29. What type of problem most commonly affects the urinary system organs?

30. Describe the changes that occur in kidney and bladder function in old age.

## Critical Thinking and Clinical Application Questions

31. Jessica has been experiencing a raging and insatiable thirst for some months. The unfortunate consequence of her increased water intake is a constant need to urinate. She is finding it difficult to get enough sleep and this is now interfering with her work. Explain her symptoms.

32. Mitchell's parents bring him to the clinic because his urine is tinged with blood. Two days before, he was diagnosed with strep throat. His face and hands are swollen. What is the probable cause of Mitchell's current kidney problem?

33. Lucy is worried about the variation in the color of her urine. Sometimes her urine is pale yellow and sometimes it is very deep yellow. Is there really a cause for concern? Explain.

34. What happens to the rate of RBC production in a patient on dialysis with total renal failure? What could be given to the patient to counteract such a problem? (Hint: You might want to check Table 9.2 on p. 356 for help with this one.)

35. Alexander is 72 and is in good health. However, he is recently having difficulties emptying his bladder. What is the likely cause?

36. Mr. Jessup, a 55-year-old man, is operated on for a cerebral tumor. About one month later, he complains that he is excessively thirsty and that he has been voiding almost continuously. A urine sample is collected, and its specific gravity is 1.001. What is your diagnosis of Mr. Jessup's problem, and how might it be related to his previous surgery?

37. Raymond is hypertensive and was recently diagnosed with impaired kidney function based on urinalysis and a blood test for creatinine. What sorts of test results would you expect, and how is hypertension related to kidney function?

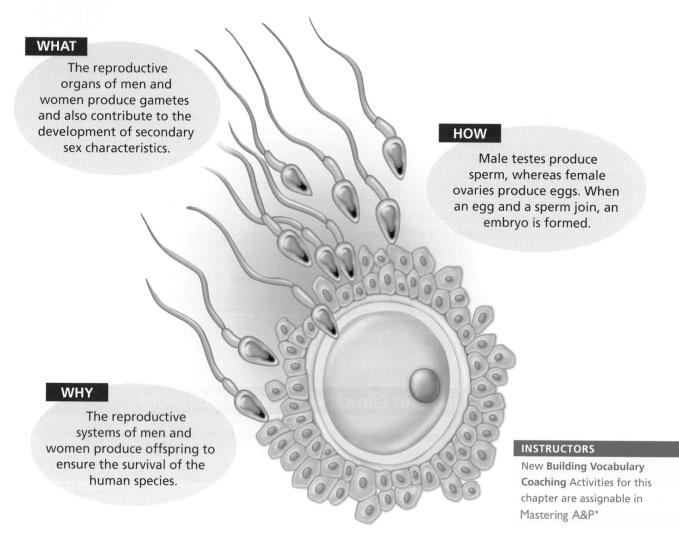

**WHAT**

The reproductive organs of men and women produce gametes and also contribute to the development of secondary sex characteristics.

**HOW**

Male testes produce sperm, whereas female ovaries produce eggs. When an egg and a sperm join, an embryo is formed.

**WHY**

The reproductive systems of men and women produce offspring to ensure the survival of the human species.

**INSTRUCTORS**

New **Building Vocabulary Coaching** Activities for this chapter are assignable in Mastering A&P®

→ **Learning Objective**

☐ **Discuss the common purpose of the reproductive system organs.**

Most organ systems of the body function almost continuously to maintain the well-being of the individual. The reproductive system, however, appears to "slumber" until *puberty*, when reproductive organs become functional. The **gonads** (go'nadz; "seeds"), or **primary sex organs**, are the *testes* in men and the *ovaries* in women. The gonads produce sex cells, or **gametes** (gam'ēts), and secrete sex hormones. The remaining reproductive system structures are **accessory reproductive organs**.

Although male and female **reproductive systems** are quite different, their joint purpose is to produce offspring. The reproductive role of the man is to manufacture male gametes called **sperm** and deliver them to the woman's reproductive tract. The woman, in turn, produces female gametes, called **ova** (singular *ovum*), or *eggs*. If the conditions are suitable, a sperm and egg fuse to produce a fertilized egg, called a *zygote*. Once fertilization has occurred, the female uterus provides a protective environment in which the *embryo*, later called the *fetus*, develops until birth.

The sex hormones play vital roles both in the development and function of the reproductive organs, and in sexual behavior and drives. These hormones also influence the growth and development of many other organs and tissues of the body.

# Anatomy of the Male Reproductive System

→ **Learning Objectives**

☐ **When provided with a model or diagram, identify the organs of the male reproductive system, and discuss the general function of each.**

☐ **Name the endocrine and exocrine products of the testes.**

☐ **Discuss the composition of semen, and name the glands that produce it.**

☐ **Trace the pathway followed by a sperm from the testis to the body exterior.**

☐ **Define** *circumcision*, *erection*, **and** *ejaculation*.

As already noted, the primary reproductive organs of the male are the paired **testes** (tes′tēz), which have both an exocrine (sperm-producing) function and an endocrine (testosterone-producing) function. The accessory reproductive structures are ducts or glands that aid in the delivery of sperm to the body exterior or to the female reproductive tract.

## Testes

Each golf ball–sized testis is approximately 4 cm (1.5 inches) long and 2.5 cm (1 inch) wide and is connected to the trunk via the **spermatic cord**, a connective tissue sheath that encloses blood vessels, nerves, and the *ductus deferens*. A fibrous connective tissue capsule, the *tunica albuginea* (tu′nĭ-kah al″bu-jin′e-ah; "white coat") surrounds each testis. Extensions of this capsule (*septa*) plunge into the testis and divide it into a large number of wedge-shaped *lobules* **(Figure 16.1)**. Each lobule contains one to four tightly coiled **seminiferous** (sem′ĭ-nif′er-us) **tubules**, the actual "sperm-forming factories." Seminiferous tubules of each lobule empty sperm into another set of tubules, the **rete** (re′te) **testis**, located to one side of the testis. Sperm travel through the rete testis to enter the first part of the duct system, the *epididymis* (ep″ĭ-did′ĭ-mis), which hugs the external surface of the testis.

Lying in the soft connective tissue surrounding the seminiferous tubules are the **interstitial** (in″ter-stish′al) **cells**, functionally distinct cells that produce androgens—the most important of which

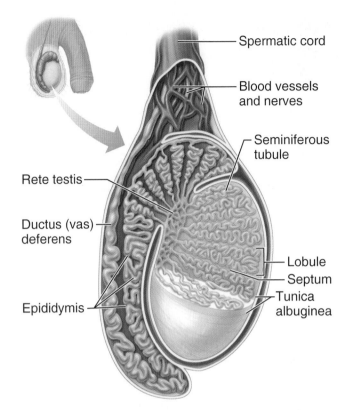

**Figure 16.1 Sagittal section of the testis and associated epididymis.**

is *testosterone*. Thus, the sperm-producing and hormone-producing functions of the testes are carried out by completely different cell populations.

## Duct System

The accessory organs forming the male duct system, which transports sperm from the body, are the *epididymis*, *ductus deferens*, and *urethra* (**Figure 16.2**, p. 566).

### Epididymis

The cup-shaped **epididymis** is a highly convoluted tube about 6 m (20 feet) long that hugs the posterior side of the testis (see Figure 16.1). The epididymis is the first part of the male duct system and provides a temporary storage site for the immature sperm. While the sperm make their way along the tortuous course of the epididymis (a trip that takes about 20 days), they mature, developing the ability to swim. When a man is sexually stimulated and **ejaculates** (*ejac* = to shoot forth), the walls of the epididymis contract to expel the sperm into the next part of the duct system, the ductus deferens.

**16**

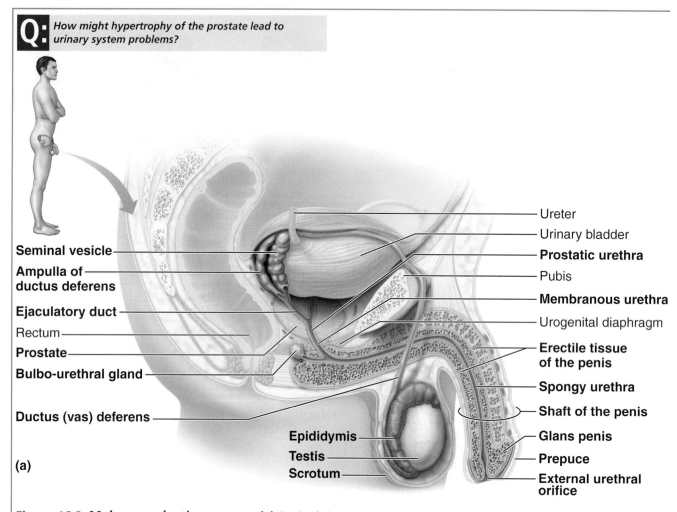

**Q:** *How might hypertrophy of the prostate lead to urinary system problems?*

Seminal vesicle

Ampulla of ductus deferens

Ejaculatory duct

Rectum

Prostate

Bulbo-urethral gland

Ductus (vas) deferens

(a)

Epididymis

Testis

Scrotum

Ureter

Urinary bladder

**Prostatic urethra**

Pubis

**Membranous urethra**

Urogenital diaphragm

**Erectile tissue of the penis**

**Spongy urethra**

**Shaft of the penis**

**Glans penis**

**Prepuce**

**External urethral orifice**

**Figure 16.2 Male reproductive organs. (a)** Sagittal view.

### Ductus Deferens

The **ductus deferens** (duk′tus def′er-enz; "carrying away"), or *vas deferens*, runs upward from the epididymis via the spermatic cord through the inguinal canal into the pelvic cavity and arches over the superior aspect of the urinary bladder. The ductus deferens then loops medially over the ureter and descends along the posterior bladder wall. The end of the ductus deferens expands as the **ampulla** and then empties into the **ejaculatory** (e-jak′u-lah-to″re) **duct**, which passes through the prostate to merge with the urethra. The main function of the ductus deferens is to propel live sperm from their storage sites—the epididymis and distal part of the ductus deferens—into the urethra. At the moment of ejaculation, the thick layers of smooth muscle in its walls create peristaltic waves that rapidly squeeze the sperm forward.

Part of the ductus deferens lies in the scrotum, a skin sac that hangs outside the body cavity and holds the testes (see Figure 16.2). Some men voluntarily opt to take full responsibility for birth control by having a *vasectomy* (vah-sek′to-me; "cutting the vas"). In this relatively minor operation, the surgeon makes a small incision into the scrotum and then cuts through and ties off the ductus deferens. Sperm are still produced, but they can no longer reach the body exterior, and eventually they deteriorate and are phagocytized. A man is sterile after this procedure, but because testosterone is

**A:** Because the prostate encircles the superior part of the urethra, prostate hypertrophy constricts the urethra in that region, impairing urination.

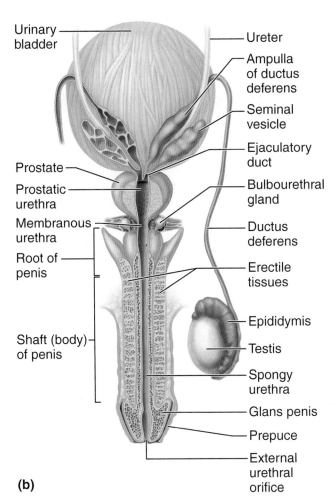

Urinary bladder

Prostate

Prostatic urethra

Membranous urethra

Root of penis

Shaft (body) of penis

Ureter

Ampulla of ductus deferens

Seminal vesicle

Ejaculatory duct

Bulbourethral gland

Ductus deferens

Erectile tissues

Epididymis

Testis

Spongy urethra

Glans penis

Prepuce

External urethral orifice

**(b)**

**Figure 16.2** *(continued)* **(b)** Frontal view; posterior aspect of penis.

still produced, he retains his sex drive and *secondary sex characteristics.*

### Urethra

The **urethra**, which extends from the base of the urinary bladder to the tip of the penis, is the terminal part of the male duct system. It has three named regions: (1) the **prostatic urethra**, surrounded by the prostate gland; (2) the **membranous urethra**, spanning the distance from the prostatic urethra to the penis; and (3) the **spongy (penile) urethra**, running within the length of the penis and opening to the body exterior via the *external urethral orifice* (see Figure 16.2). The male urethra carries both urine and sperm to the body exterior; thus, it serves two masters, the urinary and reproductive systems (recall Chapter 15). However, urine and sperm never leave the body at the same time. When ejaculation occurs and sperm enter the prostatic urethra from the ejaculatory

ducts, the bladder sphincter (internal urethral sphincter) constricts. This event not only prevents urine from passing into the urethra, but also prevents sperm from entering the urinary bladder.

### Did You Get It?

1. What are the two major functions of the testes?
2. What is the role of the seminiferous tubules?
3. Name the organs of the male duct system, in order, from the scrotum to the body exterior.

**For answers, see Appendix A.**

## Accessory Glands and Semen

The accessory glands include the paired *seminal vesicles*, the single *prostate*, and the paired *bulbourethral* (bul-bo-u-re′thral) *glands* (see Figure 16.2). These glands produce the bulk of *semen* (se′men), the sperm-containing fluid that is propelled out of the male's reproductive tract during ejaculation.

### Seminal Vesicles

The **seminal** (sem′ĭ-nul) **vesicles** are located at the base of the bladder. These large hollow glands, each 6 to 7 cm (about the shape and size of the little finger), produce about 60 percent of *seminal fluid*, the fluid portion of semen. Their thick, yellowish secretion is rich in sugar (fructose), vitamin C, prostaglandins, and other substances that nourish and activate the sperm passing through the tract. The duct of each seminal vesicle joins that of the ductus deferens on the same side to form the *ejaculatory duct* (see Figure 16.2). Thus, sperm and seminal fluid enter the urethra together during ejaculation.

### Prostate

The **prostate** is a single doughnut-shaped gland about the size of a peach pit (see Figure 16.2). It encircles the upper (prostatic) part of the urethra just inferior to the urinary bladder. Prostate fluid is milky and plays a role in activating sperm. During ejaculation, the fluid enters the urethra through several small ducts. Because the prostate is located immediately anterior to the rectum, its size and texture can be palpated (felt) by digital (finger) examination through the anterior rectal wall.

 **Homeostatic Imbalance 16.1**

The prostate has a reputation as a common cause of health problems. **Hypertrophy** of the prostate, an increase in its size independent of the body's

**16**

growth, affects nearly every older man and strangles the urethra. This troublesome condition makes urination difficult and enhances the risk of bladder infections (**cystitis**) and kidney damage. Traditional treatment has been surgery, but some newer options are becoming more popular. These include the following:

- Using drugs or microwaves to shrink the prostate

- Inserting a small inflatable balloon to compress prostate tissue away from the prostatic urethra

- Inserting a tiny needle that emits bursts of radiofrequency radiation, which incinerate excess prostate tissue

**Prostatitis**, inflammation of the prostate, is the single most common reason for a man to consult a urologist, and **prostate cancer** is the third most prevalent cancer in men. Most cases of prostate cancer are slow-growing, but it can also be a swift and deadly killer. _____**+**

### Bulbourethral Glands

The **bulbourethral glands** are tiny, pea-sized glands inferior to the prostate gland. They produce a thick, clear mucus that drains into the penile urethra. This secretion is the first to pass down the urethra when a man becomes sexually excited. It cleanses the urethra of trace acidic urine prior to ejaculation, and it serves as a lubricant during sexual intercourse.

### Semen

**Semen** is a milky white, somewhat sticky mixture of sperm and accessory gland secretions. The liquid portion acts as a transport medium for nutrients and chemicals that protect the sperm and aid their movement. Mature sperm cells are streamlined cellular "tadpoles" containing little cytoplasm or stored nutrients. The fructose in the seminal vesicle secretion provides essentially all of their energy fuel. Sperm are very sluggish under acidic conditions (below pH 6). The relative alkalinity of semen as a whole (pH 7.2–7.6) helps neutralize the acidic environment (pH 3.5–4) of the female's vagina, protecting the delicate sperm. Semen also contains antibiotic chemicals that destroy certain bacteria, the hormone relaxin, enzymes that enhance sperm motility, and substances that inhibit an immune response in the female reproductive tract.

Semen also dilutes sperm; without such dilution, sperm motility is severely impaired. The amount of semen propelled out of the male duct system during ejaculation is relatively small, only 2 to 5 ml (about a teaspoonful), but there are between 50 and 150 million sperm in each milliliter.

 **Homeostatic Imbalance 16.2**

Male infertility, the inability to conceive a child, may be due to obstructions of the duct system, hormonal imbalances, environmental estrogens, pesticides, excessive alcohol, or many other factors. One of the first series of tests done when a couple has been unable to conceive is *semen analysis*. Factors analyzed include sperm count, motility and morphology (shape and maturity), and semen volume, pH, and fructose content. A sperm count lower than 20 million per milliliter makes impregnation improbable. _____**+**

## External Genitalia

The male **external genitalia** (jen"i-tāl′e-ah) include the *scrotum* and the *penis* (see Figure 16.2). The **scrotum** (skro′tum; "pouch") is a divided sac of skin with sparse hairs that hangs outside the abdominal cavity, between the legs at the root of the penis. Under normal conditions, the scrotum hangs loosely from its attachments, providing the testes with a temperature that is below body temperature. This is a rather exposed location for a man's testes, which contain his entire genetic heritage, but apparently viable sperm cannot be produced at normal body temperature. The scrotum, which provides a temperature about 3°C (5.4°F) lower, is necessary for the production of healthy sperm. When the external temperature is very cold, the scrotum becomes heavily wrinkled as it pulls the testes closer to the warmth of the body wall. Thus, changes in scrotal surface area can maintain a temperature that favors viable sperm production.

The **penis** (pe′nis; "tail") functions to deliver sperm into the female reproductive tract. The skin-covered penis consists of a **shaft**, which ends in an enlarged tip, the **glans penis**. The skin covering the penis is loose, and it folds downward to form a sleeve of skin, the **prepuce** (pre′pūs), or **foreskin**, around the proximal end of the glans penis. Frequently in the United States, the foreskin is surgically removed shortly after birth, in a procedure called **circumcision**.

Internally, the spongy urethra (see Figure 16.2) is surrounded by three elongated areas of **erectile tissue**, a spongy tissue that fills with blood during sexual excitement. This causes the penis to enlarge and become rigid. This event, called **erection**, helps the penis serve as the male organ of *copulation* (sexual intercourse) to deliver the semen into the female reproductive tract.

### Did You Get It?

4. What is the function of the erectile tissue of the penis?
5. What is an important function of each of these two components of semen—sperm and seminal fluid?
6. Adolph, a 68-year-old gentleman, has trouble urinating and is given a rectal exam. What is his most probable condition, and what is the purpose of the rectal exam?

**For answers, see Appendix A.)**

# Male Reproductive Functions

### → Learning Objectives

☐ **Define *spermatogenesis* and *meiosis*.**

☐ **Describe the structure of a sperm, and relate its structure to its function.**

☐ **Describe the effect of FSH and LH on testis functioning.**

The chief role of the male in the reproductive process is to produce sperm and the hormone testosterone.

## Spermatogenesis

Sperm production, or **spermatogenesis** (sper″mah-to-jen′ĕ-sis), begins during puberty and continues throughout life. A man makes millions of sperm daily. Only one sperm fertilizes an egg, so it seems that nature has made sure that the human species will not be endangered for lack of sperm.

Sperm are formed in the seminiferous tubules of the testis, as noted earlier. The process is begun by primitive stem cells called **spermatogonia** (sper″mah-to-go′ne-ah), found in the outer edge, or periphery, of each tubule **(Figure 16.3)**. Spermatogonia go through rapid mitotic divisions to build up the stem cell line. From birth until puberty, all such divisions simply produce more stem cells. During puberty, however, *follicle-stimulating hormone (FSH)* is secreted in increasing amounts by the anterior pituitary gland.

### CONCEPT**LINK**

Recall that FSH is a tropic hormone that, in males, targets the testes and stimulates sperm production (Chapter 9, p. 343).

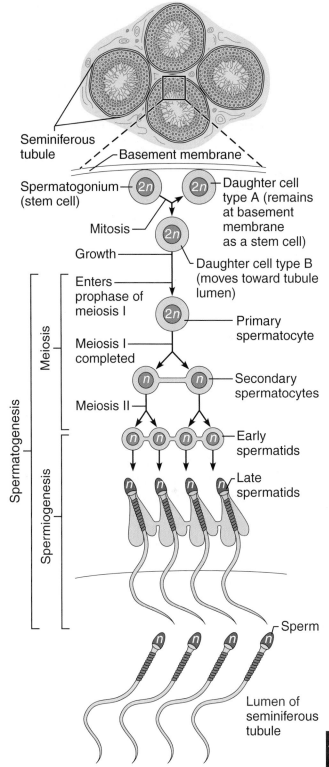

**Figure 16.3 Spermatogenesis.** Flowchart showing the relative position of the spermatogonia in the wall of the seminiferous tubule. Although the stem cells and the primary spermatocytes have the same number of chromosomes (46, designated as $2n$) as other body cells, the products of meiosis (spermatids and sperm) have only half as many (23, designated as $n$).

16

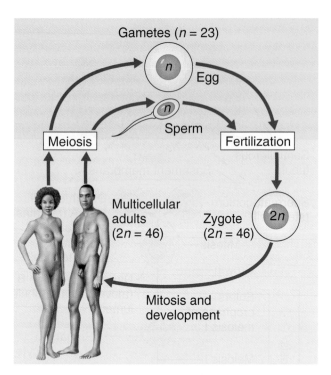

**Figure 16.4 The human life cycle.**

From puberty on, each division of a spermatogonium produces one stem cell, called a *type A daughter cell*, and another cell, called a *type B daughter cell*. The type A cell remains at the tubule periphery to maintain the stem cell population. The type B cell gets pushed toward the tubule lumen, where it becomes a **primary spermatocyte** destined to undergo *meiosis* (mi-o′sis) and form four sperm.

**Meiosis** is a special type of nuclear division that occurs for the most part only in the gonads (testes and ovaries). It differs from *mitosis* (described in Chapter 3) in two major ways. Meiosis consists of two successive divisions of the nucleus (called meiosis I and II), instead of only one division. It results in four (instead of two) daughter cells, or gametes.

In spermatogenesis, the gametes are called **spermatids** (sper′mah-tidz). Spermatids have only half as much genetic material as other body cells. In humans, this is 23 chromosomes (or the haploid number [$n$] of chromosomes) rather than the usual 46 (diploid, or $2n$). Then, when the sperm and the egg (which also has 23 chromosomes) unite, forming the fertilized egg, or zygote, the normal $2n$ number of 46 chromosomes is reestablished and is maintained in subsequent body cells by the process of mitosis **(Figure 16.4)**.

As meiosis occurs, the dividing cells (primary and then secondary spermatocytes) are pushed toward the lumen of the tubule. Thus, we can follow the progress of meiosis from the tubule periphery to the lumen. The spermatids, produced by meiosis, are *not* functional sperm. They are nonmotile and have too much excess baggage to function well in reproduction. They must undergo further changes, in which their excess cytoplasm is stripped away and a tail is formed (see Figure 16.3). In this last stage of sperm development, called **spermiogenesis** (sper″me-o-gen′ĕ-sis), all the excess cytoplasm is sloughed off, and what remains is compacted into the three regions of the mature sperm—the *head, midpiece,* and *tail* **(Figure 16.5)**. The mature sperm is a greatly streamlined cell equipped with a high rate of metabolism and a means of propelling itself, enabling it to move long distances in a short time to get to the egg. It is a prime example of the relationship between structure and function. (Remember, we talked about this in Chapter 1.)

Sperm "pack" light. The sperm head is the nucleus and contains compacted DNA, the genetic material. Anterior to the nucleus is the helmetlike **acrosome** (ak′ro-sōm), which is produced by the Golgi apparatus and is similar to a large lysosome. When a sperm comes into close contact with an egg (or more precisely, an *oocyte*), the acrosomal membrane breaks down and releases enzymes that help the sperm penetrate the capsule of follicle cells that surround the egg. *Filaments*, which form the long tail, arise from centrioles in the midpiece. Mitochondria wrapped tightly around these filaments provide the ATP needed for the whiplike movements of the tail that propel the sperm along the female reproductive tract.

The entire process from the formation of a primary spermatocyte to release of immature sperm in the tubule lumen takes 64 to 72 days. But sperm in the lumen are still unable to "swim" and so are still incapable of fertilizing an egg. They are moved by peristalsis through the tubules of the testes into the epididymis. There they undergo further maturation, which results in increased motility and fertilizing power.

## Homeostatic Imbalance 16.3

Environmental threats can alter the normal process of sperm formation. For example, some common antibiotics, such as penicillin and tetracycline, may suppress sperm formation. Radiation, lead, certain

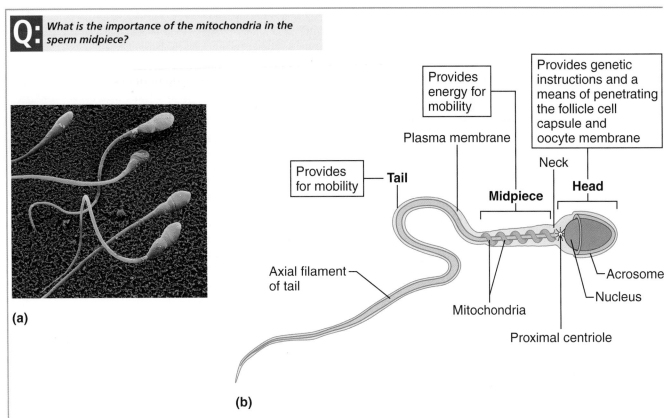

**Q:** *What is the importance of the mitochondria in the sperm midpiece?*

Provides energy for mobility

Provides genetic instructions and a means of penetrating the follicle cell capsule and oocyte membrane

Plasma membrane

Provides for mobility — **Tail**

Neck

**Midpiece**    **Head**

Axial filament of tail

Mitochondria

Acrosome

Nucleus

Proximal centriole

**(a)**

**(b)**

**Figure 16.5 Structure of sperm. (a)** Scanning electron micrograph of mature sperm (1,525×). **(b)** Diagrammatic view of a sperm.

pesticides, marijuana, tobacco, and excessive alcohol can cause production of abnormal sperm (two-headed, multiple-tailed, and so on).

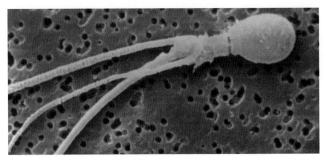

An abnormal, multi-tailed human sperm from a 66-year-old man (colorized).

## Testosterone Production

As noted earlier, the interstitial cells produce **testosterone** (tes-tos′tĕ-rōn), the most important hormonal product of the testes. During puberty, as the seminiferous tubules are being prodded by FSH to produce sperm, the interstitial cells are being activated by **luteinizing hormone (LH)**, which is also released by the anterior pituitary gland (**Figure 16.6**, p. 572). From this time on, testosterone is produced continuously (more or less) for the rest of a man's life. The rising blood level of testosterone in the young man stimulates the adolescent growth spurt, prompts his reproductive organs to develop to their adult size, underlies the sex drive, and causes the male secondary sex characteristics to appear. **Secondary sex characteristics** are features induced in nonreproductive organs by sex hormones. Male secondary sex characteristics include the following:

- Deepening of the voice as the larynx enlarges
- Increased hair growth all over the body, particularly in the axillary and pubic regions and on the face
- Enlargement of skeletal muscles to produce the heavier muscle mass typical of the male physique

**A:** They provide the ATP energy the tail needs to make the sperm motile.

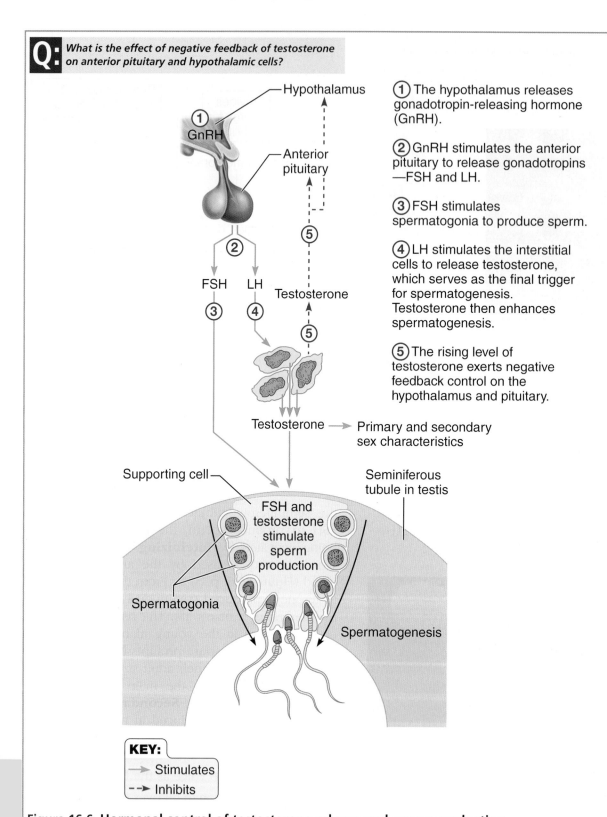

Hypothalamus

① GnRH

Anterior pituitary

②

FSH    LH

③      ④

Testosterone

⑤

Testosterone → Primary and secondary sex characteristics

Supporting cell

FSH and testosterone stimulate sperm production

Seminiferous tubule in testis

Spermatogonia

Spermatogenesis

**KEY:**
→ Stimulates
--→ Inhibits

① The hypothalamus releases gonadotropin-releasing hormone (GnRH).

② GnRH stimulates the anterior pituitary to release gonadotropins —FSH and LH.

③ FSH stimulates spermatogonia to produce sperm.

④ LH stimulates the interstitial cells to release testosterone, which serves as the final trigger for spermatogenesis. Testosterone then enhances spermatogenesis.

⑤ The rising level of testosterone exerts negative feedback control on the hypothalamus and pituitary.

**Figure 16.6 Hormonal control of testosterone release and sperm production.**

- Increased heaviness of the skeleton due to bone growth in both size and density

Because testosterone is responsible for the appearance of these typical masculine characteristics, it is often referred to as the "masculinizing" hormone.

### ✎ Homeostatic Imbalance 16.4

If testosterone is not produced in a young man, his secondary sex characteristics never appear, and his other reproductive organs remain child-like. This condition is called **sexual infantilism**. Castration of the adult male (or the inability of his interstitial cells to produce testosterone) results in a decrease in the size and function of his reproductive organs, as well as a decrease in his sex drive. **Sterility** also occurs because testosterone is necessary for the final stages of sperm production. _____+

### Did You Get It?

7. Which pituitary hormone stimulates spermatogenesis?
8. How does the final product of meiosis differ from the final product of mitosis?
9. How are nonmotile spermatids converted to functional sperm?
10. Which pituitary hormone prompts testosterone production?

**For answers, see Appendix A.**

# Anatomy of the Female Reproductive System

### → Learning Objectives

☐ Identify the organs of the female reproductive system, and discuss the general function of each.

☐ Describe the functions of the vesicular follicle and corpus luteum of the ovary.

☐ Define *endometrium, myometrium,* and *ovulation*.

☐ Indicate the location of the following regions of the female uterus: cervix, fundus, body.

The reproductive role of the female is much more complex than that of the male. Not only must she produce the female gametes (ova), but her body must also nurture and protect a developing fetus during 9 months of pregnancy. **Ovaries** (o'vah-rēz) are the primary female reproductive organs.

Like the testes, ovaries produce both an exocrine product (eggs, or *ova*) and endocrine products (estrogens and progesterone). The other organs of the female reproductive system serve as accessory structures to transport, nurture, or otherwise serve the needs of the reproductive cells and/or the developing fetus.

## Ovaries

The paired ovaries are pretty much the shape of almonds but are nearly twice as large. An internal view of an ovary reveals many tiny saclike structures called **ovarian follicles** (**Figure 16.7**, p. 574). Each follicle consists of an immature egg, called an **oocyte** (o'o-sīt; *oo* = egg), surrounded by one or more layers of very different cells called **follicle cells**. As a developing egg within a follicle begins to ripen or mature, the follicle enlarges and develops a fluid-filled central region called an *antrum*. At this stage, the follicle, called a **vesicular follicle** or **Graafian** (graf'e-an) **follicle**, is mature, and the developing egg is ready to be ejected from the ovary, an event called **ovulation**. After ovulation, the ruptured follicle is transformed into a very different-looking structure called a **corpus luteum** (kor'pus lu'te-um; "yellow body"), which eventually degenerates. Ovulation generally occurs every 28 days, but it can occur more or less frequently in some women.

The ovaries flank the uterus laterally (**Figure 16.8**, p. 575). These organs are secured to the lateral walls of the pelvis by the **suspensory ligaments** and are anchored to the uterus medially by the **ovarian ligaments**. In between, they are enclosed and held in place by a fold of peritoneum, the **broad ligament**.

## Duct System

The *uterine tubes, uterus,* and *vagina* form the duct system of the female reproductive tract (see Figure 16.8).

### Uterine Tubes

The **uterine** (u'ter-in) **tubes**, or **fallopian** (fal-lo'pe-an) **tubes**, form the initial part of the duct system. They receive the ovulated oocyte and provide a site where fertilization can occur. Each of the uterine tubes is about 10 cm (4 inches) long and extends medially from an ovary to empty into the superior region of the uterus. Like the ovaries,

**16**

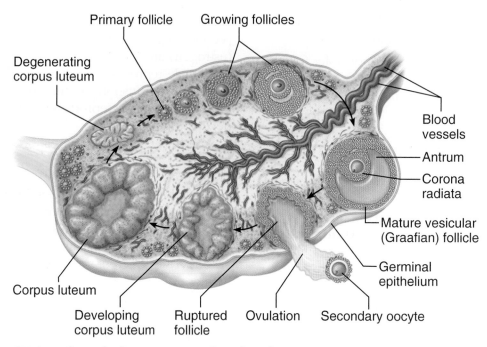

Figure 16.7 **Sagittal section of a human ovary showing the developmental stages of an ovarian follicle.**

the uterine tubes are enclosed and supported by the broad ligament.

Unlike in the male duct system, which is continuous with the tubule system of the testes, there is little or no actual contact between the uterine tubes and the ovaries. The distal end of each uterine tube expands as the funnel-shaped *infundibulum*, which has fingerlike projections called **fimbriae** (fim'bre-e) that partially surround the ovary (see Figure 16.8b). As an oocyte is expelled from an ovary during ovulation, the waving fimbriae create fluid currents that carry the oocyte into the uterine tube, where it begins its journey toward the uterus. (Many potential eggs, however, are lost in the peritoneal cavity.) The oocyte is carried toward the uterus by a combination of peristalsis and the rhythmic beating of *cilia*.

Because the journey to the uterus takes 3 to 4 days and the oocyte is viable for at most 24 hours after ovulation, the usual site of fertilization is the uterine tube. To reach the oocyte, the sperm must swim upward through the vagina and uterus to reach the uterine tubes. Because they must swim against the downward current created by the cilia, it is rather like swimming against the tide!

 **Homeostatic Imbalance 16.5**

The fact that the uterine tubes are not continuous distally with the ovaries places women at risk for infections spreading from the reproductive tract into the peritoneal cavity. **Gonorrhea** (gon"o-re'ah) and other sexually transmitted bacteria sometimes infect the peritoneal cavity in this way, causing a severe inflammation called **pelvic inflammatory disease (PID)**. Unless treated promptly, PID can cause scarring and closure of the narrow uterine tubes, which is one of the major causes of female infertility. _____✦

### Uterus

The **uterus** (u'ter-us; "womb"), located in the pelvis between the urinary bladder and rectum, is a hollow organ that functions to receive, retain, and nourish a fertilized egg. In a woman who has never been pregnant, it is about the size and shape of a pear. During pregnancy, the uterus increases tremendously in size and during the latter part of pregnancy can be felt well above the umbilicus. The uterus is suspended in the pelvis by the broad ligament and anchored anteriorly and posteriorly by the **round ligaments** and **uterosacral ligaments**, respectively (see Figure 16.8).

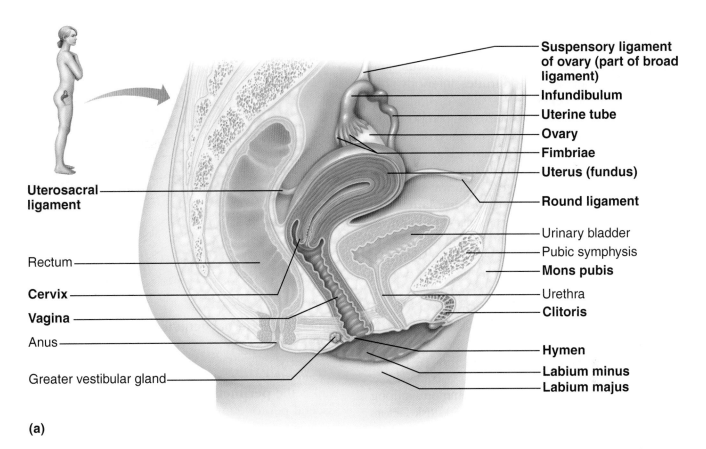

**(a)**

Uterosacral ligament

Rectum

**Cervix**

**Vagina**

Anus

Greater vestibular gland

**Suspensory ligament of ovary (part of broad ligament)**

**Infundibulum**

**Uterine tube**

**Ovary**

**Fimbriae**

**Uterus (fundus)**

**Round ligament**

Urinary bladder

Pubic symphysis

**Mons pubis**

Urethra

**Clitoris**

**Hymen**

**Labium minus**

**Labium majus**

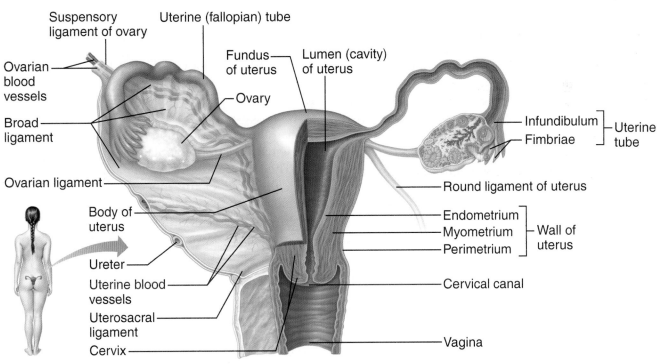

Suspensory ligament of ovary

Uterine (fallopian) tube

Ovarian blood vessels

Broad ligament

Ovarian ligament

Body of uterus

Ureter

Uterine blood vessels

Uterosacral ligament

Cervix

Fundus of uterus

Lumen (cavity) of uterus

Ovary

Infundibulum

Fimbriae

Uterine tube

Round ligament of uterus

Endometrium

Myometrium

Perimetrium

Wall of uterus

Cervical canal

Vagina

**(b)**

16

**Figure 16.8 The human female reproductive organs. (a)** Sagittal section.
(The plurals of *labium minus* and *majus* are *labia minora* and *majora*, respectively.)
**(b)** Posterior view. The posterior organ walls have been removed on the right side to
reveal the shape of the lumen of the uterine tube, uterus, and vagina.

The major portion of the uterus is referred to as the **body**. Its rounded region superior to the entrance of the uterine tubes is the **fundus**, and its narrow outlet, which protrudes inferiorly into the vagina, is the **cervix**.

The thick wall of the uterus is composed of three layers. The inner layer, or mucosa, is the **endometrium** (en-do-me′tre-um). If fertilization occurs, the fertilized egg (actually the young embryo by the time it reaches the uterus) burrows into the endometrium (in a process called **implantation**) and resides there for the rest of its development. When a woman is not pregnant, the endometrial lining sloughs off periodically, usually about every 28 days, in response to changes in the levels of ovarian hormones in the blood. This process, called *menstruation* or *menses*, is often referred to as a woman getting her "period" (discussed on p. 580).

## Homeostatic Imbalance 16.6

**Cervical cancer** is relatively common among women between the ages of 30 and 50. Risk factors include sexually transmitted diseases such as human papillomavirus (HPV), frequent cervical inflammation, multiple pregnancies, and many sexual partners. Although HPV infection is a major cause of cervical cancer, other factors may also be involved. A yearly *Pap smear* is the single most important diagnostic test for detecting this slow-growing cancer. When results are inconclusive, a test for HPV can be done from the same Pap sample or from a blood sample.

*Gardasil*, a three-dose vaccine that provides protection from the most common types of HPV-induced cervical cancer, is a recent addition to the official childhood immunization schedule. It is recommended for all 11- and 12-year-old girls but can also be given to boys and girls ages 13–26. In unexposed girls, the vaccine specifically blocks two cancer-causing kinds of HPV as well as two additional types that are not associated with cervical cancer. Whether or not this vaccine will become a requirement for school is currently decided on a state-to-state basis. _____✛

The bulky middle layer of the uterus wall is the **myometrium** (mi-o-me′tre-um), which is composed of interlacing bundles of smooth muscle (see Figure 16.8b). The myometrium plays an active role during childbirth, when it contracts rhythmically to force the baby out of the mother's body. The outermost serous layer of the uterus

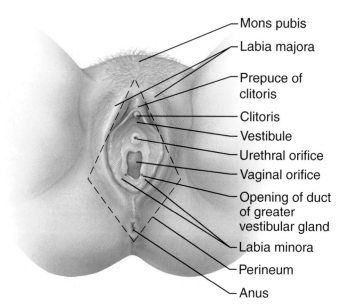

**Figure 16.9 External genitalia of the human female.**

Mons pubis
Labia majora
Prepuce of clitoris
Clitoris
Vestibule
Urethral orifice
Vaginal orifice
Opening of duct of greater vestibular gland
Labia minora
Perineum
Anus

wall is the **perimetrium** (per-ĭ-me′tre-um), or the visceral peritoneum.

### Vagina

The **vagina** (vah-ji′nah) is a thin-walled tube 8 to 10 cm (3 to 4 inches) long. It lies between the bladder and rectum and extends from the cervix to the body exterior (see Figure 16.8). Often called the *birth canal*, the vagina provides a passageway for the delivery of an infant and for the menstrual flow to leave the body. Because it receives the penis (and semen) during sexual intercourse, it is the female organ of copulation.

The distal end of the vagina is partially closed by a thin fold of the mucosa called the **hymen** (hi′men). The hymen is very vascular and tends to bleed when it is ruptured during the first sexual intercourse. However, its durability varies. In some women, it is torn during a sports activity, tampon insertion, or pelvic examination. Occasionally, it is so tough that it must be ruptured surgically if intercourse is to occur.

## External Genitalia and Female Perineum

The female reproductive structures that are located external to the vagina are the **external genitalia (Figure 16.9)**. The external genitalia, also called the **vulva**, include the *mons pubis, labia, clitoris, urethral* and *vaginal orifices*, and *greater vestibular glands*.

The **mons pubis** ("mountain on the pubis") is a fatty, rounded area overlying the pubic symphysis. After puberty, this area is covered with pubic hair. Running posteriorly from the mons pubis are two elongated hair-covered skin folds, the **labia majora** (la′be-ah ma-jo′ra), which enclose two delicate, hair-free folds, the **labia minora**. The labia majora enclose a region called the **vestibule**, which contains the external openings of the urethra,* followed posteriorly by that of the vagina.

Just anterior to the vestibule is the **clitoris** (kli′to-ris; "hill"), a small, protruding structure that corresponds to the male penis. Like the penis, it is hooded by a prepuce and is composed of sensitive erectile tissue that becomes swollen with blood during sexual excitement. The clitoris differs from the penis in that it lacks a reproductive duct.

A pair of mucus-producing glands, the **greater vestibular glands**, flank the vagina, one on each side (see Figure 16.8a). Their secretion lubricates the distal end of the vagina during intercourse. The diamond-shaped region between the anterior end of the labial folds, the anus posteriorly, and the ischial tuberosities laterally is the **perineum** (per″i-ne′um).

### Did You Get It?

11. What is the exocrine product of the ovary?
12. Which organ of the female duct system serves as an "incubator" for fetal development? What is the most common site of fertilization?
13. What name is given to an ovarian follicle that is ready to ovulate?

For answers, see Appendix A.

# Female Reproductive Functions and Cycles

Although sperm production in men continues throughout life, the production of eggs in women is quite different. The total supply of eggs that a female can release is already determined by the time she is born. In addition, a female's reproductive ability (that is, her ability to release eggs) begins during puberty, heralded by the beginning of her *menstrual cycle*, and usually ends in her fifties. The period in which a woman's reproductive

capability ends is called *menopause* (see pp. 596 and 598).

## Oogenesis and the Ovarian Cycle

→ **Learning Objectives**

☐ **Define *oogenesis*.**

☐ **Describe the influence of FSH and LH on ovarian function.**

Meiosis, the special kind of cell division that occurs in the testes to produce sperm, also occurs in the ovaries. But in this case, ova or female gametes are produced, and the process is called **oogenesis** (o″o-jen′ĕ-sis; "the beginning of an egg"). This process is shown in **Figure 16.10**, p. 578.

In the developing female fetus, **oogonia** (o″o-go′ne-ah), the female stem cells, multiply rapidly to increase their number, producing daughter cells called **primary oocytes**. These primary oocytes then push into the ovary connective tissue, where they become surrounded by a single layer of cells to ultimately form the *primary follicles*. By birth, the oogonia no longer exist, and a female's lifetime supply of primary oocytes (approximately 1 million of them) is already in place in the ovarian follicles, awaiting the chance to undergo meiosis to produce functional eggs. Because the primary oocytes remain in this state of suspended animation all through childhood, their wait is a long one—usually from 10 to 14 years.

At puberty, the anterior pituitary gland begins to release FSH (follicle-stimulating hormone), which stimulates a small number of primary follicles to grow and mature each month, and ovulation begins to occur each month. These cyclic changes that occur monthly in the ovary constitute the **ovarian cycle**. At puberty, perhaps 300,000 oocytes remain; and, beginning at this time, a small number of oocytes are activated each month. As the reproductive life of a female is about 40 years (from the age of 11 to approximately 51), and there is typically only one ovulation event per month, fewer than 500 ova out of the potential 300,000 are released during her lifetime. Again, nature has provided us with a generous oversupply of sex cells.

As a follicle prodded by FSH grows larger, it accumulates fluid in the central chamber called the *antrum* (look back at Figure 16.7), and the primary oocyte it contains replicates its chromosomes and begins meiosis. The first meiotic division produces

---

*Unlike the male urethra, which carries both urine and semen, the female urethra has no reproductive function—it is strictly a passageway for urine.

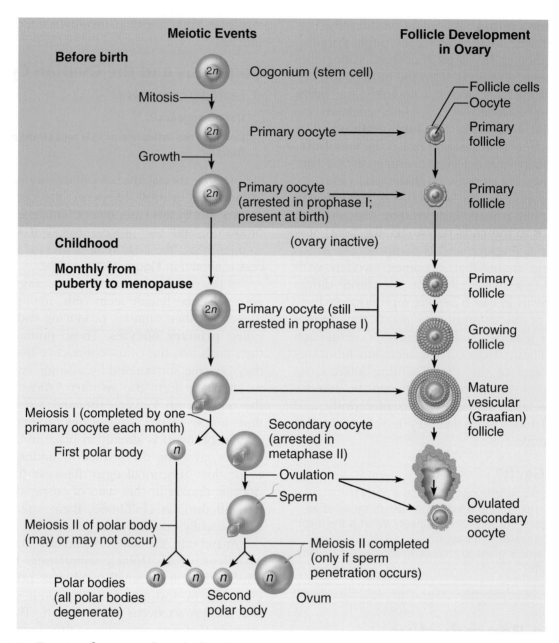

**Figure 16.10 Events of oogenesis.** Left, flowchart of meiotic events. Right, correlation with follicular development and ovulation in the ovary.

two cells that are very dissimilar in size (see Figure 16.10). The larger cell is a **secondary oocyte** and the other, very tiny cell is a **polar body**. By the time a follicle has ripened to the mature (*vesicular follicle*) stage, it contains a secondary oocyte and protrudes from the external surface of the ovary. Follicle development to this stage takes about 14 days, and ovulation (of a secondary oocyte) occurs at just about that time in response to the burstlike release of a second anterior pituitary hormone, luteinizing hormone (LH) **(Figure 16.11)**. The ovulated secondary oocyte is still surrounded by its follicle-cell capsule, now called the *corona radiata* ("radiating crown") (as shown in Figure 16.7).

Some women experience a twinge of abdominal pain in the lower abdomen when ovulation occurs. This phenomenon, called *mittelschmerz* (mit′el-shmerts; German for "middle pain"), is caused by the intense stretching of the ovarian wall during ovulation.

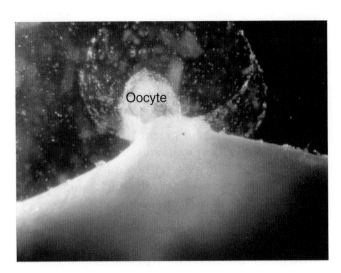

**Figure 16.11 Ovulation.** A secondary oocyte is released from a follicle at the surface of the ovary. The orange mass below the ejected oocyte is part of the ovary. The "halo" of follicle cells around the secondary oocyte is the *corona radiata*.

Generally speaking, one of the developing follicles outstrips the others each month to become the dominant follicle. Just how this follicle is selected or selects itself is not understood, but the follicle that is at the proper stage of maturity when the LH level surges is the one that ovulates its oocyte. The mature follicles that are not ovulated soon become overripe and deteriorate. In addition to playing a major role in triggering ovulation, LH also causes the ruptured follicle to change into a very different hormone-producing structure, the *corpus luteum*. (Both the maturing follicles and the corpus luteum produce hormones, as we will describe later.)

If the ovulated secondary oocyte is fertilized by a sperm in one of the uterine tubes, the oocyte quickly completes the second meiotic division that produces the **ovum** and another polar body. Once the ovum is formed, its 23 chromosomes are combined with those of the sperm to form the zygote. However, if a sperm does not penetrate the secondary oocyte, it simply deteriorates without ever completing meiosis to form a functional egg. Although meiosis in males results in four functional sperm, meiosis in females yields only one functional ovum and three tiny polar bodies. The polar bodies, produced to reduce the number of chromosomes in the developing oocyte, have essentially no cytoplasm, so they die quickly.

Another major difference between male and female gametes concerns the size and structure of these cells. Sperm are tiny and equipped with tails for locomotion. They have little nutrient-containing cytoplasm; thus, the nutrients in seminal fluid are vital to their survival. In contrast, the egg is a large, nonmotile cell, well stocked with nutrient reserves that nourish the developing embryo until it can take up residence in the uterus.

## Did You Get It?

14. Besides the one functional gamete (ovum), what other cell types are produced during oogenesis, and what happens to them?
15. Which anterior pituitary hormone promotes follicle development in the ovary?
16. Which anterior pituitary hormone causes ovulation?

**For answers, see Appendix A.**

## Hormone Production by the Ovaries

As the ovaries become active at puberty and start to produce oocytes, they also begin to produce ovarian hormones. The follicle cells of the growing and mature follicles produce **estrogens**,* which cause the appearance of secondary sex characteristics in the young woman. Such changes include the following:

- Enlargement of the accessory organs of the female reproductive system (uterine tubes, uterus, vagina, external genitals)
- Development of the breasts
- Appearance of axillary and pubic hair
- Increased deposits of fat beneath the skin in general, and particularly in the hips and breasts
- Widening and lightening of the pelvis
- Onset of *menses*, or the menstrual cycle

Estrogen also has metabolic effects. For example, it helps maintain low total blood cholesterol levels (and a high HDL level) and facilitates calcium ion uptake, which sustains bone density.

A second ovarian hormone, **progesterone**, is produced by the glandular corpus luteum (look back at Figure 16.7). As mentioned earlier, after ovulation occurs, the ruptured follicle is converted to the *corpus luteum*, which looks and acts completely different from the growing and mature fol-

---

*Although the ovaries produce several different estrogens, the most important are *estradiol*, *estrone*, and *estriol*. Of these, estradiol is the most abundant and is most responsible for estrogenic effects.

**16**

licle. Once formed, the corpus luteum produces progesterone (and some estrogen) as long as LH is still present in the blood. Generally speaking, the corpus luteum stops producing hormones by 10 to 14 days after ovulation. Except for working with estrogens to establish the menstrual cycle, progesterone does not contribute to the appearance of the secondary sex characteristics. Its other major effects are exerted during pregnancy, when it helps to maintain the pregnancy and prepare the breasts for milk production. However, the source of progesterone during pregnancy is the placenta, not the ovaries.

## Uterine (Menstrual) Cycle

### → Learning Objective

☐ Describe the phases and controls of the menstrual cycle.

Although the young embryo implants and develops in the uterus, this organ is receptive to implantation only for a very short period each month. Not surprisingly, this brief interval coincides exactly with the time when a fertilized egg would begin to implant, approximately 7 days after ovulation. The **uterine cycle**, or **menstrual cycle**, is a series of cyclic changes that the endometrium (mucosa of the uterus) goes through each month as it responds to changing blood levels of ovarian hormones.

The cyclic production of estrogens and progesterone by the ovaries is, in turn, regulated by the anterior pituitary gonadotropic hormones, FSH and LH. It is important to understand how these "hormonal pieces" fit together. Generally speaking, both female cycles (the ovarian and the uterine cycles) are about 28 days long. Ovulation typically occurs midway in the cycles, on or about day 14. **Figure 16.12** (p. 581) illustrates the events of the ovarian cycle and the uterine cycle at the same time.

The three phases of the menstrual cycle are as follows:

- **Days 1–5: Menstrual phase.** During this interval, the superficial *functional layer* of the thick endometrial lining of the uterus is sloughing off (detaching) from the uterine wall. The detached tissues and blood pass through the vagina as menstrual flow (the "period") for 3 to 5 days. The average blood loss during this period is 50 to 150 ml (or about ¼ to ½ cup). By day 5, growing ovarian follicles are beginning to produce more estrogens.

- **Days 6–14: Proliferative phase.** Stimulated by a rising level of estrogens produced by the growing follicles of the ovaries, the basal layer of the endometrium regenerates the functional layer, glands form in it, and the endometrial blood supply increases. The endometrium once again becomes velvety, thick, and well vascularized. (Ovulation occurs in the ovary at the end of this stage, in response to the sudden surge of LH in the blood.)

- **Days 15–28: Secretory phase.** A rising level of progesterone production by the corpus luteum acts on the estrogen-primed endometrium and increases its blood supply even more. Progesterone also causes the endometrial glands to grow and begin secreting nutrients into the uterine cavity. These nutrients will sustain a developing embryo (if one is present) until it has implanted. If fertilization does occur, the embryo produces a hormone very similar to LH that causes the corpus luteum to continue producing its hormones.

If fertilization does not occur, the corpus luteum begins to degenerate toward the end of this period as the LH blood level declines. Lack of ovarian hormones in the blood causes the blood vessels supplying the functional layer of the endometrium to go into spasms and kink. When deprived of oxygen and nutrients, those endometrial cells begin to die, which sets the stage for menses to begin again on day 28.

Although this explanation assumes a classic 28-day cycle, the length of the menstrual cycle is quite variable. It can be as short as 21 days or as long as 40 days. Only one interval is fairly constant in all females—the time from ovulation to the beginning of menses is almost always 14 or 15 days.

## Mammary Glands

### → Learning Objective

☐ Describe the structure and function of the mammary glands.

The **mammary glands** are present in both sexes, but they normally function only in women.

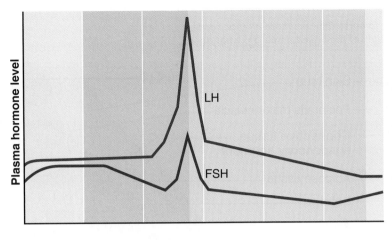

**(a) Fluctuation of gonadotropin levels:**
Fluctuating levels of pituitary gonadotropins (FSH and LH) in the blood regulate the events of the ovarian cycle.

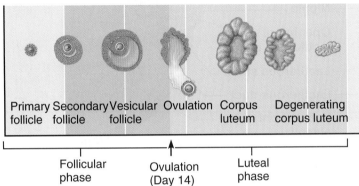

**(b) Ovarian cycle:** Structural changes in the ovarian follicles during the ovarian cycle are correlated with (d) changes in the endometrium of the uterus during the uterine cycle.

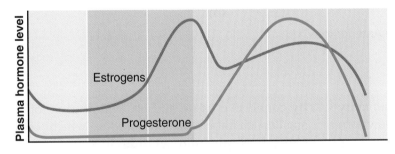

**(c) Fluctuation of ovarian hormone levels:**
Fluctuating levels of ovarian hormones (estrogens and progesterone) cause the endometrial changes of the uterine cycle. The high estrogen levels are also responsible for the LH/FSH surge in (a).

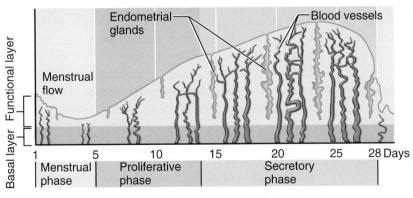

**(d) The three phases of the uterine cycle:**
- Menstrual: Shedding of the functional layer of the endometrium.
- Proliferative: Rebuilding of the functional layer of the endometrium.
- Secretory: Begins immediately after ovulation. Enrichment of the blood supply and glandular secretion of nutrients prepare the endometrium to receive an embryo.

The menstrual and proliferative phases occur before ovulation and together correspond to the follicular phase of the ovarian cycle. The secretory phase corresponds in time to the luteal phase of the ovarian cycle.

**Figure 16.12 Hormonal interactions of the female cycles.** Shown is the correlation of relative levels of anterior pituitary gonadotropins with hormonal and follicular changes of the ovary and with the menstrual cycle.

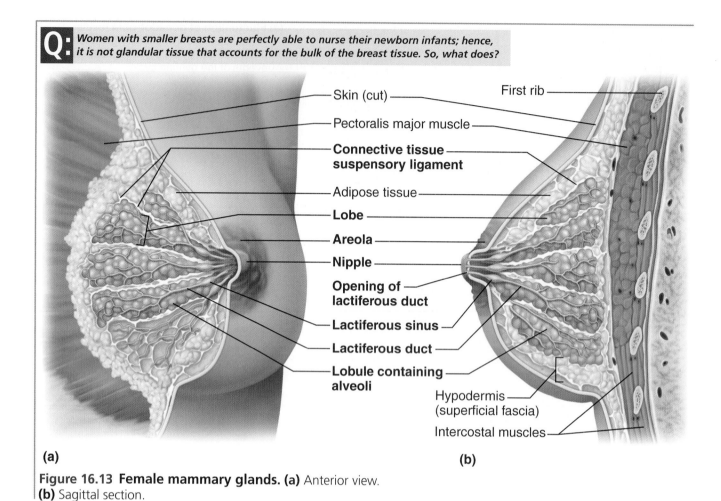

**(a)**

**(b)**

**Figure 16.13 Female mammary glands. (a)** Anterior view.
**(b)** Sagittal section.

Because the biological role of the mammary glands is to produce milk to nourish a newborn baby, they are actually important only when reproduction has already been accomplished. Stimulation by female sex hormones, especially estrogens, causes the female mammary glands to increase in size at puberty.

The mammary glands are modified *sweat glands* that are part of the skin. Each mammary gland is contained within a rounded skin-covered breast anterior to the pectoral muscles. Slightly inferior to the center of each breast is a pigmented area, the **areola** (ah-re'o-lah), which surrounds a central protruding **nipple (Figure 16.13)**.

Internally, each mammary gland consists of 15 to 25 *lobes* that radiate around the nipple. The lobes are padded and separated from one another by connective tissue and fat. Within each lobe are smaller chambers called *lobules*, which contain clusters of **alveolar glands** that produce milk when a woman is **lactating** (producing milk). Milk produced by the alveolar glands exits each lobule by passing into **lactiferous** (lak-tif'er-us) **ducts**, which open to the outside at the nipple. Just deep to the areola, each duct has a dilated region called a **lactiferous sinus**, where milk accumulates during nursing.

### Homeostatic Imbalance 16.7

**Breast cancer** is the second most common cause of death in American women; one woman in eight will develop this condition. Some 10 percent of breast cancers stem from hereditary defects, and half of these can be traced to dangerous mutations in a pair of genes (*BRCA1* and *BRCA2*). Fifty to 80 percent of women who carry

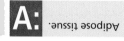

**A:** Adipose tissue.

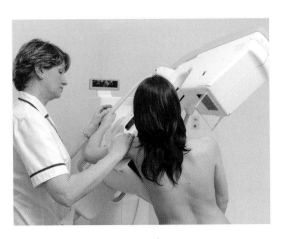

**(a) Mammogram procedure**

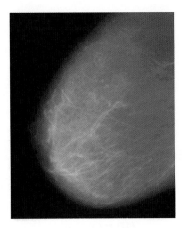

**(b) Film of normal breast**

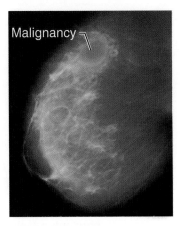

Malignancy

**(c) Film of breast with tumor**

**Figure 16.14 Mammograms.**

the altered gene develop breast cancer. With the possible exception of family history, most risk factors reflect lifelong exposure to estrogens (early menses, late menopause, estrogen replacement therapy, and others).

Breast cancer is often signaled by a change in skin texture, puckering, or leakage from the nipple. Early detection by breast self-examination and mammography is unquestionably the best way to increase a woman's chances of surviving breast cancer. Because most breast lumps are discovered by women themselves in routine monthly breast exams, this simple examination should be a priority in every woman's life. Currently the American Cancer Society recommends scheduling **mammography**—X-ray examination that detects breast cancers too small to feel (less than 1 cm)—yearly for women between 45 and 54 years old and every 2 years thereafter if the results are normal **(Figure 16.14)**. _____✚

## Did You Get It?

17. Which ovarian hormone promotes the formation of female secondary sex characteristics?
18. What happens during the proliferative stage of the uterine cycle?
19. What are three important functions of progesterone in women?
20. What problems do mutations of the *BRCA* genes cause?

For answers, see Appendix A.

# Pregnancy and Embryonic Development

→ **Learning** Objectives

☐ Define *fertilization* and *zygote*.
☐ Describe implantation.
☐ Distinguish between an embryo and a fetus.
☐ List the major functions of the placenta.

Because the birth of a baby is such a familiar event, we tend to lose sight of the wonder of this accomplishment. In every instance it begins with a single cell, the fertilized egg, and ends with an extremely complex human being consisting of trillions of cells. The development of an embryo is very complex, and the details of this process can fill a book. Our intention here is simply to outline the important events of pregnancy and embryonic development.

Let's begin by defining some terms. The term **pregnancy** refers to events that occur from the time of fertilization (conception) until birth. The pregnant woman's developing offspring is called the **conceptus** (kon-sep'tus; "that which is conceived"). Development occurs during the **gestation period** (*gestare* = to carry), which extends by convention from the last menstrual period (a date the woman is likely to remember) until birth, approximately 280 days.

From fertilization through week 8, the *embryonic period*, the conceptus is called an **embryo**, and from week 9 through birth, the *fetal period*, the conceptus is called a **fetus** ("the young in the

**16**

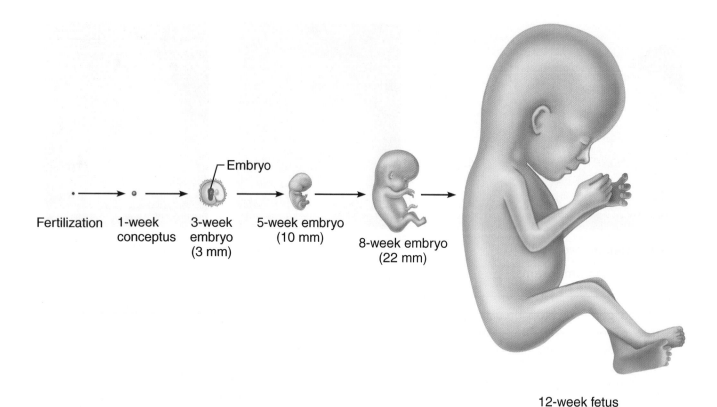

**Figure 16.15 Diagrams showing the approximate size of a human conceptus from fertilization to the early fetal stage.** Measurements indicate the length from crown to rump.

womb"). **Figure 16.15** shows the changing size and shape of the conceptus as it progresses from fertilization to the early fetal stage. At birth, the fetus becomes an infant.

## Accomplishing Fertilization

For fertilization to occur, a sperm must reach the ovulated secondary oocyte. The oocyte is viable for 12 to 24 hours after ovulation, and sperm generally retain their fertilizing power within the female reproductive tract for 24 to 48 hours after ejaculation. Consequently, for fertilization to occur, sexual intercourse must occur no more than 2 days before ovulation and no later than 24 hours after. At this point, the oocyte is approximately one-third of the way down the uterine tube. Remember that sperm are motile cells that can propel themselves by lashing movements of their tails. If sperm are deposited in a female's vagina at the approximate time of ovulation, they are attracted to the oocyte by chemicals that act as "homing devices," allowing them to locate the oocyte.

**CONCEPTLINK**

Recall the concept of chemotaxis, in which cells move toward or away from a stimulus (positive and negative chemotaxis, respectively) (Chapter 12, p. 432).

It takes 1 to 2 hours for sperm to complete the journey up the female duct system into the uterine tubes, even though they are only about 12 cm (5 inches) away. However, millions of sperm leak out of the vagina, and of those remaining, millions more are destroyed by the vagina's acidic environment. Only a few hundred to a few thousand sperm make it to the egg's vicinity.

When the swarming sperm reach the oocyte **(Figure 16.16)**, their cell surface hyaluronidase enzymes break down the "cement" that holds the follicle cells of the corona radiata together around the oocyte. Once a path has been cleared through the corona, thousands of sperm undergo the **acrosomal reaction**, in which the acrosome membranes break down, releasing enzymes that digest holes in the surrounding oocyte membrane. Then, when the membrane is adequately weakened and

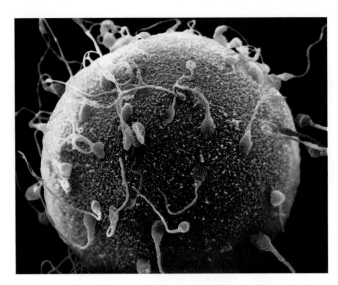

**Figure 16.16 Sperm and oocyte during fertilization.**

a single sperm makes contact with one of the oocyte's membrane receptors for sperm, the head (nucleus) of the sperm fuses with the oocyte membrane, and the snakelike sperm contents enter the oocyte cytoplasm. This is one case that does not bear out the adage "The early bird catches the worm," because sperm that are in the best position to be *the* fertilizing sperm are the ones that come along after hundreds of sperm have undergone acrosomal reactions to expose the oocyte membrane. Once a single sperm has penetrated the oocyte, the oocyte nucleus completes the second meiotic division, forming the ovum and a polar body.

Once the sperm has entered, the ovum sheds its remaining membrane surface receptors for sperm, preventing other sperm from gaining entry. In humans, of the millions of sperm ejaculated, only *one* can penetrate an oocyte. **Fertilization** occurs at the moment the genetic material of a sperm combines with that of an ovum to form a fertilized egg, or **zygote** (zi′gōt), with a complete set of 46 chromosomes. The zygote represents the first cell of the new individual.

## Events of Embryonic and Fetal Development

As the zygote journeys down the uterine tube (propelled by peristalsis and cilia), it begins to undergo rapid mitotic cell divisions—forming first two cells, then four, and so on. This early stage of embryonic development is called **cleavage** (**Figure 16.17**, p. 586). Because there is not much

time for cell growth between divisions, the daughter cells become smaller and smaller. Cleavage provides a large number of cells to serve as building blocks for constructing the embryo. Consider for a moment how difficult it would be to construct a building from one huge block of granite. If you now consider how much easier your task would be if you could use hundreds of brick-sized granite blocks, you will quickly grasp the importance of cleavage.

By the time the developing embryo reaches the uterus (about 3 days after ovulation, or on day 17 of the woman's cycle), it is a *morula*, a tiny ball of 16 cells that looks like a microscopic raspberry. The uterine endometrium is still not fully prepared to receive the embryo at this point, so the embryo floats free in the uterine cavity, temporarily using the uterine secretions for nutrition. While still unattached, the embryo continues to develop until it has about 100 cells, and then it hollows out to form a ball-like structure called a **blastocyst** (blas′to-sist). At the same time, it secretes an LH-like hormone called **human chorionic gonadotropin (hCG)**, which prods the corpus luteum of the ovary to continue producing its hormones. (If this were not the case, the functional layer of the endometrium would be sloughing off shortly in menses.) Many home pregnancy tests detect hCG in urine.

The blastocyst has two important functional areas: the **trophoblast**, which forms the large fluid-filled sphere, and the **inner cell mass**, a small cluster of cells on one side (see Figure 16.17e). By day 7 after ovulation, the blastocyst has attached to the endometrium and has eroded away the lining in a small area, embedding itself in the thick velvety mucosa. All of this is occurring even while development is continuing and the three primary germ layers are being formed from the inner cell mass (**Figure 16.18**, p. 587). The *primary germ layers* are the **ectoderm** (which will become the nervous system and the epidermis of the skin), the **endoderm** (which forms mucosae and associated glands), and the **mesoderm** (which will become virtually everything else). Implantation has usually been completed and the uterine mucosa has grown over the burrowed-in embryo by day 14 after ovulation—the day the woman would ordinarily be expecting to start her period (menses).

**16**

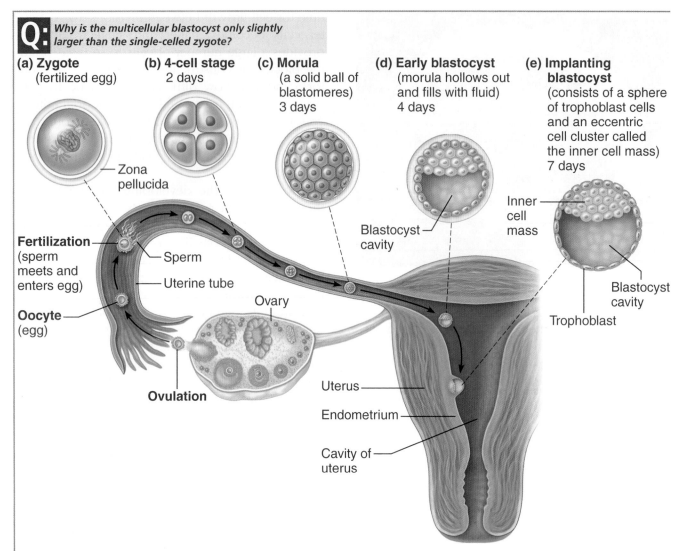

**Q:** *Why is the multicellular blastocyst only slightly larger than the single-celled zygote?*

**(a) Zygote**
(fertilized egg)

**(b) 4-cell stage**
2 days

**(c) Morula**
(a solid ball of blastomeres)
3 days

**(d) Early blastocyst**
(morula hollows out and fills with fluid)
4 days

**(e) Implanting blastocyst**
(consists of a sphere of trophoblast cells and an eccentric cell cluster called the inner cell mass)
7 days

Zona pellucida

Fertilization (sperm meets and enters egg)

Sperm

Uterine tube

Oocyte (egg)

Ovary

Ovulation

Blastocyst cavity

Inner cell mass

Blastocyst cavity

Trophoblast

Uterus

Endometrium

Cavity of uterus

**Figure 16.17 From fertilization and cleavage to implantation.** Cleavage is a rapid series of mitotic divisions that begins with the zygote and ends with the blastocyst. The zygote begins to divide about 24 hours after fertilization and continues to divide rapidly (undergo cleavage) as it travels down the uterine tube. The embryo reaches the uterus 3 to 4 days after ovulation and floats freely for another 2 to 3 days, nourished by secretions of the endometrial glands. At the late blastocyst stage, the embryo implants into the endometrium; this begins at about day 7 after ovulation.

After it is securely implanted, the trophoblast part of the blastocyst develops elaborate projections, called **chorionic villi**, which combine with the tissues of the mother's uterus to form the **placenta** (plah-sen′tah) **(Figure 16.19)**. Once the placenta has formed, the platelike embryonic body, now surrounded by a fluid-filled sac called the **amnion** (am′ne-on), is attached to the placenta by a blood vessel–containing stalk of tissue, the **umbilical cord** (see Figure 16.19). (We discussed the special features of the umbilical blood vessels and fetal circulation in Chapter 11.)

Generally by the third week, the placenta is functioning to deliver nutrients and oxygen to and remove wastes from the embryonic blood. All exchanges are made through the placental barrier. By the end of the second month of pregnancy, the placenta has also become an endocrine organ and

**A:** Because as the zygote and then its descendants divide, little or no time is provided for growth between subsequent division cycles. As a result, the cells get smaller and smaller, and the size of the cell mass stays approximately the same size as the initial zygote.

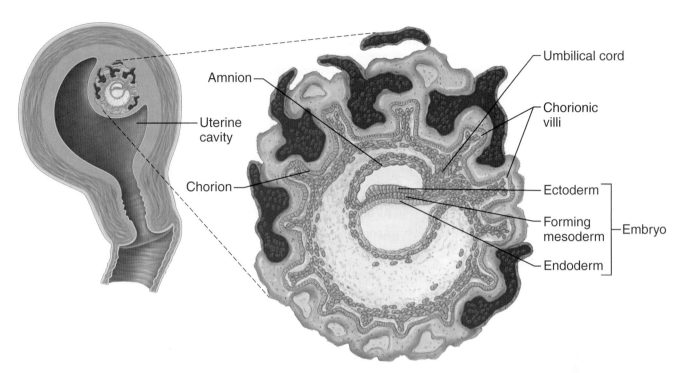

**Figure 16.18 Embryo of approximately 18 days.** At this stage of development, embryonic membranes are present.

is producing estrogens, progesterone, and other hormones that help to maintain the pregnancy. At this time, the corpus luteum of the ovary becomes inactive and degenerates.

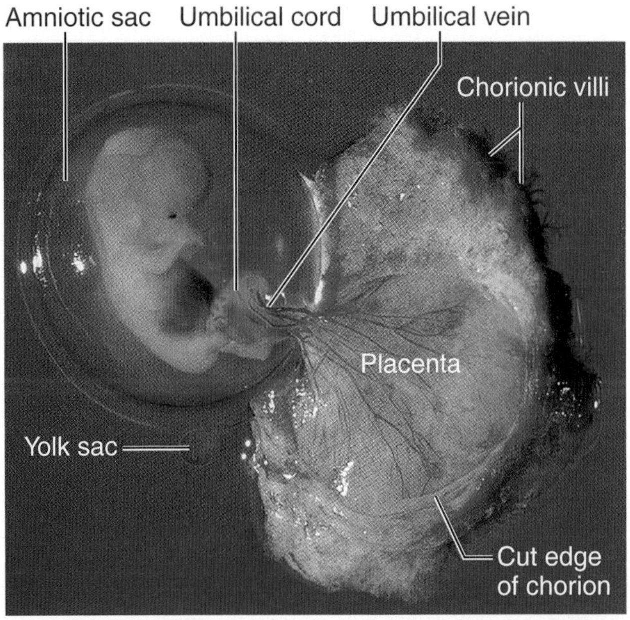

By the eighth week of embryonic development, all the groundwork has been completed. All the organ systems have been laid down, at least in rudimentary form, and the embryo looks distinctly human. Beginning in the ninth week of development, we refer to the embryo as a *fetus*. From this point on, the major activities are growth and organ specialization, accompanied by changes in body proportions. During the fetal period, the developing fetus grows from a crown-to-rump length of about 3 cm (slightly more than 1 inch) and a weight of approximately 1 g (0.03 ounce) to about 36 cm (14 inches) and 2.7 to 4.1 kg (6 to 10 pounds) or more. (Total body length at birth is about 55 cm, or 22 inches.) As you might expect with such tremendous growth, the changes in fetal appearance are quite dramatic (**Figure 16.20**, p. 589). The most significant of these changes are summarized in Table 16.1 on pp. 588–589. By approximately 270 days after fertilization, the fetus is said to be "full-term" and is ready to be born.

**Figure 16.19 The 7-week embryo.** A 7-week embryo, encased in its amniotic sac, and the chorionic villi (to the right), which cooperate with maternal uterine tissues to form the placenta.

## Did You Get It?

21. How does cleavage differ from cell divisions occurring after birth?
22. What are three roles of the placenta?

For answers, see Appendix A.

16

Table **16.1**   **Development of the Human Fetus**

| Time | | Changes/accomplishments |
|---|---|---|
| 8 weeks (end of embryonic period) | 8 weeks | Head nearly as large as body; all major brain regions present <br><br> Liver disproportionately large and begins to form blood cells <br><br> Limbs present; though initially webbed, fingers and toes are free by the end of this interval <br><br> Bone formation has begun <br><br> Heart has been pumping blood since the fourth week <br><br> All body systems present in at least rudimentary form <br><br> Approximate crown-to-rump length: 22 mm (0.9 inch); weight: 2 grams (0.07 ounce) |
| 9–12 weeks (third month) | 12 weeks | Head still dominant, but body elongating; brain continues to enlarge <br><br> Facial features present in crude form <br><br> Walls of hollow visceral organs gaining smooth muscle <br><br> Blood cell formation begins in bone marrow <br><br> Bone formation accelerating <br><br> Sex readily detected from the genitalia <br><br> Approximate crown-to-rump length at end of interval: 90 mm (9 cm) |
| 13–16 weeks (fourth month) | 16 weeks | General sensory organs are present; eyes and ears assume characteristic position and shape; blinking of eyes and sucking motions of lips occur <br><br> Face looks human, and body is beginning to outgrow head <br><br> Kidneys attain typical structure <br><br> Most bones are distinct, and joint cavities are apparent <br><br> Approximate crown-to-rump length at end of interval: 140 mm (14 cm) |
| 17–20 weeks (fifth month) | | Vernix caseosa (fatty secretions of sebaceous glands) covers body; silklike hair (lanugo) covers skin <br><br> Fetal position (body flexed anteriorly) assumed because of space restrictions <br><br> Limbs achieve near-final proportions <br><br> Quickening occurs (mother feels spontaneous muscular activity of fetus) <br><br> Approximate crown-to-rump length at end of interval: 190 mm |
| 21–30 weeks (sixth and seventh months) | | Substantial increase in weight (may survive if born prematurely at 27–28 weeks, but hypothalamus still too immature to regulate body temperature, and surfactant production by the lungs is still inadequate) <br><br> Myelination of spinal cord begins; eyes are open <br><br> Skin is wrinkled and red; fingernails and toenails are present <br><br> Body is lean and well proportioned <br><br> Approximate crown-to-rump length at end of interval: 280 mm |

Table **16.1**    *(continued)*

| Time | | Changes/accomplishments |
|---|---|---|
| 30–40 weeks (term) (eighth and ninth months) |   At birth | Bone marrow becomes sole site of blood cell formation |
| | | Testes enter scrotum in seventh month (in males) |
| | | Fat laid down in subcutaneous tissue of skin |
| | | Approximate crown-to-rump length at end of interval: 360 mm (14 inches); weight: 3.2 kg (7 pounds) |

## Effects of Pregnancy on the Mother

### → Learning Objectives

☐ **Indicate several ways that pregnancy alters or modifies the functioning of the mother's body.**

☐ **List several agents that can interfere with normal fetal development.**

The period from conception to birth can be a difficult time for the mother. Pregnancy causes not only obvious anatomical changes in the mother, but also striking changes in her physiology.

### Anatomical Changes

The ability of the uterus to enlarge during pregnancy is nothing less than remarkable. Starting as a fist-sized organ, the uterus grows to occupy most of the pelvic cavity by 16 weeks. As pregnancy continues, the uterus pushes higher and higher into the abdominal cavity (**Figure 16.21**, p. 590). As birth nears, the uterus reaches the level of the xiphoid process and occupies the bulk of the abdominal cavity. The crowded abdominal organs press superiorly against the diaphragm, which

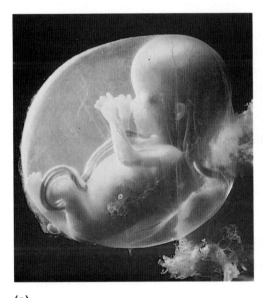

**(a)**

**(b)**

**Figure 16.20  Examples of fetal development. (a)** Fetus in month 3, about 6 cm (2.5 inches) long. **(b)** Fetus late in month 5, about 19 cm (8 inches) long.

16

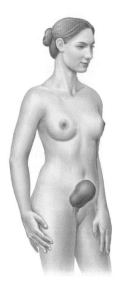

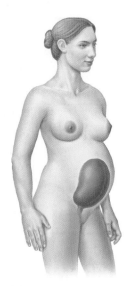

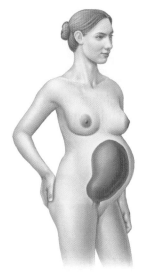

**(a) Before conception**
(Uterus is the size of a fist and resides in the pelvis.)

**(b) 4 months**
(Fundus of the uterus is halfway between the pubic symphysis and the umbilicus.)

**(c) 7 months**
(Fundus is well above the umbilicus.)

**(d) 9 months**
(Fundus reaches the xiphoid process.)

**Figure 16.21 Relative size of the uterus before conception and during pregnancy.**

intrudes on the thoracic cavity. As a result, the ribs flare, causing the thorax to widen.

The increasing bulkiness of the abdomen changes the woman's center of gravity, and many women develop an accentuated lumbar curvature (lordosis), often accompanied by backaches, during the last few months of pregnancy. Placental production of the hormone **relaxin** causes pelvic ligaments and the pubic symphysis to relax, widen, and become more flexible. This increases the width of the birth canal to ease birth passage, but it may also result in a waddling gait during pregnancy.

Good maternal nutrition is necessary throughout pregnancy if the developing fetus is to have all the building materials (proteins, calcium, iron, and the like) to form its tissues and organs. The old expression "A pregnant woman is eating for two" has encouraged many women to eat *twice* the amount of food actually needed during pregnancy, which, of course, leads to excessive weight gain. Actually, a pregnant woman needs only about 300 additional calories daily to sustain proper fetal growth. The emphasis should be on high-quality nutrition, not just more calories.

### Homeostatic Imbalance 16.8

Many potentially harmful substances can cross through the placental barrier into the fetal blood; therefore, a pregnant woman should be very aware of what she is taking into her body. Substances that may cause life-threatening birth defects (and even fetal death) include alcohol, nicotine, and many types of drugs (anticoagulants, antihypertensives, sedatives, and some antibiotics). Maternal infections, particularly German measles (rubella), may also cause severe fetal damage. Natural termination of a pregnancy before a fetus can survive on its own is called **miscarriage**. A medical termination of a pregnancy is called **abortion**. _____✚

### Physiological Changes

***Gastrointestinal System*** Many women suffer nausea, commonly called *morning sickness*, during the first few months of pregnancy, until their system adjusts to the elevated levels of progesterone and estrogens. *Heartburn* is common because the esophagus is displaced and the stomach is crowded by the growing uterus, which favors reflux of stomach acid into the esophagus. Another problem is constipation due to decline in motility of the digestive tract during pregnancy.

***Urinary System*** The kidneys have the additional burden of disposing of fetal metabolic wastes, and they produce more urine during pregnancy. Because the uterus compresses the bladder, urination becomes more frequent, more urgent, and sometimes uncontrollable. The last condition is called *stress incontinence*.

***Respiratory System*** The nasal mucosa responds to estrogens by becoming swollen and congested; thus, nasal stuffiness and occasional nosebleeds may occur. Vital capacity and respiratory rate increase during pregnancy, but residual volume declines, and many women exhibit *dyspnea* (difficult breathing) during the later stages of pregnancy.

***Cardiovascular System*** Perhaps the most dramatic physiological changes occur in the cardiovascular system. Total body water rises, and blood volume increases by 25 to 40 percent to accommodate the additional needs of the fetus. The rise in blood volume also acts as a safeguard against blood loss during birth. Blood pressure and pulse typically rise and increase cardiac output by 20 to 40 percent; this helps propel the greater blood volume around the body. Because the uterus presses on the pelvic blood vessels, venous return from the lower limbs may be impaired somewhat, resulting in varicose veins or swollen ankles and feet.

## Childbirth

### → Learning Objective

☐ **Describe how labor is initiated, and briefly discuss the three stages of labor.**

Childbirth, also called **parturition** (par″tur-ish′un; "bringing forth young"), is the culmination of pregnancy. It usually occurs within 15 days of the calculated due date (which is 280 days from the last menstrual period). The series of events that expel the infant from the uterus is referred to as **labor**.

### Initiation of Labor

Several events work together to trigger labor. During the last few weeks of pregnancy, estrogens reach their highest levels in the mother's blood. This has two important consequences: it causes the myometrium to form abundant *oxytocin receptors* (so that it becomes more sensitive to the hormone *oxytocin*), and it interferes with progesterone's quieting influence on the uterine muscle. As

a result, weak, irregular uterine contractions begin to occur. These contractions, called *Braxton Hicks contractions*, have caused many women to go to the hospital, only to be told that they were in **false labor** and sent home.

As birth nears, two more chemical signals cooperate to convert these false labor pains into the real thing. Certain cells of the fetus begin to produce oxytocin, which in turn stimulates the placenta to release *prostaglandins*. Both hormones stimulate more frequent and powerful contractions of the uterus. At this point, the increasing emotional and physical stresses activate the mother's hypothalamus, which signals for oxytocin release by the posterior pituitary. The combined effects of rising levels of oxytocin and prostaglandins initiate the rhythmic, expulsive contractions of true labor. Once the hypothalamus is involved, a positive feedback mechanism is propelled into action: stronger contractions cause the release of more oxytocin, which causes even more vigorous contractions, forcing the baby ever deeper into the mother's pelvis (**Figure 16.22**, p. 592).

#### ◄ CONCEPT**LINK**

Remember the concept of the feedback loop (see Chapter 1, p. 45). A stimulus triggers a receptor, the information is sent to the brain for processing, and a signal is sent to an effector with instructions for a response. Most of the feedback in the body is negative feedback, in which the response *decreases* the initial stimulus. Labor, however, involves positive feedback: The response (stronger contractions) actually *increases* the initial stimulus (oxytocin release) until the child is born.

Because both oxytocin and prostaglandins are needed to initiate labor, anything that interferes with production of either of these hormones will hinder the onset of labor. For example, antiprostaglandin drugs such as aspirin and ibuprofen can inhibit labor at the early stages, and such drugs are used occasionally to prevent premature births.

### Stages of Labor

The process of labor is commonly divided into three stages (**Figure 16.23**, p. 592).

***Stage 1: Dilation Stage*** The **dilation stage** is the time from the appearance of true contractions until the cervix is fully dilated by the baby's head (about 10 cm in diameter). As labor starts, regular

**16**

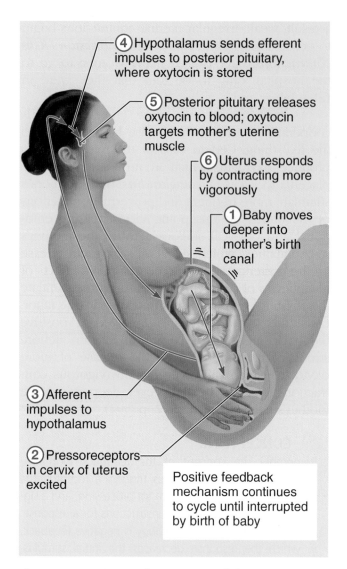

(4) Hypothalamus sends efferent impulses to posterior pituitary, where oxytocin is stored

(5) Posterior pituitary releases oxytocin to blood; oxytocin targets mother's uterine muscle

(6) Uterus responds by contracting more vigorously

(1) Baby moves deeper into mother's birth canal

(3) Afferent impulses to hypothalamus

(2) Pressoreceptors in cervix of uterus excited

Positive feedback mechanism continues to cycle until interrupted by birth of baby

**Figure 16.22 Oxytocin promotes labor contractions during birth by a positive feedback mechanism.**

but weak uterine contractions begin in the upper part of the uterus and move inferiorly toward the vagina. Gradually, the contractions become more vigorous and more rapid, and, as the infant's head is forced against the cervix with each contraction, the cervix begins to soften, becomes thinner (*effaces*), and dilates. Eventually, the amnion ruptures, releasing the amniotic fluid, an event commonly referred to as the woman's "water breaking." The dilation stage is the longest part of labor and usually lasts for 6 to 12 hours or more.

**Stage 2: Expulsion Stage** The **expulsion stage** is the period from full dilation to delivery of the infant. In this stage, the infant passes through the cervix

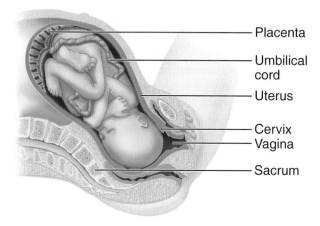

Placenta

Umbilical cord

Uterus

Cervix
Vagina

Sacrum

**(a) Dilation (of cervix)**

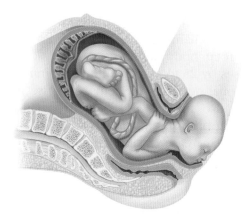

**(b) Expulsion (delivery of infant)**

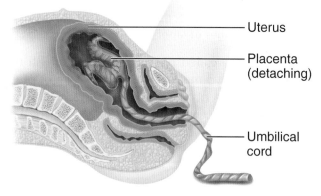

Uterus

Placenta (detaching)

Umbilical cord

**(c) Placental (delivery of the placenta)**

**Figure 16.23 The three stages of labor.**

and vagina to the outside of the body. During this stage, a mother experiencing natural childbirth (that is, undergoing labor without local anesthesia) has an increasing urge to push, or bear down, with the abdominal muscles. Although this phase can take as long as 2 hours, it is typically 50 minutes in a first birth and around 20 minutes in subsequent births.

When the infant is in the usual head-first (*vertex*) position, the skull (its largest diameter) acts as a wedge to dilate the cervix. The head-first presentation also allows the baby to be suctioned free of mucus and to breathe even before it has completely exited from the birth canal. Once the head has been delivered, the rest of the baby's body is delivered much more easily. After birth, the umbilical cord is clamped and cut. In *breech* (buttocks-first) presentations and other nonvertex presentations, these advantages are lost and delivery is often much more difficult, sometimes requiring the use of forceps or a vacuum extractor.

 ## Homeostatic Imbalance 16.9

During an extremely prolonged or difficult stage 2, a condition called **dystocia** (dis-to′se-ah) may occur. In dystocia, oxygen delivery to the infant is inadequate, leading to fetal brain damage (resulting in cerebral palsy or epilepsy) and decreased viability of the infant. To prevent these outcomes, a **cesarean** (se-zayr′e-an) **section**, also called a **C-section**, may be performed. A C-section is delivery of the infant through a surgical incision made through the abdominal and uterine walls. _____+

***Stage 3: Placental Stage***  The **placental stage**, or the delivery of the placenta, is usually accomplished within 15 minutes after birth of the infant. The strong uterine contractions that continue after birth compress uterine blood vessels, limit bleeding, and cause the placenta to detach from the uterine wall. The placenta and its attached fetal membranes, collectively called the **afterbirth**, are then easily removed by a slight tug on the umbilical cord. All placental fragments must be removed to prevent continued uterine bleeding after birth (*postpartum bleeding*).

### Did You Get It?

23. Explain how pregnancy affects a woman's respiratory and digestive processes.
24. What are the three stages of labor?

For answers, see Appendix A.

# Developmental Aspects of the Reproductive System

### → Learning Objectives

☐ Describe the importance of the presence/absence of testosterone during embryonic development of the reproductive system organs.

☐ Define *menarche* and *menopause*.

☐ List common reproductive system problems seen in adult and aging men and women.

Although the genetic sex of an individual is determined at the time of fertilization (males have X and Y sex chromosomes and females have two X sex chromosomes), the gonads do not begin to form until about the eighth week of embryonic development. Prior to this time, the embryonic reproductive structures of males and females are identical and are said to be in the *indifferent stage*. After the gonads have formed, development of the accessory structures and external genitalia begins. Whether male or female structures will form depends entirely on whether testosterone is present or absent. The usual case is that, once formed, the embryonic testes produce testosterone, and the development of the male duct system and external genitalia follows. When testosterone is not produced, as is the case in female embryos that form ovaries, the female ducts and external genitalia result.

 ## Homeostatic Imbalance 16.10

Any interference with the normal pattern of sex hormone production in the embryo results in abnormalities. For example, if the embryonic testes fail to produce testosterone, a genetic male develops the female accessory structures and external genitalia. If a genetic female is exposed to testosterone (as might happen if the mother has an androgen-producing tumor of her adrenal gland), the embryo has ovaries but develops male accessory ducts and glands, as well as a penis and an empty scrotum. Individuals with external genitalia that do not "match" their gonads are called **pseudohermaphrodites** (su″do-her-maf′ro-dītz) to distinguish them from true **hermaphrodites**, rare individuals who possess both ovarian and testicular tissues. In recent years, many pseudohermaphrodites have sought sex change operations to match their outer selves (external genitalia) with their inner selves (gonads).

*(Text continues on page 596.)*

16

# Contraception: Preventing Pregnancy

In a society such as ours, where many women opt for professional careers or must work for economic reasons, **contraception** (*contra* = against; *cept* = taking), or *birth control*, is often seen as a necessity.

The key to birth control is dependability. As shown by the red arrows in the accompanying flowchart, the birth control techniques and products currently available have many sites of action for blocking the reproductive process. Let's examine the relative advantages of a few of these methods more closely.

The most-used contraceptive product in the United States is the *birth control pill*, or simply, "the pill," a preparation taken daily that contains tiny amounts of estrogens and progestins (progesterone-like hormones), except that for the last 7 days of the 28-day cycle the tablets are hormone-free. The pill tricks the hypothalamic-pituitary control system and "lulls it to sleep" because the relatively constant blood levels of ovarian hormones make it appear that the woman is pregnant (both estrogen and progesterone are produced throughout pregnancy). Ovarian follicles do not mature, ovulation ceases, and menstrual flow is much reduced.

The pill has adverse cardiovascular effects, such as strokes, heart attacks, and blood clots in a small number of users, and there is still debate about whether it increases the incidence of ovarian, uterine, and particularly breast cancer. However, it appears that the new, very-low-dose preparations may actually help protect against ovarian and endometrial cancer and may also have reduced the incidence of seri-

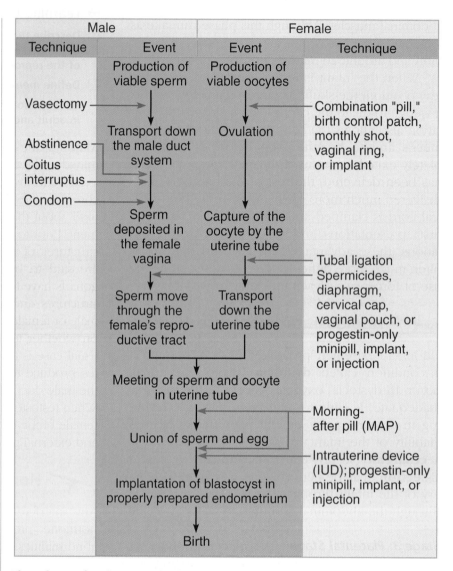

**Flowchart of points at which the events that must occur to produce a baby can be interrupted.** Techniques or products that interfere with this process are indicated by red arrows at the site of interference; they act to prevent the next step.

ous cardiovascular side effects that occurred (rarely) with earlier forms of the pill.

Presently, the pill is one of the most widely used drugs in the world; well over 50 million women use these drugs to prevent pregnancy. The incidence of failure is less than 1 percent.

A different combination hormone pill, the *morning-after pill (MAP)* or *emergency contraceptive pill (EC)*, is the therapy of choice for rape victims and is now available over the counter. Taken within 3 days of unprotected intercourse, the concentrated estrogen-progesterone combination pills "mess up" the normal hormonal signals so much that fertilization is prevented altogether or a fertilized egg is prevented from implanting.

Other hormonal approaches are progestin-only products that cause cervical mucus to become thick, blocking sperm passage and making the endometrium inhospitable to implantation. These include the *minipill* (a tablet); *implants* of a tiny silicon rod that releases the hormone over several years; and *Depo-Provera*, an injectable synthetic progesterone that lasts for 3 months. The failure rate of the implant or injection is even lower than that of the pill.

For several years, the second most used contraceptive method was the *intrauterine device (IUD)*, a plastic or metal device inserted into the uterus that prevented implantation of the fertilized egg (see flowchart). Although the failure rate of the IUD was nearly as low as that of the pill, IUDs were taken off the market in the United States because of problems with occasional contraceptive failure, uterine perforation, or pelvic inflammatory disease (PID). New IUD products that deliver sustained doses of synthetic progesterone to the endometrium are currently being recommended for women who have given birth and for women with a lower risk of developing PID in monogamous relationships.

Sterilization techniques, such as *tubal ligation* and *vasectomy* (cutting

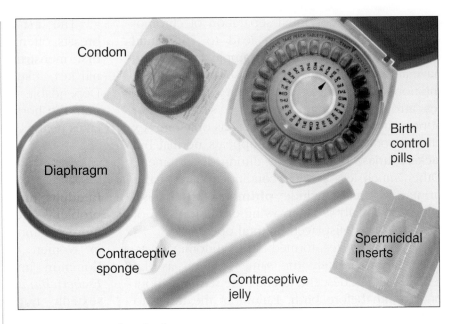

**Some contraceptive devices.**

or cauterizing the uterine tubes or ductus deferens, respectively), are nearly foolproof and are the choice of approximately 33 percent of couples of childbearing age in the United States. Both procedures can be done in the physician's office. However, these changes are usually permanent, so they are not for individuals who plan to have children at a later time.

*Coitus interruptus*, or withdrawal of the penis just before ejaculation, is simply against nature, and control of ejaculation is never assured. *Rhythm*, or *fertility awareness*, *methods* depend on avoiding intercourse during the period of ovulation or fertility. With a failure rate of 10 to 20 percent, rhythm techniques require

accurate recordkeeping for several cycles before they can be used with confidence.

*Barrier methods*, such as diaphragms, cervical caps, condoms (see photo), spermicidal foams, gels, and sponges, are quite effective, especially when used by both partners. But many people avoid them because they can reduce the spontaneity of sexual encounters.

A large number of experimental birth control drugs are now awaiting clinical trials, and other methods are sure to be developed in the near future. In the final analysis, however, the only 100 percent effective means of birth control is the age-old one—*abstinence*.

Additionally, abnormal separation of chromosomes during meiosis can lead to congenital defects of this system. For example, males who have an extra female sex chromosome (XXY) have the normal male accessory structures, but their testes atrophy, causing them to be sterile. Other abnormalities occur when a child has only one sex chromosome. An XO female appears normal but lacks ovaries; YO males die during development. Other, much less serious, conditions affect males primarily; these include **phimosis** (fi-mo′sis), which essentially is a narrowing of the foreskin of the penis, and misplaced urethral openings.

The male testes, formed in the abdominal cavity at approximately the same location as the female ovaries, descend to enter the scrotum about one month before birth. Failure of the testes to make their normal descent leads to a condition called **cryptorchidism** (krip-tor′kǐ-dizm). Because this condition results in sterility of a male (and also puts him at risk for cancer of the testes), surgery is usually performed during childhood to rectify this problem. _____✚

Because the reproductive system organs do not function until puberty, there are few problems with this system during childhood. **Puberty** is the period of life, generally between the ages of 10 and 15 years, when the reproductive organs grow to their adult size and become functional under the influence of rising levels of gonadal hormones (testosterone in men and estrogens in women). After this time, reproductive capability continues until old age in men and menopause in women. Earlier we described the secondary sex characteristics and major events of puberty, so we will not repeat these details here. It is important to remember, however, that puberty represents the earliest period of reproductive system activity.

The events of puberty occur in the same sequence in all individuals, but the age at which they occur varies widely. In boys, the event that signals puberty's onset is enlargement of the testes and scrotum, around the age of 13 years, followed by the appearance of pubic, axillary, and facial hair. Growth of the penis goes on over the next 2 years, and sexual maturation is indicated by the presence of mature sperm in the semen. In the meantime, the young man has unexpected erections and occasional nocturnal emissions ("wet dreams") as his hormones surge and hormonal controls struggle to achieve a normal balance.

The first sign of puberty in girls is budding breasts, often apparent by the age of 11 years. The first menstrual period, called **menarche** (mě-nar′ke), usually occurs about 2 years later. Dependable ovulation and fertility are deferred until the hormonal controls mature, an event that takes nearly 2 more years.

## Homeostatic Imbalance 16.11

In adults, the most common reproductive system problems are infections. Vaginal infections are more common in young and elderly women and in women whose immune resistance is low. Common infections include those caused by *Escherichia coli* (spread from the digestive tract); sexually transmitted microorganisms (such as gonorrhea, syphilis, and genital warts); and yeasts (a type of fungus). Untreated vaginal infections may spread throughout the female reproductive tract, causing pelvic inflammatory disease and sterility. Problems involving painful or abnormal menses may result from infection or hormone imbalance.

The most common inflammatory conditions in men are **urethritis**, **prostatitis**, and **epididymitis** (ep″ǐ-did-ǐ-mi′tis), all of which may follow sexual contacts in which sexually transmitted microorganisms are transmitted. **Orchitis** (or″ki′tis), inflammation of the testes, is rather uncommon but is serious because it can cause sterility. Orchitis most commonly follows sexually transmitted infection (STI) or mumps in an adult man.

As noted earlier, neoplasms represent a major threat to reproductive system organs. Tumors of the breast and cervix are the most common reproductive cancers in adult women, and prostate cancer (a common sequel to prostatic hypertrophy) is a widespread problem in adult men. _____✚

Most women reach peak reproductive abilities in their late twenties. After that, a natural decrease in ovarian function occurs. As production of estrogens declines, ovulation becomes irregular, and menstrual periods become scanty and shorter in length. Eventually, ovulation and menses cease entirely, ending childbearing ability. This event, called **menopause**, normally occurs between the ages of 46 and 54 years. A woman is said to have reached menopause when she has gone a whole year without menstruation.

# Homeostatic Relationships between the **Reproductive System** and Other Body Systems

**Endocrine System**
- Gonadal hormones exert feedback effects on hypothalamic-pituitary axis; placental hormones help to maintain pregnancy
- Gonadotropins help regulate function of gonads

**Lymphatic System/Immunity**
- Developing embryo/fetus escapes immune surveillance (not rejected)
- Lymphatic vessels drain leaked tissue fluids; transport sex hormones; immune cells protect reproductive organs from disease; IgA is present in breast milk

**Digestive System**
- Digestive organs crowded by developing fetus; heartburn, constipation common during pregnancy
- Digestive system provides nutrients needed for health

**Urinary System**
- Hypertrophy of the prostate inhibits urination; compression of bladder during pregnancy leads to urinary frequency and urgency
- Kidneys dispose of nitrogenous wastes and maintain acid-base balance of blood of mother and fetus; semen exits the body through the urethra of the male

**Muscular System**
- Androgens promote increased muscle mass
- Abdominal muscles are active during childbirth; muscles of the pelvic floor support reproductive organs and aid erection of penis/clitoris

**Nervous System**
- Sex hormones masculinize or feminize the brain and influence sex drive
- Hypothalamus regulates timing of puberty; neural reflexes regulate sexual response

**Respiratory System**
- Late stages of pregnancy impair descent of the diaphragm, causing difficult breathing
- Respiratory system provides oxygen; disposes of carbon dioxide; vital capacity and respiratory rate increase during pregnancy

**Cardiovascular System**
- Estrogens lower blood cholesterol levels and promote cardiovascular health in premenopausal women; pregnancy increases workload of the cardiovascular system
- Cardiovascular system transports needed substances to organs of reproductive system; local vasodilation is involved in erection; blood transports sex hormones

## Reproductive System

**Integumentary System**
- Male sex hormones (androgens) activate oil glands, which lubricate skin and hair; gonadal hormones stimulate characteristic fat distribution and appearance of pubic and axillary hair; estrogen increases skin hydration; enhances facial skin pigmentation during pregnancy
- Skin protects all body organs by enclosing them externally; mammary gland secretions (milk) nourish the infant

**Skeletal System**
- Androgens masculinize the skeleton and increase bone density; estrogen feminizes skeleton and maintains bone mass in women
- The bony pelvis encloses some reproductive organs; if narrow, the bony pelvis may hinder vaginal delivery of an infant

Although production of estrogens continues for a while after menopause, the ovaries eventually stop functioning as endocrine organs. When deprived of the stimulatory effects of estrogens, the reproductive organs and breasts begin to atrophy. The vagina becomes dry; intercourse may become painful (particularly if infrequent), and vaginal infections become increasingly common. Other consequences of the deficit of estrogens include irritability and other mood changes (depression in some); intense vasodilation of the skin's blood vessels, which causes uncomfortable sweat-drenching "hot flashes"; gradual thinning of the skin and loss of bone mass; and slowly rising blood cholesterol levels, which place postmenopausal women at risk for cardiovascular disorders. At one time, physicians prescribed low-dose estrogen–progesterone preparations to help women through this often difficult period and to prevent the skeletal and cardiovascular complications. These seemed like great bonuses, and until July 2002, some 14 million American women were taking some form of estrogen-containing hormone replacement therapy (HRT). Then, on July 9, the Women's Health Initiative (WHI) abruptly ended a clinical trial of 16,000 postmenopausal women, reporting that in those taking a popular estrogen–progesterone hormone combination there was an increase of 51 percent in heart disease, 24 percent in invasive breast cancer, 31 percent in stroke, and a doubling of the risk of dementia compared to those taking placebos. The backlash of this information is still spreading through physicians' offices and research labs, and it has dampened enthusiasm for HRT in both the medical community and postmenopausal women. A new and encouraging study in 2006 reported a significant drop in breast cancers, with the drop due nearly entirely to fewer women using HRT.

There is no equivalent of menopause in males. Although aging men exhibit a steady decline in testosterone secretion and a longer latent period after orgasm, a condition called *andropause*, their reproductive capability continues until death. Healthy men well into their eighties and beyond are able to father offspring.

## Did You Get It?

25. Which sex chromosome combination yields a boy—XX or XY? What hormone must be produced by an XY fetus during development to stimulate the formation of the male duct system?

26. What is cryptorchidism, and what results if it is not rectified?

27. What are the major health threats to the adult woman's reproductive system?

*For answers, see Appendix A.*

# Summary

## Anatomy of the Male Reproductive System (pp. 565–569)

1. The paired testes, the male gonads, reside in the scrotum outside the abdominopelvic cavity. Testes produce both sperm (exocrine function) and testosterone (endocrine function).

2. The male duct system includes the epididymis, ductus deferens, and urethra. Sperm mature in the epididymis. When ejaculation occurs, sperm are propelled through duct passageways to the body exterior.

3. Male accessory glands include the seminal vesicles, prostate, and bulbourethral glands. Collectively, these glands produce a fluid that activates and nourishes sperm.

4. External genitalia:
   a. Scrotum—a skin sac that hangs outside the abdominopelvic cavity and provides the proper temperature for producing viable sperm.
   b. Penis—consists of three columns of erectile tissue surrounding the urethra. Erectile tissue provides a way for the penis to become rigid so it may serve as a penetrating device during sexual intercourse.

## Male Reproductive Functions (pp. 569–573)

1. Spermatogenesis (sperm production) begins at puberty in seminiferous tubules in response to FSH. Spermatogenesis involves meiosis, a special nuclear division that halves the chromosomal number in resulting spermatids. An additional process that strips excess cytoplasm from the spermatid, called spermiogenesis, is necessary for production of functional, motile sperm.

2. Testosterone production begins at puberty in response to LH. Testosterone is produced by interstitial cells of the testes. Testosterone causes the appearance of male secondary sex characteristics and is necessary for sperm maturation.

## Anatomy of the Female Reproductive System (pp. 573–577)

1. The ovaries, the female gonads, are located against the lateral walls of the pelvis. They produce oocytes (exocrine function) and estrogen and progesterone (endocrine function).

2. The duct system:

   a. Uterine (fallopian) tubes extend from the vicinity of an ovary to the uterus. Ends form fringed fimbriae that "wave" to direct ovulated oocytes into uterine tubes. These tubes then conduct the oocyte (or embryo) to the uterus by peristalsis and ciliary action.

   b. The uterus is a pear-shaped muscular organ in which the embryo implants and develops. Its mucosa (endometrium) sloughs off each month in menses unless an embryo has become embedded in it. The myometrium contracts rhythmically during the birth of a baby.

   c. The vagina is a passageway between the uterus and the body exterior that allows a baby or the menstrual flow to leave the body. It also receives the penis and semen during sexual intercourse.

3. Female external genitalia include labia majora and minora (skin folds), clitoris, and urethral and vaginal openings.

## Female Reproductive Functions and Cycles (pp. 577–580)

1. Oogenesis (production of female sex cells) occurs in ovarian follicles, which are activated at puberty by FSH and LH to mature and eject oocytes (ovulation) on a cyclic basis. The egg (ovum) is formed only if sperm penetrates the secondary oocyte. In females, meiosis produces only one functional ovum (plus three nonfunctional polar bodies), as compared with the four functional sperm per meiosis event produced by males.

2. Hormone production: Estrogens are produced by ovarian follicles in response to FSH. Progesterone, produced in response to LH, is the main hormonal product of the corpus luteum. Estrogens stimulate development of female secondary sex characteristics.

3. The menstrual cycle involves changes in the endometrium in response to fluctuating blood levels of ovarian hormones. There are three phases:

   a. Menstrual phase. Endometrium sloughs off and bleeding occurs. Ovarian hormones are at their lowest levels.

   b. Proliferative phase. Endometrium is repaired, thickens, and becomes well vascularized in response to increasing levels of estrogens.

   c. Secretory phase. Endometrial glands begin to secrete nutrients, and lining becomes more vascular in response to increasing level of progesterone.

4. If fertilization does not occur, the phases are repeated about every 28 days.

## Mammary Glands (pp. 580–583)

1. Mammary glands are milk-producing glands found in the breasts. After the birth of a baby, they produce milk in response to hormonal stimulation.

## Pregnancy and Embryonic Development (pp. 583–593)

1. An oocyte can be fertilized up to 24 hours after release; sperm are viable within the female reproductive tract for up to 48 hours. Hundreds of sperm must release their acrosomal enzymes to break down the oocyte's plasma membrane.

2. Following sperm penetration, the secondary oocyte completes meiosis II. Then ovum and sperm nuclei fuse (fertilization), forming a zygote.

3. If fertilization occurs, embryonic development begins immediately. Cleavage, a rapid series of mitotic divisions without intervening growth, begins with the zygote and ends with a blastocyst.

4. By day 14 after ovulation, the young embryo (blastocyst) has implanted in the endometrium, and the placenta is being formed. Human chorionic gonadotropin (hCG) released by the blastocyst maintains hormone production of the corpus luteum, preventing menses, until the placenta assumes its endocrine role.

5. The placenta serves respiratory, nutritive, and excretory needs of the embryo and produces hormones of pregnancy.

6. All major organ systems have been formed by 8 weeks, and at 9 weeks the embryo is called a fetus. Growth and tissue/organ specialization are the major events of the fetal period.

7. A pregnant woman has increased respiratory, circulatory, and urinary demands placed on her system by the developing fetus. Good nutrition is necessary to produce a healthy baby.

8. Childbirth (parturition) includes a series of events called labor. It is initiated by several factors but

**16**

most importantly by rising levels of oxytocin and prostaglandins, which promote vigorous uterine contractions. The three stages of labor are dilation, expulsion, and placental stages.

## Developmental Aspects of the Reproductive System (pp. 593–598)

1. Reproductive system structures of males and females are identical during early development. Gonads begin to develop in the eighth week. The presence or absence of testosterone determines whether male or female accessory reproductive organs are formed.

2. Important congenital defects result from abnormal separation of sex chromosomes during sex cell formation.

3. The reproductive system is inactive during childhood. Reproductive organs mature and become functional for childbearing at puberty.

4. Common reproductive problems during young adulthood are infections of the reproductive tract. Neoplasms of breast and cervix are major threats to women. Prostate cancer is the most common reproductive system cancer in men.

5. During menopause, female reproductive capabilities end, and reproductive organs begin to atrophy. Hot flashes and mood changes may occur. Reproductive capacity does not appear to decline significantly in aging men.

## Review Questions

 Access additional practice questions using your smartphone, tablet, or computer: Mastering A&P®> Study Area > Practice Tests & Quizzes

### Multiple Choice

*More than one choice may apply.*

1. Which of the following can be found in semen?
   a. Sperm
   b. Relaxin
   c. Cholesterol
   d. Glucose

2. In terms of development, which of these pairs is *mismatched*?
   a. Vagina—penis
   b. Testis—ovary
   c. Labia majora—scrotum
   d. Uterine tube—ductus deferens

3. The myometrium is the muscular layer of the uterus, and the endometrium is the _____ layer.
   a. serosa
   b. adventitia
   c. submucosa
   d. mucosa

4. Which of the following is (are) true concerning human spermatids?
   a. They are the female gametes.
   b. They contain 46 chromosomes.
   c. They contain 23 chromosomes.
   d. They are motile.

5. The ductus deferens can be found in the
   a. scrotum.
   b. prostate.
   c. inguinal canal.
   d. penis.

6. Which of the following attach to the ovary?
   a. Fimbriae
   b. Ovarian ligament
   c. Suspensory ligaments
   d. Broad ligament

7. Human ova and sperm are similar in that
   a. about the same number of each is produced per month.
   b. they have the same degree of motility.
   c. they are about the same size.
   d. they have the same number of chromosomes.

8. Select the *false* statement about the cervix of the uterus.
   a. It is the superiormost part of the uterus.
   b. It projects into the vagina.
   c. Its cervical glands secrete mucus.
   d. It contains the cervical canal.

9. Each month, typically only one
   a. primary follicle is stimulated.
   b. follicle secretes estrogens.
   c. vesicular follicle undergoes ovulation.
   d. ovary is stimulated.

10. After ovulation, the ruptured follicle
    a. degenerates.
    b. becomes a corpus luteum.
    c. sloughs off as waste material.
    d. mends and produces another oocyte.

11. The outer layer of the blastocyst, which attaches to the uterine wall, is the
    a. yolk sac.          c. amnion.
    b. inner cell mass.   d. trophoblast.

12. The crown-to-rump length of an 8-week human embryo is
    a. 3 mm.    c. 22 mm.
    b. 10 mm.   d. 90 mm.

13. During human embryonic development, organogenesis occurs
    a. during the first trimester.
    b. during the second trimester.
    c. during the third trimester.
    d. just before birth.

## Short Answer Essay

14. What are the primary sex organs, or gonads, of males? What are their two major functions?

15. What is the function of seminal fluid? Name the three types of glands that help produce it.

16. The penis contains erectile tissue that becomes engorged with blood during sexual excitement. What term is used to describe this event?

17. Define *ejaculation*.

18. Why are the male gonads not found in the abdominal cavity? Where are they found?

19. How does enlargement of the prostate interfere with a man's reproductive function?

20. When does spermatogenesis begin? What causes it to begin?

21. Testosterone causes the male secondary sex characteristics to appear at puberty. Name three examples of male secondary sex characteristics.

22. What is circumcision? Does it affect secondary sex characteristics?

23. Name the female gonad, and describe its two major functions.

24. Why is the term *urogenital system* more applicable to males than females?

25. Name the structures of the female duct system, and describe the important functions of each.

26. How is the uterus anchored in the pelvis?

27. What is a follicle? What is ovulation?

28. What is mittelschmerz, and when is it most likely to occur?

29. What is the most important diagnostic method to detect cervical cancer?

30. List and describe the events of the menstrual cycle. Why is the menstrual cycle so important?

31. Define *menopause*. What does this mean to a woman?

32. Define *fertilization*. Where does fertilization usually occur? Describe the process of implantation.

33. How is body functioning of a pregnant woman altered by her pregnancy?

34. What events trigger labor?

35. Delivery of the infant occurs during which stage of labor?

36. What is the *indifferent stage* of embryonic development?

37. What are the major events of puberty?

38. How important are hereditary defects in the development of breast cancer? What are some of the methods of early detection of this cancer?

16

## Critical Thinking and Clinical Application Questions

39. A pregnant woman in substantial pain called her doctor and explained (between sobs) that she was about to have her baby "right away." The doctor calmed her and asked how she had come to that conclusion. She said that her water had broken and that her husband could see the baby's head. **(a)** Was she right to believe that birth was imminent? If so, what stage of labor was she in? **(b)** Do you think she had time to make it to the hospital 60 miles away? Why or why not?

40. Crystal had both her left ovary and her right uterine tube removed surgically at age 17 because of a cyst and a tumor in these organs. Now, at age 32, she remains healthy and is expecting her second child. How could Crystal conceive a child with just one ovary and one uterine tube, separated on opposite sides of the pelvis?

41. A sperm enters a polar body, and their nuclei fuse. Why would it be unlikely for the resulting cell to develop into a healthy embryo?

# Appendix A

## Answers to Did You Get It? Questions and Multiple Choice Review Questions

### Chapter 1

#### Did You Get It?

**1.** Anatomy and physiology are related. A given function can occur only if the corresponding structure allows it. **2.** False. They are all topics of physiology. **3.** The stomach exhibits the organ level of structural organization. Glucose is at the chemical level. **4.** These organs are part of the respiratory system. **5.** The urinary system rids the body of waste and helps regulate blood pressure. **6.** Survival also depends on the ability to maintain one's boundaries, to move, to respond to stimuli, and to reproduce. **7.** Oxygen is required for the chemical reactions that release energy from food and form ATP. Metabolism is the life function in which oxygen directly participates. **8.** The anatomical position is a standard position that serves as a reference point in descriptions of the body and its structures. In this position, the body is standing erect, feet are parallel, arms hang at the sides, and palms face forward. It's important to understand it because most descriptions of the body using anatomical terminology refer to body regions as if the body is in the anatomical position regardless of its actual position. **9.** The axillary region is the armpit. The acromial region is the point of the shoulder. **10.** Left posterior antebrachial region. **11.** To separate the thoracic and abdominal cavities, you would make a transverse section (cross section). **12.** Of these organs, the small intestine and uterus are in the abdominopelvic cavity. **13.** Joe might have appendicitis. **14.** No. We mean that they vary within a narrow and regulated range. **15.** Thirst is part of a negative feedback system. Thirst prods us to drink fluids (the response), which in turn causes the thirst sensation to decrease and end. Were it a positive feedback mechanism, we would become even more thirsty (the stimulus for drinking would increase).

#### Review Questions

**1.** d; **2.** a, c, d, e; **3.** superior, superficial, proximal, lateral, medial, posterior; **4.** 1-e, 2-c, 3-i, 4-f, 5-h, 6-a, 7-b, 8-d, 9-g; **5.** c; **6.** c; **7.** b, c, d; **8.** a, c, d, e

### Chapter 2

#### Did You Get It?

**1.** Matter is the substance of living and nonliving objects. Energy is the mover of the substance—the ability to do work. **2.** Electrical and chemical energy are both involved in transmitting messages in the body. **3.** Potential energy and kinetic energy, respectively. **4.** Every time energy changes from one form to another, some heat is given off to the environment (is lost) and is unusable. **5.** Carbon, oxygen, hydrogen, and nitrogen. **6.** An atom is the smallest particle of an element that still retains the element's properties. **7.** The atom's atomic number (#p) is 4; its atomic mass number (#n + p) is 9. **8.** Radioisotope. **9.** A molecule is two or more atoms chemically bound together. **10.** A molecule of an element is a chemical combination of two or more atoms of the same kind. In a molecule of a compound, the atoms differ. **11.** In ionic bonds, electrons are completely transferred from one atom to another. In covalent bonds, the interacting atoms share one or more electron pairs. **12.** Hydrogen bonds. **13.** It is a decomposition reaction. **14.** A reversible reaction is shown by double reaction arrows. **15.** The high heat capacity of water prevents rapid changes in body temperature. **16.** Acids are proton donors. **17.** A pH of 11 is basic. A pH of 5 is 1,000,000-fold more acidic than pH 11. **18.** All chemical reactions in the body take place in a watery environment. **19.** They conduct an electrical current when dissolved in water. **20.** The structural units of carbohydrates are monosaccharides. Those of lipids are glycerol and fatty acids. **21.** Phospholipids and cholesterol are in cellular membranes; phospholipids are the most abundant. **22.** Dehydration synthesis. **23.** A chain of amino acids. **24.** Fibrous proteins. **25.** The enzyme is able to specifically bind with its substrate(s). **26.** DNA contains the bases A, T, G, and C, and the sugar deoxyribose. RNA contains the bases A, G, C, and U, and its sugar is ribose. **27.** ATP is the immediately useful form of energy for all body cells.

#### Review Questions

**1.** b, c, e; **2.** b, c, d, f; **3.** a, b, c, d, e; **4.** c; **5.** b, c; **6.** b; **7.** a, e; **8.** c; **9.** a; **10.** a, c

### Chapter 3

#### Did You Get It?

**1.** A *cell* is the (living) unit of life. **2.** What its cells can do. **3.** Plasma membrane = external barrier that regulates what enters and leaves the cell. Cytoplasm = the area where most cell activities occur. Nucleus = control center of the cell. **4.** The generalized cell is a concept that describes organelles and functions common to all cells. **5.** Nucleoli are the sites of synthesis of ribosomes, which are important in protein synthesis. **6.** The phospholipids have both polar (heads) and nonpolar (tails) regions. Polar aligns with polar (water and other polar molecules inside and outside the cell). Nonpolar aligns with nonpolar in the membrane interior. **7.** They act as receptors, determine blood type, and play a role in cell-to-cell interactions. **8.** Communication and binding together, respectively. **9.** The cytosol is the liquid portion of the cytoplasm. Cytoplasm includes cytosol, organelles, and inclusions. **10.** Lysosomes break down ingested bacteria, worn-out organelles, and dead cells. Peroxisomes detoxify a number of harmful toxic substances and disarm free radicals. **11.** Mitochondria are the major site of ATP synthesis. The Golgi apparatus is the packaging site.

**12.** Microtubules and microfilaments are involved in cell mobility. **13.** The basis of centrioles is microtubules; that of microvilli is a core of actin filaments. **14.** Microvilli increase the cell surface area for absorption. **15.** Fibroblasts and erythrocytes. **16.** Neurons gather information and control body functions. **17.** Kinetic energy is the energy source for diffusion. **18.** The concentration gradient determines the direction that water and solutes move by diffusion. Movement is from high to low concentration of a given substance. **19.** A channel protein is an opening formed by membrane proteins for diffusion of certain small solutes. A carrier protein undergoes shape changes that allow diffusion of a specific substance through the membrane. **20.** "Down the concentration gradient" means to move from where there is a lot of something (high concentration) to where there is not a lot (low concentration). **21.** When the pump protein is phosphorylated, it changes shape. When $K^+$ binds, phosphate is released. **22.** Phagocytosis moves large particles into the cell. **23.** Receptor-mediated endocytosis. **24.** DNA is double-stranded. When it is replicated, each strand serves as a template to build a complementary strand. Thus, if the template strand is ACT, the complementary strand formed at that site is TGA. **25.** Cytokinesis is the division of the cytoplasm. If cytokinesis does not occur, the result is a binucleate cell. **26.** mRNA carries the coded information for building proteins from the DNA to the ribosome where protein synthesis occurs. tRNA delivers amino acids to the ribosome and "checks" the location by recognizing the mRNA codon with its anticodon. **27.** Transcription and translation. Proteins are synthesized during translation. Transcription is the production of mRNA using DNA as a template. **28.** Cell shape and cell arrangement (layers) are the two criteria used to classify epithelium. **29.** Exocrine glands have ducts that carry their secretion (typically a protein-containing secretion other than hormones) to a free body surface. Endocrine glands produce only hormones and are ductless glands. **30.** Can repair itself; cells have specialized cell junctions. **31.** The two hallmarks of epithelial tissue include having one free surface and being avascular. **32.** The two unique features of connective tissues include variations in blood supply and the production of a nonliving, extracellular matrix that surrounds their living cells. **33.** Collagen fibers provide strength. **34.** Skeletal muscle. **35.** It allows neurons to control structures some distance away. **36.** Nervous tissue is irritable, meaning it is able to respond to stimuli. **37.** Dense fibrous connective tissue is the basis for fibrosis, or scar formation. **38.** Epithelium and some connective tissues remain mitotic. **39.** *Neoplasm* means "new growth." It is an abnormal growth that occurs when normal control of cell division is lost; this was also described as "mitosis gone wild." **40.** Endocrine activity tends to decline with age.

## Review Questions

**1.** a; **2.** a, d; **3.** b; **4.** e; **5.** c; **6.** a, c; **7.** b; **8.** b; **9.** a; **10.** a; **11.** b, e; **12.** c

# Chapter 4

## Did You Get It?

**1.** Serous membranes line ventral body cavities closed to the exterior. Mucous membranes line body cavities open to the exterior (respiratory, digestive, urinary, and reproductive organ cavities). **2.** Parietal pleura, visceral pleura, (lung), visceral pleura, parietal pleura, parietal pericardium, visceral pericardium, (heart). **3.** Lining a fibrous capsule surrounding a joint. **4.** The skin is the epithelial membrane that covers the body surface. *Cutaneous membrane* is a synonym for skin, as is *integument*, which means "covering." The integumentary system is the skin and its derivatives (nails, hair, glands). **5.** Any three of the functions listed in Table 4.1. **6.** Keratinocytes are the most abundant cell type in the epidermis. **7.** The stratum basale. **8.** The stratum corneum. **9.** Cells begin to die as they leave the stratum granulosum. **10.** The dermal papillae in the papillary layer of the dermis are responsible for fingerprints. **11.** Just the epidermis. **12.** Melanin, carotene, and hemoglobin all contribute to skin color. **13.** The sebaceous glands, which produce oily secretions. **14.** Both are dilute salt solutions containing vitamins and wastes. Apocrine secretion also contains proteins and fatty acids. **15.** No, the nail won't regrow, because the growth region (nail matrix) is torn off. **16.** Loss of body fluids containing needed proteins and electrolytes, resulting in circulatory shock, and overwhelming infection. **17.** First-degree burns are red and swollen; only epidermis damaged. Second-degree burns damage the epidermis and the superficial part of the dermis; blisters appear, but epithelial regeneration can occur. Third-degree burns destroy the entire skin thickness and can extend into the subcutaneous tissue; the burn is gray and painless; must graft. Fourth-degree burns destroy all layers of the skin; appear charred; extend into deep tissues such as muscle or bone. **18.** ABCDE rule. **19.** UV radiation exposure (sun or tanning bed). **20.** Because stratum corneum cells are dead and thus no longer are dividing. **21.** Loss of subcutaneous fat. **22.** The sebaceous glands of the baby.

## Review Questions

**1.** c; **2.** b, e; **3.** a, b, c; **4.** a, c, e; **5.** a, c, d; **6.** c, d; **7.** c; **8.** a, c; **9.** 1-e, 2-d, 3-f, 4-b, 5-g, 6-c

# Chapter 5

## Did You Get It?

**1.** Muscles use bones as levers to bring about body movements. **2.** Red bone marrow provides a site for hematopoiesis, and yellow bone marrow is a storage site for fat. **3.** Most long bones are found in the limbs. **4.** Shaft = diaphysis; bone ends = epiphyses. **5.** Compact bones appear solid and very dense with few holes. Spongy bone areas look like the cross-beams of a house with lots of space between the bone spicules. **6.** They carry nutrients to the bone cells. **7.** Membranes or cartilage. **8.** The hormonal stimulus (PTH) maintains blood calcium homeostasis. **9.** Bones will become thinner and weaker because osteoclasts are bone-destroying cells. **10.** A fracture is a break in a bone. Compression and comminuted fractures are common in the elderly. **11.** Skull, vertebral column, and bony thorax. **12.** Eating or talking. These are the temporomandibular joints. **13.** Maxillae. **14.** Ethmoid bone. **15.** Frontal joins with the parietals at the coronal suture. Parietals join one another at the sagittal suture. **16.** Cervical, thoracic, lumbar, sacral, coccygeal. **17.** All typical cervical vertebrae are small, have holes in their transverse processes, and have a split spinous process. Lumbar vertebrae are large, blocklike vertebrae with a blunt spinous process that projects directly back. They have a large body and no foramina in their transverse processes. **18.** A true rib is attached directly to the sternum by its own costal cartilage. A false rib attaches indirectly (by the costal cartilage of a superior rib) or not at all. **19.** Vertebrae. **20.** Flat bones. **21.** The thoracic and sacral curvatures are present. **22.** At birth the newborn's spine is an arc (C-shaped), whereas an adult's spine has two additional curvatures and is S-shaped. **23.** The axial skeleton forms the body axis and protects the brain and viscera. The appendicular skeleton allows for mobility and manipulation of the external environment. **24.** The clavicle attaches medially to the sternum. **25.** The humerus forms the skeleton of the arm. **26.** Carpals are the short bones found in the wrist. **27.** Radius and ulna. **28.** Ilium, ischium, and pubis form the hip bone. Each pelvic girdle is formed by two coxal (hip) bones and the sacrum. **29.** The female pelvis is broader, lighter, has a less acute pubic angle, a wider inlet and outlet, and shorter ischial spines. **30.** Tibia and fibula form the skeleton of the leg. **31.** The talus allows toe pointing. **32.** The femur has those markings. **33.** Joints connect bones together while allowing flexibility of the body. **34.** The material between the articulating bone ends, which is connective tissue fibers in fibrous joints and cartilage in cartilaginous joints. **35.** Lining a synovial joint capsule to provide a source of lubricating fluid in the joint. **36.** The shoulder and hip joints are ball-and-socket joints. The carpometacarpal joint of the thumb is a saddle joint. **37.** Compression fracture; osteoporosis. **38.** Lower limbs and facial skeleton grow most rapidly during childhood.

## Review Questions
**1.** a, b, d; **2.** d; **3.** a, d; **4.** d; **5.** a, b, c, d, e; **6.** a, c; **7.** d; **8.** b, d, e; **9.** c; **10.** b; **11.** 1-a, b, 2-a, 3-a, 4-a, 5-b, 6-c, 7-c, 8-a, 9-c; **12.** 1-a, d, f; 2-b, e; 3-c

## Chapter 6

### Did You Get It?
**1.** Skeletal muscle fibers are long multinucleate cells with obvious striations. Cardiac cells are branching, typically uninucleate cells with less obvious striations but obvious junctions. Smooth muscle fibers are spindle-shaped uninucleate cells, nonstriated. **2.** Skeletal muscle. **3.** Striped or having bands = striated. **4.** Skeletal muscle movements can be very forceful and rapid, whereas smooth muscle movements tend to be slow and often rhythmic. **5.** The alignment of the bands on the myofilaments is responsible for the banding pattern in skeletal muscle cells. **6.** The axon ending of a motor neuron and the sarcolemma of the skeletal muscle fiber. **7.** Sodium ions enter the cells during action potential generation. **8.** Calcium stimulates the release of neurotransmitter (ACh) at the axon terminal of the motor neuron and causes the regulatory proteins blocking the myosin-binding sites on actin to move so cross bridges can form and allow contraction. **9.** Calcium ions trigger the sliding of the myofilaments. **10.** Pulling a bucket out of a well. **11.** Phosphorylation of ADP by CP, stored ATP, and ATP generated by glucose oxidation. **12.** Stored ATP. **13.** Isometric contraction. **14.** Oxygen deficit occurs when a person is not able to take in oxygen fast enough to keep the muscles supplied with all the oxygen they need when working vigorously. **15.** Resistance exercise leads to increased muscle size. **16.** Abduction. **17.** Flexion and extension. **18.** They anchor or aid the activity of the prime mover. **19.** Tibialis anterior—a muscle overlying the tibia anteriorly. Erector spinae—muscles that straighten the spine. Rectus abdominus—muscle that runs straight up the abdomen. Extensor carpi radialis longus—long muscle on the posterior of the forearm that extends the wrist and attaches at the distal radius. **20.** Circular. **21.** Frontalis raises the eyebrows. **22.** Masseter and temporalis are synergists in jaw closure. **23.** Erector spinae. **24.** It's like plywood. The various abdominal muscles run in different directions across the abdomen, making the abdominal wall very strong. **25.** Latissimus dorsi. **26.** Triceps brachii. **27.** Quadriceps on anterior thigh. **28.** The medial gluteal site and the deltoid muscle of the shoulder. **29.** Soleus and gastrocnemius. They plantar flex the foot. **30.** Nerve fibers must be myelinated. **31.** Exercise defers or reduces the natural loss in muscle mass and strength that occurs in old age.

## Review Questions
**1.** c, e; **2.** a; **3.** c; **4.** a, d; **5.** a, b, c, d; **6.** a, b; **7.** a; **8.** b, d

## Chapter 7

### Did You Get It?
**1.** CNS = brain and spinal cord. PNS = nerves that extend to and from the CNS. **2.** Astrocytes are the most numerous neuroglia. Oligodendrocytes produce myelin. **3.** Neuroglia can divide. Most neurons cannot. A criterion of cancer cells is their uncontrolled division. **4.** A tract is a bundle of nerve fibers in the CNS. A nerve is a bundle of nerve fibers in the PNS. **5.** A ganglion is a cluster of nerve cell bodies in PNS; a nucleus is a cluster of nerve cell bodies in the CNS. **6.** Dendrites conduct impulses toward the nerve cell body; the axon terminal releases neurotransmitters. **7.** The fiber that conducts at 40 m/sec. **8.** A graded potential is a local current that dies out with distance. An action potential is a current that is continuously regenerated along the length of the axon and does not die out. **9.** The synapse is the location where an impulse passes from one neuron to another or to a target cell. The synaptic cleft is the gap at the synapse that neurotransmitter must cross to transmit the signal. Chemically via the release of a neurotransmitter and binding of the neurotransmitter to the postsynaptic membrane. **10.** Dendrites. **11.** A reflex is a rapid, predictable, and involuntary response to a stimulus. **12.** Cerebral cortex, white matter, and basal nuclei. **13.** Mostly myelinated nerve fibers. **14.** The brain stem controls vital activities. **15.** The cerebellum provides precise timing for skeletal muscle activity and helps control our balance and equilibrium. **16.** Diencephalon. **17.** Ventricles. **18.** Blood-brain barrier. **19.** Arachnoid mater. **20.** Contusion. **21.** Nerve cell bodies of interneurons and motor neurons. **22.** Sensory pathways are ascending pathways. **23.** Because the leash of nerve fibers there looks like a horse's tail, the literal translation of *cauda equina.* **24.** Around each nerve fiber. **25.** Vagus nerves. **26.** Nerve plexus = complex network of nerves. **27.** Sciatic nerve of the sacral plexus. **28.** Visceral organs (smooth and cardiac muscles and glands) are served by the ANS. Skeletal muscles are served

by the somatic nervous system. **29.** The ANS has a two-motor neuron pathway from the CNS to the organ to be served. The somatic nervous system has just one motor neuron in the motor pathway. **30.** Sympathetic division. **31.** They are unable to regulate their body temperature until the hypothalamus matures. **32.** It is hypotension caused by a rapid change in position, such as getting up quickly from a reclining position. The sympathetic nervous system, which regulates blood pressure, is less efficient in old age.

## Review Questions

**1.** d; **2.** b, c; **3.** b; **4.** d; **5.** c; **6.** 1-e, 2-d, 3-f, 4-b, 5-e, 6-h, 7-a; **7.** b, c; **8.** c; **9.** b; **10.** a; **11.** b; **12.** b, c, d

## Chapter 8

### Did You Get It?

**1.** The eyelids protect the eyes. **2.** The tarsal glands, ciliary glands, conjunctiva, and lacrimal glands all contribute to lubricating the eye, though their secretions differ. **3.** Lysozyme is a component of tears that helps to destroy bacteria and protect the eye from infection. **4.** They direct the eyeball toward what you wish to view. **5.** The blind spot contains no photoreceptors; it is the site where the optic nerve leaves the eyeball. **6.** Both contain pigment, which prevents light scattering in the eye. **7.** Rods have a rodlike outer segment containing the photopigment, whereas cones have a shorter cone-shaped outer segment. Rods respond to low light conditions and produce black-and-white vision; cones need bright light and provide color vision. **8.** Refractory media include the cornea, aqueous humor, the lens, and vitreous humor. **9.** Accommodation. **10.** The optic nerves leave the eyeballs, and the medial half of the fibers of each optic nerve cross over to the opposite side, joining there with the fibers from the outside half of the opposite eye to form the optic tracts. **11.** It causes pupillary constriction in very bright light. Intense light stimulation can injure the photoreceptors. **12.** Astigmatism results from unequal curvatures on the lens surface, not from an eyeball that is too long or too short to focus the image on the retina. The unequal curvatures of astigmatism result in points of light that focus on the retina as lines, not points, leading to blurry images. **13.** External

and middle ears serve hearing only. **14.** The ossicles (hammer, anvil, and stirrup). **15.** Balance or equilibrium. **16.** Dynamic receptors located in the semicircular canals (crista ampullaris) have embedded in the gel-like cupula; static receptors located in the vestibule (maculae) have otoliths that move when the head moves, causing hairs embedded in the otolithic membrane to bend. **17.** Otoliths are tiny stones made of calcium salts that are located in the maculae of the vestibule. They respond to static equilibrium cues relative to the position of the head in space. **18.** Tympanic membrane to bones of ossicles to fluids of the cochlear scalae. **19.** Cochlear nerve (division of cranial nerve VIII). **20.** Close to the oval window. **21.** Sensorineural deafness results from damage to neural structures involved in hearing (cochlear nerve, auditory region of the brain), whereas conductive deafness results from anything that prevents sound vibrations from reaching the cochlea (earwax, fusion of the ossicles, fluid in the middle ear). **22.** Chemoreceptors, because they respond to chemicals in solution. **23.** On the tongue. **24.** Odor receptors are located at the superior aspect of the nasal cavity. Sniffing brings the air upward. **25.** Presbyopia; caused by reduced elasticity of the lens as a result of aging. **26.** Vision. **27.** Deafness of old age.

## Review Questions

**1.** d; **2.** b, c; **3.** a, c; **4.** d; **5.** b; **6.** a, c; **7.** a, b, c, d; **8.** a, c; **9.** b; **10.** a, b, c

## Chapter 9

### Did You Get It?

**1.** The endocrine system delivers its commands slowly via hormones carried by the blood. The nervous system uses rapid electrical messages that are much faster, allowing you to lift your foot off the glass more quickly. **2.** A hormone is a chemical messenger used by the endocrine system. The target organ of a hormone is the specific cells or tissues that the hormone affects. **3.** Because cAMP is not a first messenger, which is a molecule that attaches to a receptor on the cell's plasma membrane and triggers the reactions leading to synthesis of the second messenger in the cell. **4.** Endocrine organs are stimulated by hormones, by chemicals other than hormones, and by the nervous system.

**5.** Endocrine glands are ductless, and they release their hormone products directly into the intercellular fluid. Exocrine glands release their nonhormonal products to an epithelial surface via a duct. **6.** The posterior pituitary is nervous tissue that acts as a storage and releasing area for hormones sent to it by the hypothalamus. **7.** Diabetes insipidus, in which copious amounts of urine are voided as a result of hyposecretion of ADH. **8.** Tropic hormones stimulate certain endocrine organs to secrete their hormones. **9.** Functional thyroid hormone has iodine as part of its structure. **10.** Melatonin, produced by the pineal gland. **11.** The parathyroid glands reside on the posterior thyroid gland. **12.** PTH, produced by the parathyroid glands, increases the blood calcium ion level. **13.** Calcitonin, produced by the parafollicular cells of the thyroid gland, reduces the blood calcium ion level. **14.** Thymosin programs the T lymphocytes, which essentially direct our immune responses. **15.** Aldosterone stimulates the kidneys to reabsorb more sodium ions. **16.** Glucocorticoids such as cortisol. **17.** Her blood glucose level would be elevated because of her stress. When we are stressed, both glucocorticoids (adrenal cortical hormones) and epinephrine and norepinephrine (adrenal medullary hormones) are produced in increased amounts. Both sets of hormones promote a rise in the blood glucose level. **18.** Insulin stimulates cellular uptake of glucose. **19.** Estrogen. **20.** Stomach and small intestine. **21.** Placenta. **22.** Ovaries. **23.** Growth hormone decline results in muscle atrophy; estrogen decline leads to osteoporosis.

## Review Questions

**1.** d; **2.** b, c; **3.** a, c; **4.** c; **5.** a, b, d; **6.** a, b, c, d; **7.** c, d; **8.** b; **9.** b

## Chapter 10

### Did You Get It?

**1.** The liver is the major source of plasma proteins. **2.** Erythrocytes, leukocytes, and platelets. **3.** The color of blood varies with the amount of oxygen it is carrying. From most oxygen to least, the blood goes from scarlet to dull red or purple. **4.** Hemoglobin transports oxygen and a small amount of carbon dioxide in the blood. **5.** Lymphocytes are the main actors in body immunity. **6.** Infection in the body causes an increase in WBC count, thus 15,000/mm$^3$. **7.** Anemia. **8.** The hemocytoblast gives rise to all formed elements. **9.** Lack of a nucleus; therefore, they cannot carry out transcription and translation to produce proteins (enzymes and others). **10.** The stem cell (megakaryocyte) undergoes mitosis many times, forming a large multinucleate cell, which then fragments into platelets. **11.** Inactivity, leading to blood pooling, and anything that roughens or damages the lining of a blood vessel (laceration, atherosclerosis, or physical trauma). **12.** The self-antigens (agglutinogens) the RBCs bear. **13.** A transfusion reaction in which the RBCs are lysed and hemoglobin enters the bloodstream, potentially leading to kidney shutdown. **14.** O-positive. **15.** An antigen is a substance foreign to the body that activates and is attacked by the immune system. An antibody is a protein released by immune cells that binds with a specific antigen and inactivates it in some way. **16.** Fetal HbF has a greater ability to bind oxygen and binds it more strongly than adult HbA. **17.** Leukemia, pernicious anemia, and clotting disorders are particularly common in the elderly.

### Review Questions
**1.** a, b, d; **2.** a, b, c, d; **3.** d; **4.** c; **5.** a, c; **6.** c, d; **7.** b, d; **8.** c; **9.** b, c, d; **10.** a, d; **11.** d; **12.** a

## Chapter 11

### Did You Get It?
**1.** The heart is in the mediastinum between the lungs. **2.** The left ventricle has the thickest walls. This reflects its function, which is to pump blood through the whole body. **3.** Pulmonary circulation strictly serves gas exchange. Oxygen is loaded and carbon dioxide is unloaded from the blood in the lungs. Systemic circulation provides oxygen-laden blood to all body organs. **4.** Heart valves keep blood flowing in one direction through the heart. **5.** The coronary arteries supply the myocardium (cardiac muscle) with oxygen. If that circulation fails, the heart fails. **6.** The intrinsic conduction system of the heart coordinates the action of the heart chambers and causes the heart to beat faster than it would otherwise. **7.** Left

ventricle **8.** The atria are relaxing, and the ventricles are contracting. **9.** The operation of the heart valves. **10.** CO = amount of blood pumped out by each side of the heart in 1 minute. **11.** Fever increases the heart rate because the rate of metabolism of the cardiac muscle increases. **12.** Venous return. **13.** It is a vein. **14.** Blood pressure in veins is much lower than that in arteries because veins are farther along in the circulatory pathway. Hence, veins need extra measures to force blood back to the heart. **15.** Capillary walls consist only of the innermost intima layer, which is very thin. Capillaries are the exchange vessels between the blood and tissue cells; thus, thin walls are desirable. **16.** Lower limb. **17.** Upper limb. **18.** Hepatic portal vein. **19.** The pulmonary circulation is much shorter and requires a less powerful pump than the systemic circulation does. Pulmonary arteries carry oxygen-depleted/carbon dioxide–rich blood, whereas the pulmonary veins carry oxygen-rich/carbon dioxide–depleted blood. The opposite is true of the arteries and veins of the systemic circulation. **20.** Radial artery at the wrist; femoral at the groin; common carotid at the side of the neck. **21.** It decreases from heart to venae cavae. **22.** Hemorrhage reduces blood pressure initially because blood volume drops. **23.** Fluid enters the capillary bed at its venous end.

### Review Questions
**1.** c; **2.** b; **3.** d; **4.** b; **5.** c; **6.** a, c; **7.** c; **8.** a, c; **9.** a, d; **10.** b; **11.** a; **12.** b; **13.** a, b, c; **14.** b, c; **15.** b; **16.** a, b, c; **17.** b; **18.** a

## Chapter 12

### Did You Get It?
**1.** Lymphatic vessels pick up fluid and proteins leaked from the blood into the interstitial space. **2.** Lymphatic capillaries are blind-ended and not fed by arteries as blood capillaries are. They also have flaplike minivalves that make them more permeable than blood capillaries. **3.** Particularly large collections of lymph nodes occur in the axillary, inguinal, and cervical regions. **4.** The number of afferent lymphatic vessels entering the node is greater than the number of efferent vessels leaving the node at the hilum. Therefore, lymph flow stagnates somewhat. **5.** The spleen destroys worn-

out RBCs. **6.** The lymphoid tissue associated with the mucosa—tonsils in the throat, and Peyer's patches in the intestinal walls—which acts to prevent pathogens from entering the body through the mucosa. **7.** Innate defenses are nonspecific and always ready to protect the body. They include intact membranes (mucosa, skin), inflammatory response, and several protective cell types and chemicals. Adaptive defenses must be programmed and specifically target particular pathogens or antigens. **8.** Redness, heat, swelling (edema), and pain. **9.** It attaches to foreign cells, and when it is activated, membrane attack complexes (MACs) are inserted in the foreign cell's membrane and produce lesions that allow water entry and cause lysis. **10.** Viruses elicit interferon formation. **11.** An antigen is a foreign substance in the body. A self-antigen is a body protein, typically displayed in the plasma membrane, that is foreign to anyone but oneself. **12.** B lymphocytes and T lymphocytes. B lymphocytes mount a humoral response by producing antibodies. T lymphocytes mount the cellular response by activating B cells and cytotoxic T cells and stimulating the inflammatory response. **13.** In the thymus. **14.** They are antigen presenters to the T cells. **15.** Primary response—this year's new strain will be seen by the body as a different virus. **16.** This is the region that binds to the antigen. **17.** IgA, a dimer. **18.** Neutralization occurs when antibodies attach to viruses or bacterial toxins, thereby blocking the virus or toxin from injuring the body. **19.** Agglutination cross-links cell-bound antigen; precipitation cross-links soluble antigen, which then falls out of solution like clotting proteins that have been activated; opsonization is the "marking" of antigen for phagocytosis, which also makes it a bit sticky so it is easier for phagocytes to "catch." **20.** T cells have to bind to both an antigenic particle and to a self-protein on the antigen-presenting cell. **21.** The cytotoxic T cell inserts perforin (toxic chemicals) into the foreign antigen's plasma membrane, which then develops holes. Then granzymes, enzymes from the T cell's cytoplasmic granules, enter and kill the foreign cell. **22.** To slow or stop the immune response once the "enemy" has been conquered. **23.** Helper T cells "direct" both humoral and cell-mediated adaptive responses; they interact with both B cells and T cells to stimulate cell division; release cytokines to recruit other phagocytes, such as neutrophils; and enhance macrophage activity. **24.** Allograft. **25.** An overzealous immune response against an otherwise harmless substance that causes injury to the body. **26.** In anaphylactic shock, released histamine causes constriction of the bronchioles, sudden vasodilation, and fluid loss. **27.** Abnormal production or functioning of immune cells or complement. **28.** Self-proteins that were not previously exposed to the immune system appear in the circulation, or foreign antigens that resemble self-antigens arouse antibodies that attack the self-antigens.

## Review Questions

**1.** d; **2.** a; **3.** c, d; **4.** a, c; **5.** a; **6.** b, d; **7.** c; **8.** c, d; **9.** b; **10.** b, c; **11.** d; **12.** a, b, c, d; **13.** a, b

## Chapter 13

### Did You Get It?

**1.** Because the respiratory mucosa rests on thin-walled veins that warm the incoming air, mucus produced by mucous glands and goblet cells moistens the air and traps dust and bacteria. The oral mucosa performs none of these functions. **2.** Ciliated cells of the mucosa move the sheet of contaminated mucus away from the lungs and toward the throat for swallowing. **3.** Nasal cavity, pharynx, larynx, trachea, bronchi, bronchioles, alveoli. **4.** The right bronchus, because it's wider and straighter. **5.** The passageways conduct air. The elastic tissue allows lungs to recoil passively when exhaling, saving energy. **6.** Respiratory bronchiole, alveolar duct, alveolar sac, and individual alveoli. **7.** To exchange gases between the external environment and the blood—oxygen in, carbon dioxide out. **8.** Increased air pressure in the lungs as they recoil. **9.** VC is largest; TV is smallest. **10.** About 350 ml reaches the alveoli. **11.** His left lung collapsed because the pressure in the intrapleural space (normally negative) became equal to atmospheric pressure. **12.** Diffusion. **13.** As bicarbonate ion. **14.** Cyanosis is a bluish cast to the skin (in light-skinned people), mucous membranes, and nailbeds due to inadequate oxygenation of the blood. **15.** The medulla oblongata. **16.** Increased

$CO_2$, which results in a drop in blood pH. **17.** He has COPD, specifically, emphysema. **18.** The lung alveoli will collapse after every breath; IRDS.

## Review Questions

**1.** d; **2.** a, d; **3.** a, b, d; **4.** b; **5.** b, c; **6.** c; **7.** c, e

## Chapter 14

### Did You Get It?

**1.** Mouth, pharynx, esophagus, stomach, small intestine, large intestine, anus. **2.** Protein digestion begins in the stomach. **3.** Alkaline mucus protects the stomach wall from being digested. Intrinsic factor is needed for intestinal absorption of vitamin $B_{12}$. **4.** The pyloric sphincter. **5.** Villi are extensions of the mucosa of the small intestine. They increase the surface area of the small intestine tremendously for nutrient absorption. **6.** Water absorption and absorption of some vitamins made by resident bacteria. **7.** The incisors cut, the canines tear or pierce, the premolars and molars crush and grind. **8.** Starch. Saliva contains salivary amylase. **9.** Bile, produced and secreted by the liver, acts as a detergent to mechanically break up large fat masses into smaller ones for enzymatic digestion. **10.** The pancreas. **11.** Ingestion, digestion, absorption, and defecation. **12.** Mechanical breakdown breaks food down physically by squeezing and pummeling it. Digestion uses enzymes to break the chemical bonds of the food molecules and release the nutrients. **13.** This is the voluntary stage in which the chewed food is pushed into the pharynx by the tongue. **14.** The main goal of segmentation is to mix the food thoroughly, though some propulsion occurs. Peristalsis is a propulsive activity. **15.** Food went down the air passageway (trachea) instead of the digestive passage (esophagus). **16.** Pepsin, the main enzyme of the stomach, needs acidic conditions to work. **17.** Amino acids. **18.** CCK causes the pancreas to secrete enzyme-rich fluid and stimulates the gallbladder to contract and the duodenal papilla to open. **19.** Enzymes located on the microvilli of intestinal absorptive cells. **20.** Plants—fruits, vegetables, and grains. **21.** It provides fiber, which is important for moving feces along the colon and defecation. **22.** Oils are

unsaturated lipids. **23.** Most are coenzymes that work with enzymes to bring about metabolic reactions. **24.** Cellular (aerobic) respiration. **25.** Carbon dioxide and reduced coenzymes. **26.** ATP and water. **27.** Fats are used in synthesis of myelin sheaths and of cellular membranes, and as body insulation. **28.** Making glucose from noncarbohydrate molecules, such as fats and proteins. **29.** The liver combines ammonia with carbon dioxide to form urea. Urea is then eliminated from the body in urine. **30.** HDLs, because they are headed back to the liver for breakdown and elimination from the body. **31.** Large surface area relative to body volume. **32.** Vasoconstriction of blood vessels serving the skin, and shivering. **33.** It decreases body temperature. **34.** Cystic fibrosis causes excessive production of mucus, which clogs the pancreatic ducts so that pancreatic enzymes cannot reach the small intestine; as a result, fat digestion is virtually stopped, resulting in fatty stools and an inability to absorb fat-soluble vitamins. **35.** A diet low in phenylalanine. **36.** You gain weight (fat).

## Review Questions

**1.** c, d; **2.** a, b, e; **3.** c; **4.** 1-e, 2-h, 3-c, 4-b, 5-g, 6-a, 7-f, 8-d; **5.** a; **6.** a, b; **7.** b, d; **8.** d; **9.** d; **10.** a; **11.** b, d; **12.** c, d; **13.** d

## Chapter 15

### Did You Get It?

**1.** Retroperitoneal = behind, or in back of, the peritoneum. **2.** Fat helps keep the kidneys anchored against the body trunk wall. When she lost weight, the amount of fat decreased, and the kidneys fell to a lower position (ptosis), causing the ureter(s) to kink and inhibit urine flow. **3.** Cortex, medulla, pelvis. **4.** The nephron. **5.** The renal tubule carries out tubular reabsorption and secretion. The peritubular capillaries receive the fluid, nutrients, and needed ions to be returned to the general circulation. **6.** It decreases glomerular pressure. **7.** The specific gravity of urine is higher because it contains more solutes than water. **8.** The bladder. **9.** It allows the cavity of the bladder to increase in volume to store more urine when necessary. **10.** It is longer (8″ vs. 1.5″), and it transports both urine and semen. The female urethra transports urine only. **11.** Voiding or urination. **12.** Excreting nitrogenous wastes, maintaining acid-base balance, water balance, and electrolyte balance of the blood. **13.** Thirst. **14.** Aldosterone increases sodium ion reabsorption by the kidney tubules, and water follows (if able). **15.** Osmoreceptors are in the brain (hypothalamus), and they respond to changing (increasing) solute content (osmolarity) of the blood. **16.** The homeostatic pH of blood in the body varies between 7.35 and 7.45. Any blood pH below 7.35 is considered too acidic for normal body function. **17.** A weak base, because a weak base will neutralize acidity by tying up only enough hydrogen ions to maintain a physiological pH, and there is no chance of overshoot and making the pH strongly basic. **18.** It will increase, because water will be lost in sweat (and urine). **19.** Kidneys maintain acid-base balance by excreting bicarbonate ions ($HCO_3^-$) and reabsorbing or generating new $HCO_3^-$. **20.** A condition in a newborn baby boy, in which the urethral opening is on the ventral surface of the penis. **21.** Urgency is the urgent feeling that you have to urinate. Frequency is the condition in which you urinate small amounts of urine frequently. **22.** Polycystic kidney disease is a degenerative condition in which cysts (blisterlike sacs) of urine interfere with normal kidney function and lead to kidney failure; glomerulonephritis is clogging of the glomerular filter with antigen-antibody complexes.

## Review Questions

**1.** d; **2.** d; **3.** a, d; **4.** b, c, d; **5.** a, b, c, d; **6.** c; **7.** c; **8.** c

## Chapter 16

### Did You Get It?

**1.** The testes produce male gametes (sperm) and secrete sex hormones, mainly testosterone. **2.** The seminiferous tubules produce sperm. **3.** Epididymis, ductus deferens, ejaculatory duct, and urethra (prostatic, membranous, and spongy regions). **4.** To make the penis firm by allowing it to fill with blood, so that it can serve as a penetrating device during sexual activity. **5.** Sperm are male gametes that fertilize the female's "eggs." Seminal fluid serves as a transport medium for sperm and contains various substances that nourish and protect the sperm or aid their movement. **6.** His prostate is probably hypertrophied. The prostate is immediately anterior to the rectum and can be examined digitally through the anterior rectal wall. **7.** FSH stimulates spermatogenesis. **8.** The final product of mitosis is two diploid cells that are genetically identical to each other and to the mother cell. The final product of meiosis is four cells, each with half the normal number of chromosomes (haploid). Genetic variation is introduced by meiosis. **9.** Excess cytoplasm is discarded, and what remains is compacted into the head, midpiece, and tail regions. Final maturation processes in the epididymis result in increased motility. **10.** Luteinizing hormone stimulates testosterone production. **11.** Female gametes (eggs, or ova). **12.** The uterus serves as an incubator. The uterine (fallopian) tubes are the most common site of fertilization. **13.** Vesicular, or Graafian, follicle. **14.** Polar bodies are also produced. They deteriorate and die because they lack sustaining cytoplasm. **15.** FSH promotes follicle development. **16.** LH stimulates ovulation. **17.** Estrogen. **18.** The functional layer of the endometrium is rebuilt. **19.** Progesterone (1) causes the secretory phase of the menstrual cycle, (2) helps prepare the breasts for lactation, and (3) helps quiet the uterine muscle to maintain pregnancy. **20.** A large percentage (80%) of women who carry the altered gene(s) develop breast cancer. **21.** Cleavage involves successive divisions not separated by growth of the cells, so the cells get smaller with each division. **22.** The placenta produces hormones of pregnancy, delivers nutrients and oxygen to the fetus, and removes wastes from the fetal blood. **23.** During pregnancy, many women have heartburn because the uterus crowds the stomach. Constipation is another problem because mobility of the digestive tract decreases. Nasal stuffiness and difficulty breathing are common. **24.** Dilation, expulsion, and placental stages. **25.** XY; testosterone. **26.** Cryptorchidism is failure of the testes to descend into the scrotum; if uncorrected, the result is infertility. **27.** Vaginal infections (*E. coli*, STIs, and fungal infections), which may lead to PID; abnormal or painful menses.

## Review Questions

**1.** a, b; **2.** a; **3.** d; **4.** c; **5.** a, c; **6.** b, c, d; **7.** d; **8.** a; **9.** c; **10.** b; **11.** d; **12.** c; **13.** a

# Appendix B

## Word Roots, Prefixes, and Suffixes

| Word element | Meaning | Word element | Meaning | Word element | Meaning |
|---|---|---|---|---|---|
| **Prefixes** | | | | | |
| a-, an- | absence, lack | hemo- | blood | phren- | diaphragm, mind |
| ab- | away from | hist- | tissue | pod- | foot |
| acro- | extreme, extremities | homo- | same | poly- | multiple |
| ad- | toward, to | hydro- | water | post- | after, behind |
| adeno- | gland | hygro- | moisture | pre- | before, ahead |
| ambi- | around, on both sides | hyper- | excess | procto- | rectum, anus |
| | | hypo- | below, deficient | pseudo- | false |
| amyl- | starch | ileo- | ileum | psycho- | mind |
| ante- | before, preceding | in- | in, within, into | pyel- | pelvis of the kidney |
| ant-, anti- | opposed to, preventing, inhibiting | inter- | between | pyo- | pus |
| | | intra- | within, inside | pyro- | fire |
| | | juxta- | near, close to | quadri- | four |
| bi- | two | lapar- | abdomen | radio- | radiation |
| bili- | bile | laryngo- | larynx | re- | back, again |
| bio- | life | later- | side | ren- | kidney |
| brachi- | arm | leuko- | white | retro- | backward, behind |
| brady- | slow | macro- | large | rhin- | nose |
| cholecysto- | gallbladder | mal- | bad | sacro- | sacrum |
| circum- | around | mast- | breast | salpingo- | fallopian tube |
| co, con- | together | medi- | middle | sarco- | flesh |
| contra- | against, opposite | mega- | large | sclero- | hard |
| cost- | ribs | men- | month | semi- | half |
| demi- | half | mono- | single | sex- | six |
| derm- | skin | multi- | many | skeleto- | skeleton |
| dis- | from | myelo- | marrow, spinal cord | steno- | narrow |
| dors- | back | | | sub- | beneath, under |
| dys- | difficult, faulty, painful | myo- | muscle | super-, supra- | above, upon |
| | | neo- | new | | |
| electro- | electric | nitro- | nitrogen | syn- | together |
| en-, em- | in, inside | non- | not | tachy- | rapid |
| equi- | equal | ob- | before, against | thyro- | a shield |
| eryth- | red | ortho- | straight, direct | trache- | trachea |
| exo- | outside, outer layer | osteo- | bone | trans- | across, through |
| | | pan- | all, universal | tri- | three |
| extra- | outside, beyond | para- | beside, near | ultra- | beyond |
| ferr- | iron | path- | disease | un- | not, back reversal |
| fibro- | fiber | ped- | child, foot | uni- | one |
| fore- | before, in front of | per- | by, through | uretero- | ureter |
| glyco- | sugar | peri- | around | urethro- | urethra |
| hemi- | half | pharyngo- | pharynx | uro- | urine, urinary organs |

# Word Roots, Prefixes, and Suffixes *(continued)*

| Word element | Meaning | Word element | Meaning | Word element | Meaning |
|---|---|---|---|---|---|
| **Suffixes** | | | | | |
| *-able* | able to | *-graph* | an instrument used for recording data or writing | *-otomy* | to incise |
| *-algia* | pain, in a certain spot | | | *-pathy* | disease |
| | | | | *-pexy* | fixation |
| *-cele* | tumor, swelling | *-ism* | condition | *-phage* | ingesting |
| *-centesis* | surgical puncture to remove fluid | *-itis* | inflammation | *-phobia* | fear |
| | | *-ize* | to treat | *-plasty* | plastic surgery |
| *-cide* | kill, destroy | *-lith* | stone, calculus | *-plegia* | paralysis |
| *-cule* | little | *-lithiasis* | presence of stones | *-rhexis* | rupture |
| *-ectasia* | dilating, stretching | *-lysis* | loosening or breaking down | *-rrhagia* | abnormal or excessive discharge |
| *-ectomy* | cutting out, surgical removal | *-megaly* | enlargement | *-rrhea* | flow or discharge |
| | | *-meter* | instrument that measures | *-scope* | lighted instrument for visual examination |
| *-esis* | action | | | | |
| *-form* | shape | *-oid* | like, resemble | *-scopy* | to examine visually |
| *genesis, genetic* | formation, origin | *-oma* | tumor | *-stomy* | establishment of an artificial opening |
| | | *-orrhaphy* | surgical repair | | |
| *-gram* | data that are systematically recorded, a record | *-osis* | disease, condition of | *-tomy* | incision into |
| | | *-ostomy* | to form an opening or outlet | *-uria* | urine |
| | | | | *-zyme* | enzyme, ferment |

# Appendix C

## Periodic Table of the Elements*

### Periodic Table of the Elements

Representative (main group) elements — Transition metals — Representative (main group) elements

| IA | IIA | IIIB | IVB | VB | VIB | VIIB | VIIIB | VIIIB | VIIIB | IB | IIB | IIIA | IVA | VA | VIA | VIIA | VIIIA |
|---|---|---|---|---|---|---|---|---|---|---|---|---|---|---|---|---|---|
| **1 H** 1.0079 | | | | | | | | | | | | | | | | | **2 He** 4.003 |
| **3 Li** 6.941 | **4 Be** 9.012 | | | | | | | | | | | **5 B** 10.811 | **6 C** 12.011 | **7 N** 14.007 | **8 O** 15.999 | **9 F** 18.998 | **10 Ne** 20.180 |
| **11 Na** 22.990 | **12 Mg** 24.305 | | | | | | | | | | | **13 Al** 26.982 | **14 Si** 28.086 | **15 P** 30.974 | **16 S** 32.065 | **17 Cl** 35.453 | **18 Ar** 39.948 |
| **19 K** 39.098 | **20 Ca** 40.078 | **21 Sc** 44.956 | **22 Ti** 47.867 | **23 V** 50.942 | **24 Cr** 51.996 | **25 Mn** 54.938 | **26 Fe** 55.845 | **27 Co** 58.933 | **28 Ni** 58.69 | **29 Cu** 63.546 | **30 Zn** 65.38 | **31 Ga** 69.723 | **32 Ge** 72.64 | **33 As** 74.922 | **34 Se** 78.96 | **35 Br** 79.904 | **36 Kr** 83.8 |
| **37 Rb** 85.468 | **38 Sr** 87.62 | **39 Y** 88.906 | **40 Zr** 91.224 | **41 Nb** 92.906 | **42 Mo** 95.96 | **43 Tc** 98 | **44 Ru** 101.07 | **45 Rh** 102.906 | **46 Pd** 106.42 | **47 Ag** 107.868 | **48 Cd** 112.411 | **49 In** 114.82 | **50 Sn** 118.71 | **51 Sb** 121.76 | **52 Te** 127.60 | **53 I** 126.905 | **54 Xe** 131.29 |
| **55 Cs** 132.905 | **56 Ba** 137.327 | **57 La** 138.906 | **72 Hf** 178.49 | **73 Ta** 180.948 | **74 W** 183.84 | **75 Re** 186.207 | **76 Os** 190.23 | **77 Ir** 192.22 | **78 Pt** 195.08 | **79 Au** 196.967 | **80 Hg** 200.59 | **81 Tl** 204.383 | **82 Pb** 207.2 | **83 Bi** 208.980 | **84 Po** 209 | **85 At** 210 | **86 Rn** 222 |
| **87 Fr** 223 | **88 Ra** 226 | **89 Ac** 227 | **104 Rf** 267 | **105 Db** 268 | **106 Sg** 271 | **107 Bh** 272 | **108 Hs** 270 | **109 Mt** 276 | **110 Ds** 281 | **111 Rg** 280 | **112 Cn** 285 | **113 Uut** 284 | **114 Uuq** 289 | **115 Uup** 288 | **116 Uuh** 293 | **117 Uus** 292 | **118 Uuo** 294 |

Rare earth elements

Lanthanides:

| 58 Ce 140.116 | 59 Pr 140.908 | 60 Nd 144.24 | 61 Pm 145 | 62 Sm 150.36 | 63 Eu 151.964 | 64 Gd 157.25 | 65 Tb 158.925 | 66 Dy 162.5 | 67 Ho 164.93 | 68 Er 167.26 | 69 Tm 168.934 | 70 Yb 173.054 | 71 Lu 174.967 |
|---|---|---|---|---|---|---|---|---|---|---|---|---|---|

Actinides:

| 90 Th 232.038 | 91 Pa 231.036 | 92 U 238.029 | 93 Np 237.048 | 94 Pu 244 | 95 Am 243 | 96 Cm 247 | 97 Bk 247 | 98 Cf 251 | 99 Es 252 | 100 Fm 257 | 101 Md 258 | 102 No 259 | 103 Lr 262 |
|---|---|---|---|---|---|---|---|---|---|---|---|---|---|

The periodic table arranges elements according to atomic number and atomic weight into horizontal rows called *periods* and 18 vertical columns called *groups* or *families*. The elements in the groups are classified as being in either A or B classes.

Elements of each group of the A series have similar chemical and physical properties. This reflects the fact that members of a particular group have the same number of valence shell electrons, which is indicated by the number of the group. For example, group IA elements have one valence shell electron, group IIA elements have two, and group VA elements have five. In contrast, as you progress across a period from left to right, the properties of the elements change in discrete steps, varying gradually from the very metallic properties of groups IA and IIA elements to the nonmetallic properties seen in group VIIA (chlorine and others), and finally to the inert elements (noble gases) in group VIIIA. This change reflects the continual increase in the number of valence shell electrons seen in elements (from left to right) within a period.

Class B elements are referred to as *transition elements*. All transition elements are metals, and in most cases they have one or two valence shell electrons. (In these elements, some electrons occupy more distant electron shells before the deeper shells are filled.)

In this periodic table, the colors are used to convey information about the phase (solid, liquid, or gas) in which a pure element exists under standard conditions (25 degrees centigrade and 1 atmosphere of pressure). If the element's symbol is solid black, then the element exists as a solid. If its symbol is red, then it exists as a gas. If its symbol is dark blue, then it is a liquid. If the element's symbol is green, the element does not exist in nature and must be created by some type of nuclear reaction.

*Atomic weights of the elements per IUPAC Commission on Isotopic Abundances and Atomic Weights, 2007.

# Appendix D

## Key Information about Vitamins and Many Essential Minerals

| Vitamin/mineral | Primary functions | Recommended intake* | Reliable food sources | Toxicity/deficiency symptoms |
|---|---|---|---|---|
| **Fat-Soluble** | | | | |
| A (retinol, retinal, retinoic acid) | Required for ability of eyes to adjust to changes in light Protects color vision Assists cell differentiation Required for sperm production in men and fertilization in women Contributes to healthy bone Contributes to healthy immune system | RDA: Men: 900 µg/day Women: 700 µg/day UL: 3,000 µg/day | Preformed retinol: Beef and chicken liver, egg yolks, milk Carotenoid precursors: Spinach, carrots, mango, apricots, cantaloupe, pumpkin, yams | *Toxicity:* Fatigue; bone and joint pain; spontaneous abortion and birth defects of fetuses in pregnant women; nausea and diarrhea; liver damage; blurred vision; hair loss; skin disorders *Deficiency:* Night blindness, xerophthalmia; impaired growth, immunity, and reproductive function |
| D (cholecalciferol) | Regulates blood calcium levels Maintains bone health Assists cell differentiation | AI (assumes that person does not get adequate sun exposure): Adult aged 19 to 50: 5 µg/day Adult aged 50 to 70: 10 µg/day Adult aged >70: 15 µg/day UL: 50 µg/day | Canned salmon and mackerel, milk, fortified cereals | *Toxicity:* Hypercalcemia *Deficiency:* Rickets in children; osteomalacia and/or osteoporosis in adults |
| E (tocopherol) | As a powerful antioxidant, protects cell membranes, polyunsaturated fatty acids, and vitamin A from oxidation Protects white blood cells Enhances immune function Improves absorption of vitamin A | RDA: Men: 15 mg/day Women: 15 mg/day UL: 1,000 mg/day | Sunflower seeds, almonds, vegetable oils, fortified cereals | *Toxicity:* Rare *Deficiency:* Hemolytic anemia; impairment of nerve, muscle, and immune function |
| K (phylloquinone, menaquinone, menadione) | Serves as a coenzyme during production of specific proteins that assist in blood coagulation and bone metabolism | AI: Men: 120 µg/day Women: 90 µg/day | Kale, spinach, turnip greens, brussels sprouts | *Toxicity:* None known *Deficiency:* Impaired blood clotting; possible effect on bone health |
| **Water-Soluble** | | | | |
| Thiamine (vitamin B₁) | Required as enzyme cofactor for carbohydrate and amino acid metabolism | RDA: Men: 1.2 mg/day Women: 1.1 mg/day | Pork, fortified cereals, enriched rice and pasta, peas, tuna, legumes | *Toxicity:* None known *Deficiency:* Beriberi; fatigue, apathy, decreased memory, confusion, irritability, muscle weakness |
| Riboflavin (vitamin B₂) | Required as enzyme cofactor for carbohydrate and fat metabolism | RDA: Men: 1.3 mg/day Women: 1.1 mg/day | Beef liver, shrimp, milk and dairy foods, fortified cereals, enriched breads and grains | *Toxicity:* None known *Deficiency:* Ariboflavinosis; swollen mouth and throat; seborrheic dermatitis; anemia |
| Niacin, nicotinamide, nicotinic acid | Required for carbohydrate and fat metabolism Plays role in DNA replication and repair and cell differentiation | RDA: Men: 16 mg/day Women: 14 mg/day UL: 35 mg/day | Beef liver, most cuts of meat/fish/poultry, fortified cereals, enriched breads and grains, canned tomato products | *Toxicity:* Flushing, liver damage, glucose intolerance, blurred vision differentiation *Deficiency:* Pellagra; vomiting, constipation, or diarrhea; apathy |

*Abbreviations: RDA, Recommended Dietary Allowance; UL, upper limit; AI, adequate intake.

# Key Information about Vitamins and Many Essential Minerals
*(continued)*

| Vitamin/ mineral | Primary functions | Recommended intake* | Reliable food sources | Toxicity/deficiency symptoms |
|---|---|---|---|---|
| **Water-Soluble** | | | | |
| Pyridoxine, pyridoxal, pyridoxamine (vitamin $B_6$) | Required as enzyme cofactor for carbohydrate and amino acid metabolism<br>Assists synthesis of blood cells | RDA:<br>Men and women aged 19 to 50: 1.3 mg/day<br>Men aged >50: 1.7 mg/day<br>Women aged >50: 1.5 mg/day<br>UL: 100 mg/day | Chickpeas (garbanzo beans), most cuts of meat/fish/poultry, fortified cereals, white potatoes | *Toxicity:* Nerve damage, skin lesions<br>*Deficiency:* Anemia; seborrheic dermatitis; depression, confusion, and convulsions |
| Folate (folic acid) | Required as enzyme cofactor for amino acid metabolism<br>Required for DNA synthesis<br>Involved in metabolism of homocysteine | RDA:<br>Men: 400 µg/day<br>Women: 400 µg/day<br>UL: 1,000 µg/day | Fortified cereals, enriched breads and grains, spinach, legumes (lentils, chickpeas, pinto beans), greens (spinach, romaine lettuce), liver | *Toxicity:* Masks symptoms of vitamin $B_{12}$ deficiency, specifically signs of nerve damage<br>*Deficiency:* Macrocytic anemia; neural tube defects in a developing fetus; elevated homocysteine levels |
| Cobalamin (vitamin $B_{12}$) | Assists with formation of blood<br>Required for healthy nervous system function<br>Involved as enzyme cofactor in metabolism of homocysteine | RDA:<br>Men: 2.4 µg/day<br>Women: 2.4 µg/day | Shellfish, all cuts of meat/fish/poultry, milk and dairy foods, fortified cereals | *Toxicity:* None known<br>*Deficiency:* Pernicious anemia; tingling and numbness of extremities; nerve damage; memory loss, disorientation, and dementia |
| Pantothenic acid | Assists with fat metabolism | AI:<br>Men: 5 mg/day<br>Women: 5 mg/day | Meat/fish/poultry, shiitake mushrooms, fortified cereals, egg yolk | *Toxicity:* None known<br>*Deficiency:* Rare |
| Biotin | Involved as enzyme cofactor in carbohydrate, fat, and protein metabolism | RDA:<br>Men: 30 µg/day<br>Women: 30 µg/day | Nuts, egg yolk | *Toxicity:* None known<br>*Deficiency:* Rare |
| Ascorbic acid (vitamin C) | Antioxidant in extracellular fluid and lungs<br>Regenerates oxidized vitamin E<br>Assists with collagen synthesis<br>Enhances immune function<br>Assists in synthesis of hormones, neurotransmitters, and DNA<br>Enhances iron absorption | RDA:<br>Men: 90 mg/day<br>Women: 75 mg/day<br>Smokers: 35 mg more per day than RDA<br>UL: 2,000 mg | Sweet peppers, citrus fruits and juices, broccoli, strawberries, kiwi | *Toxicity:* Nausea and diarrhea, nosebleeds, increased oxidative damage, increased information of kidney stones in people with kidney disease<br>*Deficiency:* Scurvy; bone pain and fractures, depression, and anemia |
| **Major Minerals** | | | | |
| Sodium | Fluid balance<br>Acid-base balance<br>Transmission of nerve impulses<br>Muscle contraction | AI:<br>Adults: 1.5 g/day (1,500 mg/day) | Table salt, pickles, most canned soups, snack foods, cured luncheon meats, canned tomato products | *Toxicity:* Water retention, high blood pressure, loss of calcium in urine<br>*Deficiency:* Muscle cramps, dizziness, fatigue, nausea, vomiting, mental confusion |
| Potassium | Fluid balance<br>Transmission of nerve impulses<br>Muscle contraction | AI:<br>Adults: 4.7 g/day (4,700 mg/day) | Most fresh fruits and vegetables: potatoes, bananas, tomato juice, orange juice, melons | *Toxicity:* Muscle weakness, vomiting, irregular heartbeat<br>*Deficiency:* Muscle weakness, paralysis, mental confusion, irregular heartbeat |

*Abbreviations: RDA, Recommended Dietary Allowance; UL, upper limit; AI, adequate intake.

# Key Information about Vitamins and Many Essential Minerals

*(continued)*

| Vitamin/ mineral | Primary functions | Recommended intake* | Reliable food sources | Toxicity/deficiency symptoms |
|---|---|---|---|---|
| Phosphorus | Fluid balance Bone formation Component of ATP, which provides energy for our bodies | RDA: Adults: 700 mg/day | Milk/cheese/yogurt, soy milk and tofu, legumes (lentils, black beans), nuts (almonds, peanuts and peanut butter), poultry | *Toxicity:* Muscle spasms, convulsions, low blood calcium *Deficiency:* Muscle weakness, muscle damage, bone pain, dizziness |
| Chloride | Fluid balance Transmission of nerve impulses Component of stomach heartbeat acid (HCL) Antibacterial | AI: Adults: 2.3 g/day (2,300 mg/day) | Table salt | *Toxicity:* None known *Deficiency:* Dangerous blood acid-base imbalances, irregular heartbeat |
| Calcium | Primary component of bone Acid-base balance Transmission of nerve impulses Muscle contraction | AI: Adults aged 19 to 50: 1,000 mg/day Adults aged >50: 1,200 mg/day UL: 2,500 mg/day | Milk/yogurt/cheese (best-absorbed form of calcium), sardines, collard greens and spinach, calcium-fortified juices | *Toxicity:* Mineral imbalances, shock, kidney failure, fatigue, mental confusion *Deficiency:* Osteoporosis, convulsions, heart failure |
| Magnesium | Component of bone Muscle contraction Assists more than 300 enzyme systems | RDA: Men aged 19 to 30: 400 mg/day Men aged >30: 420 mg/day Women aged 19 to 30: 310 mg/day Women aged >30: 320 mg/day UL: 350 mg/day | Greens (spinach, kale, collard greens), whole grains, seeds, nuts, legumes (navy and black beans) | *Toxicity:* None known *Deficiency:* Low blood calcium, muscle spasms or seizures, nausea, weakness, increased risk of chronic diseases such as heart disease, hypertension, osteoporosis, and type 2 diabetes |
| Sulfur | Component of certain B vitamins and amino acids Acid-base balance Detoxification in liver | No DRI | Protein-rich foods | *Toxicity:* None known *Deficiency:* None known |

## Trace Minerals

| Vitamin/ mineral | Primary functions | Recommended intake* | Reliable food sources | Toxicity/deficiency symptoms |
|---|---|---|---|---|
| Selenium | Required for carbohydrate and fat metabolism | RDA: Adults: 55 µg/day UL: 400 µg/day | Nuts, shellfish, meat/fish/poultry, whole grains | *Toxicity:* Brittle hair and nails, skin rashes, nausea and vomiting, weakness, liver disease *Deficiency:* Specific forms of heart disease and arthritis, impaired immune function, muscle pain and wasting, depression, hostility |
| Fluoride | Development and maintenance of healthy teeth and bones | RDA: Men: 4 mg/day Women: 3 mg/day UL: 2.2 mg/day for children aged 4 to 8; 10 mg/day for children aged >8 | Fish, seafood, legumes, whole grains, drinking water (variable) | *Toxicity:* Fluorosis of teeth and bones *Deficiency:* Dental caries, low bone density |
| Iodine | Synthesis of thyroid hormones Temperature regulation Reproduction and growth | RDA: Adults: 150 µg/day UL: 1,100 µg/day | Iodized salt, saltwater seafood | *Toxicity:* Goiter *Deficiency:* Goiter, hypothyroidism, cretinism in infant of mother who is iodine deficient |

*Abbreviations: RDA, Recommended Dietary Allowance; UL, upper limit; AI, adequate intake.

# Key Information about Vitamins and Many Essential Minerals
*(continued)*

| Vitamin/ mineral | Primary functions | Recommended intake* | Reliable food sources | Toxicity/deficiency symptoms |
|---|---|---|---|---|
| **Trace Minerals** | | | | |
| Chromium | Glucose transport Metabolism of DNA and RNA Immune function and growth | AI: Men aged 19 to 50: 35 µg/day Men aged >50: 30 µg/day Women aged 19 to 50: 25 µg/day Women aged >50: 20 µg/day | Whole grains, brewers yeast | *Toxicity:* None known *Deficiency:* Elevated blood glucose and blood lipids, damage to brain and nervous system |
| Manganese | Assists many enzyme systems Synthesis of protein found in bone and cartilage | AI: Men: 2.3 mg/day Women: 1.8 mg/day UL: 11 mg/day adults | Whole grains, nuts, leafy vegetables, tea | *Toxicity:* Impairment of neuromuscular system *Deficiency:* Impaired growth and reproductive function, reduced bone density Impaired glucose and lipid metabolism, skin rash |
| Iron | Component of hemoglobin in blood cells Component of myoglobin in muscle cells Assists many enzyme systems | RDA: Adult men: 8 mg/day Women aged 19 to 50: 18 mg/day Women aged >50: 8 mg/day | Meat/fish/poultry (best-absorbed form of iron), fortified cereals, legumes, spinach | *Toxicity:* Nausea, vomiting and diarrhea; dizziness, confusion; rapid heartbeat, organ damage, death *Deficiency:* Iron-deficiency microcytic (small red blood cells), hypochromic anemia |
| Zinc | Assists more than 100 enzyme systems Immune system function Growth and sexual maturation Gene regulation | RDA: Men: 11 mg/day Women: 8 mg/day UL: 40 mg/day | Meat/fish/poultry (best-absorbed form of zinc), fortified cereals, legumes | *Toxicity:* Nausea, vomiting, and diarrhea; headaches, depressed immune function, reduced absorption of copper *Deficiency:* Growth retardation, delayed sexual maturation, eye and skin lesions, hair loss, increased incidence of illness and infection |
| Copper | Assists many enzyme systems Iron transport | RDA: Adults: 900 µg/day UL: 10 µg/day | Shellfish, organ meats, nuts, legumes | *Toxicity:* Nausea, vomiting, and diarrhea; liver damage *Deficiency:* Anemia, reduced levels of white blood cells, osteoporosis in infants and growing children |

*Abbreviations: RDA, Recommended Dietary Allowance; UL, upper limit; AI, adequate intake.

# Glossary

 Access a comprehensive glossary and flashcards, including audio pronunciation for selected terms: Mastering A&P® > Study Area > Additional Study Tools

Pronunciations in the text and this glossary use the following rules:

1. Accent marks follow stressed syllables. The primary stress is shown by ′, and the secondary stress by ″.

2. Unless otherwise noted, assume that vowels at the ends of syllables are long and vowels followed by consonants are short. Exceptions to this rule are indicated by a bar (¯) over the vowel, which indicates a long vowel, or a breve sign (˘) over the vowel, indicating that the vowel is short.

For example, the phonetic spelling of "thrombophlebitis" is *throm″bo-flĕ-bi′ tis*. The next-to-last syllable (*bi′*) receives the greatest stress, and the first syllable (*throm″*) gets the secondary stress. The vowel in the second syllable comes at the end of the syllable and is long. The vowel that comes at the end of the third syllable is short because it has a breve sign.

**Abdominal** *ab-dom′ ĭnal* pertaining to the anterior body trunk inferior to the ribs.

**Accommodation** (1) adaptation in response to differences or changing needs; (2) adjustment of the eye for seeing objects at close range.

**Acetylcholine (ACh)** *a″-se-til-ko′ lēn* a chemical transmitter substance released by certain nerve endings.

**Acid** a substance that liberates hydrogen ions when in an aqueous solution; proton donor; compare with *base*.

**Acid-base balance** the maintenance of proper pH in the body; involves buffers, the respiratory system, and the kidneys.

**Acidosis** *as″ ĭ-do′ sis* a condition in which the blood has an excess hydrogen ion concentration and a decreased pH; also called *ketoacidosis*.

**Acquired immune deficiency syndrome (AIDS)** immunodeficiency caused by human immunodeficiency virus (HIV) attacking T cells; symptoms include severe weight loss, night sweats, swollen lymph nodes, and opportunistic infections.

**Acromial** *ab-kro′ me-ul* pertaining to the point of the shoulder.

**Acrosome** *ak′ ro-sōm* an enzyme-containing structure covering the nucleus of the sperm.

**Actin** *ak′ tin* one of the principal contractile proteins found in muscle; makes up the thin filaments.

**Action potential** an electrical event occurring when a stimulus of sufficient intensity is applied to a neuron or muscle cell, allowing sodium ions to move into the cell and reverse the polarity.

**Active immunity** immunity produced by an encounter with an antigen; provides immunological memory.

**Active site** region on the surface of a globular protein (such as an enzyme) where it interacts with other molecules of complementary shape and charge (such as substrate).

**Active transport** net movement of a substance across a membrane against a concentration or electrical gradient; requires release and use of cellular energy.

**Adaptive defense system** branch of the immune system that targets specific antigen; involves B and T lymphocytes and antigen-presenting cells; also called *specific defense system*.

**Adenosine triphosphate (ATP)** *ab-den″o-sēn tri-fos′ fāt* the compound that is the important intracellular energy source; cellular energy.

**Adipose tissue** *ad′ ĭ-pōs* fat.

**Adrenal glands** *ab-dre′ nal glanz* hormone-producing glands located superior to the kidneys; each consists of a medulla and a cortex.

**Aerobic respiration** respiration in which oxygen is consumed and glucose is broken down entirely; water, carbon dioxide, and large amounts of ATP are the final products.

**Afferent** *af′ er-ent* carrying to or toward a center.

**Agglutinins** *ab-gloo′ tĭ-ninz* antibodies in blood plasma that cause clumping of corpuscles or bacteria.

**Agglutinogens** *ag″loo-tin′ o-jenz* (1) antigens that stimulate the formation of a specific agglutinin; (2) antigens found on red blood cells that are responsible for determining the ABO blood group classification.

**Albumin** *al-bu′ min* a protein found in virtually all animals; the most abundant plasma protein.

**Alkalosis** *al″kah-lo′ sis* a condition in which the blood has a lower hydrogen ion concentration than normal and an increased pH.

**Allergy** *al′ er-je* overzealous immune response to an otherwise harmless antigen, resulting in tissue damage; also called *hypersensitivity*.

**Alveolus** *al-ve′ o-lus* (1) a general term referring to a small cavity or depression; (2) an air sac in the lungs.

**Amino acid** *ab-me′ no* an organic compound containing nitrogen, carbon, hydrogen, and oxygen; the building block of protein.

**Amphiarthrosis** *am″fe-ar-thro′ sis* a slightly movable joint.

**Anabolism** *ab-nab′ o-lizm* the energy-requiring building phase of metabolism in which simpler substances are combined to form more complex substances.

**Anaerobic glycolysis** *an-a′ er-ōb-ik* a process in which glucose is broken down in the absence of oxygen, resulting in the formation of 2 ATP molecules and lactic acid.

**Anatomical position** reference point for regional and directional terminology; standing, feet parallel, palms facing forward.

**Anatomy** the science of the structure of living organisms.

**Anemia** *ab-ne′ me-ab* reduced oxygen-carrying capacity of the blood caused by a decreased number of erythrocytes or decreased percentage of hemoglobin in the blood.

**Antebrachial** *an″te-bra′ ke-ul* pertaining to the forearm.

**Antecubital** anterior surface of the elbow.

**Antibody** *an″tĭ-bod′ e* a specialized substance produced by the body that can provide immunity against a specific antigen.

**Antigen (Ag)** *an′tĭ-jen* any substance—including toxins, foreign proteins, or bacteria—that, when introduced to the body, is recognized as foreign and activates the immune system.

**Antigen-presenting cells (APCs)** cells that display portions of foreign antigens to T cells; include dendritic cells, macrophages, and B lymphocytes.

**Antimicrobial proteins** defensive proteins that assist the innate defenses by attacking microbes directly or by preventing their reproduction; see *complement* and *interferon*.

**Aorta** *a-or′ tab* the major systemic artery; arises from the left ventricle of the heart.

**Apocrine gland** *ap′ o-krin* the less numerous type of sweat gland. Its secretions contain water, salts, fatty acids, and proteins.

**Aponeurosis** *ap″o-nu-ro′ sis* fibrous or membranous sheet connecting a muscle and the part it moves.

**Appendicular skeleton** *ap′ en-dik′ u-lar* bones of the limbs and limb girdles that are attached to the axial skeleton.

**Aqueous humor** *a′ kwe-us hu′ mer* the watery fluid in the anterior chambers of the eye.

**Arachnoid granulation** special projections of the arachnoid mater that protrude through the dura mater; site of cerebrospinal fluid absorption into the dural venous sinuses.

**Arachnoid mater** middle layer of the meninges; has weblike extensions attach to the inner layer, the pia mater.

**Arrector pili** *ab-rek′ tor pi′ li* tiny, smooth muscles attached to hair follicles; when activated, they cause the hair to stand upright.

**Arteriole** *ar-ter′ e-ōl* minute artery.

**Arteriosclerosis** *ar-tēr′ e-o-skler-o′ sis* any of a number of proliferative and degenerative changes in the arteries leading to their decreased elasticity and hardening; end stage of atherosclerosis.

**Artery** a vessel that carries blood away from the heart.

**Arthritis** *ar-thri′ tis* inflammation of the joints.

**Articular cartilage** hyaline cartilage covering the epiphysis of a bone at a joint.

**Asthma** *az′ mah* disease or allergic response characterized by bronchial spasms and difficult breathing.

**Astigmatism** *ah-stig' mah-tizm* a visual defect resulting from irregularity in the lens or cornea of the eye causing the image to be out of focus.

**Astrocyte** *as' tro-sīt* type of CNS supporting cell; assists in exchanges between blood capillaries and neurons.

**Atherosclerosis** *a" ther-o" skler-o' sis* changes in the walls of large arteries consisting of lipid deposits on the artery walls; the early stage of arteriosclerosis and increased rigidity.

**Atlas** the first cervical vertebra; articulates with the occipital bone of the skull and the second cervical vertebra (axis).

**Atom** *at' um* the smallest part of an element; indivisible by ordinary chemical means.

**Atomic mass number** the sum of the number of protons and neutrons in the nucleus of an atom; also known as *mass number*.

**Atomic number** the number of protons in an atom.

**Atomic symbol** a one-or two-letter symbol indicating a particular element.

**Atrioventricular (AV) node** *a" tre-o-ven-trik' u-lar* a specialized mass of conducting cells located at the atrioventricular junction in the heart.

**Atrioventricular (AV) valves** two valves set between the atria and ventricles to prevent backflow; includes the mitral (bicuspid) on the left and the tricuspid on the right.

**Atrium** *a' tre-um* a chamber of the heart receiving blood from the veins; superior heart chamber.

**Atrophy** *at' ro-fe* a reduction in size or wasting away of an organ or cell resulting from disease or lack of use.

**Auricle** the external part of the ear surrounding the auditory canal; also called *pinna*.

**Autonomic nervous system** the division of the nervous system that functions involuntarily; innervates cardiac muscle, smooth muscle, and glands; also called *involuntary nervous system*.

**Axial skeleton** *ak' se-al* the skull, vertebral column, ribs, and sternum.

**Axilla** *ak-sib' lah* armpit; pertaining to axillary region.

**Axis** (1) the second cervical vertebra; has a vertical projection called the dens around which the atlas rotates; (2) the imaginary line about which a joint or structure revolves.

**Axon** *ak' son* neuron process that carries impulses away from the nerve cell body; efferent process; the conducting portion of a nerve cell.

**Axon terminal** one of multiple ends of the axon that branch from the motor neuron axon; interacts with the sarcolemma of different muscle cells to form neuromuscular junctions.

**B cells** lymphocytes that oversee humoral immunity; their descendants differentiate into antibody-producing plasma cells; also called *B lymphocytes*.

**Ball-and-socket joint** multiaxial synovial joint in which the rounded head of one bone fits into a socket (depression) on the other.

**Basal metabolic rate (BMR)** *met" ah-bol' ik* the rate at which energy is expended (heat produced) by the body per unit time under controlled (basal) conditions: 12 hours after a meal, at rest.

**Basal nuclei** *nu' kle-i* gray matter areas deep within the white matter of the cerebral hemispheres; also called *basal ganglia*.

**Base** (1) a substance that accepts hydrogen ions; proton acceptor; compare with *acid*; (2) the broad, posterosuperior aspect of the heart where large vessels enter or leave the heart; (3) the broad inferior area of each lung resting on the diaphragm.

**Basement membrane** a thin layer of extracellular material to which epithelial cells are attached in mucosa surfaces.

**Basilar membrane** the membrane in the cochlear duct that vibrates to transmit sound waves and also houses the receptor cells for hearing.

**Basophils** *ba' so-filz* white blood cells whose granules stain deep blue with basic dye; have a relatively pale nucleus and granular-appearing cytoplasm.

**Benign** *be-nīn'* not malignant.

**Bicarbonate buffer system** system composed of bicarbonate ion ($HCO_3^-$) and its salt, sodium bicarbonate ($NaHCO_3$), that resists changes in blood pH.

**Bile** a greenish yellow or brownish fluid produced in and secreted by the liver, stored in the gallbladder, and released into the small intestine.

**Blind spot** gap in vision caused by a lack of photoreceptors at the optic disc.

**Blood** liquid connective tissue composed of plasma and cells responsible for transporting substances such as nutrients and oxygen around the body.

**Blood pressure** the pressure exerted by blood against the inner walls of blood vessels.

**Blood sugar** the amount of glucose in the blood.

**Bone markings** surface features of bones where muscles, tendons, and ligaments attach, and where blood vessels and nerves pass.

**Bone remodeling** the process of repairing or maintaining bone by adding or removing bone matrix in response to damage or mechanical stress.

**Bony callus** "splint" of spongy bone that replaces fibrocartilage callus during bone remodeling.

**Bony thorax** *bōn' e tho' raks* bones of the thorax, including ribs, sternum, and thoracic vertebrae; also called *thoracic cage*.

**Brachial** *bra' ke-al* pertaining to the arm.

**Bradycardia** *brad" e-kar' de-ah* slow heart rate, usually defined as a rate under 60 beats per minute.

**Brain stem** the portion of the brain consisting of the medulla, pons, and midbrain.

**Broad ligament** large ligament formed by a fold of peritoneum that encloses the ovaries and holds them in place, and suspends the uterus from the pelvis.

**Bronchioles** *brong' ke-ōlz* the smallest conducting passages in the lungs.

**Buccal** *buk' al* pertaining to the cheek.

**Buffer** a substance or substances that help to stabilize the pH of a solution.

**Bulbourethral glands** tiny glands inferior to the prostate that produce clear mucus that neutralizes urine prior to ejaculation.

**Burn** tissue damage resulting in cell death caused by electricity, chemicals, too much heat or UV radiation. Burns vary in degree of severity. See *rule of nines*.

**Bursa** *ber' sah* a small sac filled with fluid and located at friction points, especially joints.

**Calcaneal** *kal-ka' ne-ul* pertaining to the heel of the foot.

**Calyx** *ka' liks* a cuplike extension of the pelvis of the kidney.

**Canaliculus** *kan" ah-lik' u-lus* extremely small tubular passage or channel.

**Cancer** a malignant, invasive cellular neoplasm that has the capability of spreading throughout the body or body parts.

**Carbohydrate** *kar" bo-hi' drāt* organic compound composed of carbon, hydrogen, and oxygen; includes starches, sugars, cellulose.

**Cardiac cycle** sequence of events encompassing one complete contraction and relaxation of the atria and ventricles of the heart.

**Cardiac muscle** specialized muscle of the heart with striations and intercalated discs; involuntary muscle.

**Cardiac output (CO)** the blood volume (in liters) ejected per minute by each ventricle.

**Cardiac veins** veins that drain the myocardium and empty into the coronary sinus.

**Cardiovascular system** organ system that distributes blood to all parts of the body.

**Carpal** *kar' pal* (1) one of the eight bones of the wrist; (2) pertaining to the wrist.

**Cartilage** *kar" tĭ-lij* white, semiopaque connective tissue.

**Cartilaginous joint** *kar" ti-laj' ĭ-nus* bones united by cartilage; no joint cavity is present.

**Catabolism** *kah-tab' o-lizm* the process in which living cells break down substances into simpler substances; destructive metabolism.

**Catalyst** *kat' ah-list* substance that increases the rate of a chemical reaction without itself becoming chemically changed or part of the product; see *enzyme*.

**Cataract** *kat' ah-rakt* partial or complete loss of transparency of the crystalline lens of the eye.

**Catecholamines** *kat" ē-kol' ah-mēnz* epinephrine and norepinephrine.

**Cauda equina** *kaw' da e-kwi' nah* the collection of spinal nerves at the inferior end of the vertebral canal.

**Cell** the basic biological unit of living organisms, enclosed by a limiting membrane; cells in more complex organisms contain a nucleus and a variety of organelles.

**Cell body** the part of a neuron containing the nucleus; the metabolic center of a neuron.

**Cell division** the phase of a cell's life cycle when it reproduces itself.

**Cellular immunity** *sel' u-lar ĭ-mu' nĭ-te* immunity conferred by lymphocytes called T cells; also called *cell-mediated immunity*.

**Central nervous system (CNS)** the brain and the spinal cord.

**Centriole** *sen' tre-ōl* a minute body found near the nucleus of the cell composed of microtubules; active in cell division.

**Centromere** *sen'tro-mear* button-like body holding sister chromatids together; also, the site of attachment to the mitotic spindle.

**Cephalic** *seh-fă'lik* pertaining to the head.

**Cerebellum** *ser"e-bel'um* part of the hindbrain; involved in producing smoothly coordinated skeletal muscle activity.

**Cerebral cortex** *sĕ-re'bral* outer gray matter of the cerebrum.

**Cerebral white matter** area of the cerebrum deep to the cerebral cortex containing fiber tracts carrying impulses to, from, or within the cortex.

**Cerebrospinal fluid (CSF)** the fluid produced by choroid plexuses; fills the ventricles and surrounds the central nervous system.

**Cerebrovascular accident (CVA)** a condition in which brain tissue is deprived of a blood supply, as in blockage of a cerebral blood vessel; also called a *stroke.*

**Cerebrum** *sĕ-rē'brum* the largest part of the brain; consists of right and left cerebral hemispheres.

**Cerumen** *sĕ-roo'men* earwax.

**Cervical** *ser'vĭ-kal* term referring to the neck or the necklike portion of an organ or structure.

**Cervical cancer** uncontrolled cellular growth in the cervix that is almost always associated with one or more strains of human papilloma virus (HPV).

**Cervical vertebrae** the seven vertebrae of the neck region.

**Cervix** *ser'viks* the inferior necklike portion of the uterus leading to the vagina.

**Chemical bond** an energy relationship holding atoms together; involves the interaction of electrons.

**Chemical reaction** process in which molecules are formed, changed, or broken down.

**Chemoreceptors** *ke'mo-re-sep'terz* receptors sensitive to various chemicals in solution.

**Chief cells** cells of the gastric glands that secrete inactive pepsinogen.

**Cholesterol** *ko-les'ter-ol* a steroid found in animal fats as well as in most body tissues; made by the liver.

**Chordae tendineae** tendinous chords that anchor the flaps of the closed AV valves to prevent inversion and backflow of blood.

**Chorionic villi** embryonic tissue projections that will combine with tissue of the uterine lining to form the placenta.

**Choroid** *ko'roid* the pigmented nutritive layer of the eye; part of the vascular layer.

**Choroid plexus** structure in each of the four brain ventricles that produces cerebrospinal fluid (CSF).

**Chromatid** *kro'mah-tid* one copy of the DNA after replication but before cell division; still paired with the other copy, together called *sister chromatids,* by a centromere.

**Chromatin** *kro'mah-tin* the structures in the nucleus that carry the hereditary factors (genes).

**Chromosome** *kro'ma-sōm* barlike body of tightly coiled chromatin; visible during cell division.

**Chyme** *kīm* the semifluid stomach contents consisting of partially digested food and gastric secretions.

**Cilia** *sil'e-ah* tiny, hairlike projections on cell surfaces that move in a wavelike manner.

**Ciliary body** *sil'e-er-e* smooth muscle of the vascular tunic of the eye that is connected to the lens by the ciliary zonule.

**Ciliary zonule** suspensory ligament that attaches the lens to the ciliary body in the anterior eye.

**Citric acid cycle** the aerobic pathway occurring within the mitochondria, in which energy is liberated during metabolism of carbohydrates, fats, and amino acids and $CO_2$ is produced.

**Clavicle** *klav'ĭ-kl* collar bone; part of the pectoral girdle that attaches medially to the manubrium and laterally to the scapula.

**Cleavage** *klēv'ij* an early embryonic phase consisting of rapid cell divisions without intervening growth periods.

**Cleavage furrow** indentation formed by a contractile ring of microfilaments; continues to contract until cytokinesis is complete and two daughter cells are formed. See *cytokinesis.*

**Clitoris** *kli'to-ris* a small, erectile structure in the female, homologous to the penis in the male.

**Clonal selection** *klo'nul* the process during which a B cell or T cell becomes sensitized through binding contact with an antigen.

**Clone** descendants of a single cell.

**Coccyx** *kok'siks* three to five fused bones forming the inferior section of the spine; tailbone.

**Cochlea** *kōk'le-ah* a cavity of the inner ear resembling a snail shell; houses the hearing receptor.

**Cochlear duct** the location of the spiral organ of Corti in the cochlea; contains endolymph.

**Codon** *ko'don* the three-base sequence on a messenger RNA molecule that provides the genetic information used in protein synthesis; codes for a given amino acid.

**Coenzymes** nonprotein molecules, such as vitamins, that are necessary for the function of some enzymes.

**Colon** the part of the large intestine between the cecum and the rectum.

**Common hepatic duct** the duct that drains bile from the liver.

**Compact bone** dense bone composed of osteons that makes up the outer layer of flat, short and irregular bones; major component of long bones.

**Complement** a group of plasma proteins that normally circulate in inactive forms; when activated by complement fixation, causes lysis of foreign cells and enhances phagocytosis and inflammation.

**Compound** substance composed of two or more different elements, the atoms of which are chemically united.

**Concentration gradient** a difference in amount of a substance between two areas.

**Conchae** *kong'ke* projections from the walls of the nasal cavity that increase air turbulence and the surface area of mucous membrane exposed to the air to aid in warming and moistening.

**Concussion** brain injury with reversible damage.

**Conducting zone structures** all respiratory passages that carry air to the terminal bronchioles and beyond.

**Conduction deafness** hearing loss due to interference in transmitting sound waves to the inner ear.

**Condylar joint** *kon'dĭ-ler* biaxial synovial joint in which the convex surface of one bone fits into the concave surface of another.

**Cones** one of the two types of photoreceptor cells in the retina of the eye. Provides for color vision.

**Conjunctiva** *kon"junk-ti'vah* the thin, protective mucous membrane lining the eyelids and covering the anterior surface of the eyeball.

**Connective tissue** a primary tissue; form and function vary extensively. Functions include support, storage, and protection.

**Control center** component of homeostatic control receiving sensory input and sending motor output.

**Contusion** brain injury in which the damage cannot be reversed.

**Convergence** *kon-ver'jens* turning toward a common point from different directions.

**Cornea** *kor'ne-ah* the transparent anterior portion of the eyeball.

**Coronal suture** interlocking joint formed at the intersection of the two parietal bones with the frontal bone; also called the *frontal suture.*

**Coronary arteries** arteries that branch off the aorta and supply the myocardium.

**Coronary sinus** enlarged vessel that receives blood from the cardiac veins and empties into the right atrium.

**Corpora quadrigemina** *cor'pora kwod'rĭ-jem'in-uh* four nuclei in the midbrain containing visual and hearing reflex centers.

**Corpus callosum** *kor'pus kah-lo'sum* large fiber tract of the cerebral white matter that connects the cerebral hemispheres.

**Corpus luteum** *kor'pus lu'te-um* structure formed by the ruptured follicle cells after ovulation; secretes estrogen and progesterone.

**Cortex** *kor'teks* the outer surface layer of an organ.

**Corticosteroids** hormones secreted by the adrenal cortex, including mineralocorticoids, glucocorticoids, and sex hormones.

**Covalent bond** *ko-va'lent* a bond involving the sharing of electrons between atoms.

**Coxal** pertaining to the hip.

**Cranial nerves** the 12 pairs of nerves that arise from the brain.

**Cranium** part of the skull that encloses and protects the brain.

**Creatine phosphate (CP)** *kre'ah-tēn fos'făt* a compound that transfers a phosphate group to ADP to regenerate ATP in muscle fibers.

**Crista ampullaris** receptors in the ampullae of the semicircular canals that detect dynamic equilibrium.

**Cross bridge** the link formed when a myosin head binds to the corresponding binding site on actin during contraction.

**Crural** pertaining to the anterior leg; shin.

**Cryptorchidism** *krip-tor'kĭ-dizm* a developmental defect in which one or both testes fail to descend into the scrotum.

**Cupula** *ku'pu-lah* a domelike structure; in the cristae of the inner ear, a gel-like cap covering the "hairs" of receptor cells.

**Cushing's syndrome** *koosh'ingz sin'drōm* a disease produced by excess secretion of glucocorticoids (such as cortisol); characterized by adipose tissue accumulation, weight gain, and osteoporosis.

**Cutaneous membrane** the skin; composed of epidermal and dermal layers.

**Cyanosis** *si'ah-no'sis* a bluish coloration of the mucous membranes and skin caused by deficient oxygenation of the blood.

**Cystic duct** duct that carries bile from the gallbladder.

**Cystitis** *sis-ti'tis* an inflammation of the urinary bladder.

**Cytokines** *si'to-kīnz* chemical messengers involved in immunity that enhance the immune and inflammatory responses.

**Cytokinesis** *si'to-ki-ne'sis* division of cytoplasm that occurs after the cell nucleus has divided.

**Cytoplasm** *si'to-plazm"* the substance of a cell other than that of the nucleus.

**Cytoskeleton** literally, cell skeleton; internal framework of proteins that determine cell shape. See *microtubules, intermediate filaments,* and *microfilaments.*

**Cytosol** *si'to-zol* the liquid component of cytoplasm containing water and solutes.

**Cytotoxic T cell** effector T cell that directly kills foreign cells; also called a *killer T cell.*

**Dead space volume** volume of air that never reaches the alveoli and does not participate in gas exchange.

**Decomposition reaction** a destructive chemical reaction in which complex substances are broken down into simpler ones.

**Defecation reflex** *def'ih-ka'shun* the elimination of the contents of the bowels (feces).

**Deglutition** *de"gloo-tish'un* the act of swallowing.

**Dehydration synthesis** process by which a larger molecule is synthesized from smaller ones by removal of a water molecule at each site of bond formation.

**Dendrites** *den'drītz* the branching extensions of neurons that carry electrical signals to the cell body; the receptive portion of a nerve cell.

**Deoxyribonucleic acid (DNA)** *de-ok"sĭ-ri"bo-nu-kle'ik* nucleic acid found in all living cells; carries the organism's hereditary information.

**Depolarization** *de-po"lar-i-za'shun* the loss of a state of polarity; the loss of a negative charge inside the plasma membrane.

**Dermis** *der'mis* the deep layer of the skin; composed of dense, irregular connective tissue.

**Diabetes mellitus** *di"ah-be'tēz mel'ĭ-tus* a disease caused by deficient insulin release or inadequate responsiveness to insulin, leading to inability of the body cells to use carbohydrates at a normal rate.

**Diapedesis** *di"ah-pĕ-de'sis* the passage of blood cells through intact vessel walls into the tissues.

**Diaphragm** *di'ah-fram* (1) any partition or wall separating one area from another; (2) a muscle that separates the thoracic cavity from the abdominopelvic cavity.

**Diaphysis** *di-af'ĭ-sis* elongated shaft of a long bone.

**Diarthrosis** *di"ar-thro'sis* a freely movable joint.

**Diastole** *di-as'to-le* a period (between contractions) of relaxation of the heart during which it fills with blood.

**Diastolic pressure** *di"as-to'lik* the arterial blood pressure during ventricular relaxation (diastole).

**Diencephalon** *di"en-sef'ah-lon* that part of the forebrain between the cerebral hemispheres and the midbrain including the thalamus, the third ventricle, and the hypothalamus; also called *interbrain.*

**Diffusion** *di-fu'zhun* the spreading of particles in a gas or solution with a movement toward uniform distribution of particles; also called *simple diffusion.*

**Digestive system** system that processes food into absorbable units and eliminates indigestible wastes.

**Digital** pertaining to the digits; fingers, toes.

**Distal convoluted tubule (DCT)** segment of kidney tubule between the ascending nephron loop and the collecting duct; site of some reabsorption and some secretion.

**Diverticulum** *di"ver-tik'u-lum* a pouch or sac in the walls of a hollow organ or structure.

**Dorsal ramus** branch of a spinal nerve that serves the skin and muscles of the posterior trunk.

**Dorsal root** the root at which sensory neurons enter the spinal cord.

**Dorsal root ganglion** the location of cell bodies of sensory neurons which enter the spinal cord via the dorsal root.

**Ductus arteriosus** a shunt in the fetal heart that connects the aorta and pulmonary trunk in order to bypass the immature lungs.

**Ductus deferens** *duk'tus def'er-enz* segment of the male duct system that links the epididymis to the ejaculatory duct.

**Ductus venosus** a shunt in the fetal heart that bypasses the immature liver.

**Duodenum** *doo"uh-de'num* the first part of the small intestine.

**Dura mater** *du'rah ma'ter* the outermost and toughest of the three membranes (meninges) covering the brain and spinal cord.

**Dynamic equilibrium** sense that reports on angular or rotatory movements of the head in space.

**Dyspnea** *disp'ne-ah* labored, difficult breathing.

**Eccrine gland** *ek'rin* the more numerous type of sweat gland found all over the body, but particularly abundant on the palms, soles of feet, and the forehead. Its secretions contain water, salts and metabolic wastes such as urea and uric acid.

**Edema** *ĕ-de'mah* an abnormal accumulation of fluid in body parts or tissues; causes swelling.

**Effector** *ef-fek'tor* an organ, gland, or muscle capable of being activated by nerve endings.

**Efferent** *ef'er-ent* carrying away or away from.

**Efferent arteriole** arteriole leaving the glomerulus and connecting it to the peritubular capillaries.

**Ejaculation** *e-jak"u-la'shun* the sudden ejection of semen from the penis.

**Electrical energy** energy form resulting from the movement of charged particles.

**Electrocardiogram (ECG)** *e-lek"tro-kar'de-o-gram"* a graphic record of the electrical activity of the heart.

**Electroencephalogram (EEG)** *(e-lek'tro-en-sef'ah-lo-gram)* graphic record of the electrical activity of the brain.

**Electrolyte** *e-lek'tro-līt* a substance that breaks down into ions when in solution and is capable of conducting an electric current.

**Electron (e⁻)** negatively charged subatomic particle; orbits the atomic nucleus.

**Electron transport chain** metabolic pathway within the mitochondria in which energy harvested from high-energy hydrogen atoms is used to make ATP. Final delivery of H to molecular oxygen produces water.

**Element** *el'ĕ-ment* any of the building blocks of matter; oxygen, hydrogen, carbon, for example.

**Embolus** a free-floating clot in an unbroken vessel.

**Embryo** *em'bre-o* an organism in its early stages of development; in humans, the first 2 months after conception.

**Emmetropia** the eye that focuses images correctly on the retina is said to have this "harmonious vision."

**Endocardium** *en"do-kar'de-um* the endothelial membrane lining the interior of the heart.

**Endocrine glands** *en'do-krin* ductless glands that empty their hormonal products directly into the blood.

**Endocytosis** *en"do-si-to'sis* means by which fairly large extracellular molecules or particle enter cells. See *phagocytosis, pinocytosis, receptor-mediated endocytosis.*

**Endolymph** *en'do-limf* thick fluid inside the membranous labyrinth similar to perilymph.

**Endometrium** *en-do-me'tre-um* the mucous membrane lining of the uterus.

**Endomysium** *en"do-mīs'e-um* the thin connective tissue surrounding each muscle cell.

**Endoneurium** a delicate connective tissue sheath surrounding each fiber in a nerve.

**Endoplasmic reticulum (ER)** *en"do-plas'mik rĕ-tik'u-lum* a membranous network of tubular or saclike channels in the cytoplasm of a cell.

**Energy** the ability to do work.

**Enteroendocrine cells** cells of the gastric glands that secrete locally acting hormones, such as gastrin.

**Enzyme** *en'zīm* a substance formed by living cells that acts as a catalyst in chemical reactions in the body; see *catalyst.*

**Eosinophils** *e'o-sin'o-filz* granular white blood cells whose granules readily take up a stain called eosin.

**Ependymal cell** *ĕ-pen'dĭ-mal* type of CNS supporting cell; lines the central cavities of the brain and spinal cord and circulates cerebrospinal fluid with their beating cilia.

**Epicardium** visceral layer of a serous membrane that tightly hugs the external surface of the heart and is actually part of the heart wall; also called *visceral pericardium.*

**Epidermis** *ep″ĭ-der′mis* the outer layers of the skin; an epithelium.

**Epididymis** *ep″ĭ-did′ĭ-mis* that portion of the male duct system in which sperm mature. Empties into the ductus deferens (vas deferens).

**Epiglottis** *ep″ĭ-glot′is* the elastic cartilage at the back of the throat; covers the glottis during swallowing.

**Epimysium** *ep″ĭ-mis′e-um* the sheath of fibrous connective tissue surrounding a muscle.

**Epineurium** a tough, fibrous sheath that binds together the fascicles in a nerve.

**Epiphyseal line** calcified line in the epiphysis of a long bone that contained hyaline cartilage during the period of long bone growth, which ends when calcification is complete.

**Epiphyseal plate** plate of hyaline cartilage in the epiphysis of a long bone that allows for growth in length.

**Epiphysis** *ĕ-pif′ĭ-sis* the end of a long bone.

**Epithalamus** roof of the third ventricle; composed of the pineal gland and the choroid plexus of the third ventricle.

**Epithelial membranes** membranes composed of epithelial tissue with an underlying layer of connective tissue; covering and lining membranes. See *mucous membrane* and *serous membrane.*

**Epithelium** *ep″ĭ-the′le-um* one of the primary tissues; covers the surface of the body and lines the body cavities, ducts, and vessels.

**Erection** enlargement and increased rigidity of the penis resulting when erectile tissue fills with blood.

**Erythrocytes** *ĕ-rith′ro-sīts* red blood cells.

**Erythropoietin** hormone that stimulates erythrocyte production; secreted by kidneys.

**Esophagus** muscular tube that carries food from the pharynx to the stomach.

**Estrogens** *es′tro-jenz* hormones that stimulate female secondary sex characteristics; female sex hormones.

**Eupnea** *ūp-ne′ah* easy, normal breathing.

**Exchange reaction** a chemical reaction in which bonds are both made and broken; atoms become combined with different atoms.

**Excretion** *ek-skre′shun* the elimination of waste products from the body.

**Exocrine glands** *ek′so-krin* glands that have ducts through which their secretions are carried to a body surface (skin or mucosa).

**Exocytosis** *ek″so-si-to′sis* method for the movement of substances from the cell interior to the extracellular space as a secretory vesicle fuses with the plasma membrane.

**Expiration** *eks″pĭ-ra′shun* the act of expelling air from the lungs; exhalation.

**Extracellular fluid (ECF)** fluid within the body but outside the cells; see *interstitial fluid.*

**Extracellular matrix** nonliving material in connective tissue consisting of ground substance and fibers that separate the living cells.

**Facilitated diffusion** passive transport process driven by a concentration gradient and requiring a membrane protein to act as a carrier or channel.

**Fallopian tube** *fal-lo′pe-an* see *uterine tube.*

**Fascicle** *fas′ĭ-kul* a bundle of nerve or muscle fibers bound together by connective tissue.

**Feces** *fe′sēz* material discharged from the bowel composed of food residue, secretions, and bacteria.

**Femoral** pertaining to the thigh.

**Fenestrated capillaries** unique capillaries with large pores that allow extensive exchange.

**Fertilization** *fer′tĭ-lĭ-za′shun* fusion of the nuclear material of an egg and a sperm.

**Fetus** *fe′tus* the unborn young; in humans, the period from the third month of development until birth.

**Fibrillation** *fĭ-brĭ-la′shun* irregular, uncoordinated contraction of muscle cells, particularly of the heart musculature.

**Fibrin** *fi′brin* the fibrous insoluble protein formed during the clotting of blood.

**Fibrocartilage callus** repair tissue formed during bone remodeling; "splint" composed of cartilage matrix, bony matrix, and collagen fibers.

**Fibrous joint** bones joined by fibrous tissue; no joint cavity is present.

**Fibrous pericardium** loose-fitting sac surrounding the heart that anchors the heart to surrounding structures.

**Fibrous protein** a strandlike protein that appears most often in body structures; very important in binding structures together and for providing strength in certain body tissues; also called *structural protein.*

**Fibular** pertaining to the area of the fibula, the lateral bone of the leg.

**Filtration** *fil-tra′shun* the passive process of forcing water and solutes through a membrane using a fluid pressure gradient.

**Fimbriae** *fim′bre-e* fingerlike projections that line the distal end of each uterine tube that create fluid currents to wave the oocyte inside after ovulation.

**Fissure** *fis′zher* (1) a groove or cleft; (2) the deepest depressions or inward folds on the brain.

**Flaccid** *flă′sid* soft; flabby; relaxed.

**Flagella** *flah-jel′ah* long, whiplike extensions of the cell membrane of some bacteria and sperm; serve to propel the cell.

**Follicle** *fol′lĭ-kul* (1) structure in an ovary consisting of a developing egg surrounded by follicle cells; (2) colloid-containing structure in the thyroid gland; (3) collection of lymphocytes in a lymph node.

**Follicle cells** cells that surround and help nourish a developing oocyte.

**Fontanels** *fon″tah-nelz′* the fibrous membranes in the skull where bone has not yet formed; babies' "soft spots."

**Foramen** *fo-ra′men* a hole or opening in a bone or between body cavities.

**Formed elements** cellular portion of blood.

**Fovea centralis** *fo′ve-ah sen-tră′lis* pit containing only cones lateral to the blind spot of the eye.

**Frequency** (1) the number of sound waves per unit time, measured in Hertz; (2) the need to urinate often.

**Frontal (coronal) section (plane)** a longitudinal plane that divides the body or an organ into anterior and posterior portions.

**Fused tetanus** *tet′ah-nus* a completely smooth, sustained muscle contraction resulting from rapid stimulation with no time for relaxation between stimulation events; also called *complete tetanus.*

**Gallstones** particles of hardened cholesterol or calcium salts that are occasionally formed in the gallbladder and bile ducts.

**Gamete** *gam′ēt* male or female sex cell (sperm/egg).

**Ganglion** *gang′le-on* a group of nerve cell bodies located in the peripheral nervous system.

**Gap junction** a passageway between two neighboring cells; formed by transmembrane proteins called connexons.

**Gastric glands** glands within the gastric pits of the stomach that secrete the components of gastric juice and include chief, parietal, mucous neck, and enteroendocrine cells.

**Gastric juice** fluid formed by secretions of the gastric glands, characterized by extreme acidity and pepsin, a protein-digesting enzyme.

**Gastrin** *gas′trin* a hormone that stimulates gastric secretion, especially release of hydrochloric acid.

**Gastrointestinal (GI) tract** continuous hollow muscular tube including all structures from mouth to anus.

**Gene** *jēn* biological units of heredity located in chromatin; transmits hereditary information.

**Gland** an organ specialized to secrete or excrete substances for further use in the body or for elimination.

**Glaucoma** *glaw-ko′mah* an abnormal increase of the pressure within the eye.

**Globular protein** a protein whose functional structure is basically spherical. Also referred to as *functional protein*; includes hemoglobin, enzymes, and some hormones.

**Glomerular capsule** *glo-mer′u-ler* double-walled cuplike end of a renal tubule; encloses a glomerulus; also called *Bowman's capsule.*

**Glomerular filtration** the process of forcing fluids and small particles out of the blood and into the glomerular capsule and renal tubule.

**Glomerulus** *glo-mer′u-lus* a knot of coiled capillaries in the kidney; forms filtrate.

**Glottis** *glot′is* the opening between the vocal cords in the larynx.

**Gluconeogenesis** *glu″ko-ne″o-jen′ĕ-sis* the formation of new glucose molecules from noncarbohydrate substances, such as fats and proteins.

**Glucose** *gloo′kōs* the principal sugar in the blood; a monosaccharide.

**Gluteal** *gloo′te-al* pertaining to the buttock.

**Glycerol** *glis′er-ol* a sugar alcohol; one of the building blocks of fats.

**Glycogen** *gli′ko-jen* the main carbohydrate stored in animal cells; a polysaccharide.

**Glycogenesis** *gli″ko-jen′ĕ-sis* formation of glycogen from glucose.

**Glycogenolysis** *gli″ko-jen-ol′ĭ-sis* breakdown of glycogen to glucose.

**Glycolysis** *gli-kol' ĭ-sis* breakdown of glucose to pyruvic acid; an anaerobic process.

**Goblet cells** individual cells (simple glands) that produce mucus.

**Goiter** *goy' ter* a benign enlargement of the thyroid gland.

**Golgi apparatus** *gol'je ap"uh-rat'is* membrane-bound organelle responsible for modifying, packaging, and shipping proteins produced by rough ER to the appropriate destination.

**Gonads** *go' nadz* organs producing gametes; ovaries or testes.

**Gout** gouty arthritis; caused by uric acid buildup in joints. See *arthritis*.

**Graafian follicle** *graf' e-an fol' lĭ-kul* see *vesicular follicle*.

**Graded potential** a local change in membrane potential that varies directly with the strength of the stimulus; declines with distance.

**Graded response** a response that varies directly with the strength of the stimulus.

**Gray matter** the gray area of the central nervous system; contains unmyelinated nerve fibers and nerve cell bodies.

**Gum** the tissue that surrounds the teeth like a tight collar; also called *gingiva (jin'jĭ-vah)*.

**Gyrus** *ji' rus* an outward fold of the surface of the cerebral cortex.

**Hair** flexible epithelial structure composed of keratinized cells; produced by hair follicle.

**Hair follicle** Structure with inner root sheath of epithelial tissue that produces the hair; the outer root sheath is dermal connective tissue that provides the blood supply to epithelial cells.

**Hard palate** roof of mouth formed by the fusion of the two maxillae bones.

**Haustra** pocketlike sacs caused by tone of the teniae coli (longitudinal muscle bands) of the colon.

**Haversian system** *hă-ver' zhen* see *osteon*.

**Heartburn** pain in the esophagus due to reflux of gastric juice.

**Helper T cell** the type of T lymphocyte that orchestrates cellular immunity by direct contact with other immune cells and by releasing chemicals called cytokines; also helps to mediate the humoral response by interacting with B cells.

**Hematocrit** *he-mat'o-krit* the percentage of erythrocytes to total blood volume.

**Hematoma** blood-filled swelling; bruise.

**Hematopoiesis** *hem"ah-to-po-e' sis* formation of blood cells.

**Hemocytoblasts** *he'mo-si' to-blastz* stem cells that give rise to all the formed elements of the blood.

**Hemoglobin (Hb)** *he'mo-glo'bin* the oxygen-transporting pigment of erythrocytes.

**Hemolysis** *he-mol' ĭ-sis* the rupture of erythrocytes.

**Hemostasis** the stoppage of bleeding.

**Hepatic portal vein** single vein that drains the digestive tract organs and carries nutrient-rich blood to the liver before it enters general circulation.

**Hiatal hernia** protrusion of the superior part of the stomach above the diaphragm, resulting in entry of gastric juice into the esophagus.

**High-density lipoproteins** proteins that carry cholesterol from the tissues to the liver for destruction.

**Hilum** *hi' lum* a depressed area where vessels enter and leave an organ.

**Histamine** *his'tah-mēn* a substance that causes vasodilation and increased vascular permeability.

**Homeostasis** *ho"me-o-sta'sis* a state of dynamic body equilibrium or stable internal environment of the body.

**Homeostatic imbalance** a disturbance or change in homeostasis that effects function.

**Hormones** *hor' mōnz* chemical messengers secreted by endocrine glands; responsible for specific regulatory effects on certain cells, tissues, or organs.

**Human chorionic gonadotropin (hCG)** an LH-like hormone produced by the developing embryo that tells the corpus luteum to continue producing hormones to prevent menses.

**Humerus** long bone of the upper arm; articulates with the scapula to form the shoulder joint and the radius and ulna to form the elbow joint.

**Humoral immunity** *hu' mor-al* immunity provided by antibodies released by sensitized B cells and their plasma cell progeny. Also called *antibody-mediated immunity*.

**Hyaline cartilage** cartilage connective tissue containing collagen fibers; has glassy appearance; found on ends of long bones and forms C-shaped rings in trachea.

**Hydrocephalus** a condition in which cerebrospinal fluid accumulates and puts pressure on the brain of an adult; in an infant, the skull may expand because the skull bones are not yet fused.

**Hydrogen bond** weak bond in which a hydrogen atom forms a bridge between two electron-hungry atoms. An important intramolecular bond.

**Hydrolysis** *hi-drol' ĭ-sis* the process in which water is used to split a substance into smaller particles.

**Hydrophilic** *hi'dro-fil'ik* refers to molecules, or portions of molecules, that interact with water and charged particles.

**Hydrophobic** *hi"dro-fo'bik* refers to molecules, or portions of molecules, that interact only with nonpolar molecules.

**Hyoid bone** *hi'oid* neck bone suspended by ligaments; does not articulate with any other bone.

**Hyperglycemia** *hi'per-gli-se' me-ah* high blood sugar.

**Hyperopia** *hi"per-o'pe-ah* farsightedness; distant objects are focused.

**Hyperpnea** *hy-perp'ne-ah* pattern of breathing faster and deeper during exercise.

**Hypersensitivities** allergies; overreaction of the immune system to a perceived threat, resulting in tissue damage.

**Hyperthyroidism** overactivity of the thyroid, resulting in higher-than-normal levels of thyroid hormone, causing an increased metabolic rate.

**Hypertonic** *hi"per-ton'ik* excessive, above normal, tone or tension.

**Hypertrophy** *hi-per'tro-fe* an increase in the size of a tissue or organ independent of the body's general growth.

**Hyperventilation** pattern of breathing in which excess carbon dioxide is exhaled in an effort to raise blood pH to its normal range.

**Hypodermis** adipose tissue beneath the skin. See *subcutaneous*.

**Hypoglycemia** low blood sugar.

**Hypospadias** *hi"pŏ-spa'de-as* condition occurring in male babies only where the urethral opening is located on the ventral surface of the penis.

**Hypothalamus** *hi"po-thal'ah-mus* the region of the diencephalon forming the floor of the third ventricle of the brain.

**Hypothyroidism** decreased thyroid activity, resulting in a slow metabolic rate.

**Hypotonic** *hi"po-ton'ik* below normal tone or tension.

**Hypoxia** *hi-pok'se-ah* a condition in which inadequate oxygen is available to tissues.

**Ileum** *il'e-um* the terminal part of the small intestine; between the jejunum and the cecum of the large intestine.

**Ilium** *il'e-um* large flaring bone of the hip; the most superior bone of the pelvis.

**Immune response** antigen-specific defenses mounted by activated lymphocytes (T cells and B cells).

**Immune system** the body system responsible for protecting the body from pathogens and foreign cells; involves both humoral and cell-mediated defenses.

**Immunodeficiency** *im"mu-no-de-fish'en-se* disease resulting from the deficient production or function of immune cells or certain molecules (complement, antibodies, and so on) required for normal immunity.

**Immunoglobulin (Ig)** *im"mu-no-glob'u-lin* a protein molecule, released by plasma cells, that mediates humoral immunity; an antibody.

**Inclusions** *in-klu'zhun* chemical substances stored in the cytoplasm of a cell.

**Incontinence** *in-kon'tĭ-nens* the inability to control the external urethral sphincter.

**Infarct** *in'farkt* a region of dead, deteriorating tissue resulting from a lack of blood supply.

**Inferior vena cava** large vein that carries blood from veins inferior to the diaphragm back to the right atrium.

**Inflammatory response** *in-flam'ah-tor"e* a physiological response of the body to tissue injury; includes dilation of blood vessels and increased blood vessel permeability.

**Inguinal** *in'gwĭ-nal* pertaining to the groin region.

**Innate defense system** natural defenses that protect against all foreign material, not specifically for any one antigen; also called *nonspecific defense system*.

**Inorganic compound** a compound that lacks carbon; for example, water.

**Insertion** *in-ser'shun* the movable attachment of a muscle as opposed to its origin.

**Inspiration** *in"spĭ-ra'shun* the drawing of air into the lungs; inhalation.

**Inspiratory reserve volume (IRV)** amount of air that can be forcibly inhaled in addition to tidal volume.

**Insulin** pancreatic hormone that decreases blood glucose.

**Integration** the process in which the nervous system processes and interprets sensory input and makes decisions about what should be done.

**Integumentary system** *in-teg"u-men'tuh-re* the skin and its accessory organs *(jin'jĭ-vah)*.

**Intercalated discs** *in-ter'kah-la'ted* specialized connections for communication between cardiac muscle cells containing gap junctions and desmosomes.

**Intermediate filaments** one of three types of rods in the cytoskeleton of a cell; resists pulling forces on the cell.

**Interneuron** completes the pathway between afferent and efferent neurons; also called an *association neuron.*

**Interosseous membrane** flexible fibrous connective tissue attachment connects the radius and ulna, and the tibia and fibula along their lengths.

**Interphase** the period in a cell's life cycle when it performs its usual metabolic activities and prepares for cell division by copying the DNA (genes).

**Interstitial cells** *in"ter-stish'al* cells in the testes between the seminiferous tubules that produce testosterone.

**Interstitial fluid (tissue fluid)** *in"terstish'al* the fluid between the cells; also called *extracellular fluid.*

**Intervertebral discs** *in"ter-ver'tĕ-bral* the discs of fibrocartilage between the vertebrae.

**Intrapleural pressure** pressure in the space between the pleurae; always negative compared to intrapulmonary pressure.

**Intrinsic conduction system** control system built into the heart that sets the rate of contraction by stimulating the myocardium in an atrial to ventricular direction; also called *nodal system.*

**Intrinsic factor** *in-trin'sik* a substance produced by the stomach that is required for vitamin $B_{12}$ absorption.

**Ion** *i'on* an atom with a positive or negative electric charge.

**Ionic bond** bond formed by the complete transfer of electron(s) from one atom to another (or others). The resulting charged atoms, or ions, are oppositely charged and attract each other.

**Iris** *i'ris* the pigmented, involuntary muscle that acts as the diaphragm of the eye; part of the vascular tunic of the eye.

**Irritability** *ir"ĭ-tah-bil'ĭ-te* ability to respond to a stimulus.

**Ischemia** *is-ke'me-ah* a local decrease in blood supply.

**Ischium** *is'ke-um* inferior bone of the hip; "sit bone."

**Isometric contraction** *i"so-met'rik* refers to "same length"; the muscle generates tension but does not shorten, and no movement occurs.

**Isotonic contraction** *i"so-ton'ik* refers to "same tone"; the muscle shortens as it contracts, and movement occurs.

**Isotope** *i'sĭ-tōp* different atomic form of the same element. Isotopes vary only in the number of neutrons they contain.

**Jaundice** *jawn'dis* an accumulation of bile pigments in the blood producing a yellow color of the skin.

**Jejunum** *jĕ-joo'num* the part of the small intestine between the duodenum and the ileum.

**Joint** the junction of two or more bones; an articulation.

**Juxtamedullary nephrons** nephrons that sit close to the cortex-medulla junction, with nephron loops that dip deep into the renal medulla.

**Keratin** *ker'ah-tin* a tough, insoluble protein found in tissues such as hair, nails, and epidermis of the skin.

**Ketoacidosis** low blood pH due to the metabolism of fats to produce ATP in the absence of sufficient glucose.

**Kilocalories (kcal)** unit used to measure the energy value of food.

**Kinetic energy** energy of motion.

**Kinins** *ki'ninz* group of polypeptides that dilate arterioles, increase vascular permeability, and induce pain.

**Kyphosis** *ki-fo'sis* hump formed by the abnormal curvature of the thoracic spine in a posterior direction.

**Labia** *la'be-ah* lips.

**Lacrimal** *lak'rĭ-mal* pertaining to tears.

**Lactation** *lak-ta'shun* the production and secretion of milk.

**Lacteal** *lak'te-al* special lymphatic capillaries of the small intestine that take up lipids.

**Lactic acid** *lak'tik* the product of anaerobic metabolism, especially in muscle.

**Lacuna** *lah-ku'nah* a little depression or space; in bone or cartilage, lacunae are occupied by cells.

**Lambdoid suture** interlocking joint between the occipital bone and the parietal bones.

**Lamella** concentric rings of bone matrix with lacuna between them.

**Lamina** *lam'ĭ-nah* (1) a thin layer or flat plate; (2) the portion of a vertebra between the transverse process and the spinous process.

**Larynx** *lar'inks* the cartilaginous organ located between the trachea and the pharynx; voice box.

**Leukemia** *lu-ke'me-ah* a cancerous condition in which there is an excessive production of immature leukocytes.

**Leukocyte** *lu'ko-sīt* white blood cell.

**Ligament** *lig'ah-ment* a cord of fibrous tissue that connects bones.

**Lingual tonsils** pair of tonsils at the base of the tongue.

**Lipid** *lip'id* organic compound formed of carbon, hydrogen, and oxygen; examples are neutral fats and cholesterol.

**Liver** gland of the digestive tract that produces bile and is the major metabolic gland of the body.

**Long bone** a bone that is longer than it is wide. See *humerus* and *femur.*

**Lordosis** *lor-do'sis* the abnormal curvature of the lumbar spine in an anterior direction.

**Low-density lipoproteins** proteins that carry cholesterol to the tissues.

**Lumbar** *lum'bar* the portion of the back between the thorax and the pelvis; the loin.

**Lumen** *lu'men* the space inside a tube, blood vessel, or hollow organ.

**Lungs** organs of the respiratory system responsible for air exchange with the outside and gas exchange with the blood.

**Luteinizing hormone (LH)** gonadotropic hormone released by the anterior pituitary gland.

**Lymph** *limf* the watery fluid in the lymph vessels collected from the tissue spaces.

**Lymphatic system** *lim-fat'ik* the lymphatic vessels and the lymphoid tissues and organs, including lymph nodes.

**Lymphocytes** *lim'fo-sītz* agranular white blood cells formed in the bone marrow that mature in the lymphoid tissue.

**Lymphoid organs** organs in the lymphatic system, including lymph nodes, spleen, and tonsils; see *lymphatic system.*

**Lysosomes** *li'so-sōmz* organelles that originate from the Golgi apparatus and contain strong digestive enzymes.

**Lysozyme** *li'so-zīm* an enzyme found in sweat, saliva, and tears that is capable of destroying certain kinds of bacteria.

**Macrophage** *mak'ro-fāj'* cell particularly abundant in lymphatic and connective tissues; important in the immune response as an antigen presenter to T cells and B cells.

**Maculae** *mak'u-le* static equilibrium receptors found in the vestibule.

**Main (primary) bronchi** divisions of the trachea that lead into the right and left lungs.

**Major nutrients** nutrients the body requires in significant amounts; include carbohydrates, proteins, and fats.

**Malignant** *mah-lig'nant* life-threatening; pertains to neoplasms that spread and lead to death, such as cancer.

**Malleus** *mă'le-us* lateral bone of the middle ear; also called the *hammer.*

**Mammary glands** *mam'mer-e* milk-producing glands of the breasts.

**Manubrium** *mah-nu'bre-um* the most superior of the three bones of the sternum.

**Mass movements** long, powerful, slow-moving contractions of the large intestine that move materials toward the rectum.

**Mastication** *mas"tĭ-ka'shun* the act of chewing.

**Matter** anything that occupies space and has mass.

**Maxillary bones** *mak-sil'uh-re* paired facial bones that fuse to form the upper jaw and articulate with all other facial bones except the mandible; also called *maxillae.*

**Mechanical energy** energy form directly involved in putting matter into motion.

**Mechanoreceptors** *mek"ah-no-re-sep'terz* receptors sensitive to mechanical pressures such as touch, sound, or contractions.

**Median (midsagittal) section** specific sagittal plane that lies exactly in the midline.

**Mediastinum** *me″de-as-ti′num* the region of the thoracic cavity between the lungs that houses the heart.

**Medulla** *me-dul′ah* the central portion of certain organs.

**Medulla oblongata** most inferior part of brainstem involved in visceral control.

**Medullary cavity** central cavity in the diaphysis of a long bone containing marrow.

**Megakaryocytes** large, multinucleate cells that fragment to produce platelets.

**Meiosis** *mi-o′sis* the two successive cell divisions in gamete formation producing nuclei with half the full number of chromosomes (haploid).

**Melanin** *mel′ah-nin* the dark pigment synthesized by melanocytes; responsible for skin color.

**Membranous urethra** the part of the male urethra that connects the prostatic urethra to the penis.

**Memory cell** member of T cell and B cell clones that provides for immunological memory.

**Menarche** *me-nar′ke* establishment of menstrual function; the first menstrual period.

**Meninges** *me-nin′jez* the membranes that cover the brain and spinal cord.

**Menopause** *men′o-pawz* the physiological end of menstrual cycles.

**Menstrual cycle** the periodic, cyclic discharge of blood, secretions, tissue, and mucus from the mature female uterus in the absence of pregnancy; a series of cyclic changes of the endometrium in response to changing ovarian hormone levels; also called *menstruation*, *uterine cycle*.

**Mental** pertaining to the chin.

**Merkel cell** cells associated with nerve endings that serve as touch receptors.

**Mesentery** *mes′en-ter″e* the double-layered membrane of the peritoneum that supports most organs in the abdominal cavity.

**Messenger RNA (mRNA)** long nucleotide strands that reflect the exact nucleotide sequence of the DNA gene; carry this information to the ribosome for protein synthesis.

**Metabolism** *me-tab′o-lizm* the sum total of the chemical reactions that occur in the body.

**Metacarpal** *met′ah-kar′pal* one of the five bones of the palm of the hand.

**Metastasis** *me-tas′tah-sis* the spread of cancer from one body part or organ into another not directly connected to it.

**Metatarsal** *met′ah-tar′sal* one of the five bones between the tarsus and the phalanges of the foot.

**Microfilaments** *mi′kro-fil′ah-ments* thin strands of the contractile protein actin; one of three types of rods in the cytoskeleton of a cell.

**Microglia** *mi-krog′le-ah* type of CNS supporting cell; phagocytes that ingest debris such as bacteria or dead cells.

**Microtubules** *mi″kro-too′bulz* one of three types of rods in the cytoskeleton of a cell; hollow tubes made of spherical protein that

determine the cell shape and the distribution of cellular organelles.

**Microvilli** *mi″kro-vil′i* the tiny projections on the free surfaces of some epithelial cells; increase surface area for absorption.

**Micturition** *mik′tu-rish′un* urination, or voiding; emptying the bladder.

**Midbrain** part of the brain stem containing reflex centers for vision and hearing, and fiber tracts (cerebral peduncles) that convey both ascending and descending impulses.

**Minor nutrients** vitamins and minerals; nutrients the body needs in small amounts.

**Miscarriage** termination of a pregnancy before the embryo or fetus is viable outside the uterus.

**Mitochondria** *mi″to-kon′dre-ah* the rod-like cytoplasmic organelles responsible for ATP generation.

**Mitosis** *mi-to′sis* the division of the cell nucleus; often followed by division of the cytoplasm of a cell.

**Mixed nerves** nerves containing the processes of motor and sensory neurons; their impulses travel to and from the central nervous system.

**Molecule** *mol′e-kyool* particle consisting of two or more atoms held together by chemical bonds.

**Monoclonal antibodies** *mon″o-klon′ul* pure preparations of identical antibodies that exhibit specificity for a single antigen.

**Monocyte** *mon′o-sit* large single-nucleus white blood cell; agranular leukocyte.

**Monosaccharide** *mon″o-sak′ĭ-rid* literally, one sugar; the building block of carbohydrates; examples include glucose and fructose.

**Motor division** the part of the PNS that sends motor signals from the CNS to muscles or glands; also called *efferent division*.

**Motor unit** a motor neuron and all the muscle cells it supplies.

**Mucosa** innermost layer of the gastrointestinal tract; see *mucous membrane*.

**Mucous membrane (mucosa)** membrane that forms the linings of body cavities open to the exterior (digestive, respiratory, urinary, and reproductive tracts).

**Multiple sclerosis (MS)** autoimmune disorder of the CNS where the myelin sheaths are converted into hardened scleroses in the brain and spinal cord; results in loss of control.

**Multipolar neuron** structural class of neuron with more than two processes extending from the cell body.

**Muscle fatigue** inability of a muscle to contract even while being stimulated.

**Muscle tissue** one of four main tissue types; specialized to contract (shorten) in order to produce a force that will cause movement.

**Muscle tone** sustained partial contraction of a muscle in response to stretch receptor inputs; keeps the muscle healthy and ready to react.

**Muscle twitch** a single rapid contraction of a muscle followed by relaxation.

**Muscular system** organ system consisting of skeletal muscles and their connective tissue attachments.

**Muscularis externa** smooth muscle layer of the gastrointestinal tract composed of an inner circular layer and an outer longitudinal layer.

**Myasthenia gravis** *mi″as-the′ne-ah grä′vis* an autoimmune disease affecting muscles; characterized by weakness and fatigue due to a shortage of acetylcholine receptors.

**Myelin** *mi′ĕ-lin* a white, fatty lipid substance.

**Myocardial infarction** *mi″o-kar′de-al in-fark′shun* a condition characterized by dead tissue areas in the myocardium caused by interruption of blood supply to the area.

**Myocardium** *mi″o-kar′de-um* the cardiac muscle layer of the heart wall.

**Myofibrils** *mi″o-fi′brilz* contractile organelles found in the cytoplasm of muscle cells.

**Myofilament** *mi″o-fil′ah-ment* filaments composing the myofibrils. Of two types: actin and myosin.

**Myometrium** *mi″o-me′tre-um* the thick uterine musculature.

**Myopia** *mi-o′pe-ah* nearsightedness; near objects are in focus.

**Myosin** *mi′o-sin* one of the principal contractile proteins found in muscle; makes up the thick filaments.

**Nasal** pertaining to the nose.

**Nasal septum** tissue at midline that separates the nasal cavity into right and left sides.

**Natural killer (NK) cells** unique lymphocytes that are part of the nonspecific defenses; kill cells that do not exhibit self markers.

**Neck** (1) portion of the body connecting the head to the thorax that contains the trachea and esophagus; (2) portion of a tooth connecting the crown and root.

**Negative feedback** feedback that causes the stimulus to decline or end.

**Nephron** *nef′ron* structural and functional unit of the kidney.

**Nephron loop** portion of the renal tubule between the proximal and distal convoluted tubules; also called *loop of Henle*.

**Nerve** bundle of neuronal processes (axons) outside the central nervous system.

**Nerve impulse** a self-propagating wave of depolarization; also called an *action potential*.

**Nervous system** fast-acting control system that employs nerve impulses to trigger muscle contraction or gland secretion.

**Nervous tissue** one of four main tissue types; specialized for irritability (respond to stimuli) and conductivity.

**Neuroglia** *nu-rog′le-ah* the nonneuronal tissue of the central nervous system that performs supportive and other functions; also called *glial cells* or *glia*.

**Neuromuscular junction** *nu″ro-mus′ku-lar* the region where a motor neuron comes into close contact with a skeletal muscle cell.

**Neurons** *nu′ronz* cells of the nervous system specialized to transmit messages throughout the body.

**Neurotransmitter** chemical released by neurons that may, upon binding to receptors of neurons or effector cells, stimulate or inhibit them.

**Neutral fats** dietary fats; also called *triglycerides*.

**Neutralization** *nu″tral-i-za′shun* (1) a chemical reaction that occurs between an acid and a base; (2) blockage of the harmful effects of bacterial exotoxins or viruses by the binding of antibodies to their functional sites.

**Neutron (n⁰)** *nu′tron* uncharged subatomic particle; found in the atomic nucleus.

**Neutrophils** *nu′tro-filz* the most abundant of the white blood cells.

**Nose** the only external organ of the respiratory system.

**Nuclear envelope** the double membrane barrier of the nucleus; separates the nucleoplasm from the cytoplasm.

**Nucleic acid** *nu-kle′ik* class of organic molecules that includes DNA and RNA.

**Nucleoli** *nu-kle′o-li* small spherical bodies in the cell nucleus; function in ribosome synthesis.

**Nucleotide** *nu′kle-o-tīd* building block of nucleic acids.

**Nucleus** *nu′kle-us* (1) a dense central body in most cells containing the genetic material of the cell; (2) cluster of neuronal cell bodies in the brain or spinal cord.

**Nutrient** chemical substance used by the body for normal growth, maintenance, and repair.

**Occipital** *ok-sip′ĭ-tal* pertaining to area at the back of the head.

**Olecranal** *ol-eh-kra′nel* pertaining to the posterior surface of the elbow.

**Oligodendrocyte** *ol′ĭ-go-den′dro-sīt* type of CNS supporting cell that has many cellular processes, each of which makes up a myelin sheath.

**Oocyte** *o′o-sīt* an immature egg.

**Oogenesis** *o″o-jen′ĕ-sis* the process of formation of the ova.

**Optic chiasma** *op′tik ki-as′mah* the partial crossover of fibers of the optic nerves.

**Optic disc** site where the optic nerve meets the retina; causes the blind spot.

**Oral cavity** the mouth; contains teeth and the tongue.

**Orbital** eye area.

**Organ** a part of the body formed of two or more tissues that performs a specialized function.

**Organ system** a group of organs that work together to perform a vital body function; e.g., nervous system.

**Organelles** *or″gah-nelz′* specialized structures in a cell that perform specific metabolic functions.

**Organic compound** a compound containing carbon; examples include proteins, carbohydrates, and fats.

**Organism** an individual living thing.

**Origin** the muscle attachment that is not movable or is less movable than the insertion.

**Orthostatic hypotension** temporary hypotension due to abrupt movement from a sitting or reclining position to an upright position.

**Osmoreceptor** *oz″mo-re-sep′tor* a structure sensitive to osmotic pressure or concentration of a solution.

**Osmosis** *oz-mo′sis* the diffusion of water (a solvent) through a membrane from a dilute solution into a more concentrated one.

**Ossicles** *os′sĭ-kulz* the three bones of the middle ear: hammer (malleus), anvil (incus), and stirrup (stapes); also called the *auditory ossicles*.

**Ossification** the process of bone formation.

**Osteoarthritis (OA)** most common form of arthritis; caused by wear and tear at joints that leads to articular cartilage breakdown. See *arthritis*.

**Osteoblasts** *os′te-o-blasts″* bone-forming cells.

**Osteoclasts** *os′te-o-klasts″* large cells that resorb or break down bone matrix.

**Osteocyte** *os′te-o-sīt* a mature bone cell.

**Osteon** *os′te-on* the structural and functional unit of compact bone; Haversian system.

**Osteoporosis** *os″te-o-po-ro′sis* an increased softening of the bone resulting from a gradual decrease in rate of bone formation; a common condition in older people.

**Otolith** *o′to-lith* one of the small calcified masses in the maculae of the vestibular apparatus in the inner ear.

**Otolithic membrane** gelatinous membrane in a macula with receptor "hairs" embedded in it and studded with otoliths.

**Otosclerosis** *o″to-sklĕ-ro′sis* fusion of the ossicles (bones) in the middle ear; results in conduction deafness.

**Oval window** the superior, membrane-covered opening in the middle ear wall; sound is conveyed to the oval window by the stirrup (stapes).

**Ovarian cycle** *o-va′re-an* the monthly cycle of follicle development, ovulation, and corpus luteum formation in an ovary.

**Ovarian follicles** saclike structures consisting of follicle cells and a developing oocyte.

**Ovarian ligaments** bilateral ligaments that anchor the ovaries to the uterus.

**Ovary** *o′var-e* the female sex organ in which ova (eggs) are produced.

**Ovulation** *ov″u-la′shun* the release of an ovum (or oocyte) from the ovary.

**Ovum** *o′vum* the female gamete (germ cell); an egg.

**Oxygen deficit** *ok′sĭ-jen* the volume of oxygen required after exercise to oxidize the lactic acid formed during exercise.

**Oxyhemoglobin** *ok″se-he′mo-glo″bin* hemoglobin combined with oxygen.

**P wave** wave on an ECG representing depolarization of the atria.

**Pacemaker** see *sinoatrial node*.

**Palate** *pal′et* roof of the mouth.

**Palatine tonsils** pair of tonsils flanking the oropharynx.

**Pancreas** *pan′kre-as* gland posterior to the stomach, between the spleen and the duodenum; produces both endocrine and exocrine secretions.

**Pancreatic juice** *pan″kre-at′ik* a secretion of the pancreas containing enzymes for digestion of all food categories.

**Papilla** *pah-pil′ah* small nipplelike projection.

**Papillary layer** superficial layer of dermis bordering epidermis; contains dermal papillae, the cause of fingerprints. See *dermal papillae*.

**Paranasal sinuses** air-filled spaces within the maxillae on either side of the nasal cavity that are lined with mucous membrane; help lighten the skull.

**Parasympathetic division** *par″ah-sim′pah-thet′ik* a division of the autonomic nervous system; also referred to as the *craniosacral division*.

**Parathyroid glands** *par″ah-thi′roid* small endocrine glands located on the posterior aspect of the thyroid gland.

**Parathyroid hormone (PTH)** hormone released by the parathyroid glands that regulates blood calcium level.

**Parietal cells** cells of the gastric glands that produce hydrochloric acid (HCl).

**Parietal pericardium** outermost layer of the serous pericardium that lines the interior surface of the fibrous pericardium.

**Parietal peritoneum** lining of the abdominopelvic cavity that is continuous with the serosa, or visceral peritoneum of the gastrointestinal tract.

**Parietal pleura** *ploor′ah* serous membrane layer covering the cavity in which the lung resides.

**Parotid glands** large salivary glands anterior to the ears.

**Passive immunity** short-lived immunity resulting from the introduction of "donated antibodies" obtained from an immune animal or human donor; immunological memory is not established.

**Pathogen** disease-causing microorganism (some bacteria, fungi, viruses, etc.).

**Pectoral** *pek′to-ral* pertaining to the chest.

**Pectoral girdle** composite of two bones, scapula and clavicle, that attach the upper limb to the axial skeleton; also called the *shoulder girdle*.

**Pedicle** bony extension connecting the centrum and transverse process of a vertebra.

**Pelvic** pertaining to the anterior area overlying the pelvis.

**Pelvic girdle** incomplete bony basin formed by the two coxal bones and the sacrum.

**Penis** *pe′nis* the male organ of copulation and urination.

**Pepsin** an enzyme capable of digesting proteins in an acid pH.

**Perforating canal** canal running at right angles to the central canal in an osteon. Also called *Volkmann's canal*.

**Perforating fibers** connective tissue fibers that secure the periosteum to the bone; also called *Sharpey's fibers*.

**Pericardium** *per″ĭ-kar′de-um* the membranous sac enveloping the heart.

**Perilymph** *per′ĭ-limf* plasmalike fluid filling the bony labyrinth.

**Perimysium** *per″ĭ-mis′e-um* the connective tissue enveloping bundles of muscle fibers.

**Perineum** *per′i-ne′um* that region of the body extending from the anus to the scrotum in males and from the anus to the vulva in females.

**Perineurium** coarse connective tissue wrapping that binds groups of fibers in a nerve, forming fascicles, or fiber bundles.

**Periosteum** *per-e-os′te-um* double-layered connective tissue membrane that covers and nourishes the bone.

**Peripheral nervous system (PNS)** *pĕ-rif′er-al* a system of nerves that connects the outlying parts of the body with the central nervous system.

**Peripheral resistance** the resistance to blood flow offered by the systemic blood vessels; a measure of the amount of friction encountered by blood.

**Peristalsis** *per″ĭ-stal′sis* the waves of contraction seen in tubelike organs; propels substances along the tract.

**Peritoneum** *per″ĭ-to-ne′um* the serous membrane lining the interior of the abdominal cavity and covering the surfaces of the abdominal organs.

**Peritubular capillaries** low-pressure capillaries that arise from the efferent arteriole and surround the renal tubule of each nephron to aid in reabsorption.

**Peroxisomes** *per-ok′sih-sōmz* membranous sacs in the cytoplasm containing powerful oxidase enzymes that use molecular oxygen to detoxify harmful or toxic substances, such as free radicals.

**Petechiae** *pĕ-te′ke-e* rashlike spots on the skin that indicate widespread bleeding.

**pH** the symbol for hydrogen ion concentration; a measure of the relative acidity or alkalinity of a solution.

**Phagocyte** *fag′o-sīt* cell capable of engulfing and digesting particles or cells harmful to the body using the process of phagocytosis.

**Phagocytosis** *fag″o-si-to′sis* the ingestion of solid particles by cells.

**Phalanges** *fah-lan′jēz* the bones of the finger or toe.

**Pharyngeal-esophageal phase** the second, involuntary phase of swallowing.

**Pharyngeal tonsil** *far-rin′je-al* single tonsil, also called *adenoid*, located in the nasopharynx.

**Pharyngotympanic tube** tube that connects the middle ear and the pharynx; allows pressure to be equalized on both sides of the eardrum; also called the *auditory tube* and the *eustachian tube*.

**Pharynx** *far′inks* the muscular tube extending from the posterior of the nasal cavities to the esophagus.

**Phospholipid** *fos″fo-lip′id* a modified triglyceride containing phosphorus.

**Photoreceptors** *fo″to-re-sep′torz* specialized receptor cells that respond to light energy.

**Physiology** *fiz″e-ol′o-je* the science of the functioning of living organisms.

**Pia mater** innermost meningeal layer; clings to the surface of the brain and spinal cord.

**Pinocytosis** *pin″uh-si-to′sis* the engulfing of extracellular fluid by cells.

**Pivot joint** uniaxial synovial joint in which the rounded end of one bone fits into a sleeve or ring of bone or ligaments.

**Placenta** *plah-sen′tah* the temporary organ that provides nutrients and oxygen to the developing fetus, carries away wastes, and produces the hormones of pregnancy.

**Plane joint** nonaxial synovial joint in which the two bones have flat articular surfaces; a plane joint allows only short gliding movements.

**Plantar** *plan′tar* pertaining to the sole of the foot.

**Plasma** *plaz′mah* the fluid portion of the blood.

**Plasma cell** member of a B cell clone; specialized to produce and release antibodies.

**Plasma membrane** membrane that encloses cell contents; outer limiting membrane.

**Platelet** *plāt′let* one of the irregular cell fragments of blood; involved in clotting.

**Pleura** *ploo′rah* the serous membrane covering the lung and lining the thoracic cavity.

**Plexus** network of nerves formed by joining ventral rami of spinal nerves; contains both sensory and motor fibers.

**Plicae circulares** *pli′kuh″ ser ku-lar′ēz* corkscrew-like folds within the small intestine that increase the surface area and slow the progression of food through the small intestine; also called *circular folds*.

**Pneumothorax** *nu″mo-tho′raks* the presence of air or gas in a pleural cavity.

**Polar body** a minute cell produced during meiosis in the ovary.

**Polycythemia** *pol″e-si-the′me-ah* presence of an abnormally large number of erythrocytes in the blood.

**Polymer** *pol′ ĭ″mer* a long, chainlike molecule consisting of many similar or repeated units.

**Polysaccharide** *pol″e-sak′ ĭ-rīd* literally, many sugars; a polymer of linked monosaccharides; examples include starch and glycogen.

**Pons** (1) any bridgelike structure or part; (2) the brain area connecting the medulla with the midbrain, providing linkage between upper and lower levels of the central nervous system.

**Popliteal** related to the posterior knee.

**Positive feedback** feedback that tends to cause a variable to change in the same direction as the initial change; enhances the stimulus.

**Potential energy** stored energy; energy at rest; compare to *kinetic energy*.

**Precipitation** formation of insoluble complexes that settle out of solution.

**Pressure gradient** difference in hydrostatic (fluid) pressure that drives filtration.

**Primary curvatures** spinal curvatures in the thoracic and sacral areas that are present at birth.

**Primary (essential) hypertension** high blood pressure that cannot be directly linked to any specific cause.

**Primary humoral response** the initial response of the humoral arm of the immune system to an antigen; involves clonal selection and establishes immunological memory.

**Primary motor area** area of the cerebral cortex responsible for voluntary movement; located in the precentral gyrus.

**Primary somatic sensory area** area of the cerebral cortex responsible for interpreting impulses from cutaneous sense organs and proprioceptors; located in the postcentral gyrus.

**Progesterone** ovarian hormone produced by the corpus luteum and later the placenta; responsible for maintaining the uterine lining during pregnancy, inhibiting contraction of uterine muscle, and helping to prepare the breasts for lactation.

**Proprioceptor** *pro″pre-o-sep′tor* a receptor located in a muscle or tendon; concerned with locomotion, posture, and muscle tone.

**Prostate** accessory organ of the male reproductive tract that produces a milky fluid that contributes to semen by activating sperm.

**Prostatic urethra** the segment of the urethra that passes through the prostate.

**Protein** *pro′tēn* a complex nitrogenous substance; the main building material of cells.

**Proton (p⁺)** *pro′ton* subatomic particle that bears a positive charge; located in the atomic nucleus.

**Proximal convoluted tubule (PCT)** segment of the renal tubule between the glomerular capsule and the descending nephron loop; site of most reabsorption and some secretion.

**Puberty** *pu′ber-te* the period at which reproductive organs become functional.

**Pubic** pertaining to the genital region.

**Pubic symphysis** cartilaginous joint formed at the anterior of the two hip bones where the pubis bones articulate.

**Pubis** *pu′bis* most anterior bone of the hip; the pubis from each hip bone meets to articulate at the pubic symphysis joint anteriorly. See *pubic symphysis*.

**Pulmonary circulation** system of blood vessels that carry blood to and from the lungs for gas exchange.

**Pulmonary edema** *ĕ-de′mah* a leakage of fluid into the air sacs and tissue of the lungs.

**Pulmonary pleura** *ploor′ah* serous membrane layer covering the surface of each lung; also called *visceral pleura*.

**Pulmonary trunk** large artery connecting the right ventricle with the pulmonary arteries.

**Pulp** the part of the tooth that provides nutrients and sensation; composed of blood vessels, connective tissue, and nerves.

**Pulse** the rhythmic expansion and recoil of arteries resulting from heart contraction; can be felt from the outside of the body.

**Pupil** an opening in the center of the iris through which light enters the eye.

**Purkinje fibers** *pur-kin′je* the modified cardiac muscle fibers of the conduction system of the heart that carry impulses to the myocardium.

**Pus** the fluid product of inflammation, composed of white blood cells, the debris of dead cells, and a thin fluid.

**Pyelonephritis** *pi″ĕ-lo-nĕ-fri′tis* an inflammation of the kidney pelvis and surrounding kidney tissues.

**Pyloric valve** valve formed by the pyloric sphincter between the stomach and small intestine.

**Pyramidal tracts** major motor pathways concerned with voluntary movement; connect primary motor area in the frontal lobes of each cerebral hemisphere with the spinal cord; also called *corticospinal tracts*.

**Pyrogen** *pi′ro-jen* an agent or chemical substance that induces fever.

**QRS complex** wave on an ECG representing depolarization of the ventricles.

**Radiant energy** energy of the electromagnetic spectrum, which includes heat, light, ultraviolet waves, infrared waves, and other forms.

**Radioactivity** the process of spontaneous decay seen in some of the heavier isotopes, during which particles or energy is emitted from the atomic nucleus; results in the atom becoming more stable.

**Radioisotope** *ra″de-o-i′sĭ-tōp* isotope that exhibits radioactive behavior.

**Radius** lateral forearm bone; articulates with the capitulum of the humerus at the elbow and the carpals distally.

**Real image** image formed on the retina that is reversed left to right and inverted (upside down) as a result of the refraction bending of light by the lens.

**Receptor** *re-sep′tor* (1) a peripheral nerve ending specialized for response to particular types of stimuli; (2) molecule that binds specifically with other molecules, e.g., hormones and neurotransmitters.

**Receptor-mediated endocytosis** the type of endocytosis in which engulfed particles attach to receptors on the cell surface before endocytosis occurs.

**Rectum** portion of the large intestine between the sigmoid colon and the anal canal.

**Red marrow** bone marrow that is the site of blood cell production; see *hematopoiesis*.

**Reduction** restoring broken bone ends (or a dislocated bone) to its original position.

**Reflex** automatic reaction to a stimulus.

**Refract** bend; usually refers to light.

**Regulatory T cell** type of T lymphocyte that slows or stops activity of B and T cells once the antigenic threat is ended.

**Renin** *re′nin* an enzyme released by the kidneys that is involved with raising blood pressure.

**Renin-angiotensin mechanism** system triggered by low blood pressure or changes to filtrate solute concentrations that results in angiotensin II production and rising blood pressure.

**Repolarization** restoration of the membrane potential to the initial resting (polarized) state.

**Reproductive system** organ system that functions to produce offspring.

**Residual volume** air remaining in the lungs after exhalation that cannot be voluntarily exhaled.

**Resistance exercise** refers to building muscle mass and strength; also called *isometric exercise.*

**Respiration** the process of supplying the body with oxygen and removing carbon dioxide; includes both internal and external respiration.

**Respiratory membrane (air-blood barrier)** membrane composed of the alveolar wall, the capillary wall, and their basement membranes; gases must cross this membrane for proper exchange to occur.

**Respiratory system** organ system that carries out gas exchange; includes the nose, pharynx, larynx, trachea, bronchi, and lungs.

**Respiratory zone** part of the lung that leads into the alveoli and is involved in gas exchange.

**Responsiveness** the ability to sense changes (stimuli) in the environment and then to react to them; see also *irritability.*

**Rete testis** *re′te* structures that carry sperm from the seminiferous tubules to the epididymis.

**Reticular layer** deep layer of the dermis; contains hair follicles, sweat glands, sebaceous glands and is highly vascular.

**Retina** *ret′ĭ-nah* light-sensitive layer (tunic) of the eye; contains rods and cones.

**Rheumatoid arthritis (RA)** chronic inflammatory form of arthritis caused by an autoimmune disorder in which the immune system mistakenly attacks the joints bilaterally. See *arthritis.*

**Ribonucleic acid (RNA)** *ri″bo-nu-kle′ik* the nucleic acid that contains ribose; acts in protein synthesis.

**Ribosomes** *ri′bo-sōmz* cytoplasmic organelles at which proteins are synthesized.

**Rickets** disease in which the bones of children fail to calcify properly, leading to bowing of the leg bones.

**Rods** one of the two types of photosensitive cells in the retina.

**Round ligaments** ligaments that anchor the uterus to the pelvis anteriorly.

**Round window** inferior, membrane-covered opening in the middle ear wall between the vestibule and the cochlea.

**Rule of nines** method of computing the extent of burns by dividing the body into 11 areas, each accounting for 9 percent of the total body area plus an area around the genitals representing 1 percent of the body.

**Sacral** *sa′krul* referring to the lower portion of the back, just superior to the buttocks at the base of the spine.

**Sacrum** *sa′krum* bone made up of five fused bones; forms the posterior of the pelvis.

**Saddle joint** biaxial synovial joint in which both articulating bones have both convex and concave surfaces.

**Sagittal section (plane)** *saj′ĭ-tal* a longitudinal (vertical) plane that divides the body or any of its parts into right and left portions.

**Sagittal suture** interlocking joint formed by the intersection of the two parietal bones at the midline of the skull.

**Salivary amylase** enzyme in saliva that begins the digestion of starch.

**Salivary glands** three pairs of glands that produce a mixture of mucus and serous fluid containing the enzyme salivary amylase, lysozyme, and IgA antibodies.

**Salt** ionic compound that dissociates into charged particles (other than hydrogen or hydroxyl ions) when dissolved in water.

**Sarcomere** *sar′ko-mēr* the smallest contractile unit of muscle; extends from one Z disc to the next.

**Sarcoplasmic reticulum** specialized smooth endoplasmic reticulum found in muscle cells that stores and releases calcium during muscle contraction.

**Satellite cell** type of neuroglia in the PNS; protects and cushions peripheral nerve cell bodies.

**Scapula** *skap′u-luh* triangular bone of the pectoral girdle that articulates with the clavicle and humerus.

**Schwann cell** type of neuroglia in the PNS; many Schwann cells form the myelin sheath on a single neuron.

**Sclera** *skle′rah* the firm white fibrous outer layer of the eyeball; protects and maintains eyeball shape.

**Scoliosis** *sko″le-o′sis* the abnormal curvature of the spine in a lateral direction.

**Scrotum** *skro′tum* a divided sac that suspends the testes outside the body at the root of the penis.

**Sebaceous glands** *seh-ba′shus* glands that empty their sebum secretion into hair follicles.

**Sebum** *se′bum* the oily secretion of sebaceous glands.

**Second messenger** intracellular molecule generated by binding of a chemical to a membrane receptor; mediates intracellular responses.

**Secondary curvatures** spinal curvatures in the cervical and lumbar regions that develop after birth.

**Secondary humoral response** second and subsequent responses of the humoral arm of the immune system to a previously met antigen; more rapid and more vigorous than the primary response.

**Secondary oocyte** the developing oocyte after the first meiotic division; the cell that gets ovulated.

**Secondary sex characteristics** anatomical features that develop under influence of sex hormones that are not directly involved in the reproductive process. Include male or female pattern of muscle development, bone growth, body hair, etc.

**Secretion** *se-kre′shun* (1) the passage of material formed by a cell to its exterior; (2) cell product that is transported to the cell exterior.

**Segmentation** the mixing of foodstuffs by the alternating contraction and relaxation of the circular muscle layer in the muscularis externa of the gastrointestinal tract.

**Selective permeability** characteristic exhibited by a barrier, such as a membrane, that allows some substances through and excludes others.

**Semen** *se′men* fluid mixture produced by male reproductive structures; contains sperm, nutrients, and mucus.

**Semicircular canals** circular canals of the inner ear situated in three different planes; location of dynamic equilibrium receptors.

**Semilunar valves** *sem″ĭ-lu′nar* valves that prevent blood return to the ventricles after contraction; see *pulmonary valve* and *aortic valve.*

**Seminal vesicles** *sem′ĭ-nul* glands located at the base of the bladder that contribute to semen by providing a food source for sperm.

**Seminiferous tubules** *sem′ĭ-nif′er-us* highly convoluted tubes within the testes that form sperm.

**Sensory division** the part of the PNS that sends sensory input to the CNS from sensory receptors around the body; also called *afferent division.*

**Serosa** outermost layer of the gastrointestinal tract; also called the *visceral peritoneum*.

**Serous fluid** *ser'us* a clear, watery fluid secreted by the cells of a serous membrane.

**Serous membrane** membrane that lines a cavity without an opening to the outside of the body (except for joint cavities, which have a synovial membrane); serosa.

**Serous pericardium** double-layered serous membrane deep to the fibrous pericardium; the layers secrete serous fluid to reduce friction as the heart beats.

**Serum** the fluid portion of plasma minus clotting proteins.

**Sinoatrial (SA) node** *si″no-a′tre-al* the mass of specialized myocardial cells in the wall of the right atrium; pacemaker of the heart.

**Sinus** *si′nus* (1) a mucous membrane–lined, air-filled cavity in certain cranial bones; (2) a dilated channel for passage of blood or lymph.

**Skeletal muscle** muscle composed of cylindrical multinucleate cells with obvious striations; the muscle(s) attached to the body's skeleton; also called *voluntary muscle*.

**Skeletal system** system of protection and support composed primarily of bone and cartilage.

**Skull** component of the axial skeleton containing cranium (bony enclosure for the brain) and facial bones.

**Small intestine** segment of the gastrointestinal tract between the stomach and large intestine; site where most absorption occurs.

**Smooth (visceral) muscle** muscle consisting of spindle-shaped, unstriped (nonstriated) muscle cells; involuntary muscle.

**Soft palate** portion of the roof of the mouth that is not supported by bone.

**Solute** *sol′yoot* the dissolved substance in a solution.

**Solute pump** protein carrier involved in active transport.

**Solution** a homogenous mixture of two or more components.

**Solvent** the substance present in the largest amount in a solution.

**Somatic nervous system** *so-mat′ik* a division of the peripheral nervous system; also called the *voluntary nervous system*.

**Sperm** mature male sex cell.

**Spermatic cord** a group of blood vessels, nerves, and ducts that are wrapped in a connective tissue sheath that travels through the inguinal canal.

**Spermatids** *sper′mah-tidz* gametes after the second meiotic division that still need to undergo spermiogenesis.

**Spermatogenesis** *sper″mah-to-jen′ĕ-sis* the process of sperm production in the male; involves meiosis.

**Spermatogonia** *sper″mah-to-go′ne-ah* male stem cells that will produce sperm.

**Spermiogenesis** *sper″me-o-gen′ĕ-sis* the last stage of sperm development in which all excess cytoplasm is sloughed off and the sperm takes its mature shape.

**Spinal cord** part of the CNS that provides a two-way conduction system to and from the brain; also a major reflex center.

**Spinal nerves** 31 pairs of nerves originating from the spinal cord; formed when the dorsal and ventral roots merge.

**Spinous process** bony projection from the posterior of the vertebral arch.

**Spiral organ of Corti** the location of the hearing receptors in the cochlea.

**Splanchnic nerves** *splank′nik* preganglionic sympathetic neurons that do not synapse at the sympathetic trunk ganglion, but instead synapse at a collateral ganglion anterior to the spinal cord, then innervate the visceral organs.

**Spongy bone** internal layer of bone in flat, short, and irregular bones, and in the epiphyses of long bones.

**Spongy (penile) urethra** the portion of the urethra that travels down the penis and opens to the outside of the body.

**Squamous suture** interlocking joint formed by the intersection of temporal bones with the parietal bones.

**Stapes** *sta′pēz* medial bone of the middle ear; also called the *stirrup*.

**Static equilibrium** *stat′ik e″kwĭ-lib′re-um* balance concerned with changes in the position of the head.

**Sterility** the inability to produce offspring.

**Sternum** breastbone; flat bone composed of the manubrium, body, and xiphoid process.

**Steroids** *stĕ′roidz* a specific group of chemical substances including certain hormones and cholesterol.

**Stroke volume (SV)** a volume of blood ejected by a ventricle during systole.

**Submucosa** layer of the gastrointestinal tract just deep to the mucosa that contains blood vessels, nerves, lymphoid tissue (MALT), and lymphatic vessels.

**Sudoriferous glands** *su″do-rif′er-us* the glands that produce a saline solution called sweat

**Sulcus** *sul′kus* a furrow on the brain, less deep than a fissure.

**Superior vena cava** large vein that carries blood from veins superior to the diaphragm back to the right atrium.

**Sural** *soo′ral* related to the posterior leg; calf.

**Surfactant** *sur-fak′tant* a chemical substance coating the pulmonary alveoli walls that reduces surface tension, thus preventing collapse of the alveoli after expiration.

**Suspensory ligaments** ligaments that secure the ovaries to the lateral walls of the pelvis.

**Swallowing** complex process of moving food from the mouth into the esophagus; involves the buccal and pharyngeal-esophageal phases; also called *deglutition*.

**Sympathetic division** a division of the autonomic nervous system; opposes parasympathetic functions; called the *fight-or-flight division*; also called *thoracolumbar division*.

**Synapse** *sin′aps* the region of communication between neurons, or a neuromuscular junction between a neuron and a muscle cell.

**Synaptic cleft** *sĭ-nap′tik* the fluid-filled space at a synapse between neurons.

**Synarthrosis** *sin″ar-thro′sis* an immovable joint.

**Synovial fluid** *sĭ-no′ve-al* a fluid secreted by the synovial membrane; lubricates joint surfaces and nourishes articular cartilages.

**Synovial joint** freely movable joint exhibiting a joint cavity enclosed by a fibrous capsule lined with synovial membrane.

**Synovial membrane** membrane that lines the fibrous capsule of a synovial joint.

**Synthesis reaction** chemical reaction in which larger molecules are formed from simpler ones.

**Systemic circulation** system of blood vessels that carries nutrient- and oxygen-rich blood to all body organs.

**Systole** *sis′to-le* the contraction phase of heart activity.

**Systolic pressure** *sis-tŏ′lik* the pressure generated by the left ventricle during systole.

**T cells** lymphocytes that mediate cellular immunity; include helper, cytotoxic, regulatory, and memory cells. Also called *T lymphocytes*.

**T wave** wave on an ECG representing repolarization of the ventricles.

**Tachycardia** *tak″e-kar′de-ah* an abnormal, excessively rapid heart rate; over 100 beats per minute.

**Talus** *tă′lus* ankle bone; articulates with the tibia and calcaneus; allows plantar flexion.

**Tarsal** *tahr′sal* (1) one of the seven bones that form the ankle and heel; (2) relating to the ankle.

**Tectorial membrane** *tek-to′re-al* gel-like membrane in the cochlear duct in which the "hairs" of the receptor cells are embedded.

**Tendon** *ten′dun* cord of dense fibrous tissue attaching a muscle to a bone.

**Tendon sheath** elongated bursa that wraps a tendon subject to friction.

**Testis** *tes′tis* the male primary sex organ that produces sperm.

**Testosterone** *tes-tos′tĕ-rōn* male sex hormone produced by the testes; during puberty promotes virilization and is necessary for normal sperm production.

**Thalamus** *tha′luh-mus* a mass of gray matter in the diencephalon of the brain.

**Thoracic** *tho-ras′ik* refers to the area between the neck and abdomen supported by the ribs, costal cartilages and sternum; chest.

**Thoracic vertebrae** the 12 vertebrae that are in the middle part of the vertebral column and articulate with the ribs.

**Thrombin** *throm′bin* an enzyme that induces clotting by converting fibrinogen to fibrin.

**Thrombus** *throm′bus* a fixed clot that develops and persists in an unbroken blood vessel.

**Thymus** *thi′mus* an endocrine gland active in the immune response.

**Thyroid gland** *thi′roid* one of the largest of the body's endocrine glands; straddles the anterior trachea.

**Tidal volume (TV)** amount of air inhaled or exhaled with a normal breath.

**Tight junction** area where plasma membranes of neighboring cells are tightly bound together, forming an impermeable barrier.

**Tissue** a group of similar cells specialized to perform a specific function; primary tissue types are epithelial, connective, muscle, and nervous tissues.

**Trachea** *tra'ke-ah* The windpipe; the respiratory tube extending from larynx to bronchi.

**Tracheoesophageal fistula** *tra"ke-o-ĕ-sof"ah-je' al fis'tu-lah* congenital defect in which the esophagus and trachea are connected physically; the esophagus often ends in a blind sac that does not lead to the stomach.

**Tract** a collection of nerve fibers in the CNS having the same origin, termination, and function.

**Transcription** one of the two major steps in protein synthesis; the transfer of information from DNA base sequence to the complementary messenger RNA base sequence.

**Transient ischemic attack (TIA)** the temporary restriction of blood flow to an area of the brain.

**Translation** the second major step in protein synthesis; information carried by messenger RNA is decoded and used to assemble amino acids into a protein.

**Transverse process** lateral bony projection that originates at the vertebral arch; occur in pairs, one on each side.

**Transverse section (plane)** plane that divides the body or its parts into superior and inferior parts; cross section.

**Triglycerides** *tri-glis'er-īdz* compounds composed of fatty acids and glycerol; fats and oils; also called neutral fats.

**Trigone** *tri'gon* smooth, triangular region of the bladder formed by the openings of the two ureters and the urethra.

**Trochanter** *tro-kan'ter* a large, somewhat blunt process.

**Tropic hormone** *tro'pik* a hormone that regulates the function of another endocrine organ.

**True ribs** rib pairs 1–7; are attached directly to the sternum via costal cartilage.

**True vocal cords** folds in the mucous membrane of the larynx that vibrate to produce sound; also called *vocal folds.*

**Tubercle** *tu'ber-kul* a nodule or small rounded process.

**Tuberosity** *tu"bĕ-ros'ĭ-te* a broad process, larger than a tubercle.

**Tubular reabsorption** the process of reclaiming nutrients, ions, and water that the body needs from the filtrate in the renal tubule.

**Tubular secretion** the process of actively moving substances from peritubular blood into the renal tubule for elimination from the body.

**Tunica externa** *tu'nĭ-kah eks'tern-ah* outermost layer of blood vessel walls made of fibrous connective tissue.

**Tunica intima** *tu'nĭ-kah in'ti-mah* the innermost lining of blood vessels made of endothelium on a basement membrane; the only layer forming a capillary wall.

**Tunica media** *tu'nĭ-kah me'de-ah* the middle layer of blood vessel walls containing smooth muscle and elastic fibers.

**Tympanic membrane** *tim-pan'ik* the eardrum.

**Ulna** medial bone of the forearm; articulates with the trochlea of the humerus and the carpals.

**Umbilical** pertaining to the navel area.

**Umbilical cord** *um-bil'ĭ-kul* a structure bearing arteries and veins connecting the placenta and the fetus.

**Unfused tetanus** *tet'ah-nus* a muscle contraction in which the muscle does not completely relax between stimulation events; causes an increase in force because individual twitches are added together, or summed; also called *incomplete tetanus.*

**Unipolar neuron** structural class of neuron with one process extending from the cell body.

**Urea** *u-re'ah* the main nitrogen-containing waste excreted in the urine.

**Ureters** *yer'ĕ-terz* tubes that carry urine from kidney to bladder.

**Urethra** *u-re'thrah* the canal through which urine passes from the bladder to the outside of the body.

**Urinary system** system primarily responsible for water, electrolyte, and acid-base balance and the removal of nitrogen-containing wastes from the blood.

**Urine** filtrate containing waste and excess ions excreted by the kidneys.

**Uterine cycle** see *menstrual cycle.*

**Uterine tube** the oviduct: the tube through which the ovum is transported to the uterus; also called *fallopian tube.*

**Uterosacral ligaments** bilataeral ligaments that anchor the uterus posteriorly.

**Uterus** *u'ter-us* hollow pelvic organ of the female reproductive system that functions to receive, retain, and nourish a fertilized egg.

**Uvula** *u'vu-lah* tissue tag hanging from soft palate.

**Vaccine** weakened or killed antigen injected into a person for the purpose of stimulating a primary immune response; conveys artificial active immunity; results in immunological memory.

**Vagina** *vah-ji'nah* the female copulatory organ that runs from the cervix to the outside of the body.

**Valence shell** *va'lens* the outermost energy level of an atom that contains electrons; the electrons in the valence shell determine the bonding behavior of the atom.

**Valves** structures that close to prevent backflow of blood in the heart or in large veins.

**Vascular layer** middle layer (tunic) of the eyeball wall that contains the choroid, ciliary body, and iris.

**Vascular shunt** a vessel that directly connects the arteriole and venule of a capillary bed.

**Vasoconstriction** *vās"o-kon-strik'shun* narrowing of blood vessels.

**Vein** *vān* a vessel carrying blood away from the tissues toward the heart.

**Ventral ramus** branch of a spinal nerve that serves the anterior and lateral trunk; ventral rami of spinal nerves $T_1$–$T_{12}$ form the intercostal nerves; all other ventral rami form the four nerve plexuses.

**Ventral root** the root at which motor neurons of the somatic nervous system exit the spinal cord.

**Ventricles** *ven'trĭ-kulz* (1) discharging chambers of the heart; (2) cavities within the brain.

**Venule** *vēn'ūl* a small vein.

**Vertebral arch** arch formed by connecting all bony projections posteriorly from the centrum; includes pedicles and laminae.

**Vertebral column** the spine, formed of a number of individual bones called vertebrae and two composite bones (sacrum and coccyx).

**Vesicle** *ves'ĭ-kul* a small fluid-filled sac formed by membrane.

**Vesicular breathing sounds** soft respiratory sounds produced when air fills the alveoli.

**Vesicular follicle** a mature ovarian follicle; also called *Graafian follicle.*

**Vesicular transport** an active transport process in which items too large to pass through the plasma membrane are transported in bulk via a vesicle. See *endocytosis* and *exocytosis.*

**Vestibular apparatus** name for equilibrium receptors of the inner ear; includes both static and dynamic equilibrium receptors.

**Vestibule** *ves'tĭ-būl* (1) area of the inner ear between the cochlea and semicircular canals; (2) the space in the oral cavity inside the lips and cheeks and outside the teeth and gums.

**Villi** *vil'i* fingerlike projections of the small intestinal mucosa that tremendously increase its surface area for absorption.

**Visceral pericardium** innermost layer of the serous pericardium that lines the surface of the heart, also called *epicardium.*

**Visceral peritoneum** outermost layer of the gastrointestinal tract; also called the *serosa.*

**Visceral pleura** *ploor'ah* see *pulmonary pleura.*

**Visual acuity** *ah-ku'ĭ-te* the ability of the eye to distinguish detail.

**Vital capacity (VC)** the volume of air that can be expelled from the lungs by forcible expiration after the deepest inspiration; total exchangeable air.

**Vitamins** organic compounds required by the body in very small amounts for physiological maintenance and growth.

**Vitreous humor** a gel-like substance that helps prevent the eyeball from collapsing inward by reinforcing it internally; also called *vitreous body.*

**Vulva** *vul'va* female external genitalia.

**White matter** white substance of the central nervous system; the myelinated nerve fibers.

**Xiphoid process** *zif'oyd* the most inferior of the three bones making up the sternum; begins as hyaline cartilage and gradually ossifies throughout adulthood.

**Yellow marrow** bone marrow that stores adipose (fat) tissue.

**Zygote** *zi'gōt* the fertilized ovum; produced by union of two gametes.

# Credits

## Photographic Credits

VWT smart phone, tablet, and computer: Shutterstock. Author photo: Elaine N. Marieb, Suzanne Keller, Essentials of Human Anatomy & Physiology, 12 Ed., © 2018, Pearson Education, Inc., New York, NY.

**Chapter 1** 1.1: CLS Digital Arts/Shutterstock; A Closer Look, a: Southern Illinois University/Science Source; A Closer Look, b: Laurent/Hop. Americain/Science Source; A Closer Look, c: Medical Body Scans/Science Source; A Closer Look, d: Science Source; A Closer Look, e: Scott Camazine/Science Source; 1.5a2: CNRI/Science Source; 1.5c2: James Cavallini/Science Source; 1.5a-c1: Pearson Education, Inc.; 1.5b2: Scott Camazine/Science Source; 1.8a: Pearson Education, Inc.

**Chapter 2** 2.4.1: Chip Clark/Fundamental Photographs, NYC; 2.4.2: Leslie Garland Picture Library/Alamy Stock Photo; 2.4.3: Levent Konuk/Shutterstock; 2.9b: BMJ/Shutterstock; 2.16a: Image Source/Getty Images; 2.16b: Ivaschenko Roman/Shutterstock; 2.21c: Will & Deni McIntyre/Science Source; Focus on Careers: Don Hammond/Photolibrary, Inc.

**Chapter 3** 3.7a: Mary Osborn; 3.7b: Dr. Frank Solomon; 3.7c: Mark S. Ladinsky and J. Richard McIntosh; A Closer Look, a-c: David M Philips/Science Source; 3.12b: Birgit H. Satir, Dept. of Anatomy and Structural Biology, Albert Einstein College of Medicine.; 3.18a: Biophoto Associates/Science Source; 3.18b, c, e, f: Allen Bell/Pearson Education, Inc.; 3.18d: William Karkow/Pearson Education, Inc.; 3.19a, c, g: Allen Bell/Pearson Education, Inc.; 3.19b, h: Ed Reschke/Photolibrary/Getty Images; 3.19d: William Karkow/Pearson Education, Inc.; 3.19e: Ed Reschke; 3.19f: Biophoto Associates/Science Source; 3.20a: Eric Graves/Science Source; 3.20b: Marian Rice; 3.20c: Steve Gschmeissner/Science Photo Library/Getty Images; 3.21: Biophoto Associates/Science Source; Homeostatic Imbalance 3.3: Biophoto Associates/Science Source.

**Chapter 4** 4.5: Ed Reschke/Photolibrary/Getty Images; A Closer Look: Lauren Shear/Science Source; Homeostatic Imbalance 4.2a: Mediscan/Alamy Stock Photo; Homeostatic Imbalance 4.2b: National Institute of Health; Homeostatic Imbalance 4.4: Ian Boddy/Science Source; 4.7a: Pearson Education, Inc.; 4.7b: CNRI/Science Source; 4.7c: CNRI/Photo Researchers, Inc./Science Source; 4.9a, b: Dr. P. Marazzi/Science Source; 4.9c: Medicshots/Alamy Stock Photo; 4.10b.1: Gianni Muratore/Alamy Stock Photo; 4.10b.3: John Radcliffe Hospital/Science Source; 4.10b.2: Scott Camazine/Science Source; 4.11a: Dr. P. Marazzi/Science Source; 4.11b: Girand/Science Source; 4.11c: Custom Medical Stock Photo/Alamy Stock Photo; Focus on Careers: Elena Dorfman/Pearson Education, Inc.

**Chapter 5** 5.1: Seelevel.com; 5.4c: William Krakow/Pearson Education, Inc.; Focus on Careers: Frederic Ciroues/PhotoAlto/Alamy Stock Photo; Homeostatic Imbalance 5.1: Jeff Rotman/Science Source; 5.16: Reik/AGE Fotostock; Homeostatic Imbalance 5.5a: Princess Margaret Rose Orthopaedic Hospital/Science Source; Homeostatic Imbalance 5.5b: Nordic Photos/SuperStock; Homeostatic Imbalance 5.5c: Steve Gorton/Dorling Kindersley Ltd.; 5.20b: Mark Nielsen and Shawn Miller; Closer Look, a: Lawrence Livermore National Laboratory/Science Source; Closer Look, b: Elaine N. Marieb; Closer Look, c: Frank Perry/AFP/Getty Images; Homeostatic Imbalance 5.7: CNRI/Science Photo Library/Science Source; 5.31: Scott Camazine/Science Source; 5.34a: Professor Pietro M. Motta/Science Source; 5.34b: P. Motta/Department of Anatomy, University "La Sapienza," Rome/Science Source.

**Chapter 6** Table 6.1.1: Pearson Education, Inc.; Table 6.1.2: Elaine N. Marieb; Table 6.1.3: William Karkow/Pearson Education, Inc.; 6.4b: Eric V. Grave/Science Source; 6.11a: Simon Balson/Alamy Stock Photo; 6.11b: JGI/Blend Images/AGE Fotostock.; 6.13: Phocasso/J W White/Pearson Education, Inc.; A Closer Look, 1: Corbis/SuperStock; 2: Rodney Turner KRT/Newscom; Homeostatic Imbalance 6.5: Dr. M.A. Ansary/Science Source.

**Chapter 7** 7.4: Phanie/Science Source; 7.13b: A. Glauberman/Science Source; 7.17b: From A Stereoscopic Atlas of Human Anatomy by David Bassett/Robert A. Chase; Homeostatic Imbalance 7.6: Ulrich Doering/Alamy Stock Photo; A Closer Look: AdMedia/Splash News/Newscom; Homeostatic Imbalance 7.11: Jeffrey Greenberg/Science Source; A Closer Look, a: Hank Morgan/Science Source.

**Chapter 8** 8.1: Richard Tauber/Pearson Education, Inc.; 8.4: From A Stereoscopic Atlas of Human Anatomy by David Bassett/Robert A. Chase; Homeostatic Imbalance 8.5: Arztsamui/Shutterstock; 8.7: Thalerngsak Mongkolsin/Shutterstock; Focus on Careers: Creatas/Photolibrary, Inc.; Homeostatic Imbalance 8.11: Biophoto Associates/Science Source.

**Chapter 9** Homeostatic Imbalance 9.2: Caters News/ZUMA Press/Newscom; A Closer Look: Andersen Ross/Blend Images/Getty Images; 9.6b: Michael Wiley/Pearson Education, Inc.; Homeostatic Imbalance 9.4.1: Alison Wright/Science Source; Homeostatic Imbalance 9.4.2: Ralph C. Eagle, Jr./Science Source; Homeostatic Imbalance 9.6: Biophoto Associates/Science Source; 9.11b: Pearson Education, Inc.

**Chapter 10** 10.2: Victor Eroschenko/Pearson Education, Inc.; Homeostatic Imbalance 10.1: Ingram Publishing/Photolibrary, Inc.; 10.6: Lennart Nilsson/Scanpix Sweden AB; Homeostatic Imbalance 10.3: Moredun Scientific/Science Source; Focus on Careers: BSIP SA/Alamy Stock Photo; 10.7: Jack Scanlon/Holyoke Community College.

**Chapter 11** Homeostatic Imbalance 11.2: David Campione/Science Source; 11.9: Mark Shaiken/Moment/Getty Images; 11.10a: Ed Reschke/Photolibrary/Getty Images; 11.15b: CNRI/Science Source; A Closer Look: SPL/Science Source.

**Chapter 12** Homeostatic Imbalance 12.2: Chris Bjornberg/Science Source; 12.9a: SPL/Science Source; 12.15a: Arthur J. Olson, Molecular Graphics Laboratory, Scripps Research Institute; A Closer Look: Eye of Science/Science Source.

**Chapter 13** 13.3b: CNRI/Science Source; 13.4a: Richard Tauber/Pearson Education, Inc.; 13.5b: Lisa Lee/Pearson Education, Inc.; Homeostatic Imbalance 13.7: Du Cane Medical Imaging Ltd./Science Source; A Closer Look: Stacey Green/Photographer's Choice/Getty Images.

**Chapter 14** 14.4b: From A Stereoscopic Atlas of Human Anatomy by David Bassett/Robert A. Chase; 14.5a: Robert A. Chase; A Closer Look, a: CNRI/Science Source; A Closer Look, b: Eye of Science/Science Source; 14.21b: Andrew Doran/Shutterstock; Homeostatic Imbalance 14.15: Center for Craniolfacial Anomalies; A Closer Look: Keith Srakocic/AP Images.

**Chapter 15** 15.1b: Richard Tauber/Pearson Education, Inc.; 15.2a: Karen Karbbenhoft/Pearson Education, Inc.; 15.3b: From A Stereoscopic Atlas of Human Anatomy by David Bassett/Robert A. Chase; Homeostatic Imbalance 15.3: G. Warrick/Science Source; A Closer Look: AJPhoto/Science Source; 15.11: Susanna Blavarg/Getty Images; Focus on Careers: Design Pics/Don Hammond/Getty Images.

**Chapter 16** 16.5a: Juergen Berger/Science Source; Homeostatic Imbalance 16.3: Scimat/Science Source; 16.11: C. Edleman/La Vilette/Science Source; 16.14a: Thomas/Science Source; 16.14b, c: Southern Illinois University/Science Source; 16.16: David M. Phillips/Science Source; 16.19: From A Stereoscopic Atlas of Human Anatomy by David Bassett/Robert A. Chase; 16.20a: Lennart Nilsson/Scanpix Sweden AB; 16.20b: Neil Bromhall/Science Source; A Closer Look: Susanna Price/Dorling Kindersley Ltd.

## Art Credits

All illustrations are by Imagineering STA Media Services unless noted below.

**Chapter 3** 3.2: Carla Simmons/Kristin Mount; 3.4: Tomo Narashima; 3.15: Adapted from Campbell, Neil A.; Mitchell, Lawrence G.; Reece, Jane B., Biology: Concepts and Connections, 3rd Ed., ©2000, pp. 132–133. Reprinted and Electronically reproduced by permission of Pearson Education, Inc., Upper Saddle River, New Jersey.

**Chapter 4** 4.4: Electronic Publishing Services, Inc.

**Chapter 5** 5.3: Carla Simmons/Imagineering STA Media Services; 5.7: Carla Simmons/Imagineering; 5.9–5.12: Kristin Mount.

**Chapter 6** 6.1: Raychel Ciemma; 6.2b: Carla Simmons/Kristin Mount; 6.14abcd: Electronic Publishing Services, Inc./Imagineering STA Media Services.

**Chapter 7** 7.20: Laurie O'Keefe/Kristin Mount; 7.22: Charles W. Hoffman/Kristin Mount; 7.27: Electronic Publishing Services, Inc.

**Chapter 11** 11.1: Wendy Hiller Gee/Kristin Mount; 11.2, 11.3b, 11.4: Barbara Cousins; 11.5, 11.6: Barbara Cousins/Imagineering STA Media Services; 11.7: Barbara Cousins/Kristin Mount; 11.15, 11.16, 11.24: Kristin Mount.

**Chapter 14** 14.17b: http://www.choosemyplate.gov.

**Chapter 15** 15.6: Linda McVay/Imagineering STA Media Services.

**Chapter 16** 16.3: Precision Graphics/Imagineering STA Media Services; 16.5: Precision Graphics/Imagineering STA Media Services; 16.8: Martha Blake/Kristin Mount.

## Text Credits

**Chapter 14** Table 14.2: Thompson, Janice; Manore, Melinda; Vaughan, Linda, The Science of Nutrition, 3rd Ed., ©2014, pp. C1–C12, 57. Reprinted and Electronically reproduced by permission of Pearson Education, Inc., Upper Saddle River, New Jersey.

**Appendix D** Thompson, Janice; Manore, Melinda; Vaughan, Linda, The Science of Nutrition, 3rd Ed., ©2014, pp. C1–C12, 57. Reprinted and Electronically reproduced by permission of Pearson Education, Inc., Upper Saddle River, New Jersey.

# Subject Index